T0235084

QUANTUM DISSIPATIVE SYSTEMS

Fifth Edition

QUANTUM DISSIPATIVE SYSTEMS

Fifth Edition

Ulrich Weiss

University of Stuttgart, Germany

World Scientific

NEW JERSEY · LONDON · SINGAPORE · BEIJING · SHANGHAI · HONG KONG · TAIPEI · CHENNAI · TOKYO

Published by

World Scientific Publishing Co. Pte. Ltd.
5 Toh Tuck Link, Singapore 596224
USA office: 27 Warren Street, Suite 401-402, Hackensack, NJ 07601
UK office: 57 Shelton Street, Covent Garden, London WC2H 9HE

Library of Congress Cataloging-in-Publication Data
Names: Weiss, U. (Ulrich) author.
Title: Quantum dissipative systems / Ulrich Weiss, University of Stuttgart, Germany.
Description: Fifth edition. | New Jersey : World Scientific Publishing, [2022] |
    Includes bibliographical references and index.
Identifiers: LCCN 2021034598 (print) | LCCN 2021034599 (ebook) |
    ISBN 9789811243134 (hardcover) | ISBN 9789811241499 (paperback) |
    ISBN 9789811241505 (ebook) | ISBN 9789811241512 (ebook for individuals)
Subjects: LCSH: Quantum theory. | Mathematical physics. | Thermodynamics. | Path integrals.
Classification: LCC QC174.12 .W45 2022 (print) | LCC QC174.12 (ebook) |
    DDC 530.12--dc23
LC record available at https://lccn.loc.gov/2021034598
LC ebook record available at https://lccn.loc.gov/2021034599

British Library Cataloguing-in-Publication Data
A catalogue record for this book is available from the British Library.

For any available supplementary material, please visit
https://www.worldscientific.com/worldscibooks/10.1142/12402#t=suppl

Printed in Singapore

Preface

The launch of a fifth edition of *Quantum Dissipative Systems*, twenty eight years after initial publication and nine years after publication of the fourth edition, evidences that the monograph found continuing recognition and interest. As the book has been sold out again and the topic has kept lively and exciting until now, I found it appropriate and useful to make a thorough revision of the previous edition.

The content underwent a rather extensive rewriting throughout, and it is restructured in parts. The 29 chapters of the fourth edition with the addition of new subject matter are now allocated to 32 chapters. Among the major amendments there are in Section 5.6 the implementation of an influence functional controlling the various energy flux channels within the composite system, and application of it to energy transfer in the driven TSS at strong coupling in Section 23.6, and to driven quantum Brownian motion in a corrugated potential in Chapter 32. Major further additions are (i) the discussion of anomalous dissipation resulting from momentum coupling instead of coordinate coupling in Section 3.6, and (ii) the discussion of anomalies in the thermodynamics of a quantum Brownian particle, such as negative specific heat and reentrant classicality, in Section 7.5. Part V is restructured in order to improve logical composition and line of arguments, and additional subject matter is integrated. The new content amounts to about 40 pages. To keep the overall length of the monograph moderate, several passages of lesser importance have been streamlined.

The revision of the monograph has been made with the aim to preserve the advantages of the previous editions and to improve and refine argumentation, clarity and readability wherever I considered it possible. The target readership is still the same. The level is such that anyone with a first course in quantum mechanics and thermodynamics should not have a hard time to understand the essentials. The issue of dissipative quantum mechanics is reviewed in a systematic way making the topic also very accessible to interested researchers normally working in different fields. Last but not least, active researchers in the field will find this monograph a rich and stimulating source.

Finally I would like to thank Dr. Elena Nash as the responsible editor of World Scientific Publishing for her continuous assistance and useful advice.

Stuttgart, Germany
April 2021   *Ulrich Weiss*

Preface to the Fourth Edition

The prospect of a fourth edition of *Quantum Dissipative Systems*, almost twenty years after initial publication, and four years after publication of the third edition, shows on the one hand that the field is rapidly developing and on the other hand that the book has been paid the compliment of growing use as a text since it appeared for the first time.

In preparing the fourth edition I have tried to sustain as much as possible the advantages of the previous editions while taking into account developments in the subject itself and its applications to other fields. It was a persistent intention of mine to escalate the pedagogical objectives. What has emerged is a thorough revision and a very considerable enlargement of the third edition to which about 60 pages have been added. The bibliography has undergone similar expansion, which reflects the enormous advance in this field in the four years since the third edition. Hardly a passage of the text has been left untouched with the aim of enhancing physical insights and clarity. The advanced formal techniques retained are in my opinion essential as to avoid that the reader must take anything on trust. At the same time I hope that the book is relatively easily readable by, for example, beginning graduate students in theory, or by a general readership with a background in quantum statistical physics. The major changes are of various kinds.

In Part I new sections have been added. They deal with ergodicity and treat a scheme in which stochastic non-Markovian quantum dynamics based on the unraveled influence functional is combined with semiclassical propagation within the frozen Gaussian approximation. The section on the harmonic oscillator bath with linear coupling has been augmented by subsections on the fractional Langevin equation and on the problem of a charged particle interacting with the radiation field.

The major additions in Part II are as follows. The chapter on the damped linear oscillator has been extended by a section on quantum mechanical master equations for the reduced density matrix. Further subsections dealing with radiation damping and with thermodynamic properties (e.g., internal energy, free energy and entropy) of the damped quantum oscillator have been added. The chapter on quantum Brownian free motion has been augmented by a section on the partition function and on thermodynamic properties. Further, a section on electron decoherence in a disordered conductor has been subjoined.

In Part IV, the chapter on the dissipative two-state dynamics has been rearranged and extended considerably in order to improve clarity. Now, there is a chapter on the basics and methods, and a chapter dealing with explicit results in various regimes of the parameter space. New sections on the pure dephasing regime and on decoherence resulting from $1/f$ noise have been appended.

Part V now contains a chapter on twisted partition functions for the dissipative multi-state system in the field theory limit. This includes derivation of the exact scaling solution for the partition function using properties of Jack symmetric functions and presentation of a conjecture by which the nonlinear conductance can be extracted from the the twisted partition function. This chapter complements the chapter on duality symmetry by giving a different perspective.

Errors that I have caught, or which have been pointed out to me, have been corrected. I hope that not too many new ones have crept into the revised material. I shall be very grateful to any reader who brings such deficiencies to my attention.

The selection made here represents something of a personal compromise. During recent years several topics, in particular those connected with quantum state engineering and quantum computing, have undergone enormous expansion both on the theoretical and experimental side. I am very conscious that the coverage of most recent developments in these fields is far from being complete. One area omitted deserves special mention – decoherence and relaxation in quantum gases. The importance of the field is without question, but I felt that an adequate treatment deserves a separate book.

An unforeseen event deferred the delivery by several months. I am very grateful to the responsible editor Kim Tan of World Scientific for her continuous advice and patience.

Finally, I would like to thank many colleagues and readers who made valuable remarks on previous editions. Particular thanks are due to Doron Cohen, Giuseppe Falci, Hermann Grabert, Milena Grifoni, Frank Großmann, Gert-Ludwig Ingold, Elisabetta Paladino, Gerd Schön, Udo Seifert, Jürgen Stockburger, and Andrei Zaikin for their comments and suggestions.

Stuttgart
November 2011   *Ulrich Weiss*

Preface to the Third Edition

In the fourteen years since the appearance of the first edition, the subject kept freshness. There have been interesting theoretical progress, important new applications and lots of stunning new experiments in the field.

The present edition of *Quantum Dissipative Systems* reflects two endeavors on my part: the improvement and refinement of material contained already in the second edition; the addition of new topics (and the omission of few).

The emphasis and major intensions are still the same, but there are changes, augmentations and additions. The major extensions, altogether about 60 pages, are: Chapter 3 contains a more detailed discussion of the quasiclassical Langevin equation and a subsection on Josephson flux and charge qubits. Chapter 4 gives wider space to the basics of path integration and to the treatment of an electromagnetic environment. Chapter 5 discusses the stochastic unraveling of path integrals for the reduced density matrix. Chaper 6 gives an extended discussion of the damped quantum harmonic oscillator. It includes discussion of internal energy, purity, and uncertainty. Chapter 15 presents a generalization of the Smoluchowski diffusion equation which includes quantum effects. Chapter 20 offers a broader discussion of single-charge tunneling in the weak-tunneling or Coulomb blockade regime. Chapter 21 discusses relaxation and decoherence in the spin-boson model at zero temperature. It presents analytical results for the relaxation and decoherence rate at general damping strength which cover the entire regime extending from weak to strong tunneling. Chapter 24 includes a discussion of the full counting statistics for Poissonian quantum transport and presents many analytical results available in special cases. Chapter 25 presents the scaling-invariant solution of the full counting statistics in diverse limits and discusses application to charge transport in Josephson junctions. Chapter 26 gives an extended discussion of charge transport in quantum impurity systems, including full counting statistics. It points out an intimate connection of these systems with models for coherent conductors and with others discussed in the preceding Chapters 23 – 25. The bibliography is updated.

This new edition has benefited from comments, suggestions and criticisms from many students and colleagues. Among those to whom I owe specific debt of gratitude are Holger Baur, Pino Falci, Hermann Grabert, Milena Grifoni, Yuli Nazarov, Elisabetta Paladino, Jürgen Stockburger, and Ruggero Vaia.

It is a pleasure to thank the students P. Diemand and A. Herzog for proofreading and tracking down misprints. I also wish to thank the responsible editor Kim Tan for advice and patience until completion of the third edition. Finally, I am grateful to my wife Christel for her sympathy and constant encouragement.

Stuttgart
October 2007   *Ulrich Weiss*

Preface to the Second Edition

Since the first publication of this book in 1993, there have been enormous research activities in quantum dissipative mechanics both experimentally and theoretically. For this reason, it has been highly desirable after the book has been sold out almost three years ago to undergo a number of extensions and improvements. I have been encouraged by the positive reception of this book by a large community and by many colleagues to write not simply an updated second edition. What came out now after all is almost a new book of roughly double content.

In an extensive rewriting, the 19 Chapters of the First Edition have been expanded by about one third to better meet the desires of both the newcomers to the field and the advanced readership, and I have added 7 new chapters. The most relevant extensions are as follows. In the first part, I have added a section on stochastic dynamics in Hilbert space and I have extended the discussion of relevant microscopic global models considerably. Now, there are also treated acoustic phonons with two-phonon coupling, a microscopic model for tunneling between surfaces, charging and environmental effects in normally conducting and superconducting tunnel junctions, and nonlinear quantum environments. Part II now contains an extended discussion of the damped harmonic oscillator (e.g., a study of the density of states is added), and new chapters on the thermodynamic variational approach and variational perturbation expansion method, and on the quantum decoherence problem. Part III, which deals with quantum-statistical decay, is extended by two chapters. In the new edition, the turnover theory to the energy-diffusion limited regime is discussed, and the treatment of dissipative quantum tunneling has been extended and improved. Ample space is now provided in Part IV to a thorough discussion of the dissipative two-state system. A number of new results on the thermodynamics and dynamics of this archetypal system are presented. An extensive discussion of electron transfer in a solvent, incoherent tunneling in the nonadiabatic regime, and single-charge tunneling is provided in a unified framework. Regarding dynamics, new sections on exact master equations, improved approximation schemes, and recent results on correlation functions have been written, and a new chapter on the driven dissipative two-state is included. Part V, which deals with the dissipative multi-state system, is completely rewritten. It now contains four chapters on quantum Brownian motion in a cosine potential, multi-state dynamics, duality symmetry, and tunneling of charge through an impurity in a quantum wire. Many new results available only very recently are presented. The about 460 references are suggestions for additional reading on particular subjects and are not intended as a comprehensive bibliography.

Stuttgart
December 1998   *Ulrich Weiss*

Preface to the First Edition

This book is an outgrowth of a series of lectures which I taught at the ICTP at Trieste and at the University of Stuttgart during the spring and summer of 1991. The purpose of my lectures was to present the approaches and techniques that accurately treat quantum processes in the presence of frictional influences.

The problem of *open* quantum systems has been around since the beginnings of quantum mechanics. Important contributions to this general area have been made by researchers working in fields as diverse as solid-state physics, chemical physics, biophysics, quantum measurement theory, quantum optics, nuclear and particle physics. Often, there has been used, and still is used, a language well known in one context or one field, yet sufficiently different from others that it is not altogether easy to make out the connection. Here, I offer a collection of ideas and examples rather than a comprehensive review of the topic and the history.

The central theme is the space-time functional integral or path integral formulation of quantum theory. This approach is particularly well suited for treating the quantum generalization of friction. Here we are faced to understand the behavior of a system with few quantal degrees of freedom coupled to a thermal reservoir. After integrating out the bath while keeping the system's coordinates fixed we get the influence functional describing the influence of the many bath degrees of freedom on the few relevant ones. This leads to an effective action weighting the paths of the open system in the functional integral description. Indeed, if one wishes to perform numerical computations on a rigorous level, there are no alternatives to this approach at present.

Path integration in condensed matter and chemical physics has become a growth industry in the last one or two decades. A newcomer to this thriving field may not yet be very familiar with the path integral method. Here, I do assume a knowledge of standard text book quantum mechanics and statistical mechanics augmented by a knowledge of Feynman's approach on a first introductory level. The books by Baym [1], Chandler [2], Feynman and Hibbs [3], and by Feynman [4] provide the elementary material in this regard. Further background and supplementary material on the path integral method are contained in the books by Schulman [5] and by Kleinert [6]. However, advanced mastery of these subjects is not necessary.

Some of the more sophisticated concepts, such as preparation functions, propagating functions, and correlation functions, are basic to the development as it is presented here. To cover this material at an introductory level, I make frequent use of simplified models. In this way, I can keep the mathematics relatively simple.

Many of the problems, methods, and ideas which are discussed here have become essential to the current understanding of quantum statistical mechanics. I have made a considerable effort to make the material largely self-contained. Thus, although the theoretical tools are not developed systematically and in its full beauty, the material may be useful to many graduate students to become familiar with the field and learn the methods. For the most part, I refrain from just quoting results without explaining where they come from. With regard to citations, I have tried to give references to the historical development and also to provide a selection of the very recent important ones. But the list is surely not a comprehensive bibliography.

This book exists because of the physics I learned and enjoyed from the fertile collaboration with Hermann Grabert, Peter Hänggi, Gert-Ludwig Ingold, Peter Riseborough, and Maura Sassetti. I am particularly indebted to Maura who took time off her research to weed out points of confusion and who persistently encouraged me to finish this venture. I am also grateful to my students Reinhold Egger, Manfried Milch, Jürgen Stockburger, and Dietmar Weinmann for helpful comments concerning the presentation of many subjects discussed here and for preparing the figures.

In writing this book, I have benefited from the discussion with many companions; in particular Uli Eckern, Enrico Galleani d'Agliano, Anthony J. Leggett, Hajo Leschke, Franco Napoli, Albert Schmid, Gerd Schön, Larry Schulman, Peter Talkner, Valerio Tognetti, Andrei Zaikin, and Wilhelm Zwerger, who helped me in increasing my understanding of many of the subjects which are discussed here.

It is a pleasure to thank my teacher, colleague, and friend Wolfgang Weidlich for many fruitful discussions over the years.

Finally, and most importantly, I am deeply grateful to my wife Christel and my children Ulrike, Jan, and Meike for their infinite patience and omnipresent sympathy.

Stuttgart
October 1992

Ulrich Weiss

Acknowledgements

A number of colleagues and friends have made valuable remarks on the first edition. In this regard, I am especially grateful to Theo Costi, Thorsten Dröse, Reinhold Egger, Hermann Grabert, Peter Hänggi, Gert-Ludwig Ingold, Chi Mak, Maura Sassetti, Rolf Schilling, Herbert Spohn, and Wilhelm Zwerger. In writing the new edition, discussions with Pino Falci, Igor Goychuk, Milena Grifoni, Gunther Lang, Elisabetta Paladino, and Manfred Winterstetter have been extremely profitable, and I wish to thank them for their suggestions and for tracking down misprints. I also would like to acknowledge the preparation of a number of figures by Jochen Bauer, Gunther Lang, Jörg Rollbühler, and Manfred Winterstetter. Finally, I would like to thank Mrs. Karen Yeo as the responsible Editor of World Scientific Publishing Co. for her useful advice.

Contents

Contents

II MISCELLANEOUS APPLICATIONS 153

6 Damped linear quantum mechanical oscillator 154

Contents

Contents

Contents

1 Introduction

Quantum-statistical mechanics is a very rich and checkered field. It is the theory dealing with the dynamical behavior of spontaneous quantal fluctuations.

When probing dynamical processes in complex many-body systems, one usually employs an external force which drives the system slightly or far away from equilibrium, and then measures the time-dependent response to this force. The standard experimental methods are quasielastic and inelastic scattering of light, electrons, and neutrons off a sample, and the system's dynamics is analyzed from the line shapes of the corresponding spectra. Other experimental tools are, e. g., spin relaxation experiments, study of the absorptive and dispersive acoustic behaviors, and investigation of transport properties. In such experiments, the system's response gives information about the dynamical behavior of the spontaneous fluctuations. Theoretically, the response is rigorously described in terms of time correlation functions. Therefore, time correlation functions and their Fourier transforms are at the core of interest in theoretical studies of the relaxation dynamics of nonequilibrium systems.

This book deals with the theories of open quantum systems with emphasis on phenomena in condensed matter physics. A preliminary consideration for any real dissipative quantum system is the separation of the underlying global system into a relevant subsystem - the dissipative system of interest - and the environment, of which the detailed dynamics is insignificant. In most cases of practical interest, the environment is thought to be in thermal equilibrium. The quantum statistical environment manifests itself in a fluctuating force acting on the relevant system and carrying the spectral characteristics of the heat reservoir. It is the very nature of the fluctuating force to cause decoherence and damping, and to drive everything to disorder.

While quantum mechanics was conceived as a theory for the microcosm, there is apparently no contradiction with this theory in the mesoscopic and macroscopic world. The understanding of the appearance of classical behavior within quantum mechanics is of fundamental importance. This issue is intimately connected with the understanding of decoherence. Despite the stunning success of quantum theory, there is still no general agreement on the interpretation. The main disputes circle around "measurement" and "observation".

Decoherence is the phenomenon that the superposition of macroscopically distinct states decays on a short time scale. It is omnipresent because information about quantum interference is carried away in some physical form into the surroundings. In a sense, the environmental coupling acts as a continuous measuring apparatus, leading to an incessant destruction of phase correlations. The relevance of this coupling for macroscopic systems is nowadays generally accepted by the respectable community.

This is a book of methods, techniques, and applications. The level is such that anyone with a first course in quantum mechanics and rudimentary knowledge of path integration should not find difficulties. An attempt is made to present the subject of dissipation in quantum mechanics in a unified form. A general framework is developed which can deal with weak and strong dissipation, and with all kinds of memory effects. The reader will find a presentation of the relevant ideas and theoretical concepts, and a discussion of a wide collection of microscopic models. In the models and applications, emphasis is put on condensed matter physics. I have tried to use vocabulary and notation which should be fairly familiar to scientists working in chemical and condensed matter physics.

The book is divided into five parts. The following sequence of topics is adopted. The first part of the book is devoted to the general theory of open quantum systems. In Chapter 2, I review traditional approaches, such as formulations by master equations for weak coupling, Lindblad theory, operator-valued and quasiclassical Langevin equations. I also discuss attempts to interpret the dynamics of an open quantum system in terms of a stochastic process in the Hilbert space of state vectors pertaining to the reduced system. In Chapter 3, a variety of system-plus-reservoir models are introduced. They are partly motivated by phenomenological reasoning and partly connected with microscopic models which are of relevance in condensed matter physics. The models represent both continuous and discrete quantum systems, and they cover the diverse regimes of frequency-dependent friction and noise. The criteria for ergodic behavior of the open system are discussed. Emphasis is put on models in which dissipation is linear. Also scenarios are treated in which the quantum environment is intrinsically nonlinear. Chapter 4 is devoted to the equilibrium statistical mechanics for the relevant subsystems of these models using the imaginary-time path integral approach. Chapter 5 concerns dynamics – quantum-mechanical motion, decoherence and relaxation of macroscopic systems that are far from or close to equilibrium. I discuss the Feynman-Vernon real-time influence functional approach and the relation with the Keldysh method. The concepts of preparation functions, propagating functions, and correlation functions are treated. Exact formal expressions for these quantities are derived using path sum techniques. This chapter also deals with more recent developments in which the two-time nonlocal influence functional is unravelled into a single-time stochastic influence functional. The appealing aspect is that one can derive stochastic Schrödinger equations and equivalent stochastic Liouville–von Neumann equations which are free of quantum memory effects.

Part II with Chapters 6 – 9 covers a discussion of exactly solvable damped linear quantum systems, the useful thermodynamic variational approach with extension to open nonlinear quantum systems, and different scenarios for loss of quantum coherence. I have also tried to provide a balanced survey of electron dephasing in a diffusive conductor at low temperature.

Part III deals with quantum-statistical metastability: a problem of fundamental importance in chemical physics and reaction theory. After an introduction into the problem in Chapter 10, the relevant theoretical concepts and the characteristic features of the decay are discussed in Chapters 11 to 17. The treatment mainly relies on a thermodynamic method in which the decay rate is related to the imaginary part of the analytically continued free energy of the damped system. This allows for a uniform theoretical description in the entire temperature range. The discussion extends from high temperatures where thermal activation prevails down to zero temperature where the system can only decay by quantum-mechanical tunneling out of the ground state in the metastable well. Results in analytic form are presented where available.

In Part IV, I have tried to consider the thermodynamics and dynamics of the dissipative two-state or spin-boson system on a uniform theoretical basis. The spin-boson system is the simplest nonlinear system featuring the interplay between quantum coherence, quantal and thermal fluctuations, and friction, and it is the premaster of the archetypal qubit. After an introduction into the model in Chapter 18, the discussion in Chapter 19 is focused on equilibrium properties for a general form of the system-bath coupling. In particular, the partition function is discussed and the specific heat and static susceptibility are studied. The relationship with Kondo and Ising models is explained. Chapter 20 is devoted to the issue of electron transfer in a solvent, nonadiabatic tunneling under exchange of energy, and single charge tunneling in the presence of an electromagnetic environment. Chapter 21 deals with the basics and suitable methods relevant to the dynamics of the dissipative two-state system. Different kinds of initial preparations of the system-plus-bath complex are treated and exact formal expressions for the system's dynamics in the form of series expressions and generalized master equations are derived. Various approximation schemes useful in different regimes of the parameter space are introduced and discussed. Chapter 22 is devoted to miscellaneous dynamical aspects of the dissipative two-state system. Ample space is given to the discussion of non-equilibrium and equilibrium correlation functions, and to adequate approximation schemes in the various regions of the parameter space. The effects of $1/f$ noise, sub-Ohmic, Ohmic and super-Ohmic friction and noise are discussed. Part IV closes with a chapter on the dynamics and internal energy flow of the time-periodically driven dissipative two-state system.

The last part reviews dissipative quantum transport of a quantum Brownian particle in a tilted cosine potential. In Chapter 24, I introduce the weak- and strong-tunneling representations of the respective global model. Chapters 25 and 26 provide an outline of the nonequilibrium quantum transport formalism and derivation of exact formal expressions describing the system's dynamics for factorizing and thermal initial states. Explicit results in analytic form are given in Chapter 27 for Ohmic friction in various limits. Chapter 28 contains a discussion of the duality symmetry between the weak- and strong-tunneling representations which becomes an exact self-duality in the so-called Ohmic scaling limit. I show that self-duality offers the possibility to construct the exact scaling function at zero temperature for the nonlinear mobility. Chapter 29 presents in analytic form the full counting statistics at zero temperature and the leading thermal correction in the Ohmic scaling limit. In Chapter 30, I discuss a route in which (i) the partition function on a ring with a magnetic charge at the origin is solved with the aid of Jack polynomials, and (ii) the nonlinear mobility is found via a conjectured relation with the partition function and by analytic continuation of the winding number to the physical bias.

In Chapter 31, I am concerned with quantum transport of charge in quantum impurity models. Both the weak- and strong-tunneling representations are discussed, and the close relationship of the quantum impurity model with the Brownian particle model, and with charge transport in a coherent conductor and in Josephson systems is pointed out. This allows to translate results obtained for one of these system in corresponding results for other systems which are related by a map of the parameters. The book closes with a chapter on the operation of a driven nonlinear quantum Brownian particle as a work-to-work converter, and on the various energy transport channel in this system.

Throughout the book I have tried to concentrate on models which are simple enough to be largely tractable by means of analytical methods. There are, however, important examples where numerical computations have given clues to the analytical solution of a problem. If one wishes to calculate the full dynamics of the global system, one is faced with the problem that the number of basis states is growing exponentially. Therefore, even on supercomputers, the number of reservoir modes which can be treated numerically exactly, is rather limited. When the number of bath modes is above ten or even tends to infinity, an inclusive description of the environmental effects, e.g., in terms of the influence functional method (cf. Chapters 4 and 5) is indispensable. Various numerical schemes developed within the framework of the influence functional approach are available. The most valuable numerical tool in many-body quantum theory is probably the path integral Monte Carlo simulation method. Unfortunately, in simulations of the real-time quantum dynamics, the numerical stability of long-time propagation is spoilt by the destructive interference of different paths contributing to the path sum. This so-called *dynamical sign problem* is intrinsic in real-time quantum mechanics, and is characterized by an exponential drop of the signal-to-noise ratio with increasing propagation time.

In recent years, considerable progress in reducing the sign problem has been achieved. Promising routes have been, e.g., blocking algorithms in quantum Monte Carlo simulations, iterative procedures such as the tensor-propagator approach, which relies on a maximal memory time of the bath correlation function, and techniques for stochastic unraveling of the influence functional. I have refrained from adding sections which deal with numerical methods in detail. Where appropriate, I have tried to provide relevant informations and literature.

After all, the reader may not find a comprehensive account of what interests him most. Since the number of articles in this general field has become enormous in recent years, a somewhat arbitrary choice among the various efforts is inevitable. My choice of topics is just one possibility. It reflects, to some extent, the author's personal valuation of an active and rapidly developing area in science.

PART I:

GENERAL THEORY
OF OPEN QUANTUM SYSTEMS

Often in condensed phases, a rather complex physical situation can adequately be described by a global model system consisting of only one or few relevant dynamical variables in contact with a huge environment, of which the number of degrees of freedom is very large or even infinity. If we are interested in the physical properties of the *small* relevant system alone, we have to handle this system as an *open system* which exchanges energy with its surroundings in a random manner. In the last fifty years, a great variety of different theoretical methods for open quantum systems has been developed and employed. In this book, emphasis is put on the functional integral approach to open quantum systems. Over the years, this method has turned out to be very powerful and has found broad application. Nevertheless, I find it appropriate to begin with a survey of various other formalisms. Clearly, the discussion given subsequently can not do justice to all of them. However, I hope that the interested reader will be able to get a line along the given references for deeper studies. I find it appropriate to begin with a brief discussion of the classical regime.

2 Diverse limited approaches: a brief survey

2.1 Classical Langevin equation and quantum mechanical generalization

2.1.1 Langevin equation for a classical damped system

It is our everyday experience that the motion of any macroscopic physical system comes to a stop when supply of energy is cut off. The reason for this is energy dissipation. In fact, there is no physical system that can completely be isolated from the surroundings. Therefore, energy accumulated in the system is inevitably given away to the environment until it reaches an equilibrium state. For instance, a particle in a fluid collides with the surrounding molecules. Thereby, momentum and energy of the particle is transferred to the fluid and the velocity of the particle declines. Eventually, the particle is trapped in a local minimum of the potential. The loss of energy is phenomenologically described by a smooth frictional force. Since the energy exchange goes on randomly, the systematic force is superimposed by a random force, which is zero on average. Now, a fundamental principle of classical and quantum equilibrium thermodynamics asserts that absorption of energy and fluctuations of the random force, usually called Langevin force, are closely related.

Consider for simplicity an open system with a single degree of freedom, which we may associate with the coordinate $q(t)$ of a particle with mass M. The simplest assumptions about the dissipative process one can make is that dissipation is state-independent. Then the frictional force is a *linear* functional of the history of the velocity $\dot{q}(t)$, and the Langevin force $\xi(t)$ obeys stationary Gaussian statistics. This means that $\xi(t)$ is fully characterized by the classical ensemble averages

$$\langle \xi(t) \rangle_{\mathrm{cl}} = 0 , \qquad \langle \xi(t)\xi(0) \rangle_{\mathrm{cl}} \equiv \mathcal{K}_{\mathrm{cl}}(t) . \tag{2.1}$$

Gaussian statistics means that all ensemble averages of higher order with an odd number of ξ-functions are zero, whereas those with an even number of ξ-functions can be written as a sum of combinations of factorized pair correlations $\mathcal{K}_{\mathrm{cl}}(t)$.

A classical heat bath at temperature T with zero memory time constitutes a white noise source. Then the frictional force is local in time, $F_{\mathrm{fric}}(t) = -M\gamma\dot{q}(t)$, where γ is the damping rate, and the Langevin force $\xi(t)$ is δ-correlated according to

$$\mathcal{K}_{\mathrm{cl}}(t) = 2M\gamma k_{\mathrm{B}}T \, \delta(t) . \tag{2.2}$$

A damped particle in a potential $V(q)$ is described by the Langevin equation

$$M\ddot{q}(t) + M\gamma \, \dot{q}(t) + V'[q(t)] = \xi(t) . \tag{2.3}$$

Time-local friction proportional to the velocity is usually called Ohmic because of the correspondence with a series resistor in an electrical circuit. Eq. (2.3) along with the relations (2.1) and (2.2) forms the basis of the theory of Brownian motion since the seminal studies by Einstein, Langevin and Smoluchowski. The early theoretical work on Brownian motion was reviewed in an excellent article by Chandrasekhar [7].

In many cases of practical interest, the environment is reacting with delay: the friction force depends on the history, and the noise is colored. The relevant dynamical equation with initial state at $t = 0$ is the generalized classical Langevin equation

$$M\ddot{q}(t) + M \int_0^t ds\, \gamma(t - s)\, \dot{q}(s) + V'[q(t)] = \xi(t) . \tag{2.4}$$

Since the random force $\xi(t)$ has zero mean, the effect of the reservoir on average is in the memory-friction kernel $\gamma(t)$ obeying causality, $\gamma(t) = 0$ for $t < 0$,[1]

$$\gamma(t) = \frac{1}{2\pi} \int_{-\infty}^{\infty} d\omega\, \tilde{\gamma}(\omega)\, e^{-i\omega t}, \qquad \tilde{\gamma}(\omega) \equiv \tilde{\gamma}'(\omega) + i\tilde{\gamma}''(\omega) = \int_0^{\infty} dt\, \gamma(t)\, e^{i\omega t} . \tag{2.5}$$

The damping function $\tilde{\gamma}(\omega)$ has three important properties which are directly connected with three fundamental physical principles. First, $\tilde{\gamma}(\omega)$ is analytic in the upper half-plane $\text{Im}(\omega) > 0$ and therefore satisfies Kramers-Kronig relations [cf. Eq. (6.31)]. This is an implication of causality, $\gamma(t) = 0$ for $t < 0$. Second, $\tilde{\gamma}'(\omega) > 0$. This is a consequence of the second law of thermodynamics. Third, $\tilde{\gamma}(\omega) = \tilde{\gamma}^*(-\omega)$. This holds because of the physical condition that the memory-friction kernel $\gamma(t)$ is a real function. As a result of these properties, $\gamma(t)$ has also the spectral representations

$$\gamma(t) = \Theta(t) \frac{2}{\pi} \int_0^{\infty} d\omega\, \tilde{\gamma}'(\omega) \cos(\omega t) = \Theta(t) \frac{2}{\pi} \int_0^{\infty} d\omega\, \tilde{\gamma}''(\omega) \sin(\omega t) . \tag{2.6}$$

The spectral damping function $\tilde{\gamma}(\omega)$ is connected with the force autocorrelation function $\langle \xi(t)\, \xi(s) \rangle_{\text{cl}} = \langle \xi(t - s)\, \xi(0) \rangle_{\text{cl}}$ by the Green-Kubo formula [8]

$$\tilde{\gamma}(\omega) = \frac{1}{Mk_{\text{B}}T} \int_0^{\infty} dt\, \langle \xi(t)\, \xi(0) \rangle_{\text{cl}}\, e^{i\omega t} . \tag{2.7}$$

From this we infer that the power spectrum of the classical stochastic force

$$\tilde{\mathcal{K}}_{\text{cl}}(\omega) = \int_{-\infty}^{\infty} dt\, \mathcal{K}_{\text{cl}}(t) \cos(\omega t) = \int_{-\infty}^{\infty} dt\, \langle \xi(t)\, \xi(0) \rangle_{\text{cl}} \cos(\omega t) \tag{2.8}$$

is related to the real part $\tilde{\gamma}'(\omega)$ of the damping function $\tilde{\gamma}(\omega)$ by the relation

$$\tilde{\mathcal{K}}_{\text{cl}}(\omega) = 2Mk_{\text{B}}T\, \tilde{\gamma}'(\omega) . \tag{2.9}$$

This is a version of the classical fluctuation-dissipation or Nyquist theorem. The Langevin equation (2.4) will be derived from a global system-plus-reservoir Hamiltonian in Subsection 3.1.2. Evidently, Eq. (2.4) with (2.9) is limited to the classical domain. However, we should expect that at sufficiently low temperatures all types of quantum effects will occur. Since the standard procedure of quantization draws upon the existence of a Lagrangian or a Hamiltonian function for the system, the question arises: how can one reconcile dissipation with the canonical scheme of quantization?

---

[1] Throughout, Fourier transformation has normalization and sign convention as in Eq. (2.5). Where appropriate, Fourier transforms are marked with a tilde and Laplace transforms with a hat.

2.1.2 Quantum mechanical and quasiclassical Langevin equation

Quantum Langevin equation

A natural and promising attempt towards quantum mechanics consists in a quantum mechanical generalization of the classical Langevin equation (2.4) [21, 51, 52, 81]. The quantum Langevin equation (QLE) is the Heisenberg equation of motion for the coordinate operator $\hat{q}(t)$. With the initial state at time zero, it formally coincides with the generalized classical Langevin equation (2.4),

$$M\ddot{\hat{q}}(t) + M\int_0^t ds\,\gamma(t-s)\,\dot{\hat{q}}(s) + V'[\hat{q}(t)] = \hat{\xi}(t) . \tag{2.10}$$

Here, $\hat{\xi}(t)$ is the Gaussian random force operator with symmetric autocorrelation

$$\tfrac{1}{2}\langle\,\hat{\xi}(t)\hat{\xi}(0) + \hat{\xi}(0)\hat{\xi}(t)\,\rangle_\beta = \frac{M}{\pi}\int_0^\infty d\omega\,\hbar\omega\,\tilde{\gamma}'(\omega)\coth(\tfrac{1}{2}\beta\hbar\omega)\cos(\omega t) . \tag{2.11}$$

and the nonequal-time commutator

$$[\hat{\xi}(t),\,\hat{\xi}(0)] = -i\frac{2M}{\pi}\int_0^\infty d\omega\,\hbar\omega\,\tilde{\gamma}(\omega)\sin(\omega t) . \tag{2.12}$$

Piecing Eqs. (2.11) and (2.12) together, the quantum-statistical force correlation is

$$\mathcal{K}(t) \equiv \langle\,\hat{\xi}(t)\hat{\xi}(0)\,\rangle_\beta = \frac{\hbar M}{\pi}\int_0^\infty d\omega\,\omega\tilde{\gamma}'(\omega)[\coth(\tfrac{1}{2}\beta\hbar\omega)\cos(\omega t) - i\sin(\omega t)] , \tag{2.13}$$

and the power spectrum of the random force is

$$\tilde{\mathcal{K}}(\omega) \equiv \tfrac{1}{2}\int_{-\infty}^\infty dt\,[\langle\hat{\xi}(t)\hat{\xi}(0)\rangle_\beta + \langle\hat{\xi}(0)\hat{\xi}(t)\rangle_\beta]\,e^{i\omega t} = M\hbar\omega\coth(\tfrac{1}{2}\beta\hbar\omega)\,\tilde{\gamma}'(\omega) . \tag{2.14}$$

The dynamical equation (2.10) is a Langevin equation with a linear memory-friction force and a random force operator representing additive stationary Gaussian noise. Expression (2.11) for the symmetrized force auto-correlation function, and concomitant with this, expression (2.14) for the power spectrum of the stochastic force are general implications of the quantum fluctuation-dissipation theorem (cf. Sec. 6.1). They are therefore independent of the model. On the other hand, the Gaussian property of the stochastic force operator $\hat{\xi}(t)$ strictly holds only when the reservoir is harmonic.

The QLE (2.10) can be derived for a system which is bilinearly coupled to a bath of harmonic oscillators. The derivation is given in Section 3.1. The dependences of the force operator $\hat{\xi}(t)$ on the initial conditions of the system $\hat{q}(0)$ and of the set of bath positions $\{\hat{x}_\alpha(0)\}$ and bath momenta $\{\hat{p}_\alpha(0)\}$, and subtleties of the thermal average $\langle\cdots\rangle_\beta$ in Eq. (2.11) are discussed in Subsection 3.1.4. An attempt for an axiomatic construction of quantum Langevin equations is presented in Ref. [53].

Benguria and Kac [54], and Ford and Kac [51] argued that a system which obeys the QLE (2.10) with Gaussian noise characteristics (2.11) and commutator (2.12) approaches the proper equilibrium state.

In the strictly Ohmic case, $\tilde{\gamma}(\omega) = \gamma$, damping is memoryless, $\gamma(t) = 2\gamma\,\Theta(t)\delta(t)$, whereas the spectral noise (2.14) is still colored and thus the symmetrized force auto-correlation function (2.11) continues to be not delta-correlated. The white-noise (Markovian) regime is reached in the classical limit $\hbar \to 0$, in which the commutator (2.12) vanishes and the stochastic force is delta-correlated as in Eq. (2.2).

Quasiclassical Langevin equation

The practical implementation of the generally nonlinear QLE is severely impeded because it is an operator equation in the Hilbert space of system and reservoir. A manifest approximation now is (i) to abandon the operator character of the QLE and (ii) to replace the quantum noise by colored classical noise. The so-called *quasiclassical Langevin equation* [20, 55] coincides with the classical Langevin equation (2.4),

$$M\ddot{q}(t) + M \int_0^t ds\,\gamma(t-s)\,\dot{q}(s) + V'[q(t)] \;=\; \xi(t)\,, \qquad (2.15)$$

but the power spectrum of the noise force $\xi(t)$ matches with that of quantum noise,[2]

$$\tilde{K}_{\mathrm{qucl}}(\omega) \;\equiv\; \int_{-\infty}^{\infty} dt\,\langle\,\xi(t)\,\xi(0)\,\rangle_\beta \cos(\omega t) \;=\; M\hbar\omega \coth\!\left(\frac{\hbar\omega}{2k_{\mathrm{B}}T}\right)\tilde{\gamma}'(\omega)\,. \qquad (2.16)$$

While in the QLE the operators $\hat{q}(t)$ and $\hat{p}(t)$ obey the Heisenberg uncertainty relations for all times, this feature is abandoned in the quasiclassical Langevin equation. The quasiclassical Langevin equation is an exact dynamical equation for *linear* damped quantum systems. It yields a reasonable description for systems which are nearly harmonic [56, 57]. However, the predictions of the quasiclassical Langevin equation are unreliable when the anharmonicity of the potential is decisive like, for instance, in quantum tunneling events. The conclusion of a detailed study is that use of the colored noise (2.16) in a classical Langevin equation is insufficient to render a proper description of the quantum statistical decay of a metastable state [56].

When the anharmonicity of the potential is pivotal, the most successful approach is the functional integral method. Like in the classical regime, the dissipative system is considered to interact with a complex environment, and the "complete universe" formed by the system plus environment is assumed to be energy-conserving so that it can be quantized in the standard way. For equations of motion which are linear in the bath coordinates, the environment can easily be eliminated. Thus one obtains closed equations for the damped system alone. In the path integral representation, the environment reveals itself through an influence functional which depends on the spectral properties of the environmental coupling and on temperature. A general discussion of the influence functional method is presented in Chapters 4 and 5.

---

[2]The derivation of the QLE within the path integral method, following the approach by Schmid [55], is presented in Sec. 5.3.3.

2.2 New schemes of quantization

The equation of damped motion (2.3) can not be obtained from the application of Hamilton's principle unless the Lagrangian has a particular time dependence. The use of time-dependent Lagrangians or Hamiltonians would permit us to use the standard schemes of quantization directly. Historically, the first researchers taking this path were Caldirola [9] and Kanai [10] who employed a time-dependent mass chosen in such a way that a friction term appears in the classical equation of motion. However, the commutator of the corresponding momentum and position operator decays exponentially with time, and hence the uncertainty principle is violated [11]. In a study by Schuch, explicitly time-dependent Hamiltonians were related to nonlinear Schrödinger equations, and violation of the uncertainty principle was obliterated by a noncanonical transformation of variables q and p and a nonunitary transformation of the wave function [12]. Nevertheless, it is generally accepted meanwhile that dissipation cannot be described adequately by simply employing a time-dependent mass.

Many approaches to open quantum systems were introduced over the last forty to fifty years. The variety of historical attempts falls into three main categories. One either modifies the procedure of quantization, or one trusts in heuristic stochastic Schrödinger equations for state vectors, or one starts out from the system-plus-reservoir approach.

Among the first group, Dekker [13] proposed a theory with a canonical quantization procedure for complex variables, thereby reproducing the Fokker-Planck equation for the Wigner distribution function. However, some *ad hoc* assumptions in the theory seem questionable, such as the introduction of noise sources in the canonical equations for position and momentum. Kostin [14] introduced a theory with a nonlinear Schrödinger equation. The same equation was found later by Yasue [15] using Nelson's stochastic quantization procedure [16]. However, this theory violates the superposition principle. It also yields some dubious results such as stationary undamped states. Apart from the fact that the theoretical foundations are completely unclear, these approaches can reproduce, at best, known results only for very limited cases, such as weakly damped linear systems. Therefore, all attempts of the first group can be assessed to have failed. I shall not consider them further here. The second category is discussed in some detail in Section 2.4.

The more natural and most successful approach has been to regard the dissipative system as a relevant system with a single or few significant degrees of freedom, which is in contact with (infinitely) many degrees of freedom. These additional degrees of freedom are commonly referred to as bath, or reservoir or environment. Both the relevant system and the reservoir are the constituents of an energy-conserving global system which obeys the standard rules of quantization. In this picture, friction comes about by the transfer of energy from the "small" system to the "huge" environment. Broadly speaking, the energy, once transferred, dissipates into the environment and is not given back within any physically relevant period of time.

2.3 Traditional system-plus-reservoir methods

Common approaches to open quantum systems based on system-plus-reservoir models are generally divided into two classes. Working in the Schrödinger picture, the dynamics is conventionally described in terms of generalized quantum master equations for the reduced density matrix or density operator [17]–[19]. Working in the Heisenberg picture, the description is given in terms of generalized Langevin equations for the relevant set of operators of the reduced system [20]–[22].

2.3.1 Quantum-mechanical master equations for weak coupling

The starting point of this method is the familiar Liouville equation of motion for the density operator $W(t)$ of the global system,

$$\dot{W} = -(i/\hbar)\,[\,H, W(t)\,] \equiv \mathcal{L}\,W(t)\,, \tag{2.17}$$

where H is the Hamiltonian of the total system, and where the second equality defines the Liouville operator $\mathcal{L}$. Next, assume that the Hamiltonian H and the Liouvillian $\mathcal{L}$ of the total system are decomposed as

$$H = H_\mathrm{S} + H_\mathrm{R} + H_\mathrm{I}\,, \qquad \mathcal{L} = \mathcal{L}_\mathrm{S} + \mathcal{L}_\mathrm{R} + \mathcal{L}_\mathrm{I}\,. \tag{2.18}$$

The individual parts refer to the free motion of the relevant system and of the reservoir, and to the interaction term, respectively. Employing a certain projection operator P, chosen as to project on the relevant part of the density matrix, the full density operator is reduced to an operator acting only in the space of the relevant variables,

$$\rho(t) = PW(t)\,. \tag{2.19}$$

The operator $\rho(t)$ is usually called *reduced density* operator. For systems with the Hamiltonian form (2.18), the projection operator contains a trace operation over the reservoir coordinates. By means of the projection operator P, the density operator can be decomposed into the relevant part $\rho(t)$ and the irrelevant part $(1-P)W(t)$,

$$W(t) = \rho(t) + (1-P)W(t)\,, \qquad P^2 = P\,. \tag{2.20}$$

Upon substituting the decomposition (2.20) into Eq. (2.17), and acting on the resulting equation from the left with the operator P and with the operator $1-P$, respectively, we obtain two coupled equations for the relevant part $\rho(t)$ and the irrelevant part $(1-P)W(t)$. A closed equation for $\rho(t)$ is obtained by inserting the formal integral for $(1-P)W(t)$ into the first equation. We then finally arrive at the formally exact generalized quantum master equation, the *Nakajima-Zwanzig equation* [17, 18]

$$\dot{\rho}(t) = P\mathcal{L}\,\rho(t) + \int_0^t ds\, P\mathcal{L}\,e^{(1-P)\mathcal{L}s}\,(1-P)\mathcal{L}\,\rho(t-s) \\ + P\mathcal{L}\,e^{(1-P)\mathcal{L}t}\,(1-P)W(0)\,. \tag{2.21}$$

The generalized master equation is an inhomogeneous integro-differential equation in time. It describes the dynamics of the open (damped) system in contact with the reservoir $\mathcal{R}$. Observe that the inhomogeneity in Eq. (2.21) still depends on the initial value of the irrelevant part $(1 - P)W(0)$. In applications, it is attempted to choose the projection operator in such a way that the irrelevant part of the initial state $(1 - P)W(0)$ can be disregarded. Assuming further that P commutes with $\mathcal{L}_S$, one then finds the homogeneous time-retarded quantum master equation

$$\dot{\rho}(t) = P(\mathcal{L}_S + \mathcal{L}_I)\rho(t) + \int_0^t ds\, P\mathcal{L}_I\, e^{(1-P)\mathcal{L}s}\,(1 - P)\mathcal{L}_I\rho(t - s) \,. \qquad (2.22)$$

The first (instantaneous) term describes the reversible motion of the relevant system, while the second (time-retarded) term brings on irreversibility. It includes all effects the reservoir may exert on the system, such as relaxation, decoherence and energy shifts. Eq. (2.22) is still too complicated for explicit evaluation. First, the kernel of (2.22) contains any power of $\mathcal{L}_I$. Secondly, the dynamics of ρ at time t depends on the whole history of the density matrix. In order to surmount these difficulties, one usually considers the kernel of Eq. (2.22) only to second order in $\mathcal{L}_I$. Disregarding also retardation effects, one finally arrives at the quantum master equation in Born-Markov approximation

$$\dot{\rho}(t) = P(\mathcal{L}_S + \mathcal{L}_I)\rho(t) + \int_0^t ds\, P\mathcal{L}_I\, e^{(1-P)(\mathcal{L}_S + \mathcal{L}_R)s}\,(1 - P)\mathcal{L}_I\,\rho(t) \,. \qquad (2.23)$$

Master equations of this form were successfully used to describe weak-damping phenomena, for instance in quantum optics or spin dynamics. Various excellent reviews of this sort of approach including many applications are available in Refs. [23] – [31]. While the Markov assumption can easily be dropped, the more severe limitation of this method is the Born approximation for the kernel. The truncation of the Born series at second order in the interaction $\mathcal{L}_I$ effectively restricts the application of the master equation (2.23) or of its non-Markovian generalization to weakly damped systems with relaxation times that are large compared to the relevant time scales of the reversible dynamics.

When the Born-Markov quantum master equation (2.23) is given in the energy eigenstate basis of H_S, it is usually referred to as *Redfield* equation [32, 28, 33]

$$\dot{\rho}_{nm}(t) = -i\,\omega_{nm}\,\rho_{nm}(t) - \sum_{k,\ell} R_{nmk\ell}\,\rho_{k\ell}(t) \,. \qquad (2.24)$$

The first term represents the reversible motion in terms of the transition frequencies ω_{nm}, and the second term describes relaxation. The Redfield relaxation tensor reads

$$R_{nmk\ell} = \delta_{\ell m}\sum_r \Gamma^{(+)}_{nrrk} + \delta_{nk}\sum_r \Gamma^{(-)}_{\ell rrm} - \Gamma^{(+)}_{\ell mnk} - \Gamma^{(-)}_{\ell mnk} \,. \qquad (2.25)$$

The rates are given by the Golden Rule expressions

$$\Gamma^{(+)}_{\ell mnk} = \frac{1}{\hbar^2}\int_0^\infty dt\; e^{-i\omega_{nk}t} \left\langle \tilde{H}_{I,\ell m}(t)\tilde{H}_{I,nk}(0)\right\rangle_R ,$$

$$\Gamma^{(-)}_{\ell mnk} = \frac{1}{\hbar^2}\int_0^\infty dt\; e^{-i\omega_{\ell m}t} \left\langle \tilde{H}_{I,lm}(0)\tilde{H}_{I,nk}(t)\right\rangle_R . \qquad (2.26)$$

Here, $\tilde{H}_I(t) = e^{iH_R t/\hbar} H_I e^{-iH_R t/\hbar}$ is the interaction in the interaction picture, and the angular bracket denotes thermal average of the bath degrees of freedom.

The Redfield equations (2.24) are well-established in wide areas of physics and chemistry, e.g., in nuclear magnetic resonance (NMR), in optical spectroscopy, and in laser physics. In NMR, one deals with the externally driven dynamics of the density matrix for the nuclear spin [34, 35, 36]. In optical spectroscopy, a variant of the Redfield equations are the optical Bloch equations [37, 38].

Multilevel Redfield theory has been applied to the electron transfer dynamics in condensed phase reactions by several authors [39].

Markovian reduced density matrix (RDM) theory has been also utilized in the diabatic state representation [40]. Denoting electronic-vibrational direct-product states in the diabatic representation[3] by α, β, $\cdots$, the RDM equations of motion read

$$\dot{\rho}_{\alpha\beta}(t) = -i\omega_{\alpha\beta}\rho_{\alpha\beta}(t) - \frac{i}{\hbar}\sum_{\nu}[V_{\alpha\nu}\rho_{\nu\beta}(t) - V_{\nu\beta}\rho_{\alpha\nu}(t)] - \sum_{\nu,\sigma}R_{\alpha\beta\nu\sigma}\rho_{\nu\sigma}(t). \quad (2.27)$$

The $\omega_{\alpha\beta}$ are the transition frequencies between unperturbed diabatic surfaces, the $V_{\alpha\beta}$ are the matrix elements of the diabatic interstate coupling, and the relaxation tensor $R_{\alpha\beta\nu\sigma}$ describes relaxation of the diabatic electronic-vibrational states. For electron transfer processes, the diabatic surfaces can often be taken as harmonic, which simplifies the calculation of the relaxation tensor drastically. The master equation (2.27) in the diabatic basis is especially useful when the interstate coupling is weak or moderate. In contrast to Redfield theory in the system's eigenstate basis, the computation of the Redfield tensor in the diabatic basis, Eq. (2.27), is without difficulties even for multimode vibronic-coupling systems [43]. When the RDM approach is applied to complex systems, the dimension N of the Hilbert space of the relevant system H_S is possibly very large. Since the density matrix scales with N^2 and the relaxation tensor with N^4, the computational problem may become easily nontrivial.

The so-called Monte-Carlo wave function propagation or quantum jump method discussed below in Section 2.4 is equivalent to the Born-Markov RDM approach. Most importantly, this method provides *linear* scaling of the computational costs with the number of states, and is therefore clearly preferable to direct integration of the Redfield equations when the number N of relevant states is very large.

2.3.2 Lindblad theory

Within the conception of quantum mechanics, the time evolution of the density operator of a closed system is a unitary map. If the system is open, the possible transformations are thought to be "completely positive" [44, 30], $\rho \rightarrow \sum_n O_n \rho O_n^\dagger$. Here $\{O_n\}$ is a set of linear operators on the reduced state space, restricted only by $\sum_n O_n^\dagger O_n = 1$, which guarantees that $\text{tr}\,\rho$ does not change. The most general form of generators $\mathcal{L}$, $\dot{\rho}(t) = \mathcal{L}\rho(t)$, preserving trace and complete positivity of density operators, and conveying time-directed irreversibility, is established by the Lindblad theory [44]. The

---

[3]The concepts of diabatic states are discussed in recent reviews [41, 42, 43].

Lindblad form of the quantum master equation for an open system reads

$$\frac{d\rho(t)}{dt} = -\frac{i}{\hbar}[H_{\mathrm{S}}, \rho(t)] + \frac{1}{2}\sum_j \left\{ [L_j\rho(t), L_j^\dagger] + [L_j, \rho(t)L_j^\dagger] \right\}. \qquad (2.28)$$

The first term represents the reversible dynamics of the relevant system. The Lindblad operators L_j carry the impact of the environment on the system in Born-Markov approximation. In concrete applications of Eq. (2.28), the L_j transmit emission and absorption processes. For linear dissipation, the simplest form of a Lindblad operator is a linear combination of a raising and lowering operator, or equivalently of a coordinate and a momentum operator [45],

$$L = \mu q + i\nu p, \qquad L^\dagger = \mu q - i\nu p, \qquad (2.29)$$

where μ and ν are real coefficients. Explicit temperature-dependent expressions for μ and ν are obtained for the damped harmonic quantum oscillator by matching the dissipation terms of the Lindblad master equation with corresponding expressions of this exactly solvable model (cf. Chapter 6) in the Born-Markov limit [46].

Complete positivity of reduced density operators is considered as a strict guideline by researchers working on stochastic Schrödinger equations (see Section 2.4). However, as argued by Pechukas [47, a], complete positivity of the RDM is an artefact of product initial conditions $W^{(i)} = \rho^{(i)} \otimes \rho_{\mathrm{R}}^{(i)}$ for the global density matrix. Product initial conditions are only appropriate in the weak-coupling limit. In general, reduced dynamics need not be completely positive. It is known that the Redfield-Bloch master equation may break the positivity of the density matrix. Violation of positivity may occur if memory effects are not adequately taken into account at the initial stage of the time evolution. Then some "slippage" of the initial conditions must occur before the reduced dynamics looks Markovian [47, b]. The slippage captures the effects of the actual non-Markovian evolution in a short transient regime of the order of the reservoir's memory time (see also [47, c]). The Markovian master equation holds only in a subspace in which rapidly decaying components of the density matrix are disregarded, whereas Lindblad theory requires validity for *any* reduced density matrix. These findings have been confirmed in a study of the exactly solvable damped harmonic oscillator [48]. Using the exact path integral solution, it has been shown that in general there is no exact dissipative Liouville operator describing the dynamics of the oscillator in terms of an exact master equation that is independent of the initial preparation. Exact non-stationary Liouville operators can be found only for particular preparations. Time-independent Liouville operators which are valid for arbitrary preparation of the initial state can be extracted only in the sub-space left after the fast transient components have died out. However, the Liouville operators are still not yet in Lindblad form. The Lindblad master equation is gained only after the weak-coupling limit is performed and a coarse graining in time is carried out. Discussion of this issue is given in Section 6.8. It is perfectly obvious from the study of this exactly solvable linear model that the reduced dynamics in general does not constitute a dynamical semigroup. This shows that the Lindblad master equation has no fundamental significance from a physical point of view.

Besides the Markov approximation and the weak-coupling limit, there is another severe limitation of the applicability of the above quantum master equations. These equations do not take into account relaxation contributions with time scales of the inverse Matsubara frequencies, $\sum_n a_n e^{-|\nu_n|t}$ ($\nu_n = 2\pi n/\hbar\beta$). These originate from the fluctuation dissipation theorem (see, e.g., the discussion in Chapters 6 and 17). Hence the relaxation dynamics towards the equilibrium state ρ_β is correctly described by master equations of Lindblad form only when the system's relaxation times are large compared to the thermal time $\hbar\beta$.

In conclusion, the Born-Markov quantum master equation approach provides a reasonable description in many cases, e. g., in NMR, in quantum optics, and in a variety of chemical reactions. However, this method is not applicable in many problems in solid state physics at low temperature in which neither the Born approximation is valid nor the Markov assumption holds.

2.3.3 Operator Langevin equations for weak coupling

Just as we projected the density matrix $W(t)$ of the global system onto the relevant part $\rho(t)$, we may proceed in the Heisenberg picture by projecting the operators of the global system on the set of macroscopically relevant operators. The various efforts in studying the dynamics of these operators have been described by Gardiner [50].

Let us denote the set of operators of the global system by $\{X\}$ and the set of macroscopically relevant operators governing the open system by $\{Y\}$. Now consider the operators X_i and Y_μ as elements $|X_i)$ and $|Y_\mu)$ in the Liouville space Λ.[4] At time $t = 0$, the operators $\{Y\}$ span a subspace Λ_Y of the Liouville space Λ. I use the convention that an operator in the Heisenberg representation without time argument denotes the operator at time zero. Next, it is convenient to define a time-independent projection operator which projects onto the subspace Λ_Y,

$$\mathcal{P} = \sum_{\mu,\nu} |Y_\mu) \, g_{\mu\nu} \, (Y_\nu| \; ; \qquad \mathcal{P}^2 = \mathcal{P} \, . \tag{2.30}$$

The metric $g_{\mu\nu}$ is the inverse of the scalar product $(Y_\mu|Y_\nu)$ which has to be chosen appropriately in practical calculations. For quantum statistical linear response and relaxation problems, a suitable form is the Mori scalar product [22, 49]

$$(Y_\mu|Y_\nu) \equiv \frac{1}{\beta} \int_0^\beta d\lambda \left\langle e^{-\lambda H} Y_\mu^\dagger e^{\lambda H} Y_\nu \right\rangle \, . \tag{2.31}$$

The angular brackets denote average with respect to the canonical ensemble of the global system $W_\beta = Z_\beta^{-1} e^{-\beta H}$, and $\beta = 1/k_B T$. The superoperator $\mathcal{P}$ projects onto the subspace Λ_Y according to

---

[4]See Ref. [49] for a review of the formulation of quantum mechanics in Liouville space.

$$P|X_i) = \sum_{\mu,\nu} |Y_\mu) g_{\mu\nu} (Y_\nu|X_i) .\tag{2.32}$$

Acting from the left with $\mathcal{P}$ and with $1 - \mathcal{P}$ on the Heisenberg equation of motion

$$|\dot{X}_i) = \mathcal{L}|X_i) ,\tag{2.33}$$

where $\mathcal{L}$ is the Liouville superoperator, and eliminating $(1 - \mathcal{P})|X_i)$ with the aid of the exact formal solution, it is straightforward to derive for $Y_\mu(t)$ a system of coupled integro-differential equations, which have been popularized as the Mori equations,

$$\dot{Y}_\mu(t) = i \sum_\nu Y_\nu(t) \Omega_{\nu\mu}(t) - \sum_\nu \int_0^t ds\, \gamma_{\nu\mu}(s) Y_\nu(t-s) + \xi_\mu(t) .\tag{2.34}$$

The generally temperature-dependent drift matrix $\Omega_{\nu\mu}(t)$ is given by

$$i\,\Omega_{\nu\mu}(t) = \sum_\rho g_{\nu\rho} (Y_\rho|\dot{Y}_\mu(t)) .\tag{2.35}$$

The stochastic force $\xi_\nu(t)$ is a functional of the operators of the *irrelevant* part,

$$\xi_\nu(t) = \exp[(1 - \mathcal{P})\mathcal{L}\,t]\,(1 - \mathcal{P})\dot{X}_\nu(t) .\tag{2.36}$$

Finally, the memory matrix is expressed in terms of the correlation function of the stochastic force as

$$\gamma_{\nu\mu}(t) = \sum_\rho g_{\nu\rho}(\xi_\rho|\xi_\mu(t)) .\tag{2.37}$$

The actual computation of the fluctuating force and of the memory matrix is again restricted to weak coupling. Altogether, the operator Langevin method is subject to exactly the same limitations we encountered above in the master equation approach.

In conclusion, it is important in the Mori formalism that the complete set of macro variables spans the subspace. Otherwise, the fluctuating force contains slowly varying components, and the separation of time scales is incomplete.

2.3.4 Phenomenological methods

Often, a physical or chemical system cannot be characterized by a simple model Hamiltonian of the form (2.10), or the Hamiltonian is even unknown. In such cases, it is sometimes useful to describe dissipative quantum dynamics on a phenomenological level. For instance, one may introduce a dynamical description of the system in terms of occupation probabilities $p_n(t)$ of energy levels or of spatially localized states, rather than in terms of complex probability amplitudes or wave functions, or the full reduced density matrix ρ_{nm}. Then, the relaxation dynamics of a macroscopic system is described by a *Pauli master equation* for probability $p_n(t) \equiv \rho_{nn}(t)$,

$$\dot{p}_n(t) = \sum_m [A_{nm}\, p_m(t) - A_{mn}\, p_n(t)] .\tag{2.38}$$

The first term on the r.h.s. describes the gain and the second term the loss of probability to occupy the state n. In this formulation, knowledge of the full set of transition rates $\{A_{nm}\}$ is required in order to have a complete description of the relaxation process. The transition rates may be inferred, e. g., from standard quantum mechanical perturbation theory, or from experimental data, or they may be chosen by a phenomenological ansatz. The Pauli master equation (2.38) has found widespread application to the study of population dynamics in physics, chemical kinetics, biophysics, and biology; not to forget application to the migration dynamics of social groups and to the infection dynamics in pandemics.

In the quantum coherence regime, off-diagonal matrix elements of the density matrix become relevant. Nevertheless, the full coherent dynamics of the populations can still be formulated in terms of dynamical equations for diagonal matrix elements. However, the corresponding master equation is time-nonlocal (see Section 21.2.6).

2.4 Stochastic dynamics in Hilbert space

In recent years, there have been made numerous attempts to postulate non-Hamiltonian dynamics as a fundamental modification of the Schrödinger equation in order to explain the spontaneous stochastic collapse of the wave function and the appearance of a "classical world" (cf. for a survey the article by I.-O. Stamatescu in Ref. [58]). In these approaches, the non-unitary dynamics of open quantum systems is interpreted in terms of a fundamental stochastic process in the Hilbert space of state vectors pertaining to the system. To retain the standard probability rules, the respective dynamical equations become inevitably nonlinear. The evolution of state vectors is considered as a stochastic Markov process, and the covariance matrix of the state vector is taken as the density operator. The stochastic process is usually constructed in such a way that the equation of motion of the density operator is the familiar Markovian quantum master equation in Lindblad form (2.28), e.g., the optical Bloch equations [37, 38]. First of all, approaches of this type were introduced on phenomenological grounds. Later on, a fundamental significance has been allocated to the "stochastic Schrödinger equations" by several authors, in particular to propose explanation of the omnipresent decoherence phenomena observed in real quantum systems. The above scheme does not lead to a definite stochastic representation of the dynamics of the reduced system in Hilbert space, even though the Markov approximation is made, since the stochastic process is not unambiguously determined by merely fixing the covariance. An infinity of different realizations is possible. Basically, one may distinguish two classes of stochastic models for the evolution of state vectors in the Schrödinger picture. In the first class of stochastic Schrödinger equations, the stochastic increment is a diffusion process, the so-called Wiener process. In the second class, the evolution of the state vector is represented as a stochastic process of which the realizations are piecewise deterministic paths, and the smooth segments are interrupted by stochastic sudden jump processes [59].

The *quantum-state diffusion* (QSD) method proposed by Gisin [60] and developed further by Gisin, Percival and coworkers [61, 62] belongs to the first class. The QSD method is based upon a correspondence between the solutions of the master equation for the ensemble density operator ρ and the solutions of a Langevin-Itô diffusion equation for the normalized pure state vector $|\psi>$ of an individual system of the ensemble. If the master equation has the Lindblad form (2.28), then the corresponding QSD equation is the nonlinear stochastic differential equation

$$|d\psi> = -\frac{i}{\hbar} H|\psi> dt + \sum_j \left(<L_j^\dagger> L_j - \tfrac{1}{2} L_j^\dagger L_j - \tfrac{1}{2} <L_j^\dagger><L_j> \right) |\psi> dt$$

$$+ \sum_j (L_j - <L_j>) |\psi> d\xi_j , \qquad (2.39)$$

where $<L_j> \equiv <\psi|L_j|\psi>$ is the quantum expectation.[5] The first sum describes the nonlinear drift of the state vector in the state space and the second sum the random fluctuations. The $d\xi_j$ are complex differential variables of a Wiener process satisfying

$$\langle d\xi_j \rangle = 0 , \qquad \langle d\xi_j\, d\xi_k \rangle = 0 , \qquad \langle d\xi_j^*\, d\xi_k \rangle = \delta_{j,k}\, dt . \qquad (2.40)$$

Here $\langle \cdots \rangle$ denotes the ensemble average. The respective density operator is the ensemble average of the projector onto the quantum state,

$$\rho = \langle\, |\psi><\psi|\, \rangle . \qquad (2.41)$$

A relativistic quantum state diffusion model has been proposed in Ref. [63].

Attempts have also been made to describe the stochastic evolution of the state vector in terms of a stochastic differential equation with a linear drift [64].

The Monte-Carlo wave function simulation or *quantum jump methods* proposed by Diósi [65], by Dalibard, Castin, and Mølmer [66], by Zoller and coworkers [67], and by Carmichael *et al.* [68] belong to the second class. In these related methods, the Schrödinger equation is supplemented by a non-Hermitian term and by a stochastic term undergoing a Poisson jump process. Because of the non-unitary time evolution of the state vector under a non-Hermitian Hamiltonian, the trace of the density operator is no more conserved. Conservation of probability is restored again and again by imposing stochastically chosen quantum jumps (see Refs. [66, 67, 69]). In the Monte Carlo algorithm by Mølmer *et al.* [70], the deviation of the norm δp of the wave function from unity after a certain time step is compared with a number ε, which is randomly chosen from the interval $[0, 1]$. If $\delta p > \varepsilon$, a quantum jump occurs by which the wave function is renormalized to unity. A comparison of some of the quantum jump and state diffusion models is given in Ref. [71].

Breuer and Petruccione have shown that a unique stochastic process in Hilbert space for the dynamics of the open system may be derived directly from the underlying microscopic system-plus-reservoir model [72, 73]. They employed a description of quantum mechanical ensembles in terms of probability distributions on a projective Hilbert space. In the elimination of the reservoir, they made the Markov approximation, and they employed second-order perturbation theory in the system-reservoir

---

[5]I constantly employ the ket symbol $|\cdots>$ for a pure state, and $\langle \cdots \rangle$ for the ensemble average.

coupling. They then obtained a Liouville-master functional equation for the reduced probability distribution. The Liouville part of this equation corresponds to a deterministic Schrödinger-type equation with a non-Hermitian Hamiltonian which is intrinsically nonlinear in order to preserve the norm. The master part of this equation describes gain and loss of the probabilities for individual states due to discontinuous quantum jumps. In this description, the realization of the stochastic process is very similar to those generated by the piecewise deterministic quantum jump method [66]–[68]. Therefore, the stochastic simulation algorithms of all these approaches are very similar likewise. The equation of motion for the reduced density matrix derived from the Liouville-master equation is in the Lindblad form (2.28).

In the first place, the stochastic wave function methods are computational tools with which the solution of the Born-Markov master equation is simulated by using Monte Carlo importance-sampling techniques [69]. The stochastic methods are numerically superior to the conventional integration of the master equation when the rank of the reduced density matrix is large. Over and above the computational advantage of stochastic wave function methods for Born-Markov processes, some groups are presuming to claim that the instantaneous discontinuous processes are *real* and provide a natural description of individual quantum jump events (and not only of their statistics) as observed, e.g., in experiments with single ions in radio-frequency traps [74, 75]. Against that, I wish to point out that the assignment of definite states to a subsystem is *incompatible* with standard quantum theory, and has been proven wrong, e.g., in Einstein-Rosen-Podolsky experiments. Moreover, there is no experimental indication for non-standard phenomena (e.g., spontaneous collapse) in connection with the explanation of classical properties. Hence there is no phenomenological necessity for the introduction of a stochastic equation for state vectors. Besides computational advantages in the simulation of Born-Markov processes, the quantum-state diffusion method provides an alternative approach to measurement theory. In this method, a continuous measurement process, by which the system is steadily reduced within a certain time period to an eigenstate, is an integral part of the dynamical description. In conclusion, the stochastic wave function approaches provide efficient numerical simulation schemes for quantum master equations which are of the Lindblad form. However, since the Markov and the Born approximation are made, the application of these methods to solid state physics problems is as limited as the Born-Markov quantum master equation approach.

In the sequel we move on firm ground by taking the conservative view that the Hamiltonian dynamics of a global system induces a non-unitary dynamics for a subsystem. Upon performing a reduction of the global system to the relevant subsystem, all effects of the environmental coupling are put in an influence functional. Non-Markovian generalizations of quantum state diffusion and related stochastic Schrödinger equations can be found by a stochastic unraveling of the influence functional. The related discussion is given in Section 5.4.

3 System-plus-reservoir models

A complex quantum system in nature can often clearly be divided into a subsystem of real interest with only few relevant degrees of freedom and a mostly large residual system. Such composite system is usually referred to as a system-plus-reservoir entity, in which the system corresponds to the subsystem of interest and the reservoir to the residual system. In general, the system-reservoir coupling leads to exchange of energy between system and reservoir, which, in a phenomenological classical description, usually shows up as damping of the system. Therefore, it is important to understand, how the classical Langevin equation and its quasiclassical generalization discussed above in Section 2.1, emerge from a reduction of the system-plus-reservoir entity,

For many complex quantum systems we do not have a clear understanding of the microscopic origin of damping. In certain systems, however, it is possible to track down the power spectrum of the stochastic force in the classical regime, and hence the spectral damping function $\tilde{\gamma}(\omega)$. Therefore, it is very important to establish phenomenological system-plus-reservoir models which on the one hand reduce in the classical limit to a description in terms of a Langevin equation of the form (2.4), and on the other hand allow for full quantum mechanical treatment.

The simplest model of a dissipative quantum mechanical system that one can envisage is a damped quantum mechanical linear oscillator: a central harmonic oscillator is coupled linearly via its displacement coordinate q to a fluctuating collective coordinate of a dynamical reservoir or bath. If the equilibrium state of the bath is only weakly perturbed by the central oscillator, its classical dynamics can be represented by linear equations. Then the bath can be represented as a (infinite) set of harmonic oscillators, or as a field of free bosons. Accordingly, the noise statistics of the stochastic process induced by the reservoir is strictly Gaussian. This simple system-plus-reservoir model has been introduced and discussed in a series of four papers by Ullersma [76]. Zwanzig generalized the model to the case in which the central particle moves in an anharmonic potential and studied the classical regime [77]. Caldeira and Leggett [78] were among the first who applied this model to a study of quantum mechanical tunneling of a macroscopic variable.

In the first section of this chapter, I introduce the model and I seek the relations between the parameters of the model and the quantities appearing on the classical phenomenological level. In the next section, I study the conditions that the open system is ergodic. In the subsequent sections, I introduce the versatile spin-boson model, and I discuss a number of physically important particle-plus-reservoir systems for which we can base the description to some extent on a microscopic footing. In this chapter, I can not deal with these systems in any detail. I shall outline the underlying Hamiltonians and postpone the path integral formulation of the quantum statistical mechanics until the next chapter. In the last section, I touch on the discussion of nonlinear quantum environments.

3.1 Harmonic bath with linear coordinate coupling

In this section, I first introduce the most general Hamiltonian underlying a dissipative system obeying Eq. (2.4) in the classical limit, and I discuss the important generalization to the case where the viscosity is state-dependent. After that, I determine the relation between the parameters of the global model and the phenomenological frequency-dependent friction coefficient $\tilde{\gamma}(\omega)$.

3.1.1 The Hamiltonian of the global system

Consider a system with one or few degrees of freedom which is coupled to a huge environment and imagine that the environment is represented by a bath of harmonic excitations above a stable ground state. The interaction of the system with each individual degree of freedom of the reservoir is proportional to the inverse of the volume of the reservoir. Hence, *the coupling to an individual bath mode is weak for a geometrically macroscopic environment.* Therefore, it is physically very reasonable for macroscopic composite systems to assume that the system-reservoir coupling depends *linearly* on the deflections of the individual bath modes. This feature is favorable since it allows to eliminate the bath degrees of freedom exactly. Importantly, the weak perturbation of any individual bath mode does not necessarily mean that the dissipative influence of the reservoir on the system is weak as well since the couplings of the individual bath modes add up and the number of modes can be very large.

The general form of the Hamiltonian for the global system complying with these characteristics (barring pathological cases) is (see Ref. [78], Appendix C)

$$H = H_{\mathrm{S}} + H_{\mathrm{R}} + H_{\mathrm{I}} \,. \tag{3.1}$$

Here, H_{S} is the Hamiltonian of the system that is actually of interest. For simplicity, imagine a particle of mass M moving in a generally anharmonic potential $V(q)$,

$$H_{\mathrm{S}} = \frac{p^2}{2M} + V(q) \,, \tag{3.2}$$

and suppose that the reservoir consists of a set of discrete harmonic oscillators,

$$H_{\mathrm{R}} = \sum_{\alpha} \left(\frac{p_{\alpha}^2}{2m_{\alpha}} + \frac{1}{2} m_{\alpha} \omega_{\alpha}^2 x_{\alpha}^2 \right) \,. \tag{3.3}$$

As justified above, the system-bath interaction H_{I} is assumed to be linear in the bath coordinates, but for each bath mode α the system may have individual coupling functions $F_{\alpha}(q)$,

$$H_{\mathrm{I}} = -\sum_{\alpha} x_{\alpha} F_{\alpha}(q) + \Delta V(q) \,. \tag{3.4}$$

The potential $\Delta V(q)$ is to compensate renormalization of the potential $V(q)$ caused by the coupling $-\sum_\alpha x_\alpha F_\alpha(q)$. With the interaction (3.4), the minimum of the potential surface of the global system in x_α-direction for fixed q is at $x_\alpha = F_\alpha(q)/m_\alpha\omega_\alpha^2$. Thus, the system's *effective* potential, i.e., the potential renormalized by the coupling, is

$$V_{\text{eff}}(q) = V(q) - \sum_\alpha \frac{F_\alpha^2(q)}{2m_\alpha\omega_\alpha^2} + \Delta V(q) . \tag{3.5}$$

If the coupling of the system to the reservoir may cause solely dissipation – and not in addition a renormalization of the potential $V(q)$ – we must compensate the second term in Eq. (3.5) by an appropriate choice of $\Delta V(q)$. Full compensation of the coupling-induced potential deformation is achieved if we put

$$\Delta V(q) = \sum_\alpha \frac{F_\alpha^2(q)}{2m_\alpha\omega_\alpha^2} . \tag{3.6}$$

Of particular interest is the case of *separable* system-bath interaction,

$$\sum_\alpha x_\alpha F_\alpha(q) = F(q) \sum_\alpha c_\alpha x_\alpha . \tag{3.7}$$

Under the above assumptions, the most general translational invariant Hamiltonian with separable system-bath interaction is

$$H = \frac{p^2}{2M} + V(q) + \frac{1}{2}\sum_\alpha \left[\frac{p_\alpha^2}{m_\alpha} + m_\alpha\omega_\alpha^2 \left(x_\alpha - \frac{c_\alpha}{m_\alpha\omega_\alpha^2} F(q) \right)^2 \right] . \tag{3.8}$$

Nonlinear form functions $F(q)$ are encountered, e.g., in rotational tunneling systems, in polaron systems, and in Josephson systems.

It has been argued [79] that the periodicity of a hindering potential in a *rotational tunneling system* is an exact symmetry which cannot be destroyed whatever the external influences are. The argument applies when several identical particles are tunneling at the same time, as e.g., in a rotating molecule complex. If there are N identical particles coherently tunneling [e.g., $N = 3$ for a methyl-(CH_3-)group], both the potential $V(\varphi)$ and the coupling function $F(\varphi)$, where φ is the dynamical angular variable, belong to the same symmetry group C_N, i.e., $V(\varphi) = V(\varphi + 2\pi/N)$ and $F(\varphi) = F(\varphi + 2\pi/N)$.

In a polaron system, the particle's interaction energy due to a linear lattice distortion is nonlinear in the particle's coordinate according to $F_{\mathbf{k}}(\mathbf{q}) = e^{i\,\mathbf{k}\cdot\mathbf{q}}$ [cf. Sec. 3.3].

Quasiparticle tunneling in Josephson systems is another important case. Then the coordinate q is again identified with a phase variable φ. In a phenomenological modeling of charge tunneling between superconductors the interaction term is [80]

$$H_{\text{I}} = \sin(\varphi/2) \sum_\alpha c_\alpha^{(1)} x_\alpha^{(1)} + \cos(\varphi/2) \sum_\alpha c_\alpha^{(2)} x_\alpha^{(2)} , \tag{3.9}$$

where $\{x_\alpha^{(1)}\}$ and $\{x_\alpha^{(2)}\}$ represent the sets of coordinates of two independent oscillator baths. For a discussion of the microscopic theory, see Subsection 4.2.10.

If dissipation is postulated as *strictly* linear, $F(q)$ must be linear in q,

$$F(q) = q. \tag{3.10}$$

This is the important case of state-independent dissipation. The form (3.10) describes in von Neumann's sense an ideal measurement of the particle's position by the environment. With $F(q) = q$, the potential counter term takes the form.

$$\Delta V(q) = \frac{1}{2} M \delta\omega^2 q^2, \quad \text{with} \quad \delta\omega^2 = \frac{\mu}{M} \quad \text{and} \quad \mu = \sum_\alpha \frac{c_\alpha^2}{m_\alpha \omega_\alpha^2}. \tag{3.11}$$

In the absence of the counter term $\Delta V(q)$, a harmonic potential $V(q) = \frac{1}{2} M \omega_0^2 q^2$ would be renormalized by the coupling as $V_{\text{ren}}(q) = \frac{1}{2} M (\omega_0^2 - \delta\omega^2) q^2$. When the coupling is large, the change of the potential is large as well, and if $\omega_0^2 - \delta\omega^2 < 0$, the dressed potential would have changed even qualitatively from stable to unstable.

Substituting the expression (3.10) into Eq. (3.8), the Hamiltonian takes the form

$$H = \frac{p^2}{2M} + V(q) + \frac{1}{2} \sum_\alpha \left[\frac{p_\alpha^2}{m_\alpha} + m_\alpha \omega_\alpha^2 \left(x_\alpha - \frac{c_\alpha}{m_\alpha \omega_\alpha^2} q \right)^2 \right]. \tag{3.12}$$

For later convenience, the Hamiltonian (3.12) is rewritten as

$$H = \frac{p^2}{2M} + \sum_\alpha \frac{p_\alpha^2}{2m_\alpha} + V(q, \mathbf{x}), \tag{3.13}$$

$$V(q, \mathbf{x}) = V(q) + \frac{1}{2} \sum_\alpha m_\alpha \omega_\alpha^2 \left(x_\alpha - \frac{c_\alpha}{m_\alpha \omega_\alpha^2} q \right)^2. \tag{3.14}$$

Here, $V(q, \mathbf{x})$ is the potential of the global system, and $\mathbf{x}$ represents the set of bath coordinates $\{x_\alpha\}$. The Hamiltonian (3.12) has been used to model dissipation for about forty years. Early studies were limited to a harmonic potential $V(q)$. Probably the first who showed that Eq. (3.12) leads to dissipation was Magalinskiĭ [81]. Shortly later studies include the work by Rubin [82] for classical systems, and Senitzky [20], Ford *et al.* [21], and Ullersma [76] for quantum systems. In the more recent literature, the model described by Eq. (3.12) is usually referred to as the Caldeira-Leggett model.

The coordinate coupling in the Hamiltonian (3.8) or (3.12) results in *spatial* dissipation, generally referred to as *normal* dissipation. The contrasting case of momentum coupling leads to anomalous dissipation. This issue will be discussed in Sec. 3.6.

3.1.2 The road to a Langevin equation with memory-friction

The equations of motion determined by the global Hamiltonian (3.12) read

$$M\ddot{q}(t) + V'[q(t)] + \sum_\alpha \left(c_\alpha^2 / m_\alpha \omega_\alpha^2 \right) q(t) = \sum_\alpha c_\alpha x_\alpha(t), \tag{3.15}$$

$$m_\alpha \ddot{x}_\alpha(t) + m_\alpha \omega_\alpha^2 x_\alpha(t) = c_\alpha q(t), \tag{3.16}$$

where $V'(q) = \partial V/\partial q$. The dynamical equation for the oscillator position $x_\alpha(t)$ is an ordinary second order linear differential equation with inhomogeneity $c_\alpha q(t)$. This equation is solved by standard Green function techniques. In Fourier space, an individual solution of the inhomogeneous equation (3.16) is

$$\tilde{x}_\alpha(\omega) = \tilde{\chi}_\alpha(\omega)\, c_\alpha \tilde{q}(\omega) \quad \text{with} \quad \tilde{\chi}_\alpha(\omega) = \lim_{\varepsilon \to 0^+} \frac{1}{m_\alpha} \frac{1}{\omega_\alpha^2 - \omega^2 - i\,\omega\varepsilon} . \quad (3.17)$$

The function $\tilde{\chi}_\alpha(\omega)$ is the dynamical susceptibility of an individual bath oscillator. Of particular later interest is the absorptive part $\tilde{\chi}''_\alpha(\omega) = (\pi/m_\alpha)\,\mathrm{sgn}(\omega)\,\delta(\omega_\alpha^2 - \omega^2)$. Without loss of generality, I fix the time of preparation of the initial state to $t_0 = 0$. The solution evolving from the initial state $x_{\alpha,0} = x_\alpha(0)$ and $p_{\alpha,0} = p_\alpha(0)$ then reads

$$x_\alpha(t) = x_{\alpha,0}\cos(\omega_\alpha t) + \frac{p_{\alpha,0}}{m_\alpha \omega_\alpha}\sin(\omega_\alpha t) + \frac{c_\alpha}{m_\alpha \omega_\alpha}\int_0^t ds\,\sin\left[\omega_\alpha(t-s)\right] q(s). \quad (3.18)$$

The expression (3.18) may be rewritten as a functional of the particle's velocity. Upon integrating the last term by parts, we get

$$\begin{aligned} x_\alpha(t) &= x_{\alpha,0}\cos(\omega_\alpha t) + \frac{p_{\alpha,0}}{m_\alpha \omega_\alpha}\sin(\omega_\alpha t) \\ &\quad + \frac{c_\alpha}{m_\alpha \omega_\alpha^2}\left(q(t) - \cos(\omega_\alpha t)q(0) - \int_0^t ds\,\cos\left[\omega_\alpha(t-s)\right]\dot{q}(s)\right). \end{aligned} \quad (3.19)$$

Next, we may eliminate the bath degrees of freedom in Eq. (3.15) by use of Eq. (3.19). Then, the last term of the l.h.s. in Eq. (3.15) and an occuring counter term cancel each other out. The resulting dynamical equation for $q(t)$ alone is

$$M\ddot{q}(t) + M\int_0^t ds\,\gamma(t-s)\dot{q}(s) + V'[q(t)] = \zeta(t) - M\gamma(t)q(0). \quad (3.20)$$

The memory-friction kernel $\gamma(t)$ and the force $\zeta(t)$ are given by

$$\gamma(t) = \Theta(t)\frac{1}{M}\sum_\alpha \frac{c_\alpha^2}{m_\alpha \omega_\alpha^2}\cos(\omega_\alpha t), \quad (3.21)$$

$$\zeta(t) = \sum_\alpha c_\alpha\left(x_{\alpha,0}\cos(\omega_\alpha t) + \frac{p_{\alpha,0}}{m_\alpha \omega_\alpha}\sin(\omega_\alpha t)\right). \quad (3.22)$$

The dynamical equation of motion (3.20) is a Langevin equation with memory-friction modified by the initial-slip force $-M\gamma(t)q(0)$. The random force $\zeta(t)$ depends explicitly on the set of bath variables $\{q_\alpha^{(0)}\}$ and $\{p_\alpha^{(0)}\}$ at time zero. The distribution of these variables determines the statistical properties of the force $\zeta(t)$. Further processing of the inhomogeneity in Eq. (3.20) is given in Subsection 3.1.4.

Let us next consider the reduced dynamics when the system-reservoir coupling is nonlinear in the system's coordinate. If we had started at the Hamiltonian (3.8) and had eliminated the bath dynamics, we would have arrived at the equation

$$\begin{aligned} M\ddot{q}(t) + MF'[q(t)]\int_0^t ds\,\gamma(t-s)F'[q(s)]\dot{q}(s) + V'[q(t)] \\ = F'[q(t)]\,\{\zeta(t) - M\gamma(t)q(0)\}. \end{aligned} \quad (3.23)$$

In this modified Langevin equation the damping term describes *state-dependent* memory friction, and the random force represents multiplicative noise.

Important examples of state-dependent dissipation are quasiparticle tunneling through a barrier between superconductors (contact) and strong electron tunneling through a mesoscopic junction [cf. Subsections 4.2.10 and 5.2.4].

Before we embark in the subsection after next on the statistical properties of the stochastic force, we next deal with the phenomenological modeling of friction.

3.1.3 Modeling of general frequency-dependent damping

Phenomenological modeling of friction

If the number N of bath oscillators is small, the period of time, in which the energy transferred to the bath is returned to the central system, is of the order of other relevant time scales. However, when N is about 20 or larger, the Poincaré recurrence time is found to be practically infinity. In such cases, it is appropriate to replace the sum over the discrete bath modes by a frequency integral with a continuous spectral density of the reservoir coupling. With these preliminary remarks it is now straightforward to establish the connection of the frequency-dependent damping function $\tilde{\gamma}(\omega)$ defined in Eq. (2.5) with the parameters of the Hamiltonian (3.8) or (3.12).

The Fourier transform of the retarded memory-friction kernel (3.21) is

$$\tilde{\gamma}(\omega) = \lim_{\varepsilon \to 0^+} \frac{-i\omega}{M} \sum_\alpha \frac{c_\alpha^2}{m_\alpha \omega_\alpha^2} \frac{1}{\omega_\alpha^2 - \omega^2 - i\omega\varepsilon} . \tag{3.24}$$

Next we introduce the spectral density of the environmental coupling

$$J(\omega) = \frac{\pi}{2} \sum_\alpha \frac{c_\alpha^2}{m_\alpha \omega_\alpha} \delta(\omega - \omega_\alpha) . \tag{3.25}$$

For a set of discrete modes, the spectral density consists of a sequence of δ-peaks. In order to work on a genuine heat bath, we assume that the eigenfrequencies ω_α are so dense as to form a continuous spectrum. In the continuum limit, $J(\omega)$ becomes a smooth function of ω, and the sum in Eq. (3.24) is replaced by the integral

$$\tilde{\gamma}(\omega) = \lim_{\varepsilon \to 0^+} \frac{-i\omega}{M} \frac{2}{\pi} \int_0^\infty d\omega' \frac{J(\omega')}{\omega'} \frac{1}{\omega'^2 - \omega^2 - i\omega\varepsilon} . \tag{3.26}$$

By this subtle modification, the function $\tilde{\gamma}(\omega)$ acquires a smooth real part,

$$\tilde{\gamma}'(\omega) = \tilde{\gamma}'(-\omega) = J(\omega)/M\omega . \tag{3.27}$$

Provided that $J(\omega)$ is a positive function of ω which drops to zero both in the IR and UV limit, the resulting spectral damping function $\tilde{\gamma}(\omega)$ owns all essential properties specified after Eq. (2.5).

Often it is advantageous to work in Laplace space. The Fourier and Laplace transforms of $\gamma(t)$ are related by analytic continuation,

$$\hat{\gamma}(z) \;=\; \tilde{\gamma}(\omega = iz) \,; \qquad \tilde{\gamma}(\omega) \;=\; \lim_{\varepsilon \to 0^+} \hat{\gamma}(z = -i\omega + \varepsilon) \,. \qquad (3.28)$$

With use of Eqs. (3.24) and (3.26) we obtain for $\hat{\gamma}(z)$ the expressions

$$\hat{\gamma}(z) \;=\; \frac{z}{M} \sum_{\alpha} \frac{c_{\alpha}^2}{m_{\alpha} \omega_{\alpha}^2} \frac{1}{(\omega_{\alpha}^2 + z^2)} \;=\; \frac{2}{\pi} \frac{z}{M} \int_0^{\infty} d\omega' \, \frac{J(\omega')}{\omega'} \frac{1}{\omega'^2 + z^2} \,, \qquad (3.29)$$

holding for a discrete and continuous spectral density of the coupling, respectively.

We may express $\tilde{\gamma}(\omega)$ and $J(\omega)$ in terms of the dynamical susceptibilities of the individual reservoir modes given in Eq. (3.17). The resulting expressions are

$$\tilde{\gamma}(\omega) \;=\; \frac{1}{-i\omega M} \sum_{\alpha} c_{\alpha}^2 \Big(\tilde{\chi}_{\alpha}(0) - \tilde{\chi}_{\alpha}(\omega) \Big) \,, \qquad (3.30)$$

and

$$J(\omega) \;=\; \Theta(\omega) \sum_{\alpha} c_{\alpha}^2 \, \tilde{\chi}_{\alpha}''(\omega) \,. \qquad (3.31)$$

These forms will pave the way to deal with cases in which the bath modes are effectively nonlinear. The case of a nonlinear spin bath is discussed below in Sec. 3.5.

As far as we are interested in properties of the particle alone, the dynamics is fully determined by the mass M, the potential $V(q)$, and the spectral density $J(\omega)$. We see from Eq. (3.27) that $J(\omega)$ is directly related to the absorptive part $\tilde{\gamma}'(\omega)$ of the spectral damping function, and thus determined by the damping function $\gamma(t)$,

$$J(\omega) \;=\; M\omega\tilde{\gamma}'(\omega) \;=\; M\omega \int_0^{\infty} dt \, \gamma(t) \cos(\omega t) \,. \qquad (3.32)$$

Conversely, the damping kernel $\gamma(t)$ is found from the spectral coupling $J(\omega)$ by

$$\gamma(t) \;=\; \Theta(t) \frac{1}{M} \frac{2}{\pi} \int_0^{\infty} d\omega \, \frac{J(\omega)}{\omega} \cos(\omega t) \,. \qquad (3.33)$$

The spectral density $J(\omega)$ may also be related to the Laplace transform of the damping kernel. Observing the relation (3.32), we get

$$J(\omega) \;=\; \lim_{\varepsilon \to 0^+} M\omega [\, \hat{\gamma}(\varepsilon + i\omega) + \hat{\gamma}(\varepsilon - i\omega) \,]/2 \,. \qquad (3.34)$$

At this point, an important remark is appropriate. In the case of linear normal dissipation, the spectral density $J(\omega)$, which after all determines dissipative quantum mechanics, can be extracted from the classical phenomenological equation of motion (2.4) [83, 84]. For instance, $\tilde{\gamma}'(\omega)$, and hence $J(\omega)$, may be found from a conventional molecular dynamics simulation. Therefore, the relation (3.32) plays a fundamental role in the phenomenological modeling of dissipative quantum systems.

We infer from the relation (3.29) that the spectral damping function is completely determined by the spectral density of the coupling $J(\omega)$. For any real system, the spectral density drops to zero on physical grounds as $\omega \to \infty$. The minimal high-frequency decrease required depends on the specific physical quantity under consideration.

Modeling of frequency-dependent damping

For a discussion in some detail, it is expedient to divide the total spectral density with a suitably chosen filtering function centered at $\omega = \omega_c$ into two contributions

$$J_{\text{tot}}(\omega) = J(\omega) + J_{\text{hf}}(\omega) \,. \tag{3.35}$$

The filtering function provides a high frequency cutoff for the relevant spectral density $J(\omega)$ and a low-frequency cutoff for $J_{\text{hf}}(\omega)$. The detailed form of $J_{\text{hf}}(\omega)$ is irrelevant if ω_c is chosen very large compared to the characteristic frequencies of the central system. Then the only property we need to postulate is that $J_{\text{hf}}(\omega)$ falls off sufficiently strong below ω_c, so that the mass contribution ΔM_{hf} associated with $J_{\text{hf}}(\omega)$,

$$\Delta M_{\text{hf}} = \frac{2}{\pi} \int_0^\infty d\omega \, \frac{J_{\text{hf}}(\omega)}{\omega^3} \,, \tag{3.36}$$

is finite. With this, one finds from Eq. (3.29) that $J_{\text{hf}}(\omega)$ makes to the spectral damping function in the regime $z \ll \omega_c$ the contribution $\hat{\gamma}(z) = z \Delta M_{\text{hf}}/M$. Interestingly, this term adds to the inertia mass M in the kinetic term of the dynamical equation of motion the additional mass ΔM_{hf}. In Laplace space, there holds

$$M[z^2 + z\hat{\gamma}(z)]\hat{q}(z) \;\rightarrow\; (M + \Delta M_{\text{hf}}) z^2 \hat{q}(z) \,. \tag{3.37}$$

Thus, if we are not interested in the initial slip regime $0 < t \lesssim \omega_c^{-1}$, we may treat $J_{\text{hf}}(\omega)$ in adiabatic approximation. In the sequel, if not stated differently, $J_{\text{hf}}(\omega)$ is collectively taken into account by regarding the mass M as the renormalized mass M_{r} being dressed by the reservoir's high-frequency modes, $M_{\text{r}} = M + \Delta M_{\text{hf}}$.

Next, consider the spectral density $J(\omega)$ responsible for general frequency-dependent damping and equipped with a high-frequency cutoff at ω_c, as just argued. It is convenient to assume (though it is not strictly necessary) that the function $J(\omega)$ has a power law form at low frequencies, $J(\omega) \propto \omega^s$. It will become clear later on that the dissipative influences can be classified by the power s. Negative values of s are excluded since otherwise the frequency integral in Eq. (3.33) would be infrared-divergent, which is pathological. The power-law form is assumed to hold in the frequency range $0 \leqslant \omega \lesssim \omega_c$. In physical applications, ω_c is identified with the Drude, Debye, Fermi frequency, etc., depending on the particular system.

Expedient forms of the high-frequency drop of $J(\omega)$ are a sharp cutoff, an exponential cutoff, and a smooth algebraic cutoff. In the sequel, we shall take as a basis one of the following choices

$$J(\omega) = J_0(\omega) \times \begin{cases} \Theta(1 - \omega/\omega_c)\,, \\ e^{-\omega/\omega_c}\,, \\ 1/[1 + (\omega/\omega_c)^2]^p\,, \end{cases} \quad \text{with} \quad J_0(\omega) = M\gamma_s \omega_{\text{ph}}^{1-s} \omega^s \,. \tag{3.38}$$

For the algebraic cut-off, the parameter p must be chosen sufficiently large, so that occuring frequency integrals are UV-convergent. In the power-law density $J_0(\omega)$, I have introduced for $s \neq 1$ a "phononic" reference frequency ω_{ph}, so that the coupling constant $\eta_s = M\gamma_s$ has dimension viscosity for all s. For purely dimensional reasons, one may choose $\omega_{\mathrm{ph}} = \omega_{\mathrm{c}}$. In general, however, the frequency scales ω_{ph} and ω_{c} have different physical origin. Therefore, I find it appropriate to keep both frequency scales separate. The latter scale is of different relevance in the regimes $s < 1$ and $s > 1$, as we shall see. Where convenient, I shall use the temperature scales

$$T_{\mathrm{ph}} \equiv \hbar\omega_{\mathrm{ph}}/k_{\mathrm{B}} \qquad \text{and} \qquad T_{\mathrm{c}} \equiv \hbar\omega_{\mathrm{c}}/k_{\mathrm{B}} \,. \tag{3.39}$$

In many physical situations, the frequency ω_{c}, at which the power law $J(\omega) \propto \omega^s$ is cut off, is very large compared to all other relevant frequencies of the system. Numerous results in analytic form are available in this limit.

The case $s = 1$ in Eq. (3.38) describes *Ohmic* friction. The cases $0 < s < 1$ and $s > 1$ have been dubbed *sub-Ohmic* and *super-Ohmic*, respectively [86]. In Sections 3.3 and 4.2, I shall present various microscopic models and consider the respective spectral densities of the coupling. The Ohmic case is important, e.g., for tunneling systems in a metallic environment. Importantly, Ohmic friction is virtually omnipresent in real physical systems at low temperature, as we shall see in Subsection 4.2.5. A phonon bath in d spatial dimensions corresponds to the case $s = d$ or $s = d+2$, depending on the underlying symmetry of the strain field (see Subsection 4.2.4). Irrational values of s may occur for complex environments with effectively fractal dimension, such as porous and viscoelastic media, and reservoirs with chaotic dynamics.

The Ohmic case $(s = 1)$:

For a strictly Ohmic system-bath coupling, the damping function $\gamma(t)$ is memoryless, and thus the spectral damping function is real and frequency-independent,

$$\tilde{\gamma}(\omega) = \gamma \,. \tag{3.40}$$

We see from Eq. (3.32) or from Eq. (3.34) that strict Ohmic damping is described by the model (3.12) in which the spectral density $J(\omega)$ is chosen as [78]

$$J(\omega) = \eta\omega = M\gamma\omega \tag{3.41}$$

for all frequencies ω. In the first equality, we have introduced the familiar viscosity coefficient η. The relation (3.40) or (3.41) implies damping without retardation, $\gamma(t) = 2\gamma\Theta(t)\delta(t)$.[1] This, of course, is an idealized situation. In reality, any particular spectral density $J(\omega)$ of physical origin falls off in the limit $\omega \to \infty$, because there is always a microscopic memory time setting the time scale for inertia effects in the reservoir. If $J(\omega)$ would grow steadily with ω, many physical quantities would be divergent (cf., e.g., the discussion of momentum spread in Sections 6.6 and 7.3).

In a physically natural way, the damping kernel $\gamma(t)$ in the classical equation of motion (2.4) is regularized with a Drude memory time $\tau_{\mathrm{D}} = 1/\omega_{\mathrm{D}}$,

---

[1] In the integral in Eq. (2.4) the δ-function counts only half, and thus the damping term reduces to the Ohmic form $M\gamma\dot{q}(t)$ in Eq. (2.3).

$$\gamma(t) = \gamma \omega_D \Theta(t) \exp(-\omega_D t) , \tag{3.42}$$

known as *Drude regularization*, and equivalently in the frequency domain

$$\tilde{\gamma}(\omega) = \gamma/[1 - i\omega/\omega_D] , \qquad \text{and} \qquad \hat{\gamma}(z) = \gamma/[1 + z/\omega_D] . \tag{3.43}$$

Since the imaginary part of $\tilde{\gamma}(\omega)$ differs from the real part by a factor ω/ω_D, the former is small at frequencies well below the Drude frequency.

The Drude form (3.43) emerges from a spectral density with algebraic cutoff,

$$J(\omega) = M\gamma\omega/\left(1 + \omega^2/\omega_D^2\right) . \tag{3.44}$$

The damping kernel carries memory-friction on the time scale $\tau_D = \omega_D^{-1}$. When the relevant frequencies of the system's dynamics are much lower than the Drude cutoff frequency ω_D, the spectral density of the reservoir coupling (3.44) induces Ohmic damping with effective damping strength $\gamma_{\text{eff}} = \int_0^\infty dt\, \gamma(t) = \gamma$.

The case of Ohmic friction is particularly important since it is frequently encountered in nature. Over and above it is responsible for a quantum phase transition at a critical damping strength, and it shows interesting scaling properties, as we shall see.

The sub-Ohmic case $(0 < s < 1)$:

The sub-Ohmic case is distinguished by a high density of low energy excitations. As a result, the dissipative effects are most significant in the low-temperature regime.

In the limit $s \to 0$ of the sub-Ohmic regime, the spectral density $J(\omega)$ takes the constant value $\eta_0 \omega_{\text{ph}}$. Then we have $\tilde{\gamma}'(\omega) = \eta_0 \omega_{\text{ph}}/(M|\omega|)$, and the power spectrum (2.14) of the random force takes the form $\tilde{\mathcal{K}}(\omega) = \eta_0 \hbar\omega_{\text{ph}} \coth(\beta\hbar|\omega|/2)$. Thus, in the regime $\omega \ll k_B T/\hbar$ the power spectrum is $1/f$ or flicker noise,

$$\tilde{\mathcal{K}}(\omega) = 2k_B T \eta_0 \omega_{\text{ph}}/|\omega| . \tag{3.45}$$

$1/f$ noise occurs, e.g., in a set of bistable (two-level) systems making random transitions between the two states. Discussion of this and of possible other origin of $1/f$ noise, and of the implications on decoherence is given in Section 22.5.

The super-Ohmic case $(s > 1)$:

The super-Ohmic case is characterized by a low spectral density of low-energy excitations in the environment. This feature drives the composite system usually into the weak-coupling regime, as temperature is lowered.

The case $s = 3$ describes, e.g., phonon-coupling of tunneling centers in amorphous solids. Destructive combination of two Drude kernels of the form (3.42) gives also rise to super-Ohmic friction $J(\omega) \propto \omega^3$ [87]. Upon putting

$$\gamma(t) = \Theta(t)\gamma\left(\omega_1 e^{-\omega_1 t} - \omega_2 e^{-\omega_2 t}\right) , \tag{3.46}$$

we obtain with use of Eq. (3.32)

$$J(\omega) = M\gamma\omega^3 \frac{\omega_1^2 - \omega_2^2}{(\omega^2 + \omega_1^2)(\omega^2 + \omega_2^2)} . \tag{3.47}$$

Consider next the spectral damping function $\hat{\gamma}(z)$ for general power s of the spectral coupling $J(\omega)$. Choosing for $J(\omega)$ the power law form (3.38) with a sharp high-frequency cutoff, the spectral damping function (3.29) takes the analytic form [88]

$$\hat{\gamma}(z) = \frac{2\gamma_s}{\pi s} \left(\frac{\omega_c}{\omega_{\mathrm{ph}}}\right)^{s-1} \frac{\omega_c}{z} \, {}_2F_1\left(1, \tfrac{1}{2}s; 1+\tfrac{1}{2}s; -\frac{\omega_c^2}{z^2}\right), \qquad (3.48)$$

where ${}_2F_1(a, b; c; x)$ is a hypergeometric function [89, 90]. With a linear transformation this can be written as a convergent hypergeometric series for $|z| < \omega_c$,

$$\hat{\gamma}(z) = \rho_s \left(\frac{z}{\omega_{\mathrm{ph}}}\right)^{s-1} + \frac{2\gamma_s}{\pi(s-2)} \left(\frac{\omega_c}{\omega_{\mathrm{ph}}}\right)^{s-1} \frac{z}{\omega_c} \, {}_2F_1\left(1, 1-\tfrac{1}{2}s; 2-\tfrac{1}{2}s; -\frac{z^2}{\omega_c^2}\right), \quad (3.49)$$

where $\rho_s = \gamma_s/\sin(\pi s/2)$. In the particular cases $s = 1$ and $s = 3$, one obtains

$$\hat{\gamma}(z) = \frac{2}{\pi}\gamma_1 \left[\pi/2 - \arctan(z/\omega_c)\right], \qquad\qquad s = 1, \qquad (3.50)$$

$$\hat{\gamma}(z) = \frac{2}{\pi}\frac{\gamma_3}{\omega_{\mathrm{ph}}^2} \left[\omega_c z - z^2[\pi/2 - \arctan(z/\omega_c)]\right], \qquad s = 3. \qquad (3.51)$$

The leading contribution at high frequency, $z \gg \omega_c$, is proportional to $1/z$,

$$\hat{\gamma}(z) = \frac{\varrho}{z}, \qquad \varrho = \frac{2}{\pi M}\int_0^\infty d\omega \, \frac{J(\omega)}{\omega} = \frac{2\gamma_s}{\pi s}\left(\frac{\omega_c}{\omega_{\mathrm{ph}}}\right)^{s-1}\omega_c. \qquad (3.52)$$

The first form of ϱ is generally valid, and the second holds for the sharp cutoff. In contrast, the leading terms for general s in the low-frequency regime $|z| \ll \omega_c$ are

$$\hat{\gamma}(z) = \rho_s(z/\omega_{\mathrm{ph}})^{s-1} + \mathfrak{m}_1(s)\, z + \mathfrak{m}_3(s)\, z^3/\omega_c^2 + \mathcal{O}\left(z^5\right). \qquad (3.53)$$

For the sharp cutoff in Eq. (3.38), the dimensionless mass coefficients $\mathfrak{m}_{1,3}(s)$ are

$$\mathfrak{m}_1(s) = \frac{2}{\pi(s-2)}\frac{\gamma_s}{\omega_c}\left(\frac{\omega_c}{\omega_{\mathrm{ph}}}\right)^{s-1}, \qquad \mathfrak{m}_3(s) = \frac{2}{\pi(4-s)}\frac{\gamma_s}{\omega_c}\left(\frac{\omega_c}{\omega_{\mathrm{ph}}}\right)^{s-1}. \qquad (3.54)$$

We see that $\mathfrak{m}_1(s)$ diverges at $s = 2$, and $\mathfrak{m}_3(s)$ at $s = 4$. Similar divergence occurs in the terms $\propto z^{2n-1}$ for $s = 2n$, where $n = 3, 4, \cdots$. In each case, the singularity is cancelled with a singularity of $\rho_s = \gamma_s/\sin(\pi s/2)$ in the first term of Eq. (3.53), and hence a logarithmic term emerges. For $s = 2$ and $s = 4$ one finds

$$\hat{\gamma}(z) = \frac{2}{\pi}\gamma_2 \frac{z}{\omega_{\mathrm{ph}}}\ln(\omega_c/z) + \mathfrak{m}_3(2)\, z^3/\omega_c^2, \qquad s = 2, \qquad (3.55)$$

$$\hat{\gamma}(z) = \mathfrak{m}_1(4)\, z - \frac{2}{\pi}\gamma_4\left(\frac{z}{\omega_{\mathrm{ph}}}\right)^3 \ln(\omega_c/z), \qquad s = 4. \qquad (3.56)$$

The linear term $\mathfrak{m}_1(s)z$ in Eq. (3.53) contributes to the inertial force just as the term M_{hf} in Eq. (3.37). It thus entails mass renormalization. The important point, however, is that the cases $s < 2$ and $s > 2$ are qualitatively different.

- In the regime $s > 2$ and $z \ll 1$, the linear term in z of $\hat{\gamma}(z)$ is the leading one. As the integral

$$\mathfrak{m}_1 = \frac{2}{\pi M} \int_0^\infty d\omega \, \frac{J(\omega)}{\omega^3} \tag{3.57}$$

  is IR-convergent for $s > 2$, this term dresses the inertia mass with a regular positive contribution, $M_{\mathrm{r}} = M + M_{\mathrm{bath}}$ with $M_{\mathrm{bath}} = \mathfrak{m}_1 M > 0$.

- For $s \leqslant 2$, the integral (3.57) is IR-divergent so that the total mass of the bath oscillators dragged by the particle is infinite. Hence the interpretation just given for $\mathfrak{m}_1$ ceases to hold. Nevertheless, the scaled mass $\mathfrak{m}_1(s)$ can be analytically continued to the regime $0 < s < 2$. To show this, we write $J(\omega) = J_0(\omega) + [J(\omega) - J_0(\omega)]$, where $J(\omega)$ and $J_0(\omega)$ are given in Eq. (3.38). The spectral damping function (3.29) with the term $J_0(\omega)$ just yields the first term in the series expression (3.53). With the residual contribution, the integral (3.57) is regular. Most interestingly, however, the scaled mass $\mathfrak{m}_1$ turns out negative,

$$\mathfrak{m}_1 = \frac{2}{\pi M} \int_0^\infty d\omega \, \frac{J(\omega) - J_0(\omega)}{\omega^3} = \frac{M_{\mathrm{bath}} - M_{\mathrm{bath},0}}{M} < 0 \,. \tag{3.58}$$

This relation unveils that $-\mathfrak{m}_1 M$ represents the total mass of oscillators which is missing in the actual bath relative to the reference bath without cutoff. At the critical damping strength $\gamma_s = \gamma_s^\star$, where for a sharp cutoff

$$\gamma_s^\star = (2 - s)\pi\omega_{\mathrm{c}}(\omega_{\mathrm{ph}}/\omega_{\mathrm{c}})^{s-1}/2 \,, \tag{3.59}$$

one has $\mathfrak{m}_1 = -1$, and thus the dressed mass is zero, $M_{\mathrm{r}} \equiv [1 + \mathfrak{m}_1(s)]M = 0$. At this point, the term $\propto z^3$ in Eq. (3.53) becomes relevant at low T. In the regime $\gamma_s > \gamma_s^\star$, the dressed mass is negative. This results, e.g., in thermodynamic anomalies for a free Brownian particle, as discussed below in Subsection 7.5.3.

The above qualitative behaviors do not depend on the particular cutoff function chosen. Only the s-dependent numerical factors in the series (3.53) hinge on the specific form of the cutoff function. The implications of an algebraic cutoff are compared with those of a sharp cutoff in Ref. [91].

In the most interesting regime $0 < s < 2$ and $z \ll 1$, the leading term of the damping function $\hat{\gamma}(z)$ varies as z^{s-1}. For a free particle, this leads to sub-diffusive behavior in the sub-Ohmic regime $0 < s < 1$, and to super-diffusive behavior in the super-Ohmic regime $1 < s < 2$, as we shall see in Section 7.4. In the range $s > 2$, the leading term at low z induces renormalization of the mass of the free particle. Sub-leading terms are relevant in the transient time regime (cf. Section 7.4).

Here I anticipate that in the context of quantum state engineering the weak coupling regime for a general spectral density is of interest.

3.1.4 Quantum statistical properties of the stochastic force

In Subsection 3.1.2 we have studied the reduced dynamics on the classical level. Interestingly, the quantum mechanical equations of motion in the Heisenberg picture are the same, as we have seen in Subsection 2.1.2. In the quantum regime, the random force $\zeta(t)$ given in Eq. (3.22) bears operator character via the bath operators at initial time zero, $\hat{x}_{\alpha,0}$ and $\hat{p}_{\alpha,0}$. It is pertinent to write these operators in terms of annihilation and creation operators b_α and $b_\alpha^\dagger$, respectively,

$$\hat{x}_{\alpha,0} = \left(\frac{\hbar}{2m_\alpha\omega_\alpha}\right)^{1/2} \left(b_\alpha + b_\alpha^\dagger\right) , \qquad \hat{p}_{\alpha,0} = i\left(\frac{m_\alpha\omega_\alpha\hbar}{2}\right)^{1/2} \left(b_\alpha^\dagger - b_\alpha\right) , \qquad (3.60)$$

which obey the commutation relations

$$\left[b_\alpha, b_\gamma\right] = \left[b_\alpha^\dagger, b_\gamma^\dagger\right] = 0 , \qquad \text{and} \qquad \left[b_\alpha, b_\gamma^\dagger\right] = \delta_{\alpha\gamma} . \qquad (3.61)$$

In the language of these operators, the random force operator (3.22) reads

$$\hat{\zeta}(t) = \sum_\alpha \left(\frac{\hbar c_\alpha^2}{2m_\alpha\omega_\alpha}\right)^{1/2} \left(e^{i\omega_\alpha t} b_\alpha^\dagger + e^{-i\omega_\alpha t} b_\alpha\right) . \qquad (3.62)$$

It obeys the commutator relation

$$\left[\hat{\zeta}(t), \hat{\zeta}(0)\right] = -i\hbar \sum_\alpha \frac{c_\alpha^2}{m_\alpha\omega_\alpha} \sin(\omega_\alpha t) . \qquad (3.63)$$

Suppose now that at time $t = 0$ the system and the reservoir are decoupled, with the reservoir being in thermal equilibrium. Then the density operator of the total system at initial time $t = 0$ has the factorized form

$$\hat{W}_{\text{fc}}(0) = \hat{\rho}_{\text{S}}(0) \otimes \hat{W}_{\text{R},0} . \qquad (3.64)$$

Here $\hat{\rho}_{\text{S}}(0)$ is the reduced density operator at $t = 0$, i.e., the density operator of the central system, and $\hat{W}_{\text{R},0}$ is the canonical density operator of the uncoupled bath,

$$\hat{W}_{\text{R},0} = \prod_\alpha \frac{e^{-\beta :H_{\text{R},\alpha}(\hat{p}_{\alpha,0},\hat{x}_{\alpha,0}):}}{\text{tr}\, e^{-\beta :H_{\text{R},\alpha}(\hat{p}_{\alpha,0},\hat{x}_{\alpha,0}):}} = \prod_\alpha \frac{e^{-\beta\hbar\omega_\alpha b_\alpha^\dagger b_\alpha}}{\text{tr}\, e^{-\beta\hbar\omega_\alpha b_\alpha^\dagger b_\alpha}} , \qquad (3.65)$$

with the Hamiltonian of the bath mode α in normal order

$$:H_{\text{R},\alpha}(\hat{p}_{\alpha,0},\hat{x}_{\alpha,0}): = :\frac{\hat{p}_{\alpha,0}^2}{2m_\alpha} + \frac{m_\alpha\omega_\alpha^2}{2}\hat{x}_{\alpha,0}^2: = \hbar\omega_\alpha b_\alpha^\dagger b_\alpha . \qquad (3.66)$$

The relevant averages of the bath operators with the density operator $\hat{W}_{\text{R},0}$ are

$$\langle b_\alpha\rangle_{\hat{W}_{\text{R},0}} = \langle b_\alpha^\dagger\rangle_{\hat{W}_{\text{R},0}} = \langle b_\alpha b_\gamma\rangle_{\hat{W}_{\text{R},0}} = \langle b_\alpha^\dagger b_\gamma^\dagger\rangle_{\hat{W}_{\text{R},0}} = 0 ,$$

$$\langle b_\alpha^\dagger b_\gamma\rangle_{\hat{W}_{\text{R},0}} = \delta_{\alpha\gamma}\, n(\omega_\alpha) , \qquad \langle b_\alpha b_\gamma^\dagger\rangle_{\hat{W}_{\text{R},0}} = \delta_{\alpha\gamma}\left[1 + n(\omega_\alpha)\right] , \qquad (3.67)$$

where $n(\omega)$ is the single-particle Bose distribution

$$n(\omega) = \frac{1}{e^{\beta\hbar\omega} - 1} . \tag{3.68}$$

Thus, the noise force operator $\hat{\zeta}(t)$ obeys Gaussian statistics with ensemble averages

$$\langle \hat{\zeta}(t) \rangle_{\hat{W}_{R,0}} = 0 ,$$

$$\langle \hat{\zeta}(t)\hat{\zeta}(0) \rangle_{\hat{W}_{R,0}} = \hbar \sum_\alpha \frac{c_\alpha^2}{2m_\alpha\omega_\alpha} D_{\omega_\alpha}(t) = \frac{\hbar}{\pi} \int_0^\infty d\omega \, J(\omega) D_\omega(t) . \tag{3.69}$$

The second form holds for a dense spectrum, and $D_\omega(t)$ is the equilibrium propagator

$$D_\omega(t) = n(\omega) \, e^{i\omega t} + [1 + n(\omega)] \, e^{-i\omega t} = \coth(\tfrac{1}{2}\beta\hbar\omega) \cos(\omega t) - i \sin(\omega t) . \tag{3.70}$$

The force operator $\hat{\zeta}(t)$ is fluctuating because of random emission of energy quanta $\hbar\omega$ from the bath (the term $\propto e^{i\omega t}$) and random absorption of this quantum into the bath (the term $\propto e^{-i\omega t}$). The form (3.69) with (3.70) is a version of the quantum-mechanical fluctuation-dissipation theorem [cf. Section 6.1]. Emission and absorption are related by detailed balance,

$$\frac{n(\omega)}{1 + n(\omega)} = e^{-\beta\hbar\omega} . \tag{3.71}$$

The stochastic equation of motion (3.20) differs from the standard form (2.4) by a transient term depending on the particle's initial position, $-M\gamma(t)\,q(0)$. This is owed to the factorizing initial state (3.64) with the thermal state (3.65) of the uncoupled bath [85]. In reality, due to the particle-bath coupling, the true canonical density operator of the bath is that of spatially shifted bath oscillators, with the shift depending on the particle's initial position $q_{\mathrm{i}} = q(0)$,

$$\hat{W}_{\mathrm{R}}[q_{\mathrm{i}}] = \prod_\alpha \frac{e^{-\beta : H_{\mathrm{R},\alpha}[\hat{p}_{\alpha,0}, \hat{x}_{\alpha,0} - (c_\alpha/m_\alpha\omega_\alpha^2)\,q_{\mathrm{i}}] :}}{\mathrm{tr}\, e^{-\beta : H_{\mathrm{R},\alpha}[\hat{p}_{\alpha,0}, \hat{x}_{\alpha,0} - (c_\alpha/m_\alpha\omega_\alpha^2)\,q_{\mathrm{i}}] :}} . \tag{3.72}$$

If we now define

$$\hat{\xi}(t) = \hat{\zeta}(t) - M\gamma(t)q_{\mathrm{i}} ,$$

$$= \sum_\alpha c_\alpha \left[\left(\hat{x}_{\alpha,0} - \frac{c_\alpha}{m_\alpha\omega_\alpha^2}\,q_{\mathrm{i}} \right) \cos(\omega_\alpha t) + \frac{\hat{p}_{\alpha,0}}{m_\alpha\omega_\alpha} \sin(\omega_\alpha t) \right] , \tag{3.73}$$

the shifted force $\hat{\xi}(t)$ is a real Langevin force with zero mean provided that the quantum-statistical average is taken with the shifted canonical distribution (3.72),

$$\langle \hat{\xi}(t) \rangle_{\hat{W}_{\mathrm{R}}[q_{\mathrm{i}}]} = 0 , \tag{3.74}$$

$$\mathcal{K}(t) = \langle \hat{\xi}(t)\hat{\xi}(0) \rangle_{\hat{W}_{\mathrm{R}}[q_{\mathrm{i}}]} = \hbar \sum_\alpha \frac{c_\alpha^2}{2m_\alpha\omega_\alpha} D_{\omega_\alpha}(t) = \frac{\hbar}{\pi} \int_0^\infty d\omega \, J(\omega) D_\omega(t) .$$

The second form coincides with the expression (2.13) assumed in Subsection 2.1.2. With the shift (3.73), the Langevin equation (3.20) takes the standard operator form

$$M \frac{d^2 \hat{q}(t)}{dt^2} + M \int_0^t ds\, \gamma(t-s) \frac{d\hat{q}(s)}{ds} + V'[\hat{q}(t)] = \hat{\xi}(t) . \qquad (3.75)$$

The QLE (3.75) is an operator equation that acts in the full Hilbert space of system and bath. The coupling induces entanglement over time even when the density matrix of the global system is initially in factorized form.

From the above we see that care has to be taken if one refers to statistical properties of a random force [85]. We have removed the initial transient slippage by taking the thermal average with bath modes which are shifted by the coupling to the particle.

3.1.5 Displacement correlation function

Consider as a little exercise of the preceding findings the displacement correlation function $\mathcal{C}^+(t) \equiv \langle [\hat{q}(0) - \hat{q}(t)] \hat{q}(0) \rangle_\beta$, where the position satisfies the operator Langevin equation (3.75). The linear response to the fluctuating force $\xi(t)$ is expressed in terms of the response function $\chi(t)$ as

$$\hat{q}(t) = \int_{-\infty}^t ds\, \chi(t-s) \hat{\xi}(s) . \qquad (3.76)$$

With use of Eq. (3.76) the displacement correlation function takes the form

$$\mathcal{C}^+(t) = \int_{-\infty}^\infty ds\, [\chi(-s) - \chi(t-s)] \int_{-\infty}^\infty ds_1\, \chi(-s_1) \langle \hat{\xi}(s) \hat{\xi}(s_1) \rangle_{\hat{W}_R} . \qquad (3.77)$$

With the Fourier representation (3.74) of the force correlator we then obtain

$$\mathcal{C}^+(t) = \frac{\hbar M}{\pi} \int_0^\infty d\omega\, \omega \tilde{\gamma}'(\omega) |\tilde{\chi}(\omega)|^2 \left[D_\omega(0) - D_\omega(t) \right] \qquad (3.78)$$

For linear quantum systems the quantum mechanical response function coincides with the classical one, as follows from Ehrenfest's theorem (see Section 6.2). Putting $V'(q) = M\omega_0^2 q$, we obtain from Eq. (2.4) or from Eq. (3.75)

$$\hat{\chi}(z) = \frac{1}{M} \frac{1}{\omega_0^2 + z^2 + z\hat{\gamma}(z)} , \qquad \tilde{\chi}(\omega) = \frac{1}{M} \frac{1}{\omega_0^2 - \omega^2 - i\omega\tilde{\gamma}(\omega)} , \qquad (3.79)$$

and the absorptive part of the dynamical susceptibility is found as

$$\tilde{\chi}''(\omega) \equiv \mathrm{Im} \int_0^\infty dt\, \chi(t)\, e^{i\omega t} = M\omega\tilde{\gamma}'(\omega) |\tilde{\chi}(\omega)|^2 . \qquad (3.80)$$

Thus we readily get

$$\mathcal{C}^+(t) = \frac{\hbar}{\pi} \int_0^\infty d\omega\, \tilde{\chi}''(\omega) \left\{ \coth(\tfrac{1}{2}\beta\hbar\omega)[1 - \cos(\omega t)] + i\sin(\omega t) \right\} . \qquad (3.81)$$

The displacement correlation function (3.81) is studied for the damped linear quantum oscillator in Section 6 and for the free Brownian particle in Chapter 7.

3.1.6 Thermal boson propagator and imaginary-time correlations

As an expedient groundwork we now analytically continue the above correlation functions from real time $z = t$ to Euclidean or imaginary time $z = -i\tau$.

Thermal propagator

Consider first analytic continuation of the boson function (3.70). We have

$$\mathfrak{D}_\omega(\tau) \equiv D_\omega(-i\tau) = [1 + n(\omega)]e^{-\omega\tau} + n(\omega)e^{\omega\tau} = \frac{\cosh[\omega(\frac{1}{2}\hbar\beta - \tau)]}{\sinh(\frac{1}{2}\omega\hbar\beta)}. \quad (3.82)$$

We now assume (i) that the functional form (3.82) is limited to the prinipal interval $0 \leqslant \tau < \hbar\beta$, and (ii), that this function is periodically continued outside the priniple interval. The periodic function is represented by the Fourier series

$$\mathfrak{D}_\omega(\tau) = \frac{1}{\hbar\beta} \sum_{n=-\infty}^{+\infty} \frac{2\omega}{\nu_n^2 + \omega^2} e^{i\nu_n\tau}, \quad (3.83)$$

where $\nu_n = 2\pi n/\hbar\beta$, is the bosonic Matsubara frequency. This function has symmetric cusps at $\tau = k\hbar\beta$ with peak value $\coth(\beta\hbar\omega/2)$ and minima at $\tau = (k + 1/2)\hbar\beta$ with value $1/\sinh(\beta\hbar\omega/2)$, where $k = 0, \pm1, \pm2, \cdots$. The function (3.83) satisfies the inhomogeneous differential equation

$$[-\partial^2/\partial\tau^2 + \omega^2]\mathfrak{D}_\omega(\tau) = 2\omega:\delta(\tau):, \quad (3.84)$$

where $:\delta(\tau):$ is the periodically continued δ-function,

$$:\delta(\tau): \equiv \frac{1}{\hbar\beta} \sum_{n=-\infty}^{+\infty} e^{i\nu_n\tau} = \sum_{n=-\infty}^{+\infty} \delta(\tau - n\hbar\beta). \quad (3.85)$$

This shows that $\mathfrak{D}_\omega(\tau)$ is the imaginary-time propagator in thermal equilibrium or thermal Green function of a boson with energy $\hbar\omega$.

Imaginary-time force correlation

Consider next analytic continuation of the correlation function of the Langevin force, $\mathfrak{K}(\tau) = \mathcal{K}(-i\tau) = \langle\xi(-i\tau)\xi(0)\rangle_{\hat{W}_R[\hat{q}(0)]}$. We obtain from (3.69) or from (3.74)

$$\mathfrak{K}(\tau) = \hbar \sum_\alpha \frac{c_\alpha^2}{2m_\alpha\omega_\alpha} \mathfrak{D}_{\omega_\alpha}(\tau) = \frac{\hbar}{\pi} \int_0^\infty d\omega \, J(\omega)\mathfrak{D}_\omega(\tau), \quad (3.86)$$

where $\mathfrak{D}_\omega(\tau)$ is the thermal propagator (3.83). Using for $\mathfrak{D}_\omega(\tau)$ the form (3.83), and regarding the representation (3.29) for the spectral damping $\hat{\gamma}(z)$, one obtains

$$\mathfrak{K}(\tau) = \hbar\left(\mu:\delta(\tau): -\frac{M}{\hbar\beta} \sum_{n=-\infty}^{\infty} |\nu_n|\hat{\gamma}(|\nu_n|)e^{i\nu_n t}\right), \quad (3.87)$$

where

$$\mu = \frac{2}{\pi} \int_0^\infty d\omega \, \frac{J(\omega)}{\omega} = \gamma(t \to 0^+) \,. \tag{3.88}$$

Imaginary-time displacement correlation

The imaginary-time displacement correlation function $\mathfrak{C}(\tau) = C^+(-i\tau)$ is found from the expression (3.81), holding for linear systems, as

$$\mathfrak{C}(\tau) = \frac{\hbar}{\pi} \int_0^\infty d\omega \, \tilde{\chi}''(\omega) \left[\mathfrak{D}_\omega(0) - \mathfrak{D}_\omega(\tau) \right] \,. \tag{3.89}$$

Next, we write $\mathfrak{D}_\omega(\tau)$ as the Matsubara sum (3.83) and employ the relation

$$\frac{2}{\pi} \int_0^\infty d\omega \, \frac{\omega \tilde{\chi}''(\omega)}{\omega^2 + z^2} = \mathrm{Im} \frac{1}{\pi} \int_{-\infty}^\infty d\omega \, \frac{\omega \tilde{\chi}(\omega)}{\omega^2 + z^2} = \hat{\chi}(z) \,. \tag{3.90}$$

The latter equality holds since $\tilde{\chi}(\omega)$ is analytic for $\mathrm{Im}(\omega) > 0$. We then obtain

$$\mathfrak{C}(\tau) = \frac{1}{\beta} \sum_{n \neq 0} \hat{\chi}(|\nu_n|) \left[1 - e^{i\nu_n\tau} \right] = \frac{1}{\beta} \sum_{n \neq 0} \frac{1 - e^{i\nu_n\tau}}{M[\nu_n^2 + |\nu_n|\hat{\gamma}(|\nu_n|)]} \,. \tag{3.91}$$

The second form holds for the free particle susceptibility, Eq. (3.79) with $\omega_0 = 0$.

3.1.7 Fractional Langevin equation

For spectral coupling with non-integer power s and associated spectral damping function $\hat{\gamma}(z) \propto z^{s-1}$, the generalized Langevin equation (3.75) can be cast into a fractional Langevin equation. Fractional calculus is the branch of mathematical analysis that generalizes the derivative of a function to non-integer order. We start off with Cauchy's formula for the calculation of an n-fold iterated integral,

$$D^{-n}g(t) \equiv \int_0^t d\tau_n \int_0^{\tau_n} d\tau_{n-1} \cdots \int_0^{\tau_2} d\tau_1 \, g(\tau_1) = \frac{1}{\Gamma(n)} \int_0^t d\tau \, g(\tau)(t-\tau)^{n-1} \,. \tag{3.92}$$

Evidently, the resulting integral can be analytically continued to non-integer values of n. The choice $n = -\alpha$ is the Riemann-Liouville derivative

$$D^\alpha g(t) \equiv \frac{\partial^\alpha g(t)}{\partial t^\alpha} = \frac{1}{\Gamma(-\alpha)} \int_0^t d\tau \, \frac{g(\tau)}{(t-\tau)^{1+\alpha}} \,, \qquad -1 < \alpha < 0 \,. \tag{3.93}$$

Since the integral is constrained to the regime $-1 < \alpha < 0$, it is actually a fractional antiderivative. In the fractional derivative of order α, where α is assumed in the range $n-1 < \alpha < n$, the integer derivative of order n has to be performed before the fractional antiderivative. Thus we have

$$D^\alpha g(t) = \frac{1}{\Gamma(n-\alpha)} \left(\frac{\partial}{\partial t}\right)^n \int_0^t d\tau \, \frac{g(\tau)}{(t-\tau)^{1+\alpha-n}} \,, \qquad n-1 < \alpha < n \,. \tag{3.94}$$

The key point now is that the representation (3.94) for the fractional derivative is a convolution. Taking the Laplace transform, and observing that

$$\int_0^\infty dt\, e^{-zt}\, t^{n-\alpha-1} \;=\; \Gamma(n-\alpha)\, z^{\alpha-n}\,, \qquad n-1 < \alpha < n\,, \tag{3.95}$$

we have

$$\int_0^\infty dt\, e^{-zt}\, D^\alpha g(t) \;=\; z^\alpha\, \hat{g}(z)\,. \tag{3.96}$$

Being targeted to the Langevin equation (3.75) with the friction kernel

$$M\hat{\gamma}(z) \;=\; M\rho_s\, \omega_{\text{ph}}^{1-s}\, z^{s-1}\,, \qquad 0 < s < 2\,, \tag{3.97}$$

where $\rho_s = \gamma_s/\sin(\pi s/2)$, the fractional Langevin equation reads

$$M\frac{d^2 q(t)}{dt^2} + M\rho_s\, \omega_{\text{ph}}^{1-s}\, D^{s-1}\, \dot{q}(t) + V'[q(t)] \;=\; \xi(t)\,. \tag{3.98}$$

The operator-valued form of the fractional Langevin equation (3.98) is equivalent to the generalized quantum Langevin equation (3.75) with the damping kernel (3.97).

Fractional kinetic equations were used to describe chaotic dynamics in low-dimensional systems [92], and anomalous diffusion in biology and physics [93, 94].

3.1.8 Rubin model

In Rubin's model, a heavy particle of mass M and coordinate q is bi-linearly coupled to a semi-infinite chain of harmonic oscillators with masses m and spring constants $f = m\omega_R^2/4$ [95], as sketched in Fig. 3.1. The Hamiltonian of the global model is

$$H \;=\; \frac{p^2}{2M} + V(q) + \sum_{n=1}^\infty \left(\frac{p_n^2}{2m} + \frac{f}{2}(x_{n+1} - x_n)^2 \right) + \frac{f}{2}(q - x_1)^2\,. \tag{3.99}$$

The model is not yet in the standard form (3.12) since the harmonic bath modes are coupled with each other. The harmonic chain is diagonalized with the transformation

$$x_n \;=\; \sqrt{2/\pi} \int_0^\pi dk\, \sin(kn)\, X(k)\,. \tag{3.100}$$

In the normal mode representation of the reservoir, the Hamiltonian takes the form

$$H \;=\; \frac{p^2}{2M} + V(q) + \frac{f}{2}q^2 + \int_0^\pi dk \left(\frac{P^2(k)}{2m} + \frac{m}{2}\omega^2(k)X^2(k) - c(k)X(k)q \right)\,, \tag{3.101}$$

where the eigenfrequencies $\omega(k)$ and the coupling function $c(k)$ are given by

$$\omega(k) \;=\; \omega_R \sin\left(\tfrac{1}{2}k\right)\,, \qquad c(k) \;=\; \frac{1}{2\sqrt{2\pi}}\, m\omega_R^2 \sin(k)\,. \tag{3.102}$$

The frequency ω_R is the highest frequency of the reservoir modes. The spectral density (3.25) of the coupling to the semi-infinite chain is found with Eq. (3.102) as

$$M \qquad f \qquad m$$

Figure 3.1: Pictorial sketch of the mechanical analogue of the Rubin model.

$$J_{\text{si}}(\omega) = \frac{m\,\omega_{\text{R}}^3}{16} \int_0^\pi dk\, \frac{\sin^2(k)}{\sin(\frac{1}{2}k)}\, \delta\left[\omega - \omega(k)\right] = \frac{m\,\omega_{\text{R}}}{2}\, \omega \left(1 - \frac{\omega^2}{\omega_{\text{R}}^2}\right)^{1/2} \Theta(\omega_{\text{R}} - \omega) \,.$$

If the particle is embedded in an infinite chain, the sum in Eq. (3.99) is also over negative integers n, and the resulting spectral density is $J_{\text{inf}}(\omega) = 2J_{\text{si}}(\omega)$. With this form, the damping kernel (3.33) in Laplace and real space emerges as [96, 97]

$$\hat{\gamma}_{\text{inf}}(z) = \frac{m}{M}\left(\sqrt{z^2 + \omega_{\text{R}}^2} - z\right)\,, \quad \text{and} \quad \gamma_{\text{inf}}(t) = \frac{m}{M}\,\omega_{\text{R}}\, \frac{J_1(\omega_{\text{R}} t)}{t}\,, \qquad (3.103)$$

respectively, where $J_1(z)$ is a Bessel function of the first kind. The damping kernel of the Rubin model oscillates with time, and the envelope function decays algebraically with power $\frac{3}{2}$. This particular behavior significantly differs from the exponential decay of the Drude kernel (3.42). The memory time of the Rubin kernel is of the order ω_{R}^{-1}.

Finally, consider the velocity response $\mathcal{R}(t)$ to a δ-stroke in the absence of the potential $V(q)$. In Laplace space we have with $\kappa = m/M$

$$\hat{\mathcal{R}}(z) \equiv \frac{1}{z + \hat{\gamma}_{\text{inf}}(z)} = \frac{1}{(1 - \kappa)z + \kappa\sqrt{z^2 + \omega_{\text{R}}^2}}\,, \qquad (3.104)$$

There are three interesting cases:

(1) In the limit $m/M \to 0$ with $\gamma_0 \equiv \omega_{\text{R}}\, m/M$ held fixed, the case of *memory-less* Ohmic friction is reached, $\hat{\gamma}_{\text{inf}}(z) = \gamma_0$. The response function in Laplace space then is $\hat{\mathcal{R}}(z) = 1/[z + \gamma_0]$, and thus in the time regime

$$\mathcal{R}(t) = e^{-\gamma_0 t}\,. \qquad (3.105)$$

(2) When $M = m$, we have $\hat{\mathcal{R}}(z) = 1/\sqrt{z^2 + \omega_{\text{R}}^2}$. In the time regime, the velocity response then displays oscillatory behavior of a Bessel function of order zero,

$$\mathcal{R}(t) = J_0(\omega_{\text{R}} t)\,. \qquad (3.106)$$

The memory time is of the scale $1/\omega_{\text{R}}$, and the envelope function of $\mathcal{R}(t)$ decays asymptotically as $1/\sqrt{t}$.

(3) When $M = 2m$, we have $\hat{\mathcal{R}}(z) = 2/[z + \sqrt{z^2 + \omega_{\text{R}}^2}]$. This yields in the time domain

$$\mathcal{R}(t) = \frac{J_1(\omega_{\text{R}} t)}{\omega_{\text{R}} t}\,. \qquad (3.107)$$

Interestingly, in this particular case the time-dependence of $\mathcal{R}(t)$ coincides, except for an overall factor, with that of $\gamma_{\text{inf}}(t)$.

3.1.9 Interaction of a charged particle with the radiation field

The dynamics of a one-dimensional particle of mass M and charge e in a potential $V(q)$ and interacting with the 3-d radiation field is described by the Hamiltonian

$$H = (p - eA(q)/c)^2/2M + V(q) + \sum_{\mathbf{k},\lambda} \hbar\omega_k a_{\mathbf{k},\lambda}^\dagger a_{\mathbf{k},\lambda} . \qquad (3.108)$$

The last term is the Hamiltonian of the radiation field. In dipole approximation, the vector potential is spatially constant [98]. We then have the mode decomposition

$$A(t) = \sqrt{2\pi\hbar c^2/L^3} \sum_{\mathbf{k},\lambda} [\, \mathbf{e}_q \cdot \mathbf{e}_{\mathbf{k},\lambda} (f_k^*/\sqrt{\omega_k})\, a_{\mathbf{k},\lambda}(t) + \text{h.c.}\,] , \qquad (3.109)$$

where $\mathbf{e}_q \cdot \mathbf{e}_{\mathbf{k},\lambda}$ is the projection of the polarization vector $\mathbf{e}_{\mathbf{k},\lambda}$ on the particle's straight trajectory, L^3 is the black-body volume, and f_k is the form factor of the charged particle. The form factor is unity at small k and falls to zero above a cutoff frequency ω_{D}.[2] The Heisenberg equations of motion are found from Eq. (3.108) as

$$M\dot{q}(t) = p(t) - (e/c)A(t) , \qquad \dot{p}(t) = -V'[q(t)] , \qquad (3.110)$$

$$\dot{a}_{\mathbf{k},\lambda}(t) = -i\,\omega_k a_{\mathbf{k},\lambda}(t) + i\sqrt{2\pi e^2/\hbar L^3}\, \mathbf{e}_q \cdot \mathbf{e}_{\mathbf{k},\lambda}^* (f_k/\sqrt{\omega_k})\dot{q}(t) . \qquad (3.111)$$

Eq. (3.111) is easily solved with the result

$$a_{\mathbf{k},\lambda}(t) = a_{\mathbf{k},\lambda}^{(0)} e^{-i\omega_k t} + i\sqrt{\frac{2\pi e^2}{\hbar L^3}}\, \frac{f_k}{\sqrt{\omega_k}}\, \mathbf{e}_q \cdot \mathbf{e}_{\mathbf{k},\lambda}^* \int_0^t ds\, e^{-i\omega_k(t-s)}\dot{q}(s) , \qquad (3.112)$$

where $a_{\mathbf{k},\lambda}^{(0)}$ is the free-field annihilation operator. The expansion (3.109) with (3.112) yields an explicit form for $A(t)$. Upon insertion of this expression into Eq. (3.110), we readily obtain the generalized Langevin equation

$$M\ddot{q}(t) + M\int_0^t ds\,\gamma(t-s)\dot{q}(s) + V'[q(t)] = \xi(t) . \qquad (3.113)$$

The friction kernel $\gamma(t)$ and the stochastic force $\xi(t)$ are given by [99]

$$\gamma(t) = \frac{4\pi e^2}{ML^3} \sum_{\mathbf{k},\lambda} |\mathbf{e}_q \cdot \mathbf{e}_{\mathbf{k},\lambda}|^2 |f_k|^2 \cos(\omega_k t) , \qquad (3.114)$$

$$\xi(t) = \sqrt{2\pi\hbar e^2/L^3} \sum_{\mathbf{k},\lambda} \sqrt{\omega_k} \left(i\, f_k^*\, \mathbf{e}_q \cdot \mathbf{e}_{\mathbf{k},\lambda}^* a_{\mathbf{k},\lambda}^{(0)} e^{-i\omega_k t} + \text{h.c.} \right) . \qquad (3.115)$$

The relation (3.32) with the form (3.114) yields the spectral density of the coupling,

$$J_{\mathrm{rad}}(\omega) = \frac{2\pi^2 e^2}{L^3} \sum_{\mathbf{k},\lambda} |\mathbf{e}_q \cdot \mathbf{e}_{\mathbf{k},\lambda}|^2 |f_k|^2 \omega_k\, \delta(\omega - \omega_k) = \frac{2e^2}{3c^3} \frac{\omega^3}{1 + \omega^2/\omega_{\mathrm{D}}^2} . \qquad (3.116)$$

The second form holds for the convenient choice $|f_k|^2 = 1/(1 + \omega_k^2/\omega_{\mathrm{D}}^2)$ and after execution of the continuum limit. Thus, radiation damping is super-Ohmic with power $s = 3$. In addition, we find following the reasoning given in Subsec. 3.1.4 that black body radiation equips the stochastic force (3.115) precisely with the force correlation function (2.13), where $\tilde{\gamma}'(\omega) = J_{\mathrm{rad}}(\omega)/M\omega$.

---

[2]For convenience in Sect. 6.3, the cutoff frequency is identified with the Drude frequency ω_{D}.

3.1.10 Ergodicity

A quantum statistical system is called ergodic when the ensemble average of observables coincides with the time average at asymptotic time. Expressed differently, the system is ergodic when it forgets over time its initial state and therefore relaxes to a unique equilibrium state. In order that the system is not caught in a particular state, it must be able to exchange an arbitrarily small amount of energy with the environment. This implies that the spectral density of the environmental coupling $J(\omega)$ must have nonzero weight at the smallest frequencies. In turn, an open system is *nonergodic* when the spectral density has a low-frequency cutoff.

Whether an open quantum system of interest is ergodic can be found, e.g., with a study of the asymptotic dynamics of the response function $\chi(t)$ pertinent to the Langevin equation (3.75). The two opposite cases are that the Brownian particle is confined in a potential and that it is free. In the first case, the static susceptibility χ_0 is finite, and in the second case it is infinite.

The paradigm of the first case is the damped linear oscillator of which the susceptibility in Laplace space is (cf. Section 6.2)

$$\hat{\chi}(z) = \frac{1}{M[\omega_0^2 + z^2 + z\hat{\gamma}(z)]} . \tag{3.117}$$

In general, the response function $\chi(t)$ consists of a coherent part $\chi_{\mathrm{coh}}(t)$, which is due to a complex conjugate pair of simple poles of $\hat{\chi}(z)$ in the cut z-plane located at z_{p} and z_{p}^*, and a incoherent cut contribution $\chi_{\mathrm{cut}}(t)$,

$$\chi(t) = \chi_{\mathrm{coh}}(t) + \chi_{\mathrm{cut}}(t) . \tag{3.118}$$

Since $\mathrm{Re}(z_{\mathrm{p}}) < 0$, the contribution $\chi_{\mathrm{coh}}(t)$ describes damped oscillations. Thus we have asymptotically $\lim_{t\to\infty} \chi_{\mathrm{coh}}(t) = 0$.

Next we turn to the leading cut contribution at asymptotic time. The expansion of $\hat{\chi}(z)$ at $z = 0$ is

$$\hat{\chi}(z) = \chi_0 - M\chi_0^2 z^2 - M\chi_0^2 z\hat{\gamma}(z) + \mathcal{O}\left[(z^2 + z\hat{\gamma}(z))^2/\omega_0^4\right] , \tag{3.119}$$

where $\chi_0 = 1/M\omega_0^2$ is the static susceptibility. At first we observe that integer powers of z do not contribute to $\chi(t)$. The leading relevant term of $\hat{\gamma}(z)$ for all s is the term $\rho_s(z/\omega_{\mathrm{ph}})^{s-1}$. With this, the third term of Eq. (3.119) is the leading cut contribution,

$$\hat{\chi}_{\mathrm{cut}}(z) = -M\chi_0^2 \rho_s \omega_{\mathrm{ph}}^{1-s} z^s , \tag{3.120}$$

and in the time regime[3]

$$\chi_{\mathrm{cut}}(t) = \frac{2\Gamma(1+s)}{\pi} \cos\left(\frac{\pi s}{2}\right) \frac{M\gamma_s}{\omega_{\mathrm{ph}}^{s-1}} \frac{\chi_0^2}{t^{1+s}} . \tag{3.121}$$

It is striking that the cut contribution vanishes when s is an odd integer.

---

[3]While $\hat{\gamma}(z)$ has logarithmic terms for $s = 2n$, the form (3.121) also holds for these values of s.

The important result of this analysis is that the response function drops to zero for all s. We shall see in Chapter 6 that also the equilibrium correlation functions of the damped linear quantum oscillator, e.g., the equilibrium autocorrelation function of the position, approach asymptotically a unique equilibrium state for all s.

The dissipative two-state system is another open quantum system with a finite static susceptibility. We shall find below in Chapter 21 that this nonlinear system shows similar behavior. Like the damped linear oscillator, it relaxes for general s to a unique equilibrium state.

An open system with divergent static susceptibility is the free Brownian particle. The related Langevin equation (3.75) for the velocity operator $\hat{v}(t)$ in the absence of deterministic forces yields

$$\hat{v}(t) = \mathcal{R}(t)\hat{v}(0) + \frac{1}{M}\int_0^t ds\,\mathcal{R}(t-s)\hat{\xi}(s) , \qquad (3.122)$$

where $\mathcal{R}(t) = M\dot{\chi}(t)$ describes the velocity response to a δ-shaped force,

$$\hat{\mathcal{R}}(z) = \frac{1}{z + \hat{\gamma}(z)} . \qquad (3.123)$$

In the regime $0 < s < 2$, we find with use of Eq. (3.53) $\lim_{z\to 0}\hat{\mathcal{R}}(z)z^{s-1} = \omega_{\mathrm{ph}}^{s-1}/\rho_s$ and hence the asymptotic time dependence

$$\mathcal{R}(t) = \frac{\sin(\pi s/2)}{\Gamma(s-1)}\frac{\omega_{\mathrm{ph}}}{\gamma_s}\frac{1}{(\omega_{\mathrm{ph}}t)^{2-s}} , \qquad 0 < s < 2 . \qquad (3.124)$$

When the parameter s exceeds the value 2, the dominant term in $\hat{\mathcal{R}}(z)$ is the kinetic one with the polaronic mass term included, $\lim_{z\to 0} z\,\hat{\mathcal{R}}(z) = 1/(1 + \Delta M_s/M)$. Hence the frictional effects of the damping kernel fade away in the course of time, and $\mathcal{R}(t)$ approaches asymptotically a constant, like for a free undamped particle but with change from bare to renormalized mass,

$$\mathcal{R}(t \to \infty) = \frac{M}{M + \Delta M_s} , \qquad s > 2 . \qquad (3.125)$$

Hence the mean velocity of the particle depends asymptotically on the initial velocity v_0 as $\langle v(t \to \infty)\rangle = [M/(M + \Delta M_s)]\,v_0$ when $s > 2$. Further discussion of free Brownian motion is given in Chapter 7.

From these little thoughts we deduce the following conjecture. Every open system with a finite static susceptibility is ergodic in the entire regime $s > 0$. On the other hand, an open system with divergent static susceptibility, like a free Brownian particle, is ergodic in the range $0 < s < 2$ and nonergodic for super-Ohmic friction with power $s > 2$. The findings given in this section are in correspondence with the analysis in Ref. [100].

To corroborate these findings, consider two spectral densities $\propto \omega$, the one with a high-frequency cutoff, $J_{\mathrm{hfc}}(\omega) = M\gamma\,\omega\,\Theta(\omega_{\mathrm{c}} - \omega)$, and the other with a low frequency cutoff, $J_{\mathrm{lfc}}(\omega) = M\gamma\,\omega\,\Theta(\omega-\omega_{\mathrm{c}})$. The respective spectral damping functions are found from Eq. (3.29) as

$$\begin{aligned}
\hat{\gamma}_{\mathrm{hfc}}(z) &= (2/\pi)\gamma\arctan(\omega_{\mathrm{c}}/z)\,, \\
\hat{\gamma}_{\mathrm{lfc}}(z) &= (2/\pi)\gamma\arctan(z/\omega_{\mathrm{c}})\,.
\end{aligned} \tag{3.126}$$

We have $\lim_{z\to 0}\hat{\gamma}_{\mathrm{hfc}}(z) = \gamma$ and $\lim_{z\to 0}\hat{\gamma}_{\mathrm{lfc}}(z) = (2/\pi)(\gamma/\omega_{\mathrm{c}})\,z$. In the first case, there is Ohmic friction down to the lowest frequency. In the second case, the spectral damping function ends up in mass renormalization $\Delta M/M = (2/\pi)\gamma/\omega_{\mathrm{c}}$ at time $t \gg 1/\omega_{\mathrm{c}}$. Hence the system is ergodic in the first and nonergodic in the second case.

A discussion of susceptibility, correlation functions, and transport properties of linear open systems is given in Section 6.3, and in Subsections 6.4.2 and 7.4.1. We conclude with the remark that the relation (3.120) is the essence of the Shiba relation. This relation is studied in Subsection 6.4.3 for the damped linear oscillator, and in Subsections 22.6.5, and 22.7.2 for the dissipative two-state system.

3.2 The spin-boson model

Many physical and chemical systems can be described by a generalized coordinate with which is associated an effective potential energy function with two separate minima at roughly the same energy. At thermal energy which is very small compared to the level spacing of the low-lying states in the individual wells, only the ground states of the two wells are involved in the dynamics. Then the time evolution goes off in a two-dimensional Hilbert space. For a high barrier, the two eigen states are clearly localized in the left and right well, respectively, and they are spatially well-separated. The localized states are weakly coupled through a transfer or tunneling amplitude. The two-state system (TSS) is the simplest system showing constructive and destructive quantum interference effects, e.g., clockwise oscillations of the occupation of the left and right well. There is a great amount of interest in the thermodynamical and dynamical behavior of this model due to the diversity of physical realizations, as well as to theoretical advances in recent years.

The basic element in a quantum computer is a *qubit*. Principally, it can be formed by any physical system, whose motion is effectively restricted to a two-dimensional Hilbert space. It is often convenient to regard such system as a spin-$\frac{1}{2}$ system. In any real physical situation, the spin is in contact with the environment. If the surroundings is imagined as a reservoir of bosons, the global system consisting of the spin and the boson bath is represented by the *spin-boson model*. Most interesting in this system, in particular in the context of quantum state engineering and quantum computing, is the extent to which the phase relation between the "spin-up" and "spin-down" components of the wave function is preserved.

Other examples of the situation just described include the motion of defects in crystalline solids (e.g., impurity ions in alkali halides), tunneling of light particles (e.g., hydrogen, muon, proton) in metals [101, 102, 103], the tunneling entities believed to be responsible for the anomalous low-temperature thermal and acoustic properties of oxide glasses and amorphous metals [104] and for the anomalous conductance in mesoscopic wires [105, 106], and also some types of chemical reactions involving electron transfer processes [107, 108].

3.2.1 The model Hamiltonian

I begin with considering the reduction of a spatially expanded double-well system to a two-state system. I characterize the asymmetric double well potential by a "detuning" energy $\hbar\epsilon$ between the two potential minima and by a coupling energy $\hbar\Delta_0$ representing the tunnel splitting of the symmetric system, as sketched in Fig. 3.2. Throughout the book, I shall use the convention that the right well is the lower one for positive bias, $\epsilon > 0$. The tunneling amplitude Δ_0 may be calculated using standard WKB or instanton methods (see Refs. [109, 110]).

Consider a symmetric double well potential $V(q)$ with minima at $q = \pm\frac{1}{2}q_0$ and potential energy $V(\pm\frac{1}{2}q_0) = 0$. The frequency of small oscillations in the wells is $\omega_0 = \sqrt{V''(\pm\frac{1}{2}q_0)/M}$. In the semiclassical limit, the tunnel splitting for a particle of mass M is determined by properties of the so-called instanton path $q_{\text{inst}}(\tau)$. This path describes motion in the *upside-down* potential $-V(q)$ with zero energy,

$$\frac{1}{2}M\dot{q}_{\text{inst}}^2(\tau) = V[q_{\text{inst}}(\tau)] , \tag{3.127}$$

The instanton has the boundary conditions $q_{\text{inst}}(\tau \to \pm\infty) = \pm\frac{1}{2}q_0$. It is a kink-like path centered at time zero at which the fictive particle rushes with maximal velocity through the minimum of the upside-down potential $-V(q)$. The Euclidean action of this path is

$$S_{\text{inst}} = M \int_{-\infty}^{\infty} d\tau \, \dot{q}_{\text{inst}}^2(\tau) = \int_{-q_0/2}^{q_0/2} dq \, \sqrt{2MV(q)} . \tag{3.128}$$

The instanton approaches the boundary values $\pm\frac{1}{2}q_0$ as follows,

$$q_{\text{inst}}(\tau \to \pm\infty) = \pm\frac{q_0}{2} \mp C_0 \left(\frac{S_{\text{inst}}}{2M\omega_0}\right)^{1/2} e^{-\omega_0|\tau|} . \tag{3.129}$$

The factor C_0 is a numerical constant of order unity which depends on the shape of the potential barrier. In the semiclassical limit, the tunneling matrix element Δ_0 for any particular symmetric double well is determined by the instanton action S_{inst}, the frequency ω_0, and by the numerical constant C_0 occurring in the asymptotic behavior (3.129). The semiclassical expression for the tunneling matrix element expressed in terms of instanton parameters reads[4] [109, 110]

$$\Delta_0/2 = \omega_0 C_0 \, (S_{\text{inst}}/2\pi\hbar)^{1/2} \, e^{-S_{\text{inst}}/\hbar} , \tag{3.130}$$

For the archetypal double-well potential of quartic form

---

[4]Eq. (3.130) is analogous to the decay rate (12.42) of the ground state in a metastable potential.

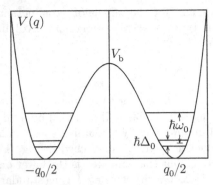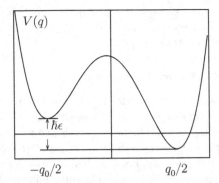

Figure 3.2: Symmetric double well potential (left) with barrier height V_b. The spacing between the first excited state and the ground state in each well is $\hbar\omega_0$, and the tunnel splitting is $\hbar\Delta_0$. The biased double well with detuning energy $\hbar\epsilon$ is sketched on the right.

$$V(q) = \tfrac{1}{2}M\omega_0^2 \left(q^2 - \tfrac{1}{4}q_0^2\right)^2 /q_0^2 , \qquad (3.131)$$

which has barrier height $V(q = 0) \equiv V_\mathrm{b} = M\omega_0^2 q_0^2/32$, the instanton trajectory reads

$$q_\mathrm{inst}(\tau) = \tfrac{1}{2}q_0 \tanh\left(\tfrac{1}{2}\omega_0\tau\right) . \qquad (3.132)$$

For this particular path, the action S_inst and the numerical factor C_0 are given by

$$S_\mathrm{inst} = M\omega_0 q_0^2/6 = 16V_\mathrm{b}/3\omega_0 , \qquad \text{and} \qquad C_0 = 2\sqrt{3} . \qquad (3.133)$$

Thus we find from Eq. (3.130) for the tunneling matrix element the expression

$$\Delta_0 = 8(2V_\mathrm{b}/\pi\hbar\omega_0)^{1/2}\omega_0 \exp\left(-16V_\mathrm{b}/3\hbar\omega_0\right) . \qquad (3.134)$$

For later convenience, we use in the sequel instead of the bare Δ_0 the renormalized tunnel matrix element Δ. The latter is dressed by a Franck-Condon factor which represents the polarization cloud of the high-frequency environmental modes in adiabatic approximation (see the discussion in Subsection 18.1.2). In the parameter regime

$$V_\mathrm{b} \gg \hbar\omega_0 \gg \hbar\Delta, \hbar|\epsilon|, k_\mathrm{B}T , \qquad (3.135)$$

where V_b is the barrier height and $\hbar\omega_0$ is the separation of the first excited state from the ground state in either well, the system will be effectively restricted to a two-dimensional Hilbert space spanned by the states $|R> \;\hat{=}\; |\uparrow>$ (right) and $|L> \;\hat{=}\; |\downarrow>$ (left) localized in the right and left well, respectively. The two-state Hamiltonian may be expressed in terms of the Pauli matrices in the pseudo-spin form[5]

---

[5]I choose spin matrix representations $\{\sigma_j\}$ and $\{\tau_j\}$ in which σ_z and τ_z are diagonal,

$$\sigma_x = \begin{pmatrix} 0 & 1 \\ 1 & 0 \end{pmatrix} , \qquad \sigma_y - \begin{pmatrix} 0 & -i \\ i & 0 \end{pmatrix} , \qquad \sigma_z = \begin{pmatrix} 1 & 0 \\ 0 & -1 \end{pmatrix} .$$

$$H_{\text{TSS}} = -\tfrac{1}{2}\hbar\Delta\,\sigma_x - \tfrac{1}{2}\hbar\epsilon\,\sigma_z = -\frac{\hbar}{2}\begin{pmatrix} \epsilon & \Delta \\ \Delta & -\epsilon \end{pmatrix}. \tag{3.136}$$

The basis is formed by the localized states $|R\!>$ and $|L\!>$ which are eigenstates of σ_z with eigenvalues $+1$ and -1, respectively. The position operator is $\hat{q} = \tfrac{1}{2}q_0\sigma_z$ and the eigenvalues $\pm\tfrac{1}{2}q_0$ of $\hat{q}$ are the positions of the localized states. In this representation, the Pauli operators may equivalently be written in the form

$$\sigma_z = |R\!><\!R| - |L\!><\!L| ,$$

$$\sigma_x = |R\!><\!L| + |L\!><\!R| , \tag{3.137}$$

$$\sigma_y = i\left(|L\!><\!R| - |R\!><\!L|\right) .$$

The localized states $|R\!>$ and $|L\!>$ are related to the eigenstates $|g\!>$ (ground) and $|e\!>$ (excited) of the Hamiltonian (3.136) by the orthogonal transformation

$$\begin{pmatrix} |R\!> \\ |L\!> \end{pmatrix} = \mathcal{R}(\varphi)\begin{pmatrix} |g\!> \\ |e\!> \end{pmatrix} \equiv \begin{pmatrix} \cos\tfrac{1}{2}\varphi & -\sin\tfrac{1}{2}\varphi \\ \sin\tfrac{1}{2}\varphi & \cos\tfrac{1}{2}\varphi \end{pmatrix}\begin{pmatrix} |g\!> \\ |e\!> \end{pmatrix}, \tag{3.138}$$

where

$$\sin\varphi = \Delta/\Delta_{\text{b}} , \qquad \cos\varphi = \epsilon/\Delta_{\text{b}} , \qquad \tan\varphi = \Delta/\epsilon . \tag{3.139}$$

The tunnel splitting energy of the biased TSS is

$$E_e - E_{\text{g}} = \hbar\Delta_{\text{b}} = \hbar\sqrt{\Delta^2 + \epsilon^2} . \tag{3.140}$$

With the orthogonal transformation

$$\mathcal{R}(-\varphi)[\sin\varphi\,\sigma_x + \cos\varphi\,\sigma_z]\mathcal{R}(\varphi) = \tau_z , \tag{3.141}$$

the TSS Hamiltonian is converted into diagonal form

$$H'_{\text{TSS}} \equiv \mathcal{R}(-\varphi)H_{\text{TSS}}\mathcal{R}(\varphi) = -\tfrac{1}{2}\hbar\Delta_{\text{b}}\,\tau_z . \tag{3.142}$$

The expectation values of the operators $\tau_{\pm} \equiv \tfrac{1}{2}(\tau_x \pm i\tau_y)$ are the off-diagonal elements or coherences of the TSS density matrix in energy representation. These are expressed in terms of the matrices σ_x, σ_y and σ_z as

$$\tau_{\pm} = \mathcal{R}(-\varphi)\tfrac{1}{2}[\cos\varphi\,\sigma_x \pm i\,\sigma_y - \sin\varphi\,\sigma_z]\mathcal{R}(\varphi) . \tag{3.143}$$

Consider the coupling to a heat bath which is sensitive to the position of the TSS and which is represented by a collective bath mode, the polarization energy $\mathfrak{E}(t)$,[6]

$$H_{\text{I}} = -\tfrac{1}{2}\sigma_z\mathfrak{E}(t) , \qquad \mathfrak{E}(t) \equiv q_0\sum_{\alpha}c_{\alpha}x_{\alpha}(t) . \tag{3.144}$$

---

[6]The potential renormalization term $\Delta V(q) = \tfrac{1}{2}\sum_{\alpha}(c_{\alpha}^2/m_{\alpha}\omega_{\alpha}^2)q^2$ does not contribute in the spin-boson model because $\Delta V(\tfrac{1}{2}q_0) = \Delta V(-\tfrac{1}{2}q_0)$.

The $x_\alpha(t)$ are the individual modes of the reservoir. A dipole-local-field interaction provides a simple physical model for this kind of coupling. As already discussed in some detail in Subsection 3.1.2, the coupling to the heat bath introduces a fluctuating force $\xi(t) = \sum_\alpha c_\alpha x_\alpha(t)$ which causes a fluctuating bias or polarization energy $\mathfrak{E}(t)$. Assuming strict Gaussian statistics, the heat bath is modeled by a bath of harmonic oscillators, Eq. (3.3). Thus the essential physics is captured by the Hamiltonian,

$$H \;=\; H_{\mathrm{TSS}} - \tfrac{1}{2}\hbar\epsilon\,\sigma_z - \tfrac{1}{2}\sigma_z\mathfrak{E}(t) + H_{\mathrm{R}}\;. \tag{3.145}$$

In the eigenbasis of H_{TSS} the Hamiltonian takes the form

$$H' \equiv \mathcal{R}(-\varphi)H\mathcal{R}(\varphi) = -\tfrac{1}{2}\hbar\Delta_{\mathrm{b}}\,\tau_z - \tfrac{1}{2}(\cos\varphi\,\tau_z - \sin\varphi\,\tau_x)\,\mathfrak{E}(t) + H_{\mathrm{R}}\;, \tag{3.146}$$

where τ_z and τ_x are the Pauli matrices in the transformed basis. In the energy representation of H_{TSS} the reservoir coupling consists of a transverse ($\propto \sin\varphi$) and a longitudinal ($\propto \cos\varphi$) coupling. Clearly, only the transverse part can flip the spin.

Upon writing the bath operators in terms of the annihilation and creation operators b_α and $b_\alpha^\dagger$ introduced in Eq. (3.60), and omitting the zero-point energy, we arrive at a form which has become known in the literature as the "spin-boson" Hamiltonian,

$$H_{\mathrm{SB}} = -\tfrac{1}{2}\hbar\Delta\,\sigma_x - \tfrac{1}{2}\hbar\epsilon\,\sigma_z - \tfrac{1}{2}\sigma_z\sum_\alpha \hbar\lambda_\alpha\left(b_\alpha + b_\alpha^\dagger\right) + \sum_\alpha \hbar\omega_\alpha b_\alpha^\dagger b_\alpha\;. \tag{3.147}$$

The environmental effects are again administered by a spectral density of the coupling,

$$G(\omega) \;=\; \sum_\alpha \lambda_\alpha^2\,\delta(\omega - \omega_\alpha) \;=\; \frac{q_0^2}{\pi\hbar}\,J(\omega)\;. \tag{3.148}$$

The second equality relates the spin-boson spectral density $G(\omega)$ to the spectral density $J(\omega)$, Eq. (3.25), of the continuous model (3.12). The latter has dimension mass times frequency squared, while the former has dimension frequency. The thermodynamics and dynamics of the spin-boson model is discussed in some detail in Part IV.

3.2.2 Flux and charge qubits: reduction to the spin-boson model

Superconducting circuits based on Josephson junctions are considered as promising candidates for quantum computing devices. Building qubits with such components seems advantageous because coupling of qubits, electrical control, and scaling to large numbers should be feasible using integrated-circuit fabrication technology.

The standard phenomenological model for a Josephson junction is the Resistively and Capacitively Shunted Junction (RCSJ) model, in which the current through the junction runs through three parallel channels:

(i) the supercurrent channel described by the Josephson relation $I_{\mathrm{s}} = I_c \sin(\psi)$, where ψ is the phase difference between the Ginsburg-Landau order parameters on the two sides of the junction. The critical current I_c is the maximum supercurrent which the junction can sustain. It depends on junction properties (cf. Subsec. 4.2.10) and is related to the Josephson coupling energy E_{J} by $E_{\mathrm{J}} = I_c\phi_0/2\pi = I_c\hbar/2e$.

(ii) The resistive channel is characterized by a generally nonlinear temperature-dependent resistance R arising from quasiparticle tunneling across the junction.

(iii) The capacitive channel carries the displacement current associated with the geometric shunt capacitance of the electrodes attached to the junction.

Within the RCSJ model, an externally applied bias current I_{ext} thus splits up into the contributions

$$I_{\text{ext}} = I_{\text{c}} \sin(\psi) + U/R + C\dot{U} \,. \tag{3.149}$$

Elimination of the voltage U using the Josephson relation $2eU = \hbar\dot{\psi}$ yields

$$C(\phi_0/2\pi)^2 \ddot{\psi} + (\phi_0/2\pi)^2 \dot{\psi}/R + E_{\text{J}} \sin\psi = (\phi_0/2\pi)I_{\text{ext}} \,. \tag{3.150}$$

This classical equation forms the basis for the $I - U$ characteristics of the current biased Josephson junction. The decay of the zero-voltage state in this model by quantum mechanicl tunneling is discussed in Sec. 17.4.

A simple design of a qubit is an rf SQUID which is formed by a superconducting ring with self-inductance L interrupted by a junction with capacitance C and charging energy $E_{\text{C}} = Q^2/2C$. The macroscopic degree of freedom is the magnetic flux ϕ threading the ring. It is related to the phase difference ψ of the Cooper pair wave function[7] across the junction (cf. Subsec. 4.2.10) by the Josephson relation

$$\psi = 2\pi(n - \phi/\phi_0) \,, \tag{3.151}$$

where $2\pi n$ (n integer) is the change of the phase of the pair wave function per cycle around the ring, and where $\phi_0 = 2\pi\hbar/2e$ is the flux quantum. The total flux ϕ through the ring consists of the externally applied flux ϕ_{ext} and the flux induced by the current I_{ext} flowing in the loop, $\phi = \phi_{\text{ext}} + LI_{\text{ext}}$.

For a qubit operating in the regime $E_{\text{J}} \gg E_{\text{C}}$, the phase ψ is virtually sharp, and hence the flux ϕ threading the ring is the relevant quantum degree of freedom (flux qubit). For a qubit operating, in contrast, in the regime $E_{\text{C}} \gg E_{\text{J}}$, the variable conjugate to ψ, i.e., the charge Q of excess Cooper pairs on the superconducting islands, $\hat{Q} = -i\,2e\,\partial/\partial\psi$, is virtually sharp (charge qubit).

Eliminating in Eq. (3.150) the phase ψ and the bias current I_{ext} in favor of the flux ϕ and the bias flux ϕ_{ext}, we obtain the second order differential equation

$$C\ddot{\phi} + \dot{\phi}/R + \partial V(\phi)/\partial\phi = 0 \,, \tag{3.152}$$

in which the potential $V(\phi)$ is a sinusoidally modulated parabola,

$$V(\phi) = (\phi - \phi_{\text{ext}})^2/2L - E_{\text{J}} \cos(2\pi\phi/\phi_0) \,. \tag{3.153}$$

Apart from the missing stochastic force, the equation of motion (3.152) is homologous to the Langevin equation of a damped particle, Eq. (2.3).

For $\phi_{\text{ext}} = 0$, the SQUID potential $V(\phi)$ is symmetric. There is only a single stable minimum located at $\phi = 0$ in the regime $\beta_{\text{L}} \equiv 4\pi^2 LE_{\text{J}}/\phi_0^2 < 3\pi/2$. As β_{L} is steadily

---

[7]We consistently use for the phase jump across a Josephson junction the variable ψ and across a normal junction the variable φ.

increased above $3\pi/2$, by and by relative minima are formed. The flux trapped in one of these minima is metastable and decays either by thermal fluctuations over the barrier, or by quantum tunneling through the barrier, as discussed in Part III.

Josephson systems have been employed since the mid 1980s in order to study new specific quantum effects, such as macroscopic quantum tunneling (MQT) of the phase (or flux) [see Part III], as well as indirect (spectroscopic) evidence for quantum superpositions of macroscopic states [112, 113].

Flux qubit

When the external flux ϕ_{ext} is tuned to a half flux quantum, the potential (3.153) becomes symmetric in the shifted phase variable $\chi = 2\pi\phi/\phi_0 - \pi$,

$$V(\chi) = (\tfrac{1}{2}\chi^2/\beta_{\text{L}} + \cos\chi)\, E_{\text{J}}\,. \tag{3.154}$$

This becomes in quartic approximation

$$V_{\text{quart}}(\chi) = [\,1 - \tfrac{1}{2}(1 - 1/\beta_{\text{L}})\chi^2 + \tfrac{1}{24}\chi^4\,]\, E_{\text{J}}\,. \tag{3.155}$$

For $\beta_{\text{L}} > 1$, the potential $V_{\text{quart}}(\chi)$ is a symmetric double well. The two degenerate minima are at $\chi = \chi_{\pm}$, where $\chi_{\pm} = \pm\sqrt{6(1 - 1/\beta_{\text{L}})}$. These states correspond to the two different senses of rotation of the supercurrent in the ring. The barrier height is

$$V_{\text{b}} = \tfrac{3}{2}(1 - 1/\beta_{\text{L}})^2 E_{\text{J}}\,. \tag{3.156}$$

When the external flux is put out of tune, $\phi_{\text{ext}} \neq \tfrac{1}{2}\phi_0$, the double well becomes asymmetric. In the regime $\beta_{\text{L}} - 1 \ll 1$, the bias energy $\hbar\epsilon \equiv V_{\text{quart}}(\chi_+) - V_{\text{quart}}(\chi_-)$ depends linearly on the external flux ϕ_{ext} as $\hbar\epsilon = 4\pi\sqrt{6(\beta_{\text{L}} - 1)}(\phi_{\text{ext}}/\phi_0 - \tfrac{1}{2})\, E_{\text{J}}$.

The condition $\beta_{\text{L}} > 1$ requires a relatively large loop, which makes the system quite vulnerable by external noise. This impairment may be overcome by using smaller superconducting loops with three or four junctions [114, 115].

The environment of flux-qubit systems usually consists of resistive elements in the circuits needed for manipulations and measurements. These elements produce voltage and current noise. In the usual case the noise is Gaussian. Friction and noise can be introduced in the relevant model by coupling the flux ϕ dynamically to a harmonic heat bath, as discussed above in Section 3.1. Under these conditions, the dynamics of the extended flux qubit setting is well described by the spin-boson model with tunable bias and tunable coupling energy.

Charge qubit

The simplest form of a charge qubit is a superconducting charge box populated by excess Cooper pair charges. It consists of a small superconducting island ("box") connected to a superconducting electrode by a tunnel junction with capacitance C_{J} and coupling energy E_{J} [116]. A control gate voltage U_{g} is coupled to the system via a gate capacitor C_{g}. When the gap energy $\hbar\Delta_{\text{g}}$ is the largest relevant energy of the system, quasiparticle tunneling is suppressed so that quasiparticle excitations on the island are negligible. Then the system is described by the Hamiltonian

$$H = 4E_{\mathrm{C}}\,(N - N_{\mathrm{g}})^2 - E_{\mathrm{J}}\cos\psi\,, \tag{3.157}$$

where $E_{\mathrm{C}} = e^2/2(C_{\mathrm{g}} + C_{\mathrm{J}})$ is the single-electron charging energy and N is the number operator of excess Cooper pair charges on the island, which is conjugate to the phase of the order parameter, $N = -i\partial/\partial\psi$. The gate voltage U_{g} controls the gate charge $2eN_{\mathrm{g}}$ according to the relation $2eN_{\mathrm{g}} = C_{\mathrm{g}}U_{\mathrm{g}}$. In the regime $E_{\mathrm{C}} \gg E_{\mathrm{J}}$, the charge on the island is almost sharp. It is then convenient to introduce a basis of charge states, parametrized by the number N of Cooper pairs on the island,

$$Q = 2e\sum_N N\,|N\rangle\langle N|\,, \qquad e^{i\psi} = \sum_N |N+1\rangle\langle N|\,. \tag{3.158}$$

In this basis, the Hamiltonian (3.157) reads

$$H = \sum_N \left[\,4E_{\mathrm{C}}(N - N_{\mathrm{g}})^2\,|N\rangle\langle N| - \tfrac{1}{2}E_{\mathrm{J}}\left(|N+1\rangle\langle N| + |N\rangle\langle N+1|\right)\right]\,. \tag{3.159}$$

When the gate voltage is tuned to the symmetry point, $U_{\mathrm{g}} = e/C_{\mathrm{g}}$, the gate charge N_{g} is $\tfrac{1}{2}$. Hence the states with $N = 0$ and $N = 1$ are degenerate at energy E_{c}, while the state with $N = 2$ is at $9E_{\mathrm{C}}$. Thus, when $k_{\mathrm{B}}T \ll 9E_{\mathrm{C}}$, the charge box reduces to an effective two-state or spin-$\tfrac{1}{2}$ quantum system (qubit) of the form (3.136). The charge states with $N = 0$ and $N = 1$ correspond to the localized states $|R\rangle$ and $|L\rangle$ of the TSS. The coupling of the states is described by E_{J}, and the bias energy is

$$\delta E_{\mathrm{ch}}(U_{\mathrm{g}}) = 4E_{\mathrm{C}}[\,1 - 2N_{\mathrm{g}}(U_{\mathrm{g}})\,]\,. \tag{3.160}$$

When the single junction is replaced by two identical junctions in a loop, each with coupling energy E_{J}^0, and when the loop is penetrated by an external flux ϕ_{ext}, the effective coupling energy of the modified device is

$$E_{\mathrm{J}}(\phi_{\mathrm{ext}}) = E_{\mathrm{J}}^0\,\cos(\pi\phi_{\mathrm{ext}}/\phi_0)\,. \tag{3.161}$$

Thus, in this setting the trapped flux ϕ_{ext} in the auxiliary ring serves to provide an independent parameter which controls the strength of the coupling energy between the two localized states. The SQUID-controlled charge qubit is thus a TSS with tunable coupling energy and tunable bias energy,

$$H = -\frac{1}{2}E_{\mathrm{J}}(\phi_{\mathrm{ext}})\,\sigma_x - \frac{1}{2}\delta E_{\mathrm{ch}}(U_{\mathrm{g}})\,\sigma_z\,. \tag{3.162}$$

Quantum manipulations of Josephson charge qubits described by the Hamiltonian (3.162) have been performed [117].

In charge qubits the most serious source of decoherence is the noise of the biasing voltage ("charge noise"). Fluctuations of the voltage entail fluctuations of the phase ψ, which are described by the phase autocorrelation function given below in Eq. (3.249) with (3.250). In charge-qubit experiments, the major source of decoherence was found to be low-frequency $1/f$ noise which originates from background charge fluctuators [118, 119, 120]. When the system is affected by $1/f$ noise in the σ_z or longitudinal

($\propto \cos \varphi$) coupling, this source of decoherence exceeds Ohmic noise at very low temperature. There are experimental indications for a connection between high-frequency Ohmic noise, which is responsible for relaxation, and low frequency $1/f$ noise, which accounts for dephasing. This can be explained with a distribution of coherent two-level background charges which is log-uniform in the tunnel splitting and linear in the bias, like the distribution of two-level tunneling systems in amorphous materials, which has been introduced in order to understand anomalous low-temperature properties [121, 122, 123]. A discussion of $1/f$ noise is given in Section 22.5.

The possibility of tuning the coupling energy and the biasing energy offers twofold advantage. First, one can adjust the gate voltage U_g to the degeneracy point $U_{g,0}$ at which the biasing charging energy vanishes, $\delta E_{ch}(U_{g,0}) = 0$. Secondly, one can tune the applied flux ϕ_{ext} such that the Josephson coupling energy has an extremum, $\partial E_J(\phi_{ext})/\partial \phi_{ext} = 0$. At this optimal point, the qubit Hamiltonian (3.162) depends on voltage fluctuations δU and flux fluctuations $\delta \phi$ as

$$H_{opt} = -\frac{1}{2}\left(E_J(\phi_{ext,0}) + \frac{1}{2}\frac{\partial^2 E_J}{\partial \phi_{ext}^2}\bigg|_{\phi_{ext,0}} \delta\phi^2 \right) \sigma_x - \frac{1}{2}\frac{\partial E_{ch}}{\partial U_g}\bigg|_{U_{g,0}} \delta U\, \sigma_z . \qquad (3.163)$$

Hence at the optimal point, the energy splitting of the two states depends only quadratically on the fluctuations $\delta\phi$ [124]. Operation of the charge qubit at this optimal point can increase the coherence time by 2-3 orders of magnitude, as shown for a charge-phase qubit ($E_c \approx E_J$) [125]. Meanwhile several types of superconducting circuits based on Josephson junctions have given evidence that control of coherent quantum state evolution is feasible. Overall, they have proven to be promising candidates for qubit implementation [117], [125]–[128]. There has been made also substantial progress in the control of the interactions between individual flux qubits operating at the optimal point while retaining quantum coherence [129].

3.3 Microscopic models

For numerous physical system-plus-environment entities, the Hamiltonian can be determined from microscopic considerations.

In unpolar crystals like semiconductors and metals, a charged particle, which may be, e.g., an electron, a heavy interstitial particle, or a defect particle, distorts the lattice in its neighborhood, and when the particle moves, the lattice distortion moves with it. In the underlying microscopic description, the particle interacts with acoustic phonons. In the conceptually simplest case, the interaction is described by a scalar deformation potential. The particle together with its attached vibrating environment is called *acoustic polaron.*

An electron in an *ionic* crystal interacts with surrounding ions and thus polarizes the lattice. The interaction lowers the energy of the electron, and when the electron moves, the polarization cloud moves with it. The accompanying cloud of vibrating displacements of the ionic sub-lattices originates from the electron's coupling to longitudinal optical phonons. The electron together with its polarized environment is called *optical polaron*.

Even without lattice vibrations, there is a rigid periodic potential $H_{\text{rig}}(\mathbf{q})$ acting on the particle at position $\mathbf{q}$. The Hamiltonian in the absence of lattice vibrations is

$$H_{\text{S}} = \frac{p^2}{2M} + H_{\text{rig}}(\mathbf{q}) . \tag{3.164}$$

One distinguishes between *small* and *large* polaron, depending on whether it is represented by a localized state in a discrete lattice (small polaron) or by an extended state in a continuous medium (large polaron).

In the *large-polaron* system, the rigid potential $H_{\text{rig}}(\mathbf{q})$ effectuates, as if the electron would have an effective mass $M_{\text{eff}} > M$. This is the usual case for ionic crystals, semiconductors and metals. Furthermore, when the electron moves, the lattice is subject to accompanying local distortions. As a result, the actual mass of the polaron is even higher than the effective mass M_{eff} for a rigid lattice. This dynamical effect is the subject of Feynman's polaron problem [4].

In the opposite *small-polaron* limit, the particle moves in a discrete tight-binding lattice. The rigid lattice potential leads to spatially localized states. The overlap of these states is represented by a transfer amplitude Δ describing transitions of the particle between the localized states. The dynamics of the particle in the absence of lattice vibrations is then described by the transport Hamiltonian

$$H_{\text{S}} = -\frac{1}{2}\hbar\Delta \sum_n \left[a_{n+1}^\dagger a_n + a_n^\dagger a_{n+1} \right] , \tag{3.165}$$

where the operators $a_n^\dagger$ and a_n create and annihilate a particle at lattice site n. To this class belong motion of a single excess electron (or hole) in a molecular crystal [130, 131], reorientation processes of dipolar defects [132], and phonon assisted transport [133] and quantum diffusion of light interstitials and defects in crystals [134]–[136].

The Hamiltonian of the particle-phonon system with interaction due to linear lattice distortions, historically introduced by Herbert Fröhlich [137], reads

$$H = H_{\text{S}} + \sum_{\mathbf{k},\lambda} \hbar\omega_{\mathbf{k},\lambda} b_{\mathbf{k},\lambda}^\dagger b_{\mathbf{k},\lambda} + \sum_{\mathbf{k},\lambda} W_{\mathbf{k},\lambda}\, e^{i\mathbf{k}\cdot\mathbf{q}} \left(b_{\mathbf{k},\lambda} + b_{-\mathbf{k},\lambda}^\dagger \right) , \tag{3.166}$$

where the $b_{\mathbf{k},\lambda}$ and $b_{\mathbf{k},\lambda}^\dagger$ are the annihilation and creation operators for the phonon normal mode with wave vector $\mathbf{k}$ and branch index λ. In order that the interaction is Hermitian, we need to have $W_{\mathbf{k},\lambda}^* = W_{-\mathbf{k},\lambda}$. The linear coupling to the phonon modes leads to a shift of the equilibrium position and to a shift of the potential energy, as we have already seen in Subsection 3.1.1 [cf. Eq. (3.5)],

$$\Delta V_{\text{relax}} = -\sum_{\mathbf{k},\lambda} |W_{\mathbf{k},\lambda}|^2 / \hbar\omega_{\mathbf{k},\lambda} . \tag{3.167}$$

In connection with optical polarons, ΔV_{relax} is called the *relaxation energy*. The relaxation energy has no effect on the dynamics of the polaron. It is a matter of convenience whether or not ΔV_{relax} is subtracted from the Hamiltonian (3.166).

Below, the form factor $W_{\mathbf{k},\lambda}$ is calculated from a microscopic consideration both for acoustic and optical polarons. In many particular cases, the form factor is practically independent of the lattice coordinates. In this case, the Hamiltonian is translational invariant. The Fröhlich Hamiltonian has been of basic importance for numerous branches of solid state physics over the last five to six decades.

In metals, the polaron is also interacting with the conduction electrons. The electromagnetic interaction of the polaron distorts the electron gas or Fermi liquid. This results in a screening cloud of virtual electron-hole excitations around the particle which is dragged along as the particle moves. The high density of these excitations leads to various singular behaviors, as we shall see. These phenomena are known as Fermi surface effects [138].

3.3.1 Acoustic polaron: one-phonon and two-phonon coupling

The interaction of the charged particle with the lattice shifts the position of the atom n from the equilibrium value $\mathbf{R}^{(n)}$ to the actual position $\mathbf{X}^{(n)}$ by the displacement vector $\mathbf{u}^{(n)}$, $\mathbf{X}^{(n)} = \mathbf{R}^{(n)} + \mathbf{u}^{(n)}$. Since the displacement $\mathbf{u}^{(n)}$ is usually small, it is convenient to expand the particle-lattice potential in a power series in the displacement. We assume that the interaction potential depends only on the distance of the particle from the host atoms, and we express this property by the short-hand notation $H_{\mathrm{latt}}(\mathbf{q} - \{\mathbf{X}\})$. We then have [139]

$$H_{\mathrm{latt}}\left(\mathbf{q} - \{\mathbf{X}\}\right) = H_{\mathrm{rig}}(\mathbf{q}) + H_{\mathrm{lin}}(\mathbf{q}) + H_{\mathrm{quadr}}(\mathbf{q}) + \cdots , \tag{3.168}$$

where

$$H_{\mathrm{rig}}(\mathbf{q}) = H_{\mathrm{latt}}(\mathbf{q} - \{\mathbf{R}\}) ,$$

$$H_{\mathrm{lin}}(\mathbf{q}) = \sum_{n} \left(\nabla_{\mathbf{X}^{(n)}} H_{\mathrm{latt}}(\mathbf{q} - \{\mathbf{X}\}) \Big|_{\mathbf{u}^{(n)}=0} \right) \cdot \mathbf{u}^{(n)}(\mathbf{q}) , \tag{3.169}$$

$$H_{\mathrm{quadr}}(\mathbf{q}) = \frac{1}{2} \sum_{n,m} \mathbf{u}^{(n)}(\mathbf{q}) \cdot \left(\nabla_{\mathbf{X}^{(n)}} \nabla_{\mathbf{X}^{(m)}} H_{\mathrm{latt}}(\mathbf{q} - \{\mathbf{X}\}) \Big|_{\mathbf{u}^{(n,m)}=0} \right) \cdot \mathbf{u}^{(m)}(\mathbf{q}) .$$

The first term is the rigid lattice potential introduced already in Eq. (3.164). The displacement is expressed in terms of the usual bosonic annihilation and creation operators. We find it convenient to choose the form [140]

$$\mathbf{u}^{(n)}(\mathbf{q}) = i \sum_{\mathbf{k},\lambda} \left(\frac{\hbar}{2V\varrho\,\omega_{\mathbf{k},\lambda}} \right)^{1/2} \mathbf{e}(\mathbf{k},\lambda) \exp\left[i\,\mathbf{k}\cdot(\mathbf{q} - \mathbf{R}^{(n)}) \right] \left(b_{\mathbf{k},\lambda} + b^{\dagger}_{-\mathbf{k},\lambda} \right) , \tag{3.170}$$

where V is the volume of the lattice, ϱ is the mass density, and $\mathbf{e}(\mathbf{k},\lambda)$ is the polarization vector. Hermiticity of the Hamiltonian requires that $-i\,\mathbf{e}^{*}(\mathbf{k},\lambda) = i\,\mathbf{e}(-\mathbf{k},\lambda)$. We choose the polarization vectors $\mathbf{e}(\mathbf{k},\lambda)$ real, and with change of sign on reversal of the $\mathbf{k}$ direction,

$$\mathbf{e}(-\mathbf{k},\lambda) = -\mathbf{e}(\mathbf{k},\lambda) . \tag{3.171}$$

For notational clearness, we use in the sequel the short form

$$k := (\mathbf{k}, \lambda) \qquad \text{and} \qquad -k := (-\mathbf{k}, \lambda) \,. \tag{3.172}$$

Next we define the scalar coupling functions

$$\kappa_k^{(n)}(\mathbf{q}) = -i\nabla_{\mathbf{X}^{(n)}} H_{\text{latt}} \left(\mathbf{q} - \{\mathbf{X}\}\right)\Big|_{\mathbf{u}^{(n)}=0} \cdot \mathbf{e}(k) \equiv \mathbf{g}^{(n)}(q) \cdot \mathbf{e}(k) \,, \tag{3.173}$$

$$\gamma_{k,k'}^{(n,m)}(\mathbf{q}) = -\mathbf{e}(k) \cdot \left(\nabla_{\mathbf{X}^{(n)}} \nabla_{\mathbf{X}^{(m)}} H_{\text{latt}} \left(\mathbf{q} - \{\mathbf{X}\}\right)\Big|_{\mathbf{u}^{(n/m)}=0}\right) \cdot \mathbf{e}(k') \,. \tag{3.174}$$

The interaction which is linear in the lattice diplacement reads

$$H_{\text{lin}}(\mathbf{q}) = -\left(\frac{\hbar}{2V\varrho}\right)^{1/2} \sum_n \sum_k \frac{\kappa_k^{(n)}(\mathbf{q})}{\sqrt{\omega_k}} \left(b_k + b_{-k}^\dagger\right) \exp\left[i\,\mathbf{k}\cdot\left(\mathbf{q} - \mathbf{R}^{(n)}\right)\right]\,, \tag{3.175}$$

and the interaction which is quadratic in the displacement is given by

$$H_{\text{quadr}}(\mathbf{q}) = \frac{\hbar}{4V\varrho} \sum_{n,m} \sum_{k,k'} \frac{\gamma_{k,k'}^{(n,m)}(\mathbf{q})}{\sqrt{\omega_k \omega_{k'}}} \left(b_k + b_{-k}^\dagger\right)\left(b_{k'} + b_{-k'}^\dagger\right)$$

$$\times \exp\left[i\,\mathbf{k}\cdot\left(\mathbf{q} - \mathbf{R}^{(n)}\right)\right] \exp\left[i\,\mathbf{k}'\cdot\left(\mathbf{q} - \mathbf{R}^{(m)}\right)\right]\,. \tag{3.176}$$

The elementary processes of the interaction $H_{\text{lin}}(\mathbf{q})$ are absorption and emission of a single acoustic phonon. In the nth order of the perturbation series in $H_{\text{lin}}(\mathbf{q})$, the polaron absorbs and emits altogether n *uncorrelated* phonons. In contrast, the elementary process of the interaction $H_{\text{quadr}}(\mathbf{q})$, which is of second order in the lattice displacement, describes simultaneous absorption and emission of two phonons. In different terms, the contribution which is quadratic in $H_{\text{lin}}(\mathbf{q})$ is a single-phonon process of second order while the first order in $H_{\text{quadr}}(\mathbf{q})$ describes a genuine two-phonon process. We shall see in Section 4.2 that the effects of acoustic phonons on the tunneling of atoms between surfaces, as occurs in a scanning-tunneling microscope, are qualitatively different from atom tunneling in the bulk. The first case implies Ohmic dissipation, while dissipation in the second case belongs to the super-Ohmic variety. The correlated two-phonon contribution (3.176) is disregarded in most studies because it is usually small compared to the contributions of $H_{\text{lin}}(\mathbf{q})$. Interestingly, as we shall see in Subsection 4.2.5, the two-phonon coupling $H_{\text{quadr}}(\mathbf{q})$ brings about Ohmic dissipation for tunneling in the bulk. However, the viscosity coefficient turns out to be strongly temperature dependent.

3.3.2 Optical polaron

In *polar* crystals, such as sodium chloride, the ions of positive charge oscillate in anti-phase with the ions of negative charge. The relative displacement of the ionic sub-lattices gives rise to polarization charges. The latter induce a polarization field which scatters the electrons. The displacement mode which causes strong polarization is the *longitudinal optical* (LO) mode [$\lambda = \ell$ in Eq. (3.166)]. The relevant polarization vector is a unit vector in $\mathbf{k}$ direction,

$$\mathbf{e}(\mathbf{k}, \ell) = \hat{\mathbf{k}} . \tag{3.177}$$

Next we assume that the LO mode is independent of the positions of the individual ions and that the frequency of the optical phonon is constant, $\omega_{\mathbf{k},\ell} = \omega_{\mathrm{LO}}$. The relative displacement of the ionic sub-lattices by the longitudinal optical mode is then obtained from Eq. (3.170) as

$$\mathbf{u}(\mathbf{q}) = i \left(\frac{\hbar}{2V \varrho \, \omega_{\mathrm{LO}}} \right)^{1/2} \sum_{\mathbf{k}} \hat{\mathbf{k}} \, e^{i \mathbf{k} \cdot \mathbf{q}} \left(b_{\mathbf{k}} + b_{-\mathbf{k}}^{\dagger} \right) . \tag{3.178}$$

The optical mode induces a polarization field $\mathbf{P}(\mathbf{q})$ which is proportional to the displacement $\mathbf{u}(\mathbf{q})$. We may write

$$\mathbf{P}(\mathbf{q}) = U_{\mathrm{pol}} \, \mathbf{u}(\mathbf{q}) . \tag{3.179}$$

where U_{pol} is a kind of polarization charge per volume, yet to be determined. Generally, the polarization charge density ρ_{pol} is a source for a polarization field,

$$\rho_{\mathrm{pol}}(\mathbf{q}) = -\nabla \cdot \mathbf{P}(\mathbf{q}) \tag{3.180}$$

where $\mathbf{P}(\mathbf{q})$ itself is a gradient field of a scattering potential, $4\pi e \, \mathbf{P}(\mathbf{q}) = -\nabla V_{\mathrm{scatt}}(\mathbf{q})$. Thus, the scattering potential $V_{\mathrm{scatt}}(\mathbf{q})$ satisfies the Poisson equation

$$\nabla^2 V_{\mathrm{scatt}}(\mathbf{q}) = 4\pi e \, \rho_{\mathrm{pol}}(\mathbf{q}) = -4\pi e \, U_{\mathrm{pol}} \, \nabla \cdot \mathbf{u}(\mathbf{q}) , \tag{3.181}$$

where the electron has charge $-e$. The solution of Eq. (3.181) reads

$$V_{\mathrm{scatt}}(\mathbf{q}) = -4\pi e \, U_{\mathrm{pol}} \left(\frac{\hbar}{2V \varrho \, \omega_{\mathrm{LO}}} \right)^{1/2} \sum_{\mathbf{k}} \frac{e^{i \mathbf{k} \cdot \mathbf{q}}}{|\mathbf{k}|} \left(b_{\mathbf{k}} + b_{-\mathbf{k}}^{\dagger} \right) . \tag{3.182}$$

The electron's interaction with the modes of the polar crystal, Eq. (3.182), is exactly of the Fröhlich form (3.166) [4, 137, 140]. Comparing Eq. (3.182) with the coupling term in Eq. (3.166), the form factor of the longitudinal optical polaron $W_{\mathbf{k},\ell}$ reads

$$W_{\mathbf{k},\ell} = -4\pi e \, U_{\mathrm{pol}} \left(\frac{\hbar}{2V \varrho \, \omega_{\mathrm{LO}}} \right)^{1/2} \frac{1}{|\mathbf{k}|} . \tag{3.183}$$

Finally, the charge density U_{pol} can be deduced from the interaction potential between two electrons at fixed distance $|\mathbf{q}|$. The interaction energy originates from the relaxation of the ionic environment in which the charges are embedded. For a finite distance $\mathbf{q}$ of the two charges, we find from Eq. (3.167) the relaxation contribution

$$V_{\mathrm{relax}}(\mathbf{q}) = -2 \sum_{\mathbf{k}} \frac{|W_{\mathbf{k},\ell}|^2}{\hbar \omega_{\mathrm{LO}}} \, e^{i \mathbf{k} \cdot \mathbf{q}} . \tag{3.184}$$

With insertion of Eq. (3.183) we obtain

$$V_{\text{relax}}(\mathbf{q}) = -\frac{(4\pi e\, U_{\text{pol}})^2}{\varrho\, \omega_{\text{LO}}^2} \int \frac{d^3 k}{(2\pi)^3} \frac{e^{i\mathbf{k}\cdot\mathbf{q}}}{|\mathbf{k}|^2} \,, \tag{3.185}$$

which in real space is the Coulomb potential

$$V_{\text{relax}}(\mathbf{q}) = -\gamma\, e^2/|\mathbf{q}| \qquad \text{with} \qquad \gamma = 4\pi\, U_{\text{pol}}^2/(\varrho\, \omega_{\text{LO}}^2)\,. \tag{3.186}$$

The relaxation potential $V_{\text{relax}}(\mathbf{q})$ represents the contribution of the optical phonons to the dielectric screening of the Coulomb potential. This is the difference in screening between the cases of low and high frequencies. Thus we have

$$\frac{e^2}{\epsilon_0 |\mathbf{q}|} = \frac{e^2}{|\mathbf{q}|}\left(\frac{1}{\epsilon_\infty} - \gamma\right) \qquad \text{with} \qquad \gamma = \frac{1}{\epsilon_\infty} - \frac{1}{\epsilon_0}\,. \tag{3.187}$$

Here ϵ_∞ is the high-frequency (optical) and ϵ_0 the static dielectric constant. The polarization charge density U_{pol} emerges from Eqs. (3.186) and (3.187) as

$$U_{\text{pol}}^2 = \frac{\varrho\, \omega_{\text{LO}}^2}{4\pi}\left(\frac{1}{\epsilon_\infty} - \frac{1}{\epsilon_0}\right)\,. \tag{3.188}$$

One directly sees that U_{pol} is completely determined by measurable quantities. In the literature, it is customary to use the dimensionless polaron coupling constant

$$\alpha \equiv \left(\frac{1}{\epsilon_\infty} - \frac{1}{\epsilon_0}\right)\frac{e^2}{\hbar\omega_{\text{LO}}}\left(\frac{M\omega_{\text{LO}}}{2\hbar}\right)^{1/2}\,. \tag{3.189}$$

With the parameter α, the form factor for optical polarons (3.183) is

$$W_{\mathbf{k},\ell} = -\left(\frac{4\pi\alpha}{V}\right)^{1/2}\left(\frac{\hbar}{2M\omega_{\text{LO}}}\right)^{1/4}\frac{\hbar\omega_{\text{LO}}}{|\mathbf{k}|}\,. \tag{3.190}$$

With the coupling parameter α given by Eq. (3.189), the self-energy of the polaron can be put in a simple form [4]. Typical values of α run from about 1 to 20.

3.3.3 Interaction with fermions (normal and superconducting)

As an implication of the Pauli principle, the response of the noninteracting electron gas to a time-dependent local potential shows interesting characteristic features. The transient perturbations of the Fermi gas or Fermi liquid lead to a screening cloud of virtual electron-hole excitations which turns out to be very singular in character. The singular behavior can be seen, e. g., in the energy dependence of soft X-ray absorption or emission of metals (see, e.g., the review by Mahan [141], and by Ohtaka and Tanabe [142]). The singular response of the electron gas is also reflected in the electrical resistivity when dilute magnetic impurities are dissolved in a non-magnetic metallic host crystal. The understanding of these peculiar properties constitutes the crux of the Kondo problem (cf. the review by Tsvelik and Wiegmann [143]). Closely related to the above phenomena, quantum diffusion of charged interstitials in normally conducting metals also exhibits singular behavior, as shown first by Kondo.

The singular transient response of the conduction electrons results from the high density of electron-hole excitations around the Fermi surface.[8] When the particle moves, it drags behind it a screening cloud of electron-hole pairs. These excitations have bosonic character. As we shall see, they can adequately be represented by an Ohmic spectral density of a virtual coupling to a bosonic reservoir. This particular coupling leads, e.g., to self-trapping at zero temperature above a critical value of the coupling strength [144] – [149], and to anomalous temperature dependence such as the increase of electron transfer rates and of the diffusion coefficient with decreasing temperature [138]). This issue is studied in Subsec. 20.2.5 and Sec. 27.1. Ohmic friction fades away when the fermionic reservoir changes from the normally conducting to the superconducting state. We shall discuss the respective properties of the coupling in some detail in Subsection 4.2.8.

In this subsection I introduce the underlying Hamiltonians for the normal and the superconducting state of the fermions. I postpone the derivation of the effective action of the particle until Subsection 4.2.8. The description of the fermionic effects on the charged interstitial is based on the Hamiltonian

$$H = H_S(\mathbf{q}) + H_R + H_I(\mathbf{q}) , \qquad (3.191)$$

where H_S is the Hamiltonian of the system in the absence of the environment, and H_R and H_I are the terms due to the reservoir and the interaction, respectively. It is convenient to formulate H_R and H_I in the scheme of second quantization.

For fermions in the *normally conducting* state, the reservoir Hamiltonian has the standard free fermion form

$$H_R^{(\mathrm{nc})} = \sum_{\mathbf{k},\sigma} \hbar\omega_\mathbf{k}\, c_{\mathbf{k}\sigma}^\dagger c_{\mathbf{k}\sigma} , \qquad (3.192)$$

where $c_{\mathbf{k}\sigma}^\dagger$ is the creation operator for a conduction electron with wave vector $\mathbf{k}$ and spin polarization σ. The energy $\hbar\omega_\mathbf{k}$ is measured relative to the Fermi energy E_F.

The charged interstitial couples to local fluctuations of the electronic density via a potential $V(\mathbf{r})$. For a point particle at position $\mathbf{q}$, the interaction term reads

$$H_I^{(\mathrm{nc})}(\mathbf{q}) = \int_{L^3} d^3\mathbf{r}\, V(\mathbf{r} - \mathbf{q})\Psi^\dagger(\mathbf{r})\Psi(\mathbf{r}) . \qquad (3.193)$$

Upon using the normal mode expansion $\Psi^\dagger(\mathbf{r}) = L^{-3/2}\sum_{\mathbf{k},\sigma} e^{-i\,\mathbf{k}\cdot\mathbf{r}}\, c_{\mathbf{k},\sigma}^\dagger$, where L^3 is the quantization volume, one obtains

$$H_I^{(\mathrm{nc})}(\mathbf{q}) = \sum_{\mathbf{k},\mathbf{k}',\sigma,\sigma'} < \mathbf{k},\sigma|\,V\,|\mathbf{k}',\sigma' > e^{i\,(\mathbf{k}'-\mathbf{k})\cdot\mathbf{q}}\, c_{\mathbf{k}\sigma}^\dagger c_{\mathbf{k}'\sigma'} . \qquad (3.194)$$

The potential field $V(\mathbf{r})$ depends on the particular case. For a charged particle in a metal, it is the Coulomb potential, while in a superconductor, it is a screened Coulomb potential. Other choices, such as dipolar or multipolar couplings for neutral particles, are also pertinent.

---

[8]The Fermi surface effects are reviewed by Kondo in Ref. [138].

When the fermions are in the *superconducting* state, the reservoir is made up of BCS quasiparticles,

$$H_R^{(sc)} = \sum_{\mathbf{k},\alpha} \hbar\Omega_\mathbf{k}\,\gamma_{\mathbf{k}\alpha}^\dagger \gamma_{\mathbf{k}\alpha} \,. \tag{3.195}$$

Here, $\gamma_{\mathbf{k}\alpha}^\dagger$ is the creation operator for a quasi-particle with wave vector $\mathbf{k}$, spin polarization α, and excitation energy $\hbar\Omega_\mathbf{k}$, where

$$\hbar\Omega_\mathbf{k} = \hbar\left(\omega_\mathbf{k}^2 + \Delta_g^2\right)^{1/2}, \tag{3.196}$$

and $\hbar\Delta_g$ is the temperature dependent gap energy of the superconducting state. The operators $\gamma_{\mathbf{k}\alpha}^\dagger$ are connected with $c_{\mathbf{k}\sigma}$ and $c_{\mathbf{k}\sigma}^\dagger$ by Bogoliubov relations [150]

$$\gamma_{\mathbf{k}\uparrow}^\dagger = u_\mathbf{k} c_{\mathbf{k}\uparrow}^\dagger - v_\mathbf{k} c_{-\mathbf{k}\downarrow}\,, \qquad \gamma_{\mathbf{k}\downarrow}^\dagger = u_\mathbf{k} c_{\mathbf{k}\downarrow}^\dagger + v_\mathbf{k} c_{-\mathbf{k}\uparrow}\,. \tag{3.197}$$

Since the screening of the electrons in the superconducting state is the same as in the normal state [151], the interaction is the same as in Eq. (3.194). Next I wish to express the interaction as a function of the quasi-particle operators. The inversion of Eq. (3.197) gives

$$c_{\mathbf{k}\alpha}^\dagger = u_\mathbf{k}\gamma_{\mathbf{k}\alpha}^\dagger + \sum_\beta \rho_{\alpha\beta} v_\mathbf{k}\gamma_{-\mathbf{k}\beta}\,, \tag{3.198}$$

where the matrix ρ is

$$\rho = \begin{pmatrix} 0 & -1 \\ 1 & 0 \end{pmatrix}. \tag{3.199}$$

Insertion of Eq. (3.198) into Eq. (3.194) yields

$$H_I(\mathbf{q}) = \sum_{\mathbf{k},\mathbf{k}',\alpha,\alpha'} <\mathbf{k},\alpha|V|\mathbf{k}',\alpha'> e^{i(\mathbf{k}'-\mathbf{k})\cdot\mathbf{q}}$$

$$\times \left\{ u_\mathbf{k} u_{\mathbf{k}'}\gamma_{\mathbf{k}\alpha}^\dagger\gamma_{\mathbf{k}'\alpha'} + v_\mathbf{k} v_{\mathbf{k}'}\sum_{\sigma,\sigma'}\rho_{\alpha\sigma}\rho_{\alpha'\sigma'}\gamma_{-\mathbf{k}\sigma}\gamma_{-\mathbf{k}'\sigma'}^\dagger \right.$$

$$\left. + u_\mathbf{k} v_{\mathbf{k}'}\sum_{\sigma'}\rho_{\alpha'\sigma'}\gamma_{\mathbf{k}\alpha}^\dagger\gamma_{-\mathbf{k}'\sigma'}^\dagger + u_{\mathbf{k}'} v_\mathbf{k}\sum_\sigma \rho_{\alpha\sigma}\gamma_{-\mathbf{k}\sigma}\gamma_{\mathbf{k}'\alpha'} \right\}. \tag{3.200}$$

The first two terms of the interaction describe scattering of quasiparticles. The other terms are irrelevant at low temperature because they create and annihilate two quasiparticles, respectively. The scattering contribution can be written in the form

$$H_I^{(sc)}(\mathbf{q}) = \sum_{\mathbf{k},\mathbf{k}',\alpha,\alpha'} M(\mathbf{k},\alpha|\mathbf{k}',\alpha') \, e^{i(\mathbf{k}'-\mathbf{k})\cdot\mathbf{q}}\,\gamma_{\mathbf{k}\alpha}^\dagger\gamma_{\mathbf{k}'\alpha'}\,, \tag{3.201}$$

Using the relation $\gamma_i\gamma_j^\dagger = -\gamma_j^\dagger\gamma_i$ for $i \neq j$, one finds [150]

$$M(\mathbf{k},\alpha|\mathbf{k}',\alpha') = u_\mathbf{k} u_{\mathbf{k}'} <\mathbf{k},\alpha|V|\mathbf{k}',\alpha'> - v_\mathbf{k} v_{\mathbf{k}'}\sum_{\sigma,\sigma'}\rho_{\sigma'\alpha'}\rho_{\sigma\alpha} <-\mathbf{k}',\sigma'|V|-\mathbf{k},\sigma>\,.$$

The term $\sum_{\sigma\sigma'}\rho_{\sigma'\alpha'}\rho_{\sigma\alpha} <-\mathbf{k}',\sigma'|V|-\mathbf{k},\sigma>$ is essentially the matrix element of V in which the polarizations and momenta of the electrons have opposite signs.

Accordingly, the sign of this term depends on the behavior of the interaction under time-reversal. Thus we may write

$$M(\mathbf{k}, \alpha | \mathbf{k}', \alpha') = [u_{\mathbf{k}} u_{\mathbf{k}'} - \zeta\, v_{\mathbf{k}} v_{\mathbf{k}'}] < \mathbf{k}, \alpha |\, V\, | \mathbf{k}', \alpha' > , \qquad (3.202)$$

where $\zeta = \pm 1$. The square bracket in Eq. (3.202) is called the coherence factor of the transition. One finds with use of the definitions

$$u_{\mathbf{k}} v_{\mathbf{k}} = \frac{\Delta_g}{2\Omega_{\mathbf{k}}} ; \quad u_{\mathbf{k}}^2 = \frac{1}{2}\left(1 + \frac{\omega_{\mathbf{k}}}{\Omega_{\mathbf{k}}}\right) ; \quad v_{\mathbf{k}}^2 = \frac{1}{2}\left(1 - \frac{\omega_{\mathbf{k}}}{\Omega_{\mathbf{k}}}\right) , \qquad (3.203)$$

$$|M(\mathbf{k}, \sigma | \mathbf{k}', \sigma')|^2 = \frac{1}{2}\left\{1 + \frac{\omega_{\mathbf{k}} \omega_{\mathbf{k}'}}{\Omega_{\mathbf{k}} \Omega_{\mathbf{k}'}} - \zeta \frac{\Delta_g^2}{\Omega_{\mathbf{k}} \Omega_{\mathbf{k}'}}\right\} |< \mathbf{k}, \sigma |V|\mathbf{k}', \sigma' >|^2 . \qquad (3.204)$$

If the interaction does not break time-reversal symmetry, which is the case of most interest, we have $\zeta = +1$. Since to each value of $\Omega_{\mathbf{k}}$ there belong the values $\pm\omega_{\mathbf{k}}$, the term $\omega_{\mathbf{k}} \omega_{\mathbf{k}'}/\Omega_{\mathbf{k}} \Omega_{\mathbf{k}'}$ cancels out when Eq. (3.204) is summed over these two values.

Our subsequent study of the dissipative influences of the fermionic environment, which is given in in Subsection 4.2.8, will be based on the above microscopic Hamiltonians. For normally conducting electrons, these are Eq. (3.191) with Eq. (3.192) and Eq. (3.194), whereas in the superconducting case the underlying model is defined by Eq. (3.191) with Eq. (3.195) and Eq. (3.201).

3.3.4 Superconducting tunnel junction

A superconducting tunnel junction consists of two superconducting electrodes which are weakly coupled through a normally conducting barrier. The phase difference of the superconducting state across the barrier represents a collective variable which determines the dynamics of the device. The quasiparticle degrees of freedom together with the electromagnetic surroundings constitute a heat bath which causes phase fluctuations. A well-known simple phenomenological description of this scenario is rendered by the RCSJ model introduced in Subsec. 3.2.2. In this model, a resistive current and a displacement current are added to the tunneling current of Cooper pairs. Alternatively, one may start from the microscopic theory, as shown by Ambegaokar *et al.* [152], and by Larkin and Ovchinnikov [153]. We shall see that this leads to a model Hamiltonian equivalent to the form (3.8) with linear or nonlinear Ohmic friction, depending on the magnitude of the phase difference across the junction. The description of the bulk superconductor is based on the grand canonical Hamiltonian [154]

$$H_{\text{bulk}} = \int d^3\mathbf{r}\, \Psi_\sigma^\dagger(\mathbf{r}) \left[-\frac{\hbar^2}{2m}\left(\nabla - \frac{i e}{\hbar}\mathbf{A}\right)^2 - \mu + V(\mathbf{r})\right] \Psi_\sigma(\mathbf{r}) \qquad (3.205)$$

$$- \frac{1}{2}\int d^3\mathbf{r}\, \Psi_\sigma^\dagger(\mathbf{r})\, \Psi_{-\sigma}^\dagger(\mathbf{r})\, g(\mathbf{r})\, \Psi_{-\sigma}(\mathbf{r})\, \Psi_\sigma(\mathbf{r}) + \frac{1}{8\pi}\int d^3\mathbf{r}\,(\mathbf{h} - \mathbf{h}_{\text{ext}})^2 .$$

Here, summation over spin polarizations is implied. The vector field $\mathbf{A}$ represents the vector potential, and μ is the chemical potential. The Hamiltonian describes conduction electrons in a potential $V(\mathbf{r})$. This may account for impurities and boundaries.

The second line describes an effective attractive BCS interaction of strength $g(\mathbf{r})$. In the last line, the magnetic field contribution is added. Here, $\mathbf{h}$ is the magnetic field related to the vector potential $\mathbf{A}$, and $\mathbf{h}_{\text{ext}}$ is the externally applied magnetic field.

A tunnel junction with superconducting electrodes to the left (L) and to the right (R) of the barrier is conveniently described by the Hamiltonian

$$H = H_{\text{bulk,L}} + H_{\text{bulk,R}} + H_{\text{T}} + H_{\text{Q}} . \qquad (3.206)$$

The coupling of the leads is due to the transfer of electrons through the barrier and due to the Coulomb interaction. The former is described by the tunneling term

$$H_{\text{T}} = \int_{\mathbf{r} \in R} d^3\mathbf{r} \int_{\mathbf{r}' \in L} d^3\mathbf{r}' \left(T_{\mathbf{r}\mathbf{r}'} \Psi^{\dagger}_{R,\sigma}(\mathbf{r}) \Psi_{L,\sigma}(\mathbf{r}') + \text{h.c.} \right) . \qquad (3.207)$$

The spatial range of the tunneling amplitude $T_{\mathbf{r}\mathbf{r}'}$ is limited to the vicinity of the barrier. As long as we are interested in frequencies much smaller than the plasma frequency, the Coulomb interaction across the barrier effectively behaves as a capacitive interaction depending on the Cooper pair charges Q_L and Q_R stored on the electrodes and on the capacitor with capacitance C. The latter is determined by the geometry and by properties of the insulating barrier. Then the Coulomb term reads

$$H_{\text{Q}} = (Q_L - Q_R)^2 /8C , \qquad (3.208)$$

$$Q_{L(R)} = e \int_{\mathbf{r} \in \text{ins}} d^3\mathbf{r} \, \Psi^{\dagger}_{L(R),\sigma}(\mathbf{r}) \, \Psi_{L(R),\sigma}(\mathbf{r}) . \qquad (3.209)$$

The model (3.206) also describes inhomogeneous systems, such as normally conducting domains $[g(\mathbf{r}) = 0]$ and constrictions, as well as superconducting rings with a weak link in which the electrodes L and R are joined up in a loop.

The relevant quantum statistical properties of the system can be extracted from the generating functional

$$Z(\xi) = \text{tr}_{\mathbf{A},\Psi} \left\{ T_\tau \exp \left(-\frac{1}{\hbar} \int_0^{\hbar\beta} d\tau \left[H(\tau) - \hbar\xi(\tau)B(\tau) \right] \right) \right\} , \qquad (3.210)$$

where $B(\tau)$ is the quantity of interest and $\xi(\tau)$ is the usual source term. Here, T_τ is the time ordering operator along the imaginary-time axis τ (cf. Chapter 4). The trace is taken over the fermion fields Ψ and the vector potential $\mathbf{A}$. The respective imaginary-time effective action is discussed in Subsection 4.2.10.

3.4 Charging and environmental effects in tunnel junctions

With the rapid progress in micro-fabrication techniques, one can nowadays fabricate metallic tunnel junctions with capacitances C of 10^{-15} F or less. The charging energy $E_C = e^2/2C$ of an electron for such a small capacitance is larger than the thermal energy for temperatures of $1\,\text{K}$ or smaller. Therefore, in this low temperature regime electron transport through ultrasmall tunnel junctions is strongly influenced by the charging energy. Charging effects are also relevant in semiconductor nanostructures, e.g., quantum dots in a 2-D electron gas with typical capacitances of 10^{-15} F. In the still smaller molecular electron transfer systems, the charging energy can be so large that single-electron tunneling becomes observable even at room temperatures. On the one hand, the systems are small, on the other hand, they are still large enough that they can be connected to macroscopic current and voltage sources or probes. Just for this reason, the systems are sensitive to the electromagnetic environment.

In the sequel, we introduce the basics for the quantum mechanical treatment of charging effects. The concepts presented in this section and continued in Sec. 20.3 are generally important for metallic, semiconductor, and molecular systems. Here we concentrate the attention on metallic systems with a high electron density of states. A collection of research papers in the field of single-charge tunneling has been published as a special issue of Zeitschrift für Physik B [155]. A series of nine tutorial articles by renowned experts which deal with particular aspects of the field are compiled in the book *Single Charge Tunneling* [156]. We now introduce the global system for single-electron tunneling and present examples of electromagnetic environments. For a further study, we refer the interested reader to the review by Ingold and Nazarov [157].

Consider now first a local view at the tunnel junction, in which interaction with the electromagnetic environment is ignored.

For an ideal current-biased junction at $T = 0$ a charge can only tunnel if the balance of the electrostatic energy before and after the charge transfer is positive,

$$\Delta E_C \equiv Q^2/2C - (Q - e)^2/2C = e[U_J - e/2C] > 0. \qquad (3.211)$$

Here U_J is the voltage directly applied at the junction. This condition is satisfied if $Q > e/2$, which in turn implies that the applied voltage U_J across the junction must be larger than the critical voltage $U_{crit} = e/2C$. When the junction is current-biased with current I, the charge Q on the junction capacitor increases until the threshold charge $e/2$ is reached. At $Q = e/2$, the charge e tunnels through the junction thereby leaving a charge $Q = -e/2$ on the capacitor. After this, a new charging cycle begins. Altogether, the junction voltage performs sawtooth single electron tunneling (SET) oscillations [158, 154] with a frequency $\nu_{SET} = I/e$. [9]

Alternatively, we may drive the current through a normal- or superconducting junction with an ideal voltage source directly applied at the junction. Pursuant to the above argument, current flow is blocked in the regime $-e^*/2C < U_J < e^*/2C$, where e^* is the charge e in the normal state and $2e$ in the superconducting state. This is the *Coulomb blockade* phenomenon for a single junction. With an applied voltage larger than $e/2C$, one finds for a normal state junction at $T = 0$ an Ohmic (linear) current-voltage characteristics, which is shifted by the Coulomb gap $e/2C$, $I(U_J) = (U_J - e/2C)/R_T$ (see also Subsec. 20.3.2).

---

[9] By a similar argument, analogous so-called Bloch oscillations occur for a superconducting junction device with fundamental frequency $\nu_{Bloch} = I/2e$.

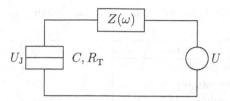

Figure 3.3: An ultrasmall tunnel junction with capacitance C and tunnel resistance R_T coupled to an ideal voltage source U via the external impedance $Z(\omega)$ which models frequency-dependent losses.

3.4.1 The global system for single electron tunneling

To begin with, it is useful to establish that the phase jump $\varphi(t)$ across the junction is linked to the history of the voltage U_J applied at the junction by the relation

$$\varphi(t) = e \int_0^t ds \, U_J(s)/\hbar \, . \qquad (3.212)$$

For applied voltage $U_J(t)$, the charge on the junction is $Q(t) = CU_J(t)$. The charge $Q(t)$ is conjugate to the variable $(\hbar/e)\varphi(t)$. In the quantum regime, both are operators satisfying the commutation relation $[\hat{\varphi}(t), \hat{Q}(t)] = i\,e$.

As the single-electron device is attached to leads, it is influenced, in reality, by stray capacitances, inductances, and resistances, as well as by quantum and thermal fluctuations. The simplest model in which we can study these influences on charge tunneling is a single junction in series with a general impedance $Z(\omega)$ and connected to an ideal voltage source U, as sketched in Fig. 3.3. The impedance $Z(\omega)$ represents the frequency response of the electromagnetic environment.

In a phenomenological modeling inspired by the harmonic oscillator reservoir, the electromagnetic environment is formed by LC-circuits with inductances L_α and capacitances C_α. To derive an "electromagnetic" Hamiltonian for the junction-plus-environment complex, we rely on the familiar correspondence between electrical and mechanical quantities assembled in Table I [157]. Using these equivalences, we get for a single voltage-biased LC-circuit the Hamiltonian

$$H = \frac{Q^2}{2C} + \frac{1}{2L} \left(\frac{\hbar}{2e} \right)^2 (\varphi - eUt/\hbar)^2 \, . \qquad (3.213)$$

The term $\varphi - eUt/\hbar$ in the magnetic field energy is because the phase differences at the capacitor and inductor should add up to $(e/\hbar)\,Ut$. The variables

$$\widetilde{Q}(t) = Q(t) - CU \, , \qquad \widetilde{\varphi}(t) = \varphi(t) - eUt/\hbar \, . \qquad (3.214)$$

represent the fluctuations of charge and phase about the mean values CU and $eUt/\hbar$. The commutator of the fluctuation operators is

$$\left[\hat{\widetilde{\varphi}}(t), \hat{\widetilde{Q}}(t) \right] = i\,e \, . \qquad (3.215)$$

Table I. Correspondence between mechanical and electrical quantities

Mechanical quantity	Electrical quantity
mass M	capacitance C
momentum p	charge Q
velocity $v = p/M$	voltage $U = Q/C$
coordinate x	phase φ
$[\hat{x}, \hat{p}] = i\hbar$	$[\hat{\varphi}, \hat{Q}] = ie$
spring constant f	inverse inductance $1/L$
harmonic oscillator	LC-circuit

Except for an irrelevant energy shift, the Hamiltonian in the variables (3.214) reads

$$H = \frac{\tilde{Q}^2}{2C} + \frac{1}{2L}\left(\frac{\hbar}{2e}\tilde{\varphi}\right)^2 . \tag{3.216}$$

According to this preparatory work, the electromagnetic environment consists of the capacitance of the junction, a set of LC-circuits, and the bilinear circuit-junction couplings. The Hamiltonian modeling the environmental coupling is

$$H_{\text{env}} = \frac{\tilde{Q}^2}{2C} + \sum_\alpha \left[\frac{q_\alpha^2}{2C_\alpha} + \left(\frac{\hbar}{e}\right)^2 \frac{1}{2L_\alpha}(\tilde{\varphi} - \varphi_\alpha)^2\right] . \tag{3.217}$$

The correspondence of the mechanical Hamiltonian (3.12) for $V(q) = 0$ with the electrodynamical Hamiltonian (3.217) becomes apparent, if we put $c_\alpha = m_\alpha \omega_\alpha^2$, which does not restrict generality, and employ the equivalences listed in Table I. Using the relations $\tilde{Q} = (\hbar/e)C\dot{\tilde{\varphi}}$ and $q_\alpha = (\hbar/e)C_\alpha \dot{\varphi}_\alpha$, the equations of motion emerge as

$$C\ddot{\tilde{\varphi}} + \sum_\alpha \frac{1}{L_\alpha}\tilde{\varphi} = \sum_\alpha \frac{1}{L_\alpha}\varphi_\alpha , \tag{3.218}$$

$$C_\alpha \ddot{\varphi}_\alpha + C_\alpha \omega_\alpha^2 \varphi_\alpha = \tilde{\varphi}/L_\alpha . \tag{3.219}$$

After elimination of the bath variable φ_α, the reduced dynamics is described by a Langevin equation for φ similar to that of a free Brownian particle (cf. Subsec. 3.1.2),

$$C\ddot{\tilde{\varphi}}(t) + \int_0^t ds\, Y(t-s)\dot{\tilde{\varphi}}(s) = \frac{e}{\hbar}I_{\text{noise}}(t) , \tag{3.220}$$

with a random current $I_{\text{noise}}(t)$ analogous to Eq. (2.13), and with[10]

$$Y(t) = \Theta(t)\sum_\alpha \frac{1}{L_\alpha}\cos(\omega_\alpha t) = \frac{1}{2\pi}\int_{-\infty}^{\infty} d\omega\, \tilde{Y}^*(\omega)\, e^{-i\omega t} . \tag{3.221}$$

---

[10]We use for $\tilde{Z}(\omega)$ and $\tilde{Y}(\omega)$ the engineering standard. These functions are integrated into our quantum mechanical convention as complex conjugate quantities [cf. footnote to Eq. (2.5).]

The Fourier transform of the memory-friction kernel $Y(t)$ is the admittance

$$\tilde{Y}^*(\omega) = \lim_{\varepsilon \to 0^+} \sum_\alpha \frac{1}{L_\alpha} \frac{-i\,\omega}{\omega_\alpha^2 - (\omega + i\,\varepsilon)^2} = \lim_{\varepsilon \to 0^+} \frac{2}{\pi} \int_0^\infty d\omega' \frac{J(\omega')}{\omega'} \frac{-i\,\omega}{\omega'^2 - (\omega + i\,\varepsilon)^2} . \quad (3.222)$$

With regard to a dense spectrum, we have put in the second form the spectral density

$$J(\omega) = \frac{\pi}{2} \sum_\alpha \frac{\omega_\alpha}{L_\alpha} \delta(\omega - \omega_\alpha) . \quad (3.223)$$

Again, the quantum Langevin equation for the phase operator $\hat{\tilde{\varphi}}$ coincides with the classical one. The noise current operator $\hat{I}_{\text{noise}}(t)$ obeys stationary Gaussian statistics with equilibrium averages found along the proceeding in Subsection 3.1.4,

$$\langle \hat{I}_{\text{noise}}(t) \rangle_{\hat{W}_{\text{R}}} = 0 ,$$

$$\langle \hat{I}_{\text{noise}}(t)\, \hat{I}_{\text{noise}}(0) \rangle_{\hat{W}_{\text{R}}} = \frac{\hbar}{\pi} \sum_\alpha \frac{\omega_\alpha}{L_\alpha} D_{\omega_\alpha}(t) = \frac{\hbar}{\pi} \int_0^\infty d\omega\, \omega \tilde{Y}'(\omega) D_\omega(t) . \quad (3.224)$$

Here, we have used the relation $J(\omega) = \omega\, \tilde{Y}'(\omega) = \omega \operatorname{Re} \tilde{Y}(\omega)$.

Next we move from the current to the dimensionless charge autocorrelator

$$Q_Q(t) \equiv (2\pi/e)^2 \langle\, [\hat{Q}(0) - \hat{Q}(t)]\hat{Q}(0) \rangle_{\hat{W}_{\text{R}}} . \quad (3.225)$$

We get from Eq. (3.224) by integrating twice over time

$$Q_Q(t) = \int_0^\infty d\omega\, \frac{G_Q(\omega)}{\omega^2} \Big(\coth(\tfrac{1}{2}\beta\hbar\omega)[\, 1 - \cos(\omega t)\,] + i \sin(\omega t) \Big). \quad (3.226)$$

With the resistance quantum $R_{\text{K}} = 2\pi\hbar/e^2$, the spectral density $G_Q(\omega)$ is found as

$$G_Q(\omega) = 2 R_{\text{K}} \omega \tilde{Y}'(\omega) . \quad (3.227)$$

The phase displacement correlation function $Q_\varphi(t) \equiv \langle\, [\hat{\varphi}(0) - \hat{\varphi}(t)]\hat{\varphi}(0) \rangle_{\hat{W}_{\text{R}}}$ is also of particular interest. In accordance with the Langevin equation (3.220), the phase fluctuation $\hat{\varphi}(t)$ is a linear functional of the current fluctuation $\hat{I}_{\text{noise}}(t)$,

$$\hat{\varphi}(t) = (e/\hbar) \int_{-\infty}^t dt'\, \chi(t - t')\hat{I}_{\text{noise}}(t) , \quad (3.228)$$

with the dynamical susceptibiliy

$$\tilde{\chi}(\omega) = \frac{\tilde{Z}_t^*(\omega)}{-i\,\omega} , \qquad \text{with} \qquad \tilde{Z}_t(\omega) = \frac{1}{i\,\omega C + \tilde{Y}(\omega)} . \quad (3.229)$$

Hence the total impedance $\tilde{Z}_t(\omega)$ consists of the impedance $\tilde{Z}(\omega) = 1/\tilde{Y}(\omega)$ due to the electromagnetic environment with the capacitance C in parallel connection. We then obtain for the correlator $Q_\varphi(t)$ upon employing Eqs. (3.228) and (3.224) the expression [cf. analogy with Eqs. (3.76)–(3.81) and (3.74)][11]

---

[11] The spectral coupling densities $G_Q(\omega)$ and $G_\varphi(\omega)$ are in correspondence with the spectral density $G(\omega)$ introduced in Eq. (3.148). Likewise, the correlation functions $Q_Q(t)$ and $Q_\varphi(t)$ are in correspondence with the pair interaction $Q(t)$ introduced in Eq. (18.42). This function is central in the charge representation of dissipative tight-binding models.

$$Q_\varphi(t) = \int_0^\infty d\omega \, \frac{G_\varphi(\omega)}{\omega^2} \Big(\coth(\tfrac{1}{2}\beta\hbar\omega)[\,1 - \cos(\omega t)\,] + i\sin(\omega t) \Big) \qquad (3.230)$$

with the spectral density

$$G_\varphi(\omega) = (e^2/\pi\hbar)\,\omega^2\tilde{\chi}''(\omega) = 2\omega\,\tilde{Z}_t'(\omega)/R_K\,. \qquad (3.231)$$

An alternative approach is presented in Subsection 4.2.11. There we calculate the imaginary-time correlation function $W_\varphi(\tau) = Q_\varphi(z = -i\,\tau)$ upon performing average of the phase correlator $e^{i\,[\varphi(0) - \varphi(\tau)]}$ with the Gaussian weight $\exp[-S_{\text{eff},0}(\varphi)/\hbar]$, where $S_{\text{eff},0}(\varphi)$ is the effective Euclidean action of a LC-circuit environment.

Apparently, any particular harmonic electromagnetic environment can be modeled by a suitable choice of the parameters L_α and C_α in the Hamiltonian (3.217). Any real physical systems is ultimately described by a continuous spectral density of the coupling. Hence, sums over α are replaced in the end by spectral integrals.

So far we have treated the junction merely as a capacitor. Next, we allow for tunneling of charge through the junction by adding the tunneling Hamiltonian

$$H_T' = \sum_{k,k',\sigma} T_{k,k'}\, c_{k',\sigma}^\dagger c_{k,\sigma}\, e^{-i\varphi} + \text{h.c.}\,. \qquad (3.232)$$

The operators $c_{k,\sigma}$ and $c_{k,\sigma}^\dagger$ are annihilation and creation operators for fermionic quasiparticles with wave vector k, energy E_k, and spin polarization σ. The wave vectors k and k' pertain to quasiparticles in the left and right electrode of the tunnel barrier, respectively.

Just as $e^{-i\,p_0\hat{q}/\hbar}$ is the momentum translational operator, the unitary operator $e^{-i\hat{\varphi}}$ is the charge translational operator, as follows from Table I on page 62,

$$e^{i\hat{\varphi}}\,\hat{Q}\,e^{-i\hat{\varphi}} = \hat{Q} - e\,. \qquad (3.233)$$

The junction model is completed by adding a Hamiltonian describing the quasi-particles in the left and right electrode, respectively,

$$H_{\text{qp}}' = \sum_{k,\sigma} E_k\, c_{k,\sigma}^\dagger c_{k,\sigma} + \sum_{k',\sigma} E_{k'}\, c_{k',\sigma}^\dagger c_{k',\sigma}\,. \qquad (3.234)$$

Next, we perform time dependent unitary transformations of H_T' and H_{qp}' in order to replace in Eq. (3.232) the phase φ by the phase $\tilde{\varphi} = \varphi - eUt/\hbar$. This is achieved with the unitary operator acting on quasiparticle operators in the left electrode,

$$\mathcal{U} = \prod_{k,\sigma} \exp\Big(i\,\frac{e}{\hbar}\, U t\, c_{k,\sigma}^\dagger c_{k,\sigma} \Big)\,, \qquad (3.235)$$

yielding the tunneling Hamiltonian

$$H_T = \mathcal{U}^\dagger H_T' \mathcal{U} = \sum_{k,k',\sigma} T_{k,k'}\, c_{k',\sigma}^\dagger c_{k,\sigma}\, e^{-i\tilde{\varphi}} + \text{h.c.}\,. \qquad (3.236)$$

The unitary transformation has no effect on H_{env}, whereas it shifts the quasiparticle energies in the left electrode by $e\,U$,

$$H_{\mathrm{qp}} = \mathcal{U}^\dagger H'_{\mathrm{qp}} \mathcal{U} - i\hbar \mathcal{U}^\dagger \partial \mathcal{U}/\partial t$$
$$= \sum_{\mathbf{k},\sigma} [\, E_{\mathbf{k}} + eU \,]\, c^\dagger_{\mathbf{k},\sigma} c_{\mathbf{k},\sigma} + \sum_{\mathbf{k}',\sigma} E_{\mathbf{k}'}\, c^\dagger_{\mathbf{k}',\sigma} c_{\mathbf{k}',\sigma} \, . \tag{3.237}$$

The transformed Hamiltonian of the tunnel junction with the quasiparticles in the leads and with the electromagnetic environment is

$$H = H_{\mathrm{T}} + H_{\mathrm{qp}} + H_{\mathrm{env}} \, . \tag{3.238}$$

The charge degrees of freedom in H_{env} are coupled with the quasiparticles in H_{qp} via the tunneling Hamiltonian H_{T}. This coupling results in a nonlinear current-voltage characteristics of the tunnel junction, as we discuss in Sec. 20.3.

For weak tunneling, the charge on the junction is nearly sharp. It is then natural to choose as basis the discrete charge representation

$$Q = e \sum_N N |N\rangle\langle N| \, , \qquad e^{i\varphi} = \sum_N |N+1\rangle\langle N| \, . \tag{3.239}$$

In this limit, the phase fluctuations described by the correlation function $Q_\varphi(t)$, Eq. (3.230), are large. In Section 20.3, single-electron tunneling is considered to lowest order in the tunneling coupling. There we shall see that the energy-dependence of the Golden Rule tunneling rate is governed by the probability density $P(E)$ for exchange of energy E with the environment. The function $P(E)$ will turn out as the Fourier transform of $e^{-Q_\varphi(t)}$.

On the other hand, for strong tunneling, the jump of the phase at the junction is nearly sharp, whereas the charge fluctuations at the junction are large. Then, it is favorable to switch over to the discrete phase representation

$$\varphi = 2\pi \sum_n n |n\rangle\langle n| \, , \qquad e^{i 2\pi Q/e} = \sum_n |n\rangle\langle n+1| \, . \tag{3.240}$$

Now, the probability for exchange of energy with the environment will turn out as the Fourier transform of the function $e^{-Q_Q(t)}$, where $Q_Q(t)$ is given in Eq. (3.226).

These two different representation are dual to each other. We shall discuss the charge-phase duality in the context of Cooper pair tunneling in Subsection 28.3.

The full weak- and strong-tunneling series expansions of charge transport across a weak link and a weak constriction, respectively, will be discussed in Section 31.1.

3.4.2 Resistor, inductor, and transmission lines

A significant case is, when the environmental impedance is described by an Ohmic resistor, $Z(\omega) = R$. Then we obtain from Eqs. (3.229) and (3.231)

$$G_\varphi(\omega) = 2\alpha \frac{\omega}{1 + (\omega/\omega_{\mathrm{D}})^2} \, . \tag{3.241}$$

Here we have introduced a dimensionless resistance α and a cutoff frequency ω_{D},

$$\alpha = R/R_{\mathrm{K}} = R\,e^2/2\pi\hbar\,, \qquad \omega_{\mathrm{D}} = 1/RC = E_C/\pi\alpha\hbar\,, \qquad (3.242)$$

and $E_C = e^2/2C$ is the charging energy. The spectral density is Ohmic at low frequencies, and the junction capacitance provides a cutoff of the Drude form (3.44).

When the impedance $Z(\omega)$ is purely inductive, $Z(\omega) = i\,\omega L$, the real part of the total impedance (3.229) is given by $Z_{\mathrm{t}}'(\omega) = (\pi/2C)\,\delta(\omega - \omega_{\mathrm{L}})$, where $\omega_{\mathrm{L}} = 1/\sqrt{LC}$ is the resonance frequency of the LC circuit. Hence the junction is effectively coupled to this single environmental mode,

$$G_\varphi(\omega) = (E_C/\hbar)\,\omega_{\mathrm{L}}\,\delta(\omega - \omega_{\mathrm{L}})\,. \qquad (3.243)$$

When a resistor is put in series with an inductor, $Z(\omega) = R + i\,\omega L$, the resonance at frequency ω_{L} is broadened and the quality factor $Q_{\mathrm{qual}} \equiv \omega_{\mathrm{D}}/\omega_{\mathrm{L}} = L/C$ becomes finite. Upon employing Eqs. (3.229) and (3.231), we get the spectral density

$$G_\varphi(\omega) = 2\alpha\,\frac{\omega}{[\,1 - (\omega/\omega_{\mathrm{D}})^2 Q_{\mathrm{qual}}^2\,]^2 + (\omega/\omega_{\mathrm{D}})^2}\,. \qquad (3.244)$$

This expression includes the Ohmic case (3.241) in the limit $Q_{\mathrm{qual}} \to 0$, and the single mode case (3.243) in the limit $Q_{\mathrm{qual}} \to \infty$. Thus, upon tuning the quality factor, the characteristics of the environment can be varied between resonant and resistive.

So far, we have considered impedances which are built by one or two lumped circuit elements. A more realistic model adapted to real experiments should include the coaxial leads attached to a junction as a resistive transmission line with distributed resistors, inductors, and capacitors. A corresponding model for $Z(\omega)$ can be thought of as a ladder of discrete building blocks, in which the inductor and the resistor are in series along a section of the leg, and the capacitor is on the attached rung. Putting the resistance, inductance, and capacitance per building block as R_0, L_0, and C_0, one finds for an infinitely long transmission line the impedance [157]

$$Z(\omega) = \sqrt{(R_0 + i\,\omega L_0)/(i\,\omega C_0)}\,. \qquad (3.245)$$

There are two important limiting cases:

- A *LC* transmission line has the purely resistive impedance $Z_{\mathrm{LC}} = \sqrt{L_0/C_0}$. The corresponding spectral density $G_\varphi(\omega)$ is Ohmic with Drude cutoff, Eq. (3.241), with $\alpha = \sqrt{L_0/C_0}/R_{\mathrm{K}}$ and $\omega_{\mathrm{D}} = \sqrt{C_0/L_0}/C$.

- A *RC* transmission line has impedance $Z_{\mathrm{RC}}(\omega) = \sqrt{R_0/(i\,\omega C_0)} \propto \omega^{-1/2}$. With $\tilde{Z}_{\mathrm{t}} = 1/[\,i\,\omega C + 1/Z_{\mathrm{RC}}(\omega)\,]$, the spectral density (3.231) is found as

$$G_\varphi(\omega) = 2\alpha\,\frac{\tilde{\omega}^{1/2}\,\omega^{1/2}}{1 + (2\omega/\omega_{\mathrm{c}})^{1/2} + \omega/\omega_{\mathrm{c}}}\,, \qquad (3.246)$$

where $\alpha = R_0/R_{\mathrm{K}}$, $\tilde{\omega} = 1/(2R_0C_0)$, and $\omega_{\mathrm{c}} = C_0/(C^2 R_0)$. Thus in the frequency range $\omega \ll \omega_{\mathrm{c}}$, the spectral density modeling the coupling of the junction to a *RC* transmission line is sub-Ohmic with power $s = \frac{1}{2}$, $G_\varphi(\omega) = 2\alpha\,\tilde{\omega}^{1/2}\,\omega^{1/2}$, and the parameters α and $\tilde{\omega}$ do not depend on the junction parameter C.

I shall study the effect of these diverse electromagnetic environments on single-electron tunneling in Section 20.3.

3.4.3 Charging effects in Josephson junctions

There are two kinds of charge carriers traversing the barrier in tunnel junctions: Cooper pairs and fermionic quasiparticles.[12] As the coupling between Cooper pairs and quasiparticles is small, the tunneling of these entities can be treated separately.

In Subsec. 3.4.1, the concept of a dynamical phase instead of a coordinate turned out to be very useful to describe quasiparticle tunneling through a normal junction. Clearly the same concept applies for quasiparticle tunneling through a superconducting junction, and thus the same methodology can be applied. The important difference however is that the density of states of quasiparticles in a superconductor close to the gap strongly depends on energy, as described below in Subsec. 4.2.9. This is in striking contrast to the constant density of states around the Fermi energy for a normal junction. At temperature well below the critical temperature of the superconductor and applied voltage well below the gap voltage, the tunneling of quasiparticles can be disregarded.

The phase concept plays an even clearer role in the superconducting case. Since the charge of a Cooper pair is twice the quasiparticle charge, we write the Josephson phase as $\psi = 2\varphi$, where φ is again defined in Eq. (3.212). Tunneling of Cooper pairs through the barrier gives rise to the potential [cf. Eq. (3.153)]

$$H_{\mathrm{J}} = -E_{\mathrm{J}} \cos(2\varphi) \qquad (3.247)$$

Here, ψ is the phase leap of the Cooper pair wave function across the junction. The coupling energy depends linearly on the critical current, $E_{\mathrm{J}} = (\hbar/2e)I_{\mathrm{c}}$. The operator relation $e^{i\,2\hat{\varphi}}\hat{Q}\,e^{-i\,2\hat{\varphi}} = \hat{Q} - 2e$ points up that tunneling of a Cooper pair changes the charge on the junction by $2e$. Again, we assume that the junction has capacitance C and is coupled to an ideal voltage source U via a general impedance $Z(\omega)$, as sketched in Fig. 3.3. There are two observations:

- As the Josephson phase ψ is 2φ and the cooper pair charge is $2e$, the Hamiltonian of the environment is the same as given in Eq. (3.217).

- Since the Cooper pairs form a condensate, there are no extra degrees of freedom. Hence a Hamitonian analogous to H_{qp} in Eq. (3.234) is absent.

In short, the Hamiltonian of the Josephson junction coupled to the electromagnetic surroundings is

$$H = \frac{Q^2}{2C} + \sum_{\alpha} \left[\frac{q_{\alpha}^2}{2C_{\alpha}} + \left(\frac{\hbar}{e}\right)^2 \frac{1}{2L_{\alpha}} \left(\varphi - eUt/\hbar - \varphi_{\alpha}\right)^2 \right] - E_{\mathrm{J}} \cos(2\varphi) . \qquad (3.248)$$

It is added for completeness that the charge q_{α} is conjugate to $(\hbar/e)\varphi_{\alpha}$, and the charging energy for Cooper pairs is $E_{\mathrm{C}} = 2e^2/C$.

---

[12] The effective action governing the equilibrium properties of the junction including quasiparticle and Cooper pair tunneling is considered below in Subsection 4.2.10.

In the regime $E_J \ll E_C$ discussed until now the charge on the junction is quite sharp. Hence the charge representation (3.158) is appropriate, and the coupling H_J may be treated as a perturbation. The electromagnetic coupling leads to phase fluctuations with the autocorrelation function $Q_\psi(t) = \langle [\psi(0) - \psi(t)]\psi(0) \rangle_{\hat{W}_R}$ given by

$$Q_\psi(t) = \int_0^\infty d\omega \, \frac{G_\psi(\omega)}{\omega^2} \Big(\coth(\tfrac{1}{2}\beta\hbar\omega)[1 - \cos(\omega t)] + i\sin(\omega t) \Big) \qquad (3.249)$$

with the spectral density of the coupling

$$G_\psi(\omega) = 2\omega \, Z'_t(\omega)/R_Q . \qquad (3.250)$$

Here, $Z_t(\omega)$ is the total impedance of the circuit, as specified in Eq. (3.229). The reference value is the resistance quantum for Cooper pairs, $R_Q = R_K/4 = 2\pi\hbar/4e^2$. Observe the close resemblance with the normal junction, Eq. (3.230) with (3.231). The golden rule rate in the regime $E_J \ll E_C$ is studied below in Subsec. 20.3.4.

In the opposite limit of large Josephson coupling, $E_J \gg E_C$, the phase is rather sharp, and the environmental influences produce charge fluctuations. The appropriate phase representation and the relevant correlation functions of the charge fluctuations are presented in Sec. 28.3. There, a striking and meanwhile well-known phase-charge duality transformation is studied. This symmetry property establishes profound relations between tunneling of Cooper pairs in the weak-coupling limit and tunneling of the phase in the strong-coupling limit.

3.5 Nonlinear quantum environments

Generally, the low-energy physics of an intricate system which is coupled to complex surroundings can be studied upon using an effective Hamiltonian. The reduction usually leads to one of few canonical forms. For instance, a double well may be reduced to a two-state system, and a reservoir mode with a whole ladder of roughly equidistant excited states may be represented by a harmonic oscillator. The opposite extreme case is a nonlinear bath mode with only one or few accessible excited states. This mode behaves like a spin. Evidently, a distribution of such modes forms a bath of spins. As opposed to a harmonic bath, a spin bath features saturation at elevated temperature. Every nonlinear bath is situated between these two idealized cases.

Consider a composite Hamiltonian analogous to Eq. (3.1) consisting of a system part, a nonlinear reservoir part, and an interaction part with a counter term,

$$H = \frac{p^2}{2M} + V(q) + H_R(\{\mathbf{x}\}) - F(q)\sum_\alpha c_\alpha x_\alpha + \frac{1}{2}F^2(q)\sum_\alpha c_\alpha^2 \tilde{\chi}_\alpha(0) . \qquad (3.251)$$

The counter-term is defined in terms of the static susceptibility $\tilde{\chi}_\alpha(\omega = 0)$ of the nonlinear mode x_α. It would reduce to the previous form (3.6) with (3.7) for a linear bath. It is straight again to write down the equations of motion and to eliminate the reservoir coordinates. This is most conveniently performed in frequency space upon introducing the dynamical susceptibility of the reservoir mode x_α. We then find an equation of motion of the form (3.20) in which the Fourier transform of the damping kernel has the form (3.30). The corresponding expression for the spectral density of the coupling is as in Eq. (3.31),

$$J(\omega) = \Theta(\omega) \sum_\alpha c_\alpha^2 \, \tilde{\chi}_\alpha''(\omega) \, . \tag{3.252}$$

The derivation reveals that the form (3.252) is generally valid for any composite system in which the coupling is *linear* in the bath coordinate. For a nonlinear bath, the dynamical susceptibility $\tilde{\chi}(\omega)$ usually differs drastically from the harmonic form (3.17). In general, it depends on $\hbar$ and on temperature. Hence the spectral density of the coupling for a nonlinear environment is temperature-dependent as well.

A "nanomagnet", i.e., a mono-domain particle, may contain as many as 10^8 magnetically ordered spins. Such a giant spin, denoted by $\mathbf{S}$, is usually coupled to surrounding nuclear spins and to paramagnetic impurity spins. The global Hamiltonian for a central spin bi-linearly coupled to a spin bath in a magnetic field and with internal interactions has the canonical form

$$H\left(\mathbf{S}, \{\sigma^\alpha\}\right) = H_{\mathbf{S}} + \frac{1}{S} \sum_\alpha E_\alpha \mathbf{S} \cdot \sigma^\alpha + \sum_\alpha \mathbf{h}^\alpha \cdot \sigma^\alpha + \tfrac{1}{2} \sum_{\alpha,\alpha';i,j} V_{ij}^{\alpha,\alpha'} \sigma_i^\alpha \sigma_j^{\alpha'} \, .$$

Many different forms for the giant spin Hamiltonian $H_{\mathbf{S}}$ have been discussed in the literature (cf. the book [159] and the review [160], and references therein). A simple example of the giant spin Hamiltonian $H_{\mathbf{S}}$ is the biaxial (easy axis/easy plane) form

$$H_{\mathbf{S}} = \left(-K_\| S_z^2 + K_\perp S_x^2\right)/S + g\mu_{\mathrm{B}} H_z S_z \, , \tag{3.253}$$

in which tunneling between the classical minima at $S_z = \pm S$ is accomplished by the symmetry-breaking transverse $K_\perp$-term, and we have added a magnetic field term. The low-energy physics of the Hamiltonian (3.253) can be understood in terms of the spin-$\tfrac{1}{2}$ Hamiltonian (3.136). The passage from the form (3.253) to the form (3.136) for $S \gg 1$ has been studied using WKB methods [162] and instanton methods [163]. Single molecule magnets like $Mn_{12}ac$ are conveniently described by similar spin Hamiltonians [164].

For giant spin systems, there are two scenarios. In the usual case, the coupling between the giant spin and the nuclear spins is very large. Then the nuclear spins are slaved to the giant spin, and the spectrum of the spin environment is drastically modified. The coupling to the nuclear spins may introduce three different effects [160]. Transitions of the giant spin may cause a change of the phase in the nuclear bath state, which then reacts back upon the giant spin and induces phase randomization ("topological decoherence"). Secondly, transitions of the giant spin may be hindered by a mismatch between the initial and final bath state ("orthogonality blocking"). This would lead to a Franck-Condon type dressing factor (cf. Subsection 18.1.4). Thirdly, the spread of coupling energies of the nuclear spins leads to a spread of bias energies which may bring the initial state of the tunneling giant spin out of resonance with the final state. This would result in a Landau-Zener type suppression of tunneling transitions ("degeneracy blocking").

The strong-coupling case contrasts with the usual weakly-coupled harmonic oscillator environment. It is impossible to integrate out the spin bath using the functional averaging method described in the next chapter, and one has to study the composite system directly, e.g., the instanton trajectory in the full system-bath space [165].

In the other scenario, the coupling to an individual bath spin is weak. In this case, the treatment is like that of a boson bath. The resulting spectral density $J_{\text{spin}}(\omega)$ has the form (3.252), in which $\tilde{\chi}_\alpha(\omega)$ is the dynamical susceptibility for a spin degree. Consider as example a global system with a system part $H_S(p,q)$, a spin-$\frac{1}{2}$ bath with excitation frequencies $\{\omega_\alpha\}$, a bilinear coupling term and a counter term,

$$H = H_S(p,q) - \tfrac{1}{2}\sum_\alpha \hbar\omega_\alpha \sigma_x^\alpha - \tfrac{1}{2}\sum_\alpha c_\alpha q_0^{(\alpha)}\sigma_z^\alpha q + \tfrac{1}{2}\sum_\alpha c_\alpha^2 \tilde{\chi}_\alpha(0)q^2 \ . \qquad (3.254)$$

The coupling terms in Eq. (3.12) and (3.254) correspond to each other at $T=0$ if we identify the length $\frac{1}{2}q_0^{(\alpha)}$ with the position spread of the equivalent bath oscillator with eigenfrequency ω_α in the ground state, $\frac{1}{2}q_0^{(\alpha)} = \sqrt{\hbar/2m_\alpha\omega_\alpha}$ (cf. Subsection 6.4.3).

The spin correlation function is $\text{Re}\langle \sigma_z^\alpha(t)\sigma_z^\alpha(0)\rangle = \cos(\omega_\alpha t)$. Taking the Fourier transform and anticipating the fluctuation dissipation theorem (6.24), we obtain

$$\tilde{\chi}''_{\alpha,\,\text{spin}}(\omega) = (\pi/\hbar)\tanh(\beta\hbar\omega_\alpha/2)\delta(\omega - \omega_\alpha) \ . \qquad (3.255)$$

With the equivalent oscillator susceptibility, $\tilde{\chi}_\alpha(\omega) = (\frac{1}{2}q_0^{(\alpha)})^2\tilde{\chi}_{\alpha,\,\text{spin}}(\omega)$, we find

$$J_{\text{spin}}(\omega) = \Theta(\omega)\sum_\alpha c_\alpha^2\,\tilde{\chi}''_\alpha(\omega) = \frac{\pi}{2}\sum_\alpha \frac{c_\alpha^2}{m_\alpha\omega_\alpha}\tanh(\beta\hbar\omega_\alpha/2)\,\delta(\omega-\omega_\alpha) \ . \qquad (3.256)$$

The same form is found by calculation of the influence function (see Section 4.2) of a spin bath up to second order in c_α [166]. Because of the non-Gaussian nature of the spin-bath, the cumulant expansion does not break off at this order. Cumulants of higher order in c_α are not considered here. We see from Eq. (3.256) that, to order c_α^2, the spectral densities of a spin-$\frac{1}{2}$ bath and a boson bath are related by

$$J_{\text{spin}}(\omega) = J_{\text{boson}}(\omega)\tanh(\beta\hbar\omega/2) \ . \qquad (3.257)$$

The two reservoirs yield the same spectral coupling function at zero temperature. At finite temperature, the spin bath has a smaller effect on the system owing to the possibility for saturation of the populations of excited bath states. The striking counter-example to a spin bath is when the energy gap between neighboring excited states of a nonlinear bath mode is decreasing with increasing energy. In this case, we find enhancement of $J(\omega)$ compared with the spectral density of a harmonic bath.

The findings of this section can be summarized as follows. For weak coupling of the individual reservoir degrees of freedom to the system, the elimination of a nonlinear bath is similar to that of a linear bath, and the effects of a nonlinear environment are captured, to second order in the coupling, by a temperature-dependent spectral density of the coupling.

3.6 Anomalous linear dissipation

So far we have considered dissipation for systems in which the potential $V(q)$ is anharmonic, and the interaction depends on the coordinates of the reservoir and the system. Now I consider the case in which the system reservoir interaction depends on the momenta. Here I choose a Hamiltonian with bilinear momentum coupling,

$$H = \frac{p^2}{2M} + V(q) + \sum_\alpha \left[\frac{1}{2m_\alpha}(p_\alpha - d_\alpha p)^2 + \frac{1}{2}m_\alpha \omega_\alpha^2 x_\alpha^2 \right] . \tag{3.258}$$

Mixed couplings $q\, p_\alpha$ and $p\, x_\alpha$ are already included in the Hamiltonian (3.12) and (3.258), respectively, as one sees by applying a canonical transformation which interchanges coordinates and momenta of the bath oscillators.

It has been pointed out by Leggett [83] that the $q\, x_\alpha$-coupling leads to *normal* dissipation and the $p\, p_\alpha$-coupling to *anomalous* dissipation. For state-independent memory-less friction, the rate of dissipated energy is proportional to $\dot{q}^2$ for normal dissipation and proportional to $\dot{p}^2$ for anomalous dissipation. Models of the type (3.258) have been studied in recent years by different groups [167]. Anomalous dissipation may occur in LC circuits with a nonlinear capacitance, and with a linear Ohmic resistance in parallel [83]. Here, the charge takes the role of the coordinate, and flux through the inductance corresponds to the momentum.

For a harmonic potential, $V(q) \to V_{\rm h}(q) = \frac{1}{2}M\omega_0^2 q^2$, the Hamiltonians (3.12) and (3.258) can be transferred into one another by means of a canonical transformation in which coordinates and momenta are exchanged,

$$q \to -p/M\omega_0 , \qquad p \to M\omega_0 q , \qquad x_\alpha \to -p_\alpha/m_\alpha \omega_\alpha , \qquad p_\alpha \to m_\alpha \omega_\alpha x_\alpha . \tag{3.259}$$

At this mapping, one finds the correspondence

$$c_\alpha \triangleq M\omega_0\, \omega_\alpha\, d_\alpha . \tag{3.260}$$

With this, the spectral density of the coupling $J_q(\omega)$ defined in Eq. (3.25) reads

$$J_q(\omega) = M^2\omega_0^2\, J_p(\omega) , \qquad \text{with} \qquad J_p(\omega) = \frac{\pi}{2}\sum_\alpha \frac{\omega_\alpha}{m_\alpha} d_\alpha^2\, \delta(\omega - \omega_\alpha) . \tag{3.261}$$

To be specific, here I consider the case in which the nonlinear element of the Hamiltionian depends on the coordinate q. This includes the possibility of tunneling. The coupled equations of motion determined by the Hamiltonian (3.258) read

$$\begin{aligned}
\dot{q} &= p/M + p\sum_\alpha d_\alpha^2/m_\alpha - \sum_\alpha d_\alpha p_\alpha/m_\alpha , & \dot{p} &= -V'(q) , \\
\dot{x}_\alpha &= p_\alpha/m_\alpha - p\, d_\alpha/m_\alpha , & \dot{p}_\alpha &= -m_\alpha \omega_\alpha^2\, x_\alpha .
\end{aligned} \tag{3.262}$$

Next we eliminate the x_α, p_α and p in Fourier space. We find that the Fourier transform $\tilde{q}(\omega)$ of $q(t)$ obeys the unconventional Langevin equation

$$-M\omega^2\, \tilde{q}(\omega) + \left[1 - i\,\omega\, \tilde{\gamma}_q(\omega)/\omega_0^2\right] [\partial V(q)/\partial q](\omega) = \tilde{\zeta}(\omega) . \tag{3.263}$$

The Langevin force $\zeta(t)$ depends on the reservoir's initial state analogous to (3.22), and $\tilde{\gamma}_q(\omega) = \hat{\gamma}(-i\,\omega)$ coincides with the standard spectral damping function (3.29),

$$\hat{\gamma}_q(z) = M\omega_0^2 \sum_\alpha \frac{d_\alpha^2}{m_\alpha} \frac{z}{\omega_\alpha^2 + z^2} = M\omega_0^2 \frac{2}{\pi} \int_0^\infty d\omega \, \frac{J_p(\omega)}{\omega} \frac{z}{\omega^2 + z^2} . \tag{3.264}$$

Here, ω_0 is a characteristic frequency of the nonlinear system. This yields the a damped equation of motion in the unorthodox form

$$M\ddot{q}(t) + \frac{1}{\omega_0^2} \int_{-\infty}^t ds\, \gamma_q(t-s) \frac{\partial}{\partial s} V'[q(s)] + V'[q(t)] = \zeta(t) . \tag{3.265}$$

For Ohmic anomalous dissipation, the equation (3.265) can be formally cast into the normal form (2.3), but at the expense of a position-dependent friction coefficient,

$$M\ddot{q} + \eta(q)\,\dot{q} + V'(q) = \zeta(t) , \qquad \text{with} \qquad \eta(q) = \gamma V''(q)/\omega_0^2 . \tag{3.266}$$

In the region around a stable minimum, where $V''(q) \to V_h''(q) = M\omega_0^2$, the equation (3.266) reduces to the standard damped equation of motion,

$$M\ddot{q} + M\gamma\,\dot{q} + V'(q) = \zeta(t) . \tag{3.267}$$

But in the case of tunneling through a barrier, the pseudo-friction coefficient $\eta(q)$ is negative over major parts of the tunneling region. This already indicates that the sign of the effect of tunneling is opposite to that in the normal case [168, 83].

Alternatively, we may write the equation of motion (3.265) in the usual form (2.4),

$$\mathbf{K}q\,(t) + V'[q(t)] = \xi(t) , \tag{3.268}$$

where $\mathbf{K}$ is a linear operator of the integro-differential type with Fourier transform

$$\tilde{K}(\omega) = -M\omega^2 - i\,M\omega\tilde{\gamma}_p(\omega) , \tag{3.269}$$

where $\tilde{\gamma}_p(\omega) = \hat{\gamma}_p(-i\,\omega)$ is the spectral damping function for anomalous dissipation,

$$\hat{\gamma}_p(z) = -z\,\frac{z\hat{\gamma}_q(z)/\omega_0^2}{1 + z\hat{\gamma}_q(z)/\omega_0^2} = -z + \frac{z}{1 + z\hat{\gamma}_q(z)/\omega_0^2} . \tag{3.270}$$

With this, the kinetic spectral function $\tilde{K}(\omega)$ takes the form

$$\tilde{K}(\omega) = -M\omega^2/[\,1 - i\,\omega\tilde{\gamma}_q(\omega)/\omega_0^2\,] . \tag{3.271}$$

Applications of anomalous dissipation are made to quantum tunneling on page 82 and to the position and momentum dispersions of the damped harmonic oscillator on page 180. The general conclusions are that anomalous dissipation usually effectuates opposite behaviors compared to those in the normal case. As an example, anomalous dissipation enhances quantum tunneling through a barrier, while normal dissipation suppresses it. Equally, anomalous dissipation increases cordinate dispersion, while normal dispersion reduces it. All in all, in presence of both normal and anomalous dissipation, there are competing effects of reversely acting influences. These may lead to nonmonotonic behaviors, which in the absence of one of the two would be monotonic. The case of mixed dissipation seems to be of some practical interest [167].

4 Imaginary-time approach and equilibrium dynamics

In a canonical ensemble – the entirety of all microstates with fixed particle number and volume – the system is kept at equilibrium by contact with a heat reservoir at temperature T, and thus only the energy fluctuates. The equilibrium characteristics of a canonical ensemble of quantum systems governed by the Hamiltonian $\hat{H}$ is determined by properties of the canonical density operator $\hat{W}_\beta = Z^{-1} e^{-\beta \hat{H}}$, where $Z = \text{tr}\, e^{-\beta \hat{H}} = \sum_k e^{-\beta E_k}$ is the partition function. The source of the imaginary-time path integral approach to equilibrium thermodynamics is the fact that the canonical density operator $e^{-\beta \hat{H}}$ is related to the time-evolution operator $e^{-i\hat{H}t/\hbar}$ by analytic continuation of time, $t \to z = -i\hbar\beta$, known in field theory as *Wick rotation* from Minkowskian to Euclidean field theory. Feynman taught us that the coordinate matrix elements of $e^{-i\hat{H}t/\hbar}$ and $e^{-\beta \hat{H}}$ can be written as a sum over histories of real- and imaginary-time paths in configuration space, respectively. From a contemporary point of view, path integrals represent not only an approach alternative to canonical quantization of classical mechanics, but are also basic to the foundation and interpretation of quantum mechanics. Besides that, they form a perfect basis for numerical simulations of quantum thermodynamics and dynamics by means of Monte-Carlo methods.

4.1 General concepts

4.1.1 Density matrix and reduced density matrix

In quantum-statistical mechanics, the state of a system at a particular time is not perfectly known. When one has incomplete information about a system, one usually appeals to the concept of probability. Typically, the incomplete information about a system presents itself in quantum mechanics as a *statistical mixture*: the state of the system may be either the state $|\psi_1>$ with a probability p_1, or the state $|\psi_2>$ with a probability p_2, etc. There holds $0 \leqslant p_1, p_2, \cdots, p_k, \cdots \leqslant 1$, and $\sum_k p_k = 1$. A system in a mixed state is specified by the density operator

$$\hat{W}(t) = \sum_k p_k |\psi_k(t)><\psi_k(t)| . \tag{4.1}$$

We have $\text{tr}\, \hat{W} = \sum_k p_k = 1$, whereas $\text{tr}\, \hat{W}^2 = \sum_k p_k^2 \leqslant 1$. The equal sign holds for a pure state, $p_k = \delta_{k,i}$. In the orthonormal basis $\{|\varphi_n>\}$, the density matrix reads

$$W_{n,m} = \sum_k p_k <\varphi_n|\psi_k><\psi_k|\varphi_m> . \tag{4.2}$$

The diagonal matrix elements $W_{n,n}$ of $\hat{W}$ in the $\{|\varphi_n>\}$ basis are

$$W_{n,n} = \sum_k p_k |<\varphi_n|\psi_k>|^2 , \tag{4.3}$$

where $| <\varphi_n|\psi_k> |^2$ is a real positive number with the following physical interpretation: if the state of the system is $|\psi_k>$, it is the probability of finding, in a measurement, this system in the state $|\varphi_n>$. Because of the uncertainty of the state $|\psi_k>$ before the measurement, $W_{n,n}$ represents the average probability of finding the system in the state $|\varphi_n>$. For this reason, the diagonal matrix element $W_{n,n}$ is called the *population* of the state $|\varphi_n>$. The off-diagonal element

$$W_{n,m} = \sum_k p_k <\varphi_n|\psi_k> <\psi_k|\varphi_m> \qquad (4.4)$$

is the average of the cross terms. While $W_{n,n}$ is a sum of real positive numbers, $W_{n,m}$ is a sum of complex numbers. When $W_{n,m}$ is nonzero, coherence between the states $|\varphi_n>$ and $|\varphi_m>$ persists. For this reason, the off-diagonal elements of the density matrix are dubbed *coherences*. When $W_{n,m}$ is zero, there is no interference between the states $|\varphi_n>$ and $|\varphi_m>$. It is easy to prove that

$$W_{n,n}\, W_{m,m} \geqslant |W_{n,m}|^2 \ . \qquad (4.5)$$

We remark that the distinction between populations and coherences depends on the basis $\{|\varphi_n>\}$ chosen in the state space. The density matrix concept for coupled few-state systems in $SU(n)$ representation is neatly presented in Ref. [169].

Canonical density operator and partition function

A fundamental quantity in quantum thermodynamics is the normalized canonical density operator in which the energy eigenstates $|\psi_k>$ are thermally occupied,

$$\hat{\mathsf{W}}^{(\mathrm{eq})} \equiv \hat{\mathsf{W}}_\beta = e^{-\beta\hat{H}}/Z = Z^{-1}\sum_k e^{-\beta E_k}\, |\psi_k><\psi_k| \ . \qquad (4.6)$$

The quantity Z is the quantum statistical partition function, $Z = \mathrm{tr}\, e^{-\beta\hat{H}}$.

Reduced density operator

The density operator of the global system acts in the full state space $\mathcal{H} = \mathcal{H}_{\mathrm{S}} \otimes \mathcal{H}_{\mathrm{R}}$. Within the density matrix representation, expectation values are expressed as

$$\langle A \rangle = \mathrm{tr}\,\{\hat{W}\hat{A}\} \qquad (4.7)$$

For observables of which the respective operators act only in the system's space $\mathcal{H}_{\mathrm{S}}$, the trace with respect to the reservoir in Eq. (4.7) can be separated. This suggests to introduce a *reduced* density operator $\hat{\rho}$ acting only in the system's space $\mathcal{H}_{\mathrm{S}}$,

$$\hat{\rho} = \mathrm{tr}_{\mathrm{R}}\,\hat{W} \ . \qquad (4.8)$$

The operation in Eq. (4.8) is the *partial trace* with respect to the reservoir's degrees of freedom. The essence of the reduced density operator $\hat{\rho}$ is that all physical predictions about measurements bearing only on the system S are captured by this operator,

$$\langle A^{(\mathrm{S})} \rangle = \mathrm{tr}_{\mathrm{S}}\,\mathrm{tr}_{\mathrm{R}}\{\hat{W}\hat{A}^{(\mathrm{S})}\} = \mathrm{tr}_{\mathrm{S}}\{\hat{\rho}\hat{A}^{(\mathrm{S})}\} \ . \qquad (4.9)$$

For a normalized reduced density matrix, $\mathrm{tr}\,\hat{\rho} = 1$, we have

$$0 < \mathrm{tr}\,\hat{\rho}^2 \leqslant 1 \ . \qquad (4.10)$$

The maximum value $\mathrm{tr}\,\hat{\rho}^2 = 1$ indicates that the system is in a pure quantum state, whereas small values, $\mathrm{tr}\,\hat{\rho}^2 \ll 1$, describe nearly classical scenarios.

4.1.2 Imaginary-time path integral

To begin with, we sketch the derivation of the path integral for the imaginary time propagator $\langle q''| e^{-\tau\hat{H}/\hbar}|q'\rangle$ for the simple Hamiltonian $\hat{H} = \hat{p}^2/2M + V(\hat{q})$. The starting point of the derivation is a group property of the imaginary-time evolution,

$$\langle q''| e^{-\tau\hat{H}/\hbar}|q'\rangle = \langle q''| e^{-\tau\hat{H}/N\hbar}\, 1\, e^{-\tau\hat{H}/N\hbar}\, 1\, e^{-\tau\hat{H}/N\hbar} \cdots e^{-\tau\hat{H}/N\hbar}|q'\rangle . \qquad (4.11)$$

Upon insertion of the completeness relation of the orthonormal position eigenstates

$$1 = \int_{-\infty}^{\infty} dq\, |q\rangle\langle q| , \qquad (4.12)$$

the thermal propagator $\langle q''| e^{-\tau\hat{H}/\hbar}|q'\rangle$ becomes a multiple integral representing succession of N short-time propagators over imaginary time intervals of length $\varepsilon = \tau/N$,

$$\langle q''| e^{-\tau\hat{H}/\hbar}|q'\rangle = \lim_{N\to\infty} \int_{-\infty}^{\infty} dq_{N-1} \cdots dq_1 \prod_{n=1}^{N} \langle q_n| e^{-\varepsilon\hat{H}/\hbar}|q_{n-1}\rangle , \qquad (4.13)$$

where $q_N = q''$ and $q_0 = q'$, and where we eventually take the limit $N \to \infty$. The short-time propagator with initial condition $\langle q_n|q_{n-1}\rangle = \delta(q_n - q_{n-1})$ reads

$$\langle q_n| e^{-\varepsilon\hat{H}/\hbar}|q_{n-1}\rangle = \sqrt{M/2\pi\hbar\varepsilon}\; e^{-S_n/\hbar} , \qquad (4.14)$$

where $S_n = \varepsilon\{M[(q_n - q_{n-1})/\varepsilon]^2/2 + V(q_n)\}$ is the Euclidean short-time action. With the form (4.14), the limit $N \to \infty$ in Eq. 4.13) is well defined,

$$\langle q''| e^{-\tau\hat{H}/\hbar}|q'\rangle = \lim_{N\to\infty} \int_{-\infty}^{\infty} \frac{dq_1\, dq_2 \cdots dq_{N-1}}{(2\pi\hbar\varepsilon/M)^{N/2}}\, e^{-S_N/\hbar} , \qquad (4.15)$$

where $S_N = \sum_{n=1}^{N} S_n$. As $N \to \infty$, S_N turns into the Euclidean action integral[1]

$$\lim_{N\to\infty} S_N \to S[q(\cdot)] = \int_0^\tau d\tau' \left(\frac{M}{2}\dot{q}^2(\tau') + V[q(\tau')]\right) . \qquad (4.16)$$

With the short-hand notation for the multiple-integral measure

$$\int \mathcal{D}q(\cdot) \cdots = \lim_{N\to\infty} \int_{-\infty}^{\infty} \frac{dq_1\, dq_2 \cdots dq_{N-1}}{(2\pi\hbar\varepsilon/M)^{N/2}} \cdots , \qquad (4.17)$$

the propagator takes the suggestive form of a path sum or path integral

$$\langle q''| e^{-\tau\hat{H}/\hbar}|q'\rangle = \int_{q(0)=q'}^{q(\tau)=q''} \mathcal{D}q(\cdot)\, e^{-S[q(\cdot)]/\hbar} . \qquad (4.18)$$

As already emphasized by Feynman [3], the concept of the sum over all paths, like the concept of an ordinary integral, is independent of a special definition of the path

---

[1] Throughout the book, the center dot in "$q(\cdot)$" is used to express that the path q is a functional of time (here imaginary-time), while $q(\tau)$ means the value the path takes at a particular time τ. The notation $q(\tau)$ for the path – though standard in the literature – is not used here since the path $q(\cdot)$ and the value $q(\tau)$ it takes at a particular time will often appear close together.

sum. It may be chosen according to convenience. In the representation (4.15), the path is a zigzag of straight line segments. Evaluation of path integrals by Fourier series is discussed in Subsection 4.3.4. The notion of a path integral can be extended to discrete systems for which no classical analogue exists. The tight-binding (TB) position operator has a discrete spectrum. Hence the paths piecewise stay for some time at localized TB positions with a countable number of sudden jumps between these states. The respective path sum is discussed in Chapters 19, 21 and 25.

This concludes the brief introduction to path integrals. The reader interested in getting familiarized with this field is referred to Refs. [3]–[6].

Let us first consider a one-dimensional system described by a coordinate q and a real-time Lagrangian (with regard to correlation functions, we add a source term)

$$\mathcal{L}_S[q(t), \dot{q}(t)] = \tfrac{1}{2} M \dot{q}^2(t) - V[q(t)] - \mathcal{J}(t) q(t) . \tag{4.19}$$

For simplicity, we assume that there is no velocity-dependent force. The canonical density matrix $\mathbb{W}_\beta(q'', q')$ of the system, Eq. (4.6), may be expressed in terms of the eigenstates $\psi_n(q)$ and eigenvalues E_n of the Hamiltonian H_S. We then have

$$< q'' | \mathbb{W}_\beta | q' > \equiv \mathbb{W}_\beta(q'', q') = Z_S^{-1} \sum_n \psi_n(q'') \psi_n^*(q') \exp(-\beta E_n) , \tag{4.20}$$

According to Eq. (4.18), this can be written as the imaginary-time path integral [3, 4]

$$\mathbb{W}_\beta(q'', q') = Z_S^{-1} \int_{q(0)=q'}^{q(\hbar\beta)=q''} \mathcal{D}q(\cdot) \exp\left\{ - \int_0^{\hbar\beta} d\tau\, \mathbb{L}_S[q(\tau), \dot{q}(\tau)]/\hbar \right\} . \tag{4.21}$$

The path sum in Eq. (4.21) covers all imaginary-time paths $q(\tau)$ which leave position q' at time $\tau = 0$ and arrive at position q'' at thermal time $\tau = \hbar\beta$. As a result of the analytical continuation $t \to -i\tau$, and hence

$$i \int dt\, \mathcal{L}_S[q(t), \dot{q}(t)] \to - \int d\tau\, \mathbb{L}_S[q(\tau), \dot{q}(\tau)] , \tag{4.22}$$

the classical mechanics in imaginary time is governed by the Euclidean Lagrangian

$$\mathbb{L}_S[q(\tau), \dot{q}(\tau)] = \tfrac{1}{2} M \dot{q}^2(\tau) + V[q(\tau)] + \mathcal{J}(\tau) q(\tau) . \tag{4.23}$$

which comes with the reverse sign in the potential term.[2]

Reduced density operator

The reduced density operator in thermal equilibrium is defined as the partial trace of the canonical density operator $\hat{\mathbb{W}}_\beta$ of the total system with respect to the reservoir,

$$\hat{\rho}_\beta \equiv \mathrm{tr}_R \hat{\mathbb{W}}_\beta . \tag{4.24}$$

Our main goal in this chapter is to derive for the composite systems discussed in the preceding chapter the path integral expressions for the respective reduced density

---

[2]In real-time quantities, the overdot denotes differentiation with respect to real time t, whereas in *Euclidean* quantities, the overdot denotes differentiation with respect to the imaginary time τ.

matrices in thermal equilibrium. The essential element of the imaginary-time path integral for the reduced system (often called open system) is the effective Euclidean action. This is found by integrating out the reservoir coordinates in the path integral for the equilibrium density matrix W_β of the composite system.

In the next section, we calculate the effective Euclidean actions for the various composite systems introduced above. After that, we study the partition function in the semiclassical limit, and for a composite harmonic system. The chapter is concluded with a treatment of quantum statistical expectation values in phase space.

4.2 Effective action and equilibrium density matrix

We now generalize the discussion to systems coupled to a thermal bath. The equilibrium density matrix of the "universe" (system plus environment) may be written as

$$W_\beta(q'', \mathbf{x}''; q', \mathbf{x}') = Z_{\text{tot}}^{-1} \sum_{\{n\}} \Psi_{\{n\}}(q'', \mathbf{x}'') \Psi_{\{n\}}^*(q', \mathbf{x}') \exp(-\beta E_{\{n\}}) . \qquad (4.25)$$

The N-component vector $\mathbf{x}$ stands for the coordinates $x_1, x_2, \ldots, x_N$ of the reservoir, $\Psi_{\{n\}}$ denotes the eigenstates of the Hamiltonian of the total system, and $\{n\}$ is the full set of quantum numbers. Again, Z_{tot} ensures that $\text{tr}\,\hat{W}_\beta = 1$. We shall assume that the forces in the Hamiltonian of the global system are velocity-independent. Then the canonical density matrix (4.25) can be written as a path integral of standard form,

$$W_\beta(q'', \mathbf{x}''; q', \mathbf{x}') = Z_{\text{tot}}^{-1} \int_{q(0)=q'}^{q(\hbar\beta)=q''} \mathcal{D}q(\cdot) \int_{\mathbf{x}(0)=\mathbf{x}'}^{\mathbf{x}(\hbar\beta)=\mathbf{x}''} \mathcal{D}\mathbf{x}(\cdot)\, e^{-\mathcal{S}[q(\cdot),\mathbf{x}(\cdot)]/\hbar} . \qquad (4.26)$$

The path sum is over all paths taken by the coordinates $q(\tau)$ and $\mathbf{x}(\tau)$ with endpoints $q(0) = q'$, $\mathbf{x}(0) = \mathbf{x}'$, $q(\hbar\beta) = q''$ and $\mathbf{x}(\hbar\beta) = \mathbf{x}''$. The Euclidean action $\mathcal{S}$ receives contributions from the system (S), the reservoir (R), and the interaction (I),

$$\mathcal{S} = \mathcal{S}_S + \mathcal{S}_R + \mathcal{S}_I = \int_0^{\hbar\beta} d\tau \, (\mathbb{L}_S + \mathbb{L}_R + \mathbb{L}_I) . \qquad (4.27)$$

From here on we focus our interest on a reduced description in which the reservoir coordinates are eliminated. To this end, we introduce the *reduced* equilibrium density operator $\hat{\rho}_\beta = \text{tr}_R \hat{W}_\beta$ describing the open system, which in position space is

$$\rho_\beta(q'', q') = \int_{-\infty}^{+\infty} d\mathbf{x}'\, W_\beta(q'', \mathbf{x}'; q', \mathbf{x}') , \qquad (4.28)$$

and the so-called *reduced* partition function,

$$Z(\hbar\beta) = Z_{\text{tot}}(\hbar\beta)/Z_R(\hbar\beta) . \qquad (4.29)$$

Here Z_{tot} is the partition function of the composite system, and Z_R is the partition function of the reservoir. For a bath of N harmonic oscillators there is

$$Z_R \equiv \prod_{\alpha=1}^{N} Z_R^{(\alpha)} = \prod_{\alpha=1}^{N} \frac{1}{2\sinh(\beta\hbar\omega_\alpha/2)} . \qquad (4.30)$$

With Eqs. (4.26)-(4.29), the reduced equilibrium density matrix can be written as

$$\rho_\beta(q'', q') = \frac{1}{Z} \int_{q(0)=q'}^{q(\hbar\beta)=q''} \mathcal{D}q(\cdot) \exp\{-S_S[q(\cdot)]/\hbar\} \, \mathbb{F}[q(\cdot)] \,. \qquad (4.31)$$

The Euclidean or imaginary-time *influence functional*

$$\mathbb{F}_{\mathrm{infl}}[q(\cdot)] \equiv \exp\{-S_{\mathrm{infl}}[q(\cdot)]/\hbar\} = \frac{1}{Z_R} \oint \mathcal{D}x(\cdot) \exp\{-S_{R,I}[q(\cdot), x(\cdot)]/\hbar\} \qquad (4.32)$$

with

$$S_{R,I}[q, x] = S_R[x] + S_I[q, x] \qquad (4.33)$$

captures the effects of the environmental coupling on the equilibrium properties of the open system. The trace operation in Eq. (4.28) appears in the path integral (4.32) as the sum over all *periodic* paths with period $\hbar\beta$ taken by the coordinate x. The factor Z_R^{-1} normalizes $\mathbb{F}_{\mathrm{infl}}[q(\cdot)]$ such that $\mathbb{F}_{\mathrm{infl}}[q(\cdot)] = 1$ when the coupling $S_I[q, x]$ is switched off, and the factor Z^{-1} in Eq. (4.31) normalizes $\hat\rho_\beta$ as $\mathrm{tr}_S \hat\rho_\beta = 1$. Next, we calculate the Euclidean influence action $S_{\mathrm{infl}}[q(\cdot)]$ for particular systems.

4.2.1 Open system with bilinear coordinate coupling to a harmonic reservoir

For the model described by the Hamiltonian (3.12), the Euclidean Lagrangian $\mathbb{L}_S$ is given in Eq. (4.23) with $\mathcal{J} = 0$, and with

$$\begin{aligned}
\mathbb{L}_R &= \sum_\alpha \frac{m_\alpha}{2}(\dot{x}_\alpha^2 + \omega_\alpha^2 x_\alpha^2) \,, \\
\mathbb{L}_I &= \sum_\alpha \left(-c_\alpha x_\alpha q + \frac{c_\alpha^2 q^2}{2m_\alpha \omega_\alpha^2}\right) \,.
\end{aligned} \qquad (4.34)$$

The stationary paths of the total action (4.27), which we denote by $\bar{q}$ and by $\bar{x}_\alpha$, obey the Euclidean classical equations of motion

$$M\ddot{\bar{q}} - \frac{\partial V(\bar{q})}{\partial \bar{q}} + \sum_\alpha c_\alpha \left(\bar{x}_\alpha - \frac{c_\alpha}{m_\alpha \omega_\alpha^2}\bar{q}\right) = 0 \,, \qquad (4.35)$$

$$m_\alpha \ddot{\bar{x}}_\alpha - m_\alpha \omega_\alpha^2 \bar{x}_\alpha + c_\alpha \bar{q} = 0 \,. \qquad (4.36)$$

Since the Lagrangian is a quadratic form in x and $\dot{x}$, the multiple functional integration of the periodic path $x(\tau)$ can be done in closed form. To proceed, we choose periodic continuation of the paths $q(\tau)$ and $x(\tau)$ outside the range $0 \leqslant \tau < \hbar\beta$ by writing them as Fourier series

$$\begin{aligned}
x_\alpha(\tau) &= \frac{1}{\hbar\beta} \sum_{n=-\infty}^{+\infty} x_{\alpha,n} \, e^{i\nu_n \tau} \,, \\
q(\tau) &= \frac{1}{\hbar\beta} \sum_{n=-\infty}^{+\infty} q_n \, e^{i\nu_n \tau} \,,
\end{aligned} \qquad (4.37)$$

where $x_{\alpha,n} = x_{\alpha,-n}^*$, $q_n = q_{-n}^*$, and $\nu_n = 2\pi n/\hbar\beta$ is a bosonic Matsubara frequency. Insertion of the series expressions (4.37) into Eq. (4.33) with Eq. (4.34) yields

$$S_{R,I}[q, \mathbf{x}] = \sum_{\alpha} \frac{1}{\hbar\beta} \sum_{n=-\infty}^{+\infty} \frac{m_\alpha}{2} \left(\nu_n^2 |x_{\alpha,n}|^2 + \omega_\alpha^2 \left| x_{\alpha,n} - \frac{c_\alpha}{m_\alpha \omega_\alpha^2} q_n \right|^2 \right). \tag{4.38}$$

Next, we split $x_{\alpha,n}$ into the classical part $\bar{x}_{\alpha,n}$ and the quantum fluctuation part $y_{\alpha,n}$,

$$x_{\alpha,n} = \bar{x}_{\alpha,n} + y_{\alpha,n} = \frac{c_\alpha}{m_\alpha(\nu_n^2 + \omega_\alpha^2)} q_n + y_{\alpha,n}. \tag{4.39}$$

In the second form, we have used the solution of Eq. (4.36) in Fourier space. Since $\bar{x}_\alpha(\tau)$ is a stationary point of the action $S_{R,I}[q, \mathbf{x}]$, the term linear in the deviation $y_{\alpha,n}$ is absent, and we get pure quadratic forms in $\mathbf{y}$ and q,

$$S_{R,I}[q, \bar{\mathbf{x}} + \mathbf{y}] = S_R[\mathbf{y}] + S_{infl}[q], \tag{4.40}$$

$$S_R[\mathbf{y}] = \sum_{\alpha} \frac{1}{\hbar\beta} \sum_{n=-\infty}^{+\infty} \frac{m_\alpha}{2} (\nu_n^2 + \omega_\alpha^2) |y_{\alpha,n}|^2 = \sum_{\alpha} \int_0^{\hbar\beta} d\tau \frac{m_\alpha}{2} (\dot{y}_\alpha^2 + \omega_\alpha^2 y_\alpha^2),$$

$$S_{infl}[q] = \sum_{\alpha} \frac{c_\alpha^2}{2m_\alpha} \frac{1}{\hbar\beta} \sum_{n=-\infty}^{+\infty} \left(\frac{|q_n|^2}{\omega_\alpha^2} - \frac{|q_n|^2}{\nu_n^2 + \omega_\alpha^2} \right). \tag{4.41}$$

The first term in the round bracket of $S_{infl}[q]$ originates from the potential counter term $\propto q^2$ in $\mathbb{L}_I$. The path sum of the $\hbar\beta$-periodic quantum fluctuations $\mathbf{y}(\tau)$ directly yields the partition function of the reservoir,

$$Z_R = \oint \mathcal{D}\mathbf{y}(\cdot) \exp\left\{ -S_R[\mathbf{y}(\cdot)]/\hbar \right\}. \tag{4.42}$$

This term cancels the factor Z_R^{-1} in Eq. (4.32). Hence the thermal influence functional $\mathbb{F}_{infl}[q(\cdot)]$ and the influence action $S_{infl}[q(\cdot)]$ are found as

$$\mathbb{F}_{infl}[q(\cdot)] = e^{-S_{infl}[q(\cdot)]/\hbar} \quad \text{with} \quad S_{infl}[q(\cdot)] = \frac{M}{2} \frac{1}{\hbar\beta} \sum_{n=-\infty}^{+\infty} \xi_n |q_n|^2,$$

$$\xi_n = \frac{1}{M} \sum_{\alpha} \frac{c_\alpha^2}{m_\alpha \omega_\alpha^2} \frac{\nu_n^2}{(\nu_n^2 + \omega_\alpha^2)} = \frac{2}{M\pi} \int_0^\infty d\omega \frac{J(\omega)}{\omega} \frac{\nu_n^2}{\nu_n^2 + \omega^2}. \tag{4.43}$$

In the second form of ξ_n, we have introduced the spectral density (3.25). The Fourier coefficient ξ_n of the influence action can be expressed in terms of the spectral damping functions $\tilde{\gamma}(\omega)$ and $\hat{\gamma}(z)$ defined in Eq. (3.26) and (3.28), respectively, as

$$\xi_n = |\nu_n| \tilde{\gamma}(i|\nu_n|) = |\nu_n| \hat{\gamma}(|\nu_n|). \tag{4.44}$$

Consequently, the effective action of the reduced system in Matsubara representation for general spectral coupling is

$$S_{eff}[q(\cdot)] = \frac{M}{2} \frac{1}{\hbar\beta} \sum_{n} [\nu_n^2 + |\nu_n| \hat{\gamma}(|\nu_n|)] |q_n|^2 + S_V(q_n), \tag{4.45}$$

where $S_V(q_n) = \int_0^{\hbar\beta} d\tau V[q(\tau)]$ is the potential contribution to the action.

The influence action (4.43) is a convolution in imaginary-time representation,

$$S_{\text{infl}}[q(\cdot)] = \int_0^{\hbar\beta} d\tau \int_0^\tau d\tau'\, k(\tau - \tau')\, q(\tau)\, q(\tau')\,, \tag{4.46}$$

$$k(\tau) = \frac{M}{\hbar\beta} \sum_{n=-\infty}^{+\infty} \xi_n\, e^{i\nu_n\tau}\,. \tag{4.47}$$

The kernel $k(\tau)$ has the symmetry $k(\tau) = k(\hbar\beta - \tau)$. With use of the property

$$\int_0^{\hbar\beta} d\tau\, k(\tau) = M\xi_0 = 0 \tag{4.48}$$

the influence action (4.46) can be cast into the form

$$S_{\text{infl}}[q(\cdot)] = -\frac{1}{2} \int_0^{\hbar\beta} d\tau \int_0^\tau d\tau'\, k(\tau - \tau')\, [q(\tau) - q(\tau')]^2\,. \tag{4.49}$$

This action is *time-nonlocal* and therefore free of potential renormalization.
Alternatively, we may express the nonlocal action (4.49) with the kernel

$$K(\tau) = \mu :\delta(\tau): - k(\tau) = \frac{M}{\hbar\beta} \sum_{n=-\infty}^{+\infty} \zeta_n\, e^{i\nu_n\tau}\,, \tag{4.50}$$

where $:\delta(\tau):$ is the periodically continued δ-function (3.85),

$$S_{\text{infl}}[q(\cdot)] = \frac{1}{2} \int_0^{\hbar\beta} d\tau \int_0^\tau d\tau'\, K(\tau - \tau')\, [q(\tau) - q(\tau')]^2\,. \tag{4.51}$$

With the particular choice

$$\mu = \sum_\alpha \frac{c_\alpha^2}{m_\alpha \omega_\alpha^2} = \frac{2}{\pi} \int_0^\infty d\omega\, \frac{J(\omega)}{\omega} = \lim_{t\to 0^+} M\gamma(t) = M\gamma(0^+)\,, \tag{4.52}$$

the Fourier coefficient ζ_n of the kernel $K(\tau)$ is

$$\zeta_n = \frac{\mu}{M} - \xi_n = \frac{1}{M} \sum_\alpha \frac{c_\alpha^2}{m_\alpha} \frac{1}{\nu_n^2 + \omega_\alpha^2} = \frac{2}{\pi M} \int_0^\infty d\omega\, J(\omega) \frac{\omega}{\nu_n^2 + \omega^2}\,. \tag{4.53}$$

The Matsubara sum (4.50) is just a spectral integral of the boson propagator (3.83),

$$K(\tau) = \frac{1}{\pi} \int_0^\infty d\omega\, J(\omega)\, \mathfrak{D}_\omega(\tau) = \frac{1}{\pi} \int_0^\infty d\omega\, J(\omega) \frac{\cosh[\omega(\hbar\beta/2 - \tau)]}{\sinh(\omega\hbar\beta/2)}\,. \tag{4.54}$$

The second form holds in the regime $0 \leqslant \tau < \hbar\beta$. The kernel $K(\tau)$ is related to the correlation function $\mathfrak{K}(\tau)$ of the stochastic force $\xi(\tau) = \sum_\alpha c_\alpha x_\alpha(\tau)$, Eq. (3.86), by

$$K(\tau) = \mathfrak{K}(\tau)/\hbar\,. \tag{4.55}$$

The function $K(\tau)$ may also be expressed with the real-time memory-friction kernel $\gamma(t)$. Upon substituting the expression (3.32) for $J(\omega)$ into the second form of Eq. (4.54), reversing the order of integrations, and performing the ω-integral, we get

$$K(\tau) = \frac{M}{\hbar\beta} \int_0^\infty dt\, \gamma(t)\, \frac{\partial}{\partial t} \frac{\sinh(\nu_1 t)}{\cosh(\nu_1 t) - \cos(\nu_1 \tau)}\,, \qquad (4.56)$$

where ν_1 is the lowest Matsubara frequency. For Ohmic dissipation, $J(\omega) = M\gamma\omega$ and $\xi_n = \gamma|\nu_n|$, the kernels have the simple analytic form

$$k(\tau) = -K(\tau) = -\frac{\eta}{\pi} \frac{(\pi k_B T/\hbar)^2}{\sin^2(\pi k_B T\tau/\hbar)}\,, \qquad 0 < \tau < \hbar\beta\,. \qquad (4.57)$$

Another form of the influence action $S_{\text{infl}}[q(\cdot)]$ is found if we agree to periodically continue the path $q(\tau)$ outside the range $0 \leqslant \tau < \hbar\beta$ according to $q(\tau + n\hbar\beta) = q(\tau)$. By extension of the integration regime, the expression (4.51) is transformed into [78]

$$S_{\text{infl}}[q(\cdot)] = \frac{1}{4} \int_0^{\hbar\beta} d\tau \int_{-\infty}^\infty d\tau'\, K_0(\tau - \tau')\, [q(\tau) - q(\tau')]^2\,, \qquad (4.58)$$

where $K_0(\tau)$ is the kernel $K(\tau)$ at zero temperature,

$$K_0(\tau) = \sum_\alpha \frac{c_\alpha^2}{2m_\alpha\omega_\alpha} e^{-\omega_\alpha|\tau|} = \frac{1}{\pi} \int_0^\infty d\omega\, J(\omega)\, e^{-\omega|\tau|}\,. \qquad (4.59)$$

While $K(\tau - \tau')$ in Eq. (4.51) is acting only within the principal interval of length $\hbar\beta$, the function $K_0(\tau - \tau')$ in Eq. (4.58) arranges also correlations between the path in the principal interval and the periodically continued path.

In summary, the reduced equilibrium density matrix takes the concise form

$$\rho_\beta(q'', q') = \frac{1}{Z} \int_{q(0)=q'}^{q(\hbar\beta)=q''} \mathcal{D}q(\cdot)\, e^{-S_{\text{eff}}[q(\cdot)]/\hbar}\,. \qquad (4.60)$$

The effective action $S_{\text{eff}}[q(\cdot)] = S_S[q(\cdot)] + S_{\text{infl}}[q(\cdot)]$, and generalization to many dimensions, describes any system with linear dissipation. It is the essential component in the path integral approach towards the quantum statistics and thermodynamics of open quantum systems. Taking for $S_{\text{infl}}[q(\cdot)]$ the form (4.46), one has

$$S_{\text{eff}}[q(\cdot)] = \int_0^{\hbar\beta} d\tau\, \left[\tfrac{1}{2}M\dot{q}(\tau)^2 + V[q(\tau)] \right] + \int_0^{\hbar\beta} d\tau \int_0^\tau d\tau'\, k(\tau - \tau')\, q(\tau)q(\tau')\,. \qquad (4.61)$$

In a perturbative treatment of the potential $V(q)$, it is pertinent to introduce the effective action without the potential contribution,

$$S_{\text{eff},0}[q(\cdot)] = \frac{M}{2} \frac{1}{\hbar\beta} \sum_{n=-\infty}^\infty \lambda_n |q_n|^2\,, \qquad (4.62)$$

where

$$\lambda_n = \nu_n^2 + \xi_n = \nu_n^2 + |\nu_n|\hat{\gamma}(|\nu_n|)\,. \qquad (4.63)$$

Consider finally the displacement correlation function $\mathfrak{C}(\tau) = \langle [q(0) - q(\tau)]q(0)\rangle_\beta$. Here, $\langle \cdots \rangle_\beta$ denotes average with the weight function $\exp[-S_{\text{eff},0}[q]/\hbar]$. With the quadratic form (4.62) one obtains

$$\mathfrak{C}(\tau) = \frac{1}{\beta} \sum_{n \neq 0} \hat{\chi}(|\nu_n|) \left[1 - e^{i\nu_n\tau}\right] \quad \text{with} \quad \hat{\chi}(z) = \frac{1}{M[z^2 + z\hat{\gamma}(z)]} . \tag{4.64}$$

The function $\hat{\chi}(z)$ is the Laplace transform of the susceptibility $\chi(t)$ of a free Brownian particle. In fact, the expression (4.64) coincides with the earlier result (3.91) found along different line. Quantum Brownian free motion will be discussed in Chapter 7.

Momentum dissipation

So far, we have been concerned with the influence action related to spatial dissipation. With regard to momentum dissipation resulting from the global Hamiltonian (3.258), the relevant influence action is again of the form (4.46) with (4.47), but with the modified spectral damping function (3.270) in the Matsubara kernel, $\xi_n \to \hat{\xi}_p(|\nu_n|)$,

$$\hat{\xi}_p(|\nu_n|) = |\nu_n|\hat{\gamma}_p(|\nu_n|) \quad \text{with} \quad \hat{\gamma}_p(z) = -z + \frac{z}{1 + z\hat{\gamma}_q(z)/\omega_0^2} . \tag{4.65}$$

With this substitution in (4.45), the effective action for momentum dissipation is

$$S_{p,\text{eff}}[q(\cdot)] = \frac{M}{2} \frac{1}{\hbar\beta} \sum_n \frac{\nu_n^2}{1 + |\nu_n|\hat{\gamma}_q(|\nu_n|)/\omega_0^2} |q_n|^2 + S_V(q_n) , \tag{4.66}$$

Since the value of the effective action $S_{p,\text{eff}}[q(\cdot)]$ for any function q_n is less than it would be in the absence of dissipation ($\hat{\gamma}_q(z) \to 0$), it is clear that momentum dissipation decreases the effective action [83]. Consequently, anomalous dissipation arising from system-reservoir momentum coupling effectuates exponential increase of the tunneling rate. This is in complete agreement with the findings on page 180 that momentum dissipation increases coordinate dispersion instead of reducing it.

In the Ohmic case $\hat{\gamma}_q(z) \to \gamma_q \equiv \omega_0^2/\bar{\gamma}$, there is the obvious inequality

$$\nu_n^2/[1 + |\nu_n|/\bar{\gamma}] \geqslant \bar{\gamma}|\nu_n| - \bar{\gamma}^2 . \tag{4.67}$$

yielding the estimate

$$S_{p,\text{eff}}[q(\cdot)] \geqslant \frac{M}{2} \frac{1}{\hbar\beta} \sum_n \bar{\gamma}|\nu_n|q_n^2 + \int_0^{\hbar\beta} d\tau \left[V[q(\tau)] - M\bar{\gamma}^2 q(\tau)^2/2 \right] . \tag{4.68}$$

This is just the action for a particle in the potential $V_p(q) = V(q) - \frac{1}{2}M\bar{\gamma}^2 q(\tau)^2$ subjected to normal Ohmic dissipation with friction coefficient $\bar{\eta} = M\bar{\gamma}$, but without the kinetic term $\frac{1}{2}M\nu_n^2 q_n^2$. In the case of a metastable well located at the origin, the barrier of the effective potential $V_p(q)$ is shallower than that of the original potential.

In Sec. (17.3), the action (4.68) is calculated in analytic form for the periodic bounce in the cubic metastable potential $V(q) = \frac{1}{2}M\omega_0^2 q^2(1 - q/q_{\text{ex}})$. The effective potential $V_p(q)$ is metastable in the range $\bar{\gamma} < \omega_0$, but it has a smaller well frequency, $\omega_p^2 = \omega_0^2 - \bar{\gamma}^2$, and a smaller barrier width or tunneling distance, $q_{\text{ex},p} = (1 - \bar{\gamma}^2/\omega_0^2)q_{\text{ex}}$. The thermal crossover energy above which the bounce does not exist anymore is $k_\text{B}T_p = \hbar\omega_p^2/2\pi\bar{\gamma}$. With the exact solution (17.27), and with the modified potential parameters, one finds that the bounce action for anomalous Ohmic dissipation in the WKB rate expression (17.7) has the lower bound

$$S_{\text{B},p}(T) \geqslant \frac{2\pi}{9} M\bar{\gamma} q_{\text{ex},p}^2 \left[1 - (T/T_p)^2 \right] , \qquad T \leqslant T_p . \tag{4.69}$$

4.2.2 State-dependent memory friction

The analysis given in the previous subsection can easily be generalized to nonlinear state-dependent friction with all types of memory effects provided that the system-bath coupling is linear in the bath degrees of freedom. For the global system described by the Hamiltonian (3.8), the reduced density matrix in thermal equilibrium adopts again the form (4.60), but now the influence action is

$$
\begin{aligned}
S_{\mathrm{infl}}[q(\cdot)] &= \frac{1}{2} \int_0^{\hbar\beta} d\tau \int_0^\tau d\tau' \, K(\tau - \tau') \, [\, F[q(\tau)] - F[q(\tau')]\,]^2 \\
&= \int_0^{\hbar\beta} d\tau \int_0^\tau d\tau' \, k(\tau - \tau') \, F[q(\tau)] F[q(\tau')] \,.
\end{aligned}
\tag{4.70}
$$

A nonlinear form function $F(q)$ applies, e.g., in rotational tunneling systems, as the system's bath coupling must obey the symmetries of the hindering potential [79] [cf. the discussion following Eq. (3.8)], and for quasiparticle tunneling between superconductors. For the phenomenological coupling (3.9), the influence action reads

$$
S_{\mathrm{infl}}[\varphi(\cdot)] = \sum_{m=1}^2 \int_0^{\hbar\beta} d\tau \int_0^\tau d\tau' \, k^{(m)}(\tau - \tau') \, F^{(m)}[\varphi(\tau)] \, F^{(m)}[\varphi(\tau')] \,,
\tag{4.71}
$$

where $F^{(1)}[\varphi(\tau)] = \sin[\varphi(\tau)/2]$ and $F^{(2)}[\varphi(\tau)] = \cos[\varphi(\tau)/2]$. The action (4.71) can be brought into the form (4.191) which applies to quasiparticle tunneling in a Josephson junction [80] (cf. Subsec. 4.2.10).

Another important example of state-dependent friction is when the central system interacts with individual bath modes within a spatially restricted region [170, 171]. The case in which the system interacts with localized modes is discussed within the context of decoherence in Section 9.3. The reasoning of this subsection applies accordingly to a weakly coupled nonlinear bath considered in Sec. 3.5.

4.2.3 Spin-boson model

In the model (3.145) and the equivalent spin-boson model (3.147), the respective paths $q(\tau)$ and $\sigma(\tau)$, where $q(\tau) = \frac{1}{2} q_0 \, \sigma(\tau)$, are piecewise constant, and the transitions between positions $\pm\frac{1}{2}q_0$ and ± 1, respectively, are sudden jumps. Fig. 4.1 shows a spin path with 8 moves (spin flips, or kinks, or instantons). A path of $2m$ alternating flips with centers $\{s_j\}$ in chronological order (m sequential anti-kink–kink pairs) reads

$$
\sigma^{(m)}(\tau) = 1 + 2 \sum_{j=1}^{2m} (-1)^j \, \Theta(\tau - s_j) \,.
\tag{4.72}
$$

When the spin initially is in the state $\sigma = \pm 1$, and the bath has relaxed to the corresponding shifted canonical state $W_{\mathrm{R}}[\pm 1]$, and then the spin suddenly jumps to the state $\sigma = \mp 1$, the bath is reorganized from the shifted state $W_{\mathrm{R}}[\pm 1]$ to the shifted state $W_{\mathrm{R}}[\mp 1]$. In the classical regime, the shifted canonical state is

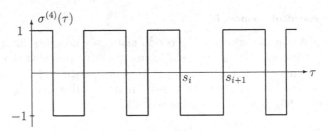

Figure 4.1: The spin path (4.72) consisting of eight jumps at s_i, $i = 1, \cdots, 8$.

$$W_{\mathrm{R}}(\sigma) = \prod_\alpha \sqrt{\frac{\beta \, m_\alpha}{2\pi}} \, w_\alpha \exp\left[-\frac{\beta \, m_\alpha \omega_\alpha^2}{2} \left(x_\alpha - \frac{c_\alpha}{m_\alpha \omega_\alpha^2} \frac{q_0}{2} \sigma \right)^2 \right] . \qquad (4.73)$$

The reorganisation of the classical bath releases the *classical* reorganisation energy Λ_{cl}. This is the difference of the interaction energy (3.144), $H_{\mathrm{I}}(\sigma) = -\frac{1}{2} q_0 \, \sigma \sum_\alpha c_\alpha x_\alpha$, averaged with the two shifted canonical states $W_{\mathrm{R}}(\sigma = \pm 1)$,

$$\begin{aligned}
\Lambda_{\mathrm{cl}} &\equiv \langle H_{\mathrm{I}}(\sigma = \mp 1)\rangle_{W_{\mathrm{R}}(\pm 1)} - \langle H_{\mathrm{I}}(\sigma = \mp 1)\rangle_{W_{\mathrm{R}}(\mp 1)} \\
&= \frac{q_0^2}{2} \sum_\alpha \frac{c_\alpha^2}{m_\alpha \omega_\alpha^2} = \frac{q_0^2}{\pi} \int_0^\infty d\omega \, \frac{J(\omega)}{\omega} = \hbar \int_0^\infty d\omega \, \frac{G(\omega)}{\omega} .
\end{aligned} \qquad (4.74)$$

The last form holds for the continuous spin-boson spectral density $G(\omega)$ defined in Eq.(3.148). The reorganization energy of the bath plays a crucial role in the Marcus theory of electron transfer in a solvent. This theory is discussed in Section 20.1.

Consider next the fluctuations of the polarization energy $\mathfrak{E}(\tau) = q_0 \sum_\alpha c_\alpha x_\alpha(\tau)$ in imaginary time. We obtain with Eq. (3.86) and with the spectral density (3.148)

$$\mathcal{X}(\tau) \equiv \frac{1}{\hbar^2}\langle \delta \mathfrak{E}(\tau) \delta \mathfrak{E}(0)\rangle_{\tilde{W}_0} = \frac{q_0^2}{\hbar^2} \mathfrak{K}(\tau) = \int_0^\infty d\omega \, G(\omega) \mathfrak{D}_\omega(\tau) , \qquad (4.75)$$

where $\mathfrak{D}_\omega(\tau)$ is the thermal boson propagator (3.82) or (3.83). Upon writing the action (4.51) in terms of the spin path $\sigma(\tau)$, and using the relation (4.55), the Euclidean influence functional (4.43) with (4.51) takes the form

$$\mathbb{F}[\sigma(\cdot)] = \exp\left\{ -\frac{1}{8} \int_0^{\hbar\beta} d\tau \int_0^\tau d\tau' \, \mathcal{X}(\tau - \tau') \, [\sigma(\tau) - \sigma(\tau')]^2 \right\} . \qquad (4.76)$$

For the spin path (4.72), the time integrations can be done. We then get a representation in terms of repulsive kink-kink and attraktive kink-antikink interactions $\pm \mathcal{W}(\tau)$, where $-\mathcal{W}(\tau)$ is the second integral of the kernel $\mathcal{X}(\tau)$, $\ddot{\mathcal{W}}(\tau) = -\mathcal{X}(\tau)$. For a path with m consecutive kink–anti-kink pairs, the influence functional takes the form

$$\mathbb{F}_m[\{s_j\}] = \exp\left\{ \sum_{j=2}^{2m} \sum_{i=1}^{j-1} (-1)^{i+j} \, \mathcal{W}(s_j - s_i) \right\} , \qquad (4.77)$$

$$\mathcal{W}(\tau) = \int_0^\infty d\omega \, \frac{G(\omega)}{\omega^2} \left[\mathfrak{D}_\omega(0) - \mathfrak{D}_\omega(\tau) \right] .$$

Eq. (4.76) is the state representation of the influence function, whereas Eq. (4.77) is usually referred to as charge representation. The fictitious charges ± 1 are at the kink and antikink positions. The function $\mathcal{W}(\tau)$ describes the interaction between two likely charges at imaginary-time distance τ. The exponent of the influence function (4.77) is the sum of all interactions of the $2m$ charges with alternating sign in chronological order.

4.2.4 Acoustic polaron and defect tunneling: one-phonon coupling

Consider a particle which is coupled to acoustic phonons via the interaction $H_{\mathrm{lin}}(\mathbf{q})$ given in Eq. (3.175). Since the global Hamiltonian is a mixed quadratic form in the phonon variables, the thermal average of the phonon degrees of freedom can be done exactly. Rather than perform the average in the path sum representation, here we choose, for a change, to proceed directly in the operator formulation. The influence functional which describes the effects of the coupling H_{lin} is formally defined by

$$\mathbb{F}_{\mathrm{lin}}[\mathbf{q}(\cdot)] \equiv \exp\left[-\mathcal{S}_{\mathrm{lin}}[\mathbf{q}(\cdot)]/\hbar\right] = \left\langle T_\tau \exp\left(-\int_0^{\hbar\beta} d\tau\, \tilde{H}_{\mathrm{lin}}[\mathbf{q}(\tau)]/\hbar\right)\right\rangle_\beta. \quad (4.78)$$

Here T_τ is the imaginary-time ordering operator, $\langle\cdots\rangle_\beta$ denotes the thermal average with respect to $\exp(-\sum_k \beta\hbar\omega_k b_k^\dagger b_k)$, where again $k = (\mathbf{k}, \lambda)$, and $\tilde{H}_{\mathrm{lin}}[\mathbf{q}(\tau)]$ is the electron-phonon interaction (3.175) in the interaction picture (denoted by a tilde),

$$\tilde{H}_{\mathrm{lin}}[\mathbf{q}(\tau)] = \exp(\hat{H}_{\mathrm{R}}\tau/\hbar)\, \hat{H}_{\mathrm{lin}}[\mathbf{q}(\tau)] \exp(-\hat{H}_{\mathrm{R}}\tau/\hbar). \quad (4.79)$$

In imaginary time we have

$$\tilde{b}_k(\tau) = b_k\, e^{-\omega_k\tau}, \qquad \tilde{b}_k^\dagger(\tau) = b_k^\dagger\, e^{\omega_k\tau}. \quad (4.80)$$

Next, the formal expression (4.78) is expanded into a power series in $\tilde{H}_{\mathrm{lin}}[\mathbf{q}]$,

$$\begin{aligned}
\mathbb{F}_{\mathrm{lin}}[\mathbf{q}(\cdot)] = 1 + \sum_{n=1}^\infty \left(\frac{-1}{\hbar}\right)^n \int_0^{\hbar\beta} d\tau_n \cdots \int_0^{\tau_3} d\tau_2 \int_0^{\tau_2} d\tau_1 \\
\times \left\langle \tilde{H}_{\mathrm{lin}}[\mathbf{q}(\tau_n)] \cdots \tilde{H}_{\mathrm{lin}}[\mathbf{q}(\tau_1)]\right\rangle_\beta,
\end{aligned} \quad (4.81)$$

and *Wick's theorem* for thermodynamic averages (see, e.g., Ref. [172]) is employed. The resulting cumulant expansion then yields the influence action. Since the interaction (3.175) is linear in the phonon field, the averages of all odd powers of $\tilde{H}_{\mathrm{lin}}[\mathbf{q}]$ vanish. Because of the Gaussian statistics, also the cumulants of fourth and of higher even order in $\tilde{H}_{\mathrm{lin}}[\mathbf{q}]$ are zero. Hence the only non-vanishing cumulant contraction is the second order term $\langle\tilde{H}_{\mathrm{lin}}[\mathbf{q}(\tau)]\tilde{H}_{\mathrm{lin}}[\mathbf{q}(\tau')]\rangle_\beta$, and thus the influence action is

$$\mathcal{S}_{\mathrm{lin}}[\mathbf{q}(\cdot)] = -\frac{1}{\hbar}\int_0^{\hbar\beta} d\tau \int_0^\tau d\tau'\, \left\langle\tilde{H}_{\mathrm{lin}}[\mathbf{q}(\tau)]\tilde{H}_{\mathrm{lin}}[\mathbf{q}(\tau')]\right\rangle_\beta. \quad (4.82)$$

With use of the form (3.175) for $H_{\mathrm{lin}}[\mathbf{q}(\tau)]$, the contraction may be written in terms of the boson imaginary-time propagator (3.82) as

$$\left\langle \tilde{H}_{\text{lin}}[\mathbf{q}(\tau)] \tilde{H}_{\text{lin}}[\mathbf{q}(\tau')] \right\rangle_\beta = \sum_{n,m} \sum_{\mathbf{k},\lambda} \frac{\hbar \kappa_k^{(n)}[\mathbf{q}(\tau)] \kappa_k^{(m)}[\mathbf{q}(\tau')]}{2 V \varrho \, \omega_k} \, e^{-i\mathbf{k}\cdot[\mathbf{R}^{(n)} - \mathbf{R}^{(m)}]}$$

$$\times \exp\left(i\,\mathbf{k}\cdot[\mathbf{q}(\tau) - \mathbf{q}(\tau')]\right) \mathfrak{D}_{\omega_k}(\tau - \tau') . \qquad (4.83)$$

Quantum diffusion of light interstitials and defect tunneling in solids usually take place by a succession of incoherent tunneling transitions between localized states in the rigid multi-well potential of the host crystal. At thermal energy small compared with the level spacing of the low-lying states in the individual wells, only the lowest state in each local minimum is relevant to the dynamics [173].[3]

Consider the case of a defect particle which tunnels between two spatially separated interstitial positions located at $\mathbf{q}_1 = \frac{1}{2}\mathbf{q}_0$ and $\mathbf{q}_2 = -\frac{1}{2}\mathbf{q}_0$. We then have

$$\mathbf{q}(\tau) = \frac{1}{2}\mathbf{q}_0\,\sigma(\tau) , \qquad (4.84)$$

where $\sigma(\tau) = \pm 1$, as sketched in Fig. 4.1. The values $\pm\frac{1}{2}\mathbf{q}_0$ are the two energetically accessible interstitial positions of the defect. For this formation, the contraction (4.83) takes the form

$$\left\langle \tilde{H}_{\text{lin}}[\sigma(\tau)] \tilde{H}_{\text{lin}}[\sigma(\tau')] \right\rangle_\beta = \sum_{n,m} \sum_{\mathbf{k},\lambda} \frac{\hbar \kappa_k^{(n)}(\sigma) \kappa_k^{(m)}(\sigma')}{2 V \varrho \, \omega_k} \, e^{-i\mathbf{k}\cdot[\mathbf{R}^{(n)} - \mathbf{R}^{(m)}]} \mathfrak{D}_{\omega_k}(\tau - \tau')$$

$$\times \left[\cos^2\left(\tfrac{1}{2}\mathbf{k}\cdot\mathbf{q}_0\right) + \sin^2\left(\tfrac{1}{2}\mathbf{k}\cdot\mathbf{q}_0\right) \sigma(\tau)\sigma(\tau') \right] . \quad (4.85)$$

Here we have used the notation

$$\kappa_k^{(n)}(\sigma) \equiv \kappa_k^{(n)}(\mathbf{q} = \tfrac{1}{2}\sigma\mathbf{q}_0) . \qquad (4.86)$$

In defect tunneling, the energy shift which is caused by the linear coupling $H_{\text{lin}}(\mathbf{q})$ plays an ancillary role [cf. the discussion below Eq. (3.4)]. Therefore we may restrict the attention to the change of the action when the particle hops. With subtraction of the contribution resulting from a constant path σ, we have

$$\mathbb{S}_{\text{lin}}[\sigma(\cdot)] = \frac{1}{\hbar} \int_0^{\hbar\beta} d\tau \int_0^\tau d\tau' \left\langle \tilde{H}_{\text{lin}}[\sigma] \tilde{H}_{\text{lin}}[\sigma] - \tilde{H}_{\text{lin}}[\sigma(\tau)] \tilde{H}_{\text{lin}}[\sigma(\tau')] \right\rangle_\beta . \quad (4.87)$$

Putting $[\kappa_k^{(n)}(+)]^2 = [\kappa_k^{(n)}(-)]^2$, the action may be written in the form (4.76),

$$\mathbb{S}_{\text{lin}}[\sigma(\cdot)]/\hbar = \frac{1}{8} \int_0^{\hbar\beta} d\tau \int_0^\tau d\tau' \, \mathcal{X}_{\text{lin}}(\tau - \tau') \left[\sigma(\tau) - \sigma(\tau') \right]^2 , \quad (4.88)$$

where the kernel $\mathcal{X}_{\text{lin}}(\tau)$ is defined in terms of the integral representation

$$\mathcal{X}_{\text{lin}}(\tau) = \int_0^\infty d\omega \, G_{\text{lin}}(\omega) \mathfrak{D}_\omega(\tau) . \qquad (4.89)$$

Here, $\mathfrak{D}_\omega(\tau)$ is the phonon propagator (3.82), and the spectral density reads

---

[3]The dissipative two- and multi-state system are discussed in Part IV and Part V, respectively.

$$G_{\text{lin}}(\omega) = \sum_{\mathbf{k},\lambda} \frac{\delta(\omega - \omega_{\mathbf{k},\lambda})}{2V\varrho\,\hbar\omega_{\mathbf{k},\lambda}} \left\{ \cos^2\left(\tfrac{1}{2}\mathbf{k}\cdot\mathbf{q}_0\right) \Big| \sum_n e^{i\mathbf{k}\cdot\mathbf{R}^{(n)}} \left[\kappa_{\mathbf{k},\lambda}^{(n)}(+) - \kappa_{\mathbf{k},\lambda}^{(n)}(-)\right] \Big|^2 \right.$$
$$\left. + \sin^2\left(\tfrac{1}{2}\mathbf{k}\cdot\mathbf{q}_0\right) \Big| \sum_n e^{i\mathbf{k}\cdot\mathbf{R}^{(n)}} \left[\kappa_{\mathbf{k},\lambda}^{(n)}(+) + \kappa_{\mathbf{k},\lambda}^{(n)}(-)\right] \Big|^2 \right\}. \quad (4.90)$$

Consider next the Debye model for the phonon degrees of freedom. For three space dimension, linear dispersion $\omega_{\mathbf{k},\lambda} = v_\lambda|\mathbf{k}|$, and with the same Debye frequency ω_{D} for the longitudinal and transversal phonon branches, we have

$$\frac{1}{V}\sum_{\mathbf{k},\lambda}\cdots \;\rightarrow\; \frac{1}{2\pi^2}\sum_\lambda \frac{1}{v_\lambda^3}\int\frac{d\Omega}{4\pi}\int_0^\infty d\omega_{\mathbf{k},\lambda}\,\omega_{\mathbf{k},\lambda}^2\,\Theta(\omega_{\text{D}} - \omega_{\mathbf{k},\lambda})\cdots. \quad (4.91)$$

Let N denote the number of atoms per volume V. Then the summation (4.91) yields

$$\omega_{\text{D}} = \left(\frac{2\pi^2 N}{V}\right)^{1/3}\bar{v} \quad\text{with}\quad \frac{1}{\bar{v}^3} = \frac{1}{3}\left(\frac{1}{v_\ell^3} + \frac{2}{v_t^3}\right). \quad (4.92)$$

The low-frequency expansion of the spectral density $G_{\text{lin}}(\omega)$ is found upon Taylor expansion of the exponential and trigonometric functions in Eq. (4.90),

$$G_{\text{lin}}(\omega) = \left(2\alpha_1\omega + 2\alpha_3\frac{\omega^3}{\omega_{\text{D}}^2} + 2\alpha_5\frac{\omega^5}{\omega_{\text{D}}^4} + \mathcal{O}(\omega^7)\right)\Theta(\omega_{\text{D}} - \omega). \quad (4.93)$$

The coupling parameter of the Ohmic contribution $G_{\text{lin}}(\omega) \propto \omega$ is given by

$$\alpha_1 = \frac{1}{8\pi^2\varrho\hbar}\sum_\lambda \frac{1}{v_\lambda^3}\int\frac{d\Omega}{4\pi}\Big|\sum_n\left(\kappa_{\mathbf{k},\lambda}^{(n)}(+) - \kappa_{\mathbf{k},\lambda}^{(n)}(-)\right)\Big|^2. \quad (4.94)$$

The dimensionless coupling coefficients α_3 and α_5 are conveniently expressed in terms of the symmetrized and anti-symmetrized components of the deformation potential tensor ($\mathfrak{P}$ has dimension energy, κ has dimension force)

$$\mathfrak{P}_{\text{s/a}} = \frac{1}{2}\sum_n \mathbf{R}^{(n)} \otimes \left[\mathbf{g}^{(n)}(+) \pm \mathbf{g}^{(n)}(-)\right], \quad (4.95)$$

where $\mathbf{g}^{(n)}(\sigma)$ is the Kanzaki force defined in Eq. (3.173). Equation (4.90) yields [139]

$$\alpha_3 = \frac{\omega_{\text{D}}^2}{2\pi^2\varrho\hbar}\sum_\lambda \frac{1}{v_\lambda^5}\int\frac{d\Omega}{4\pi}\Big|\hat{\mathbf{k}}\cdot\mathfrak{P}_{\text{a}}\cdot\mathbf{e}(\hat{\mathbf{k}},\lambda)\Big|^2, \quad (4.96)$$

and

$$\alpha_5 = \frac{\omega_{\text{D}}^4}{32\pi^2\varrho\hbar}\sum_\lambda \frac{1}{v_\lambda^7}\int\frac{d\Omega}{4\pi}\left\{\Big|\sum_n\left(\hat{\mathbf{k}}\cdot\mathbf{R}^{(n)}\right)^2\left(\kappa_{\mathbf{k},\lambda}^{(n)}(+) - \kappa_{\mathbf{k},\lambda}^{(n)}(-)\right)\Big|^2\right.$$
$$\left. + 4\left(\hat{\mathbf{k}}\cdot\mathbf{q}_0\right)^2\left(\Big|\hat{\mathbf{k}}\cdot\mathfrak{P}_{\text{s}}\cdot\mathbf{e}(\hat{\mathbf{k}},\lambda)\Big|^2 - \Big|\hat{\mathbf{k}}\cdot\mathfrak{P}_{\text{a}}\cdot\mathbf{e}(\hat{\mathbf{k}},\lambda)\Big|^2\right)\right\}, \quad (4.97)$$

where $\hat{\mathbf{k}}$ is a unit vector. The amplitude factors $\big|\sum_n(\mathbf{R}^{(n)})^{(\nu-1)/2}\cdots\big|$ in the coefficients α_ν are moments of the positions of the atoms. Therefore, we may regard α_1 as the coefficient of a monopole force, α_3 as the coefficient of a dipole force, etc. [174].

For a crystalline environment, the *monopole* force is absent because of the symmetry of the interstitial positions. Therefore, the coupling coefficient α_1 is zero. Hence dissipation by acoustic phonons is super-Ohmic instead of Ohmic.

The tensor character of $\mathfrak{P}$ in Eqs. (4.96) and (4.97) may lead to strong selection rules for the coefficients α_3 and α_5. The particular case depends on the symmetry of the interstitial sites [134, 136, 139].

In bcc lattices, such as Nb and Fe, the octahedral and tetrahedral interstitial sites have tetragonal symmetry. Hence there is no selection rule. Denoting the eigenvalues of the deformation potential tensor $\mathfrak{P}_a$ in Eq. (4.95) by p_1 and p_3, where p_1 is double and p_3 belongs to the tetragonal axis, we obtain for an isotropic elastic medium

$$\alpha_3 = \frac{\omega_D^2 (p_1 - p_3)^2}{2\pi^2 \varrho \hbar} \left(\frac{1}{10} \frac{1}{v_t^5} + \frac{1}{15} \frac{1}{v_\ell^5} \right) . \tag{4.98}$$

For niobium, we have $\alpha_3 \approx 1.3 \, (p_1 - p_3)^2/[eV]^2$ (Ref. [139]), and tabulated values of p_1 and p_3 can be found in Ref. [175]. In bcc metals, the coefficient α_5 is estimated to be smaller than or at most of the same order of magnitude as the coefficient α_3.

In fcc lattices, such as Cu and Al, the octa- and tetrahedral sites have cubic symmetry. Hence the deformation potential tensor is degenerate, $p_1 = p_2 = p_3 \equiv p$. As a result of this symmetry, the dipole force is absent, and thus $\alpha_3 = 0$. Furthermore, there follows from Eq. (4.97) that only the longitudinal phonon branch contributes to the coefficient α_5. One finds [139]

$$\alpha_5 = \frac{\omega_D^4 q_0^2}{24\pi^2 \varrho \hbar v_\ell^7} p^2 . \tag{4.99}$$

For tetrahedral sites in Al, Eq. (4.99) gives the numerical value $\alpha_5 = 0.1 \, p^2/[eV]^2$, and for octahedral sites the value $\alpha_5 = 0.2 \, p^2/[eV]^2$.

For tunneling centers in amorphous solids, the monopole force term is again missing for symmetry reasons, i.e., the coupling parameter α_1 of the Ohmic spectral density vanishes again. However, there are generally no selection rules for the dipole force. It is convenient to introduce the coupling strength as in Eq. (4.93),

$$G_{\text{lin}}(\omega) = 2\alpha_3 \frac{\omega^3}{\omega_D^2} , \qquad \text{where} \qquad \alpha_3 = \frac{\omega_D^2}{2\pi^2 \varrho \hbar} \left(\frac{p_\ell^2}{v_\ell^5} + \frac{2p_t^2}{v_t^5} \right) . \tag{4.100}$$

The elastic energies or deformation potentials p_ℓ and p_t describe again the coupling to longitudinal and transversal acoustic phonons, respectively.

For acoustic phonons in d dimensions, the spectral density of states per unit volume at low frequencies is

$$\sum_{\mathbf{k},\lambda} \delta(\omega - \omega_{\mathbf{k},\lambda}) \equiv D_{\text{ac}}(\omega) = 2D_{\text{ac},0} \left(\frac{\omega}{\omega_D} \right)^{d-1} , \tag{4.101}$$

where $D_{\text{ac},0}$ is a normalization constant. Next, we assume a frequency-dependent coupling of the power-law form [$\lambda_\alpha \to \lambda(\omega)$ in Eq. (3.148)],

$$\lambda(\omega) = \lambda_0 \, (\omega/\omega_D)^\mu \qquad \text{for} \qquad \omega \lesssim \omega_D . \tag{4.102}$$

Here, the parameter λ_0 is proportional to the deformation potential. Then the spectral density $G_{\text{lin}}(\omega)$ in Eq. (3.148) takes the power-law form [cf. Eq. (3.38)]

$$G_{\text{lin}}(\omega) \equiv D_{\text{ac}}(\omega)\lambda^2(\omega) = 2\alpha_s \omega_{\text{D}}^{1-s}\omega^s, \qquad (4.103)$$

where $\alpha_s = D_{\text{ac},0}\lambda_0^2/\omega_{\text{D}}$. The power s of the spectral algebraic law is

$$s = d - 1 + 2\mu. \qquad (4.104)$$

The relation (4.104) connects the power s of the power-law form (4.103) for $G_{\text{lin}}(\omega)$ with the dimension d of the lattice, and with the power μ of the frequency-dependent coupling function $\lambda(\omega)$. In the absence and presence of a cubic symmetry, we have $\mu = 1/2$ and $\mu = 3/2$, respectively.

A concluding thought is appropriate. When the polaron moves coherently in the crystal, the two-state assumption breaks down. In this case, it is reasonable to eliminate the dependence of the expression (4.83) on the individual positions $\mathbf{R}^{(n)}$ of the atoms by averaging over many atoms. Then the angular integrals in Eq. (4.83) are easily carried out. We then find

$$\begin{aligned}
S_{\text{lin}}[\mathbf{q}(\cdot)] &= \frac{3}{\pi} \int_0^\infty d\omega \, \frac{J(\omega)}{k^2(\omega)} \int_0^{\hbar\beta} d\tau \int_0^\tau d\tau' \, \mathfrak{D}_\omega(\tau - \tau') \\
&\quad \times \left\{ 1 - \frac{\sin[k(\omega)|\mathbf{q}(\tau) - \mathbf{q}(\tau')|]}{k(\omega)|\mathbf{q}(\tau) - \mathbf{q}(\tau')|} \right\},
\end{aligned} \qquad (4.105)$$

where $\mathfrak{D}_\omega(\tau)$ is the thermal propagator (3.82) introduced in Subsection 3.1.6. The function $k(\omega)$ is the dispersion relation solved for $k = |\mathbf{k}|$, and the spectral density $J(\omega)$ is related to the function $G(\omega)$ given in Eq. (4.93) by Eq. (3.148), where q_0 is a characteristic length. For $k|\mathbf{q}(\tau) - \mathbf{q}(\tau')| \ll 1$ in the relevant integration regime, we may expand the sine in Eq. (4.105). Then, the expression (4.105) reduces to the standard form given in Eq. (4.51).

Generally, the influence action (4.51) overestimates the dissipative influences of the environment as compared with the more accurate expression (4.105). The influence action (4.105) has been used repeatedly in studies of the ground-state energy and the effective mass of the acoustic polaron.[4]

Particular interest has been devoted over about fifty years to the investigation of the possibility of a phonon-induced self-trapping phase transition. In recent years, this question has been answered in the negative. For a clarification of this issue we refer to a review by Gerlach and Löwen [178].

4.2.5 Acoustic polaron: two-phonon coupling

We now turn to the discussion of the two-phonon coupling (3.176). The influence functional of the two-phonon process $F_{2\text{ph}}[\mathbf{q}]$ is formally given by the series expansion

---

[4]Cf., e.g., Refs. [176, 177]. See also the review in Ref. [178].

(4.81) in which $\tilde{H}_{\text{lin}}[\mathbf{q}(\tau)]$ is replaced by $\tilde{H}_{\text{2ph}}[\mathbf{q}(\tau)]$. With use of Eqs. (4.80), (3.67), and (3.82), the thermal average of the first order takes the form

$$\left\langle \tilde{H}_{\text{2ph}}[\mathbf{q}(\tau)] \right\rangle_\beta = \frac{\hbar}{4V\varrho} \sum_{n,m} \sum_{\mathbf{k},\lambda} \frac{\gamma_{\mathbf{k},-\mathbf{k}}^{(n,m)}[\mathbf{q}(\tau)]}{\omega_k} e^{i\,\mathbf{k}\cdot[\mathbf{R}^{(m)}-\mathbf{R}^{(n)}]} \coth(\beta\hbar\omega_k/2) . \quad (4.106)$$

Now assume that the particle-lattice potential in Eq. (3.168) or Eq. (3.174) is a sum of pair potentials. Furthermore, at low temperature umklapp processes may be disregarded. Then one finds [139]

$$\sum_{n,m} \gamma_{\mathbf{k},-\mathbf{k}}^{(n,m)}[\mathbf{q}(\tau)] e^{i\,\mathbf{k}\cdot[\mathbf{R}^{(m)}-\mathbf{R}^{(n)}]} = 0 , \quad (4.107)$$

and thus

$$\langle \tilde{H}_{\text{2ph}}[\mathbf{q}(\tau)] \rangle_\beta = 0 . \quad (4.108)$$

The computation of the second cumulant is straightforward. With use of relations for Gaussian thermal averages involving four phonon operators , e.g.,

$$\langle b_k b_{-k'}^\dagger b_{-k''}^\dagger b_{k'''} \rangle_\beta = \langle b_k b_{-k'}^\dagger \rangle_\beta \langle b_{-k''}^\dagger b_{k'''} \rangle_\beta + \langle b_k b_{-k''}^\dagger \rangle_\beta \langle b_{-k'}^\dagger b_{k'''} \rangle_\beta , \quad (4.109)$$

and of Eqs. (4.108), (3.67), (3.82), and with the relation $\gamma_{k,k'}^{(n,m)} = \gamma_{-k,-k'}^{(n,m)}$, we find

$$\left\langle \tilde{H}_{\text{2ph}}[\mathbf{q}(\tau)]\tilde{H}_{\text{2ph}}[\mathbf{q}(\tau')] \right\rangle_\beta = \sum_{n,m} \sum_{n',m'} \sum_{k,k'} \frac{\hbar^2 \gamma_{k,k'}^{(n,m)}[\mathbf{q}(\tau)]\gamma_{k,k'}^{(n',m')}[\mathbf{q}(\tau')]}{2(2V\varrho)^2 \omega_k \omega_{k'}} \quad (4.110)$$

$$\times \; e^{i\,\mathbf{k}\cdot[\mathbf{R}^{(n')}-\mathbf{R}^{(n)}]} e^{i\,\mathbf{k'}\cdot[\mathbf{R}^{(m')}-\mathbf{R}^{(m)}]} e^{i\,(\mathbf{k}+\mathbf{k'})\cdot[\mathbf{q}(\tau)-\mathbf{q}(\tau')]}$$

$$\times \; \Big\{ 2\Theta(\omega_k - \omega_{k'})\left[n(\omega_{k'}) - n(\omega_k) \right] \mathfrak{D}_{\omega_k-\omega_{k'}}(\tau-\tau')$$

$$+ \; [1 + n(\omega_k) + n(\omega_{k'})] \, \mathfrak{D}_{\omega_k+\omega_{k'}}(\tau-\tau') \Big\} .$$

In the sequel, we restrict again the attention to defect tunneling between two interstitial sites. With use of the parametrization (4.84) we may write

$$\text{Re } e^{i\,(\mathbf{k}+\mathbf{k'})\cdot[\mathbf{q}(\tau)-\mathbf{q}(\tau')]} = \cos^2[\tfrac{1}{2}(\mathbf{k}+\mathbf{k'})\cdot\mathbf{q}_0] + \sin^2[\tfrac{1}{2}(\mathbf{k}+\mathbf{k'})\cdot\mathbf{q}_0]\,\sigma(\tau)\sigma(\tau') . \quad (4.111)$$

As in the preceding analysis of linear phonon coupling, we are not interested in the constant energy shift induced by the coupling for $\sigma = \sigma'$. Therefore, as before in Eq.(4.87), we subtract this term from the expression (4.110). With use of Eq. (4.111) we then obtain the influence action in the form

$$\mathbb{S}_{\text{2ph}}[\sigma(\cdot)]/\hbar = \frac{1}{8} \int_0^{\hbar\beta} d\tau \int_0^\tau d\tau' \, \mathcal{X}_{\text{2ph}}(\tau-\tau')[\,\sigma(\tau) - \sigma(\tau')\,]^2 \quad (4.112)$$

with the kernel

$$\mathcal{X}_{\text{2ph}}(\tau) = \int_0^\infty d\omega\, G_{\text{2ph}}(\omega)\mathfrak{D}_\omega(\tau) . \quad (4.113)$$

The spectral density resulting from two-phonon coupling is given by

$$G_{2\text{ph}}(\omega) = \frac{1}{2(2V\varrho)^2} \sum_{k,k'} \left\{ 2\delta(\omega - \omega_k + \omega_{k'})\frac{n(\omega_{k'}) - n(\omega_k)}{\omega_k \omega_{k'}} \right. \qquad (4.114)$$

$$\left. + \delta(\omega - \omega_k - \omega_{k'})\frac{1 + n(\omega_k) + n(\omega_{k'})}{\omega_k \omega_{k'}} \right\} f(k, k'),$$

where

$$f(k, k') = \cos^2\left[\tfrac{1}{2}(\mathbf{k} + \mathbf{k}') \cdot \mathbf{q}_0\right] \left| \sum_{n,m} e^{i\mathbf{k}\cdot\mathbf{R}^{(n)}} e^{i\mathbf{k}'\cdot\mathbf{R}^{(m)}} [\gamma_{k,k'}^{(n,m)}(+) - \gamma_{k,k'}^{(n,m)}(-)] \right|^2$$

$$+ \sin^2\left[\tfrac{1}{2}(\mathbf{k} + \mathbf{k}') \cdot \mathbf{q}_0\right] \left| \sum_{n,m} e^{i\mathbf{k}\cdot\mathbf{R}^{(n)}} e^{i\mathbf{k}'\cdot\mathbf{R}^{(m)}} [\gamma_{k,k'}^{(n,m)}(+) + \gamma_{k,k'}^{(n,m)}(-)] \right|^2,$$

and where we used the notation

$$\gamma_{k,k'}^{(n,m)}(\sigma) \equiv \gamma_{k,k'}^{(n,m)}(\mathbf{q} = \tfrac{1}{2}\sigma\mathbf{q}_0). \qquad (4.115)$$

Next, we expand the expression (4.114) in powers of ω. To linear order in ω, we find

$$G_{2\text{ph}}(\omega) = 2\alpha_{2\text{ph}}\omega + \mathcal{O}\left[(\omega/\omega_D)^3\right]. \qquad (4.116)$$

The prefactor $\alpha_{2\text{ph}}$ is the dimensionless expression

$$\alpha_{2\text{ph}} = \frac{\hbar\beta}{2(2V\varrho)^2} \sum_{k,k'} \delta(\omega_k - \omega_{k'})\frac{n(\omega_k)[1 + n(\omega_{k'})]}{\omega_k \omega_{k'}} f(k, k'). \qquad (4.117)$$

The contributions $\propto \omega^3, \omega^5, \cdots$ are of minor interest since they can be merged with the corresponding terms of $G_{\text{lin}}(\omega)$ in the series (4.93). This would result in small temperature dependent contributions to the coefficients α_3, α_5, etc.

Turning to the Debye model (4.91) and expanding the trigonometric functions in $\mathbf{k}$ and $\mathbf{k}'$, we obtain the low temperature expansion for α_{quadr}. We find

$$\alpha_{2\text{ph}} = \kappa_1\left(\frac{k_B T}{\hbar\omega_D}\right)^6 + \kappa_2\left(\frac{k_B T}{\hbar\omega_D}\right)^8 + \mathcal{O}\left[\left(\frac{k_B T}{\hbar\omega_D}\right)^{10}\right]. \qquad (4.118)$$

The coupling parameter κ_1 is given by

$$\kappa_1 = \frac{9\pi^2\omega_D^6}{370\varrho^2} \sum_{\lambda,\lambda'} \frac{1}{v_\lambda^5 v_{\lambda'}^5} \int\frac{d\Omega}{4\pi} \int\frac{d\Omega'}{4\pi}$$

$$\times \left| \sum_{n,m} [\gamma_{\hat{\mathbf{k}},\lambda}^{(n,m)}(+) - \gamma_{\hat{\mathbf{k}},\lambda}^{(n,m)}(-)] (\hat{\mathbf{k}}\cdot\mathbf{R}^{(n)})(\hat{\mathbf{k}}'\cdot\mathbf{R}^{(m)}) \right|^2. \qquad (4.119)$$

It is also straightforward to calculate the coefficient κ_2. Since the resulting expression is quite lengthy, it is not given here, and we refer to Ref. [139].

For tetrahedral interstitial sites in bcc crystals, the coefficient κ_1 is estimated to be about 10^4. On the other hand, for cubic symmetry of the interstitial sites, there holds $\gamma^{(n)}(+) = \gamma^{(n)}(-)$, and therefore the coefficient κ_1 vanishes. The dimensionless coefficient κ_2 is estimated to be about 10^6.

In summary, the spectral density due to the two-phonon process (4.116) is Ohmic. However, the viscosity coefficient varies strongly with T as T^6 or T^8, depending on the particular symmetry of the interstitial sites. We anticipate that coupling to electron-hole excitations in metals leads also to an Ohmic form of the spectral density, but with a temperature-independent viscosity (see the discussion in Subsection 4.2.8).

Correlated n-phonon processes of higher order, $n \geqslant 3$, which arise from the terms of order $\mathbf{u}^n$ in the expansion (3.168), yield again an Ohmic contribution to the spectral density. The respective viscosity coefficient behaves as T^{2n+2} in the absence of cubic symmetry, and as T^{2n+4} for cubic symmetry. Therefore, at low temperatures, multi-phonon processes with $n > 2$ merely result in a weak temperature-dependent renormalization of the parameters κ_1 and κ_2. The two-phonon process has been studied in Refs. [135, 179, 181, 182, 139].

In conclusion, even if one has, for some model, in leading order of the coupling a super-Ohmic form for $G(\omega)$, in most (if not all) cases an Ohmic spectral density contribution is found if one considers higher orders of the coupling. However, the respective viscosity will turn out to be strongly *temperature-dependent*.

4.2.6 Tunneling between surfaces: one-phonon coupling

One of the most interesting developments in surface science in the last thirty years is the possibility to manipulate atoms and molecules at a surface on an atomic scale by a scanning tunneling microscope (STM) [183]. In the atomic switch realized by Eigler *et al.* [184], a Xe atom has been reversibly transferred between a Ni surface and a tungsten tip. Thus, scanning-tunneling microscopy can be used not only for imaging, but also to test fundamental aspects of quantum mechanics [185]. The crystalline environment can have strong influence on the transfer of the atom. Louis and Sethna have emphasized that linear coupling of acoustic phonons to atoms tunneling between surfaces results in Ohmic dissipation, as opposed to the bulk case where dissipation is super-Ohmic [174]. With the groundwork already done in Subsection 4.2.4, the case of surface tunneling is easily understood.

The potential for the surface-tip system of a STM or atomic-force microscope (AFM) has a double well shape. Consider the parameter regime (3.135) which allows us to truncate the Hilbert space to two states. The tunneling atom exerts a force on the surface which changes by an amount $\Delta \mathbf{F}$ when the atom hops from the surface to the tip. In response to the atom being on the surface or on the tip, the surface atoms will switch their equilibrium positions. Since the displacement is small, we may consider the interaction of the tunneling atom with the surface in linear order in the displacement (one-phonon coupling). The atom on the surface or tip together with the relaxation of the atoms on the surface can thus be modeled by the spin-boson Hamiltonian (3.147). The possibility of tuning many parameters is very appealing.

By application of an electrical field and by shift of the tip position, the bias energy and the tunneling coupling can be varied in a wide range.

All we need to connect tunneling between surfaces with defect tunneling in the bulk is to identify the force $\Delta\mathbf{F}$ with the change of the sum of the Kanzaki forces $\mathbf{g}^{(n)}$ when the atom moves from the surface $(+)$ to the tip $(-)$. We have

$$\Delta\mathbf{F} = \sum_n \left[\mathbf{g}^{(n)}(+) - \mathbf{g}^{(n)}(-) \right] , \qquad (4.120)$$

where the Kanzaki forces are defined in Eq. (3.173). The link missing yet is easily gathered from the findings in Subsection 4.2.4, which for the present purpose are Eq. (4.93) with Eq. (4.94). Thus, the tip-plus-atom system coupled to acoustic phonons is represented by an Ohmic spectral density[5]

$$G(\omega) = \left[2K\omega + \mathcal{O}(\omega^3) \right] \Theta(\omega_{\mathrm{D}} - \omega) \qquad (4.121)$$

with the dimensionless coupling parameter

$$K = \frac{1}{8\pi^2 \varrho\hbar} \sum_\lambda \frac{1}{v_\lambda^3} \int \frac{d\Omega}{4\pi} \left| \Delta\mathbf{F} \cdot \mathbf{e}(\hat{\mathbf{k}}, \lambda) \right|^2 . \qquad (4.122)$$

While for defect tunneling in the bulk of a solid the monopole force is absent and hence super-Ohmic dissipation prevails, atom tunneling between a tip and a surface has lower symmetry, which results in a nonzero monopole force $\Delta\mathbf{F}$ on the surface and hence in Ohmic dissipation. Assuming a point force $\Delta\mathbf{F}/2$ on a semi-infinite isotropic medium, the Ohmic coupling parameter K is estimated as [174]

$$K = \frac{\delta}{8\pi^2 \varrho\hbar v_t^3} (\Delta\mathbf{F})^2 , \qquad (4.123)$$

where v_t is the tranversal sound velocity and the numerical constant δ varies between 0.3 and 0.8, depending on the Lamé constants of the medium. By variation of the distance between the tip and the surface, the force $\Delta\mathbf{F}$ (and thus the coupling parameter K) can be tuned besides the tunneling coupling Δ and the bias ϵ. Hence, atomic tunneling in a STM or AFM seems to be a prominent test ground for studying *macroscopic quantum tunneling* (MQT) and *macroscopic quantum coherence* (MQC) phenomena in different parameter regimes. In conclusion, we have studied the effects of phonons on the tunneling of an atom between two surfaces, and we have found, because of the lower symmetry compared with tunneling in the bulk, Ohmic dissipation.

4.2.7 Optical polaron

Consider an electron interacting with longitudinal optical phonons in a polar crystal as described by the Hamiltonian (3.166) with Eq. (3.190). In the interaction picture, the interaction term reads

---

[5]We deliberately denote the coupling parameter by K in order to emphasize the similarity with the Kondo parameter K occuring for a fermionic environment [see Eqs. (4.158) and (18.17)].

$$\tilde{H}_{\rm I}[\mathbf{q}(\tau)] = \sum_{\mathbf{k}} W_{\mathbf{k},\ell}\, e^{i\,\mathbf{k}\cdot\mathbf{q}(\tau)} \left(e^{-\omega_{\rm LO}\tau}\, b_{\mathbf{k}} + e^{\omega_{\rm LO}\tau}\, b_{\mathbf{k}}^{\dagger} \right) . \tag{4.124}$$

The influence functional $\mathbb{F}_{\rm LO}[\mathbf{q}(\cdot)]$ is calculated following the lines presented in the two preceding subsections. The formal expression for $\mathbb{F}_{\rm LO}[\mathbf{q}(\cdot)]$ is expanded into a power series in $\tilde{H}_{\rm I}$, as in Eq. (4.81), and subsequently *Wick's theorem* for thermodynamic Gaussian averages is applied. For the simple form (4.124) the only non-vanishing cumulant contraction is again the two-point function $\langle \tilde{H}_{\rm I}[\mathbf{q}(\tau)]\tilde{H}_{\rm I}[\mathbf{q}(\tau')]\rangle_{\beta}$, and thus the influence action is again given by [cf. Eq. (4.82)]

$$S_{\rm LO}^{({\rm E})}[\mathbf{q}(\cdot)] = -\frac{1}{\hbar}\int_{0}^{\hbar\beta} d\tau \int_{0}^{\tau} d\tau' \langle \tilde{H}_{\rm I}[\mathbf{q}(\tau)]\tilde{H}_{\rm I}[\mathbf{q}(\tau')]\rangle_{\beta} . \tag{4.125}$$

The contraction takes the form

$$\langle \tilde{H}_{\rm I}[\mathbf{q}(\tau)]\tilde{H}_{\rm I}[\mathbf{q}(\tau')]\rangle_{\beta} = \sum_{\mathbf{k}} |W_{\mathbf{k},\ell}|^{2}\, e^{i\,\mathbf{k}\cdot[\mathbf{q}(\tau)-\mathbf{q}(\tau')]} \tag{4.126}$$

$$\times \left(e^{-\omega_{\rm LO}(\tau-\tau')}\langle b_{\mathbf{k}} b_{\mathbf{k}}^{\dagger}\rangle_{\beta} + e^{\omega_{\rm LO}(\tau-\tau')}\langle b_{\mathbf{k}}^{\dagger} b_{\mathbf{k}}\rangle_{\beta} \right) .$$

Next, observe that the square bracket in Eq. (4.126) is just the thermal boson propagator $\mathfrak{D}_{\omega_{\rm LO}}(\tau-\tau')$, as follows from inspection of the expression (3.82). Thus we have

$$\langle \tilde{H}_{\rm I}[\mathbf{q}(\tau)]\tilde{H}_{\rm I}[\mathbf{q}(\tau')]\rangle_{\beta} = \sum_{\mathbf{k}} |W_{\mathbf{k},\ell}|^{2}\, e^{i\,\mathbf{k}\cdot[\mathbf{q}(\tau)-\mathbf{q}(\tau')]}\, \mathfrak{D}_{\omega_{\rm LO}}(\tau-\tau') . \tag{4.127}$$

The effective action of a longitudinal optical polaron has a kinetic and an influence contribution,

$$S_{\rm eff}[\mathbf{q}(\cdot)] = \frac{M}{2}\int_{0}^{\hbar\beta} d\tau\, \dot{\mathbf{q}}^{2}(\tau) + S_{\rm LO}[\mathbf{q}(\cdot)] . \tag{4.128}$$

With use of the form (3.190) for $W_{\mathbf{k},\mathrm{l}}$ the influence action is found to read

$$S_{\rm LO}[\mathbf{q}(\cdot)] = -\frac{\hbar}{V}\frac{4\pi\alpha\omega_{\rm LO}^{2}\sqrt{\hbar}}{(2M\omega_{\rm LO})^{1/2}}\int_{0}^{\hbar\beta} d\tau \int_{0}^{\tau} d\tau'\, \mathfrak{D}_{\omega_{\rm LO}}(\tau-\tau') \sum_{\mathbf{k}} \frac{e^{i\,\mathbf{k}\cdot[\mathbf{q}(\tau)-\mathbf{q}(\tau')]}}{|\mathbf{k}|^{2}} . \tag{4.129}$$

Finally, in the continuum limit of a three-dimensional isotropic medium, the influence action takes the form

$$S_{\rm LO}[\mathbf{q}(\cdot)] = -\alpha\hbar \left(\frac{\hbar}{2M\omega_{\rm LO}} \right)^{1/2} \omega_{\rm LO}^{2}\int_{0}^{\hbar\beta} d\tau \int_{0}^{\tau} d\tau'\, \frac{\mathfrak{D}_{\omega_{\rm LO}}(\tau-\tau')}{|\mathbf{q}(\tau)-\mathbf{q}(\tau')|} . \tag{4.130}$$

The polaron problem represents one of the simplest examples of the interaction of a particle with a field. With the influence action (4.130) however, we are left with a considerably complicated non-Gaussian path integral. Fortunately, one may tackle the polaron path integral with a variational approach (cf. Subsection 8.2). The variational method has been successfully utilized in diverse applications and gives reliable results for arbitrary strength of the coupling.

4.2.8 Heavy particle in a metal

We now turn our attention to a study of the response of a noninteracting Fermi liquid to a local time-dependent perturbation. The above elimination procedure adequate for bosons is generally not suitable for fermions. The anti-commutation relations for fermions involve difficulties which are absent in the case of bosons [186]. Fortunately, the low-energy excitations of the Fermi liquid have bosonic character. Therefore, it is possible to circumvent these difficulties and give a consistent formulation following previous lines. To determine the effective action for a heavy particle interacting with conduction electrons, we have to evaluate the influence functional (4.32) for the Hamiltonian (3.191). It is convenient to work out $\mathbb{F}[\mathbf{q}(\cdot)]$ using the operator formulation for the fermionic degrees of freedom. We have

$$
\begin{aligned}
\mathbb{F}[\mathbf{q}(\cdot)] &= Z_\mathrm{R}^{-1} \mathrm{tr}_\mathrm{R} \left\{ e^{-\beta H_\mathrm{R}} T_\tau \exp\left(-\int_0^{\hbar\beta} d\tau\, \tilde{H}_\mathrm{I}[\mathbf{q}(\tau)]/\hbar \right) \right\} \\
&\equiv \left\langle T_\tau \exp\left(-\int_0^{\hbar\beta} d\tau\, \tilde{H}_\mathrm{I}[\mathbf{q}(\tau)]/\hbar \right) \right\rangle_\beta ,
\end{aligned}
\tag{4.131}
$$

where T_τ is the imaginary-time ordering operator for fermion fields. In the second form, $\langle \cdots \rangle_\beta$ denotes thermal average with weight $\exp(-\beta H_\mathrm{R})$. The tilde indicates again that we are in the interaction representation [cf. Eq. (4.79)]. We proceed by expanding $\mathbb{F}[\mathbf{q}]$ into a power series in $\tilde{H}_\mathrm{I}$ as given in Eq. (4.81).

Consider first *normally conducting* electrons. We find from Eq. (3.194)

$$
\tilde{H}_\mathrm{I} = \sum_{\mathbf{k},\mathbf{k}',\sigma,\sigma'} <\mathbf{k},\sigma| V |\mathbf{k}',\sigma'>\, e^{i(\mathbf{k}'-\mathbf{k})\cdot\mathbf{q}(\tau)} \, e^{-(\omega_{\mathbf{k}'}-\omega_{\mathbf{k}})\tau} \, c_{\mathbf{k}\sigma}^\dagger c_{\mathbf{k}'\sigma'} .
\tag{4.132}
$$

Working in perturbation theory, all terms with odd powers of $\tilde{H}_\mathrm{I}$ vanish because $\langle \tilde{H}_\mathrm{I}[\mathbf{q}(\tau)]\rangle_\beta = 0$. The lowest-order term ($n = 2$) in the series for $\mathbb{F}[q(\cdot)]$ analogous to Eq. (4.81) is written in terms of the two-time electron-hole contraction

$$
\mathcal{H}_2[\mathbf{q}(\tau),\mathbf{q}(\tau')] \equiv \langle \tilde{H}_\mathrm{I}[\mathbf{q}(\tau)]\, \tilde{H}_\mathrm{I}[\mathbf{q}(\tau')]\rangle_\beta
\tag{4.133}
$$

as

$$
\mathbb{F}_2[\mathbf{q}(\cdot)] = \int_0^{\hbar\beta} d\tau \int_0^\tau d\tau'\, \mathcal{H}_2[\mathbf{q}(\tau),\mathbf{q}(\tau')]/\hbar^2 .
\tag{4.134}
$$

Next, we use the expression (4.132) and employ for thermal averages the relations

$$
\langle c_{\mathbf{k}\sigma}^\dagger c_{\mathbf{k}'\sigma'}\rangle_\beta = \delta_{\sigma\sigma'}\delta_{\mathbf{k}\mathbf{k}'} f_\mathbf{k} , \qquad \langle c_{\mathbf{k}\sigma} c_{\mathbf{k}'\sigma'}^\dagger\rangle_\beta = \delta_{\sigma\sigma'}\delta_{\mathbf{k}\mathbf{k}'}(1 - f_\mathbf{k}) .
\tag{4.135}
$$

Here, $f_\mathbf{k}$ is the Fermi distribution function,

$$
f_\mathbf{k} \equiv f(\omega_\mathbf{k}) = \frac{1}{\exp(\beta\hbar\omega_\mathbf{k}) + 1} ,
\tag{4.136}
$$

and the energy $\hbar\omega_\mathbf{k}$ is measured relatively to the Fermi energy. We then get

$$\mathcal{H}_2[\mathbf{q}(\tau),\mathbf{q}(\tau')] \;=\; \sum_{\mathbf{k},\mathbf{k'},\sigma,\sigma'} |<\mathbf{k},\sigma|V|\mathbf{k'},\sigma'>|^2 \, e^{i(\mathbf{k'}-\mathbf{k})\cdot[\mathbf{q}(\tau)-\mathbf{q}(\tau')]}$$

$$\times \; e^{-(\omega_{\mathbf{k'}}-\omega_{\mathbf{k}})(\tau-\tau')} f_{\mathbf{k}}(1-f_{\mathbf{k'}}) \, . \tag{4.137}$$

The expression (4.134) with Eq. (4.137) is a density response function of the non-interacting Fermi gas. The function $\mathcal{H}_2[\mathbf{q}(\tau),\mathbf{q}(\tau')]$ may be interpreted in terms of an electron-hole pair injected at imaginary time τ' and removed at a later imaginary time τ. This is indicated by diagram (a) in Fig. 4.2. All those diagrams representing multiple self-energy insertions arising from uncorrelated electron-hole loops, e. g., the two shown in (b) of Fig. 4.2, can be summed up and expressed as an exponential of the simple diagram (a). As we shall see immediately, the spectral density of the particle's coupling to the boson-like electron-hole excitations implicit in Eq. (4.134) is of the Ohmic form $J(\omega) \propto \omega$, this being due to the constant density of states around the Fermi surface. Contractions of higher order in the cumulant expansion for the action resulting from Eq. (4.131) [cf. Eq. (4.81)], e.g., diagram (c) in Fig. 4.2, correspond to coherent excitation of two or more electron-hole pairs. Since they represent convolutions of the densities of pairs, they come with additional powers of ω at low frequency in comparison to diagram (a). Therefore, contributions of type (c) to the response of the electron gas can be disregarded at low temperature.

Within the approximation of uncorrelated electron-hole pairs, we thus obtain the influence functional in the form

$$\mathbb{F}[\mathbf{q}(\cdot)] \;=\; \exp\left(\int_0^{\hbar\beta} d\tau \int_0^{\tau} d\tau' \, \mathcal{H}_2[\mathbf{q}(\tau),\mathbf{q}(\tau')]/\hbar^2 \right) \, . \tag{4.138}$$

The evaluation of the expression (4.137) is simple when the impurity potential is a contact potential, $<\mathbf{k},\sigma|V|\mathbf{k'},\sigma'> = \delta_{\sigma\sigma'} V_0$. The coupling strength V_0 is usually referred to as *deformation potential*. This form is also appropriate for reasonably short-ranged potentials, as the Fermi functions effectively restrict $\hbar|\omega_{\mathbf{k}}|$ and $\hbar|\omega_{\mathbf{k'}}|$ in Eq. (4.137) to small values compared to the Fermi energy E_{F}.

Computation of the angular integrals for an isotropic medium ($d = 3$) gives

$$\mathcal{H}_2[\Delta q(\tau,\tau')] \;=\; 2V_0^2 \sum_{\mathbf{k},\mathbf{k'}} \frac{\sin[k\Delta q(\tau,\tau')]}{k\Delta q(\tau,\tau')} \frac{\sin[k'\Delta q(\tau,\tau')]}{k'\Delta q(\tau,\tau')}$$

$$\times \; e^{-(\omega_{k'}-\omega_k)(\tau-\tau')} f_k(1-f_{k'}) \, , \tag{4.139}$$

with

$$\Delta q(\tau,\tau') \equiv |\mathbf{q}(\tau)-\mathbf{q}(\tau')| \, . \tag{4.140}$$

Next, we perform the continuum limit

$$\sum_{\mathbf{k}} \cdots \;\rightarrow\; \hbar\varrho \int d\omega \, \mathcal{N}(\omega) \cdots \, , \tag{4.141}$$

and count the energy relative to the Fermi energy E_{F}. The quantity $\varrho\mathcal{N}(\omega)$ is the number of states per energy interval $\hbar \, d\omega$, where ϱ is the density of states at the Fermi energy and has dimension inverse energy. Thus by definition

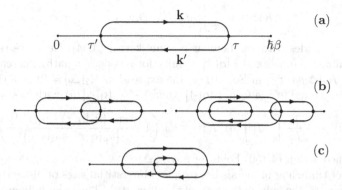

Figure 4.2: Some virtual electron-hole diagrams appearing in the expansion in $\tilde{H}_{\mathrm{I}}[\mathbf{q}(\tau)]$ are shown. The oval (thick) lines represent electron and hole propagators, respectively. Diagrams (a) and (b) are included in the expression (4.138), while diagram (c) is neglected.

$$\mathcal{N}(\omega = 0) = 1 . \tag{4.142}$$

As we are interested in the low energy excitations around the Fermi surface, we may replace k and k' in the fractions of Eq. (4.139) by the Fermi wave number k_{F}. At this point it is convenient to absorb various factors into the "viscosity" coefficient

$$\eta = 4\pi V_0^2 \varrho^2 \hbar k_{\mathrm{F}}^2 / 3 , \tag{4.143}$$

which has the usual dimension mass times frequency. Thus we may write

$$\mathcal{H}_2[\Delta q(\tau, \tau')] = \frac{3\hbar}{2k_{\mathrm{F}}^2} \frac{\sin^2[k_{\mathrm{F}}\Delta q(\tau, \tau')]}{[k_{\mathrm{F}}\Delta q(\tau, \tau')]^2} K(\tau - \tau') , \tag{4.144}$$

$$K(\tau) = \frac{\eta}{\pi} \int_{-\infty}^{\infty} d\omega' \int_{-\infty}^{\infty} d\omega'' \, \mathcal{N}(\omega')\mathcal{N}(\omega'') \exp[(\omega'' - \omega')\tau] f(\omega'') \, f(-\omega') . \tag{4.145}$$

With the substitution $\omega'' = \omega' - \omega$, the kernel $K(\tau)$ takes the form

$$K(\tau) = \frac{1}{\pi} \int_{-\infty}^{\infty} d\omega \, \frac{\exp(-\omega\tau)}{1 - \exp(-\omega\hbar\beta)} J(\omega) \tag{4.146}$$

with

$$J(\omega) = \eta \left[1 - e^{-\omega\hbar\beta} \right] \int_{-\infty}^{\infty} d\omega' \, \mathcal{N}(\omega')\mathcal{N}(\omega' - \omega) f(\omega' - \omega) f(-\omega') . \tag{4.147}$$

Use of the explicit form of the Fermi function yields

$$J(\omega) = \frac{\eta}{2} \int_{-\infty}^{\infty} d\omega' \, \mathcal{N}(\omega')\mathcal{N}(\omega' - \omega) \left(\tanh(\tfrac{1}{2}\beta\hbar\omega') - \tanh[\tfrac{1}{2}\beta\hbar(\omega' - \omega)] \right) , \tag{4.148}$$

which shows that $J(\omega)$ is antisymmetric in ω. The antisymmetric part of the fraction in Eq. (4.146) is the phonon propagator (3.82) times a factor $\tfrac{1}{2}$. Thus we obtain

$$K(\tau) = \frac{1}{\pi} \int_0^\infty d\omega \, J(\omega) \mathfrak{D}_\omega(\tau) \, . \tag{4.149}$$

This form coincides with the familiar bosonic kernel (4.54). Next, we require again that the influence functional $\mathbb{F}[\mathbf{q}(\cdot)]$ is unity for a constant path. Therefore, we subtract from $\mathcal{H}_2[\Delta q(\tau, \tau')]$ in Eq. (4.144) the expression $\mathcal{H}_2[\Delta q = 0]$. We then find the influence functional in the form $\mathbb{F}[\mathbf{q}(\cdot)] = \exp\{-\mathcal{S}_{\text{infl}}[\mathbf{q}(\cdot)]/\hbar\}$ with the action

$$\mathcal{S}_{\text{infl}}[\mathbf{q}(\cdot)] = \frac{3}{2\,k_F^2} \int_0^{\hbar\beta} d\tau \int_0^\tau d\tau' \, K(\tau - \tau') \left(1 - \frac{\sin^2[k_F|\mathbf{q}(\tau) - \mathbf{q}(\tau')|]}{[k_F|\mathbf{q}(\tau) - \mathbf{q}(\tau')|]^2} \right) . \tag{4.150}$$

The influence action (4.150) holds for any $k_F\Delta q$.

For defect tunneling in metals between two interstitial sites of distance q_0, we may pass on again to the spin path $\sigma(\tau)$, $q(\tau) = \frac{1}{2}q_0\,\sigma(\tau)$. Then the influence action can be written in the spin-boson form, as in Eq. (4.76) with Eq. (4.75),

$$\mathcal{S}_{\text{infl}}[\sigma(\cdot)] = \frac{\hbar}{8} \int_0^{\hbar\beta} d\tau \int_0^\tau d\tau' \, \mathcal{X}(\tau - \tau') \left[\sigma(\tau) - \sigma(\tau')\right]^2 \tag{4.151}$$

with

$$\mathcal{X}(\tau) = \int_0^\infty d\omega \, G(\omega) \, \mathfrak{D}_\omega(\tau) \, . \tag{4.152}$$

The spectral density $G(\omega)$ is related to $J(\omega)$ as

$$G(\omega) = \frac{3}{(k_F q_0)^2} \left(1 - \frac{\sin^2(k_F q_0)}{(k_F q_0)^2} \right) \frac{q_0^2}{\pi\hbar} \, J(\omega) \, . \tag{4.153}$$

This reduces in the regime $k_F q_0 \ll 1$ to the relation (3.148).

For conduction electrons, the density of states around the Fermi level is constant,

$$\mathcal{N}(\omega) = 1 \, . \tag{4.154}$$

If we disregard, in addition, the finite width of the conduction band, the ω'-integral in Eq. (4.148) can easily be performed. We then get

$$J(\omega) = \eta\omega \, , \tag{4.155}$$

which is just the Ohmic form (3.41).

The expression (4.143) is the viscosity in Born approximation, $\eta \propto V_0^2$, resulting from the sea of electron-hole excitations. In a non-perturbative treatment, Sassetti *et al.* [187] showed by calculation of the single-particle propagator for a contact potential that the friction coefficient may be expressed in terms of the s-wave scattering phase shift δ_0 at the Fermi energy,

$$\eta = 4\hbar \, k_F^2 \sin^2 \delta_0 / 3\pi \, , \qquad \text{where} \qquad \tan \delta_0 = -\pi V_0 \varrho \, . \tag{4.156}$$

The same form is also found by calculation of the force auto-correlation function of a heavy particle in a free electron gas. For a spin-independent spherically symmetric finite-range potential, the partial wave analysis yields the viscosity [188]

$$\eta = \frac{4\hbar}{3\pi} k_F^2 \sum_{\ell=0}^{\infty} (\ell + 1) \sin^2(\delta_{\ell+1} - \delta_\ell) , \qquad (4.157)$$

where δ_ℓ represents the scattering phase shift of the ℓth partial wave.

As intermediate result, we emphasize that the electronic excitations near to the Fermi energy collectively behave as if they were bosons with an Ohmic spectral density of the coupling, regardless of whether the impurity interaction is a contact or a finite-range potential. Similar conclusions were drawn by Chang and Chakravarty [189] who studied Dyson's equation for the fermion propagator in presence of the impurity within a real-time formulation. They solved the problem by employing the long-time approximation of the free fermion propagator, a method originally proposed by Nozières and De Dominicis [190]. Furthermore, it has been shown via bosonization of the fermion operators [191] that the low-energy excitations of a fermionic bath can be mapped onto a bosonic bath. Importantly, the scattering phase of a contact potential depends on the regularization scheme employed. Only the leading Born term is found to be universal, i.e., does not depend on the particular regularization prescription. For a discussion of this point we refer to Ref. [191].

With the Ohmic form (4.155), the spin-boson spectral density $G(\omega)$ is Ohmic too,

$$G(\omega) = 2K\omega . \qquad (4.158)$$

The dimensionless damping strength K is usually called Kondo parameter for reasons which become clear shortly. For interstitials in metals with a tunneling distance q_0 of the order of the lattice constant, we have $k_F q_0 \ll 1$. In this important limit we then find from (4.153) with (4.155)

$$K = \frac{\eta q_0^2}{2\pi\hbar} + \mathcal{O}[(k_F q_0)^4] . \qquad (4.159)$$

The expression (4.159) coincides with a corresponding result found by Sols and Guinea [192] in a linear response calculation. Schönhammer [193] has shown by taking all the scattering phase shifts into account that the relation (4.159) is generally valid in the limit $k_F q_0 \ll 1$ for any spherically symmetric potential independent of the strength of the coupling constant.

Use of Eq. (4.155) with (4.156) in Eq. (4.153) yields

$$K = (2/\pi^2) \sin^2 \delta_0 \left[1 - \sin^2(k_F q_0)/(k_F q_0)^2 \right] . \qquad (4.160)$$

This relation has the limiting forms

$$K = 2(V_0 \rho)^2 [1 - \sin^2(k_F q_0)/(k_F q_0)^2] , \qquad \delta_0 \ll \pi/2 , \qquad (4.161)$$

$$K = \frac{2}{3\pi^2} (k_F q_0)^2 \sin^2 \delta_0 + \mathcal{O}[(k_F q_0)^4] , \qquad k_F q_0 \ll 1 . \qquad (4.162)$$

The nature of the electronic screening cloud around an impurity in a metal has been the subject of numerous studies. Anderson [194] considered the overlap of the ground state $|\phi>$ of the fermions in the absence of the impurity with the corresponding state

$|\psi>$ in the presence of the impurity and found that the overlap integral for a state with N conduction electrons behaves as

$$<\psi|\phi> \; \propto \; N^{-K_+} \,, \tag{4.163}$$

where K_+ is a positive number. This relation, known as Anderson's *orthogonality theorem*, indicates that the overlap tends to zero as the system size tends to infinity. For a contact potential, one finds

$$K_+ \; = \; (\delta_0/\pi)^2 \,, \tag{4.164}$$

where δ_0 is the s-wave phase for electron scattering off the impurity.

Consider next a version of the orthogonality theorem which concerns the overlap of the ground state of the fermions with the impurity at position $\mathbf{q}_1$, denoted by $|\psi_1>$, and the ground state of the fermions with the impurity at position $\mathbf{q}_2$, denoted by $|\psi_2>$. Kondo has found that these states obey a similar orthogonality theorem [138],

$$<\psi_1|\psi_2> \; \propto \; N^{-K} \,. \tag{4.165}$$

Interestingly, the overlap parameter K introduced by Kondo coincides with the dimensionless damping strength K in the spectral density (4.158). It is is a function of the phase shifts and the distance $q_0 = |\mathbf{q}_1 - \mathbf{q}_2|$. A general calculation of K as a function of the distance is rather difficult [195, 196, 193], since each partial wave centered at $\mathbf{q}_1$ mixes with *all* partial waves centered at $\mathbf{q}_2$. Even when the impurity potential is spherically symmetric, the problem lacks spherical symmetry.

For s-wave scattering and spin degeneracy, one finds [195]

$$K \; = \; \frac{2}{\pi^2}\left\{ \arctan\left(\frac{\sqrt{1-x(q_0)}\,\tan\delta_0}{\sqrt{1+x(q_0)\tan^2\delta_0}} \right) \right\}^2 \,. \tag{4.166}$$

The phase shift δ_0 is given in Eq. (4.156). Equation (4.166) gives the bound $K \leqslant 1/2$. The function $x(q_0)$ depends on the space dimension d,

$$x(q_0) \; = \; \sin^2(k_\mathrm{F}q_0)/(k_\mathrm{F}q_0)^2 \qquad \text{for} \qquad d=3 \,, \tag{4.167}$$

$$x(q_0) \; = \; \sin^2(k_\mathrm{F}q_0) \qquad\qquad \text{for} \qquad d=1 \,. \tag{4.168}$$

For $d=3$, the expression (4.166) coincides in the limits $\delta_0 \ll \pi/2$ and $k_\mathrm{F}q_0 \ll 1$ with the expressions (4.161) and (4.162), respectively.

The *short-distance* form of K, Eq. (4.159), does not depend on the spatial dimension d [193], as distinguished from the large distance behavior which depends on d. In the one-dimensional case, the function $x(q_0)$ is oscillating around the mean value $\frac{1}{2}$. For $d \geqslant 2$, the function $x(q_0)$ goes to zero as $k_\mathrm{F}q_0 \to \infty$. Putting $x(q_0) = 0$ in Eq. (4.166), we obtain $K = 2(\delta_0/\pi)^2$, and thus $K = 2K^+$.

For s-wave scattering, the overlap parameter K has the upper bound $1/2$ for arbitrary distances as follows from Eq. (4.166). In the general case of an extended potential, K is *not* bounded in principle by this value in two or three dimensions, since the exponent K_+ is not bounded [193]. Experimentally, K turns out to be small, typically about 0.1 for charged interstitials in metals. The parameter K is small because surrounding charges are effectively screening the impurity potential.

For the Ohmic spectral density (4.158), the kernel $\mathcal{X}(\tau)$ in the retarded action (4.151) can be calculated in analytic form. The integral expression (4.152) is

$$\mathcal{X}(\tau) = 2K \frac{(\pi/\hbar\beta)^2}{\sin^2(\pi\tau/\hbar\beta)} . \tag{4.169}$$

The influence action (4.151) with the kernel (4.169) conveys the effects of a normal-state metallic environment on interstitial tunneling at thermal energies well below the Fermi energy.

4.2.9 Heavy particle in a superconductor

It is straightforward to generalize the discussion of the preceding subsection to the *superconducting* state. The underlying Hamiltonian is given by Eqs. (3.195) and (3.201) with (3.204). According to the BCS theory, the electronic spectrum is modified and the coherence factor is as specified in Eq. (3.204). The influence action is found again in the form (4.150), and with the TSS path $q(\tau) = q_0\,\sigma(\tau)/2$ in the form (4.151), respectively. Furthermore, the kernel $\mathcal{K}(\tau)$ takes again the form (4.152), but now the spectral density of the coupling (4.148) is adjusted to BCS-quasiparticle excitations,

$$\begin{aligned}
G_{\mathrm{qp}}(\omega) = \; & K \int_{-\infty}^{\infty} d\omega'\, \mathcal{N}_{\mathrm{qp}}(\omega')\,\mathcal{N}_{\mathrm{qp}}(\omega'-\omega) \left\{ 1 - \Delta_{\mathrm{g}}^2/[\,\omega'(\omega'-\omega)\,] \right\} \\
& \times \left\{ \tanh(\tfrac{1}{2}\beta\hbar\omega') - \tanh[\,\tfrac{1}{2}\beta\hbar(\omega'-\omega)\,] \right\} .
\end{aligned} \tag{4.170}$$

Here $\mathcal{N}_{\mathrm{qp}}(\omega)$ is the density of states of BCS quasiparticles, and the coherence factor is $1 - \Delta_{\mathrm{g}}^2/[\omega'(\omega'-\omega)]$, which holds for time-reversal-invariant interaction. In the BCS theory, the density of states is given by

$$\mathcal{N}_{\mathrm{qp}}(\omega) = \Theta(|\omega| - \Delta_{\mathrm{g}}) \frac{|\omega|}{\sqrt{\omega^2 - \Delta_{\mathrm{g}}^2}} . \tag{4.171}$$

In the low- frequency range $\omega \ll \Delta_{\mathrm{g}}$, the function $G_{\mathrm{qp}}(\omega)$ takes the Ohmic form

$$G_{\mathrm{qp}}(\omega) = 2K_{\mathrm{qp}}\,\omega \tag{4.172}$$

with the temperature-dependent dimensionless Ohmic coupling constant

$$K_{\mathrm{qp}} = 2K f(\Delta_{\mathrm{g}}) = 2K/[\,1 + \exp(\beta\hbar\Delta_{\mathrm{g}})\,] . \tag{4.173}$$

Thus, K_{qp} is downsized by the gap and even drops to zero at $T = 0$. The function $G_{\mathrm{qp}}(\omega)$ is discontinuous at the excitation threshold $\omega = 2\Delta_{\mathrm{g}}$, with a jump of height

$$\Delta G_{\mathrm{qp}} = 2K\pi\Delta_{\mathrm{g}} \tanh\left(\frac{\hbar\Delta_{\mathrm{g}}}{2k_{\mathrm{B}}T}\right) . \tag{4.174}$$

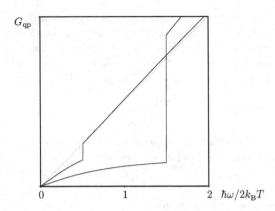

Figure 4.3: Quasiparticle spectral density G_{qp} as a function of $\hbar\omega/2k_BT$. The solid curves correspond to $\hbar\Delta_g/k_BT = 0.5$ and $\hbar\Delta_g/k_BT = 1.5$, respectively. The dotted straight line represents the spectral density in the normally conducting state.

Above the threshold, $G_{qp}(\omega)$ increases with increasing frequency and approaches in the regime $\omega \gg \Delta_g$ the Ohmic form $G_{qp}(\omega) = 2K\omega$. Fig. 4.3 shows plots of $G_{qp}(\omega)$ for different temperatures.

At zero temperature, the integral in Eq. (4.170) can be expressed in terms of the complete elliptic integral of the second kind $E(\pi/2, k)$ [89]. We then get

$$G_{qp}(\omega) = \Theta\left(\omega - 2\Delta_g\right) 2K\omega\, E\left(\pi/2, \sqrt{1 - 4\Delta_g^2/\omega^2}\right) , \qquad (4.175)$$

where $\Delta_g = \Delta_g(T = 0)$. With this, the kernel $\mathcal{X}_0(\tau) = \int_0^\infty d\omega\, G_{qp}(\omega)\mathfrak{D}_\omega(\tau)|_{T=0}$ can be expressed with the modified Bessel functions $K_0(z)$ and $K_1(z)$ in the form

$$\mathcal{X}_0(\tau) = 2K\Delta_g^2\left[K_1^2(\Delta_g|\tau|) + K_0^2(\Delta_g|\tau|) \right] . \qquad (4.176)$$

The limiting behavior of $\mathcal{X}_0(\tau)$ is

$$\mathcal{X}_0(\tau) = 2K \begin{cases} \dfrac{1}{\tau^2} + \Delta_g^2\left[\ln(\Delta_g|\tau|/2) - C_E\right]^2 , & \text{for} \quad \Delta_g|\tau| \ll 1 , \\[2mm] \dfrac{\pi}{|\tau|}\Delta_g\, e^{-2\Delta_g|\tau|} , & \text{for} \quad \Delta_g|\tau| \gg 1 , \end{cases} \qquad (4.177)$$

where C_E is the Euler constant. The exponential drop of the kernel at long time differs qualitatively from the algebraic decay $\propto 1/\tau^2$ in the normally conducting state at zero temperature [cf. Eq. (4.169)]. The former is due to the gap above the ground state in the excitation spectrum of the superconducting environment.

4.2.10 Effective action of a Josephson junction

To find the effective action of a junction, we start out from the microscopic Hamiltonian (3.206) put down in Subsec. 3.3.4 and follow the procedure outlined by Ambegaokar *et al.* [152]. In the first step, the quartic interactions in Eqs. (3.205) and (3.208) are eliminated in the functional integral representation of the expression (3.210) by means of a Gaussian identity known as Hubbard-Stratonovich transformation. In this method, new fields are introduced in exchange for the quartic pair interaction and for the quartic Coulomb interaction, respectively. The former can be identified with the complex order parameter fields $\Delta_L(\mathbf{r}, \tau)$ and $\Delta_R(\mathbf{r}, \tau)$, the latter with the (real) voltage field $U(\tau)$. The grand merit of the Hubbard-Stratonovich transformation is that the resulting action is quadratic in the fermion field. Hence the trace with respect to this field can readily be performed. The partition function takes the form

$$Z = \int \mathcal{D}^2 \Delta_L(\mathbf{r}, \tau) \, \mathcal{D}^2 \Delta_R(\mathbf{r}, \tau) \, \mathcal{D}U(\tau) \, \mathcal{D}\mathbf{A}(\mathbf{r}, \tau) \, e^{-\mathcal{A}[\Delta_L, \Delta_R, U, \mathbf{A}]/\hbar}, \qquad (4.178)$$

with the effective action

$$\mathcal{A} = \hbar \operatorname{tr} \ln \hat{\underline{G}}^{-1} + \frac{1}{2} \int_0^{\hbar\beta} d\tau \, C U^2(\tau) + \frac{1}{8\pi} \int_0^{\hbar\beta} d\tau \int d^3\mathbf{r} \, [\mathbf{h}(\mathbf{r}, \tau) - \mathbf{h}_{\text{ext}}]^2$$

$$+ \int_0^{\hbar\beta} d\tau \left\{ \int_{\mathbf{r}\in L} d^3\mathbf{r} \, g_L^{-1}(\mathbf{r}) \, |\Delta_L(\mathbf{r}, \tau)|^2 + \int_{\mathbf{r}\in R} d^3\mathbf{r} \, g_R^{-1}(\mathbf{r}) \, |\Delta_R(\mathbf{r}, \tau)|^2 \right\}. \qquad (4.179)$$

Here, $\hat{\underline{G}}$ is a 4×4 matrix Green function in the space spanned by (L) and (R) and in the Nambu pseudo-spin space.[6] In the sequel, we indicate matrices in Nambu space by carets, and matrices describing the (L) and (R) superconductors by underlines. The inverse of the Green function is given by

$$\hat{\underline{G}}^{-1}(\mathbf{r}, \tau; \mathbf{r}', \tau') = \begin{pmatrix} \hat{G}_L^{-1}(\mathbf{r}, \tau; \mathbf{r}', \tau') & \hat{T}_{\mathbf{r}\mathbf{r}'} \, \delta(\tau - \tau') \\ \hat{T}_{\mathbf{r}\mathbf{r}'}^\dagger \, \delta(\tau - \tau') & \hat{G}_R^{-1}(\mathbf{r}, \tau; \mathbf{r}', \tau') \end{pmatrix}. \qquad (4.180)$$

The diagonal elements in the L-R space are

$$\hat{G}_{L/R}^{-1}(\mathbf{r}, \tau; \mathbf{r}', \tau') = \left\{ \hbar \frac{\partial}{\partial \tau} \hat{1} + \left[-\frac{\hbar^2}{2m} \left(\nabla - \frac{ie}{\hbar c} \mathbf{A} \hat{\tau}_3 \right)^2 - \mu + i e \, U_{L/R}(\tau) \right] \hat{\tau}_3 \right.$$

$$\left. + \hat{\Delta}_{L/R}(\mathbf{r}, \tau) \right\} \delta(\mathbf{r} - \mathbf{r}') \, \delta(\tau - \tau'), \qquad (4.181)$$

where $\hat{\tau}_3$ is the Pauli matrix in the Nambu space, and where $U(\tau) = U_L(\tau) - U_R(\tau)$. The order parameter field in Nambu space reads

$$\hat{\Delta}_{L/R}(\mathbf{r}, \tau) = \begin{pmatrix} 0 & \Delta_{L/R}(\mathbf{r}, \tau) \\ \Delta_{L/R}^*(\mathbf{r}, \tau) & 0 \end{pmatrix} = |\Delta_{L/R}| \, e^{-i\varphi_{L/R}\hat{\tau}_3} \, \hat{\tau}_1. \qquad (4.182)$$

The two superconductors (L) and (R) are coupled by the transfer matrix

$$\hat{T}_{\mathbf{r}\mathbf{r}'} = \begin{pmatrix} T_{\mathbf{r}\mathbf{r}'} & 0 \\ 0 & -T_{\mathbf{r}\mathbf{r}'}^* \end{pmatrix}. \qquad (4.183)$$

---

[6]The two-dimensional Nambu space facilitates a compact formulation of BCS superconductivity.

It is useful to perform a gauge transformation with the aim of making the off-diagonal elements in Eq. (4.182) real. This is achieved with a transformation induced by the unitary matrix (do not confuse with the voltage U)

$$\hat{\underline{U}} = \begin{pmatrix} e^{i\varphi_{\mathrm{L}}(\mathbf{r},\tau)\hat{\tau}_3/2} & 0 \\ 0 & e^{i\varphi_{\mathrm{R}}(\mathbf{r},\tau)\hat{\tau}_3/2} \end{pmatrix} . \tag{4.184}$$

Then the diagonal element of $\hat{\underline{G}}'^{-1} = \hat{\underline{U}}\,\hat{\underline{G}}^{-1}\hat{\underline{U}}^{-1}$ in the L-space becomes

$$\hat{\underline{G}}_{\mathrm{L}}'^{-1}(\mathbf{r},\tau;\mathbf{r}',\tau') = \left\{\hbar\frac{\partial}{\partial\tau}\hat{1} - i\hbar\left(\mathbf{v}_{\mathrm{s,L}}\cdot\nabla\right)\hat{1} + \hat{\Delta}_{\mathrm{L}}(\mathbf{r},\tau) \right. \tag{4.185}$$

$$\left. + \left[-\frac{\hbar^2\nabla^2}{2m} - \mu + \frac{m}{2}\mathbf{v}_{\mathrm{s,L}}^2 - i\left(\frac{\hbar}{2}\frac{\partial\varphi_{\mathrm{L}}}{\partial\tau} - eU_{\mathrm{L}}(\tau)\right)\right]\hat{\tau}_3\right\}\delta(\mathbf{r}-\mathbf{r}')\,\delta(\tau-\tau') .$$

In the transformation, the left-right off-diagonal elements pick up a phase,

$$\hat{T}'_{\mathbf{r}\mathbf{r}'} = \begin{pmatrix} T_{\mathbf{r}\mathbf{r}'}\,e^{i\psi(\tau)/2} & 0 \\ 0 & -T^*_{\mathbf{r}\mathbf{r}'}\,e^{-i\psi(\tau)/2} \end{pmatrix} , \tag{4.186}$$

where $\psi(\tau)$ is the phase difference across the junction plane. Here we assumed that $T_{\mathbf{r}\mathbf{r}'}$ differs essentially from zero only for $\mathbf{r}$, $\mathbf{r}'$ near to the right and left of the barrier, respectively. Above we have introduced the gauge-invariant superfluid velocity

$$\mathbf{v}_{\mathrm{s,L/R}} = -\frac{\hbar}{2m}\left(\nabla\varphi_{\mathrm{L/R}} + \frac{2e}{\hbar c}\mathbf{A}\right) . \tag{4.187}$$

In Eq. (4.185), the order parameter matrix in Nambu space $\hat{\Delta}_{\mathrm{L}}$ has *real* matrix elements instead of the complex ones in Eq. (4.182).

The action $\mathcal{A}$ depends on the independent collective variables $|\Delta_{\mathrm{L/R}}|$, $\mathbf{A}$, U and ψ. The partition function is a functional integral over all these variables. For bulk superconductors, the quantities $|\Delta_{\mathrm{L/R}}|$, $\mathbf{A}$, and U have insignificant uncertainties, so that we may disregard fluctuations of these variables. This simplifies the further discussion of Z considerably. We proceed by observing the following three points:

- For uncoupled homogeneous bulk superconductors, the extremal condition $\delta\mathcal{A}/\delta|\Delta_{\mathrm{L/R}}| = 0$ reproduces the gap equation of the BCS theory, which yields the mean field value $|\Delta_{\mathrm{L/R}}^{\mathrm{BCS}}|$. Since the second variation $\delta^2\mathcal{A}/(\delta|\Delta_{\mathrm{L/R}}|)^2$ is proportional to the density N_0 of free electron states at E_{F} and proportional to the third power of the BCS coherence length, fluctuations of $|\Delta_{\mathrm{L/R}}|$ about $|\Delta_{\mathrm{L/R}}^{\mathrm{BCS}}|$ are suppressed except in a narrow temperature range just below the transition temperature T_{c}.

- Expansion of the action to second order in the superfluid velocity gives

$$\mathcal{A} = \mathcal{A}_0 + \int_0^{\hbar\beta} d\tau \int d^3\mathbf{r}\,\left[\frac{m}{2}\rho_{\mathrm{s}}\mathbf{v}_{\mathrm{s}}^2 + \frac{1}{8\pi}\left[\mathbf{h}(\mathbf{r},\tau) - \mathbf{h}_{\mathrm{ext}}\right]^2\right] , \tag{4.188}$$

where ρ_s is the superfluid density [152]. From this one finds that the least action condition $\delta\mathcal{A}/\delta\mathbf{A}(\mathbf{r},\tau)=0$ reproduces the familiar law $\mathbf{j}=\mathbf{j}_{\text{ext}}+\mathbf{j}_s$, where $\mathbf{j}_{\text{ext}}$ is the current density related to the external field, and where $\mathbf{j}_s$ is the supercurrent,

$$\mathbf{j}_s = \frac{e\hbar}{2m}\rho_s\left(\nabla\psi+\frac{2e}{\hbar c}\mathbf{A}\right)\,. \tag{4.189}$$

In a bulk superconductor, fluctuations of $\mathbf{j}$ are suppressed, as follows from the study of the second variation of $\mathcal{A}$ with respect to the vector potential $\mathbf{A}$.

• The action is extremal when the phase difference across the junction satisfies the mean field relation $\dot{\psi}=2e\,U/\hbar$. Again, we may disregard deviations from the mean field since fluctuations are suppressed by bulk energies.

Hence the only fluctuating variable left is the phase jump $\psi(\tau)$ across the junction. In a SQUID geometry, in which the two superconductors are joined in a loop, the relevant fluctuating variable is the total flux ϕ through the loop. The flux ϕ is related to the phase variable ψ by the Josephson relation (3.151). When the quantities $|\Delta_{\text{L/R}}|$, U and $\mathbf{j}$ are fixed to their mean values, the multiple functional integral (4.178) is reduced to a single one over the phase variable ψ. The partition function of a junction then reads

$$Z = \oint\mathcal{D}\psi(\cdot)\exp\left\{-\frac{C}{2\hbar}\int_0^{\hbar\beta}d\tau\left(\frac{\hbar}{2e}\frac{\partial\psi}{\partial\tau}\right)^2-\frac{1}{\hbar}\mathcal{A}_{\text{T}}[\psi(\cdot)]\right\}\,, \tag{4.190}$$

where $\mathcal{A}_{\text{T}}[\psi(\cdot)]$ represents the tunneling contribution to the action. Working out $\mathcal{A}_{\text{T}}[\psi(\cdot)]$ to second order in the (averaged) tunnel matrix element from the first term on the right hand side of (4.179), one finds[7] [152]–[154]

$$\begin{aligned}
\mathcal{A}_{\text{T}} = 2\int_0^{\hbar\beta}d\tau\int_0^{\hbar\beta}d\tau'&\left\{\alpha(\tau-\tau')\left(1-\cos\frac{\psi(\tau)-\psi(\tau')}{2}\right)\right.\\
&\left.+\,\beta(\tau-\tau')\cos\frac{\psi(\tau)+\psi(\tau')}{2}\right\}\,.
\end{aligned} \tag{4.191}$$

For later convenience, we have extracted a factor 2 in front of the integrals, and we have added a ψ-independent term in order that the first contribution to $\mathcal{A}_{\text{T}}$ vanishes for a constant path, $\psi(\tau)=\psi(\tau')$. The kernels $\alpha(\tau)$ and $\beta(\tau)$ are given in terms of the diagonal $(1,1)$ and off-diagonal $(1,2)$ components of the Green function $\hat{G}$ in Nambu space, usually denoted by $G(\tau,\mathbf{p})$ and $F(\tau,\mathbf{p})$,

$$\begin{aligned}
\alpha(\tau) &= -\frac{|T_{\text{t}}|^2}{\hbar}\int\frac{d^3\mathbf{p}_{\text{L}}}{(2\pi\hbar)^3}\int\frac{d^3\mathbf{p}_{\text{R}}}{(2\pi\hbar)^3}\,G_{\text{L}}(\tau,\mathbf{p}_{\text{L}})\,G_{\text{R}}(-\tau,\mathbf{p}_{\text{R}})\,,\\
\beta(\tau) &= -\frac{|T_{\text{t}}|^2}{\hbar}\int\frac{d^3\mathbf{p}_{\text{L}}}{(2\pi\hbar)^3}\int\frac{d^3\mathbf{p}_{\text{R}}}{(2\pi\hbar)^3}\,F_{\text{L}}(\tau,\mathbf{p}_{\text{L}})\,F_{\text{R}}(-\tau,\mathbf{p}_{\text{R}})\,.
\end{aligned} \tag{4.192}$$

---

[7]This expression is the leading term in the cumulant expansion.

Here we have assumed that the tunneling coupling T_t is independent of momentum near to the Fermi momentum p_F. The Green functions are given by

$$
\int \frac{d^3\mathbf{p}}{(2\pi\hbar)^3} \begin{pmatrix} G \\ F \end{pmatrix} (\tau, \mathbf{p}) = -N_0 \int_{-\infty}^{\infty} d\omega \, \mathcal{N}_{\mathrm{qp}}(\omega) \begin{pmatrix} 1 \\ \Delta_g/\omega \end{pmatrix}
$$
$$
\times e^{-\omega\tau} \left[f(-\omega)\Theta(\tau) - f(\omega)\Theta(-\tau) \right].
$$
(4.193)

Here $\mathcal{N}_{\mathrm{qp}}(\omega)$ is the density of states of BCS quasiparticles (4.171), $f(\omega)$ denotes the Fermi distribution function and Δ_g is the mean field gap frequency $\Delta_g = \Delta^{\mathrm{BCS}}/\hbar$. For notational simplicity, we now assume that $\mathcal{N}_{\mathrm{qp}}^{(\mathrm{L})}(\omega) = \mathcal{N}_{\mathrm{qp}}^{(\mathrm{R})}(\omega) = \mathcal{N}_{\mathrm{qp}}$ and $\Delta_{g,\mathrm{L}} = \Delta_{g,\mathrm{R}} = \Delta_g$. Next, we insert the expression (4.193) into Eq. (4.192) and assume that both superconductors are described by the same mean field parameters. We then get

$$
\begin{pmatrix} \alpha(\tau) \\ \beta(\tau) \end{pmatrix} = \frac{|T_t|^2}{\hbar} N_0^2 \int_{-\infty}^{\infty} d\omega' \int_{-\infty}^{\infty} d\omega'' \, \mathcal{N}_{\mathrm{qp}}(\omega') \mathcal{N}_{\mathrm{qp}}(\omega'') \begin{pmatrix} 1 \\ \Delta_g^2/\omega'\omega'' \end{pmatrix}
$$
$$
\times e^{(\omega''-\omega')|\tau|} f(\omega') f(-\omega').
$$
(4.194)

These expressions are similar to the previous form (4.145). Next, we follow the steps described below Eq. (4.145). We then find in analogy to the expression (4.54)

$$
\begin{pmatrix} \alpha(\tau) \\ \beta(\tau) \end{pmatrix} = \frac{1}{\pi} \int_0^{\infty} d\omega \begin{pmatrix} J_\alpha(\omega) \\ J_\beta(\omega) \end{pmatrix} \mathfrak{D}_\omega(\tau).
$$
(4.195)

Here, $\mathfrak{D}_\omega(\tau)$ is the boson propagator (3.82), and $J_{\alpha(\beta)}(\omega)$ is the spectral density

$$
\begin{pmatrix} J_\alpha(\omega) \\ J_\beta(\omega) \end{pmatrix} = \frac{\pi}{2} \frac{|T_t|^2}{\hbar} N_0^2 \int_{-\infty}^{\infty} d\omega' \, \mathcal{N}_{\mathrm{qp}}(\omega') \mathcal{N}_{\mathrm{qp}}(\omega' - \omega) \begin{pmatrix} 1 \\ \frac{\Delta_g^2}{\omega'(\omega'-\omega)} \end{pmatrix}
$$
$$
\times \left\{ \tanh(\tfrac{1}{2}\beta\hbar\omega') - \tanh[\tfrac{1}{2}\beta\hbar(\omega'-\omega)] \right\},
$$
(4.196)

which has dimension energy.

The physics of the two terms in Eq. (4.191) is quite different. The α-term describes dissipation because of quasiparticle tunneling. The β-term describes tunneling of Cooper pairs (tunneling). In this term, the kernel $\beta(\tau - \tau')$ appears in combination with $\cos[(\psi(\tau) + \psi(\tau'))/2]$. Assuming that $\psi(\tau)$ varies slowly on the time scale on which the kernel $\beta(\tau)$ decays, the β-term in the action (4.191) reduces to an ordinary potential term, the "washboard potential" action

$$
\mathcal{A}_T^{(\beta)}[\psi(\cdot)] = -I_c \frac{\hbar}{2e} \int_0^{\hbar\beta} d\tau \, \cos\psi(\tau),
$$
(4.197)

where

$$I_c = -\frac{4e}{\hbar} \int_{-\hbar\beta/2}^{\hbar\beta/2} d\tau\, \beta(\tau) = -\frac{8e}{\pi\hbar} \int_0^\infty d\omega\, \frac{J_\beta(\omega)}{\omega} \tag{4.198}$$

is the critical current of the junction. The nonlocal correction in the β-term represents the nonlocal supercurrent found by Wertheimer [197]. This contribution is usually referred to as the "$\cos\psi$" term or quasi-particle pair interference current [152, 153]. In order to keep the discussion simple, we shall ignore time-nonlocal current corrections, apart from a contribution to the effective capacitance (see below).

When the superconducting tunnel junction is biased by an externally applied current I_{ext}, there is an additional potential contribution $\propto \psi$ which tilts the washboard. At this level of approximation, the effective action of a current biased tunnel junction with capacitance C is

$$\mathcal{A}_{\text{eff}}[\psi(\cdot)] = \int_0^{\hbar\beta} d\tau \left[\frac{C}{2} \left(\frac{\hbar}{2e} \frac{\partial\psi}{\partial\tau} \right)^2 + V(\psi) \right] + \mathcal{A}_{\text{T}}^{(\alpha)}[\psi],$$

$$\mathcal{A}_{\text{T}}^{(\alpha)}[\psi(\cdot)] = 2 \int_0^{\hbar\beta} d\tau \int_0^{\hbar\beta} d\tau'\, \alpha(\tau-\tau') \left(1 - \cos\frac{\psi(\tau)-\psi(\tau')}{2} \right), \tag{4.199}$$

where $V(\psi)$ is the tilted washboard potential

$$V(\psi) = -\frac{\hbar}{2e} I_c \cos\psi - \frac{\hbar}{2e} I_{\text{ext}} \psi. \tag{4.200}$$

The potential energy of the total flux ϕ in a SQUID geometry is given in Eq. (3.153).

The first term in $\mathcal{A}_{\text{eff}}$ accounts for the charging energy, and the potential $V(\psi)$ for the coupling and a bias term. The term $\mathcal{A}_{\text{T}}^{(\alpha)}$ describes state-dependent damping through quasiparticle tunneling in superconducting junctions across the barrier, or through single electrons in normal junctions. The trigonometric dependence on the phase difference describes discrete quasiparticle or single electron tunneling rather than continuous flow of charge. When the phase fluctuations in thermal equilibrium are weak, we may expand the trigonometric function in the α-term. Then we recover the standard quadratic form (4.51),

$$\mathcal{A}_{\text{T}}^{(\alpha)}[\psi(\cdot)] = \frac{1}{2} \int_0^{\hbar\beta} d\tau \int_0^\tau d\tau'\, \alpha(\tau-\tau') \left[\psi(\tau)-\psi(\tau')\right]^2. \tag{4.201}$$

Observing the analogy $\psi(\tau)/2\pi \,\hat{=}\, q(\tau)/q_0$ we see that the correspondence of the dissipation kernel for quasiparticle tunneling with that of the phenomenological model discussed in Subsection 4.2.1 is $4\pi^2\alpha(\tau) \,\hat{=}\, q_0^2 K(\tau)$. Accordingly, the correspondence of the spectral densities is $4\pi^2 J_\alpha(\omega) \,\hat{=}\, q_0^2 J(\omega)$.

Important limiting cases are:

(i) When the junction is formed by ideal BCS superconductors, the excitation spectrum has a gap, and thus $J_\alpha(\omega)$ has a behavior similar to the one of $G_{\text{qp}}(\omega)$ sketched in Fig. 4.3. At zero temperature, we find [cf. Eq. (4.176)]

$$\alpha(\tau) = \frac{\hbar^2 \Delta_{\mathrm{g,L}} \Delta_{\mathrm{g,R}}}{4\pi e^2 R_{\mathrm{N}}} K_1\left(\Delta_{\mathrm{g,L}}|\tau|\right) K_1\left(\Delta_{\mathrm{g,R}}|\tau|\right) , \tag{4.202}$$

where

$$R_{\mathrm{N}}^{-1} = 4\pi N_{0,\mathrm{L}} N_{0,\mathrm{R}} \frac{|T|^2}{\hbar^2} \frac{e^2}{\hbar} \tag{4.203}$$

denotes the normal state conductance of the tunnel junction. For equal gap frequencies, $\Delta_{\mathrm{g,L}} = \Delta_{\mathrm{g,R}} = \Delta_{\mathrm{g}}$, the limiting cases are

$$\alpha(\tau) = \frac{\hbar^2}{4\pi e^2 R_{\mathrm{N}}}
\begin{cases}
\dfrac{1}{\tau^2} & \text{for} \quad \Delta_{\mathrm{g}}|\tau| \ll 1 , \\[2ex]
\dfrac{\pi \Delta_{\mathrm{g}}}{2|\tau|} e^{-2\Delta_{\mathrm{g}}|\tau|} & \text{for} \quad \Delta_{\mathrm{g}}|\tau| \gg 1 .
\end{cases} \tag{4.204}$$

If the phase varies slowly with time on the time scale $1/\Delta_{\mathrm{g}}$, we may expand the term $\psi(\tau) - \psi(\tau')$ in (4.201) in the relative variable $\xi = \tau - \tau'$. We then get

$$\mathcal{A}_{\mathrm{T}}^{(\alpha)}[\psi(\cdot)] = \frac{1}{2} \int_0^{\hbar\beta} d\tau \left(\frac{\partial\psi}{\partial\tau}\right)^2 \int_0^{\hbar\beta/2} d\xi\, \xi^2 \alpha(\xi) . \tag{4.205}$$

Hence on this condition, the α-term yields a kinetic contribution which leads to an increase of the effective capacitance. With use of the form (4.202), the additional contribution is found to read

$$\delta C^{(\alpha)} = \frac{3\pi}{32} \frac{1}{\Delta_{\mathrm{g}} R_{\mathrm{N}}} . \tag{4.206}$$

The renormalization of the capacitance is similar to the mass renormalization in a mechanical analogue for super-Ohmic dissipation ($s > 2$) discussed in Section 3.1.3.

(ii) In a normal junction ($\Delta_{\mathrm{g}} = 0$), the kernel $\alpha(\tau)$ is in Ohmic form (4.169),

$$\alpha(\tau) = \frac{\hbar^2}{4\pi e^2 R_{\mathrm{N}}} \frac{(\pi/\hbar\beta)^2}{\sin^2(\pi\tau/\hbar\beta)} . \tag{4.207}$$

However even in this case, quasiparticle tunneling differs from an Ohmic resistor by the trigonometric dependence of the action $\mathcal{A}_{\mathrm{T}}^{(\alpha)}[\psi(\cdot)]$ on the phase in Eq. (4.199), instead of a quadratic dependence for a resistor, Eq. (4.201). The difference reflects different physics [198]. In the former case we have a discrete transfer of charge in units of e, while in the latter case charge is flowing continuously.

(iii) In a non-ideal junction we may have pair-breaking effects, spatial variation of the order parameter, or locally non-perfect energy gaps which lead to a finite subgap conductance even at $T = 0$ [154]. These effects may be covered either by smearing the density of states $\mathcal{N}_{\mathrm{qp}}(\omega)$ in Eq. (4.196), or by writing the kernel $\alpha(\tau)$ as a linear combination of the expressions (4.202) and (4.207).

Given the action $\mathcal{A}_{\mathrm{eff}}[\psi(\cdot)]$, one may ask whether ψ is an extended variable defined in the range $-\infty \leqslant \psi \leqslant \infty$, or a compact variable restricted either to the interval $0 < \psi \leqslant 2\pi$ or to $0 < \psi \leqslant 4\pi$. In the first case, the minima of the potential $U(\psi)$ are distinguishable, and $Q = (2e/i)\partial/\partial\psi$ is the usual charge operator with continuous eigenvalues, while in the other two cases Q is a quasi-charge operator with discrete eigenvalues similar to quasi-momentum of a particle in a periodic potential [154]. All these cases are possible, and which case applies depends on the experimental conditions. If charge transport is dominated by Cooper pair tunneling, the 2π-periodic potential $V(\psi)$ determines the symmetry of the problem and it is convenient to choose a basis of corresponding Bloch states. If quasiparticle tunneling is relevant, the symmetry is reduced to 4π periodicity, which is the symmetry of $\mathcal{A}_{\mathrm{T}}^{(\alpha)}[\psi]$. Finally, if there is a continuous flow of charge, as in the Ohmic case, or in a SQUID, or when the junction is coupled to an external circuit, Bloch states are inappropriate and we have to choose a basis of continuous charge states. When ψ is defined on a ring, the path sum of the partition function generally includes a summation over winding numbers. In many cases it is convenient to work with the restricted space of discrete charges and allow for a continuous change of charges perturbatively [154].

When quasiparticle tunneling is hampered by the superconducting gap, higher-order processes may become relevant, e.g., correlated tunneling of two electrons across a junction with a normal and a superconducting electrode. A diagram taking into account two diagonal (G) propagators on the normal metal side and two off-diagonal (F) propagators on the superconductor side describes Andreev scattering across the interface. The resulting action is similar in form to the quasiparticle contribution $\mathcal{A}_{\mathrm{T}}^{(\alpha)}$ in Eq. (4.199) except that it is of order $|T|^4$ and that the factor $\frac{1}{2}$ in the argument of the cosine function is missing because, instead of the charge e, the charge $2e$ is transferred in the Andreev scattering process. This implies that dissipation by Andreev reflection is very similar to dissipation by quasiparticle or single-electron tunneling [199, 200]. If the normal electrode is made of a dirty metal, the average over many impurities must be performed which results in a Cooperon propagator. The respective analysis for different junction geometries is given in Ref. [201].

4.2.11 Electromagnetic environment

In analogy to the derivation of the influence action (4.43) with (4.44) for the mechanical oscillator model Hamiltonian (3.12), it is straightforward to calculate the effective action connected with the Hamiltonian (3.217), which represents besides the capacitive energy an electromagnetic environment formed by LC-circuits. The resulting action, which corresponds to the expression (4.62) with (4.63), is ($R_{\mathrm{K}} = 2\pi\hbar/e^2$)

$$S_{\mathrm{eff},0}[\varphi] \;=\; \frac{R_{\mathrm{K}}}{4\pi} \frac{1}{\beta} \sum_n \Big[|\nu_n|^2 C + |\nu_n| \hat{Y}(|\nu_n|) \Big] |\varphi_n|^2 \,, \qquad (4.208)$$

where $\hat{Y}(z)$ is the Laplace transform of the admittance function $Y(t)$ and φ_n is the Fourier coefficient of the $\hbar\beta$-periodic phase function $\varphi(\tau) = \sum_n e^{i\nu_n\tau}\varphi_n/\hbar\beta$.

Consider next the phase correlator $\langle e^{i[\varphi(0)-\varphi(\tau)]}\rangle_\beta$ in imaginary time $0 \leqslant \tau < \hbar\beta$. Here, $\langle\cdots\rangle_\beta$ denotes average with the weight function $\exp[-S_{\mathrm{eff},0}[\varphi]/\hbar]$. With the form (4.208), we get

$$\left\langle e^{i\,[\varphi(0)-\varphi(\tau)]} \right\rangle_\beta \;=\; e^{-\langle[\varphi(0)-\varphi(\tau)]\varphi(0)\rangle_\beta} \;=\; e^{-\mathcal{W}_\varphi(\tau)} \;. \qquad (4.209)$$

The Gaussian average (4.209) yields for $\mathcal{W}_\varphi(\tau)$ the expression

$$\mathcal{W}_\varphi(\tau) \;=\; \frac{2\pi}{\hbar\beta} \sum_{n\neq 0} \frac{\hat{Z}_\mathrm{t}(|\nu_n|)}{|\nu_n| R_\mathrm{K}} \left[1 - e^{i\nu_n\tau} \right] \quad \text{with} \quad \hat{Z}_\mathrm{t}(\nu_n) = \frac{1}{\nu_n C + \hat{Y}(\nu_n)} \;. \qquad (4.210)$$

Observing the equivalence of the expression (3.91) to the expression (3.89), the Matsubara sum in Eq. (4.210) may be transformed into the frequency integral

$$\mathcal{W}_\varphi(\tau) \;=\; \int_0^\infty d\omega\, \frac{G_\varphi(\omega)}{\omega^2} \left[\mathfrak{D}_\omega(0) - \mathfrak{D}_\omega(\tau) \right] \;, \qquad (4.211)$$

where [cf. Eq. (3.231) with (3.229)]

$$G_\varphi(\omega) \;=\; 2\omega\, \mathrm{Re}\, \frac{1}{R_\mathrm{K}[-i\,\omega C + Z^{-1}(\omega)]} \;. \qquad (4.212)$$

and where $\mathfrak{D}_\omega(\tau)$ is the thermal boson propagator (3.82).

For a resistive environment described by an Ohmic impedance R, the phase correlation function $\mathcal{W}_\varphi(\tau)$ takes in the regime $\tau \gg RC$ the form

$$\mathcal{W}_\varphi(\tau) \;=\; \frac{R}{R_\mathrm{K}} \frac{2\pi}{\hbar\beta} \sum_{n\neq 0} \frac{1}{|\nu_n|} \left[1 - e^{i\nu_n\tau} \right] \;. \qquad (4.213)$$

We shall employ these expressions in Section 20.3 and in Subsection 31.1.4. There we discuss quantum transport of charge through a weak link and a weak constriction under influence of an electromagnetic environment.

4.3 Partition function of the open system

4.3.1 General path integral expression

The quantum statistical quantities characterizing a system in thermal equilibrium, such as internal energy, entropy, pressure, specific heat, and susceptibility, can be calculated directly if the partition function Z is known as a function of temperature, volume, external field, etc. They are obtained by differentiation of Z with respect to these control parameters. Since the partition function is the Laplace transform of the density of states, it also carries information about the energy spectrum (see the discussion of the density of states of the damped harmonic oscillator in Sect. 6.5).

The key quantity in quantum thermodynamics of open quantum systems is the reduced partition function introduced in Eq. (4.29), $Z(\hbar\beta) = Z_\mathrm{tot}(\hbar\beta)/Z_\mathrm{R}(\hbar\beta)$. In the path sum representation, the sum over periodic paths represents the trace operation. In Eq. (4.32) we have traced out the bath to obtain the influence functional $e^{-S_\mathrm{infl}[q(\cdot)]/\hbar}$. Tracing out in addition the system coordinates, one arrive at the reduced partition function. Precisely put, the path sum representation for $Z(\hbar\beta)$ is

$$Z(\hbar\beta) \;=\; \oint \mathcal{D}q(\cdot)\, e^{-S_\mathrm{eff}[q(\cdot)]/\hbar} \;=\; \oint \mathcal{D}q(\cdot)\, e^{-\{S_\mathrm{S}[q(\cdot)]+S_\mathrm{infl}[q(\cdot)]\}/\hbar} \;. \qquad (4.214)$$

The symbol $\oint$ indicates that the path sum is over all periodic paths of period $\hbar\beta$, and $S_{\text{eff}}[q] = S_S[q(\cdot)] + S_{\text{infl}}[q(\cdot)]$ is the effective Euclidean action of the open system.

If not stated otherwise, we shall restrict our attention in the remainder of this chapter to the simple phenomenological model discussed previously in Subsection 4.2.1. The exact formal expression (4.214) with the influence action (4.51) finds application in an enormous variety of open quantum systems. We may use it as a starting point, e. g., for the calculation of thermodynamic properties, or, in problems connected with tunneling, for the calculation of the quantum statistical decay of metastable states. The respective discussion is postponed until Part III.

The factor $\exp\{-S_{\text{eff}}[q(\cdot)]/\hbar\}$ represents the weight of an individual periodic path $q(\cdot)$ contributing to the reduced partition function. Alas, the path sum can be carried out exactly only when the exponent $S_{\text{eff}}[q(\cdot)]$ is a quadratic form of the path $q(\cdot)$. Therefore, I find it useful to present also a variety of different approaches which yield reasonable approximations in particular regions of the parameter space. Among them are the semiclassical approximation and the variational approach to quantum-statistical mechanics. The discussion of the latter approach is deferred until Chapter 8, which is after the discussion of the damped harmonic oscillator.

4.3.2 Semiclassical approximation

In the semiclassical regime, $S_{\text{eff}}[q(\cdot)]/\hbar \gg 1$, the path sum (4.60) or (4.214) is dominated by paths which in function space are close to the stationary points of the action. The first variation of the action vanishes for the *extremal* path $\bar{q}(\tau)$ obeying $\partial S_{\text{eff}}[q(\cdot)]/\partial q|_{\bar{q}(\cdot)} = 0$. The extremal path satisfies the equation of motion

$$-M\ddot{\bar{q}}(\tau) + V'[\bar{q}(\tau)] + \int_0^{\hbar\beta} d\tau'\, k(\tau - \tau')\, \bar{q}(\tau') = 0 . \qquad (4.215)$$

The inertia term has reverse sign, because the motion is in imaginary time. The time-nonlocal third term originates from the projection of the original energy-conserving path $\{\bar{q}(\tau), \bar{x}(\tau)\}$ in the $(N + 1)$-dimensional $\{q, x\}$ coordinate space [see Eqs. (4.35) and (4.36)] onto the q-axis. When applied to the partition function, the paths solving Eq. (4.215) must be $\hbar\beta$-periodic , $\bar{q}(\tau + \hbar\beta) = \bar{q}(\tau)$.

To account for the paths in the vicinity of $\bar{q}(\tau)$, we expand the action $S_{\text{eff}}^{(\text{E})}[q(\cdot)]$ about the stationary action up to second order in the deviations. Putting

$$q(\tau) = \bar{q}(\tau) + y(\tau) , \qquad (4.216)$$

we find

$$S_{\text{eff}}[q(\cdot)] = S_{\text{eff}}[\bar{q}(\cdot)] + \frac{M}{2}\int_0^{\hbar\beta} d\tau\, y(\tau)\mathbf{\Lambda}[\bar{q}(\tau)]\, y(\tau) + \mathcal{O}(y^3) . \qquad (4.217)$$

Here, $\mathbf{\Lambda}[\bar{q}(\tau)]$ is a linear operator acting in the space of $\hbar\beta$-periodic functions,

$$\mathbf{\Lambda}[\bar{q}(\tau)]y(\tau) \equiv \left(-\frac{\partial^2}{\partial\tau^2} + \frac{1}{M}V''[\bar{q}(\tau)]\right) y(\tau) + \frac{1}{M}\int_0^{\hbar\beta} d\tau'\, k(\tau - \tau')\, y(\tau') . \qquad (4.218)$$

Thus, Z becomes a sum of periodic fluctuation paths $y(\tau)$ with Gaussian weight,

$$Z = e^{-S_{\text{eff}}[\bar{q}(\cdot)]/\hbar} \oint \mathcal{D}y(\cdot) \exp\left(-\frac{M}{2\hbar} \int_0^{\hbar\beta} d\tau\, y(\tau)\Lambda[\bar{q}(\tau)]y(\tau)\right). \tag{4.219}$$

Execution of all the Gaussian integrals in a suitable representation yields

$$Z = \frac{N}{\sqrt{D[\bar{q}(\cdot)]}} \exp\left\{-S_{\text{eff}}[\bar{q}(\cdot)]/\hbar\right\}. \tag{4.220}$$

Here, N is a normalization factor, and $D[\bar{q}(\cdot)]$ is the determinant of the fluctuation operator $\Lambda[\bar{q}(\cdot)]$ on the space of $\hbar\beta$-periodic functions,

$$D[\bar{q}(\cdot)] \equiv \det(\Lambda[\bar{q}(\cdot)]). \tag{4.221}$$

The yet undetermined normalization factor N is universal in the sense that it neither depends on the specific form of the potential nor on the specific form of the dissipative mechanism. This is because the normalization factor N is fixed by the kinetic part $-\partial^2/\partial\tau^2$ of $\Lambda[\bar{q}(\cdot)]$ (see Subsection 4.3.4). Therefore, in ratios of partition functions the normalization factor N drops out. Thus we may write

$$Z = Z_0 \left(\frac{D_0}{D[\bar{q}(\cdot)]}\right)^{1/2} \exp\left\{-S_{\text{eff}}[\bar{q}(\cdot)]/\hbar\right\}. \tag{4.222}$$

Here we have chosen for convenience, $Z_0 = N/\sqrt{D_0}$ as the partition function of the undamped harmonic oscillator. Here D_0 is the determinant of the respective Gaussian fluctuation operator Λ_0. With the form (4.222), the remaining computational problem is reduced to the calculation of a ratio of determinants.

The semiclassical form (4.222) is a valid expression when all fluctuation modes can be treated in Gaussian approximation and when all eigenvalues of $\Lambda[\bar{q}(\cdot)]$ are positive. The proper treatment of conflicting modes, which are those with negative eigenvalues and with zero or quasi-zero eigenvalues, is given in Part III.

4.3.3 Partition function of the damped linear oscillator

Consider a harmonic potential with the minimum located at $q = 0$, $V(q) = \frac{1}{2}M\omega_0^2 q^2$, where ω_0 denotes the frequency of oscillations about the minimum of the well. For this potential, the imaginary-time equation of motion (4.215) admits only the trivial periodic solution in which the particle sits for ever at the barrier top of the upside-down potential $-V(q)$. Hence we have $\bar{q}(\tau) = 0$ and $S_{\text{eff}}[\bar{q} = 0] = 0$. The eigenvalues of the related fluctuation operator

$$\Lambda y(\tau) \equiv \left(-\frac{\partial^2}{\partial\tau^2} + \omega_0^2\right) y(\tau) + \frac{1}{M} \int_0^{\hbar\beta} d\tau'\, k(\tau - \tau')y(\tau') \tag{4.223}$$

on the space of $\hbar\beta$-periodic functions are found with use of Eqs. (4.47) and (4.44) as

$$\Lambda_n = \nu_n^2 + \omega_0^2 + |\nu_n|\hat{\gamma}(|\nu_n|), \tag{4.224}$$

where $\nu_n = 2\pi n/\hbar\beta$ $(n = 0, \pm 1, \pm 2, \cdots)$ is a bosonic Matsubara frequency, and where $\hat{\gamma}(z)$ is the Laplace transform of the friction kernel $\gamma(t)$. With the expression (4.224), the determinant of the fluctuation operator $\mathbf{\Lambda}$ takes the form

$$D = \prod_{n=-\infty}^{\infty} \Lambda_n = \omega_0^2 \prod_{n=1}^{\infty} \Lambda_n^2 = \omega_0^2 \prod_{n=1}^{\infty} \left[\nu_n^2 + \omega_0^2 + |\nu_n|\hat{\gamma}(|\nu_n|) \right]^2 . \qquad (4.225)$$

For the undamped oscillator, the fluctuation determinant is

$$D_0 = \omega_0^2 \prod_{n=1}^{\infty} \left(\nu_n^2 + \omega_0^2 \right)^2 , \qquad (4.226)$$

and the Matsubara representation of the corresponding partition function is

$$Z_0 = \frac{1}{2\sinh(\beta\hbar\omega_0/2)} = \frac{1}{\beta\hbar\omega_0} \prod_{n=1}^{\infty} \frac{\nu_n^2}{\omega_0^2 + \nu_n^2} . \qquad (4.227)$$

With use of Eqs. (4.224) - (4.226), and with $S_{\text{eff}}^{(E)}[\bar{q} = 0] = 0$, the semiclassical partition function (4.222) of the damped oscillator is found as

$$Z = Z_0 \sqrt{\frac{D_0}{D}} = \frac{1}{\beta\hbar\omega_0} \prod_{n=1}^{\infty} \frac{\nu_n^2}{\omega_0^2 + \nu_n^2 + \nu_n\hat{\gamma}(\nu_n)} . \qquad (4.228)$$

Finally, since the action of any linear quantum system is a quadratic functional of the path, semiclassics is exact for all these systems. Thus the Matsubara representation (4.228) is the exact partition function of the damped linear quantum oscillator for any form $\hat{\gamma}(\nu_n)$ of frequency-dependent dissipation.

4.3.4 Functional measure in Fourier space

Lagrangians which consist of a kinetic term $\frac{1}{2}M\dot{q}^2$ and a general potential $V(q)$ are usually referred to as standard Lagrangians. Open systems described by Eqs. (4.27), (4.23) and (4.34) also belong to this class, whereas magnetic systems are non-standard. Consider now the normalized functional measure for standard Lagrangians in Fourier space. Upon expanding the $\hbar\beta$-periodic fluctuation $y(\tau)$ into Fourier series

$$y(\tau) = \sum_{n=-\infty}^{\infty} y_n e^{i\nu_n\tau} , \qquad (4.229)$$

the exponent of the Gaussian weight function becomes

$$\frac{M}{2\hbar} \int_0^{\hbar\beta} d\tau \, y(\tau)\mathbf{\Lambda} \, y(\tau) = \frac{M\beta}{2} \sum_{n=-\infty}^{\infty} \Lambda_n y_n y_{-n} . \qquad (4.230)$$

The fluctuation operator $\mathbf{\Lambda}$ is defined in Eq. (4.223). Apparently, the expression (4.228) is directly found from (4.219) if we define the functional measure as

$$\oint \mathcal{D}y(\cdot) \cdots \equiv \int_{-\infty}^{\infty} \frac{dy_0}{\sqrt{2\pi\hbar^2\beta/M}} \prod_{n=1}^{\infty} \left[\int_{-\infty}^{\infty} \int_{-\infty}^{\infty} \frac{dy_n \, dy_{-n}}{2\pi/(iM\beta\nu_n^2)} \right] \cdots$$

$$= \int_{-\infty}^{\infty} \frac{dy_0}{\sqrt{2\pi\hbar^2\beta/M}} \prod_{n=1}^{\infty} \left[\int_{-\infty}^{\infty} \int_{-\infty}^{\infty} \frac{d\mathrm{Re}\, y_n \, d\mathrm{Im}\, y_n}{\pi/(M\beta\nu_n^2)} \right] \cdots . \tag{4.231}$$

The measure (4.231) in quantum statistical path integrals is generally valid for Gaussian fluctuation modes in systems with standard Lagrangians. Modifications are necessary for modes with zero or quasi-zero eigenvalues (cf. Chapters 16 and 17).

4.3.5 Partition function of the damped linear oscillator revisited

In Subsection 4.3.3 we have calculated the reduced partition function (4.228) via determination of the eigenvalues of the fluctuation modes. Complementary understanding of structure and specific form of the reduced partition function may be achieved by direct reduction of the partition function Z_{tot} of the full system-plus-reservoir complex to the reduced partition function [202]. We rigorously have

$$Z = Z_{\mathrm{tot}}/Z_{\mathrm{R}} , \tag{4.232}$$

where Z_{R} is the partition function of the harmonic reservoir given in Eq. (4.30).

To this purpose we start out from the Hamiltonian (3.12) with $V(q) = \frac{1}{2}M\omega_0^2 q^2$. First, we introduce mass weighted coordinates

$$u_0 = M^{1/2} q , \qquad u_\alpha = m_\alpha^{1/2} x_\alpha \qquad (\alpha = 1, 2, \cdots, N) . \tag{4.233}$$

With these, the Hamiltonian of $N + 1$ coupled stable oscillators takes the form

$$H^{(0)} = \frac{1}{2} \sum_{\alpha=0}^{N} \sum_{\alpha'=0}^{N} \left[\delta_{\alpha\alpha'} \dot{u}_\alpha \dot{u}_{\alpha'} + U_{\alpha\alpha'}^{(0)} u_\alpha u_{\alpha'} \right] . \tag{4.234}$$

The force constant matrix $\mathbf{U}^{(0)}$ is made up of the scaled spring coefficients of the coupled oscillators,

$$\mathbf{U}^{(0)} = \begin{pmatrix} \overline{\omega}_0^2 & C_1 & C_2 & \cdots & C_N \\ C_1 & \omega_1^2 & 0 & \cdots & 0 \\ C_2 & 0 & \omega_2^2 & \cdots & 0 \\ \vdots & \vdots & \vdots & \ddots & \vdots \\ C_N & 0 & 0 & \cdots & \omega_N^2 \end{pmatrix} . \tag{4.235}$$

Here we have put

$$\overline{\omega}_0^2 \equiv \omega_0^2 + \sum_{\alpha=1}^{N} \frac{C_\alpha^2}{\omega_\alpha^2} , \qquad C_\alpha \equiv -\frac{c_\alpha}{(m_\alpha M)^{1/2}} , \qquad (\alpha = 1, 2, \cdots, N) . \tag{4.236}$$

Denoting the eigenfrequencies of the stable harmonic system by μ_α $(\alpha = 0, 1, \cdots, N)$, the Hamiltonian (4.234) in normal mode representation reads

$$H^{(0)} = \frac{1}{2} \sum_{\alpha=0}^{N} \left(\dot{\rho}_\alpha^2 + \mu_\alpha^2 \rho_\alpha^2 \right), \tag{4.237}$$

and the partition function of the global harmonic system may be written in the form

$$Z_{\text{tot}}^{(0)} = \prod_{\alpha=0}^{N} \frac{1}{2 \sinh(\beta \hbar \mu_\alpha / 2)} = \prod_{\alpha=0}^{N} \left(\frac{1}{\beta \hbar \mu_\alpha} \prod_{n=1}^{\infty} \frac{\nu_n^2}{\mu_\alpha^2 + \nu_n^2} \right). \tag{4.238}$$

In the second equality we have used again the infinite product representation of the hyperbolic sine function.

Consider next the dynamical matrix $\mathbf{A}^{(0)}$ for $\hbar\beta$-periodic functions in imaginary time associated with the harmonic Hamiltonian (4.234),

$$\mathbf{A}^{(0)}(\nu_n) \equiv \mathbf{U}^{(0)} + \nu_n^2 \mathbf{1} = \begin{pmatrix} \Omega_0^2 & C_1 & C_2 & \cdots & C_N \\ C_1 & \Omega_1^2 & 0 & \cdots & 0 \\ C_2 & 0 & \Omega_2^2 & \cdots & 0 \\ \vdots & \vdots & \vdots & \ddots & \vdots \\ C_N & 0 & 0 & \cdots & \Omega_N^2 \end{pmatrix}. \tag{4.239}$$

The coefficients on the diagonal are

$$\Omega_0^2 = \bar{\omega}_0^2 + \nu_n^2, \quad \text{and} \quad \Omega_\alpha^2 = \omega_\alpha^2 + \nu_n^2. \tag{4.240}$$

In the eigenvalue representation of the dynamical matrix $\mathbf{A}^{(0)}$, the determinant is

$$\det \mathbf{A}^{(0)}(\nu_n) = \prod_{\alpha=0}^{N} \left(\mu_\alpha^2 + \nu_n^2 \right). \tag{4.241}$$

On the other hand, we may calculate $\det \mathbf{A}^{(0)}(\nu_n)$ with the representation (4.239),

$$\det \mathbf{A}^{(0)}(\nu_n) = \left(\Omega_0^2 - \sum_{\alpha=1}^{N} \frac{C_\alpha^2}{\Omega_\alpha^2} \right) \prod_{\alpha'=1}^{N} \Omega_{\alpha'}^2. \tag{4.242}$$

This expression is transformed by means of the relations (4.236) and (4.240) into

$$\det \mathbf{A}^{(0)}(\nu_n) = \left(\omega_0^2 + \nu_n^2 + \frac{\nu_n^2}{M} \sum_{\alpha=1}^{N} \frac{c_\alpha^2}{m_\alpha \omega_\alpha^2} \frac{1}{(\omega_\alpha^2 + \nu_n^2)} \right) \prod_{\alpha'=1}^{N} \left(\omega_{\alpha'}^2 + \nu_n^2 \right)$$
$$= \left[\omega_0^2 + \nu_n^2 + |\nu_n| \hat{\gamma}(|\nu_n|) \right] \prod_{\alpha=1}^{N} \left(\omega_\alpha^2 + \nu_n^2 \right). \tag{4.243}$$

In the second form, we have used the representation (3.29) for the Laplace transform $\hat{\gamma}(\nu_n)$ of the damping kernel $\gamma(t)$. Upon equating (4.241) with (4.243), we get

$$\prod_{\alpha=0}^{N} \left(\mu_\alpha^2 + \nu_n^2 \right) = \left[\omega_0^2 + \nu_n^2 + |\nu_n| \hat{\gamma}(|\nu_n|) \right] \prod_{\alpha=1}^{N} \left(\omega_\alpha^2 + \nu_n^2 \right). \tag{4.244}$$

Putting $n = 0$ in Eq. (4.244) and observing that $\nu_0 = 0$, we obtain a simple relation between the set of eigenfrequencies $\{\mu_\alpha\}$ of the composite system, the system frequency ω_0, and the set of reservoir frequencies $\{\omega_\alpha\}$,

$$\det \mathbf{U}^{(0)} = \prod_{\alpha=0}^{N} \mu_\alpha^2 = \omega_0^2 \prod_{\alpha=1}^{N} \omega_\alpha^2 . \tag{4.245}$$

With use of the relations (4.244) and (4.245), the expression (4.238) takes the form

$$Z_{\text{tot}} = \prod_{\alpha=1}^{N} \left\{ \frac{1}{\beta\hbar\omega_\alpha} \prod_{n=1}^{\infty} \frac{\nu_n^2}{\omega_\alpha^2 + \nu_n^2} \right\} \frac{1}{\beta\hbar\omega_0} \prod_{n=1}^{\infty} \frac{\nu_n^2}{\omega_0^2 + \nu_n^2 + \nu_n \hat{\gamma}(\nu_n)} . \tag{4.246}$$

The upshot of this analysis now is that the N-fold product of the curly bracket term is the partition function Z_R of the reservoir, Eq. (4.30), and the residual set of factors is the product representation (4.228) of the reduced partition function Z of the damped quantum oscillator. Thus indeed, the partition function of the global system emerges just in the factorized form

$$Z_{\text{tot}} = Z_R \, Z . \tag{4.247}$$

In conclusion, we have shown by two different lines of reasoning that the partition function of the damped system is given by the expression (4.228).

4.4 Quantum statistical expectation values in phase space

We may ask ourselves if there is a phase space representation of quantum statistical expectation values[8] for general operator functions $\hat{A}(\hat{p}, \hat{q})$. A classical particle is described by a probability density function in phase space $f^{(cl)}(p, q)$. The average of a function of momentum and position can then be expressed as

$$\langle A \rangle_{\text{cl}} = \iint \frac{dp\, dq}{2\pi\hbar} A(p, q) f^{(cl)}(p, q) . \tag{4.248}$$

Because of the uncertainty principle, one cannot define a true phase space probability distribution for a quantum mechanical particle. Nevertheless, "quasi-probability distribution functions" which bear resemblance to classical phase space distribution functions are very useful as a calculational tool. In addition, they elucidate the nature of quantum mechanics. Within the framework of a quasi-probability distribution $f^{(qm)}(p, q)$, the quantum statistical average takes the form

$$\langle A \rangle_{\text{qm}} \equiv \text{tr}\,(\hat{A}\hat{\rho}) = \iint \frac{dp\, dq}{2\pi\hbar} A(p, q) f^{(qm)}(p, q) . \tag{4.249}$$

First of all, the question arises whether an operator function $\hat{A}(\hat{p}, \hat{q})$ which is defined in terms of a certain ordering prescription can unambiguously be represented as

---

[8]There is a neat monograph by W. P. Schleich on quantum physics in phase space [203].

an ordinary function in phase space. One positive answer is the generalized Weyl correspondence. This is discussed in the following subsection.

Since the pioneering work by Weyl [204], the association of distribution functions in phase space with operator ordering rules for operators $\hat{p}$ and $\hat{q}$ has been the subject of many studies (cf. the review article [205]). The quasi-probability distribution which is appropriate to the Weyl correspondence between operator-valued and ordinary functions is the Wigner distribution $f^{(W)}(p, q)$ [206].

4.4.1 Generalized Weyl correspondence

We now aim at establishing a correspondence between quantum mechanical operators $\hat{A}(\hat{p}, \hat{q})$ and ordinary functions $A(p, q)$ in phase space. As a specific example, we choose the case of a point-like particle in one space dimension. The generalization to a coordinate space with d dimensions and curvature is discussed in Ref. [207]. We start with the expansion of the operator $\hat{A}(\hat{p}, \hat{q})$ in position and momentum eigenstates,

$$\hat{A}(\hat{p}, \hat{q}) = \int dp'\, dp''\, dq'\, dq'' \, |p''><p''|q''><q''|\hat{A}(\hat{p}, \hat{q})|q'><q'|p'><p'| \, . \quad (4.250)$$

Next, we consider a one-parameter family of transformations of the phase space variables with Jacobian unity,

$$\begin{aligned} p'' &= p + \tfrac{1}{2}(1-\gamma)\eta \, ; & p' &= p - \tfrac{1}{2}(1+\gamma)\eta \, , \\ q'' &= q + \tfrac{1}{2}(1+\gamma)y \, ; & q' &= q - \tfrac{1}{2}(1-\gamma)y \, . \end{aligned} \quad (4.251)$$

The range of the parameter γ is $-1 \leqslant \gamma \leqslant 1$. With use of the relations

$$<q|q'> = \delta(q-q') \, ; \quad <p|p'> = \delta(p-p') \, ; \quad <q|p> = (2\pi\hbar)^{-1/2}e^{i\,pq/\hbar} \, , \quad (4.252)$$

we then obtain

$$\hat{A}(\hat{p}, \hat{q}) = \iint \frac{dp\, dq}{2\pi\hbar} \, A_\gamma(p, q)\hat{\Delta}_\gamma(p, q) \, , \quad (4.253)$$

where

$$A_\gamma(p, q) = \int dy \, e^{-i\,py/\hbar}<q + \tfrac{1}{2}(1+\gamma)y \,|\, \hat{A}(\hat{p}, \hat{q}) \,|\, q - \tfrac{1}{2}(1-\gamma)y> \, , \quad (4.254)$$

$$\hat{\Delta}_\gamma(p, q) = \int d\eta \, e^{-i\,q\eta/\hbar} \, |p + \tfrac{1}{2}(1-\gamma)\eta><p - \tfrac{1}{2}(1+\gamma)\eta| \, . \quad (4.255)$$

The function $A_\gamma(p, q)$ is the generalized Weyl transform of the operator-valued function $\hat{A}(\hat{p}, \hat{q})$. The correspondence is one-to-one and is denoted by the symbol $\longleftrightarrow$. The functional form of $A_\gamma(p, q)$ depends on the ordering of the operators $\hat{p}$ and $\hat{q}$ in $\hat{A}(\hat{p}, \hat{q})$, and on the parameter γ. By calculation one finds, e.g., the correspondence

$$\hat{p}^m \hat{F}(\hat{q})\hat{p}^n \quad \longleftrightarrow \quad \left(p - i\hbar \frac{1-\gamma}{2} \frac{\partial}{\partial q}\right)^m \left(p + i\hbar \frac{1+\gamma}{2} \frac{\partial}{\partial q}\right)^n F(q) \, . \quad (4.256)$$

Hence the ordinary function in phase space usually differs from the corresponding operator function. It is useful to define a one-parameter family of operator ordering schemes in such a way that the functional form of the operator function coincides exactly with the ordinary function in phase space. Denoting this family symbolically by $\mathcal{Z}_\gamma\{\hat{A}(\hat{p},\hat{q})\}$, we thus have the direct correspondence

$$\mathcal{Z}_\gamma\{\hat{A}(\hat{p},\hat{q})\} \quad\longleftrightarrow\quad A(p,q) . \tag{4.257}$$

By definition, $\mathcal{Z}_\gamma$ is an operator which orders the operators $\hat{p}$ and $\hat{q}$ in an operator function $\hat{A}(\hat{p},\hat{q})$ regardless of the commutation relations such that the correspondence (4.257) holds. There follows with Eq. (4.256) that the operator corresponding to the transform $p^m F(q)$ is given by

$$\mathcal{Z}_\gamma\{\hat{p}^m \hat{F}(\hat{q})\} =: \sum_{\ell=0}^{m} \binom{m}{\ell} \left(\frac{1+\gamma}{2}\hat{p}\right)^{m-\ell} \hat{F}(\hat{q}) \left(\frac{1-\gamma}{2}\hat{p}\right)^{\ell} . \tag{4.258}$$

With this form, we can find $\mathcal{Z}_\gamma\{\hat{A}(\hat{p},\hat{q})\}$ for any operator function $\hat{A}(\hat{p},\hat{q})$ of which the Weyl transform can be represented in the form of a power series in the momentum, $A(p,q) = \sum_j F_j(q)\, p^j$,

$$\mathcal{Z}_\gamma\{\hat{A}(\hat{p},\hat{q})\} =: \sum_{j}\sum_{\ell=0} \binom{j}{\ell} \left(\frac{1+\gamma}{2}\hat{p}\right)^{j-\ell} \hat{F}_j(\hat{q}) \left(\frac{1-\gamma}{2}\hat{p}\right)^{\ell} . \tag{4.259}$$

Equation (4.258) or (4.259) defines the ordering prescription for a whole family of functions parametrized by a continuous parameter γ in the range $-1 \leqslant \gamma \leqslant 1$. The special value $\gamma = 0$ represents the case of Weyl ordering. For example, if we put $m = 2$ in Eq. (4.258), we have

$$\mathcal{Z}_0\{\hat{p}^2 \hat{F}(\hat{q})\} =: \frac{1}{4}\left(\hat{p}^2\hat{F}(\hat{q}) + 2\hat{p}\hat{F}(\hat{q})\hat{p} + \hat{F}(\hat{q})\hat{p}^2\right) . \tag{4.260}$$

The cases $\gamma = 1$ and $\gamma = -1$ represent the antistandard ($\hat{p}\hat{q}$) and standard ($\hat{q}\hat{p}$) ordering prescription, respectively. With use of (4.255) and (4.252), we get

$$< q''|\hat{\Delta}_\gamma(p,x)|q' > \;=\; \delta(x-q)\, e^{i\,p(q''-q')/\hbar} , \tag{4.261}$$

where

$$q \;=\; \tfrac{1}{2}(q''+q') - \tfrac{1}{2}\gamma(q''-q') . \tag{4.262}$$

As γ is tuned from $+1$ to -1, the position q runs from the initial point q' to the end point q''. Insertion of (4.261) in the coordinate representation of (4.253) and observance of the correspondence (4.257) results in the expression

$$< q''|\mathcal{Z}_\gamma\{\hat{A}(\hat{p},\hat{q})\}|q' > \;=\; (2\pi\hbar)^{-1}\int dp\, A(p,q)\, e^{i\,p(q''-q')/\hbar} . \tag{4.263}$$

Inversion of this relation yields

$$A(p,q) = \int dy \, e^{-i\,py/\hbar} < q + \tfrac{1}{2}(1+\gamma)y \,|Z_\gamma\{\hat{A}(\hat{p},\hat{q})\}|\, q - \tfrac{1}{2}(1-\gamma)y > . \quad (4.264)$$

Thus, the generalized Weyl transform $A(p,q)$ is calculated by regarding the matrix element $< q''|Z_\gamma\{\hat{A}(\hat{p},\hat{q})\}|q' >$ as a function of q and y, and then taking the Fourier transform with respect to y. For symmetric (Weyl) ordering, $\gamma = 0$, we have

$$A(p,q) = \int dy \, e^{-i\,py/\hbar} < q + \tfrac{1}{2}y|Z_0\{\hat{A}(\hat{p},\hat{q})\}|q - \tfrac{1}{2}y > . \quad (4.265)$$

In summary of the results obtained so far, we have found for the operator ordering prescription (4.258) a unique representation in phase space.

4.4.2 Generalized Wigner function and expectation values

Wigner's function is a phase space representation of the reduced density matrix $\rho_\beta(q'', q')$ in thermal equilibrium. Assume that $\rho_\beta(q'', q')$ is expressed in terms of the variables $y = q'' - q'$ and $q = \tfrac{1}{2}(q'' + q')$. Then, Wigner's function $f^{(W)}(p,q)$ is the Fourier transform with respect to the relative coordinate y,

$$f^{(W)}(p,q) = \int dy \, e^{-i\,py/\hbar} \rho_\beta \left(q + \tfrac{1}{2}y, \, q - \tfrac{1}{2}y\right) . \quad (4.266)$$

The quasi-probability distribution $f^{(W)}(p,q)$ is the appropriate density function in phase space for operator functions with Weyl ordering.

If we aim at the phase space representation of quantum statistical expectation values for general operator functions $Z_\gamma\{\hat{A}(\hat{p},\hat{q})\}$, we have to make a subtle generalization of Wigner's function (4.266),

$$f^{(\gamma)}(p,\bar{q}) \equiv \int dy \, e^{-i\,py/\hbar} \, \rho_\beta \left(\bar{q} + \tfrac{1}{2}(1-\gamma)y, \, \bar{q} - \tfrac{1}{2}(1+\gamma)y\right) . \quad (4.267)$$

Here we have again $y = q'' - q'$, but we use as second variable

$$\bar{q} = \tfrac{1}{2}(q'' + q') + \tfrac{1}{2}\gamma(q'' - q') , \quad (4.268)$$

which differs from the expression (4.262) by the substitution $\gamma \to -\gamma$. The reason for this subtle change will become clear shortly. The generalized Wigner function at $\gamma = 0$, $f^{(0)}(p,\bar{q})$, coincides with $f^{(W)}(p,q)$. The expression (4.267) is inverted to yield

$$\rho_\beta(q'', q') = (2\pi\hbar)^{-1} \int dp \, f^{(\gamma)}(p,\bar{q}) \, e^{i\,p(q''-q')/\hbar} . \quad (4.269)$$

The quasi-probability distributions $f^{(W)}(p,q)$ and $f^{(\gamma)}(p,\bar{q})$ have properties similar to the classical density function $f(p,q)$ in phase space,

$$\frac{1}{2\pi\hbar} \int dp \, f^{(W)}(p,q) = \frac{1}{2\pi\hbar} \int dp \, f^{(\gamma)}(p,q) = < q|\hat{\rho}_\beta|q > \equiv P(q) ,$$

$$\int dq \, f^{(W)}(p,q) = \int d\bar{q} \, f^{(\gamma)}(p,\bar{q}) = < p|\hat{\rho}_\beta|p > \equiv \tilde{P}(p) . \quad (4.270)$$

The diagonal elements $P(q)$ and $\tilde{P}(p)$ represent the probability for finding the damped particle in thermal equilibrium at position q and with momentum p, respectively. Although the quasi-probability distribution $f^{(W)}(p,q)$ $[f^{(\gamma)}(p,\bar{q})]$ satisfies (4.270), it does not describe the probability for finding the particle at position q $[\bar{q}]$ with momentum p, because it can become negative for some values of p and q $[\bar{q}]$.

We are now ready to consider the phase space representation of the quantum statistical average of the operator $\mathcal{Z}_\gamma\{\hat{A}(\hat{p},\hat{q})\}$,

$$\langle A_\gamma \rangle =: \left\langle \mathcal{Z}_\gamma\{\hat{A}(\hat{p},\hat{q})\} \right\rangle = \int dq'' \int dq' <q''|\mathcal{Z}_\gamma\{\hat{A}(\hat{p},\hat{q})\}|q'> \rho_\beta(q',q'') \, . \tag{4.271}$$

Because $\rho_\beta(q',q'')$, instead of $\rho_\beta(q'',q')$, appears in Eq. (4.271), we used Eq. (4.268) in Eq. (4.269), instead of Eq. (4.262). As a result, the Wigner function related to $\rho_\beta(q',q'')$ is a function of q instead of $\bar{q}$. Substituting (4.263) and (4.269) we find

$$\langle A_\gamma \rangle = \iint \frac{dp'\, dq}{2\pi\hbar} \iint \frac{dp\, dy}{2\pi\hbar} A(p,q) f^{(\gamma)}(p',q) \, e^{i\, y(p-p')/\hbar} \, , \tag{4.272}$$

where we have interchanged q'' with q' in Eq. (4.269) and observed Eqs. (4.268) and (4.262). Since the y integral gives $\delta(p-p')$, we find in the end

$$\langle A_\gamma \rangle = \iint \frac{dp\, dq}{2\pi\hbar} A(p,q) f^{(\gamma)}(p,q) \, . \tag{4.273}$$

Thus, when the quasi-probability distribution in Eq. (4.249) is chosen to be $f^{(\gamma)}(p,q)$, the correspondence between $A(p,q)$ and $\hat{A}(\hat{p},\hat{q})$ is that given by Eq. (4.257). Notice that the expression (4.273) applies to the family of operator ordering prescriptions $\mathcal{Z}_\gamma\{\hat{A}(\hat{p},\hat{q})\}$ among which Weyl ordering, standard ordering, and anti-standard ordering are special cases. The main criterion for the choice of the parameter γ of the quasi-probability distribution function for a particular problem is convenience.

The phase space representation of quantum statistical expectation values is especially convenient if the density matrix in coordinate representation is already known. Then the remaining problem is limited to integrations of ordinary functions. We will make use of the result (4.273) in Section 8.2.

5 Real-time approach and nonequilibrium dynamics

5.1 Statement of the problem and general concepts

In the preceding chapter, I have formulated quantum statistical mechanics for a number of systems in contact with a thermal reservoir using the imaginary-time or Euclidean path integral representation. I have also shown some techniques useful in the approximate evaluation of path integrals. In this chapter we are concerned with the description of non-equilibrium time-dependent phenomena. The general formalism will be outlined with the simple phenomenological model introduced in Section 3.1. The proceeding will also illuminate the appropriate modifications in the passage from imaginary to real time for the various microscopic models introduced previously. Thus I take them up only briefly in the summary at the end of the chapter. Since we are interested in the evolution of mixed states, the proper vehicle is the density matrix representation.

Generic nonequilibrium problems are as follows: (i) The system is prepared in a nonequilibrium inital state and one is interested in the relaxation dynamics of a physical observable O at later time. (ii) The system is initially in the canonical equilibrium state and then time-dependent perturbations are switched on. The quantity of interest in both cases is the average value $\langle O(t) \rangle = \text{tr}\{\hat{\rho}(t)\hat{O}\}$ as function of time.

In the next section, I present the Feynman-Vernon unfluence functional method, which deals with the evolution of the open system from a product initial state of the composite system, and work out in the classical and semiclassical limits diverse physical implications arising from the influence functional. In the third section, the influence functional concept is generalized for any form of correlated initial states. The discussion includes the discussion of different initial preparations and representations with complex-time, real-time and closed contour path integrals In the remaining sections I discuss the semiclassical regime, in which we end up with a Langevin-type description of the quantum stochastic process, the stochastic unraveling of the influence functional, and the combination of semiclassics with stochastic unraveling. This will turn out as a promising route to handle non-Markovian dynamics. Finally, an approach covering energy transfer between the various dissipative channels is presented.

Generally, the analysis of this chapter will show that the Hamiltonian dynamics of the global system induces a non-unitary dynamics for the reduced density matrix.

Let us examine the possibility of obtaining closed formal expressions for the dynamics of a dissipative quantum system. We assume that the underlying global system is governed by the Hamiltonian (3.12). Again, it is convenient to perform the reduction of the composite system to the open system within the functional integral representation, a technique introduced by Feynman and Vernon [208] already in 1963.

Our starting point is the density operator of the composite system-plus-reservoir complex in Heisenberg representastion,

$$\hat{W}(t) = e^{-i\hat{H}t/\hbar} \, \hat{W}(0) \, e^{i\hat{H}t/\hbar} \,. \tag{5.1}$$

In coordinate representation, we have

$$< q_\mathrm{f}, \mathbf{x}_\mathrm{f}| \, \hat{W}(t) \, |q'_\mathrm{f}, \mathbf{x}'_\mathrm{f} > \; = \; \int dq_\mathrm{i} \, dq'_\mathrm{i} \, d\mathbf{x}_\mathrm{i} \, d\mathbf{x}'_\mathrm{i} \, K(q_\mathrm{f}, \mathbf{x}_\mathrm{f}, t; q_\mathrm{i}, \mathbf{x}_\mathrm{i}, 0) \tag{5.2}$$

$$\times \; < q_\mathrm{i}, \mathbf{x}_\mathrm{i}| \, \hat{W}(0) \, |q'_\mathrm{i}, \mathbf{x}'_\mathrm{i} > K^*(q'_\mathrm{f}, \mathbf{x}'_\mathrm{f}, t; q'_\mathrm{i}, \mathbf{x}'_\mathrm{i}, 0) \,.$$

Here, $\mathbf{x}_\mathrm{i/f}$ represents the set of bath oscillator positions $(x_{\mathrm{i/f},1}, \ldots, x_{\mathrm{i/f},N})$ in the initial/final state, and $K(q_\mathrm{f}, \mathbf{x}_\mathrm{f}, t; q_\mathrm{i}, \mathbf{x}_\mathrm{i}, 0) \; = \; < q_\mathrm{f}, \mathbf{x}_\mathrm{f}| \, e^{-i\hat{H}t/\hbar} |q_\mathrm{i}, \mathbf{x}_\mathrm{i} >$ is a propagation amplitude. The amplitudes K and K^* have the path sum representation

$$K(q_\mathrm{f}, \mathbf{x}_\mathrm{f}, t; q_\mathrm{i}, \mathbf{x}_\mathrm{i}, 0) \; = \; \int \mathcal{D}q(\cdot) \, \mathcal{D}\mathbf{x}(\cdot) \; e^{i \, S[q(\cdot), \mathbf{x}(\cdot)]/\hbar} \,,$$
$$K^*(q'_\mathrm{f}, \mathbf{x}'_\mathrm{f}, t; q'_\mathrm{i}, \mathbf{x}'_\mathrm{i}, 0) \; = \; \int \mathcal{D}q'(\cdot) \, \mathcal{D}\mathbf{x}'(\cdot) \; e^{-i \, S[q'(\cdot), \mathbf{x}'(\cdot)]/\hbar} \,. \tag{5.3}$$

The path sums in Eq. (5.3) cover all paths with endpoints

$$q(0) \; = \; q_\mathrm{i} \,, \qquad q(t) \; = \; q_\mathrm{f} \,, \qquad q'(0) \; = \; q'_\mathrm{i} \,, \qquad q'(t) \; = \; q'_\mathrm{f} \,, \tag{5.4}$$
$$\mathbf{x}(0) \; = \; \mathbf{x}_\mathrm{i} \,, \qquad \mathbf{x}(t) \; = \; \mathbf{x}_\mathrm{f} \,, \qquad \mathbf{x}'(0) \; = \; \mathbf{x}'_\mathrm{i} \,, \qquad \mathbf{x}'(t) \; = \; \mathbf{x}'_\mathrm{f} \,. \tag{5.5}$$

The action S is composed of the actions of system, reservoir, and interaction,

$$S \; = \; S_\mathrm{S} + S_\mathrm{R} + S_\mathrm{I} \; = \; \int_0^t dt' \, \left(\mathcal{L}_\mathrm{S}(t') + \mathcal{L}_\mathrm{R}(t') + \mathcal{L}_\mathrm{I}(t') \right) \,, \tag{5.6}$$

where [we add a source term in $\mathcal{L}_\mathrm{S}(t)$, as in Eq. (4.19)]

$$\mathcal{L}_\mathrm{S}(t) \; = \; \frac{1}{2}M\dot{q}^2(t) - V[q(t)] - \mathcal{J}(t)q(t) \,,$$
$$\mathcal{L}_\mathrm{R}(t) \; = \; \frac{1}{2}\sum_\alpha m_\alpha \left[\dot{x}_\alpha^2(t) - \omega_\alpha^2 x_\alpha^2(t) \right] \,, \tag{5.7}$$
$$\mathcal{L}_\mathrm{I}(t) \; = \; \sum_\alpha \left(c_\alpha x_\alpha(t)q(t) - \frac{1}{2}\frac{c_\alpha^2}{m_\alpha \omega_\alpha^2} q^2(t) \right) \,.$$

The expression (5.2) describes the dynamics of the system-plus-bath complex as a whole. However, in most cases of interest the only information we wish to have is the system's dynamics under the reservoir's influence. Then the quantity we are really interested in is the reduced density operator $\hat{\rho}(t) = \mathrm{tr}_\mathrm{R}\hat{W}(t)$ [cf. Eq. (4.8)], or in coordinate representation the reduced density matrix (RDM) [11, 24, 23]

$$\rho(q_\mathrm{f}, q'_\mathrm{f}; t) \; \equiv \; \int d\mathbf{x}_\mathrm{f} < q_\mathrm{f}, \mathbf{x}_\mathrm{f}| \, \hat{W}(t) \, |q'_\mathrm{f}, \mathbf{x}_\mathrm{f} >$$

$$= \; \int dq_\mathrm{i} \, dq'_\mathrm{i} \, d\mathbf{x}_\mathrm{f} \, d\mathbf{x}_\mathrm{i} \, d\mathbf{x}'_\mathrm{i} \, K(q_\mathrm{f}, \mathbf{x}_\mathrm{f}, t; q_\mathrm{i}, \mathbf{x}_\mathrm{i}, 0) \tag{5.8}$$

$$\times \; < q_\mathrm{i}, \mathbf{x}_\mathrm{i}| \, \hat{W}(0) \, |q'_\mathrm{i}, \mathbf{x}'_\mathrm{i} > K^*(q'_\mathrm{f}, \mathbf{x}_\mathrm{f}, t; q'_\mathrm{i}, \mathbf{x}'_\mathrm{i}, 0) \,.$$

5.2 Feynman-Vernon method for product initial state

Assume that at time $t = 0$ the system and the bath are uncoupled, and the bath is in thermal equilibrium. Then the density operator of the composite system is factorized at time $t = 0$ according to

$$\hat{W}_{\text{fc}}(0) \; = \; \hat{\rho}(0) \otimes \hat{W}_{\text{R},0} \; = \; \hat{\rho}(0) \otimes Z_{\text{R}}^{-1} e^{-\beta \hat{H}_{\text{R}}} \,, \tag{5.9}$$

where $\hat{\rho}$ is the reduced density operator. The product initial state (5.9) is free of correlations between system and thermal bath.

Another choice of interest is that the system is initially in a diagonal state and the bath in the shifted canonical state (3.72) induced by the system-reservoir coupling,

$$<q_{\text{i}}|\hat{W}_{\text{sh}}(0)|q_{\text{i}}> \; = \; <q_{\text{i}}|\hat{\rho}(0)|q_{\text{i}}> \hat{W}_{\text{R}}[q_{\text{i}}] \; = \; <q_{\text{i}}|\hat{\rho}(0)|q_{\text{i}}> Z_{\text{R}}^{-1} e^{-\beta[\hat{H}_{\text{R}} + \hat{H}_{\text{I}}(q_{\text{i}})]} \,. \tag{5.10}$$

The first case (class A in Subsec. 21.1.2) corresponds to the situation in which the system is suddenly prepared in a particular state and evolutes from this state before the bath has relaxed to the shifted thermal distribution. This case is important in the adiabatic regime in which the bath cut-off ω_{c} is of the order of the characteristic system frequencies or smaller. It might be relevant, e.g., in electron transfer reactions.

In the second case (class B in Subsec. 21.1.2), the initial state is prepared by holding the system for sufficiently long time at position q_{i}, e.g., by a suitably chosen bias field, so that the bath has come into equilibrium with it. At time zero the bias is turned off, and the composite system starts out of this state.

In this section we choose the product initial state (5.9). The initial state (5.10) is chosen in Sec. 5.6 in connection with energy transfer inside the composite system. We indicate in advance that the difference in the latter case is the absence of the slip action in the expression (5.31), as we can see in Eq. (5.123).

5.2.1 Influence functional

Now assume that the system-bath coupling is suddenly switched on at time $t = 0^+$, and then consider the dynamics of $\rho(t)$ for $t \geqslant 0$. With Eqs. (5.3) – (5.7) and with (5.9), the expression (5.8) can be written as

$$\rho(q_{\text{f}}, q_{\text{f}}'; t) \; = \; \int dq_{\text{i}} \, dq_{\text{i}}' \, J_{\text{FV}}(q_{\text{f}}, q_{\text{f}}', t; q_{\text{i}}, q_{\text{i}}', 0) \, \rho(q_{\text{i}}, q_{\text{i}}'; 0) \,. \tag{5.11}$$

The *propagating function* J_{FV} describes the evolution of the reduced density matrix. It has the path sum representation with the boundary points indicated on the l.h.s,

$$J_{\text{FV}}(q_{\text{f}}, q_{\text{f}}', t; q_{\text{i}}, q_{\text{i}}', 0) \; = \; \int \mathcal{D}q(\cdot) \, \mathcal{D}q'(\cdot) \, e^{i\{ S_{\text{S}}[q(\cdot)] - S_{\text{S}}[q'(\cdot)] \}/\hbar} \, \mathcal{F}_{\text{FV}}[q(\cdot), q'(\cdot)] \,, \tag{5.12}$$

The Feynman-Vernon influence functional $\mathcal{F}_{\text{FV}}[q(\cdot), q'(\cdot)]$ includes all influences of the environment on the propagation of the system [208]. It may be written as

$$\mathcal{F}_{\text{FV}}[q(\cdot), q'(\cdot)] = \int d\mathbf{x}_{\text{f}} \, d\mathbf{x}_{\text{i}} \, d\mathbf{x}_{\text{i}}' \, W_{\text{R},0}(\mathbf{x}_{\text{i}}, \mathbf{x}_{\text{i}}') \, \mathbb{F}[q(\cdot); \mathbf{x}_{\text{f}}, \mathbf{x}_{\text{i}}, t] \, \mathbb{F}^*[q'(\cdot); \mathbf{x}_{\text{f}}, \mathbf{x}_{\text{i}}', t] \,. \tag{5.13}$$

Here, $W_{R,0}(\mathbf{x}_i, \mathbf{x}'_i)$ represents the thermal initial state of the bath, and $\mathbb{F}[q(\cdot); \mathbf{x}_f, \mathbf{x}_i, t]$ is the propagation amplitude of the bath under motion $q(\cdot)$ of the system,

$$
\begin{aligned}
W_{R,0}(\mathbf{x}_i, \mathbf{x}'_i) &= \frac{1}{Z_R} \int_{\bar{\mathbf{x}}(0)=\mathbf{x}_i}^{\bar{\mathbf{x}}(\hbar\beta)=\mathbf{x}_f} \mathcal{D}\bar{\mathbf{x}}(\cdot)\, e^{-S_R[\bar{\mathbf{x}}(\cdot)]/\hbar} , \\
\mathbb{F}[q(\cdot); \mathbf{x}_f, \mathbf{x}_i, t] &= \int_{x(0)=x_i}^{x(t)=x_f} \mathcal{D}\mathbf{x}(\cdot)\, e^{i\{S_R[\mathbf{x}(\cdot)]+S_I[\mathbf{x}(\cdot),q(\cdot)]\}/\hbar} .
\end{aligned}
\tag{5.14}
$$

For the harmonic reservoir model described by $\mathcal{L}_R$, Eq. (5.7), the canonical density matrix $W_{R,0}$, the amplitude $\mathbb{F}$, and the reservoir's partiton function Z_R are in product form, in which each factor represents the contribution of an individual bath oscillator,

$$
\begin{aligned}
W_{R,0}(\mathbf{x}_i, \mathbf{x}'_i) &= \prod_\alpha W_R^{(\alpha)}(x_{i,\alpha}, x'_{i,\alpha}) , \\
\mathbb{F}[q(\cdot); \mathbf{x}_f, \mathbf{x}_i, t] &= \prod_\alpha \mathbb{F}_\alpha[q(\cdot); x_{f,\alpha}, x_{i,\alpha}, t] , \\
Z_R &= \prod_\alpha Z_R^{(\alpha)} = \prod_\alpha \frac{1}{2\sinh(\beta\hbar\omega_\alpha/2)} .
\end{aligned}
\tag{5.15}
$$

The canonical density matrix $W_R^{(\alpha)}(x_{i,\alpha}, x'_{i,\alpha})$ and the amplitude $\mathbb{F}_\alpha[q(\cdot); x_{f,\alpha}, x_{i,\alpha}]$ of oscillator α are Gaussian path sums in imaginary and real time, respectively,

$$
W_R^{(\alpha)}(x_{i,\alpha}, x'_{i,\alpha}) = \frac{1}{Z_R^{(\alpha)}} \int_{\bar{x}_\alpha(0)=x'_{i,\alpha}}^{\bar{x}_\alpha(\hbar\beta)=x_{i,\alpha}} \mathcal{D}\bar{x}_\alpha(\cdot) \exp\left\{ -\frac{m_\alpha}{2\hbar} \int_0^{\hbar\beta} d\tau\, [\dot{\bar{x}}_\alpha^2(\tau) + \omega_\alpha^2 \bar{x}^2(\tau)] \right\} ,
$$

$$
\mathbb{F}_\alpha[q(\cdot); x_{f,\alpha}, x_{i,\alpha}, t] = \int_{x_\alpha(0)=x_{i,\alpha}}^{x_\alpha(t)=x_{f,\alpha}} \mathcal{D}x_\alpha(\cdot)
\tag{5.16}
$$

$$
\times \exp\left\{ \frac{i m_\alpha}{2\hbar} \int_0^t ds\, \left[\dot{x}_\alpha^2(s) - \omega_\alpha^2 \left(x_\alpha(s) - \frac{c_\alpha}{m_\alpha\omega_\alpha^2} q(\cdot) \right)^2 \right] \right\} .
$$

The path sums in Eq. (5.16) can be carried out exactly [208, 88]. Alternatively, one may utilize that the resulting exponents are just the extremal actions connected with the imaginary- and real-time Lagrangians in Eq. (5.16), and the prefactors may be found from the condition $\operatorname{tr}\hat{W}_R^{(\alpha)} = 1$, and from the fixed-point property $\mathbb{F}[0; \mathbf{x}, \mathbf{x}', t] = \int d\mathbf{x}''\, \mathbb{F}[0; \mathbf{x}, \mathbf{x}'', t_1]\mathbb{F}[0; \mathbf{x}'', \mathbf{x}', t - t_1]$. In any case one gets

$$
W_R^{(\alpha)}(x_{i,\alpha}, x'_{i,\alpha}) = \frac{1}{Z_R^{(\alpha)}} \sqrt{\frac{B_{\alpha,\text{th}}}{2\pi\hbar}}
\tag{5.17}
$$

$$
\times \exp\left\{ -[A_{\alpha,\text{th}}(x_{i,\alpha}^2 + x_{i,\alpha}'^2) - B_{\alpha,\text{th}} x_{i,\alpha} x'_{i,\alpha}]/\hbar \right\} ,
$$

$$
\mathbb{F}_\alpha[q(\cdot); x_{f,\alpha}, x_{i,\alpha}, t] = \sqrt{\frac{B_\alpha(t)}{2\pi i\hbar}} e^{i\phi_\alpha[q(\cdot); x_{f,\alpha}, x_{i,\alpha}, t]/\hbar} ,
\tag{5.18}
$$

with the action

$$\phi_\alpha[q(\cdot); x_{f,\alpha}, x_{i,\alpha}, t] = \tag{5.19}$$

$$A_\alpha(t)(x_{i,\alpha}^2 + x_{f,\alpha}^2) - B_\alpha(t)x_{i,\alpha}x_{f,\alpha} - \frac{\mu_\alpha}{2}\int_0^t ds\, q^2(s)$$

$$+ \int_0^t ds\, \left\{ x_{i,\alpha}c_\alpha \frac{\sin[\omega_\alpha(t-s)]}{\sin(\omega_\alpha t)} + x_{f,\alpha}c_\alpha\frac{\sin(\omega_\alpha s)}{\sin(\omega_\alpha t)} \right\} q(s)$$

$$- \frac{c_\alpha^2}{m_\alpha\omega_\alpha}\int_0^t ds_2\, \frac{\sin[\omega_\alpha(t-s_2)]}{\sin(\omega_\alpha t)}q(s_2)\int_0^{s_2}ds_1\sin(\omega_\alpha s_1)\, q(s_1)\,.$$

The auxiliary functions occurring in Eqs. (5.17) – (5.19) are

$$\begin{aligned}
A_\alpha(t) &= (m_\alpha\omega_\alpha/2)\cot(\omega_\alpha t)\,, & B_\alpha(t) &= m_\alpha\omega_\alpha/\sin(\omega_\alpha t)\,, \\
A_{\alpha,\text{th}} &= (m_\alpha\omega_\alpha/2)\coth(\omega_\alpha\hbar\beta)\,, & B_{\alpha,\text{th}} &= m_\alpha\omega_\alpha/\sinh(\omega_\alpha\hbar\beta)\,.
\end{aligned} \tag{5.20}$$

With the expressions (5.15), (5.17) and (5.18) in Eq. (5.13), and subsequent integrations over $\mathbf{x}_{i,\alpha}$, $\mathbf{x}'_{i,\alpha}$ and $\mathbf{x}_{f,\alpha}$ by successive completion of the squares, one readily obtains the Feynman-Vernon influence functional as a time-nonlocal Gaussian form in the paths $q(\cdot)$ and $q'(\cdot)$ with prefactor unity. The functional is conveniently written as

$$\mathcal{F}_{\text{FV}}[q(\cdot), q'(\cdot)] = e^{-\mathcal{S}_{\text{FV}}[q(\cdot), q'(\cdot)]/\hbar}\,, \tag{5.21}$$

with the influence action

$$\mathcal{S}_{\text{FV}}[q(\cdot), q'(\cdot)] = \int_0^t ds_2\int_0^{s_2}ds_1[q(s_2) - q'(s_2)][L(s_2 - s_1)\,q(s_1) - L^*(s_2 - s_1)\,q'(s_1)]$$

$$+ i\frac{\mu}{2}\int_0^t ds\,[q^2(s) - q'^2(s)]\,. \tag{5.22}$$

The time-local action term in Eq. (5.22) stems from the potential counter term $\Delta V(q)$ in Eq. (3.11) and in the interaction $\mathcal{L}_I(t)$ given in Eq. (5.7). The coefficient μ is

$$\mu = \sum_\alpha \mu_\alpha = \sum_\alpha \frac{c_\alpha^2}{m_\alpha\omega_\alpha^2} = \frac{2}{\pi}\int_0^\infty d\omega\,\frac{J(\omega)}{\omega} = M\gamma(0^+)\,, \tag{5.23}$$

where $\gamma(t)$ is the damping kernel (3.33) in the related Langevin equation.

The real-time kernel $L(t)$ is related to the imaginary-time kernel $K(\tau)$, Eq. (4.54), in the thermal influence action (4.51) by analytic continuation, $L(t) = K(z = it)$,

$$\begin{aligned}
L(t) &\equiv L'(t) + iL''(t) = \sum_\alpha \frac{c_\alpha^2}{2m_\alpha\omega_\alpha}\frac{\cosh[\omega_\alpha(\frac{1}{2}\hbar\beta - it)]}{\sinh(\frac{1}{2}\omega_\alpha\hbar\beta)}\,, \\
&= \frac{1}{\pi}\int_0^\infty d\omega\, J(\omega)\,[\coth(\omega\hbar\beta/2)\cos\omega t - i\sin\omega t]\,.
\end{aligned} \tag{5.24}$$

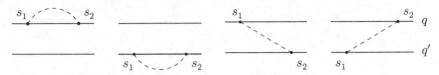

Figure 5.1: Graphical representation of the four contributions to the influence function $\mathcal{S}_{\mathrm{FV}}[q, q']$. The dashed line is the propagator $L(s_2 - s_1)$ in the first and third graph, and $L^*(s_2 - s_1)$ in the second and fourth graph.

Comparison of the expression (2.13) with (5.24), (3.144), and (3.74) shows that $L(t)$ is directly related to the correlation $\mathcal{K}(t) = \langle \hat{\xi}(t)\hat{\xi}(0)\rangle_\beta$ of the random force $\hat{\xi}(t)$,

$$L(t) = \mathcal{K}(t)/\hbar. \tag{5.25}$$

Hence the *bath correlation function* $L(t)$ meets the *fluctuation-dissipation* theorem.

To illustrate the effects of the influence action, we divide the expression (5.22) into the self interaction and the cross interaction terms,

$$\mathcal{S}_{\mathrm{FV}}[q(\cdot), q'(\cdot)] = \mathcal{S}_{\mathrm{FV, self}}[q(\cdot)] + \mathcal{S}^*_{\mathrm{FV, self}}[q'(\cdot)] + \mathcal{S}_{\mathrm{FV, cross}}[q(\cdot), q'(\cdot)],$$

$$\mathcal{S}_{\mathrm{FV, self}}[q(\cdot)] = \int_0^t ds_2 \int_0^{s_2} ds_1\, q(s_2)\, L(s_2 - s_1)\, q(s_1) + i\frac{\mu}{2}\int_0^t ds\, q^2(s), \tag{5.26}$$

$$\mathcal{S}_{\mathrm{FV, cross}}[q(\cdot), q'(\cdot)] = -\int_0^t ds_2 \int_0^{s_2} ds_1\, [q'(s_2)L(s_2-s_1)\, q(s_1) + q(s_2)\, L^*(s_2-s_1)\, q'(s_1)].$$

Overall, $\mathcal{S}_{\mathrm{FV}}[q(\cdot), q'(\cdot)]$ consists of four contributions, as diagrammatically sketched in Fig. 5.1. The first two graphs describe the self-interactions of the paths $q(\cdot)$ and $q'(\cdot)$, respectively, and the other two depict the cross interactions between the paths. The exponentiation of $\mathcal{S}_{\mathrm{FV}}[q(\cdot), q'(\cdot)]$ in $\mathcal{F}_{\mathrm{FV}}[q(\cdot), q'(\cdot)]$ gives all kinds of irreducible and reducible diagrams with any number of exchanged lines made up of the four fundamental contributions.

For a linearly responding bath, all influences of the reservoir exerted on the central system are fully captured by the autocorrelation function $\langle \xi(t)\xi(0)\rangle_\beta$. Hence a conventional molecular dynamics simulation may be used to determine this correlation function in the classical limit, which then defines $J(\omega)$ via the relations (2.7) and (3.27). The quantum mechanical response of the bath is then obtained by calculation of the integral expression (5.24).

With the determination of the Feynman-Vernon influence functional (5.21) for the model (5.6) we have all pieces together: the dynamics of the reduced density matrix is given by the relation (5.11) with (5.12), and the double functional integral in (5.12) is over paths $q(t')$ and $q'(t')$ with the endpoint constraints given in (5.4).

Consider now an ergodic open system, which forgets about its initial state in the course of time, as discussed in Section 3.1.10. The system relaxes to the thermal equilibrium state $\rho_\beta(q_{\mathrm{f}}, q_{\mathrm{f}}')$ independently of the particular initial state $\rho(q_{\mathrm{i}}, q_{\mathrm{i}}'; 0)$ chosen

Thus, upon using the relation (5.11), there holds

$$\rho_\beta(q_{\rm f}, q_{\rm f}') = \lim_{t\to\infty} \rho(q_{\rm f}, q_{\rm f}'; t) = \lim_{t\to\infty} \int dq_{\rm i}\, dq_{\rm i}'\, J_{\rm FV}(q_{\rm f}, q_{\rm f}', t; q_{\rm i}, q_{\rm i}', 0)\, \rho(q_{\rm i}, q_{\rm i}'; 0)\,. \qquad (5.27)$$

This is a remarkably useful relation: instead of calculating the thermal equilibrium state of the damped system with the imaginary-time path integral (4.60), one obtains the equilibrium density matrix $\rho_\beta(q_{\rm f}, q_{\rm f}')$ also from the real-time propagation of the density matrix at asymptotic time.

The dynamical road (5.27) towards the calculation of the equilibrium density matrix has considerable advantage over the imaginary-time method in many specific physical situations. The detailed discussion is deferred until Section 22.9.

5.2.2 Friction and quantum noise

Physical insight is gained by dividing the action into real and imaginary part,

$$\begin{aligned}
S_{\rm FV}[q(\cdot), q'(\cdot)] &= \int_0^t ds_2 \int_0^{s_2} ds_1\, [q(s_2) - q'(s_2)] L'(s_2 - s_1)\, [q(s_1) - q'(s_1)] \\
&\quad + i \int_0^t ds_2 \int_0^{s_2} ds_1\, [q(s_2) - q'(s_2)] L''(s_2 - s_1)[q(s_1) + q'(s_1)] \\
&\quad + i\frac{\mu}{2} \int_0^t ds\, [q^2(s) - q'^2(s)]\,, \qquad (5.28)
\end{aligned}$$

With the relation between $L''(t)$ and the damping kernel (3.33),

$$L''(t) = M\,\dot\gamma(t)/2\,, \qquad\qquad t > 0\,, \qquad (5.29)$$

one may perform partial integration in the second term of Eq. (5.28). Then a resulting boundary term and the potential counter term in Eq. (5.28) cancel each other out.

At this point, it is expedient to pass on to the path combinations

$$r(t) = \tfrac{1}{2}[q(t) + q'(t)]\,, \qquad \text{and} \qquad y(t) = q(t) - q'(t)\,. \qquad (5.30)$$

The path $r(\cdot)$ counts propagation along the diagonal of the RDM, and is therefore termed *quasiclassical* path. The path $y(\cdot)$ is book-keeping the system's visits of off-diagonal states while propagating. This path describes quantum fluctuations. With these, the influence action (5.28) is divided into noise, friction and slip action [88],

$$S_{\rm FV}[r(\cdot), y(\cdot)] = S^{(N)}[y(\cdot)] + i\left(S^{(F)}[r(\cdot), y(\cdot)] + S^{(\rm slip)}[r_{\rm i}, y(\cdot)]\right)\,, \qquad (5.31)$$

where

$$\begin{aligned}
S^{(N)}[y(\cdot)] &= \int_0^t ds_2 \int_0^{s_2} ds_1\, y(s_2)\, L'(s_2 - s_1)\, y(s_1)\,, \\
S^{(F)}[r(\cdot), y(\cdot)] &= M \int_0^t ds_2 \int_0^{s_2} ds_1\, y(s_2)\gamma(s_2 - s_1)\dot r(s_1) \qquad (5.32) \\
S^{(\rm slip)}[r_{\rm i}, y(\cdot)] &= M r_{\rm i} \int_0^t ds\, \gamma(s) y(s)\,.
\end{aligned}$$

Figure 5.2: Graphical representation of the two contributions to $\mathcal{S}_{\text{FV}}[r(\cdot), y(\cdot)]$. The left diagram is the self-interaction of the off-diagonal path $y(\cdot)$ and represents $\mathcal{S}^{(\text{N})}[y(\cdot)]$. The exchanged line is the propagator $L'(s_2 - s_1)$. The right diagram illustrates the correlations between $\dot{r}(\cdot)$ and $y(\cdot)$ in $\mathcal{S}^{(\text{F})}[r(\cdot), y(\cdot)]$. The exchanged line represents the friction kernel $\gamma(s_2 - s_1)$.

Eq. (5.31) with (5.32) is the influence action for the product initial state (5.9). The noise action $\mathcal{S}^{(\text{N})}[y(\cdot)]$ conveys quantum noise, the friction action $\mathcal{S}^{(\text{F})}[r(\cdot), y(\cdot)]$ carries friction, and the slip action $\mathcal{S}^{(\text{slip})}[r_{\text{i}}, y(\cdot)]$ covers the initial slip force in Eq. (3.20).

At this point it is appropriate to indicate that the slip action is missing for the shifted canonical initial state (5.10) of the bath, as will be derived below in Subsec. 5.6 and stated in Eq. (5.123).

Next, the system's action is expressed with the paths $r(\cdot)$ and $y(\cdot)$,

$$\Sigma_{\text{S}}[r(\cdot), y(\cdot)] \equiv S_{\text{S}}[q(\cdot)] - S_{\text{S}}[q'(\cdot)] \tag{5.33}$$

$$= \int_0^t ds \left[M\dot{r}(s)\dot{y}(s) - V[r(s) + \tfrac{1}{2}y(s)] + V[r(s) - \tfrac{1}{2}y(s)] \right].$$

Finally, the propagating function (5.12) in $\{r, y\}$-representation is found to read

$$J_{\text{FV}}(r_{\text{f}}, y_{\text{f}}, t; r_{\text{i}}, y_{\text{i}}, 0) = \int \mathcal{D}r(\cdot)\, \mathcal{D}y(\cdot)\, e^{i\, \Sigma_{\text{tot}}[r(\cdot), y(\cdot)]/\hbar} \tag{5.34}$$

$$\Sigma_{\text{tot}}[r(\cdot), y(\cdot)] = \Sigma_{\text{S}}[r(\cdot), y(\cdot)] - \mathcal{S}^{(\text{F})}[r(\cdot), y(\cdot)] - \mathcal{S}^{(\text{slip})}[r_{\text{i}}, y(\cdot)] + i\,\mathcal{S}^{(\text{N})}[y(\cdot)]$$

Here, the boundaries of the paths $r(\cdot)$ and $y(\cdot)$ are as indicated in the arguments of J_{FV}. The actions $\mathcal{S}^{(\text{N})}[y]$ and $\mathcal{S}^{(\text{F})}[r, y]$ are sketched diagrammatically in Fig. 5.2. Exponentiation of the action takes into account graphs with any number of the two fundamental diagrams stringed together as well as nested in each other.

In summary, the time evolution of the reduced density matrix from a product initial state may be calculated from Eq. (5.11) either with use of the $\{q, q'\}$-representation (5.12) with (5.21) and (5.26) for the propagating function, or with use of the $\{r, y\}$-representation (5.34) with (5.32) and (5.33). Apparently, the double-path sum complicates the nonequilibrium dynamics in comparison with the equilibrium dynamics in imaginary time, Eq. (4.60). In actual calculations, it is usually expedient to choose the $\{r, y\}$-representation, as friction and quantum noise are separated. This we study next, before turning to the dynamics for general initial conditions in Sec. 5.3.1.

The coupling of the system to the reservoir manifests itself in two different ways. On the one hand, the influence action $\mathcal{S}^{(\text{F})}[r, y]$ introduces friction in the system, as we shall see right away. On the other hand, the noise action $\mathcal{S}^{(\text{N})}[y]$ provides a Gaussian stochastic force, which random pumps energy back and forth between system and reservoir, and thus brings loss of coherence into the system.

5.2.3 Damping and decoherence

In the classical limit, the dynamics of the reduced system is determined by the paths for which the action (5.34) under variation of the paths $y(\cdot)$ and $r(\cdot)$ with fixed boundaries is stationary. The resulting equations of motion for $0 \leqslant s \leqslant t$ are

$$M\ddot{r}(s) + M\int_0^s ds_1\,\gamma(s-s_1)\dot{r}(s_1) + M\gamma(s)r_{\mathrm{i}}$$
$$+ \frac{\partial}{\partial y}[\,V(r+\tfrac{1}{2}y) - V(r-\tfrac{1}{2}y)\,] - i\int_0^t ds_1\,L'(s-s_1)y(s_1) = 0\,, \tag{5.35}$$

$$M\ddot{y}(s) - M\int_s^t ds_1\,\ddot{y}(s_1)\gamma(s_1-s) + \frac{\partial}{\partial r}[\,V(r+\tfrac{1}{2}y) - V(r-\tfrac{1}{2}y)\,] = 0\,. \tag{5.36}$$

Assuming that the system is initially and finally in a diagonal state,

$$y(0) = y(t) = 0\,, \tag{5.37}$$

the only solution of Eq. (5.36) is the trivial path $y(s) = 0$ for $0 \leqslant s \leqslant t$. This expresses that the system propagates along diagonal states of the RDM. With the omission of fluctuations, the resulting equation of motion is that of a classical particle subjected to memory-friction and to the force $-dV(r)/dr - M\gamma(s)r_{\mathrm{i}}$,

$$M\ddot{r}(s) + M\int_0^s ds_1\,\gamma(s-s_1)\dot{r}(s_1) + V'[r(s)] + M\gamma(s)r_{\mathrm{i}} = 0\,. \tag{5.38}$$

Pictorially, friction is brought along by the right diagram in Fig. 5.2.

Quantum interference is a phenomenon related to the fact that, in general, transition amplitudes gathered up from different paths the system takes have different phases. The interference term of two different path contributions is characterized by a phase factor $e^{\pm i\varphi}$. In the presence of bath coupling, the phase φ is fluctuating. Then the relevant object is the statistical average of $e^{\pm i\varphi}$, which is the "decoherence" factor $\langle e^{\pm i\varphi}\rangle$ [210, 211]. The extinction of quantum coherence for the pair of paths $q(t)$ and $q'(t)$ depends on the distance $y(t)$, and is determined by the noise action $\mathcal{S}^{(\mathrm{N})}[y(\cdot)]$,

$$\left\langle e^{\pm i\,\varphi[y(\cdot)]}\right\rangle = e^{-\mathcal{S}^{(\mathrm{N})}[y(\cdot)]/\hbar}\,. \tag{5.39}$$

The noise functional acts as a Gaussian filter for detours to off-diagonal states, and thus quenches quantum fluctuations. Physically, the environment is continuously measuring the position of the system. It therefore weakens quantum interference between different position eigenstates and pushes the system towards classical behavior. Diagrammatically, $\mathcal{S}^{(\mathrm{N})}[y(\cdot)]$ is represented by the left diagram in Fig. 5.2.

To elucidate in some more detail the way how $\mathcal{S}^{(\mathrm{N})}[y(\cdot)]$ works, consider the particular case $L'(t) = (2\eta/\hbar\beta)\,\delta(t)$ holding for Ohmic coupling, $J(\omega) = \eta\omega$ and high temperature, as follows from the expression (5.24). In this white-noise regime, the interference of two localized states with spatial separation q_0 decoheres with time as

$$\langle e^{-i\varphi} \rangle = e^{-\gamma_{\mathrm{dec}} t} , \qquad \text{where} \qquad \gamma_{\mathrm{dec}} = \eta q_0^2 k_{\mathrm{B}} T / \hbar^2 . \qquad (5.40)$$

The rate γ_{dec} is the inverse time scale for decoherence or dephasing. The issue of pure dephasing is taken up again in Section 22.4.

It is interesting to compare the decoherence rate γ_{dec} with the Ohmic damping rate

$$\gamma_{\mathrm{damp}} = \eta / M . \qquad (5.41)$$

The ratio of the two rates is

$$\gamma_{\mathrm{dec}} / \gamma_{\mathrm{damp}} = q_0^2 M k_{\mathrm{B}} T / \hbar^2 = q_0^2 / \lambda_{\mathrm{th}}^2 , \qquad (5.42)$$

where $\lambda_{\mathrm{th}} = \hbar / p_{\mathrm{th}} = \hbar / \sqrt{M k_{\mathrm{B}} T}$ is a thermal wave length. The ratio (5.42) relates the range of spatial coherence with the thermal wave length of the particle. For a typical macroscopic situation, $M = 1\,\mathrm{g}$, $T = 300\,\mathrm{K}$, $q_0 = 1\,\mathrm{mm}$, the ratio $\gamma_{\mathrm{dec}} / \gamma_{\mathrm{damp}}$ is 10^{38}. Thus, for practically all macroscopic systems, friction is completely negligible on the ultra-short time scale where quantum coherence is significant. The ratio between the inverse time scales for decoherence and friction, Eq. (5.42), is generally valid for all systems which are in contact with an Ohmic heat bath with thermal energy large compared to the relevant internal energy scales of the system.[1] Suppression of quantum coherence for interfering paths is discussed in Chapter 9.

5.2.4 Quasiclassical Langevin equation

We now go beyond the deterministic description by considering the Gaussian fluctuations $y(s)$ about the minimal action path. We are interested in the probability distribution $w(r_{\mathrm{f}}; t) = \rho(r_{\mathrm{f}}, y_{\mathrm{f}} = 0, t)$ evolving from the initial distribution $w(r_{\mathrm{i}}; 0)$. We choose for the evolution of $w(r_{\mathrm{f}}; t)$ in correspondence with Eq. (5.34) the ansatz

$$w(r_{\mathrm{f}}; t) = \int dr_{\mathrm{i}} \, w(r_{\mathrm{i}}; 0) \int \mathcal{D}r(\cdot) \, \mathcal{D}y(\cdot) \, e^{\, i \, \Sigma^{(\mathrm{sc})} [r(\cdot), y(\cdot)] / \hbar} , \qquad (5.43)$$

in which the yet to be determined action $\Sigma^{(\mathrm{sc})}$ is a quadratic form in $y(\cdot)$. The functional integral extends over all paths $y(t')$ and $r(t')$ which obey $y(0) = y(t) = 0$, $r(0) = r_{\mathrm{i}}$ and $r(t) = r_{\mathrm{f}}$. We proceed with the observation that the noise action factor $p[y(\cdot)] = e^{-S^{(\mathrm{N})}[y(\cdot)]}$ in Eq. (5.34) is a Gaussian filter function which suppresses large quantum fluctuations $y(\cdot)$ in the semiclassical limit. Hence it is adequate to expand the potential term in the system action about $y = 0$ and to keep the first order only,

$$-V(r + \tfrac{1}{2}y) + V(r - \tfrac{1}{2}y) = -V'(r)\, y + \mathcal{O}(y^3) . \qquad (5.44)$$

With this reduction in the system action $\Sigma_{\mathrm{S}}[r(\cdot), y(\cdot)]$, and with partial integration of the kinetic term, and with disregard of the initial slip contribution in $S^{(\mathrm{F})}[r(\cdot), y(\cdot)]$, the action $\Sigma_{\mathrm{tot}}[r(\cdot), y(\cdot)]$, given in Eq. (5.34), reduces to the semiclassical action

---

[1] See Sections 21.3 and 11.4 for the discussion of decoherence in the dissipative two-state system.

$$\Sigma^{(sc)}[r(\cdot), y(\cdot)] \;=\; -\int_0^t ds_2\, y(s_2) \Big(M\ddot{r}(s_2) + Mr_i\gamma(s_2) + V'[r(s_2)]$$

$$+ M\int_0^{s_2} ds_1\, \gamma(s_2 - s_1)\,\dot{r}(s_1) \Big)$$

$$+ \frac{i}{2}\int_0^t ds_2 \int_0^t ds_1\, y(s_2)L'(s_2 - s_1)y(s_1)\,. \qquad (5.45)$$

The path probability in the interval from $r(t)$ to $r(t) + \mathcal{D}r(t)$,

$$W[r(\cdot)]\,\mathcal{D}r(\cdot) \;\equiv\; \int \mathcal{D}y(\cdot)\, e^{i\,\Sigma^{(sc)}[r,y(\cdot)]/\hbar}\, \mathcal{D}r(\cdot) \qquad (5.46)$$

is a Gaussian functional integral in the fluctuation $y(\cdot)$, which can be evaluated by completion of the square. Omitting an irrelevant normalization constant, we obtain

$$W[r(\cdot)]\,\mathcal{D}r(\cdot) = \exp\left(-\frac{1}{2\hbar}\int_0^t ds_2 \int_0^t ds_1\, \xi[r(s_2)]\, L'^{-1}(s_2 - s_1)\, \xi[r(s_1)] \right) \mathcal{D}r(\cdot)\,. \quad (5.47)$$

Here, $L'^{-1}(s_2-s_1)$ is the inverse of $L'(s_2-s_1)$ and the functional $\xi[r(s)]$ is defined by

$$\xi[r(s)] \;\equiv\; M\ddot{r}(s) + M\int_0^s ds_1\, \gamma(s - s_1)\,\dot{r}(s_1) + Mr_i\gamma(s) + V'[r(s)]\,. \qquad (5.48)$$

The Gaussian weight function (5.47) suggests to introduce a measure of path integration $\mathcal{D}\xi(\cdot)$ associated with the $\xi(\cdot)$ fluctuations. The functional Jacobian of the transformation $\mathcal{D}r(\cdot) = \mathcal{J}[r(\cdot)]^{-1}\mathcal{D}\xi(\cdot)$ is given by

$$\mathcal{J}[r(\cdot)] \;=\; \det\left[\left(M\frac{\partial^2}{\partial t^2} + V''(r(t)) \right) \delta(t - s) + M\int_0^t ds\, \gamma(t - s)\frac{\partial}{\partial s} \right]\,. \qquad (5.49)$$

Schmid [55] has argued by discretization of the differential operator on the sliced time axis that the Jacobian is a constant independent of the choice of the potential $V(r)$.[2] Thus, we may consider $\xi(t)$ as an independent stochastic variable. As $\xi(t)$ in fact does not depend on the path $r(t)$, we may regard Eq. (5.48) as a Langevin equation with memory-friction, and with systematic, initial slip, and colored-noise forces,

$$M\ddot{r}(t) + M\int_0^t ds\, \gamma(t - s)\dot{r}(s) + V'[r(t)] + Mr_i\gamma(t) \;=\; \xi(t)\,. \qquad (5.50)$$

The fluctuating force $\xi(t)$ undergoes the Gaussian stochastic process

$$W[\xi(\cdot)]\,\mathcal{D}\xi(\cdot) \;=\; \exp\left(-\frac{1}{2\hbar}\int_0^t ds_2 \int_0^t ds_1\, \xi(s_2)L'^{-1}(s_2 - s_1)\xi(s_1) \right) \mathcal{D}\xi(\cdot)\,. \quad (5.51)$$

Average with this Gaussian weight function yields the statistical properties

---

[2]The criticism stated in Ref. [6], p.1300, on the slicing in Ref. [55] does not spoil our reasoning.

$$\langle \xi(t) \rangle \;=\; 0 \,, \tag{5.52}$$

$$\mathcal{K}_{\mathrm{qucl}}(t) \;\equiv\; \mathrm{Re}\langle \xi(t)\xi(0)\rangle_{\mathrm{qucl}} = \hbar L'(t) = \frac{\hbar}{\pi}\int_0^\infty d\omega\, J(\omega)\coth(\beta\hbar\omega/2)\cos(\omega t) \,.$$

The power spectrum of the correlation function $\mathcal{K}_{\mathrm{qucl}}(t)$ is found to read

$$\tilde{\mathcal{K}}_{\mathrm{qucl}}(\omega) \;\equiv\; \int_{-\infty}^{\infty} dt\, \mathcal{K}_{\mathrm{qucl}}(t)\,\cos(\omega t) \;=\; M\hbar\omega\coth(\beta\hbar\omega/2)\,\tilde{\gamma}'(\omega)\,, \tag{5.53}$$

where we have used the relation (3.27). The noise characteristics of the Langevin force derived here agrees with that anticipated in Subsec. 2.1.2, and consolidated in Subsec. 3.1.4. The relation (5.53) is the quantum mechanical version of Kubo's "Second Fluctuation-Dissipation Theorem". As $\hbar \to 0$, it becomes equivalent to the classical FDT of the second kind, Eq. (2.9).

In conclusion, in the semiclassical limit, the Brownian particle can be described by a Langevin equation in which the quantum mechanical stochastic nature of the Langevin force is given by the expressions (5.52). From the above derivation it is clear that the quasiclassical Langevin equation is exact for a Brownian particle with Gaussian statistics when the external force is linear in the coordinate. The results of the quasiclassical Langevin equation are also reliable when the anharmonic part of the potential can be treated as a perturbation. On this level of approximation, squeezing of thermal and quantum fluctuations by time-dependent external forces has been studied in Ref. [225]. For a discussion of the validity of the quasiclassical Langevin equation for anharmonic potentials we refer to Refs. [55] and [56], and to the remarks given in Subsection 2.1.2.

General quantum kinetic equations were derived in Ref. [226]. For weak dissipation, they can be reduced, upon employing a generalized Born approximation, to relaxation equations for the diagonal and off-diagonal components of the reduced density matrix, as familiar from nuclear magnetic relaxation theory [32]. In the classical limit, one is led to the energy-diffusion version of the Fokker-Planck equation studied below in Section 11.4. On the other hand, when coarse grained, the kinetic equation gives the full Fokker-Planck equation capturing the relevant physics in the classical limit.

The quasiclassical Langevin equation is easily generalized to state-dependent damping [152, b]. Particular cases are described by the action given in Eq. (4.71) or Eq. (4.191), and the relevant quasiclassical Langevin equation is of the form (3.20). Important examples in mesoscopic physics are quasiparticle tunneling between superconductors and strong electron tunneling through small junctions or metallic grains. When the fluctuations of the electronic tunneling current are strong, fluctuations of the conjugate phase variable φ are weak, and therefore the phase dynamics can well be described in terms of a Langevin equation for the phase variable φ with a state-dependent stochastic force $\xi_1(t)\cos\varphi + \xi_2(t)\sin\varphi$, where $\xi_1(t)$ and $\xi_2(t)$ are two independent stochastic variables obeying Gaussian statistics with a correlator of the form (5.52) with Eq. (5.53). For a recent application of a quasiclassical Langevin equation with state-dependent noise to "strong electron tunneling", see Ref. [227].

5.3 Influence functional method for general initial states

5.3.1 General initial states and preparation function

Often in experiments, the "small" system is prepared, say at $t = 0$, by first letting it equilibrate with the bath and then measuring certain dynamical variables of it. The measurement leads to a reduction of the equilibrium density matrix of the system-plus-reservoir complex. Initial states of this type are generated by a linear transformation of the equilibrium density matrix of the global system. We write

$$< q_i, \mathbf{x}_i | \hat{W}(0) | q_i', \mathbf{x}_i' > = \int d\bar{q} \, d\bar{q}' \, \lambda(q_i, q_i'; \bar{q}, \bar{q}') < \bar{q}, \mathbf{x}_i | \hat{W}_\beta | \bar{q}', \mathbf{x}_i' > , \qquad (5.54)$$

where $\hat{W}_\beta = e^{-\beta \hat{H}}/Z$. The linking function $\lambda(q, q'; \bar{q}, \bar{q}')$ is a *preparation function* which depends on the specific measurement apparatus at $t = 0$ [88].

Initial states of the form (5.54) are not in the product form (5.9). Thus we cannot apply the procedure described in Section 5.2. Before we give the corresponding generalization for correlated initial states, we briefly discuss specific initial preparations which may be experimentally relevant.

Let the effect of the measuring device be described by operators $\hat{O}_j, \hat{O}_j{}'$ which act on the system only and leave the reservoir unaffected. Then the preparation function is in the form

$$\lambda(q_i, q_i'; \bar{q}, \bar{q}') = \sum_j < q_i | \hat{O}_j | \bar{q} > < \bar{q}' | \hat{O}_j' | q_i' > . \qquad (5.55)$$

Consider briefly three important cases. First, if we are interested in the response of a system in thermal equilibrium to an external force switched on at time $t = 0$, we have $\hat{W}(0) = \hat{W}_\beta$. Then the preparation function $\lambda(q_i, q_i'; \bar{q}, \bar{q}')$ takes the simple form

$$\lambda_\beta(q_i, q_i'; \bar{q}, \bar{q}') = \delta(q_i - \bar{q}) \, \delta(q_i' - \bar{q}') . \qquad (5.56)$$

The second important example is that we perform initially a measurement of a dynamical variable of the system. In accordance with the principles of quantum measurement theory (see, for instance, Refs. [212, 58, 213]), the effect of an "observation" is to project the density matrix that described the system just prior to observation onto the manifold corresponding to the measured interval of eigenvalues. For an ideal measurement at time $t = 0^-$, the initial density matrix is then given by

$$\hat{W}(0) = \hat{P} \hat{W}_\beta \hat{P} , \qquad (5.57)$$

where $\hat{P}$ is the projection operator corresponding to the measurement. For example, the measurement of the position of a particle with the outcome q_0 with uncertainty Δq is described by the projection operator

$$\hat{P}_q = \int_{q_0 - \Delta q/2}^{q_0 + \Delta q/2} dq \, | q > < q | . \qquad (5.58)$$

133

Thirdly, the structure factor measured in scattering experiments is closely related to the Fourier transform of the position autocorrelation function of the scatterer in thermal equilibrium [214]. Now, an equilibrium correlation function, say $\langle A(t)\, B(0)\rangle$, may be formally looked upon as the expectation value of $\hat{A}$ at time t for a system starting out from the initial state $\hat{W}(0) = \hat{B}\hat{W}_\beta$. In this case, the specific form of the preparation function follows directly from Eq. (5.54). For the thermal coordinate autocorrelation function $C_{qq}(t) = \text{tr}\,\{q(t)q(0)W_\beta\}$ the preparation function reads

$$\lambda_q(q_{\mathrm{i}}, q_{\mathrm{i}}'; \bar{q}, \bar{q}') \;=\; q_{\mathrm{i}}\, \lambda_\beta(q_{\mathrm{i}}, q_{\mathrm{i}}'; \bar{q}, \bar{q}')\,, \tag{5.59}$$

while for the coordinate-momentum correlation function in thermal equilibrium $C_{qp}(t) = \text{tr}\,\{\hat{q}(t)\,\hat{p}(0)\hat{W}_\beta\}$ the preparation function is

$$\lambda_p(q_{\mathrm{i}}, q_{\mathrm{i}}'; \bar{q}, \bar{q}') \;=\; \frac{\hbar}{i}\frac{\partial}{\partial q_{\mathrm{i}}}\lambda_\beta(q_{\mathrm{i}}, q_{\mathrm{i}}'; \bar{q}, \bar{q}')\,. \tag{5.60}$$

The two cases correspond to $\hat{W}(0) = \hat{q}\,\hat{W}_\beta$ and $\hat{W}(0) = \hat{p}\,\hat{W}_\beta$, respectively. These expressions are not proper density matrices since they are not positive definite. However, they belong to the class of initial conditions specified by the relation (5.55).

5.3.2 Complex-time path integral for propagating function

For the general class of initial states discussed in the preceding section, the dynamics is characterized by a *preparation function* $\lambda(q_{\mathrm{i}}, q_{\mathrm{i}}'; \bar{q}, \bar{q}')$ and by a *propagating function* $J(q_{\mathrm{f}}, q_{\mathrm{f}}', t; q_{\mathrm{i}}, q_{\mathrm{i}}'; \bar{q}, \bar{q}')$. The propagation function describes the time evolution of the reduced system starting from an initial state procured by the preparation function. The time evolution of the prepared initial state is described by the relation

$$\rho(q_{\mathrm{f}}, q_{\mathrm{f}}'; t) \;=\; \int dq_{\mathrm{i}}\, dq_{\mathrm{i}}'\, d\bar{q}\, d\bar{q}'\; J(q_{\mathrm{f}}, q_{\mathrm{f}}', t; q_{\mathrm{i}}, q_{\mathrm{i}}'; \bar{q}, \bar{q}')\, \lambda(q_{\mathrm{i}}, q_{\mathrm{i}}'; \bar{q}, \bar{q}')\,. \tag{5.61}$$

Insertion of (5.54) into (5.8) and comparison with (5.61) yields for J the expression

$$J(q_{\mathrm{f}}, q_{\mathrm{f}}'; t; q_{\mathrm{i}}, q_{\mathrm{i}}'; \bar{q}, \bar{q}') \;=\; \int dx_{\mathrm{i}}\, dx_{\mathrm{i}}'\, dx_{\mathrm{f}}\; K(q_{\mathrm{f}}, x_{\mathrm{f}}, t; q_{\mathrm{i}}, x_{\mathrm{i}}, 0)$$
$$\times\; <\bar{q}, x_{\mathrm{i}}|W_\beta|\bar{q}', x_{\mathrm{i}}'>\, K^*(q_{\mathrm{f}}', x_{\mathrm{f}}, t; q_{\mathrm{i}}', x_{\mathrm{i}}', 0)\,. \tag{5.62}$$

With use of the functional integral expressions (4.26) for the thermal density matrix and (5.3) for the propagators we may write the propagating function (5.62) as

$$J(q_{\mathrm{f}}, q_{\mathrm{f}}'; t; q_{\mathrm{i}}, q_{\mathrm{i}}'; \bar{q}, \bar{q}') \;=\; Z^{-1}\int \mathcal{D}q(\cdot)\, \mathcal{D}q'(\cdot)\, \mathcal{D}\bar{q}(\cdot)\, \mathcal{F}_{\mathrm{ct}}[q(\cdot), q'(\cdot), \bar{q}(\cdot)]$$
$$\times\; e^{i\,\{S_{\mathrm{S}}[q(\cdot)] - S_{\mathrm{S}}[q'(\cdot)]\}/\hbar}\; e^{-S_{\mathrm{S}}^{(\mathrm{E})}[\bar{q}(\cdot)]/\hbar}\,, \tag{5.63}$$

where Z is the reduced partition function. The real-time paths $q(t')$ and $q'(t')$ and the imaginary-time path $\bar{q}(\tau)$ have the boundary points

$$q(0) = q_{\mathrm{i}}, \quad q'(0) = q'_{\mathrm{i}}, \quad \bar{q}(0) = \bar{q}',$$
$$q(t) = q_{\mathrm{f}}, \quad q'(t) = q'_{\mathrm{f}}, \quad \bar{q}(\hbar\beta) = \bar{q}. \tag{5.64}$$

The complex-time (ct) influence functional is formally given by

$$\mathcal{F}_{\mathrm{ct}}[q(\cdot), q'(\cdot), \bar{q}(\cdot)] = \int d\mathbf{x}_{\mathrm{f}}\, d\mathbf{x}_{\mathrm{i}}\, d\mathbf{x}'_{\mathrm{i}}\, \frac{1}{Z_{\mathrm{R}}} \int \mathcal{D}\mathbf{x}(\cdot)\, \mathcal{D}\mathbf{x}'(\cdot)\, \mathcal{D}\bar{\mathbf{x}}(\cdot) \tag{5.65}$$
$$\times\ e^{-\{S_{\mathrm{R}}^{(\mathrm{E})}[\bar{\mathbf{x}}] + S_{\mathrm{I}}^{(\mathrm{E})}[\bar{\mathbf{x}}, \bar{q}(\cdot)]\}/\hbar}\ e^{i\{S_{\mathrm{R}}[\mathbf{x}] + S_{\mathrm{I}}[\mathbf{x}, q(\cdot)] - S_{\mathrm{R}}[\mathbf{x}'] - S_{\mathrm{I}}[\mathbf{x}', q'(\cdot)]\}/\hbar}\ .$$

The functional integrations are evaluated for paths $\mathbf{x}(s)$, $\mathbf{x}'(s')$, $\bar{\mathbf{x}}(\tau)$ which form a closed path on a contour in the complex-time plane with the endpoint conditions

$$\mathbf{x}(t) = \mathbf{x}'(t) = \mathbf{x}_{\mathrm{f}}, \quad \mathbf{x}(0) = \bar{\mathbf{x}}(\hbar\beta) = \mathbf{x}_{\mathrm{i}}, \quad \bar{\mathbf{x}}(0) = \mathbf{x}'(0) = \mathbf{x}'_{\mathrm{i}}. \tag{5.66}$$

Next, we proceed as in Section 5.2 by writing the expression (5.65) in the form

$$\mathcal{F}_{\mathrm{ct}}[q(\cdot), q'(\cdot), \bar{q}(\cdot)] = \prod_{\alpha=1}^{N} \frac{1}{Z_R^{(\alpha)}} \int dx_{\mathrm{f},\alpha}\, dx_{\mathrm{i},\alpha}\, dx'_{\mathrm{i},\alpha} \tag{5.67}$$
$$\times\ \mathbb{F}_{\alpha}[q(\cdot), x_{\mathrm{f},\alpha}, x_{\mathrm{i},\alpha}]\, \mathbb{F}_{\alpha}^{*}[q'(\cdot), x_{\mathrm{f},\alpha}, x'_{\mathrm{i},\alpha}]\, \mathbb{F}_{\alpha}^{(\mathrm{E})}[\bar{q}(\cdot), x_{\mathrm{i},\alpha}, x'_{\mathrm{i},\alpha}]\ .$$

Here the amplitude $\mathbb{F}_{\alpha}$ is the path sum expression given in Eq. (5.16), and $\mathbb{F}_{\alpha}^{(\mathrm{E})}$ is the respective path sum in the imaginary-time interval $0 \leqslant \tau < \hbar\beta$. The resulting analytic expression for $\mathbb{F}_{\alpha}$ is put down in Eq. (5.18). The analytic expression for $\mathbb{F}_{\alpha}^{(\mathrm{E})}$ is found from that for $\mathbb{F}_{\alpha}$ by the substitution $t \to -i\,\hbar\beta$.

With these forms, the integrand of the expression (5.67) is a Gaussian in the variables $x_{\mathrm{f},\alpha}, x_{\mathrm{i},\alpha}$ and $x'_{\mathrm{i},\alpha}$. The evaluation of the Gaussian integrals is straightforward by successively completing the square. In the end one finds [88]

$$\mathcal{F}_{\mathrm{ct}}[q(\cdot), q'(\cdot), \bar{q}(\cdot)] = e^{-\mathcal{S}_{\mathrm{ct}}[q(\cdot), q'(\cdot), \bar{q}(\cdot)]/\hbar}, \tag{5.68}$$

where

$$\mathcal{S}_{\mathrm{ct}}[q(\cdot), q'(\cdot), \bar{q}(\cdot)] = \mathcal{S}_{\mathrm{FV}}[q(\cdot), q'(\cdot)] + \frac{\mu}{2} \int_{0}^{\hbar\beta} d\tau\, \bar{q}^{2}(\tau)$$
$$- \int_{0}^{\hbar\beta} d\tau \int_{0}^{\tau} d\tau'\, K(\tau - \tau')\, \bar{q}(\tau)\, \bar{q}(\tau') \tag{5.69}$$
$$- i \int_{0}^{\hbar\beta} d\tau \int_{0}^{t} ds\, L^{*}(s - i\tau)\, \bar{q}(\tau)\, [q(s) - q'(s)]\ .$$

Here, $\mathcal{S}_{\mathrm{FV}}[q(\cdot), q'(\cdot)]$ is the Feynman-Vernion action (5.22), and $K(\tau)$ and $L(t)$ are defined in Eqs. (4.54) and (5.24), respectively.

The triple path integral (5.63) with Eqs. (5.68) and (5.69) is the exact formal solution for the propagating function $J(q_{\mathrm{f}}, q'_{\mathrm{f}}; t; q_{\mathrm{i}}, q'_{\mathrm{i}}; \bar{q}, \bar{q}')$.

The results obtained in this section shed light on the evolution of a general initial state. The dynamics is specified by a preparation function $\lambda(q_i, q_i'; \bar{q}, \bar{q}')$ and by a propagating function $J(q_f, q_f', t; q_i, q_i'; \bar{q}, \bar{q}')$. The preparation function fixes the particular initial state $\rho_0 = \rho(t = 0)$ within the preparation class through the relation

$$\rho_0(q_i, q_i') = \int d\bar{q}\, d\bar{q}'\, \lambda(q_i, q_i'; \bar{q}, \bar{q})\, \rho_\beta(\bar{q}, \bar{q}')\,, \tag{5.70}$$

where $\rho_\beta(\bar{q}, \bar{q}')$ is the thermal reduced density matrix given in Eq. (4.60). The relation (5.70) is obtained by tracing out the reservoir coordinates in Eq. (5.54). It also follows from Eq. (5.61) in the limit $t \to 0$ with the initial condition [cf. Eq. (5.62)]

$$\lim_{t \to 0} J(q_f, q_f', t; q_i, q_i'; \bar{q}, \bar{q}') = \delta(q_f - q_i)\, \delta(q_f' - q_i')\rho_\beta(\bar{q}, \bar{q}')\,. \tag{5.71}$$

The preparation function $\lambda(q_i, q_i'; \bar{q}, \bar{q}')$ represents the quantum analogue of the initial classical phase space distribution. Note that it is not uniquely specified through the relation (5.70). It is rather determined by the initial density matrix of the total system through relation (5.54). In quantum mechanics, the propagating function $J(q_f, q_f', t; q_i, q_i'; \bar{q}, \bar{q}')$ has taken over the role of the classical conditional probability. From the considerations above it is clear that the propagating function is determined by the equilibrium properties of the reduced system. This is in correspondence with the classical stationary conditional probability.

5.3.3 Real-time path integral for propagating function

In the formal solution (5.63) with (5.68) and (5.69) of the propagating function, one faces difficulties which arise from the correlations of the forward and backward real-time paths $q(s)$ and $q'(s)$ with the imaginary-time path $\bar{q}(\tau)$. The complications caused by this coupling can be simplified to some extent in actual calculations for *ergodic* open systems, as we explain in the sequel.

We start with the observation that the propagating function (5.62) has the initial condition (5.71). With use of the relation (5.61) we then obtain

$$\lim_{t \to 0} \rho(q_f, q_f'; t) = \int d\bar{q}\, d\bar{q}'\, \lambda(q_f, q_f'; \bar{q}, \bar{q}')\, \rho_\beta(\bar{q}, \bar{q}')\,. \tag{5.72}$$

Suppose now that the system has been prepared at some large negative time $-|t_p|$ in a particular initial state and recall that the ergodic open system relaxes to thermal equilibrium independently of the chosen initial preparation (cf. Section 3.1.10 and the discussion given at the end of Subsection 5.2.1). Then, the system will have equilibrated with the reservoir at time $t = 0$, if it had started out in the infinite past from some non-equilibrium initial state. Therefore, we may choose in the infinite past without loss of generality a product initial state of the form

$$\lim_{t_p \to -\infty} \hat{W}(t_p) = \hat{\rho}_p \otimes e^{-\beta \hat{H}_R}/Z_R\,. \tag{5.73}$$

We then have at time $t = 0^-$ in correspondence with the relation (5.27)

$$\rho(\bar{q}, \bar{q}'; 0^-) = \rho_\beta(\bar{q}, \bar{q}') = \lim_{t_p \to -\infty} \int dq_p \, dq'_p \, J_{\text{FV}}(\bar{q}, \bar{q}', 0^-; q_p, q'_p; t_p) \, \rho_p(q_p, q'_p), \quad (5.74)$$

where the propagating function J_{FV} is given by Eq. (5.12) with (5.22) and (5.4).

After having determined the correlated equilibrium state of the damped particle at time $t = 0^-$, it is straightforward to describe the evolution of the RDM. We have

$$\rho(q_f, q'_f; t) = \int dq_i \, dq'_i \, d\bar{q} \, d\bar{q}' \, \lambda(q_i, q'_i; \bar{q}, \bar{q}') \int dq_p \, dq'_p \, \rho_p(q_p, q'_p)$$
$$\times \lim_{t_p \to -\infty} J(q_f, q'_f, t; q_i, q'_i, 0^+; \bar{q}, \bar{q}', 0^-; q_p, q'_p, t_p). \quad (5.75)$$

The propagating function J is given by the real-time double path integral

$$J(q_f, q'_f, t; q_i, q'_i, 0^+; \bar{q}, \bar{q}', 0^-; q_p, q'_p, t_p) = \int \mathcal{D}q \, \mathcal{D}q' \, e^{i(S_{\text{S}}[q] - S_{\text{S}}[q'])/\hbar} \, e^{-S_{\text{rt}}[q, q']/\hbar} \quad (5.76)$$

with the real-time (rt) influence action

$$S_{\text{rt}}[q(\cdot), q'(\cdot)] = \int_{t_p}^t ds_2 \int_{t_p}^{s_2} ds_1 \, [q(s_2) - q'(s_2)] \, [L(s_2 - s_1) \, q(s_1) - L^*(s_2 - s_1) \, q'(s_1)]$$
$$+ i\frac{\mu}{2} \int_{t_p}^t ds \, [q^2(s) - q'^2(s)]. \quad (5.77)$$

The path sum in Eq. (5.76) is over paths $q(\cdot)$ and $q'(\cdot)$ with the boundary points

$$\begin{aligned}
q(t_p) &= q_p, & q(0^-) &= \bar{q}, & q(0^+) &= q_i, & q(t) &= q_f, \\
q'(t_p) &= q'_p, & q'(0^-) &= \bar{q}', & q'(0^+) &= q'_i, & q'(t) &= q'_f.
\end{aligned} \quad (5.78)$$

Observe that the paths $q(\cdot)$ and $q'(\cdot)$ are discontinuous at time $t' = 0$.

It is again expedient to pass on to the $\{r, y\}$-representation

$$\rho(r_f, y_f; t) = \int dr_i \, dy_i \, d\bar{r} \, d\bar{y} \, \lambda(r_i, y_i; \bar{r}, \bar{y}) \int dr_p \, dy_p \, \rho_p(r_p, y_p)$$
$$\times \lim_{t_p \to -\infty} J(r_f, y_f, t; r_i, y_i, 0^+; \bar{r}, \bar{y}, 0^-; r_p, y_p, t_p), \quad (5.79)$$

with the propagating function

$$J(r_f, y_f, t; r_i, y_i, 0^+; \bar{r}, \bar{y}, 0^-; r_p, y_p, t_p) = \int \mathcal{D}r(\cdot) \, \mathcal{D}y(\cdot) \, e^{i \Sigma_{\text{tot}}[r(\cdot), y(\cdot)]/\hbar}, \quad (5.80)$$

$$\Sigma_{\text{tot}}[r(\cdot), y(\cdot)] = \Sigma_{\text{S}}[r(\cdot), y(\cdot)] - S^{(\text{F})}[r(\cdot), y(\cdot)] - S^{(\text{slip})}[r_i - \bar{r}, y(\cdot)] + i \, S^{(\text{N})}[y(\cdot)].$$

137

The actions $\Sigma_{\rm S}$, $\mathcal{S}^{({\rm N})}$ and $\mathcal{S}^{({\rm F})}$ are given in Eqs. (5.33) and (5.32), but now with time integrations starting at $t_{\rm p} < 0$. The boundary points of the paths $r(\cdot)$ and $y(\cdot)$ are

$$
\begin{aligned}
r(t_{\rm p}) &= r_{\rm p}\,, & r(0^-) &= \bar{r}\,, & r(0^+) &= r_{\rm i}\,, & r(t) &= r_{\rm f}\,, \\
y(t_{\rm p}) &= y_{\rm p}\,, & y(0^-) &= \bar{y}\,, & y(0^+) &= y_{\rm i}\,, & y(t) &= y_{\rm f}\,.
\end{aligned}
\tag{5.81}
$$

The slip action $\mathcal{S}^{({\rm slip})}[r_{\rm i}-\bar{r}, y(\cdot)]$ is a boundary term resulting from partial integration. It is given in Eq. (5.32), but now with $r_{\rm i} - \bar{r}$ in place of $r_{\rm i}$.

In the limit $t_{\rm p} \to -\infty$, the propagating function (5.80) becomes independent of the variable $r_{\rm p}$. In this limit, in addition, its dependence on the variable $y_{\rm p}$ develops into a δ–distribution centered at $y_{\rm p} = 0$. We then have

$$
\int dr_{\rm p}\, \rho_{\rm p}(r_{\rm p}, y_{\rm p} = 0) \;=\; {\rm tr}\,\rho_{\rm p} \;=\; 1\,.
\tag{5.82}
$$

Thus the memory of the reduced density matrix $\rho(r_{\rm f}, y_{\rm f}; t)$ on the particular initial distribution $\rho_{\rm p}(r_{\rm p}, y_{\rm p})$ dies out, as $t_{\rm p} \to -\infty$ in Eq. (5.79). As a result, $\rho(r_{\rm f}, y_{\rm f}; t)$ is in fact independent of the particular initial state chosen in the infinite past.

It is worth closing this subsection with the remark that we ended up with a pure real-time path integral for the propagating function for the class of preparations discussed in Section 5.2. In this formulation, the effects of the system-reservoir correlations of the equilibrium state at time $t = 0^+$ are carried by the interactions between the positive and negative time branches of the functional integral expression (5.76). These interactions are included in the influence action (5.77). The real-time approach has been applied successfully, e. g., to the calculation of equilibrium correlation functions for dissipative two-state systems [223] and to the study of universal low temperature properties [224] (see Chapter 21).

5.3.4 Closed time contour representation

The propagating functions discussed in Sec. 5.2, and Subsections 5.3.2 and 5.3.3 may be written in compact form with a path $\tilde{q}(z)$ defined on a closed time contour. The system's Minkowskian action for the path $\tilde{q}(z)$ reads

$$
\mathcal{S}_{\rm S}[\tilde{q}(\cdot)] \;=\; \int dz \left[\frac{M}{2}\left(\frac{\partial \tilde{q}}{\partial z}\right)^2 - V(\tilde{q}) \right].
\tag{5.83}
$$

The influence action for the path $\tilde{q}(z)$, with $L(z)$ given in Eq. (5.24), is

$$
\mathcal{S}_{\rm infl}[\tilde{q}(\cdot)] \;=\; \int_{z>z'} dz \int dz'\, L(z - z')\, \tilde{q}(z)\, \tilde{q}(z') \;+\; i\,\frac{\mu}{2} \int dz\, \tilde{q}^2(z)\,.
\tag{5.84}
$$

The constraint $z > z'$ means time ordering along the path $\tilde{q}(z)$.

Contour path in complex-time

Consider a contour $\mathcal{C}_{\rm I}$ and a contour $\mathcal{C}_{\rm II}$ as sketched in Fig. 5.3 top and bottom,

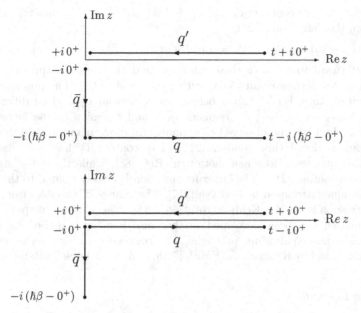

Figure 5.3: Schematic depiction of the contours $\mathcal{C}_I$ (top) and $\mathcal{C}_{II}$ (bottom). The complex time runs in the direction indicated by the arrows, and the path sequence is $\bar{q}$, q, q' for $\mathcal{C}_I$ and $\mathcal{C}_{II}$. The filled circles indicate the endpoint conditions (5.64). The infinitesimal shifts of the path segments q and q' away from the real axis serve to have proper convergence behavior of the factors $e^{iS_s[q]/\hbar}$ and $e^{-iS_s[q']/\hbar}$. The Feynman-Vernon contour $\mathcal{C}_{FV}$ for a product initial state differs from the contour $\mathcal{C}_{I\,(II)}$ by omission of the path segment $\bar{q}(\tau)$.

respectively. The path $\tilde{q}(z)$ with z on the contour $\mathcal{C}_I$ is

$$
\tilde{q}(z) \;=\; \begin{cases} q'(s) & \text{for} \quad z = s + i\,0^+\,, & 0 \leqslant s \leqslant t\,, \\[4pt] \bar{q}(\tau) & \text{for} \quad z = -i\,\tau\,, & 0 \leqslant \tau \leqslant \hbar\beta\,, \\[4pt] q(s) & \text{for} \quad z = -i\,(\hbar\beta - 0^+) + s\,, & 0 \leqslant s \leqslant t\,, \end{cases} \qquad (5.85)
$$

and on the contour $\mathcal{C}_{II}$

$$
\tilde{q}(z) \;=\; \begin{cases} q(s) & \text{for} \quad z = s - i\,0^+\,, & 0 \leqslant s \leqslant t\,, \\[4pt] q'(s) & \text{for} \quad z = s + i\,0^+\,, & 0 \leqslant s \leqslant t\,, \\[4pt] \bar{q}(\tau) & \text{for} \quad z = -i\,\tau\,, & 0 \leqslant \tau \leqslant \hbar\beta\,. \end{cases} \qquad (5.86)
$$

The propagating function for the path $\mathcal{C}_I$ ($\mathcal{C}_{II}$) now reads

$$
J(q_f, q'_f; t; q_i, q'_i; \bar{q}, \bar{q}') \;=\; Z^{-1} \int_{\mathcal{C}_I\,(\mathcal{C}_{II})} \mathcal{D}\tilde{q}(\cdot)\; e^{\,i S_s[\tilde{q}(\cdot)]/\hbar - S_{\mathrm{infl}}[\tilde{q}(\cdot)]/\hbar}\,. \qquad (5.87)
$$

The equivalence of the contours $\mathcal{C}_{\mathrm{I}}$ and $\mathcal{C}_{\mathrm{II}}$ follows with use of symmetry relations resulting from the definition (5.24),

$$L[t - i(\hbar\beta - \tau)] = L(t - i\tau)^* = L(-t - i\tau), \qquad L(-i\tau) = K(\tau). \qquad (5.88)$$

It is straightforward to see with these relations that the concise expression (5.87) is equivalent to the former result (5.63) with (5.68) and (5.69). The propagation of a product initial state, Eq. (5.12), is defined on a contour $\mathcal{C}_{\mathrm{FV}}$ which differs from $\mathcal{C}_{\mathrm{I}}$ or $\mathcal{C}_{\mathrm{II}}$ by omission of the path segment $\bar{q}(\tau)$ and omission of the factor Z^{-1}. Schwinger was the first who observed that nonequilibrium quantum field dynamics proceeds along a closed time contour [215]. The contour $\mathcal{C}_{\mathrm{I}}$ has been discussed and applied to quantum Brownian motion in Ref. [88], while $\mathcal{C}_{\mathrm{II}}$ is the standard Kadanoff-Baym contour [216]. The latter formulation is closely related to the Green function technique introduced by Keldysh [217]. The change of variables from (q, q') to (r, y) corresponds to the Keldysh rotation [218]. The closed-time-path Green function approach has been developed further and has been applied to various equilibrium and non-equilibrium problems. For reviews we refer to the work by Chou *et al.* [219], and by Rammer and Smith [220], and to the book by Rammer [221].

Contour path in real time

Consider next a contour $\mathcal{C}_{\mathrm{III}}$ and a contour $\mathcal{C}_{\mathrm{IV}}$, as illustrated in Fig. 5.4 top and bottom, respectively. The path $\tilde{q}(z)$ with z along the contour $\mathcal{C}_{\mathrm{III}}$ is given by

$$\tilde{q}(z) = \begin{cases} q'(s) & \text{for} \quad z = s + i0^+ , & t_{\mathrm{p}} \leqslant s \leqslant t , \\ q(s) & \text{for} \quad z = -i(\hbar\beta - 0^+) + s , & t_{\mathrm{p}} \leqslant s \leqslant t , \end{cases} \qquad (5.89)$$

and the path with z along the contour $\mathcal{C}_{\mathrm{IV}}$ by

$$\tilde{q}(z) = \begin{cases} q'(s) & \text{for} \quad z = s + i0^+ , & t_{\mathrm{p}} \leqslant s \leqslant t , \\ q(s) & \text{for} \quad z = s - i0^+ , & t_{\mathrm{p}} \leqslant s \leqslant t . \end{cases} \qquad (5.90)$$

For these closed time contours the propagating function J takes the compact form

$$J(q_{\mathrm{f}}, q_{\mathrm{f}}'; t; q_{\mathrm{i}}, q_{\mathrm{i}}', 0^+; \bar{q}, \bar{q}', 0^-; q_{\mathrm{p}}, q_{\mathrm{p}}', t_{\mathrm{p}}) = \int_{\mathcal{C}_{\mathrm{III}}(\mathcal{C}_{\mathrm{IV}})} \mathcal{D}\tilde{q}(\cdot)\, e^{i\,\mathcal{S}_{\mathrm{S}}[\tilde{q}(\cdot)]/\hbar - \mathcal{S}_{\mathrm{infl}}[\tilde{q}(\cdot)]/\hbar} , \qquad (5.91)$$

where $\mathcal{S}_{\mathrm{S}}[\tilde{q}(\cdot)]$ is defined in Eq. (5.83) and $\mathcal{S}_{\mathrm{infl}}[\tilde{q}(\cdot)]$ in Eq. (5.84). With the relations (5.88), the compact expression (5.91) both for the contour $\mathcal{C}_{\mathrm{III}}$ and $\mathcal{C}_{\mathrm{IV}}$ is in correspondence with the expression (5.76) with (5.77).

For an ergodic system, the expression (5.91) for the propagating function is exactly equivalent to the former expression (5.87). The crucial point behind the flexibility to employ either path integration along the contours $\mathcal{C}_{\mathrm{I}}$ or $\mathcal{C}_{\mathrm{II}}$ depicted in Fig. 5.3, or along the contours $\mathcal{C}_{\mathrm{III}}$ or $\mathcal{C}_{\mathrm{IV}}$ shown in Fig. 5.4 is that for ergodic open systems the bath correlation function $L(t)$ drops sufficiently fast at asymptotic time. As a result, differing path contributions depending on the initial state at time t_{p} vanish as $t_{\mathrm{p}} \to -\infty$. An explicit demonstration of the equivalence has been given for specific models such as the damped harmonic oscillator [222] and the spin-boson system [223].

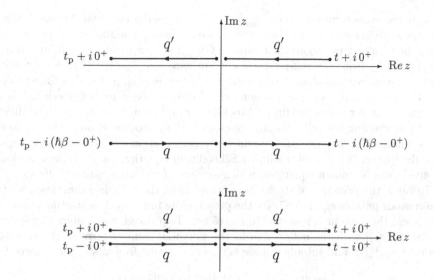

Figure 5.4: Schematic representation of the contours $\mathcal{C}_{\mathrm{III}}$ (top) and $\mathcal{C}_{\mathrm{IV}}$ (bottom). The complex time runs again in the direction indicated by the arrows, and the path sequence is q, q' both for $\mathcal{C}_{\mathrm{III}}$ and $\mathcal{C}_{\mathrm{IV}}$. The filled circles symbolize the endpoint conditions (5.78).

5.4 Stochastic unraveling of influence functionals

Exact numerical techniques such as quantum Monte Carlo (QMC) methods keep struggling with the so-called dynamical sign problem. The numerical load is due to the fact that complex-valued probability amplitudes with varying phases rather than probabilities have to be summed up in a quantum mechanical computation.

Path integral representations of the reduced density matrix are intimately connected to stochastic processes for pure quantum states. It has been shown that there is a close connection between influence functionals and non-Markovian generalizations of quantum state diffusion [228] as well as Langevin-like stochastic dynamics of the reduced density matrix [229].

Quantum state diffusion (QSD) has been established as an alternative approach to quantum dissipation in the perturbative weak-coupling regime [66, 67, 61, 62]. In this method, pure quantum states are propagated stochastically. This yields an appealing picture of individual quantum trajectories for open systems and facilitates effective numerical calculations (see also Section 2.4). Extension of this method to the case of non-perturbative, non-Markovian dynamics paves the way to overcome many of the above-mentioned numerical limitations. A corresponding generalization of quantum state diffusion has been proposed by Strunz, Diósi, and Gisin [228]. In their approach, the cross term $\exp\{-\mathcal{S}_{\mathrm{FV,\,cross}}[q(\cdot), q'(\cdot)]/\hbar\}$ of the Feynman-Vernon influence functional (5.21) with (5.26) is unraveled into a stochastic influence functional. However, the resulting propagation amplitude $G[q(\cdot)]$ still experiences quantum memory effects arising from the self-interaction term $\exp\{-\mathcal{S}_{\mathrm{FV,\,self}}[q(\cdot)]/\hbar\}$. As a result of that, the respective stochastic Schrödinger equation is equipped with a time-nonlocal functional derivative, for which a general solution strategy, even numerically, is not known, except for restrictive limits. Stockburger and Grabert overcame these difficulties by unraveling the full influence functional into a stochastic one [230]. Following these lines, they derived exact c-number representations of non-Markovian quantum dissipation in terms of stochastic Schrödinger equations or equivalent stochastic Liouville-von Neumann equations, which are free of quantum memory effects.

To catch the principle of stochastic unraveling of the influence functional we take the contour path integral (5.87) for the propagating function J as starting point, and we choose the contour $\mathcal{C}_{\mathrm{I}}$ specified in Eq. (5.85). To proceed, we combine the system's action $S_{\mathrm{S}}[\tilde{q}(\cdot)]$ with the action due to the potential counter term in the controlled action $S_{\mathrm{contr}}[\tilde{q}(\cdot)]$ and subjoin a noise action originating from a stochastic force $\xi(z)$,

$$
\begin{aligned}
S_{\mathrm{contr}}[\tilde{q}(\cdot)] &= S_{\mathrm{S}}[\tilde{q}(\cdot)] - \tfrac{1}{2}\mu \int_{\mathcal{C}_{\mathrm{I}}} dz\, \tilde{q}^2(z)\,, \\
S_{\xi}[\tilde{q}(\cdot)] &= S_{\mathrm{contr}}[\tilde{q}(\cdot)] + \int_{\mathcal{C}_{\mathrm{I}}} dz\, \xi(z)\tilde{q}(z)\,,
\end{aligned}
\tag{5.92}
$$

where the path $\tilde{q}(z)$ is specified in Eq. (5.85). For given force $\xi(z)$, the time evolution of the propagating function J_{ξ} and ultimately of the density matrix ρ_{ξ} is given by

$$
J_{\xi}(q_{\mathrm{f}}, q'_{\mathrm{f}}; t; q_{\mathrm{i}}, q'_{\mathrm{i}}; \bar{q}, \bar{q}') = \frac{1}{Z} \int_{\mathcal{C}_{\mathrm{I}}} \mathcal{D}\tilde{q}(\cdot)\, e^{i S_{\xi}[\tilde{q}(\cdot)]/\hbar}\,.
\tag{5.93}
$$

The propagating function $J(q_{\mathrm{f}}, q'_{\mathrm{f}}; t; q_{\mathrm{i}}, q'_{\mathrm{i}}; \bar{q}, \bar{q}')$, as well as the reduced density matrix $\rho(q_{\mathrm{f}}, q_{\mathrm{i}}; t)$, is the thermal average of the individual noise realizations,

$$
J(q_{\mathrm{f}}, q'_{\mathrm{f}}; t; q_{\mathrm{i}}, q'_{\mathrm{i}}; \bar{q}, \bar{q}') = \big\langle J_{\xi}(q_{\mathrm{f}}, q'_{\mathrm{f}}; t; q_{\mathrm{i}}, q'_{\mathrm{i}}; \bar{q}, \bar{q}') \big\rangle_{\beta}\,.
\tag{5.94}
$$

For Gaussian statistics, the average of the noise action factor is an exponential expressed by the autocorrelation function of the stochastic force $\xi(t)$,

$$
\Big\langle \exp\Big(\frac{i}{\hbar}\int_{\mathcal{C}_{\mathrm{I}}} dz\, \xi(z)\tilde{q}(z)\Big)\Big\rangle_{\beta} = \exp\Big(-\frac{1}{\hbar^2} \int_{\mathcal{C}_{\mathrm{I}}} dz \int_{z>z'} dz'\, \tilde{q}(z)\langle \xi(z)\xi(z')\rangle_{\beta}\, \tilde{q}(z')\Big)\,.
\tag{5.95}
$$

From this we see that the noise-averaged propagating function (5.94) with $z > z'$ along $\mathcal{C}_{\mathrm{I}}$ coincides with the former expression (5.87), if we identify [cf. Eq. (5.25)]

$$
\langle \xi(z)\xi(z')\rangle_{\beta} = \hbar L(z - z')\,.
\tag{5.96}
$$

For a factorizing initial state, the Feynman-Vernon contour $\mathcal{C}_{\mathrm{FV}}$ [cf. Fig. 5.3] applies. Then we can replace the noise force $\xi(z)$ with complex-time argument z by two noise forces parameterized with real time s, one along the forward path, $\xi_1(s) = \xi(s - i\hbar\beta)$, and the other along the backward path, $\xi_2(s) = \xi(s)$. If we assume that the random forces have correlations

$$\langle \xi_{1/2}(s)\xi_{1/2}(0)\rangle_\beta = \hbar L(s), \quad \text{and} \quad \langle \xi_1(s)\xi_2(0)\rangle_\beta = \hbar L^*(s), \quad (5.97)$$

then the stochastic influence functional (5.93) for the contour $\mathcal{C}_{\mathrm{FV}}$ is unraveling the reduced dynamics according to

$$\rho(q_{\mathrm{f}}, q_{\mathrm{f}}'; t) = \int dq_{\mathrm{i}} \int dq_{\mathrm{i}}' \Big\langle G_{\xi_1}(q_{\mathrm{f}}, q_{\mathrm{i}}; t)\, G^*_{\xi_2}(q_{\mathrm{f}}', q_{\mathrm{i}}'; t) \Big\rangle_\beta \rho(q_{\mathrm{i}}, q_{\mathrm{i}}'; 0). \quad (5.98)$$

Here $G_{\xi_j}(q_{\mathrm{f}}, q_{\mathrm{i}}; t)$ is the stochastic propagator

$$G_{\xi_j}(q_{\mathrm{f}}, q_{\mathrm{i}}; t) = \int_{q(0)=q_{\mathrm{i}}}^{q(t)=q_{\mathrm{f}}} \mathcal{D}q(\cdot)\, e^{i S_\xi[q(\cdot)]/\hbar}, \quad (5.99)$$

which is free of memory load. Next, it is expedient to introduce the linear combinations $\zeta(s) = \frac{1}{2}[\xi_1(s) + \xi_2^*(s)]$ and $\nu(s) = [\xi_1(s) - \xi_2^*(s)]/\hbar$. The correlations of the random forces $\zeta(s)$ and $\nu(s)$ are found with Eqs. (5.24) and (5.29) to read

$$\langle \zeta(s)\zeta(0)\rangle_\beta = \hbar L'(s), \qquad \langle \nu(s)\nu(0)\rangle_\beta = 0,$$
$$\langle \zeta(s)\nu(0)\rangle_\beta = 2i\,\Theta(s)\, L''(s) = i M \dot{\gamma}(s). \quad (5.100)$$

Since the propagator of the forward path differs from that of the backward path, the propagation of a stochastic sample of the density matrix with initial state $\rho_{\mathrm{ini}} = |\psi_1(0)><\psi_2(0)|$ is determined by two different stochastic Schrödinger equations,

$$i\hbar|\dot{\psi}_1(t)> = \left\{ H_{\mathrm{S}} - [\zeta(t) + \tfrac{1}{2}\hbar\nu(t)]\, q + \tfrac{1}{2}\mu q^2 \right\} |\psi_1(t)>,$$
$$i\hbar|\dot{\psi}_2(t)> = \left\{ H_{\mathrm{S}} - [\zeta^*(t) - \tfrac{1}{2}\hbar\nu^*(t)]\, q + \tfrac{1}{2}\mu q^2 \right\} |\psi_2(t)>. \quad (5.101)$$

These equations are local in time and their solution renders the propagation amplitudes (5.99). All effects of the quantum mechanical memory are contained in the averaging of the product of stochastic propagators in Eq. (5.98).

The other correlations $\langle \zeta(s)\zeta^*(0)\rangle_\beta$, $\langle \zeta(s)\nu^*(0)\rangle_\beta$ and $\langle \nu(s)\nu^*(0)\rangle_\beta$ are left undetermined by the model. They can be chosen according to computational convenience. The noise force $\zeta(s)$ can be chosen real, at which the noise force $\nu(s)$ is imaginary.

The dynamics governed by the stochastic Schrödinger equations (5.101) can be rewritten as a stochastic Liouville-von Neumann (SLN) equation in which the noise terms are in the form of a commutator $[\cdot, \cdot]$ and an anti-commutator $\{\cdot, \cdot\}$,

$$i\hbar\dot{\rho}_\xi(t) = [H_{\mathrm{S}} + \tfrac{1}{2}\mu q^2, \rho_\xi(t)] - \zeta(t)[q, \rho_\xi(t)] - \tfrac{1}{2}\hbar\nu(t)\{q, \rho_\xi(t)\}. \quad (5.102)$$

The linear Schrödinger equations (5.101) and the SLN equation (5.102) are formally exact. Monte Carlo sampling of $\rho_\xi(t)$ for the realizations of $\zeta(s)$ and $\nu(s)$ with statistical weight (5.100) will eventually converge to the exact time evolution of the RDM $\rho(t)$.

The stochastic linear Schrödinger equations (5.101), as well as other stochastic linear Schrödinger equations, do not preserve the norm of the states, since the noise force $\nu(t)$ is complex. The norm $\operatorname{tr} \rho(t)$ is preserved, however, after stochastic sampling. The diffusive spread of the sample trace slows down numerical convergence and limits practical applications. This flaw can be softened or even eliminated by subtraction of a suitably chosen guide trajectory $\bar{q}(t)$ in the anti-commutator term of Eq. (5.102), $q \to q - \bar{q}(t)$. The function $\bar{q}(t)$ may be any causal function of $\zeta(s)$ and of $\nu(s)$. Turning to the normalized density matrix $\hat{\rho}_\xi(t) = \rho_\xi(t)/\operatorname{tr} \rho_\xi(t)$, the factor $\operatorname{tr} \rho_\xi(t) = \exp[i \int_0^t ds\, \nu(s)\bar{q}(s)]$ must be included in the Gaussian integration measure of the stochastic forces. The linear shift leads to noise forces with the same variance but displaced mean values. Upon completing the square in the exponent one obtains the original Gaussian probability measure for shifted variables. This shift can be compensated by a corresponding shift of the stochastic force $\zeta(t)$ in the SLN equation, $\zeta(t) \to \tilde{\zeta}(t) = \zeta(t) - M \int_0^t ds\, \dot{\gamma}(t-s)\bar{q}(s)$. Upon performing integration by parts in the second term, the SLN equation for the normalized density $\hat{\rho}_\xi(t)$ takes the form

$$
\begin{aligned}
i\hbar\dot{\hat{\rho}}_\xi(t) &= \left[H_S + \tfrac{1}{2}\mu(q - \bar{q}(t))^2, \hat{\rho}_\xi(t) \right] - \tfrac{1}{2}\hbar\nu(t)\left\{ q - \bar{q}(t), \hat{\rho}_\xi(t) \right\} \\
&\quad - \left(\zeta(t) - M \int_0^t ds\, \gamma(t-s)\dot{\bar{q}}(s) - M\gamma(t)\bar{q}(0) \right) \left[q, \hat{\rho}_\xi(t) \right].
\end{aligned}
\tag{5.103}
$$

The random force $\zeta(t)$ has vanishing mean, and noise correlations are as in Eq. (5.100). The SLN equation (5.103) conserves the normalization of $\hat{\rho}_\xi(t)$ for each sample. A natural choice for the guide trajectory is $\bar{q}(t) = \operatorname{tr}\{q\hat{\rho}(t)\} = <\psi_2(t)|q|\psi_1(t)>$.

In the classical limit, terms $\propto q - \bar{q}$ become negligibly small. Then the dynamics of $q(t)$ resulting from Eq. (5.103) is described by the Langevin equation (3.20).

The implementation of a numerical simulation scheme based on the LVN equation (5.103) is discussed in Ref. [231]. The transition from linear QSD approaches to stochastic processes with normalized quantum states proved to be a decisive step towards robust numerical simulations [230, 231].

5.5 Non-Markovian dissipative dynamics in the semiclassical limit

5.5.1 Van Vleck and Herman-Kluk propagator

Shortly after the discovery of Schrödinger's equation in 1927, Van Vleck wrote down in 1928 a complex-valued function which nowadays is called the semi-classical or quasi-classical propagator of quantum mechanics [232]. In d space dimensions it reads

$$
K_{\text{sc}}(\mathbf{q}_f, t; \mathbf{q}_i, 0) = \sum_\alpha \frac{\sqrt{|C_\alpha(\mathbf{q}_f, t; \mathbf{q}_i, 0)|}}{(2\pi i\hbar)^{d/2}}\, e^{-i n_\alpha \pi/2}\, \exp[i\, S_{\text{cl},\alpha}(\mathbf{q}_f, t; \mathbf{q}_i, 0)/\hbar]. \tag{5.104}
$$

Here α labels the classical path from $(\mathbf{q}_i, 0)$ to $(\mathbf{q}_f, t)$, and $S_{\text{cl},\alpha}(\mathbf{q}_f, t; \mathbf{q}_i, 0)$ is the related classical action. The density $C_\alpha(\mathbf{q}'', t; \mathbf{q}', 0)$ is the so-called Van Vleck determinant

$$C_\alpha(\mathbf{q}'', t; \mathbf{q}', 0) = \det \frac{-\partial^2 S_{\text{cl},\alpha}(\mathbf{q}'', t; \mathbf{q}', 0)}{\partial q_i'' \, \partial q_j'} . \tag{5.105}$$

The root in Eq. (5.104) is taken on the absolute value of the Van Vleck determinant. The determinant is singular at conjugate or focal points along the path α, and n_α is the number of conjugate points along the classical trajectory α from $\mathbf{q}_i$ to $\mathbf{q}_f$ [233, 5]. The phase factor $e^{-i n_\alpha \pi/2}$ was absent in Van Vleck's original work.

The path integral for the quantum mechanical propagator

$$K(\mathbf{q}_f, t; \mathbf{q}_i, 0) \equiv\, <\mathbf{q}_f| e^{-i H t/\hbar} |\mathbf{q}_i> \,= \int_{\mathbf{q}(0)=\mathbf{q}_i}^{\mathbf{q}(t)=\mathbf{q}_f} \mathcal{D}\mathbf{q}(\cdot) \, \exp\{i\, S[\mathbf{q}(\cdot)]/\hbar\} , \tag{5.106}$$

or its discretized version analogous to Eq. (4.15), can be worked out in the neighborhood of the classical path, if we take only the second variation of the action into acount. The resulting Gaussian integral over the quadratic fluctuations yields exactly the Van Vleck propagator [234]. The semiclassical propagator satisfies a fixed-point property on stationary phase approximation (spa) of the integral,

$$K_{\text{sc}}(\mathbf{q}'', t''; \mathbf{q}', t') = \int_{\text{spa}} d^n\mathbf{q}\, K_{\text{sc}}(\mathbf{q}'', t''; \mathbf{q}, t)\, K_{\text{sc}}(\mathbf{q}, t; \mathbf{q}', t') . \tag{5.107}$$

Above all, it satisfies the time-dependent Schrödinger equation with a potential correction $\mathcal{V}_{\text{qu}}$ of order $\hbar^2$, $(i\hbar\partial/\partial t - H - \mathcal{V}_{\text{qu}})K_{\text{sc}}(\mathbf{q}, t; \mathbf{q}_i, 0) = 0$. The compensating potential is $\mathcal{V}_{\text{qu}}(\mathbf{q}) = \hbar^2 [\Delta_\mathbf{q}\sqrt{C(\mathbf{q})}]/[2M\sqrt{C(\mathbf{q})}]$ [234].

The semiclassical propagator (5.104) requires to determine classical trajectories $\mathbf{q}_{\text{cl}}(s)$ with prescribed values at time zero and at time t, $\mathbf{q}_{\text{cl}}(0) = \mathbf{q}_i$ and $\mathbf{q}_{\text{cl}}(t) = \mathbf{q}_f$. These boundary conditions numerically cost root searches. This drawback is avoided in the time-dependent semiclassical initial value formalism introduced by Heller [235] and justified later by Herman and Kluk [236]. The Herman-Kluk propagator propagates Gaussian non-spreading or frozen wave packets according to the laws of classical mechanics. In one space dimension, it is given by the phase space integral

$$K_{\text{HK}}(q_f, t; q_i, 0) = \int \frac{dp_i\, dx_i}{2\pi\hbar} <q_f|g_\sigma(x_t, p_t)> R_t(x_i, p_i)\, e^{i\, S_t(x_i, p_i)/\hbar} <g_\sigma(x_i, p_i)|q_i> .$$

Here $S_t(x_i, p_i)$ is the action at time t for classical paths evolving from the initial phase space point x_i, p_i, and $<g_\sigma(x_i, p_i)|q_i>$ and $<q_f|g_\sigma(x_t, p_t)>$ are complex-valued Gaussian wave packets of fixed width σ centered around the initial phase space point x_i, p_i and the time-evolved phase space point x_t, p_t, respectively,

$$<q|g_\sigma(x, p)> = (\pi\sigma^2)^{-1/4}\, e^{-(q-x)^2/2\sigma^2}\, e^{i\,p(q-x)/\hbar} . \tag{5.108}$$

The width σ may, in principle, be chosen arbitrarily. The momentum p_t and position x_t at time t evolve from p_i and p_i according to Hamilton's equations of motion. The factor R_t of the HK integrand is expressed in terms of stability matrix elements as

$$R_t(x_i, p_i) = \sqrt{\frac{1}{2}\left(\frac{\partial x_t}{\partial x_i} + \frac{\partial p_t}{\partial p_i} - i\frac{\hbar}{\sigma^2}\frac{\partial x_t}{\partial p_i} + i\frac{\sigma^2}{\hbar}\frac{\partial p_t}{\partial x_i}\right)} . \tag{5.109}$$

In view of doubts about its validity, the HK propagator was proven to be the leading term in a uniform semiclassical expansion [237]. The crucial advantage of the HK propagator compared to the Van Vleck propagator (5.104) is that the former constitutes a classical initial value problem which is free of root searches for trajectories.

5.5.2 Semiclassical dissipative dynamics

Recently, it has been proposed to combine the stochastic quantum dynamics presented in the preceding section with semiclassical propagation based on the frozen Gaussian approximation [238]. In this approach the density matrix $\rho(q_f, q_i; t)$ involves three Monte Carlo integrations, two over the forward and backward phase spaces of the HK propagators and the residual one over the noise distribution. The classical path in the HK propagator is found from the quasiclassical dynamics of the frozen Gaussians under the transformed versions of Eqs. (5.101). The non-dispersive Gaussians are, up to an irrelevant phase factor, coherent states $|\alpha>_j = e^{-|\alpha_j^2|/2} e^{\alpha_j \hat{a}^\dagger} |0>$ with

$$\alpha_j = \frac{1}{\sqrt{2}\sigma} \left(q_j + i \frac{\sigma^2}{\hbar} p_j \right) , \qquad j = 1, 2 . \tag{5.110}$$

The classical equations of motion derived from Eqs. (5.101) read

$$\dot{\alpha}_1 = \frac{1}{\sqrt{2}\sigma} \left\{ \frac{p_1}{M} + \frac{\sigma^2}{i\hbar} \left[V'(q_1) + \mu q_1 - \zeta + M \int_0^t dt' \, \dot{\gamma}(t - t') \bar{q}(t') - \tfrac{1}{2}\hbar\nu \right] \right\} ,$$

$$\dot{\alpha}_2 = \frac{1}{\sqrt{2}\sigma} \left\{ \frac{p_2}{M} + \frac{\sigma^2}{i\hbar} \left[V'(q_2) + \mu q_2 - \zeta^* + M \int_0^t dt' \, \dot{\gamma}(t - t') \bar{q}(t') + \tfrac{1}{2}\hbar\nu^* \right] \right\} .$$

These equations have to be solved for real q_j and p_j. If we disregard the ν-fluctuations we find upon elimination of p exactly the Langevin equation (3.20) with the force correlation $\langle \zeta(s)\zeta(0)\rangle_\beta = \hbar L'(s)$.

The guide trajectory $\bar{q}(t)$ is again found by the requirement that the ν-dependent terms leave the trace of the sample density matrix $\rho_\xi(t)$ invariant. This leads to a relation which includes the pair of trajectories α_1 and α_2 as

$$\bar{q}(t) = \sqrt{\sigma^2/2} \left[\alpha_1(t) + \alpha_2^*(t) \right] . \tag{5.111}$$

In the numerical algorithm, it is therefore expedient to combine the integrations over the two HK phase spaces and the space of noise trajectories $\zeta(t), \nu(t)$ into a joint Monte Carlo sampling scheme, where new initial phase space points and a new noise trajectory are chosen for each sample. Wave packet dynamics in a Morse potential has been studied with this method over many oscillation periods in Ref. [238]. The analysis showed that dissipative dynamics is accurately described even when non-Markovian effects are significant. As a perspective, systematic quantum corrections to the HK propagator [237, 239] may be implemented in the numerical scheme.

5.6 Energy transfer between system, reservoir and interaction

Towards conception of consistent thermodynamics for nonequilibrium scenarios in complex quantum systems with general system-reservoir coupling, deep understanding and precise control of energy exchange between the various dissipative channels would be an important step. The question how to conceive consistent thermodynamics under such conditions has attracted a lot of interest [240, 241]. The issue how to clearly define and precisely allocate, e.g., internal energies and dissipated heat, is still a subject of controversy. Substantial progress to identify heat unambiguously has recently been made for classical systems at strong coupling [242].

The natural setting to study energy exchange and dissipation is an externally driven quantum system coupled to one or to several heat reservoirs. When the central system is externally driven, the time-dependent force is continuously pumping energy into the system, and thus drives it out of equilibrium. Such scenario is mostly studied in the weak-coupling limit, in which perturbative methods can be used. On the other hand, strong coupling with the environment makes the non-equilibrium dynamics more intricate since the composite system can dissipate energy through different channels. In particular regions of the parameter space, the amount of energy dissipated into the interaction H_I can be as large as those amounts dissipated into the reservoir H_R and into the system H_S. All these issues can be treated with an extension of the real-time influence functional method. This is set out in the remainder of this section.

The most common approach to the quantum-statistical energy transfer dynamics is to measure the energy of interest twice, and then iterate the protocol to reconstruct statistical information. The central quantity bearing the entire statistics of the heat transfer process is the chararacteristic function $G(\varrho, t)$, defined as Fourier transform of the normalized probability distribution function $P(\mathcal{E}, t)$ of the transferred energy $\mathcal{E}$,

$$G(\varrho, t) = \int_{-\infty}^{\infty} d\mathcal{E}\, P(\mathcal{E}, t)\, e^{i\,\varrho\mathcal{E}/\hbar} = \sum_{n=0}^{\infty} \frac{(i\varrho/\hbar)^n}{n!} \langle \mathcal{E}^n(t) \rangle. \tag{5.112}$$

It represents the moment generating function (MGF) of the transfer statistics,

$$\langle \mathcal{E}^n(t) \rangle = (-i\,\hbar\,\partial/\partial\varrho)^n\, G(\varrho, t)\big|_{\varrho=0}, \tag{5.113}$$

while the function $\ln G(\varrho, t)$ is the cumulant generating function (cf. Subsec. 25.1.2),

$$\langle \mathcal{E}^n(t) \rangle_c = (-i\,\hbar\,\partial/\partial\varrho)^n\, \ln G(\varrho, t)\big|_{\varrho=0}, \tag{5.114}$$

The first moment, $n = 1$, is the average amount of energy or heat transferred,

$$\langle \mathcal{E}(t) \rangle = -i\,\hbar\,\partial G(\varrho, t)/\partial\varrho\big|_{\varrho=0}. \tag{5.115}$$

A major advantage of choosing the MGF as starting point is that it can be written as the trace of a generalized density operator [243, 244]. This gives us access to the path integral representation of the MGF [245]. In addition, we put the generalisation that one may record besides the reservoir energy also the interaction energy [246],

$$G_{\kappa_f,\kappa_i,\kappa}(\varrho,t) \;=\; Z_R^{-1} \tag{5.116}$$

$$\times\ \mathrm{tr}\left[e^{i\varrho(\hat{H}_R+\kappa_f\hat{H}_I)/\hbar}\,\hat{U}(t,0)\,e^{-i\,\varrho(\hat{H}_R+\kappa_i\hat{H}_I)/\hbar}\,\hat{\rho}(0)\,e^{-\beta(\hat{H}_R+\kappa\hat{H}_I)}\,\hat{U}^{\dagger}(t,0)\right].$$

The trace is over all degrees of freedom of system and reservoir, and $\hat{U}(t,0)$ is the unitary evolution operator of the composite system with the time-dependent generator $\hat{H}(t) = \hat{H}_S(t) + \hat{H}_R + \hat{H}_I$ with the interaction $\hat{H}_I = -\sum_\alpha c_\alpha \hat{x}_\alpha \hat{q} + \Delta V(\hat{q})$, where $\Delta V(\hat{q}) = \frac{1}{2}\sum_\alpha (c_\alpha^2/2m_\alpha\omega_\alpha^2)\hat{q}^2$. When energy is injected into or withdrawn from the composite system by an externally applied time-dependent force $F_{\mathrm{ext}}(t)$, the system Hamiltonian $\hat{H}_S$ becomes explicitly time-dependent,

$$\hat{H}_S(t) = \hat{p}^2/2M + V(\hat{q}) - F_{\mathrm{ext}}(t)\,\hat{q}\,. \tag{5.117}$$

The quantities κ, κ_i and κ_f are control parameters, each set to zero or unity, in order to deal with different protocols on the same footing. The choices $\kappa = 0$ and 1 represent the unshifted and shifted canonical bath states (5.9) and (5.10). The choices $\kappa_i = 0,\,1$ and $\kappa_f = 0,\,1$ represent measurement of the bath energy alone or together with the coupling energy H_I at start time zero and at end time t, respectively.

Now we assume that the system starts out at time zero from the diagonal state $<q_i|\hat{\rho}(0)|q_i'> = \delta(q_i - q_i')\,p_S(q_i)$. The position representation of the MGF (5.116) can be written analogous to the expression (5.11) with (5.12) as the double path integral

$$G_{\kappa_f,\kappa_i,\kappa}(\varrho,t) \;=\; \int dq_f\,dq_i\,p_S(q_i) \int_{q(0)=q_i}^{q(t)=q_f} \mathcal{D}q(\cdot) \int_{q'(0)=q_i}^{q'(t)=q_f} \mathcal{D}q'(\cdot)$$

$$\times\ e^{i\{S_S[q(\cdot)]-S_S[q'(\cdot)]\}/\hbar}\,\mathcal{F}_{\kappa_f,\kappa_i,\kappa}[q(\cdot),q'(\cdot);\varrho]\,. \tag{5.118}$$

The functional $\mathcal{F}_{\kappa_f,\kappa_i,\kappa}[q(\cdot),q'(\cdot);\varrho]$ carries all effects of the system-bath coupling both on the reduced dynamics and on the various energy transfer measurement protocols. It is a generalization of the one considered in Ref. [245] to include also the interaction energy H_I in the measurement protocol [246],

$$\mathcal{F}_{\kappa_f,\kappa_i,\kappa}[q(\cdot),q'(\cdot);\varrho] \;=\; \frac{1}{Z_R}\int dx_f'\,dx_f\,dx_i''\,dx_i'\,dx_i$$

$$\times\ <x_f'|\,e^{i\,\varrho[\hat{H}_R+\kappa_f\hat{H}_I(q_f)/\hbar}\,|x_f> \mathbb{F}[q(\cdot),x_f,x_i]$$

$$\times\ <x_i|\,e^{-i\,\varrho[\hat{H}_R+\kappa_i\hat{H}_I(q_i)]/\hbar}|x_i''> \tag{5.119}$$

$$\times\ <x_i''|\,e^{-\beta[\hat{H}_R+\kappa\hat{H}_I(q_i)]}|x_i'> \mathbb{F}^*[q'(\cdot),x_f',x_i']\,.$$

The shifted canonical initial state is given as an extension of the expression (5.17) by

$$<x_\alpha|\,e^{-\beta[\hat{H}_{R,\alpha}+\kappa\hat{H}_{I,\alpha}(q_i)]}|y_\alpha> \;=\; \sqrt{B_{\alpha,\mathrm{th}}/2\pi\hbar} \tag{5.120}$$

$$\times\ \exp\left\{-\frac{1}{\hbar}\left[A_{\alpha,\mathrm{th}}\left[\left(x_\alpha - \frac{\kappa\,c_\alpha}{m_\alpha\omega_\alpha^2}q_i\right)^2 + \left(y_\alpha - \frac{\kappa\,c_\alpha}{m_\alpha\omega_\alpha^2}q_i\right)^2\right]\right.\right.$$

$$\left.\left. - B_{\alpha,\mathrm{th}}\left(x_\alpha - \frac{\kappa\,c_\alpha}{m_\alpha\omega_\alpha^2}q_i\right)\left(y_\alpha - \frac{\kappa\,c_\alpha}{m_\alpha\omega_\alpha^2}q_i\right)\right]\right\},$$

The ϱ-dependent amplitudes ensue from this by analytical continuation $\beta \to \pm i\, \varrho/\hbar$,

$$<x_\alpha|\, e^{\mp i\, \varrho [\hat{H}_{\mathrm{R},\alpha} + \kappa_{\mathrm{i/f}} \hat{H}_{\mathrm{I},\alpha}(q_{\mathrm{i/f}})]/\hbar}|y_\alpha> \; = \; \sqrt{\frac{B_\alpha(\varrho)}{\pm 2\pi i\hbar}} \tag{5.121}$$

$$\times \, \exp\left\{ \pm \frac{i}{\hbar}\left[A_\alpha(\varrho)\left[\left(x_\alpha - \frac{\kappa_{\mathrm{i/f}}\, c_\alpha}{m_\alpha \omega_\alpha^2}q_{\mathrm{i/f}}\right)^2 + \left(y_\alpha - \frac{\kappa_{\mathrm{i/f}}\, c_\alpha}{m_\alpha \omega_\alpha^2}q_{\mathrm{i/f}}\right)^2 \right] \right. \right.$$

$$\left. \left. - \, B_\alpha(\varrho)\left(x_\alpha - \frac{\kappa_{\mathrm{i/f}}\, c_\alpha}{m_\alpha \omega_\alpha^2}q_{\mathrm{i/f}}\right)\left(y_\alpha - \frac{\kappa_{\mathrm{i/f}}\, c_\alpha}{m_\alpha \omega_\alpha^2}q_{\mathrm{i/f}}\right)\right]\right\} \, ,$$

and the real-time amplitude $\mathbb{F}_\alpha[q(\cdot), x_{\mathrm{f},\alpha}, x_{\mathrm{i},\alpha}]$ is given in Eq. (5.18) with (5.19).

As the five amplitudes in (5.119) are mixed Gaussian forms, it is straightforward to integrate out all the bath variables by successive completion of the squares. The resulting functional of the system paths emerges as the Feynman-Vernon functional times the functional accounting for energy transfer. It is convenient to express these with the quasiclassical path $r(\cdot)$ and the fluctuation path $y(\cdot)$ introduced in Eq. (5.30),

$$\mathcal{F}_{\kappa_{\mathrm{f}}, \kappa_{\mathrm{i}}, \kappa}[r(\cdot), y(\cdot); \, \varrho] = e^{-S_{\mathrm{FV}, \kappa}[r(\cdot), y(\cdot)]/\hbar}\, e^{-S_{\kappa_{\mathrm{f}}, \kappa_{\mathrm{i}}, \kappa}[r(\cdot), y(\cdot); \, \varrho]/\hbar} \tag{5.122}$$

The first factor is the Feynman-Vernon functional with modified slip action (5.32),

$$\mathcal{S}_{\mathrm{FV}, \kappa}[r(\cdot), y(\cdot)] = \mathcal{S}^{(N)}[y(\cdot)] + i\left(\mathcal{S}^{(F)}[r(\cdot), y(\cdot)] + (1 - \kappa)\mathcal{S}^{(\mathrm{slip})}[r_{\mathrm{i}}, y(\cdot)] \right) \, , \tag{5.123}$$

For the shifted thermal initial state (3.72), $\kappa = 1$, the slip action is removed, and the annoying initial-slip force occurring, e.g., in Eqs. (5.38), (5.50) and (3.20) is absent.

The action $\mathcal{S}_{\kappa_{\mathrm{f}}, \kappa_{\mathrm{i}}, \kappa}[r(\cdot), y(\cdot); \, \varrho]$ carries the various energy exchange processes of the composite system. It is expedient to give this action in velocity representation in order to gather up all boundary terms. By means of partial integration, one obtains

$$\mathcal{S}_{\kappa_{\mathrm{f}}, \kappa_{\mathrm{i}}, \kappa}[r(\cdot), y(\cdot); \, \varrho] = \tag{5.124}$$

$$i \int_0^t ds_2 \int_0^{s_2} ds_1 \left\{ \left[\, 4\dot{r}(s_2)\dot{r}(s_1) - \dot{y}(s_2)\dot{y}(s_1)\,\right] R_1(s_2 - s_1, \varrho) \right.$$

$$\left. + \, 2\left[\, \dot{r}(s_2)\dot{y}(s_1) - \dot{y}(s_2)\dot{r}(s_1)\,\right] R_2(s_2 - s_1, \varrho)\right\}$$

$$+ \, i \int_0^t ds \left\{ (1 - \kappa_{\mathrm{i}})2r_{\mathrm{i}}[\, 2R_1(s, \varrho)\dot{r}(s) - R_2(s, \varrho)\dot{y}(s)\,]\right.$$

$$- \, (1 - \kappa_{\mathrm{f}})2r_{\mathrm{f}}[\, 2R_1(t - s, \varrho)\dot{r}(s) + R_2(t - s, \varrho)\dot{y}(s)\,]$$

$$\left. - \, (\kappa - \kappa_{\mathrm{i}})r_{\mathrm{i}}[\, R''(s) - R''(s + \varrho)\,][\, 2\dot{r}(s) + \dot{y}(s)\,]\right\}$$

$$+ \, i\, (\kappa - \kappa_{\mathrm{i}})r_{\mathrm{i}}[\, (1 - \kappa_{\mathrm{i}})2r_{\mathrm{i}}R''(\varrho) - (1 - \kappa_{\mathrm{f}})2r_{\mathrm{f}}[\, R''(t + \varrho) - R''(t)\,]\,]$$

$$- \, i\, (1 - \kappa_{\mathrm{f}})(1 - \kappa_{\mathrm{i}})4r_{\mathrm{i}}r_{\mathrm{f}}R_1(t, \varrho) + 2i\, [\, (1 - \kappa_{\mathrm{i}})^2 r_{\mathrm{i}}^2 + (1 - \kappa_{\mathrm{f}})^2 r_{\mathrm{f}}^2\,]\, R_1(0, \varrho) \, .$$

The double integral contribution pertains entirely to the reservoir, while both the unilateral and ambiliteral boundary terms are mixtures of reservoir and interaction contributions. These may cancel each other out, depending on the choice of the parameters κ_{f}, κ_{i} and κ.. The occurring correlation functions are

$$
\begin{aligned}
R_1(\tau, \varrho) &= -\frac{1}{\pi} \int_0^\infty d\omega \, \frac{J(\omega)}{\omega^2} \frac{\sinh[\frac{1}{2}\omega(\hbar\beta + i\,\varrho)]}{\sinh(\frac{1}{2}\omega\hbar\beta)} \sin(\tfrac{1}{2}\omega\varrho) \cos(\omega\tau), \\
R_2(\tau, \varrho) &= \frac{i}{\pi} \int_0^\infty d\omega \, \frac{J(\omega)}{\omega^2} \frac{\cosh[\frac{1}{2}\omega(\hbar\beta + i\,\varrho)]}{\sinh(\frac{1}{2}\omega\hbar\beta)} \sin(\tfrac{1}{2}\omega\varrho) \sin(\omega\tau),
\end{aligned} \tag{5.125}
$$

The function $R(\tau)$ is the second integral of the function (5.24), $\ddot{R}(\tau) = L(\tau)$,

$$
R''(\tau) \equiv \operatorname{Im} R(\tau) = \frac{1}{\pi} \int_0^\infty d\omega \, \frac{J(\omega}{\omega^2} \sin(\omega\tau) . \tag{5.126}
$$

The nth moment $\langle \mathcal{E}^{(n)}_{\kappa_{\mathrm{f}}, \kappa_{\mathrm{i}}, \kappa}(t) \rangle$ of the probability distribution function $P(\mathcal{E}, t)$ is found from the relation (5.113) as

$$
\begin{aligned}
\langle \mathcal{E}^{(n)}_{\kappa_{\mathrm{f}}, \kappa_{\mathrm{i}}, \kappa}(t) \rangle &\equiv \int dr_{\mathrm{f}} \int dr_{\mathrm{i}} \, p_{\mathrm{S}}(r_{\mathrm{i}}) \int_{r(0)=r_{\mathrm{i}}}^{r(t)=r_{\mathrm{f}}} \mathcal{D}r(\cdot) \int_{y(0)=0}^{y(t)=0} \mathcal{D}y(\cdot) \\
&\quad \times \; e^{i\,S_{\mathrm{S}}[r(\cdot), y(\cdot)]/\hbar} e^{-S_{\mathrm{FV}, \kappa}[r(\cdot), y(\cdot)]/\hbar} \, \varepsilon^{(n)}_{\kappa_{\mathrm{f}}, \kappa_{\mathrm{i}}, \kappa}[r(\cdot), y(\cdot)] ,
\end{aligned} \tag{5.127}
$$

where

$$
\varepsilon^{(n)}_{\kappa_{\mathrm{f}}, \kappa_{\mathrm{i}}, \kappa}[r(\cdot), y(\cdot)] = (-i\,\hbar\,\partial/\partial\varrho)^n \, e^{-S_{\kappa_{\mathrm{f}}, \kappa_{\mathrm{i}}, \kappa}[r(\cdot), y(\cdot), \varrho]/\hbar}\Big|_{\varrho=0} . \tag{5.128}
$$

For $n = 0$, the expression (5.127) reduces to the expression (5.118) at $\varrho = 0$, which is the normalization condition of the RDM, $G(0, t) = \operatorname{tr}[\hat{\rho}(t) \, e^{-\beta(\hat{H}_{\mathrm{R}} + \kappa\hat{H}_{\mathrm{I}})}]/Z_{\mathrm{R}} = 1$. As a result, only terms of the functional $\varepsilon^{(n)}_{\kappa_{\mathrm{f}}, \kappa_{\mathrm{i}}, \kappa}[r(\cdot), y(\cdot)]$ which depend on the boundary values of $r(\cdot)$ actually contribute to $\langle \mathcal{E}^{(n)}_{\kappa_{\mathrm{f}}, \kappa_{\mathrm{i}}, \kappa}(t) \rangle$.

In a full description of the energy transport processes, knowledge of all moments of the energy transferred between the various dissipative channels is essential. The functional for the mean transferred energy found from Eq. (5.128) with (5.124) is

$$
\begin{aligned}
\varepsilon_{\kappa_{\mathrm{f}}, \kappa_{\mathrm{i}}, \kappa}[r(\cdot), y(\cdot)] &= \int_0^t ds_2 \, \dot{r}(s_2) \int_0^{s_2} ds_1 \, [2Y''(s_2 - s_1)\dot{r}(s_1) - i\,Y'(s_2 - s_1)\dot{y}(s_1)] \\
&\quad - (1 - \kappa_{\mathrm{f}})r_{\mathrm{f}} \int_0^t ds \, [2Y''(t - s)\dot{r}(s) - i\,Y'(t - s)\dot{y}(s)] \\
&\quad + (1 - \kappa)r_{\mathrm{i}} \int_0^t ds \, 2Y''(s)\dot{r}(s) - 2(1 - \kappa)(1 - \kappa_{\mathrm{f}}) \, r_{\mathrm{f}}r_{\mathrm{i}} \, Y''(t) \\
&\quad + [(1 - \kappa)^2 - (\kappa_{\mathrm{i}} - \kappa)^2] \Delta V(r_{\mathrm{i}}) + (1 - \kappa_{\mathrm{f}})^2 \, \Delta V(r_{\mathrm{f}}) . \tag{5.129}
\end{aligned}
$$

The function $Y(\tau)$ is first integral of the bath correlation function $L(\tau)$, $\dot{Y}(\tau) = L(\tau)$,

$$
Y(\tau) = \frac{1}{\pi} \int_0^\infty d\omega \, \frac{J(\omega}{\omega} \left[\coth(\tfrac{1}{2}\omega\hbar\beta) \sin(\omega\tau) + i \, \cos(\omega\tau)\right] . \tag{5.130}
$$

In the last line of Eq. (5.129) we have used the relation $\Delta V(q) = Y''(0)q^2$

The functional integral representation (5.127) with (5.129) for the mean transferred energy ($n = 1$) is formally exact. It generally holds for any composite quantum system with bilinear system-bath coupling. But except in special cases it cannot be evaluated exactly. However, it may serve as starting point for implementation of powerful numerical schemes, e.g., the ones proposed in Ref. [238] and dealt with in Subsec. 5.5.2.

The amounts of energy transferred from the externally driven central system to the reservoir and to the reservoir-plus-interaction as function of time and system parameters, both for the uncorrelated canonical state (5.9) and the shifted canonical state (5.10), $\kappa = 0, 1$, respectively, could be of particular interest. There holds, e.g.,

$$\langle \mathcal{E}_{0,0,\kappa}(t) \rangle \equiv \langle E_{R,R}(t) \rangle_\kappa = \langle H_R(t) \rangle_\kappa - \langle H_R(0) \rangle_\kappa \,,$$

$$\langle \mathcal{E}_{1,1,\kappa}(t) \rangle \equiv \langle E_{RI,RI}(t) \rangle_\kappa = \langle H_R(t) + H_I(t) \rangle_\kappa - \langle H_R(0) + H_I(0) \rangle_\kappa \,, \tag{5.131}$$

while the energy transferred to the interaction alone is $\langle \mathcal{E}_{1,1,\kappa}(t) \rangle - \langle \mathcal{E}_{0,0,\kappa}(t) \rangle$. All these observables are related to specific measurement protocols.

For completeness, consider next an energy balance equation holding for the partial energies of the driven composite system. These are

$$\langle \mathcal{E}_S(t) \rangle_\kappa = \mathrm{tr}\left[\hat{H}_S(t)\hat{U}(t,0)\,\hat{\rho}(0)\,e^{-\beta[\hat{H}_R + \kappa \hat{H}_I]}\,\hat{U}^\dagger(t,0) \right] \,,$$

$$\langle \mathcal{E}_R(t) \rangle_\kappa = \mathrm{tr}\left[\hat{H}_R\,\hat{U}(t,0)\,\hat{\rho}(0)\,e^{-\beta[\hat{H}_R + \kappa \hat{H}_I]}\,\hat{U}^\dagger(t,0) \right] \,, \tag{5.132}$$

$$\langle \mathcal{E}_I(t) \rangle_\kappa = \mathrm{tr}\left[\hat{H}_I\,\hat{U}(t,0)\,\hat{\rho}(0)\,e^{-\beta[\hat{H}_R + \kappa \hat{H}_I]}\,\hat{U}^\dagger(t,0) \right] \,.$$

Here, $\hat{U}(t,0)$ is the evolution operator of the composite system, as in Eq. (5.116). Now, the mean energies $\langle \mathcal{E}_R(t) \rangle_\kappa$ and $\langle \mathcal{E}_I(t) \rangle_\kappa$ can be written in the functional integral form (5.127) with (5.129), whereas this is not directly possible for $\langle \mathcal{E}_S(t) \rangle_\kappa$. Fortunately, the partial energies are linked to the excess energy $\langle \mathcal{E}_{exc}(t) \rangle_\kappa$, pumped into the system by the external drive in Eq. (5.117), by the energy balance equation

$$\langle \mathcal{E}_S(t) \rangle_\kappa + \langle \mathcal{E}_R(t) \rangle_\kappa + \langle \mathcal{E}_I(t) \rangle_\kappa = \langle \mathcal{E}_{exc}(t) \rangle_\kappa \equiv - \int_0^t ds\, \dot{F}_{ext}(s) \langle r(s) \rangle_\kappa \,. \tag{5.133}$$

Preferably, one may determine $\langle \mathcal{E}_S(t) \rangle_\kappa$ from the balance relation (5.133), rather than from direct calculation of the expression (5.132).

The method put forward in this section is applied in Sec. 23.6 to energy transfer in the dissipative two-state system or spin-boson model. Building on the expressions (5.127) and (5.129), the balance relation (5.133) is directly proven for the dissipative two-state system in Subsec. 23.6.2.

Finally, we would like to mention that the generalization of the above method to two or many heat reservoirs at different temperatures coupled to the central system is straightforward. The appropriate extension essentially is that then are two or many influence actions in the second exponent of Eq. (5.127).

5.7 Brief summary and outlook

So far we have presented the general real-time functional integral formalism for the density matrix by means of the phenomenological system-plus-reservoir model described by the Hamiltonian (3.12). In the first section we have familiarized ourselves with the general formalism by studying the evolution of a product initial state. Then we have addressed the important question of the preparation of initial states. We have introduced a preparation function which describes the reduction of the equilibrium density matrix of the global system due to a measurement of certain variables of the system carried out at time $t = 0$. The subsequent evolution of the system is determined by a propagating function.

In the third section, we have studied the propagating function and found the exact formal solution in the compact form of a path integral, in which the path's progression in time is along a certain contour in the complex-time plane. We then have discussed a real-time formulation of the propagating function in which the system is evolving since the infinite past and then is subjected to a measurement of its position at time zero, and again at time t. The appropriate modifications required for the various microscopic models introduced in Section 3.3 are easy to understand, and the implementation is left to the reader. At this, the key point is that the combined action $-iS_S[\tilde{q}] + S_{\text{infl}}[\tilde{q}]$ in Eq. (5.87) or (5.91) for the various microscopic models results from the corresponding imaginary-time action $S_{\text{eff}}[q]$ by the substitution $\tau = iz$ and $q(\tau) = \tilde{q}(z)$.

In the fourth section we have dealt with the stochastic unreaveling of the influence functional and discussed the close relationship with stochastic Liouville-von Neuman equations. In the remaining sections, the semiclassical dissipative dynamics and the energy transfer in composite quantum systems are covered.

The methods of this chapter will be applied below *inter alia* to the dissipative two-state system and to quantum transport in multi-state systems.

PART II:

MISCELLANEOUS APPLICATIONS

So far I have employed functional integral methods in order to establish exact formal expressions which describe the thermodynamics and dynamics of damped quantum systems. Now we move to new grounds and consider the physical properties of exactly solvable damped linear systems, i.e. the damped quantum oscillator and quantum Brownian particle. Subsequently, I discuss a variational method which makes it possible for us to treat nonlinear damped quantum systems in much the same way as linear systems. I conclude this part with a brief general discussion of quantum decoherence.

6 Damped linear quantum mechanical oscillator

The interplay of friction and quantum-statistical fluctuations with impact on thermodynamics and dynamics can be studied quantitatively on a microscopic basis for *linear* systems. Of particular interest is a harmonic oscillator subjected to linear (state-independent) friction of arbitrary frequency-dependence.

The understanding of the quantum-mechanical damped oscillator is very important because of its universal relevance. This model applies for every macroscopic quantum system which is slightly displaced from a stable local potential minimum. Applications are, e.g., current oscillations in circuits with impedances (cf. Sect. 3.4.3).

The problem of a damped harmonic quantum mechanical oscillator has been studied intensively since the early works in the 1960s [20, 21, 76]. A satisfactory understanding in the classical regime and in the weak-coupling limit has been reached in the 1970s [23]–[27]. Substantial progress beyond the limitations of the weak-coupling approach was made in the 1980s, in particular with phenomenological methods [247], with study of the microscopic dynamics [248], and with Feynman-Vernon path integral techniques [249]. These and related works [55, 57, 209], [250]–[253] disclosed unexpected behaviors at low temperature, such as, e.g., nonexponential decay of correlation functions. Haake and Reibold derived from the microscopic dynamics for general memory friction and factorizing initial state an exact second order partial differential equation for the Wigner function, and they transformed it into the corresponding quantum master equation for the reduced density operator [248]. Later, the master equation was rederived by Hu, Paz, and Zhang with the Feynman-Vernon path integral method [254]. Grabert, Schramm, and Ingold determined with the Feynman-Vernon method along the lines given in Sections 4.2 and 5.2 the exact time evolution of the reduced density matrix for general memory friction and general initial preparation [88]. Subsequently, Karrlein and Grabert showed that the time-dependent Liouville operator of the respective quantum master equation depends on the initial preparation, and they determined it for thermal and factorizing initial conditions, and for different types of damping [48].

Consider a central harmonic oscillator with eigenfrequency ω_0 bilinearly coupled to a harmonic oscillator bath and subjected to a time-dependent external force $F_{\text{ext}}(t)$,

$$H(t) = \frac{p^2}{2M} + \frac{M\omega_0^2 q^2}{2} - qF_{\text{ext}}(t) + \sum_{\alpha=1}^{N} \left[\frac{p_\alpha^2}{2m_\alpha} + \frac{m_\alpha\omega_\alpha^2}{2}\left(x_\alpha - \frac{c_\alpha}{m_\alpha\omega_\alpha^2}q\right)^2 \right]. \quad (6.1)$$

Clearly, the Hamiltonian of the composite system can be diagonalized directly, as we have already discussed in Section 4.3.5. Since the action of the global system is quadratic, path integrals for thermodynamical and dynamical quantities can be calculated exactly. Above all, the Hamiltonian (6.1) gives *linear* equations of motion, which

154

can be solved in analytic form. Because of the linearity, a damped linear oscillator is a system simple enough that it allows for an exact study of the quantum mechanical stochastic process in the entire range of parameters by using only phenomenological considerations. This is possible because the quantum mechanical equation for $q(t)$ is in the form of a classical Langevin equation with quantum mechanical colored noise.

Before embarking on the specific model, we consider an important theorem which is generally valid for any particular linear and nonlinear quantum system.

6.1 Fluctuation-dissipation theorem

The fluctuation-dissipation theorem (FDT) is a cornerstone of linear response theory. The theorem relates relaxation of a weakly perturbed system to the spontaneous fluctuations in thermal equilibrium. Well-known special cases of the FDT in the classical regime are the Einstein relation [255], which connects the diffusion constant with the viscosity of a Brownian particle, and the Johnson-Nyquist formula [256, 257], which relates thermal current fluctuations in an electrical circuit to the impedance.

To illustrate how the FDT ensues from principles of quantum statistical mechanics, consider a system, say a particle in a potential $V(q)$, which is in the canonical equilibrium state. Imagine that at a particular time a weak time-dependent external force $\delta F_{\mathrm{ext}}(t)$ is switched on. By the perturbation, the particle is thrown out of equilibrium. With the additional force, the particle's Hamiltonian is time-dependent,

$$\mathcal{H}(t) \;=\; H + \delta H(t) \;=\; H - q\,\delta F_{\mathrm{ext}}(t)\,. \tag{6.2}$$

The time evolution of the average position $\langle q(t)\rangle$ for a mixed initial state is carried by the density operator $\rho(t)$ according to (in the sequel, we deliberately write operators without the hat symbol)

$$\langle q(t)\rangle \;=\; \mathrm{tr}\left\{\rho(t)\,q\right\}, \tag{6.3}$$

where $\mathrm{tr}\left\{\rho(t)\right\} = 1$. Here, $\mathrm{tr}\left\{\cdots\right\}$ denotes the usual quantum mechanical trace operation. The density matrix $\rho(t)$ obeys the equation of motion

$$i\hbar\,\dot{\rho}(t) \;=\; [\,\mathcal{H}(t),\,\rho(t)\,] \;=\; [\,H,\,\rho(t)\,] + [\,\delta H(t),\,\rho(t)\,]\,, \tag{6.4}$$

where $[\,A, B\,] = AB - BA$. We assume that the system is in the thermal equilibrium state in the infinite past, and then the perturbation $\delta H(t)$ is turned on,

$$\begin{aligned}
\lim_{t\to-\infty} \rho(t) &\;=\; \rho_{\mathrm{eq}} \;=\; e^{-\beta H}/\mathrm{tr}\left\{e^{-\beta H}\right\}, \\
\lim_{t\to-\infty} \langle q(t)\rangle &\;=\; \langle q\rangle_{\mathrm{eq}} \;=\; \mathrm{tr}\left\{e^{-\beta H}q\right\}/\mathrm{tr}\left\{e^{-\beta H}\right\},
\end{aligned} \tag{6.5}$$

For weak external perturbation $\delta H(t)$, the average displacement of the particle from the equilibrium position is a linear functional of the history of the time-dependent external force $\delta F_{\mathrm{ext}}(t)$,

$$\delta\langle q(t)\rangle \;\equiv\; \langle q(t)\rangle - \langle q\rangle_{\mathrm{eq}} \;=\; \int_{-\infty}^{+\infty} dt'\,\chi(t-t')\,\delta F_{\mathrm{ext}}(t')\,. \tag{6.6}$$

The response function $\chi(t)$ is a retarded temperature Green function, $\chi(t) = 0$ for $t < 0$, called generalized susceptibility. To determine $\chi(t)$, we put

$$\rho(t) = \rho_{eq} + \delta\rho(t) . \tag{6.7}$$

The change of the density matrix $\delta\rho(t)$ caused by the perturbation obeys the equation

$$i\hbar \, \partial\delta\rho(t)/\partial t = [\delta H(t), \rho_{eq}] + [H, \delta\rho(t)] . \tag{6.8}$$

This equation is easily solved and yields

$$\delta\rho(t) = \frac{1}{i\hbar} \int_{-\infty}^{t} dt' \, e^{-i H(t-t')/\hbar} [\delta H(t'), \rho_{eq}] \, e^{i H(t-t')/\hbar} . \tag{6.9}$$

Upon inserting the expression (6.9) into $\delta\langle q(t)\rangle = \mathrm{tr}\{\delta\rho(t)\,q\}$, and comparing with the expression (6.6), we find that $\chi(t)$ is connected with the antisymmetric part of the position auto-correlation function in thermal equilibrium by the relation

$$\chi(t) = \Theta(t) \langle [\, q(t)q(0) - q(0)q(t)\,]\rangle_\beta /2 . \tag{6.10}$$

The function $\chi(t)$ describes the linear response of the equilibrated system to an external force. The step function implicit in Eq. (6.10) ensures *causality*. Since we did not use specific properties of the system, the expression (6.10) is a general result for any quantum system within linear response.

Generally, the position autocorrelation function $C^\pm(t)$ is defined with a subtraction which serves that $\lim_{t\to\infty} C^\pm(t) = 0$, so that the Fourier transform is well-defined,

$$
\begin{aligned}
C^+(t) &\equiv C_{qq}(t) &&\equiv \langle q(t)q(0)\rangle_\beta - \langle q(t)\rangle_\beta\langle q(0)\rangle_\beta , \\
C^-(t) &\equiv C_{qq}(-t) &&\equiv \langle q(0)q(t)\rangle_\beta - \langle q(0)\rangle_\beta\langle q(t)\rangle_\beta .
\end{aligned} \tag{6.11}
$$

For a harmonic potential centered at $q = 0$, the subtraction is zero. The function $C^+(t)$ is linked to the displacement correlation function $\mathcal{C}^+(t)$ discussed in Subsec. 3.1.5 by the relation $C^+(t) = \langle q^2\rangle_\beta - \mathcal{C}^+(t)$. Since $C^{+*}(t) = C^-(t)$, we may decompose $C^\pm(t)$ into a real symmetric part and an imaginary odd part,

$$C^\pm(t) = S(t) \pm i\, A(t) . \tag{6.12}$$

The real part $S(t)$ is the symmetrized correlation function

$$S(t) = S(-t) = \langle [\, q(t)q(0) + q(0)q(t)\,]\rangle_\beta /2 , \tag{6.13}$$

and the imaginary part is the expectation value of the commutator,

$$A(t) = -A(-t) = \langle [\, q(t)q(0) - q(0)q(t)\,]\rangle_\beta /2i . \tag{6.14}$$

According to Eqs. (6.10) and (6.14), the response function $\chi(t)$ is related to $A(t)$ by

$$\chi(t) = -(2/\hbar)\Theta(t)A(t) . \tag{6.15}$$

It is expedient to specify this relation in Fourier space. With the Fourier transforms

$$\tilde{C}^{\pm}(\omega) \equiv \int_{-\infty}^{\infty} dt\, C^{\pm}(t)\, e^{i\omega t} = \tilde{S}(\omega) \pm i\,\tilde{A}(\omega)\,,$$

$$\tilde{\chi}(\omega) \equiv \int_{0}^{\infty} dt\, \chi(t)\, e^{i\omega t}\,, \tag{6.16}$$

the relation (6.15) is rephrased in Fourier space as

$$\tilde{\chi}''(\omega) = \frac{i}{\hbar}\,\tilde{A}(\omega) = \frac{1}{2\hbar}\left(\tilde{C}^{+}(\omega) - \tilde{C}^{-}(\omega)\right)\,, \tag{6.17}$$

where $\tilde{\chi}''(\omega) \equiv \operatorname{Im}\chi(\omega)$ is the absorptive part of the dynamical susceptibility.

With this groundwork it is straight to establish the fluctuation-dissipation theorem. This can be done in two different ways. In the first, one utilizes that the canonical operator $e^{-\beta H}$ is an imaginary-time translational operator, $e^{-\beta H}q(0)\,e^{\beta H} = q(i\,\hbar\beta)$, and then employs invariance of the trace under cyclic permutation. This yields

$$\operatorname{tr}\{e^{-\beta H}q(0)\,q(t)\} = \operatorname{tr}\{q(i\,\hbar\beta)\,e^{-\beta H}q(t)\} = \operatorname{tr}\{e^{-\beta H}q(t)\,q(i\,\hbar\beta)\}\,. \tag{6.18}$$

Since the correlations in thermal equilibrium are time translation invariant,

$$\langle q(t)q(t')\rangle_{\beta} = \langle q(t-t')q(0)\rangle_{\beta}\,, \tag{6.19}$$

we find from Eq. (6.18) the relation

$$C^{-}(t) = C^{+}(t - i\,\hbar\beta)\,. \tag{6.20}$$

In Fourier space, this result turns into the central relation

$$\tilde{C}^{-}(\omega) = e^{-\beta\hbar\omega}\,\tilde{C}^{+}(\omega)\,. \tag{6.21}$$

In the alternative second route, we write the correlation functions (6.11) in energy representation and then switch to the respective Fourier transforms. This yields

$$\tilde{C}^{+}(\omega) = \frac{2\pi\hbar}{Z}\sum_{n,m} e^{-\beta E_{m}} <n|q|m><m|q|n> \delta(E_{m} - E_{n} + \hbar\omega)\,,$$

$$\tilde{C}^{-}(\omega) = \frac{2\pi\hbar}{Z}\sum_{n,m} e^{-\beta E_{n}} <n|q|m><m|q|n> \delta(E_{m} - E_{n} + \hbar\omega)\,. \tag{6.22}$$

Due to the constraint $E_{n} = E_{m} + \hbar\omega$, the expressions (6.22) are again linked together by the relation (6.21).

Insertion of the central relation (6.21) into Eq. (6.17) yields the fluctuation-dissipation theorem discovered first by Callen and Welton in 1951 [258],

$$\tilde{\chi}''(\omega) = \frac{1}{2\hbar}\left(1 - e^{-\beta\hbar\omega}\right)\tilde{C}^{+}(\omega)\,. \tag{6.23}$$

We can also relate the Fourier transform of the symmetrized autocorrelation function to $\tilde{\chi}''(\omega)$. Upon using the relation $\tilde{S}(\omega) = \frac{1}{2}[\tilde{C}^{+}(\omega) + \tilde{C}^{-}(\omega)]$ and Eq. (6.21), we find

$$\tilde{S}(\omega) = \hbar\coth(\beta\hbar\omega/2)\,\tilde{\chi}''(\omega)\,. \tag{6.24}$$

There are two important limiting cases of the fluctuation-dissipation theorem (6.24):

(i) At very low temperatures, $k_B T \ll \hbar\omega$, there are pure quantum fluctuations,

$$\tilde{S}(\omega) = \hbar \, \tilde{\chi}''(|\omega|) \,. \tag{6.25}$$

According to Eq. (6.25), the function $\tilde{S}(\omega)$ is non-analytic at $\omega = 0$, and thus the position autocorrelation function decays algebraically with time at $T = 0$.

(ii) In the high temperature regime, $k_B T \gg \hbar\omega$, the spectral function $\tilde{S}(\omega)$ takes the classical ($\hbar$-independent) form ,

$$\tilde{S}(\omega) = 2k_B T \, \tilde{\chi}''(\omega)/\omega \,. \tag{6.26}$$

Special cases of this formula are the Einstein relation, which relates the diffusion coefficient D to the mobility μ, $D = k_B T \, \mu$ (see Section 7.3 for linear systems and Subsection 26.3.3 for a nonlinear system), and the Johnson-Nyquist formula, which relates the power spectrum of the current fluctuations $\tilde{S}_{\mathrm{II}}(\omega) = \int_{-\infty}^{\infty} dt \, \langle I(t) I(0) \rangle_\beta \, e^{i\omega t}$ in an electrical circuit to the admittance $\tilde{Y}(\omega) = \tilde{Z}(\omega)/R_K^2$, where $\tilde{Z}(\omega)$ is the impedance of the circuit, and $R_K = 2\pi\hbar/e^2$ is the resistance quantum, $\tilde{S}_{\mathrm{II}}(\omega) = 2k_B T \, \mathrm{Re} \, \tilde{Y}(\omega)$. The quantum version of the Johnson-Nyquist formula is

$$\tilde{S}_{\mathrm{II}}(\omega) = \hbar\omega \coth\left(\frac{\hbar\omega}{2k_B T}\right) \mathrm{Re} \, \tilde{Y}(\omega) = 2\left(\frac{\hbar\omega}{2} + \frac{\hbar\omega}{e^{\beta\hbar\omega} - 1}\right) \mathrm{Re} \, \tilde{Y}(\omega) \,. \tag{6.27}$$

The quantum version was already anticipated by Nyquist in his 1928 paper [257], but devoid of the zero-point energy term in the second equality.

It is appropriate to emphasize that at no step of the derivation we used particular properties of a linear system. Thus, the FDT relations (6.23) and (6.24) are generally valid for any open quantum system in thermal equilibrium. Therefore, they have broad implications far beyond the particular linear model systems discussed here.

6.2 Stochastic modeling

The essential features of linear dissipative quantum systems can be gathered straightaway from a simple stochastic modeling [247] which is based upon the following three principles:

1. The mean values obey the classical equations of motion (*Ehrenfest's theorem*).

2. The response function and the equilibrium autocorrelation function are related by the *fluctuation-dissipation theorem*.

3. The stochastic process is a *stationary Gaussian process*.

While the first two principles are generally valid, the assumption of a strictly Gaussian stochastic process is limited to linear systems. For *linear* systems, the response to an external perturbation is *linear* for arbitrary strength of the perturbation. Thus, all we said in the previous subsection for an infinitesimal perturbation $\delta F_{\rm ext}(t)$ is generally valid for an external force $F_{\rm ext}(t)$ of any strength in Eq. (6.1).

By virtue of Ehrenfest's theorem, the average position $\langle q(t) \rangle$ of a damped quantum oscillator obeys the equation of motion

$$\langle \ddot{q}(t) \rangle + \int_{-\infty}^{t} dt' \, \gamma(t - t') \langle \dot{q}(t') \rangle + \omega_0^2 \langle q(t) \rangle \;=\; \frac{1}{M} F_{\rm ext}(t) \, . \tag{6.28}$$

To be general, we have permitted memory-friction. The response of $\langle q(t) \rangle$ to a general force $F_{\rm ext}(t)$ is given by [cf. Eq. (6.6)]

$$\langle q(t) \rangle \;=\; \int_{-\infty}^{+\infty} dt' \, \chi(t - t') F_{\rm ext}(t') \, , \tag{6.29}$$

where $\chi(t)$ is the causal response function, $\chi(t) = 0$ for $t \leqslant 0$. Causality requires that the dynamical susceptibility

$$\tilde{\chi}(\omega) \;=\; \int_{-\infty}^{\infty} dt \, \chi(t) \, e^{i\omega t} \tag{6.30}$$

is an analytic function of ω in the upper half of the complex ω-plane, $\mathrm{Im}\, \omega > 0$. This is the prerequisite underlying the remarkable Kramers-Kronig dispersion relations

$$\tilde{\chi}'(\omega) \;=\; \fint_{-\infty}^{\infty} \frac{d\nu}{\pi} \frac{\tilde{\chi}''(\nu)}{\nu - \omega} \, , \qquad \tilde{\chi}''(\omega) \;=\; -\fint_{-\infty}^{\infty} \frac{d\nu}{\pi} \frac{\tilde{\chi}'(\nu)}{\nu - \omega} \, , \tag{6.31}$$

where the dash denotes prinipal value. These relations are fundamental in nearly all areas of physics, as they express dispersion in terms of absorption and vice versa.

For a *linear* quantum system, the Ehrenfest equation (6.28) agrees exactly with the corresponding classical equation of motion. This implies that the quantum mechanical and the classical response functions are identical for linear systems, $\chi(t) = \chi_{\rm cl}(t)$.

For the linear equation of motion (6.28), the dynamical susceptibility reads

$$\tilde{\chi}(\omega) \equiv \tilde{\chi}'(\omega) + i\, \tilde{\chi}''(\omega) \;=\; \frac{1}{M} \frac{1}{\omega_0^2 - \omega^2 - i\omega\tilde{\gamma}(\omega)} \, , \tag{6.32}$$

where $\tilde{\gamma}(\omega) = \int_0^\infty dt\, \gamma(t)\, e^{i\omega t}$ is the frequency-dependent damping coefficient. To complete the picture, we also give the response function in Laplace representation,

$$\chi(t) \;=\; \frac{1}{2\pi i} \int_{-i\infty+c}^{i\infty+c} dz \, \hat{\chi}(z) \, e^{zt} \, , \tag{6.33}$$

$$\hat{\chi}(z) \;\equiv\; \int_0^\infty dt \, e^{-zt} \, \chi(t) \;=\; \frac{1}{M} \frac{1}{\omega_0^2 + z^2 + z\hat{\gamma}(z)} \, . \tag{6.34}$$

The constant c in the integral expression (6.33) must be chosen such that all singularities of $\hat{\chi}(z)$ are lying to the left of the integration path.

By virtue of the fluctuation-dissipation theorem (6.23), the position autocorrelation function in thermal equilibrium[1]

---

[1] Note that there is no subtraction in $C_{qq}(t)$ since the equilibrium average $\langle q \rangle_\beta$ is zero.

$$C^+(t) \equiv C_{qq}(t) \equiv \langle q(t)q(0)\rangle_\beta \tag{6.35}$$

is determined by the absorptive part of the dynamical susceptibility,

$$C^+(t) = \frac{\hbar}{\pi} \int_{-\infty}^{\infty} d\omega \, \tilde{\chi}''(\omega) \frac{e^{-i\omega t}}{1 - e^{-\beta\hbar\omega}}, \tag{6.36}$$

$$\tilde{\chi}''(\omega) = M\omega\tilde{\gamma}'(\omega)\tilde{\chi}(\omega)\tilde{\chi}^*(\omega) = \frac{1}{M} \frac{\omega\tilde{\gamma}'(\omega)}{[\omega^2 - \omega_0^2 - \omega\tilde{\gamma}''(\omega)]^2 + \omega^2\tilde{\gamma}'^2(\omega)}. \tag{6.37}$$

For frequency-independent damping, $\tilde{\gamma}(\omega) = \gamma$, the expression (6.37) is essentially a superposition of two Lorentzians of width $\gamma/2$ centered at $\omega = \pm\sqrt{\omega_0^2 - \gamma^2/4}$. In the limit $\gamma \to 0$, the Lorentzians reduce to δ-functions located at $\omega = \pm\omega_0$.

From the above, we draw an important conclusion: as the dynamical susceptibility $\tilde{\chi}(\omega)$ is purely classical for any linear system, quantum mechanics of linear systems is completely determined by the fluctuation-dissipation theorem (6.24).

In a stationary stochastic process, the time correlations for dynamical variables do *not* depend on absolute time. Time translation invariance of correlations is reflected by the property (6.19). Thus, once we have gained knowledge about the position correlation function $C_{qq}(t)$, some other pair correlation functions can be found by differentiation. Upon relating the momentum to the time derivative of the position and using Eq. (6.19), we obtain the simple relations

$$\begin{aligned}
C_{pq}(t) &\equiv \langle p(t)q(0)\rangle_\beta &=& M\,\dot{C}_{qq}(t)\,, \\
C_{qp}(t) &\equiv \langle q(t)p(0)\rangle_\beta &=& -M\,\dot{C}_{qq}(t)\,, \\
C_{pp}(t) &\equiv \langle p(t)p(0)\rangle_\beta &=& -M^2\,\ddot{C}_{qq}(t)\,.
\end{aligned} \tag{6.38}$$

It follows from property (iii) that correlation functions with an odd number of variables (position or momentum) vanish whereas correlation functions with an even number of variables can be written as a sum of combinations of factorized pair correlation functions. Within each pair, the original order of the variables is preserved [259]. For illustration, we give an example,

$$\langle q(t)p(t')p(t'')q(0)\rangle_\beta = \langle q(t)p(t')\rangle_\beta \langle p(t'')q(0)\rangle_\beta \tag{6.39}$$

$$+ \langle q(t)p(t'')\rangle_\beta \langle p(t')q(0)\rangle_\beta + \langle q(t)q(0)\rangle_\beta \langle p(t')p(t'')\rangle_\beta\,.$$

Thus, every correlation function with an even number of dynamical variables can be expressed in terms of the position autocorrelation function and of its time derivatives. In this way we can determine the complete quantum mechanical stochastic process of the damped harmonic system.

6.3 Susceptibility

6.3.1 Strictly Ohmic and Drude damping

For *Ohmic* or frequency-independent damping, the dynamical susceptibility (6.32) has two poles in the lower ω half-plane at $\omega = -i\lambda_1^{(0)}$ and at $\omega = -i\lambda_2^{(0)}$, where

$$\lambda_{1,2}^{(0)} = \begin{cases} \frac{1}{2}\gamma \pm i\omega_\gamma\,, & 0 < \gamma < 2\omega_0\,, \\ \frac{1}{2}\gamma \pm \gamma_\omega\,, & \gamma > 2\omega_0\,. \end{cases} \tag{6.40}$$

Here $\omega_\gamma = \sqrt{\omega_0^2 - \frac{1}{4}\gamma^2}$ and $\gamma_\omega = \sqrt{\frac{1}{4}\gamma^2 - \omega_0^2}$. In Laplace space, the susceptibility is

$$\hat{\chi}(z) = \frac{1}{M}\frac{1}{\omega_0^2 + z^2 + \gamma z} = \frac{1}{M}\frac{1}{(z + \lambda_1^{(0)})(z + \lambda_2^{(0)})}\,. \tag{6.41}$$

The response function (6.33) undergoes damped oscillations in the underdamped regime $0 < \gamma < 2\omega_0$, and biexponentiell decay in the overdamped regime $\gamma > 2\omega_0$,

$$\chi(t) = \Theta(t)\begin{cases} \dfrac{1}{M\omega_\gamma}\sin(\omega_\gamma t)\,e^{-\gamma t/2}\,, & 0 < \gamma < 2\omega_0\,, \\ \dfrac{1}{2M\gamma_\omega}\left(e^{-(\gamma/2-\gamma_\omega)t} - e^{-(\gamma/2+\gamma_\omega)t}\right)\,, & \gamma > 2\omega_0\,. \end{cases} \tag{6.42}$$

Drude damping

When $\hat{\gamma}(z)$ is rational in z, the characteristic equation determining the poles of $\hat{\chi}(z)$ is algebraic. For Drude damping, Eq. (3.43), the susceptibility (6.34) takes the form

$$\hat{\chi}(z) = \frac{1}{M}\frac{z + \omega_D}{(\omega_0^2 + z^2)(\omega_D + z) + \gamma\omega_D z} = \frac{1}{M}\frac{z + \omega_D}{(z + \lambda_1)(z + \lambda_2)(z + \lambda_3)}\,, \tag{6.43}$$

with the Vieta relations

$$\lambda_1 + \lambda_2 + \lambda_3 = \omega_D\,, \quad \lambda_1\lambda_2 + \text{cycl.} = \omega_0^2 + \gamma\omega_D\,, \quad \lambda_1\lambda_2\lambda_3 = \omega_0^2\omega_D\,. \tag{6.44}$$

The λ_k consistently have positive real parts. For $\gamma < \gamma_{cr}$, one is real, and the other two are complex conjugate, whereas for $\gamma > \gamma_{cr}$ all three are real. The critical damping up to first order in $1/\omega_D$ is $\gamma_{cr} = 2\omega_0[1 - \omega_0/\omega_D]$. In the regime $\gamma < \gamma_{cr}$, we may write

$$\lambda_{1,2} = \tfrac{1}{2}\Gamma \pm i\Omega\,, \quad \text{with} \quad \Omega = \sqrt{\Omega_0^2 - \tfrac{1}{4}\Gamma^2}\,,$$

$$\lambda_3 = \Omega_D \equiv \omega_D - \Gamma\,. \tag{6.45}$$

Expanding Ω_0 and Γ up to terms of order $1/\omega_D^2$, we obtain

$$\Omega_0 = \left(1 + \frac{\gamma}{2\omega_D} + +\frac{7\gamma^2}{8\omega_D^2}\right)\omega_0\,,$$

$$\Gamma = \left(1 + \frac{\gamma}{\omega_D} + \frac{2\gamma^2 - \omega_0^2}{\omega_D^2}\right)\gamma\,, \tag{6.46}$$

Gathering up the three pole contribution of $\hat\chi(z)$ in the contour integral (6.33), the response function takes in the underdamped regime $\gamma < \gamma_{\rm cr}$ the analytic form

$$\chi(t) = \Theta(t)\frac{1}{M[\Omega^2 + (\Omega_{\rm D} - \frac{1}{2}\Gamma)^2]}$$
$$\times \left\{\Gamma(e^{-\Omega_{\rm D}t} - e^{-\Gamma t/2}\cos\Omega t) + [\Omega + (\Omega_{\rm D}^2 - \Gamma^2/4)/\Omega]\,e^{-\Gamma t/2}\sin\Omega t\right\}. \tag{6.47}$$

In the transition from $\gamma < \gamma_{\rm cr}$ to $\gamma > \gamma_{\rm cr}$, the complex conjugate poles move to the real axis. As a result, the response function changes from (6.47) to triple exponential decay.

6.3.2 Radiation damping

The interaction of the radiation field with a charged particle is described by the super-Ohmic spectral density $J(\omega) = (2e^2/3c^3)\,\omega^3/(1 + \omega^2/\omega_{\rm D}^2)$, as we have found in Subsec. 3.1.9. With this form, the integral representation (3.29) yields the spectral damping function

$$\hat\gamma(z) = \gamma_3\frac{z}{z + \omega_{\rm D}} \qquad \text{with} \qquad \gamma_3 = \frac{2e^2\omega_{\rm D}^2}{3c^3 M}. \tag{6.48}$$

The resulting dynamical susceptibility is

$$\hat\chi(z) = \frac{1}{M}\frac{z + \omega_{\rm D}}{(\omega_0^2 + z^2)(\omega_{\rm D} + z) + \gamma_3 z^2} = \frac{1}{M}\frac{z + \omega_{\rm D}}{(z + \lambda_1)(z + \lambda_2)(z + \lambda_3)}. \tag{6.49}$$

There are two complex conjugate poles, and a real one,

$$\lambda_{1,2} = \tfrac{1}{2}\Gamma \pm i\sqrt{\Omega_0^2 - \tfrac{1}{4}\Gamma^2};, \qquad \lambda_3 = \Omega_{\rm D} \equiv \omega_{\rm D} + \gamma_3 - \Gamma. \tag{6.50}$$

Taking into accout terms up to order $1/\omega_{\rm D}^2$, one obtains

$$\Omega_0 = \left(1 - \frac{\gamma_3}{2\omega_{\rm D}} + \frac{3\gamma_3^2}{8\omega_{\rm D}^2}\right)\omega_0, \qquad \Gamma = \frac{\omega_0^2}{\omega_{\rm D}^2}\gamma_3, \tag{6.51}$$

As distinct from Eq. (6.46), the $1/\omega_{\rm D}$-term in Ω_0 is negative. This is, because the linear part $\gamma_3 z/\omega_{\rm D}$ of $\hat\gamma(z)$ yields the mass contribution $\Delta M = (\gamma_3/\omega_{\rm D})M$. The added mass results in a negative frequency shift $\Delta\omega = -\frac{1}{2}(\Delta M/M)\,\omega_0 = -\frac{1}{2}(\gamma_3/\omega_{\rm D})\omega_0$. This is precisely the leading contribution to $\Omega_0 - \omega_0$ in Eq. (6.51).

Since the susceptibilities (6.49) and (6.43) match each other, the response function for radiation damping coincides with that for Drude damping, Eq. (6.47). The only distinction is in the relations of Ω_0, Γ and $\Omega_{\rm D}$ with the original parameters. In contrast to Ohmic or Drude damping, the dynamics for radiation damping is consistently underdamped, $\Gamma/2 < \Omega_0$. With the above correspondences, the dynamical and thermodynamical results presented below for Drude-regularized Ohmic friction also hold for radiation damping.

6.4 The position autocorrelation function

Consider now the position autocorrelation function (6.36) in some detail. First of all, the function $C^+(t)$ may be analytically continued to complex time $z = t - i\tau$ in the strip $0 < \tau < \hbar\beta$. The Euclidean or imaginary-time correlation function is real,

$$\mathbb{C}(\tau) \equiv C^+(z = -i\tau) . \tag{6.52}$$

There follows from Eq. (6.20) or directly from (6.36) the relation $\mathbb{C}(0) = \mathbb{C}(\hbar\beta)$.

If we agree that $\mathbb{C}(\tau)$ is periodically continued outside of the principal interval $0 \leqslant \tau < \hbar\beta$, it can be written as Fourier series

$$\mathbb{C}(\tau) = \frac{1}{\hbar\beta} \sum_{n=-\infty}^{+\infty} c_n \, e^{i\nu_n\tau} , \tag{6.53}$$

where the frequencies ν_n are the bosonic Matsubara frequencies

$$\nu_n = \frac{2\pi}{\hbar\beta} n , \qquad n = \pm 1, \pm 2, \cdots . \tag{6.54}$$

The Fourier coefficient c_n can be related to the dynamical susceptibility $\tilde{\chi}(\omega)$. Use of Eqs. (6.52) and (6.36) yields

$$
\begin{aligned}
c_n &= \int_0^{\hbar\beta} d\tau \, \mathbb{C}(\tau) \, e^{-i\nu_n\tau} \\
&= \frac{\hbar}{\pi} \int_{-\infty}^{\infty} d\omega \, \frac{\tilde{\chi}''(\omega)}{\omega + i\nu_n} = \frac{\hbar}{\pi} \, \mathrm{Im} \int_{-\infty}^{\infty} d\omega \, \frac{\omega \, \tilde{\chi}(\omega)}{\omega^2 + \nu_n^2} .
\end{aligned}
\tag{6.55}
$$

As $\tilde{\chi}(\omega)$ is analytic in the upper half-plane, there follows by contour integration[2]

$$c_n = c_{-n} = \hbar\tilde{\chi}(\omega = i\,|\nu_n|) = \hbar\hat{\chi}(z = |\nu_n|) , \tag{6.56}$$

and finally with the explicit form (6.32),

$$c_n = \hbar\,\hat{\chi}(|\nu_n|) = \frac{\hbar}{M} \, \frac{1}{\omega_0^2 + \nu_n^2 + |\nu_n|\,\hat{\gamma}(|\nu_n|)} . \tag{6.57}$$

Back to real time, the symmetric and antisymmetric parts of $C^+(t)$ in Eq. (6.12) can be written by means of Eqs. (6.24) and (6.17) as

$$S(t) = \frac{\hbar}{2\pi} \int_{-\infty}^{\infty} d\omega \, \tilde{\chi}''(\omega) \coth(\beta\hbar\omega/2) \, e^{-i\omega t} , \tag{6.58}$$

$$A(t) = \frac{\hbar}{2\pi i} \int_{-\infty}^{\infty} d\omega \, \tilde{\chi}''(\omega) \, e^{-i\omega t} , \tag{6.59}$$

and, according to the relation (6.15), the response function has the representation

---

[2] We recall that the Fourier transform is denoted by a tilde and the Laplace transform by a hat.

$$\chi(t) \;=\; \Theta(t)\frac{2}{\pi}\int_0^\infty d\omega\,\tilde{\chi}''(\omega)\sin(\omega t) \;. \tag{6.60}$$

Consider next the initial value $S(0)$, which is the position dispersion $\langle q^2\rangle_\beta$ in thermal equilibrium. This quantity may be written as an integral representation as well as a Matsubara sum, as follows from Eq. (6.58) and from Eq. (6.53) with (6.57),

$$S(0) = \langle q^2\rangle_{\rm eq} = \frac{\hbar}{2\pi}\int_{-\infty}^\infty d\omega\,\coth(\tfrac{1}{2}\beta\hbar\omega)\,{\rm Im}\,\tilde{\chi}(\omega) \;=\; \frac{1}{\beta}\sum_{n=-\infty}^\infty \hat{\chi}(|\nu_n|) \;. \tag{6.61}$$

The integral matches the Matsubara sum because $\tilde{\chi}(\omega)$ is regular in the upper half-plane, ${\rm Im}\,\omega > 0$, and because $\coth(\tfrac{1}{2}\beta\hbar\omega)$ has simple poles at $\omega = i\,|\nu_n|$,

$$\coth(\beta\hbar\omega/2) \;=\; \frac{1}{\hbar\beta}\sum_{n=-\infty}^\infty \frac{2\omega}{\omega^2 + \nu_n^2} \;. \tag{6.62}$$

A closer look at the position dispersion $\langle q^2\rangle_\beta$ will be given below in Sec. 6.6.

In the classical limit, the position autocorrelation function takes the form

$$C_{\rm cl}^+(t) \;=\; S_{\rm cl}(t) \;=\; \frac{1}{\beta\pi}\int_{-\infty}^\infty d\omega\,\frac{\tilde{\chi}''(\omega)}{\omega}\,e^{-i\omega t} \;. \tag{6.63}$$

Analytic expressions for the functions $S(t)$, $A(t)$ and $\chi(t)$ can be derived for special forms of the damping function $\tilde{\gamma}(\omega)$.

6.4.1 Ohmic friction

For strictly Ohmic friction, $\hat{\gamma}(|\nu_n|) = \gamma$, the response function $\chi(t)$ is given in (6.42), and similar form holds for $A(t)$. As regards the symmetrized position correlation function $S(t)$ for $t > 0$, the integration contour in Eq. (6.58) must be closed in the lower half of the complex-ω-plane. This yields two contributions,

$$S(t) \;=\; S_1(t) + S_2(t) \;. \tag{6.64}$$

The first is the contribution from the poles of $\tilde{\chi}''(\omega)$ located at $\omega = -i\,\gamma/2 \pm \omega_\gamma$. In the underdamped regime, $0 < \gamma < 2\omega_0$, the frequency $\omega_\gamma = \sqrt{\omega_0^2 - \gamma^2/4}$ is real,

$$S_1(t) \;=\; \frac{\hbar}{2M\omega_\gamma}\,\frac{\left[\sinh(\beta\hbar\omega_\gamma)\cos(\omega_\gamma t) + \sin(\beta\hbar\gamma/2)\sin(\omega_\gamma|t|)\right]}{\cosh(\beta\hbar\omega_\gamma) - \cos(\beta\hbar\gamma/2)}\,e^{-\gamma|t|/2} \;. \tag{6.65}$$

The second term is the contribution from the infinite sequence of simple poles of the function $\coth(\beta\hbar\omega/2)$ located at $\omega = -i\,\nu_n$ $(n = 1, 2, \ldots)$,

$$S_2(t) \;=\; -\frac{2\gamma}{M\beta}\sum_{n=1}^\infty \frac{\nu_n\,e^{-\nu_n|t|}}{(\omega_0^2 + \nu_n^2)^2 - \gamma^2\nu_n^2} \;. \tag{6.66}$$

The sum can alternatively be written, with a partial fraction decomposition, as a linear combination of four hypergeometric functions [88].

In the Ohmic case, the Matsubara sum (6.61) for the initial value is

$$S(0) = \frac{1}{M\beta} \sum_{n=-\infty}^{\infty} \frac{1}{\omega_0^2 + \nu_n^2 + \gamma|\nu_n|} .$$ (6.67)

The interested reader may verify for himself via partial fraction decomposition that $S_1(0) + S_2(0) = S(0)$ is satisfied indeed. In the regime $k_B T \gg \hbar\omega_0$, the term $S_1(0)$ grows linearly with T, whereas the second term behaves as $S_2(0) \propto 1/T^2$. Hence the ratio $|S_2(0)|/S_1(0)$ decreases as $1/T^3$, as the classical domain is approached,

$$|S_2(0)|/S_1(0) = \zeta(3)(2\gamma/\omega_0)(\hbar\omega_0/2\pi k_B T)^3 .$$ (6.68)

Here $\zeta(3)$ is a Riemann number. In addition, when $k_B T > \hbar\gamma/4\pi$, i.e., $\nu_1 > \gamma/2$, the function $S_2(t)$ drops to zero faster than $S_1(t)$. For these two reasons, the contribution $S_2(t)$ is negligibly small at high temperature for all t. In the asymptotic high temperature limit, $S_1(t)$ becomes independent of Planck's constant, while $S_2(t)$ drops to zero. The resulting classical correlation function reads

$$S(t) \xrightarrow{\hbar \to 0} S_{cl}(t) = \frac{1}{M\beta\omega_0^2} \left[\cos(\omega_\gamma t) + \left(\frac{\gamma}{2\omega_\gamma} \right) \sin(\omega_\gamma |t|) \right] e^{-\gamma|t|/2} .$$ (6.69)

Conversely, as temperature is lowered, quantum effects appear in $S_1(t)$ and above all the term $S_2(t)$ becomes increasingly important.

Historically, the leading quantum corrections were calculated with weak-coupling theories like the quantum master equation method [23, 25, 27, 30]. However, weak-coupling theories fail to describe the temperature regime $k_B T \lesssim \hbar\gamma/4\pi$, since the weak-coupling assumption is not valid anymore.

In the temperature regime $0 \leqslant k_B T < \hbar\gamma/4\pi$, the decay of $S(t)$ at long time is determined by the contribution of the lowest Matsubara frequency in Eq. (6.66) which is $\propto e^{-\nu_1 t}$, and not anymore by the term $S_1(t) \propto e^{-\gamma t/2}$.

As temperature falls to zero, the Matsubara sum of exponentials in Eq. (6.66) merges into an integral, which is a linear combination of exponential integrals,

$$S_2(t) = -\frac{\hbar\gamma}{M\pi} \int_0^\infty d\nu \frac{\nu e^{-\nu|t|}}{(\omega_0^2 + \nu^2)^2 - \gamma^2\nu^2} .$$ (6.70)

Asymptotic evaluation of the integral (6.70) for time $t \gg 1/\omega_0$, γ/ω_0^2 yields algebraic decay. The leading contribution is

$$S_2(t) = -\frac{\hbar\gamma}{\pi M\omega_0^4} \frac{1}{t^2} = -\frac{M\hbar\gamma}{\pi} \frac{\chi_0^2}{t^2} .$$ (6.71)

In the last form, we have introduced the static susceptibility $\chi_0 = 1/M\omega_0^2$.

The long-time tail of the correlation function $S(t)$ with power law decrease $\propto t^{-2}$ is characteristic for Ohmic friction. At low T, the algebraic decay $\propto 1/t^2$ occurs at intermediate times before the exponential decay $\propto e^{-\nu_1 t}$ sets in [249].

The function $S(t)$ for the Drude model (3.43) is found along the same lines with use of Eq. (6.43). Since a high-frequency cutoff does not change the low-frequency behavior of the dynamical susceptibility, the $1/t^2$ tail in Eq. (6.71) also holds for Ohmic friction with a Drude cutoff.

6.4.2 Non-Ohmic spectral density

The damping function $\hat{\gamma}(z)$ is an analytic function of z for odd s, and is nonanalytic at $z = 0$ for all other values of s, as follows from the integral (3.29) with (3.38). As a result, the Laplace transform $\hat{\chi}(z)$ in Eq. (6.34) has a branch point at $z = 0$ for $s \neq 1 + 2n$. The complex z-plane is cut along the negative real axis, and in the cut plane $\hat{\chi}(z)$ is single-valued. The principal branch of z^s is taken in the cut plane.

Coherent contribution

The two complex conjugate poles of $\hat{\chi}(z)$ for zero friction move away from the positions $z = \pm i\,\omega_0$ when friction is switched on. Depending on the particular values chosen for the parameters s, γ_s and ω_{ph}. they may or may not lie in the cut plane. The locations of the poles can not be given in analytic form for general s. If they lie in the cut plane, they lead to damped oscillatory contributions $A_{\mathrm{coh}}(t)$, $\chi_{\mathrm{coh}}(t)$ and $S_{\mathrm{coh}}(t)$. Generally the amplitudes of $S_{\mathrm{coh}}(t)$ are temperature-dependent.

Cut contribution

In addition to the coherent contribution of the simple poles in the cut plane there is the incoherent cut contribution

$$\chi_{\mathrm{cut}}(t) = \frac{1}{\pi} \, \mathrm{Im} \int_0^\infty d\nu \, \hat{\chi}(\nu\, e^{-i\pi}) \, e^{-\nu t} \ . \tag{6.72}$$

The leading asymptotic contribution from the cut expressed in terms of the static susceptibility χ_0 is

$$\chi_{\mathrm{cut}}(t) = \frac{M\chi_0^2}{\pi} \, \mathrm{Im} \int_0^\infty d\nu \, \nu \, \hat{\gamma}(\nu\, e^{-i\pi}) \, e^{-\nu t} \ . \tag{6.73}$$

Interestingly, this form holds for any *linear* and *nonlinear* system with a finite static susceptibility.

Next we observe that contributions to $z\hat{\gamma}(z)$ with positive integer powers of z, e.g. mass renormalization terms, do not contribute to the long-time tail of $\chi_{\mathrm{cut}}(t)$. We readily find with the expression (3.53) and with $\gamma_s = \rho_s / \sin(\pi s/2)$ in the asymptotic regime $t \gg (\rho_s \omega_{\mathrm{ph}}/\omega_0^2)^{1/s}/\omega_{\mathrm{ph}}$ for all $s > 0$ the universal behavior

$$\chi_{\mathrm{cut}}(t) = \frac{2}{\pi} \Gamma(1+s)\cos(\tfrac{1}{2}\pi s) M\gamma_s \omega_{\mathrm{ph}}^2 \chi_0^2 \frac{1}{(\omega_{\mathrm{ph}} t)^{1+s}} \ . \tag{6.74}$$

For odd s the prefactor vanishes and thus the algebraic tail is absent.

Consider next the position autocorrelation function $C^+(t)$ at $T = 0$ and positive asymptotic times. There follows from Eq. (6.37) with (3.32) at low positive frequency the relation $\tilde{\chi}''(\omega) = \chi_0^2 J(\omega)$. With this, we obtain from Eq. (6.52)

$$C^+(t) = \frac{\hbar \chi_0^2}{\pi} \int_0^\infty d\omega \, J(\omega) \, e^{-i\omega t} \ . \tag{6.75}$$

With the spectral density $J(\omega) = M\gamma_s \omega_{\mathrm{ph}}^{1-s} \omega^s$, we then obtain

$$C^+(t) \;=\; \frac{\Gamma(1+s)}{\pi}\, M\hbar\gamma_s\omega_{\rm ph}^2\chi_0^2\, \frac{e^{-i\pi(1+s)/2}}{(\omega_{\rm ph}t)^{1+s}}\,. \tag{6.76}$$

The real part of $C^+(t)$ is the symmetrized correlation function

$$S(t) \;\equiv\; {\rm Re}\,C^+(t) \;=\; -\frac{\Gamma(1+s)\sin(\tfrac{1}{2}\pi s)}{\pi}\, M\hbar\gamma_s\omega_{\rm ph}^2\, \frac{\chi_0^2}{(\omega_{\rm ph}|t|)^{1+s}}\,. \tag{6.77}$$

The imaginary part, with the relation (6.15), is in line with the expression (6.74). The expression (6.76) would also result, if we had computed the thermal correlation function $\mathbb{C}(\tau)$, Eq. (6.53), for large τ in the limit $T \to 0$, and had substituted $\tau \to i\,t$.

6.4.3 Shiba relation

The algebraic decay laws (6.71), (6.74) and (6.77) holding at $T = 0$ and asymptotic time can be written in a catchy form which will turn out to be universal. To this aim, we measure length in units of the position spread of the undamped oscillator in the ground state, $\sqrt{\langle q^2\rangle}|_{T=0,\gamma_s=0} = \sqrt{\hbar/2M\omega_0} = \tfrac{1}{2}q_0$, and we introduce a dimensionless friction coefficient $\delta_s \equiv M\gamma_s q_0^2/2\pi\hbar = \gamma/\pi\omega_0$ and scaled correlation functions $S_{\rm sp}(t)$ and $\chi_{\rm sp}(t)$, complying with the spin correlation functions introduced below in Subsec. 21.2.4, $S(t) = q_0^2 S_{\rm sp}(t)/4$ and $\chi(t) = q_0^2\chi_{\rm sp}(t)/4$, and the scaled static susceptibility $\chi_{\rm sp,0} = 4\chi_0/q_0^2$. Thde universal forms are

$$\begin{aligned}
\lim_{t\to\infty} S_{\rm sp}(t)\, t^{1+s} &= -2\delta_s\Gamma(1+s)\sin(\pi s/2)\omega_{\rm ph}^{1-s}(\hbar\chi_{\rm sp,0}/2)^2\,, \\
\lim_{t\to\infty} \chi_{\rm sp}(t)\, t^{1+s} &= 2\delta_s\Gamma(1+s)\cos(\pi s/2)\omega_{\rm ph}^{1-s}(\hbar\chi_{\rm sp,0}/2)^2\,.
\end{aligned} \tag{6.78}$$

The corresponding expressions in frequency space are the *generalized Shiba relations*

$$\begin{aligned}
\lim_{\omega\to 0^\pm} \tilde{S}_{\rm sp}(\omega)/|\omega|^s &= 2\pi\delta_s\omega_{\rm ph}^{1-s}(\hbar\chi_{\rm sp,0}/2)^2\,, \\
\lim_{\omega\to 0^\pm} \hbar\,{\rm sgn}(\omega)\tilde{\chi}''_{\rm sp}(\omega)/|\omega|^s &= 2\pi\delta_s\omega_{\rm ph}^{1-s}(\hbar\chi_{\rm sp,0}/2)^2\,.
\end{aligned} \tag{6.79}$$

The standard Shiba relations pertain to the Ohmic case $s = 1$. They have been discovered first for the Anderson model [260].

We anticipate that the Shiba relations (6.79), and the analogues (6.78) in the time regime, are universally valid for any linear or nonlinear open system with spectral density of the form (3.38) which has a finite linear or nonlinear static susceptibility at zero temperature. The individual systems differ only by the associated static susceptibility $\chi_{\rm sp,0}$. For the linear oscillator model, we have $\chi_{\rm sp,0} = 2/\hbar\omega_0$.

The Shiba relations in the dissipative two-state system and in the fermionic Toulouse model are discussed in Subsections 22.6.4 and 22.7.2.

6.5 Partition function and implications

6.5.1 Partition function

The partition function $Z(\beta)$ of the damped linear quantum oscillator has been calculated with the imaginary-time path integral method in Section 4.3. There we obtained the partition function of the composite system in the factorized form $Z_{\text{tot}} = Z_{\text{R}}Z$, where Z_{R} and Z are the partition functions of the reservoir and of the damped oscillator, respectively. The reduced partition function was found in Subsec. 4.3.3 to read

$$Z(\beta) = \frac{1}{\beta\hbar\omega_0} \prod_{n=1}^{\infty} \frac{\nu_n^2}{\omega_0^2 + \nu_n^2 + \nu_n\hat{\gamma}(\nu_n)} . \qquad (6.80)$$

This expression holds for arbitrary frequency-dependent damping. In order that the infinite product is convergent, we must have $\lim_{z\to\infty}\hat{\gamma}(z) \to 0$. Hence the strict Ohmic case is excluded. For the Drude-regularized damping function (3.43), the partition function of the damped oscillator takes the form

$$Z(\beta) = \frac{1}{\beta\hbar\omega_0} \prod_{n=1}^{\infty} \frac{\nu_n^2 (\nu_n + \omega_{\text{D}})}{(\nu_n + \lambda_1)(\nu_n + \lambda_2)(\nu_n + \lambda_3)} . \qquad (6.81)$$

The λ_j ($j = 1, 2, 3$) satisfy the Vieta relations (6.44). With the infinite product form of the gamma function [89, 90] the expression (6.81) can be concisely written as

$$Z(\beta) = \frac{\beta\hbar\omega_0}{4\pi^2} \frac{\Gamma(\beta\hbar\lambda_1/2\pi)\Gamma(\beta\hbar\lambda_2/2\pi)\Gamma(\beta\hbar\lambda_3/2\pi)}{\Gamma(\beta\hbar\omega_{\text{D}}/2\pi)} . \qquad (6.82)$$

The analytic expression (6.82) holds for arbitrary ω_0, γ, ω_{D} and T.

When ω_{D} is very large compared to ω_0 and γ, and to the thermal frequency ν_1, the partition function of the damped quantum oscillator is found from Eq. (6.82) as

$$Z(\beta) = \frac{1}{\beta\hbar\omega_0} \left(\frac{2\pi}{\beta\hbar\omega_{\text{D}}}\right)^{\beta\hbar(\omega_{\text{D}}-\Omega_{\text{D}})/2\pi} \Gamma\left(1 + \frac{\beta\hbar\lambda_1^{(0)}}{2\pi}\right)\Gamma\left(1 + \frac{\beta\hbar\lambda_2^{(0)}}{2\pi}\right) , \qquad (6.83)$$

where $\lambda_{1,2}^{(0)} = \frac{1}{2}\gamma \pm i\,(\omega_0^2 - \frac{1}{4}\gamma^2)^{1/2}$. With the expressions for λ_1, λ_2, and λ_3 given in Subsection 6.3.2, the expression (6.82) also holds for radiation damping.

6.5.2 Internal energy, free energy, and entropy

The analytic expressions (6.82) and (6.83) facilitate direct calculation of thermodynamic quantities, such as the free energy $F = -\ln Z/\beta$ and the internal energy $U = -\partial \ln Z/\partial\beta$, and derivatives of these quantities, for instance the entropy $S = -\partial F/\partial T$ and the specific heat $c_{\text{U}} = \partial U/\partial T = -k_{\text{B}}\,\beta^2\,\partial U/\partial\beta$.

The free energy $f(\omega, \beta)$ and internal energy $u(\omega, \beta)$ of an individual oscillator with eigenfrequency ω at inverse temperature β are

$$f(\omega, \beta) = \ln[2\sinh(\tfrac{1}{2}\beta\hbar\omega)]/\beta , \qquad u(\omega, \beta) = \tfrac{1}{2}\hbar\omega\coth(\tfrac{1}{2}\beta\hbar\omega) , \qquad (6.84)$$

as follows from the partition function $z(\omega, \beta) = 1/[2\sinh(\frac{1}{2}\beta\hbar\omega)]$. In the eigenfrequency representation (4.238) of the system-plus-reservoir complex the total free energy and total internal energy thus take the forms

$$F_{\text{tot}} = \sum_{\alpha=0}^{N} f(\mu_\alpha, \beta), \qquad U_{\text{tot}} = \sum_{\alpha=0}^{N} u(\mu_\alpha, \beta), \qquad (6.85)$$

To divide the internal energy into the reservoir's and the reduced system's part, we start out from the representation (4.246) of the product form $Z^{(\text{tot})} = Z_{\text{R}} Z$. Thus we may write $F_{\text{tot}} = F_{\text{R}} + F$ and $U_{\text{tot}} = U_{\text{R}} + U$, where F (U) is the free (internal) energy of the open system, and F_{R} (U_{R}) is the free (internal) energy of the reservoir,

$$F_{\text{R}} = \sum_{\alpha=1}^{N} f(\omega_\alpha, \beta), \qquad U_{\text{R}} = \sum_{\alpha=1}^{N} u(\omega_\alpha, \beta). \qquad (6.86)$$

Consider now the internal and free energy of the open system. We obtain from the expression (6.80) for general frequency-dependent friction $\hat{\gamma}(z)$

$$U = \frac{1}{\beta} \sum_{n=-\infty}^{\infty} \frac{\nu_n}{2} \frac{\partial}{\partial \nu_n} \ln\left[\nu_n^2 \hat{\chi}(|\nu_n|)\right], \qquad (6.87)$$

Taking different view, we may write the internal energy U as a spectral integral of internal energies $u(\omega, \beta)$ of fictitious oscillators with eigenfrequency ω,

$$U = \int_0^\infty d\omega\, \xi_{\text{osc}}(\omega) u(\omega, \beta). \qquad (6.88)$$

Here $\xi_{\text{osc}}(\omega)$ is the change of the density of bath oscillators by the system-bath coupling [262, 263]. With the set of eigenvalues $\{\mu_\alpha\}$ of the composite system introduced in Eq. (4.237), and with the set of eigenvalues $\{w_\alpha\}$ of the uncoupled bath, we have

$$\xi_{\text{osc}}(\omega) = \sum_{\alpha=0}^{N} \delta(\omega - \mu_\alpha) - \sum_{\alpha=1}^{N} \delta(\omega - \omega_\alpha). \qquad (6.89)$$

By analogy with the equality (6.61), we may write the Matsubara sum (6.87) as the frequency integral (6.88) with the spectral density (see also Refs. [99, 261])

$$\xi_{\text{osc}}(\omega) = \frac{1}{\pi} \operatorname{Im} \frac{\partial \ln[\tilde{\chi}(\omega)]}{\partial \omega}. \qquad (6.90)$$

Following the lines put down in Ref. [262], one finds that the expressions (6.89) and (6.90) are in correspondence.

Observing that the expression (6.90) is a partial derivative, and that $\tilde{\gamma}'(\omega)$ is positive, we find for all s the sum rule

$$\Sigma_{\text{osc}} \equiv \int_0^\infty d\omega\, \xi_{\text{osc}}(\omega) = 1. \qquad (6.91)$$

In correspondence with Eq. (6.88), the free energy has the integral representation

$$F = \int_0^\infty d\omega\, \xi_{\text{osc}}(\omega) f(\omega, \beta) . \tag{6.92}$$

These remarkable expressions for the internal and free energy of the open system are exact. In sum, all the environmental influences are in the dynamical susceptibility.

Both for Drude regularized Ohmic friction and for radiation damping, Eq. (6.43), the density $\xi_{\text{osc}}(\omega)$ is a linear combination of four Lorentzians,

$$\xi_{\text{osc}}(\omega) = \frac{1}{\pi}\left(\frac{\lambda_1}{\omega^2 + \lambda_1^2} + \frac{\lambda_2}{\omega^2 + \lambda_2^2} + \frac{\lambda_3}{\omega^2 + \lambda_3^2} - \frac{\omega_D}{\omega^2 + \omega_D^2} \right) , \tag{6.93}$$

and the internal energy U is found to be a superposition of four digamma functions,

$$U = \frac{1}{\beta} + \frac{\hbar}{2\pi}\left(\omega_D\, \psi(1 + \beta\hbar\omega_D/2\pi) - \sum_{i=1}^{3} \lambda_i\, \psi(1 + \beta\hbar\lambda_i/2\pi) \right) . \tag{6.94}$$

In the remainder of this subsection we restrict ourselves to the Ohmic case. When the thermal energy $k_B T$ is large compared to the energy scales of the damped oscillator, we may expand the round bracket in the expression (6.94) in a power series in β. The resulting high temperature series for the internal energy is

$$U = k_B T \left[1 + \frac{1}{12} \frac{\gamma\omega_D + \omega_0^2}{\omega_0^2} \frac{1}{\vartheta^2} + \mathcal{O}\left(\frac{\gamma\omega_D^2}{\omega_0^3} \frac{1}{\vartheta^3} \right) \right] . \tag{6.95}$$

Here we have introduced the scaled temperature

$$\vartheta = k_B T/\hbar\omega_0 . \tag{6.96}$$

Evidently, the internal energy of the damped oscillator approaches in the classical limit the equipartition value $k_B T$ as required.

For large cutoff frequency, $\omega_D \gg \omega_0, \gamma, \nu_1$, we obtain from (6.83) or from (6.94)

$$U = \frac{\hbar\gamma}{2\pi}\left[1 + \ln\left(\frac{\beta\hbar\omega_D}{2\pi} \right) \right] + \frac{1}{\beta} - \frac{\hbar}{2\pi}\sum_{i=1}^{2} \lambda_i^{(0)}\, \psi\left(1 + \frac{\beta\hbar\lambda_i^{(0)}}{2\pi} \right) . \tag{6.97}$$

Consider next the asymptotic low-temperature regime. As $T \to 0$, the sum in Eq. (6.87) turns into an integral. The ground-state energy $E_0 = \lim_{\beta\to\infty} U(\beta)$ is

$$E_0 \equiv \hbar\varepsilon_0 = \frac{\hbar}{2\pi}\int_0^\infty d\nu\, \ln\left(\frac{\omega_0^2 + \nu^2 + \nu\hat{\gamma}(\nu)}{\nu^2} \right) . \tag{6.98}$$

This form holds for arbitrary frequency-dependent damping. For Drude regularized Ohmic friction and for radiation damping, we find either from Eq. (6.98) or from the asymptotic limit of the expression (6.94)

$$\hbar\varepsilon_0 = \hbar\left[\lambda_1 \ln(\omega_D/\lambda_1) + \lambda_2 \ln(\omega_D/\lambda_2) + \lambda_3 \ln(\omega_D/\lambda_3) \right]/2\pi . \tag{6.99}$$

In the regime $\omega_D \gg \omega_0, \gamma$, the ground state energy takes the form

$$\hbar\varepsilon_0 = \frac{\hbar\omega_0}{2}\left[(1 - \alpha^2)\, g(\alpha) + \frac{2\alpha}{\pi}\left[1 + \ln(\omega_D/\omega_0) \right] + \mathcal{O}\left(\frac{\gamma}{\omega_D} \right) \right] , \tag{6.100}$$

where
$$g(\alpha) = \frac{2}{\pi} \frac{\omega_0}{\lambda_1^{(0)} - \lambda_2^{(0)}} \left[\ln(\lambda_1^{(0)}/\omega_0) - \ln(\lambda_2^{(0)}/\omega_0) \right] . \tag{6.101}$$

With the explicit forms (6.40) for $\lambda_{1,2}^{(0)}$, and with $\alpha = \gamma/2\omega_0$, we obtain

$$g(\alpha) = \begin{cases} \dfrac{1}{\sqrt{1-\alpha^2}} \left(1 - \dfrac{2}{\pi} \arcsin \alpha \right) , & \text{for} \quad \alpha < 1 , \\[2ex] \dfrac{2}{\pi} \dfrac{\ln\left(\alpha + \sqrt{\alpha^2 - 1} \right)}{\sqrt{\alpha^2 - 1}} , & \text{for} \quad \alpha > 1 . \end{cases} \tag{6.102}$$

We see from Eqs. (6.100) - (6.102) that the vacuum energy increases monotonously with α and with ω_D. For strong damping $\alpha \gg 1$, we have $\varepsilon_0 = (\gamma/2\pi) \ln(\omega_D/\gamma)$.

With use of the asymptotic series of the digamma function [90], the low temperature expansion of the internal energy is found to read

$$U = \hbar\varepsilon_0 + a_1 \hbar\omega_0 \vartheta^2 + a_2 \hbar\omega_0 \vartheta^4 + \mathcal{O}(\vartheta^6) , \tag{6.103}$$

where
$$a_1 = \frac{\pi}{6} \frac{\gamma}{\omega_0} , \qquad a_2 = \frac{\pi^3}{15} \frac{\gamma}{\omega_0} \left(3 - \frac{\gamma^2}{\omega_0^2} \right) . \tag{6.104}$$

The corresponding expansion for the free energy $F = -k_B T \ln Z$ is

$$F = \hbar\varepsilon_0 - a_1 \hbar\omega_0 \vartheta^2 - \tfrac{1}{3} a_2 \hbar\omega_0 \vartheta^4 + \mathcal{O}(\vartheta^6) . \tag{6.105}$$

The leading thermal variation of U and $F \propto T^2$ at low T is a characteristic feature of Ohmic dissipation. This contribution can be rewritten in terms of the static susceptibility $\chi_0 = 1/M\omega_0^2$ as

$$\Delta U = -\Delta F = (\pi/6\hbar)\gamma M \chi_0 (k_B T)^2 . \tag{6.106}$$

Most interestingly, this form universally holds at low T for any linear and nonlinear Ohmic system with nonzero static susceptibility at $T = 0$ (see Section 22.9).

Finally, the low temperature series for the entropy $S = -\partial F/\partial T$ arises from Eq. (6.105) as
$$S/k_B = 2a_1 \vartheta + \tfrac{4}{3} a_2 \vartheta^3 + \mathcal{O}(\vartheta^5) . \tag{6.107}$$

Hence the entropy vanishes at zero temperature in accordance with the Third Law of thermodynamics.

6.5.3 Specific heat and Wilson ratio

The specific heat of the damped system is the difference between the specific heat of system-plus-reservoir entity and the specific heat of the uncoupled reservoir alone,

$$C = C_{S+R} - C_R . \tag{6.108}$$

With the above results for the internal energy U it is only a small move towards C. Expressed in terms of the specific heat of a single oscillator with eigenfrequency ω,

$$c_{\mathrm{ho}}(\omega,\beta) = k_{\mathrm{B}} \left(\frac{\beta\hbar\omega/2}{\sinh(\beta\hbar\omega/2)} \right)^2 , \tag{6.109}$$

the specific heat is compactly written as

$$C = \int_0^\infty d\omega\, \xi_{\mathrm{osc}}(\omega) c_{\mathrm{ho}}(\omega,\beta) . \tag{6.110}$$

In the classical limit $T \to \infty$, this expression correctly yields the value $C_\infty = k_{\mathrm{B}}$, as follows with the sum rule (6.91). As a byproduct, there follows that the specific heat is zero for all s at absolute temperature $T = 0$. This will come out differently for the the specific heat of the free Brownian particle, as discussed below in Subsec. 7.5.3.

Drude damping

In the asymptotic high temperature limit, $\nu_1 \gg \omega_{\mathrm{D}} \gg \omega_0, \gamma$, we obtain from Eq. (6.95)

$$C/k_{\mathrm{B}} = 1 - \frac{1}{12} \frac{\gamma\omega_{\mathrm{D}} + \omega_0^2}{\omega_0^2} \frac{1}{\vartheta^2} . \tag{6.111}$$

In contrast, when ω_{D} is very large compared to ω_0, γ, and ν_1, Eq. (6.97) yields

$$C/k_{\mathrm{B}} = 1 - \frac{\beta\hbar\gamma}{2\pi} + \sum_{i=1}^{2} \left(\frac{\beta\hbar\lambda_i^{(0)}}{2\pi} \right)^2 \psi'\left(1 + \frac{\beta\hbar\lambda_i^{(0)}}{2\pi}\right). \tag{6.112}$$

This form reduces in the regime $\omega_{\mathrm{D}} \gg \nu_1 \gg \omega_0, \gamma$ to the high temperature series

$$C/k_{\mathrm{B}} = 1 - \frac{\gamma}{2\pi\omega_0} \frac{1}{\vartheta} - \frac{1}{24} \frac{2\omega_0^2 - \gamma^2}{\omega_0^2} \frac{1}{\vartheta^2} . \tag{6.113}$$

The major difference to the series expressions (6.111) is the $1/\vartheta$-term as the leading quantum contribution when ω_{D} is asymptotically large.

Finally, the asymptotic low-temperature series of the specific heat is found from Eq. (6.103) as

$$C/k_{\mathrm{B}} = 2a_1\vartheta + 4a_2\vartheta^3 + \mathcal{O}(\vartheta^5) . \tag{6.114}$$

General frequency-dependent damping

Consider next the specific heat for general spectral coupling $J_{\mathrm{lf}}(\omega) = M\gamma_s\omega_{\mathrm{ph}}^{1-s}\omega^s$. The associated leading term of the spectral function $\xi_{\mathrm{osc}}(\omega)$ at low frequency reads

$$\xi_{\mathrm{osc}}(\omega) = (Ms\gamma_s\chi_0/\pi)\,(\omega/\omega_{\mathrm{ph}})^{s-1} . \tag{6.115}$$

With this, the specific heat at low T is found from Eq. (6.110) as

$$C/k_{\mathrm{B}} = \frac{s\Gamma(2+s)\zeta(1+s)}{\pi} M\gamma_s\omega_{\mathrm{ph}}\chi_0 \left(\frac{k_{\mathrm{B}}T}{\hbar\omega_{\mathrm{ph}}} \right)^s . \tag{6.116}$$

Hence the asymptotic low temperature dependence is T^s for C, and T^{1+s} for U and F.

Wilson ratio

The asymptotic low temperature dependence of the specific heat can be written in a universally valid form. To this end we introduce, as in Subsection 6.4.3, a scaled static susceptibility $\chi_{\mathrm{sp},0}$ according to $\chi_0 = \frac{1}{4}q_0^2\chi_{\mathrm{sp},0}$, and a dimensionless friction coefficient $\delta_s = M\gamma_s q_0^2/2\pi\hbar$. Further, we define the so-called generalized Wilson ratio

$$R_s \equiv \lim_{T\to 0} \frac{4C(T)/k_{\mathrm{B}}}{\chi_{\mathrm{sp},0}(\hbar\omega_{\mathrm{ph}})^{1-s}(k_{\mathrm{B}}T)^s} \ . \tag{6.117}$$

Thus the specific heat at low T, Eq. (6.116), is characterized by the Wilson ratio

$$R_s = 2s\Gamma(2+s)\zeta(1+s)\,\delta_s \ . \tag{6.118}$$

This reduces in the Ohmic case to

$$R_1 = 2\delta_1\pi^2/3 \ . \tag{6.119}$$

Interestingly, the Wilson ratio (6.118) holds for any dissipative system which has finite linear or nonlinear static susceptibility at zero temperature [cf. Section 22.9].

The power laws in T of the internal and free energy, and of the specific heat indicate that there is now energy gap above the ground state. This conjecture is confirmed in the following subsection by a direct analysis of the energy spectrum.

6.5.4 Spectral density of states

Generally, the partition function provides access to the spectrum of the quantum mechanical system. More specifically, the reduced partition function $Z(\beta)$ may be regarded as the Laplace transform of a kind of density of states $\varrho(\varepsilon)$,

$$Z(\beta) = \int_0^\infty d\varepsilon\, \varrho(\varepsilon)\, e^{-\beta\hbar\varepsilon} \ . \tag{6.120}$$

The density function $\varrho(\epsilon)$ is the inverse Laplace transform of the partition function,

$$\varrho(\varepsilon) = \frac{\hbar}{2\pi i} \int_{c-i\infty}^{c+i\infty} d\beta\, Z(\beta)\, e^{\beta\hbar\varepsilon} \ . \tag{6.121}$$

Here c is chosen such that the poles of $Z(\beta)$ lie to the left of the integration path. To be consistent with the conception of a density, $\varrho(\varepsilon)$ must be a positive function of ε.

Equipped with the integral expression (6.121) with (6.82), one can now study how the sharp lines of the discrete spectrum of the undamped oscillator,

$$\varrho_0(\varepsilon) = \sum_{n=0}^\infty \delta\!\left(\varepsilon - (n + \tfrac{1}{2})\,\omega_0\right) , \tag{6.122}$$

broaden and merge together, as damping is turned on and raised. Direct numerical computation of the integral (6.121) is difficult since the integrand is rapidly oscillating. Fortunately, valuable insights can be obtained by analytic methods.

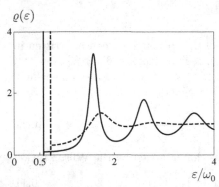
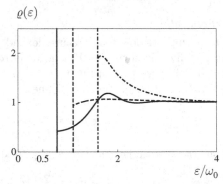

Figure 6.1: The density of states $\varrho(\varepsilon)$ is plotted versus ε/ω_0 for $\omega_{\rm D} = 10\,\omega_0$ for weak and strong damping. The left plot shows the cases $\alpha = 0.1$ (full curve) and $\alpha = 0.3$ (dashed curve), and the right plot the cases $\alpha = 0.4$ (full curve), $\alpha = 0.9$ (dashed curve) and $\alpha = 1.8$ (dashed-dotted curve). Both the ground-state energy and the height of the jump increase monotonously with α. The respective values of the ground state energies are $\varepsilon_0/\omega_0 = 0.576$, 0.721, 0.790, 1.108, and 1.607, respectively.

Drude damping

First, it is immediately clear from the low-T expansion of the free energy, Eq. (6.105), that the ground state remains as a separate δ-function at the frequency ε_0,

$$\varrho(\varepsilon) = \delta(\varepsilon - \varepsilon_0) + \Theta(\varepsilon - \varepsilon_0)\,\Phi(\varepsilon)\,. \tag{6.123}$$

The δ-peak represents the ground state. It is shifted with increasing damping strength towards higher frequency, as specified by the expression (6.99). The function $\Phi(\varepsilon)$ represents the density of states above the ground state. The expansion of $\Phi(\varepsilon)$ about ε_0 for $\varepsilon > \varepsilon_0$ is found from the asymptotic expansion of F, Eq. (6.105), as

$$\Phi(\varepsilon) = \frac{1}{\omega_0}\left\{ a_1 + \frac{a_1^2}{2}\frac{\varepsilon - \varepsilon_0}{\omega_0} + \left(\frac{a_1^3}{12} + \frac{a_2}{6}\right)\left(\frac{\varepsilon - \varepsilon_0}{\omega_0}\right)^2 + \mathcal{O}\!\left[\left(\frac{\varepsilon - \varepsilon_0}{\omega_0}\right)^3\right]\right\}\,. \tag{6.124}$$

First, from this we see that the gap in the energy spectrum above the ground state is washed out by the dissipative coupling. The function $\Phi(\varepsilon)$ makes a jump at $\varepsilon = \varepsilon_0$ of height $\pi\alpha/3\omega_0$ ($\alpha = \gamma/2\omega_0$). The leading thermal enhancement $\propto T^2$ of the internal energy is a direct implication of the nonzero density of states above ϵ_0. Secondly, the derivative of $\Phi(\varepsilon)$ at $\varepsilon = \varepsilon_0$ is positive for all α, whereas the curvature of $\Phi(\varepsilon)$ at $\varepsilon = \varepsilon_0$ changes sign, when $a_1^3/2 + a_2 = 0$, which is at $\alpha = 0.88$ (see also Fig. 6.1).

The density $\Phi(\varepsilon)$ well above ε_0 is given by the sum of contributions collected from the infinite series of simple poles in the expression (6.82) [264],

$$\Phi(\varepsilon) = \frac{1}{\omega_0} + \sum_{k=1}^{3}\sum_{n=1}^{\infty} R_{n,k}\, e^{-2\pi n\varepsilon/\lambda_k}\,. \tag{6.125}$$

The residues are

$$R_{n,1} = \frac{(-1)^{n-1}}{(n-1)!} \frac{\Gamma(-n\lambda_2/\lambda_1)\Gamma(-n\lambda_3/\lambda_1)}{\Gamma(-n\omega_D/\lambda_1)} \frac{\omega_0}{\lambda_1^2} , \qquad (6.126)$$

and the expressions for $R_{n,2}$ and $R_{n,3}$ result by cyclic permutation of the indices.

The analysis yields that the series (6.124) and (6.125) have an overlapping region of convergence. Further, for $\alpha > 0$, the contribution $\Phi(\varepsilon)$ is an absolutely continuous function of ε in the interval $\varepsilon_0 < \varepsilon < \infty$. Therefore, apart from the ground state, there are no other discrete states embedded in the continuum. In the underdamped regime, which is $\Gamma < 2\Omega_0$ in Eq. (6.45), the density of states shows roughly equidistant resonances with exponentially reduced amplitudes towards higher frequencies. The transition from the bumpy to the smooth behavior occurs roughly at the critical damping $\Gamma = 2\Omega_0$. For $\Gamma > 2\Omega_0$, the spectral density increases initially from the threshold value $\Phi(\epsilon_0) = \pi\alpha/3\omega_0$ to a maximum and then decreases monotonously to the classical value $\Phi_{cl} = 1/\omega_0$ reached at high frequency $\epsilon \gg \omega_0$. The characteristic features of the density of states in the underdamped and overdamped regime are shown in Fig. 6.1.

Experimentally, the density of states can be found by measuring the absorbed microwave power of the externally driven oscillator.

Non-Ohmic damping

Finally, consider the spectral density of states for non-Ohmic spectral coupling $J(\omega) \propto \omega^s$. In this general case the ground state contributes again a δ-peak $\delta(\epsilon - \epsilon_0)$. However, now the spectral density of states slightly above the ground state energy $\hbar\epsilon_0$ has power-law form $\Phi(\varepsilon) \propto (\varepsilon - \varepsilon_0)^{s-1}$. Hence the density of states above the ground state grows smoothly with increasing frequency from zero in the super-Ohmic case, whereas it is singular at the threshold for sub-Ohmic environmental coupling.

6.6 Mean square of position and momentum

6.6.1 General expressions for linear damping

The equilibrium dispersion of the position of the damped oscillator can be written both as a frequency integral and as a Matsubara sum,[3] as anticipated in Eq. (6.61),

$$\langle q^2 \rangle = \frac{\hbar}{\pi} \int_0^\infty d\omega \, \tilde{\chi}''(\omega) \coth(\beta\hbar\omega/2) = \frac{1}{M\beta} \sum_{n=-\infty}^{\infty} \frac{1}{\omega_0^2 + \nu_n^2 + |\nu_n|\hat{\gamma}(|\nu_n|)} . \qquad (6.127)$$

The dispersion (6.127) is conveniently split into the classical dispersion $\langle q^2 \rangle_{cl}$ and the quantum mechanical dispersion $\langle q^2 \rangle_{qm}$. There holds in the sum representation

$$\langle q^2 \rangle = \langle q^2 \rangle_{cl} + \langle q^2 \rangle_{qm} ,$$

$$\langle q^2 \rangle_{cl} = \frac{1}{M\beta\omega_0^2} , \qquad \langle q^2 \rangle_{qm} = \frac{2}{M\beta} \sum_{n=1}^{\infty} \frac{1}{\omega_0^2 + \nu_n^2 + \nu_n\hat{\gamma}(\nu_n)} . \qquad (6.128)$$

---

[3]For simplicity, in the sequel the label "eq" is omitted both for $\langle q^2 \rangle$ and $\langle p^2 \rangle$.

Consider next the momentum spread $\langle p^2 \rangle$. Relying on the expressions (6.38) with (6.36), the mean value of p^2 in thermal equilibrium is found as

$$\langle p^2 \rangle = \frac{\hbar M^2}{\pi} \int_0^\infty d\omega\, \omega^2 \tilde{\chi}''(\omega) \coth(\beta\hbar\omega/2) . \tag{6.129}$$

Alternatively, one may start from the Fourier series (6.53) with (6.57) of $\langle q(\tau)q(0)\rangle_\beta$, which has cusps at $\tau = m\hbar\beta$ $(m = 0, \pm 1, \cdots)$. Thus the periodically continued momentum correlation function has δ-function singularities at $\tau = m\hbar\beta$,

$$\langle p(\tau)p(0) \rangle = -\hbar M :\delta(\tau): + \langle p(\tau)p(0)\rangle_{\text{reg}} ,$$

$$\langle p(\tau)p(0) \rangle_{\text{reg}} = \frac{M}{\beta} \sum_{n=-\infty}^{\infty} \frac{\omega_0^2 + |\nu_n|\hat{\gamma}(|\nu_n|)}{\omega_0^2 + \nu_n^2 + |\nu_n|\hat{\gamma}(|\nu_n|)}\, e^{i\nu_n\tau} . \tag{6.130}$$

Here $:\delta(\tau):$ is the periodically continued δ-function (3.85). Clearly, only the regular part $\langle p(\tau)p(0)\rangle_{\text{reg}}$ is the analytic continuation of the real-time momentum autocorrelation function in the strip $0 < \text{Re}\,\tau < \hbar\beta$. As a result, the Matsubara representation of the momentum dispersion is found as

$$\langle p^2 \rangle = \lim_{\tau \to 0^+} \langle p(\tau)p(0)\rangle_{\text{reg}} = \frac{M}{\beta} \sum_{n=-\infty}^{\infty} \frac{\omega_0^2 + |\nu_n|\hat{\gamma}(|\nu_n|)}{\omega_0^2 + \nu_n^2 + |\nu_n|\hat{\gamma}(|\nu_n|)} . \tag{6.131}$$

The dispersions of position and momentum can also be related to the partition function $Z(\beta)$ of the damped linear oscillator. The product representation (6.80) discloses that the dispersions (6.127) and (6.131) may be written as

$$\langle q^2 \rangle = -\frac{1}{M\beta\omega_0} \frac{d}{d\omega_0} \ln Z(\beta) , \tag{6.132}$$

$$\langle p^2 \rangle = -\frac{M}{\beta} \left(\omega_0 \frac{d}{d\omega_0} + 2\gamma \frac{d}{d\gamma} \right) \ln Z(\beta) . \tag{6.133}$$

In the second relation we have employed the representation $\hat{\gamma}(z) = \gamma f(z)$. The expressions of this subsection are generally valid for any form of linear memory-friction.

6.6.2 Ohmic and Drude damping

Ohmic damping

In the Ohmic case $\hat{\gamma}(|\nu_n|) = \gamma$, the Matsubara sum in (6.127) may be decomposed as

$$\langle q^2 \rangle = \frac{1}{M\beta\omega_0^2} + \frac{2}{M\beta(\lambda_2^{(0)} - \lambda_1^{(0)})} \sum_{n=1}^{\infty} \left(\frac{1}{\nu_n + \lambda_1^{(0)}} - \frac{1}{\nu_n + \lambda_2^{(0)}} \right) , \tag{6.134}$$

where $\lambda_1^{(0)}$ and $\lambda_2^{(0)}$ are the characteristic frequencies defined in Eq. (6.40). The Matsubara sum in Eq. (6.134) is a linear combination of digamma functions,

$$\langle q^2 \rangle = \frac{1}{M\beta\omega_0^2} + \frac{\hbar}{M\pi(\lambda_2^{(0)} - \lambda_1^{(0)})} \left[\psi\left(1 + \lambda_2^{(0)}/\nu \right) - \psi\left(1 + \lambda_1^{(0)}/\nu \right) \right] , \tag{6.135}$$

where again $\nu \equiv \nu_1 = 2\pi/\hbar\beta$. In the absence of damping, this reduces to the familiar expression $\langle q^2 \rangle = (\hbar/2M\omega_0) \coth(\beta\hbar\omega_0/2)$.

In the high-temperature regime $k_B T \gg \hbar|\lambda_{1,2}^{(0)}|/2\pi$, the quantum mechanical contribution $\langle q^2 \rangle_{qm}$ is found from Eq.(6.135) with the Taylor expansion of $\psi(1+z)$ as

$$\langle q^2 \rangle_{qm} = \frac{1}{12}\frac{\hbar^2}{Mk_B T}\left[1 + \mathcal{O}\left(\frac{\hbar\gamma}{k_B T}\right)\right] . \tag{6.136}$$

This shows again that damping becomes irrelevant at high temperature.

On the other hand, in the low temperature limit the asymptotic expansion of the digamma functions applies. Then the term $\langle q^2 \rangle_{cl}$ is cancelled by a counter term stemming from $\langle q^2 \rangle_{qm}$. The resulting series is

$$\langle q^2 \rangle = \frac{\hbar}{2M\omega_0}\left\{g(\alpha) + \frac{2\pi}{3}\frac{\gamma}{\omega_0}\left(\frac{k_B T}{\hbar\omega_0}\right)^2 + \mathcal{O}\left[\left(\frac{k_B T}{\hbar\omega_0}\right)^4\right]\right\} . \tag{6.137}$$

The function $g(\alpha)$ is given in Eq. (6.102). As $g(0) = 1$, the expression (6.137) reduces in the limit $\alpha \to 0$ to the dispersion of the undamped oscillator in the ground state.

At finite T, the dispersion of the position is enhanced compared with the zero temperature value. The thermal contributions grow algebraically with T for a damped system. The leading T^2 law at low T is again a signature of Ohmic friction. For a spectral density $J(\omega \to 0) \propto \omega^s$, the leading thermal enhancement grows as T^{1+s}.

For large friction $\gamma \gg \omega_0$ and temperature in the range

$$k_B T \gg \hbar\omega_0^2/2\pi\gamma , \tag{6.138}$$

the leading quantum mechanical contribution to the coordinate dispersion is

$$\langle q^2 \rangle_{qm} = \frac{\hbar}{\pi M\gamma}\left(\psi(1 + \beta\hbar\gamma/2\pi) - \psi(1)\right) . \tag{6.139}$$

Interestingly enough, the quantum mechanical part of the coordinate dispersion does not depend in leading order on properties of the potential for strong friction and temperature in the regime (6.138). At temperature $T \gg \hbar\gamma/2\pi k_B$, the expression (6.139) reduces to the form (6.136).

In the temperature regime $\hbar\omega_0^2/2\pi\gamma \ll k_B T \ll \hbar\gamma/2\pi$, the leading quantum contribution to the coordinate dispersion is found from the expression (6.139) as

$$\langle q^2 \rangle_{qm} = \frac{\hbar}{\pi M\gamma}\ln\left(\frac{\beta\hbar\gamma}{2\pi}\right) . \tag{6.140}$$

At this stage, the following preview seems appropriate. For large friction, the dynamics of a classical Brownian particle in phase space reduces to the propagation in position space. The corresponding dynamics is well described by the Smoluchowski diffusion equation, which is discussed below in Sec. 11.3. In the temperature range (6.138), quantum effects are important. However, the time evolution of the position distribution can still be described with a diffusion equation. The corresponding quantum Smoluchowski equation (QSE) is discussed in Subsec. 15.3. We shall see that the size of quantum effects in the QSE is controlled by the strength of $\langle q^2 \rangle_{qm}$.

Drude damping

For strictly Ohmic damping, the frequency integral (6.129) as well as the Matsubara sum (6.131) are logarithmically UV-divergent. The arising divergence is an indication that care has to be exercised with the Markov assumption. The conception of a memoryless reservoir is unphysical for the momentum dispersion. There is always a "microscopic" time scale below which the inertia of the environment becomes relevant. In its simplest form, this is represented by a high-frequency cutoff in $\tilde{\gamma}(\omega)$. By this, the thermal average of $\langle p^2 \rangle$ is regularized. For the Drude model (3.43), one finds

$$
\begin{aligned}
\langle q^2 \rangle &= \frac{1}{M\beta} \sum_{n=-\infty}^{+\infty} \frac{1}{\omega_0^2 + \nu_n^2 + \gamma\omega_D |\nu_n|/(\omega_D + |\nu_n|)} , \\
\langle p^2 \rangle &= \frac{M}{\beta} \sum_{n=-\infty}^{+\infty} \frac{\omega_0^2 + \gamma\omega_D |\nu_n|/(\omega_D + |\nu_n|)}{\omega_0^2 + \nu_n^2 + \gamma\omega_D |\nu_n|/(\omega_D + |\nu_n|)} .
\end{aligned}
\tag{6.141}
$$

Partial fraction decomposition and subsequent summation yields

$$
\langle q^2 \rangle = \frac{1}{\beta M \omega_0^2} + \frac{\hbar}{M\pi} \sum_{j=1}^{3} a_j \psi(1 + \lambda_j/\nu) ,
\tag{6.142}
$$

$$
\langle p^2 \rangle = M^2 \omega_0^2 \langle q^2 \rangle + \Pi^2 ,
\tag{6.143}
$$

$$
\Pi^2 = \frac{M\gamma\hbar\omega_D}{\pi} \sum_{j=1}^{3} b_j \psi(1 + \lambda_j/\nu) ,
\tag{6.144}
$$

where

$$
a_1 = \frac{\lambda_1 - \omega_D}{(\lambda_1 - \lambda_2)(\lambda_1 - \lambda_3)} , \qquad b_1 = \frac{\lambda_1}{(\lambda_1 - \lambda_2)(\lambda_1 - \lambda_3)} ,
\tag{6.145}
$$

and where a_2, a_3 and b_2, b_3 are defined by cyclic permutation of the indices.

We see with use of Eq. (6.45) with Eq. (6.46) that the expression (6.142) for $\langle q^2 \rangle$ differs from the previous result (6.135) by terms of order ω_0/ω_D and γ/ω_D.

For asymptotically high temperature $T \gg \hbar\omega_D/k_B$, we may expand all three digamma functions in Eq. (6.144) about unity. We then obtain

$$
\Pi^2 = \frac{M\hbar\gamma}{12} \frac{\hbar\omega_D}{k_B T} \left\{ 1 + \mathcal{O}\left(\frac{\hbar\omega_D}{k_B T}\right) \right\} .
\tag{6.146}
$$

Hence the term Π^2 is negligibly small in this temperature regime, and thus the expression (6.143) has the proper classical limit, $\langle p^2 \rangle \to M^2 \omega_0^2 \langle q^2 \rangle \to M k_B T$.

In the temperature range $\hbar\omega_D/k_B \gg T \gg \hbar\lambda_{1,2}/k_B$, the leading contribution to Π^2 comes from the $j = 3$ term in Eq. (6.144). This yields

$$
\Pi^2 = (M\hbar\gamma/\pi) \ln(\beta\hbar\omega_D) .
\tag{6.147}
$$

For very low temperature $T \ll \hbar\lambda_{1,2}/k_B$, we may use for the three digamma functions in Eq. (6.144) the respective asymptotic expansions. We then find with Eq. (6.101)

$$
\Pi^2 = (M\hbar\gamma/\pi) \ln(\omega_D/\omega_0) - M\hbar\omega_0 \alpha^2 g(\alpha) - \tfrac{1}{3}\pi M\hbar\gamma (k_B T/\hbar\omega_0)^2 ,
\tag{6.148}
$$

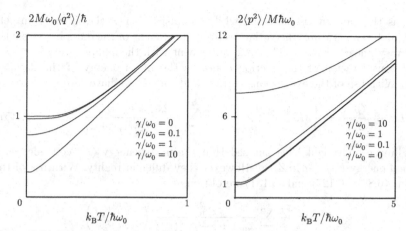

Figure 6.2: The normalized equilibrium variances of position and momentum are plotted versus temperature for the Drude model. The cutoff frequency is $\omega_D = 10\,\omega_0$. The damping strength γ/ω_0 varies between 0 and 10.

where terms of order ω_0/ω_D, γ/ω_D, and $(k_B T/\hbar\omega_0)^4$ have been disregarded. Now, Eq. (6.148) combines with Eq. (6.137) to give for $\langle p^2 \rangle$ at zero temperature

$$\langle p^2 \rangle_{T=0} = M^2 \omega_0^2 (1 - 2\alpha^2) \langle q^2 \rangle_{T=0} + (2\alpha M \hbar\omega_0/\pi) \ln(\omega_D/\omega_0) . \qquad (6.149)$$

Since, in a sense, the reservoir is continuously measuring the position of the particle, the position spread becomes smaller, while momentum spread and mean kinetic energy become larger, as damping is increased.[4] For large damping $\gamma \gg \omega_0$ we have

$$\langle q^2 \rangle_{T=0} = (2\hbar/M\pi\gamma) \ln(\gamma/\omega_0) , \qquad \langle p^2 \rangle_{T=0} = (M\hbar\gamma/\pi) \ln(\omega_D/\gamma) . \qquad (6.150)$$

The expression (6.149) reduces at $\alpha = 0$ to the momentum dispersion in the ground state of the undamped system, $\langle p^2 \rangle_{T=0} = M\hbar\omega_0/2$. In Fig. 6.3 (left diagram) the normalized dispersions $2M\omega_0\langle q^2 \rangle/\hbar$ (dashed-dotted line) and $2\langle p^2 \rangle/M\hbar\omega_0$ (dashed line), and the normalized uncertainty $2\sqrt{\langle q^2 \rangle \langle p^2 \rangle}/\hbar$ (full line) in the ground state are plotted as functions of the damping parameter α. The uncertainty reaches a plateau fairly above $\alpha = 2$. On the plateau there is only weak logarithmic dependence on α.

The low temperature series for $\langle q^2 \rangle$ and $\langle p^2 \rangle$ may be calculated from the integral expressions (6.127) and (6.129) by writing $\coth(\beta\hbar\omega/2) = 1 + 2/(e^{\beta\hbar\omega} - 1)$, and taking the low-frequency series for $\tilde{\chi}''(\omega)$. Since $\tilde{\chi}''(\omega)$ is odd in ω, the low temperature series is a power series in T^2. The leading term of $\langle q^2 \rangle$ is T^2, while that of $\langle p^2 \rangle$ is T^4.

The dispersions $\langle q^2 \rangle$ and $\langle p^2 \rangle$ of the harmonic oscillator with Drude-regularized Ohmic damping are depicted as functions of temperature in Fig. 6.2. The findings are as one would expect intuitively. For fixed temperature, the dispersion of the position

---

[4]If the reservoir would couple to the momentum of the oscillator, coordinate and momentum dispersion would perform role reversal. This is discussed on the next page.

decreases as the damping is increased while the dispersion of the momentum gets larger. At high temperature, the variances are independent of the damping strength, and they vary linearly with T. This is in agreement with the equipartition law.

A final consideration of this section concerns the mean energy of the damped oscillator. With use of the expressions (6.127) and (6.131) we have

$$\langle E \rangle \equiv \frac{\langle p^2 \rangle}{2M} + \frac{1}{2} M \omega_0^2 \langle q^2 \rangle = \frac{1}{\beta} + \frac{1}{\beta} \sum_{n=1}^{\infty} \frac{2\omega_0^2 + \nu_n \hat{\gamma}(\nu_n)}{\nu_n^2 + \nu_n \hat{\gamma}(\nu_n) + \omega_0^2} . \qquad (6.151)$$

At first glimpse one might have guessed that the mean energy $\langle E \rangle$ coincides with the internal energy $U = -\partial \ln Z / \partial \beta$. However they differ in reality. With use of the expressions (6.87), (6.127) and (6.131) we obtain

$$\begin{aligned} U - \langle E \rangle &= \frac{1}{\beta} \sum_{n=1}^{\infty} \frac{\nu_n^2}{\omega_0^2 + \nu_n^2 + \nu_n \hat{\gamma}(\nu_n)} \left(-\frac{\partial \hat{\gamma}(\nu_n)}{\partial \nu_n} \right) \\ &= \int_0^{\infty} d\omega \, u(\omega, \beta) \, \mathrm{Re} \left\{ \frac{M \omega \tilde{\chi}(\omega)}{\pi} \frac{\partial \tilde{\gamma}(\omega)}{\partial \omega} \right\} . \end{aligned} \qquad (6.152)$$

As a result, the specific heats $\partial U / \partial T$ and $\partial \langle E \rangle / \partial T$ differ, barring the strictly Ohmic case $\hat{\gamma}(z) = \gamma$. Since $Z = Z_{\mathrm{tot}} / Z_{\mathrm{R}}$ [cf. Eq. (4.247)], the system-bath coupling is fully included in the internal energy U and in the specific heat $\partial U / \partial T$. The above discrepancy shows that this interaction is only partially taken into account in $\langle E \rangle$ and in $\partial \langle E \rangle / \partial T$. The differences originate from the non-Ohmic part of the friction kernel. For the Drude-regularized damping kernel (3.43) one directly finds using the expressions(6.94) and (6.142)–(6.144) that both $U - \langle E \rangle$ and the difference of the specific heats $\partial U / \partial T - \partial \langle E \rangle / \partial T$ can be expressed in terms of polygamma functions.

Anomalous dissipation

We have just seen that normal dissipation resulting from the harmonic CL-model (6.1) has the effect of decreasing $\langle q^2 \rangle$ and increasing $\langle p^2 \rangle$. This is directly visible, e.g., in Fig. 6.2 on page 179. Interestingly, anomalous dissipation resulting from the model (3.258) with bilinear momentum coupling has the reverse effect. Writing the coordinate dispersion for moment coupling analogous to Eq. (6.127),

$$\langle q^2 \rangle_p = \frac{1}{M\beta} \sum_n \frac{1}{\omega_0^2 + \nu_n^2 + |\nu_n| \hat{\gamma}_p(|\nu_n|)} \qquad (6.153)$$

and using for $\hat{\gamma}_p(z)$ the expression (4.65), one gets

$$\langle q^2 \rangle_p = \frac{1}{M\beta\omega_0^2} \sum_n \frac{\omega_0^2 + |\nu_n| \hat{\gamma}_q(|\nu_n|)}{\omega_0^2 + \nu_n^2 + |\nu_n| \hat{\gamma}_q(|\nu_n|)} = \frac{1}{M^2 \omega_0^2} \langle p^2 \rangle_q \qquad (6.154)$$

The relations of the dispersions $\langle q^2 \rangle_q$ and $\langle p^2 \rangle_q$ of the CL-model with the dispersions $\langle q^2 \rangle_p$ and $\langle p^2 \rangle_p$ for momentum coupling are also immediately clear considering that the Hamiltonians (6.1) and (3.258) match each other for a harmonic potential under the canonical transformation (3.259) with constraint (3.260). Thus, there holds in addition to the relation (6.154) also the relation

$$\langle p^2 \rangle_p = M^2 \omega_0^2 \langle q^2 \rangle_q \, . \tag{6.155}$$

The important implication of these relations is that normal dissipation suppresses tunneling, whereas anomalous dissipation increases it (see Sec. 3.6 for some details).

6.7 Equilibrium density matrix

6.7.1 Derivation of the action

For a Gaussian process, the normalized equilibrium density matrix is completely specified by the position dispersion $\langle q^2 \rangle$ and the momentum dispersion $\langle p^2 \rangle$. Therefore, the correct form for any linear dissipative mechanism can be inferred from the equilibrium density matrix of the undamped oscillator, but with dispersions of the damped quantum oscillator,

$$\rho_\beta(q, q') \equiv\, <q|\hat\rho_\beta|q'> \,= \frac{1}{\sqrt{2\pi\langle q^2 \rangle}} \exp\left(-\frac{(q + q')^2}{8\langle q^2 \rangle} - \frac{\langle p^2 \rangle}{2\hbar^2}(q - q')^2 \right) \, . \tag{6.156}$$

Let us now see how the exponent in Eq. (6.156) emerges from the Euclidean action $S_{\text{eff}}[q(\cdot)]$ in Eq. (4.60). For a harmonic system we have

$$S_{\text{eff}}[q(\cdot)] = \int_0^{\hbar\beta} d\tau \left(\frac{M}{2}\dot q^2 + \frac{1}{2}M\omega_0^2 q^2 \right) + \int_0^{\hbar\beta} d\tau \int_0^\tau d\tau' \, k(\tau - \tau')q(\tau)q(\tau') \, . \tag{6.157}$$

The Fourier expansion method is somewhat complicated by the fact that the end-points of the path $q(\tau)$ are different and therefore the periodically continued path

$$q(\tau) = \frac{1}{\hbar\beta} \sum_{n=-\infty}^{\infty} q_n \, e^{i\nu_n \tau} \tag{6.158}$$

involves periodically repeated jumps from q back to q' and periodically repeated cusps, both at times $\tau = m\hbar\beta$ $(m = 0, \pm 1, \cdots)$. It is convenient to write

$$q(\tau) = q^{(1)}(\tau) + q^{(2)}(\tau) \, , \tag{6.159}$$

where $q^{(1)}(\tau)$ makes jumps, and $q^{(2)}(\tau)$ has cusps which result in jumps of the velocity,

$$q^{(1)}(0^- + m\hbar\beta) - q^{(1)}(0^+ + m\hbar\beta) = q - q' \, , \tag{6.160}$$

$$\dot q^{(2)}(0^- + m\hbar\beta) = -\dot q^{(2)}(0^+ + m\hbar\beta) = v \, . \tag{6.161}$$

At times $0 < \tau < \hbar\beta$, the equation of motion for the extremal path reads

$$-M\ddot{q}(\tau) + M\omega_0^2 q(\tau) + \int_{0^+}^{\tau} d\tau' \, k(\tau - \tau')q(\tau') \;=\; 0 \,. \tag{6.162}$$

Taking into account the jump conditions (6.160) and (6.161), we have in Fourier space

$$
\begin{aligned}
\left[\nu_n^2 + \omega_0^2 + |\nu_n|\hat{\gamma}(|\nu_n|)\right] q_n^{(1)} &= i\nu_n(q - q') \,, \\
\left[\nu_n^2 + \omega_0^2 + |\nu_n|\hat{\gamma}(|\nu_n|))\right] q_n^{(2)} &= 2v \,,
\end{aligned}
\tag{6.163}
$$

where v is determined by the requirement

$$q^{(2)}(\tau = m\hbar\beta) \;=\; \frac{1}{\hbar\beta} \sum_{n=-\infty}^{\infty} \frac{2v}{\nu_n^2 + |\nu_n|\hat{\gamma}(|\nu_n|) + \omega_0^2} \;=\; (q + q')/2 \,. \tag{6.164}$$

The Matsubara sum can be expressed in terms of $\langle q^2 \rangle$. This yields

$$v \;=\; \hbar(q + q')/4M\langle q^2 \rangle \,. \tag{6.165}$$

With Eqs. (6.163) $-$ (6.165) the Fourier coefficient of $q(\tau)$ is found to read

$$
\begin{aligned}
q_n \;=\; &\left(-\frac{1}{i\nu_n} + \frac{1}{i\nu_n} \frac{\omega_0^2 + |\nu_n|\hat{\gamma}(|\nu_n|)}{\omega_0^2 + \nu_n^2 + |\nu_n|\hat{\gamma}(|\nu_n|)} \right)(q - q') \\
&+ \frac{1}{\nu_n^2 + \omega_0^2 + |\nu_n|\hat{\gamma}(|\nu_n|)} \frac{\hbar}{2M\langle q^2 \rangle}(q + q') \,.
\end{aligned}
$$

With this there holds

$$
\begin{aligned}
q(0^+) &= q' \,, & q(0^-) &= q \,, \\
\dot{q}(0^\pm) &= \frac{\langle p^2 \rangle}{M\hbar}(q - q') \pm \frac{\hbar}{4M\langle q^2 \rangle}(q + q') \,.
\end{aligned}
\tag{6.166}
$$

As the extremal path satisfies the equation of motion (6.162), the related action (6.157) can be written as

$$S_{\mathrm{eff}}[q(\cdot)] \;=\; \int_{0^+}^{\hbar\beta - 0^+} d\tau \, \frac{M}{2} \frac{d}{d\tau}\Big(q(\tau)\dot{q}(\tau) \Big) \;=\; \frac{M}{2}\Big(q(0^-)\dot{q}(0^-) - q(0^+)\dot{q}(0^+) \Big) \,. \tag{6.167}$$

Here we have used $q(\hbar\beta - 0^+) = q(0^-)$. Use of the relations (6.166) finally yields

$$S_{\mathrm{eff}}[q(\cdot)]/\hbar \;=\; \frac{1}{8\langle q^2 \rangle}(q + q')^2 + \frac{\langle p^2 \rangle}{2\hbar^2}(q - q')^2 \,, \tag{6.168}$$

which is with an extra minus sign just the exponent in the expression (6.156).

6.7.2 Purity

After having substantiated the explicit form (6.156) of the equilibrium density matrix of the damped oscillator, the next step is to consider the purity

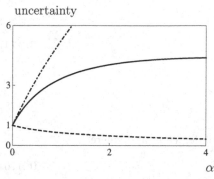
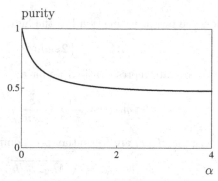

Figure 6.3: In the left figure the normalized variances $2M\omega_0\langle q^2\rangle/\hbar$ (dashed curve) and $2\langle p^2\rangle/M\hbar\omega_0$ (dashed-dotted curve), and the normalized uncertainty $2\sqrt{\langle q^2\rangle\langle p^2\rangle}/\hbar$ (full curve) of the ground state are plotted versus the damping parameter α. In the right figure, the purity pur(α) at zero temperature is plotted versus α. The Drude cutoff is $\omega_D = 100\,\omega_0$.

$$\mathrm{pur}\,(\alpha) \equiv \mathrm{tr}\,\hat{\rho}_\beta^2 = \int dq\,dq'\,\rho_\beta(q,q')\,\rho_\beta(q',q) = \frac{\hbar}{2\sqrt{\langle p^2\rangle\langle q^2\rangle}}\,. \tag{6.169}$$

For a pure quantum state, $\hat{\rho} = |\psi><\psi|$, the purity is unity. The expressions for $\langle q^2\rangle_{T=0}$ and $\langle p^2\rangle_{T=0}$ derived above, e.g., Eqs. (6.137) and (6.149), yield that already very weak damping makes the ground state rather impure. Accordingly, pure quantum states never exist in reality. On the other hand, for γ somewhat above ω_0 the purity reaches a plateau which is still of the order of one even for large α. There, the dependence of the purity on α is only logarithmic [cf. Fig 6.3]. This indicates that the damped oscillator stays quantum mechanical even for very large damping.

Let us now study what the effective mass and effective frequency of a fictional linear quantum oscillator with a discrete equidistant spectrum would be, if it had the same coordinate and momentum dispersions as specified in the preceding chapter. The respective statistical operator with normalization $\mathrm{tr}\,\hat{\rho}_\beta = 1$ is

$$\hat{\rho}_\beta = \tilde{Z}^{-1}(\beta)\,e^{-\beta\hat{H}_{\mathrm{eff}}}\,, \tag{6.170}$$

$$\hat{H}_{\mathrm{eff}} = \frac{1}{2M_{\mathrm{eff}}}\,\hat{p}^2 + \frac{1}{2}\,M_{\mathrm{eff}}\omega_{\mathrm{eff}}^2\hat{q}^2\,. \tag{6.171}$$

Here the parameters ω_{eff} and M_{eff} are fixed by the requirement that Eq. (6.170) has the coordinate representation (6.156). The density matrix can be represented as

$$\hat{\rho}_\beta = \frac{1}{\tilde{Z}(\beta)}\sum_{n=0}^{\infty} e^{-\beta E_n}\,|n><n|\,, \tag{6.172}$$

where $|n>$ is an eigenstate of the effective Hamiltonian

$$\hat{H}_{\mathrm{eff}}\,|n> = \hbar\omega_{\mathrm{eff}}(n + \tfrac{1}{2})\,|n>\,. \tag{6.173}$$

Now the partition function has the familiar form

$$\tilde{Z}(\beta) = [2\sinh(\beta\hbar\omega_{\text{eff}}/2)]^{-1} . \tag{6.174}$$

In coordinate representation, we have

$$< q|n > = \frac{1}{\pi^{1/4}} \left(\frac{c}{2^n n!}\right)^{1/2} H_n(cq) e^{-c^2 q^2/2} , \tag{6.175}$$

where the $H_n(x)$ are Hermitian polynomials of degree n, and where

$$c = \sqrt{M_{\text{eff}}\omega_{\text{eff}}/\hbar} . \tag{6.176}$$

With Eqs. (6.172), (6.173), (6.175) and with the generating function of the Hermitian polynomials, the coordinate representation of the density matrix is found as [4]

$$< q|\hat{\rho}_\beta|q' > = \frac{c}{\tilde{Z}\sqrt{2\pi\sinh\Omega}} \exp\left\{-\frac{c^2\left[(q^2 + q'^2)\cosh\Omega - 2qq'\right]}{2\sinh\Omega}\right\} , \tag{6.177}$$

where $\Omega = \omega_{\text{eff}}\hbar\beta$. Upon equating Eq. (6.156) with Eq. (6.177) one finds that the effective frequency ω_{eff} and mass M_{eff} are connected with $\langle q^2 \rangle$ and $\langle p^2 \rangle$ by

$$\omega_{\text{eff}} = \frac{1}{\hbar\beta} \ln\frac{\sqrt{\langle p^2\rangle\langle q^2\rangle} + \hbar/2}{\sqrt{\langle p^2\rangle\langle q^2\rangle} - \hbar/2} , \qquad M_{\text{eff}} = \sqrt{\langle p^2\rangle/\langle q^2\rangle}/\omega_{\text{eff}} . \tag{6.178}$$

At high temperatures, the effective frequency ω_{eff} and mass M_{eff} are close to their bare values ω_0 and M, whereas at low temperatures, they are strongly different. Substituting the zero temperature values for $\langle q^2 \rangle$ and $\langle p^2 \rangle$, we find

$$\hbar\omega_{\text{eff}} = 2 k_B T \operatorname{arcoth}\sqrt{g(\alpha)\left[(1 - 2\alpha^2)g(\alpha) + (4\alpha/\pi)\ln(\omega_D/\omega_0)\right]} , \tag{6.179}$$

where the function $g(\alpha)$ is given in Eq. (6.102). Observe that ω_{eff} vanishes linearly with T as $T \to 0$. Hence the discrete energy levels in Eq. (6.173) become very narrowly spaced near zero temperature. In fact, the occupation probabilities $p_n = \tilde{Z}^{-1}\exp(-\beta E_n)$ remain finite in the limit $T \to 0$. From this we infer that the ground state of the fictional oscillator is not a pure state but a mixture.

We conclude with the remark that the partition function of the damped oscillator Z and the partition function of the fictional oscillator $\tilde{Z}$ drastically differ at low temperatures for $\alpha > 0$. They approach each other as temperature is increased and coincide in the classical limit.

6.8 Quantum master equations for the reduced density matrix

So far we have considered selected thermal averages, e.g., the mean squares of displacement and momentum. For a complete characterization of the statistical properties of the central oscillator, knowledge of the reduced density matrix (RDM) or the Wigner function is essential. Of particular interest are the corresponding dynamical equations. The path integral expression (5.63) with (5.65) for the propagating function of the damped oscillator has been calculated in analytic form in Ref. [88]. For a given preparation function, the reduced density matrix is then given by the integral expression (5.61). Starting out from these expressions, Karrlein and Grabert have studied various quantum master equations for the RDM for different types of the initial preparation. They found that the RDM is described in terms of the "quasiclassical" coordinate $r = \frac{1}{2}(q + q')$ and the "fluctuation" coordinate $y = q - q'$ by the master equation [48]

$$\frac{\partial}{\partial t}\rho(y, r; t) = \mathcal{L}\left(y, r, \frac{\partial}{\partial y}, \frac{\partial}{\partial r}; t\right)\rho(y, r; t) \tag{6.180}$$

with the time-dependent Liouville operator

$$\mathcal{L} = \frac{i\hbar}{M}\frac{\partial^2}{\partial y \partial r} - \frac{iM}{\hbar}\gamma_q(t)ry - \gamma_p(t)y\frac{\partial}{\partial y} - \frac{iM}{\hbar}D_q(t)y\frac{\partial}{\partial r} - \frac{M^2}{\hbar^2}D_p(t)y^2 . \tag{6.181}$$

The functions $\gamma_q(t)$ and $\gamma_p(t)$ are time-dependent drift coefficients and the functions $D_q(t)$ and $D_p(t)$ are time-dependent diffusion coefficients. The analysis in Ref. [48] showed that these functions depend on the preparation function. Hence there is no unique exact Liouville operator.

6.8.1 Thermal initial condition

Consider now initial conditions specified by the preparation function (cf. Sec. 5.3.1)

$$\lambda(y_i, r_i, \bar{y}, \bar{r}) = f(y_i, r_i)\lambda_\beta(y_i, r_i, \bar{y}, \bar{r}) = f(y_i, r_i)\delta(\bar{y} - y_i)\delta(\bar{r} - r_i) . \tag{6.182}$$

In this case the diffusion and drift coefficients are expressed in terms of the symmetrized part $S(t)$ and the antisymmetrized part $A(t) = -(\hbar/2M)G(t)$ of the equilibrium position autocorrelation function $C^+(t) \equiv \langle q(t)q(0)\rangle_\beta = S(t) - i(\hbar/2M)G(t)$ discussed in Sections 6.3 and 6.4. One finds [48]

$$\gamma_q(t) = \frac{\dot{G}(t)\ddot{S}(t) - \ddot{G}(t)\dot{S}(t)}{\dot{G}(t)S(t) - G(t)\dot{S}(t)}, \qquad \gamma_p(t) = \frac{G(t)\ddot{S}(t) - \ddot{G}(t)S(t)}{\dot{G}(t)S(t) - G(t)\dot{S}(t)}, \tag{6.183}$$

$$D_q(t) = \gamma_q(t)\langle q^2\rangle - \langle p^2\rangle/M^2 , \qquad D_p(t) = \gamma_p(t)\langle p^2\rangle/M^2 .$$

The dispersions $\langle q^2\rangle$ and $\langle p^2\rangle$ are given in Eqs. (6.127) and (6.131). By virtue of the assignments

$$y \to [q, \cdot] , \qquad r \to \{q, \cdot\}/2 , \qquad \partial/\partial y \to (i/2\hbar)\{p, \cdot\} , \qquad \partial/\partial r \to (i/\hbar)[p, \cdot] , \tag{6.184}$$

the master equation (6.180) with (6.183) can be written in the form

$$\dot{\rho}(t) = -(iM/\hbar)\gamma_q(t)\left[q, \{q, \rho(t)\}/2 + (i/\hbar)\langle q^2\rangle[p, \rho(t)]\right]$$

$$- (i/\hbar)\gamma_p(t)\left[q, \{p, \rho(t)\}/2 - (i/\hbar)\langle p^2\rangle[q, \rho(t)]\right] \tag{6.185}$$

$$- (i/M\hbar)\left[p, \{p, \rho(t)\}/2 - (i/\hbar)\langle p^2\rangle[q, \rho(t)]\right] ,$$

where $[u, v] = uv - vu$ is the commutator and $\{u, v\} = uv + vu$ is the anti-commutator.

The master equation (6.185) can be transformed into a generalized Fokker-Planck equation for the Wigner function $w(p, q; t) = \int dy\, \rho(y, q; t)\, e^{-iyp/\hbar}$. With use of the correspondences

$$[q, \cdot] \to -\frac{\hbar}{i}\frac{\partial}{\partial p}, \quad \{q, \cdot\} \to 2q, \quad [p, \cdot] \to \frac{\hbar}{i}\frac{\partial}{\partial q}, \quad \{p, \cdot\} \to 2p, \quad (6.186)$$

we then obtain the generalized Fokker-Planck equation

$$
\dot{w}(p, q; t) = \left\{ \frac{\partial}{\partial p} M \gamma_q(t) \left(q + \frac{\partial}{\partial q}\langle q^2 \rangle \right) \right.
$$
$$
\left. + \left(\frac{\partial}{\partial p}\gamma_p(t) - \frac{\partial}{\partial q}\frac{1}{M} \right) \left(p + \frac{\partial}{\partial p}\langle p^2 \rangle \right) \right\} w(p, q; t) . \quad (6.187)
$$

To see what we have obtained so far, consider now the classical limit. The classical position autocorrelation function $S_{\mathrm{cl}}(t)$ is given by the integral expression (6.63). Now, the derivative $\dot{S}_{\mathrm{cl}}(t)$ is directly related to the function $G(t)$ as $G(t) = -M\beta\dot{S}_{\mathrm{cl}}(t)$. Further we have $\langle q^2 \rangle_{\mathrm{cl}} = 1/M\beta\omega_0^2$ and $\langle p^2 \rangle_{\mathrm{cl}} = M^2\omega_0^2\langle q^2 \rangle_{\mathrm{cl}}$. With these expressions the generalized Fokker-Planck equation (6.187) takes exactly the form found by Adelman [265]. He derived this equation from the generalized classical Langevin equation (2.4) with (2.9). The dynamical equation (6.187) with (6.183) represents the exact quantum mechanical generalization of Adelman's Fokker-Planck equation.

6.8.2 Product initial state

In many works it has been assumed for simplicity that the initial density matrix is in factorized form [208, 209, 248, 254]. The time evolution is then given by the expression (5.11) with (5.12). The resulting master equation is again in the form (6.180) with (6.181), but now the drift and diffusion functions are given by

$$
\gamma_q(t) = \frac{\ddot{G}^2(t) - \dot{G}(t)\dddot{G}(t)}{\dot{G}^2(t) - G(t)\ddot{G}(t)}, \quad \gamma_p(t) = \frac{G(t)\dddot{G}(t) - \dot{G}(t)\ddot{G}(t)}{\dot{G}^2(t) - G(t)\ddot{G}(t)} \Big],
$$

$$
D_q(t) = \frac{\hbar}{M}\left(\frac{1}{2}\ddot{K}_q(t) - K_p(t) + \gamma_q(t)K_q(t) + \frac{1}{2}\gamma_p(t)\dot{K}_q(t) \right), \quad (6.188)
$$

$$
D_p(t) = \frac{\hbar}{M}\left(\frac{1}{2}\dot{K}_p(t) + \gamma_p(t)K_p(t) + \frac{1}{2}\gamma_q(t)\dot{K}_q(t) \right).
$$

$$
K_q(t) = \frac{1}{M}\int_0^t dt_2 \int_0^t dt_1\, G(t_2)L'(t_2 - t_1)G(t_1),
$$

$$
K_p(t) = \frac{1}{M}\int_0^t dt_2 \int_0^t dt_1\, \dot{G}(t_2)L'(t_2 - t_1)\dot{G}(t_1), \quad (6.189)
$$

where $L'(t)$ is the real part of the kernel $L(t)$, Eq. (5.24). The master equation (6.187) with (6.188) and (6.189) corresponds to that by Haake and Reibold [248] who derived it from the microscopic dynamics. Few years later, Hu, Paz and Zhang confirmed the findings of Haake and Reibold by means of the Feynman-Vernon method [254].

In case of a product initial state, the sudden switch-on of the interaction with the bath at time zero produces initial slips which may persist over times of the order of the relaxation time. As a result, the master equation for a product initial state does not reduce in the classical limit to a master equation that is equivalent to Adelman's Fokker-Planck equation.

6.8.3 Approximate time-independent Liouville operators

Let us now assume that the functions $G(t)$ and $S(t)$ are exponential forms,

$$G(t) = a_1 e^{-\lambda_1 t} + a_2 e^{-\lambda_2 t} , \qquad S(t) = b_1 e^{-\lambda_1 t} + b_2 e^{-\lambda_2 t} , \qquad (6.190)$$

where λ_1 and λ_2 are complex conjugate or real frequencies, and where the coefficients a_1, a_2 and b_1, b_2 are complex conjugate pairs or real, respectively. Under this assumption, the drift coefficients $\gamma_p(t)$ and $\gamma_q(t)$ and diffusion coefficients $D_q(t)$ and $D_p(t)$ given in Eq. (6.183) become time-independent,

$$\gamma_q = \lambda_1\lambda_2 , \qquad\qquad \gamma_p = \lambda_1 + \lambda_2 ,$$
$$D_q = \lambda_1\lambda_2 \langle q^2 \rangle - \langle p^2 \rangle/M^2 , \qquad D_p = (\lambda_1 + \lambda_2)\langle p^2 \rangle/M^2 . \qquad (6.191)$$

Then the Wigner transform of the density matrix obeys the Fokker-Planck equation (6.187) with time-independent coefficients.

Unfortunately, the forms (6.190) can never be met jointly, as we shall immediately. Hence there is no exact time-independent Liouville operator.

Ohmic damping

For strictly Ohmic damping, the function $G(t)$ is exactly of the form (6.190), as we may infer from Subsec. 6.3.1. However, the strictly Ohmic case is excluded because the variance of momentum would diverge, as discussed in Subsec. 6.6.2. In addition, the form (6.190) for $S(t)$ would only hold if the Matsubara series (6.66) would be disregarded.

Drude damping

For Drude regularized damping, $\hat{\gamma}(z) = \gamma\omega_D/(z + \omega_D)$, the momentum dispersion is regularized, but unfortunately both $G(t)$ and $S(t)$ obtain additional contributions $\propto e^{-\Omega_D t}$, where Ω_D is given in Eq. (6.45). These contribution as well as the Matsubara contribution (6.66) to $S(t)$ must be disregarded in order that the Liouville operator becomes time-independent. According to the findings in Subsecs. 6.3.1 and 6.4.1, this happens under conditions $\omega_D \gg \omega_0$, γ and $k_B T \gg \hbar\omega_0$, $\hbar\gamma$. Only then can the dynamics be described by a time-independent Liouville operator of the form (6.181) with the coefficients (6.191). Explicit expressions for the variances $\langle q^2 \rangle$ and $\langle p^2 \rangle$ are discussed in Subsec. 6.6.2. The resulting approximate master equation is

$$\dot{\rho}(t) = -(i/\hbar)\, M\lambda_1\lambda_2 \left[q,\, \tfrac{1}{2}\{q,\, \rho(t)\} + (i/\hbar)\langle q^2\rangle\, [p,\, \rho(t)]\, \right]$$
$$-(i/\hbar)(\lambda_1 + \lambda_2)\left[q,\, \tfrac{1}{2}\{p,\, \rho(t)\} - (i/\hbar)\langle p^2\rangle\, [q,\, \rho(t)]\, \right] \qquad (6.192)$$
$$-(i/M\hbar)\left[p,\, \tfrac{1}{2}\{p,\, \rho(t)\} - (i/\hbar)\langle p^2\rangle\, [q,\, \rho(t)]\, \right]\, ,$$

first derived by Haake and Reibold [248].

The classical regime with strictly Ohmic friction is reached by first taking the high temperature limit and subsequently the limit $\omega_D \to \infty$. This yields

$$\gamma_q = \omega_0^2\,, \qquad \gamma_p = \gamma\,, \qquad D_q = 0\,, \qquad D_p = \gamma k_B T/M\,. \qquad (6.193)$$

The master equation (6.192) with (6.193) can be transformed into a Fokker-Planck equation for the Wigner function $w(p, q; t)$,

$$\dot{w}(p, q; t) = \left\{ -\frac{1}{M}\frac{\partial}{\partial q}p + \frac{\partial}{\partial p}\left[\frac{\partial V(q)}{\partial q} + \gamma p\right] + \gamma k_B T M\frac{\partial^2}{\partial p^2} \right\} w(p, q; t)\,. \qquad (6.194)$$

Here we have put $M\omega_0^2 q = \partial V(q)/\partial q$. Writing Eq. (6.194) in position and velocity space, it becomes the Klein-Kramers equation derived by Kramers already in 1940 [266]. We shall study Kramers' flux solution of this equation in Section 11.2.

For weak damping and arbitrary frequency dependence, the contribution $S_2(t)$ resulting from the poles of the expression (6.62) is negligibly small. The relevant poles of $\hat{G}(z)$ and $\hat{S}_1(z)$ are at $z = -\lambda_{1,2}$, where $\lambda_{1,2} = \tfrac{1}{2}\gamma' \pm i\,(\omega_0 + \tfrac{1}{2}\gamma'')$, and where $\gamma' + i\,\gamma'' = \hat{\gamma}(z = -i\,\omega_0)$. Thus the detailed frequency-dependence of $\tilde{\gamma}(\omega)$ is irrelevant. With these poles, the functions $G(t)$ and $S(t)$ read as in Eq. (6.190). As a result one ends up again with a quantum master equation of the form (6.192).

6.8.4 Connection with Lindblad theory

We see from the derivation that the time-independent quantum master equation (6.192) holds only for times $t > t_0$, where $t_0 > 1/\omega_D$, $\hbar\beta/2\pi$. During the initial interval $0 < t < t_0$, the fast components decay, and afterwards the density matrix is reduced to a subspace in which the fast components are absent. The Markovian master equation (6.192) holds only for density matrices within this subspace.

On the other hand, Lindblad theory supposes validity for any reduced density matrix. The Lindblad master equation (2.28) preserves the positivity of density operators for all times. The time-independent Liouville operators derived in the previous subsection are not of Lindblad form, as can be seen with use of findings by Sandulescu and Scutaru [267]. This is not disquieting because, as we have seen in Subsection 2.3.2, a slippage of fast components must occur before the dynamics effectively is Markovian. Therefore, complete positivity postulated by Lindblad theory is not necessary, as was emphasized by Pechukas [47].

Further coarse-graining was studied in Ref. [48] for weak damping $\gamma \ll \omega_0$. It was found that in the time regime $\omega_0^{-1} \ll t \ll \gamma^{-1}$ the above time-independent Liouville operators can be cast in fact into Lindblad form.

7 Quantum Brownian free motion

Historically, the Langevin equation (2.3) was used to analyze the irregular motion (*Zitterbewegung*) of a heavy particle moving in a thermally equilibrated molecular medium. This phenomenon is known as *Brownian motion*, named after Robert Brown, who observed the random motion of pollen grains immersed in a fluid. In many cases, a systematic external force is absent. Then we are left with the problem of free Brownian motion. Often, the memory times of the fluctuating forces are negligibly short compared to the time scale over which one observes the Brownian particle. The assumption of memoryless friction is usually referred to as Markov limit.

In this chapter I consider quantum Brownian free motion for general frequency-dependent damping $\tilde{\gamma}(\omega)$. I begin with an important remark: since the equation of motion is linear, the susceptibility is devoid of $\hbar$, and the particle behaves as a classical one. The Planck constant enters solely via the fluctuation-dissipation theorem, as already emphasized after Eq. (6.37). A brief discussion of genuine quantum effects resulting from state-dependent dissipation is postponed to Subsection 9.3.1. Intrinsic quantum effects are limited to nonlinear systems, i.e., a damped particle in an anharmonic potential, and they crucially affect linear and nonlinear transport properties, as we shall see in Part V.

The simplest microscopic model for a free Brownian particle with bare mass M_0 is described by the Hamiltonian (3.12) with $V(q) = 0$ and $M = M_0$. The classical equation of motion reads

$$M_0\,\ddot{q}(t) + M_0 \int_{-\infty}^{t} dt'\,\gamma(t - t')\,\dot{q}(t') \;=\; \xi(t)\,. \tag{7.1}$$

Here $\xi(t)$ is the fluctuating force with properties as discussed in Section 3.1.2. As emphasized previously, the dynamical equation (7.1) holds exactly also in the quantum regime, but with the condition that the stochastic force obeys the quantum mechanical force auto-correlation (2.13). Additional light is shed on the underlying physics by introducing new coordinates y_α and masses μ_α for the bath oscillators [268],

$$y_\alpha \;=\; \frac{m_\alpha\,\omega_\alpha^2}{c_\alpha}\,x_\alpha\;; \qquad \mu_\alpha \;=\; \frac{c_\alpha^2}{m_\alpha\,\omega_\alpha^4}\,. \tag{7.2}$$

Expressed with these variables, the Hamiltonian (3.12) with $V(q) = 0$ is

$$H \;=\; \frac{p^2}{2\,M_0} + \frac{1}{2} \sum_{\alpha=1}^{N} \mu_\alpha \left(\dot{y}_\alpha^2 + \omega_\alpha^2\,(y_\alpha - q)^2 \right). \tag{7.3}$$

We see that the model describes a particle of mass M_0 with all of the reservoir's effective masses μ_α attached with springs to its position coordinate q. This form makes the translational invariance of the model evident.

7.1 Spectral density, damping function and mass renormalization

In this section, I briefly discuss renormalization of the mass of the Brownian particle induced by the environmental coupling.

With the substitution (7.2) the spectral density of the coupling takes the form

$$J_{\text{tot}}(\omega) = \frac{\pi}{2} \sum_{\alpha=1}^{N} \mu_{\alpha} \omega_{\alpha}^{3} \delta(\omega - \omega_{\alpha}) . \tag{7.4}$$

Any desired frequency dependence of the spectral coupling $J(\omega)$ can be modelled by an appropriate spectral distribution of the oscillator masses μ_{α}.

If $\sum_{\alpha} \mu_{\alpha}$ is finite, the total energy associated with the Hamiltonian (7.3) is the sum of an internal energy, which is independent of the total momentum, and the kinetic energy of the center of mass. The latter is the square of the total momentum divided by twice the mass of the total system. The total mass is the renormalized mass

$$M_{\text{r}} = M_0 + \sum_{\alpha=1}^{N} \mu_{\alpha} = M_0 + \frac{2}{\pi} \int_{0}^{\infty} d\omega \, \frac{J_{\text{tot}}(\omega)}{\omega^3} . \tag{7.5}$$

In the second form, we have taken the continuum limit. Clearly, the notion of a total mass M_{tot} is only meaningful when the sum or rather the integral in Eq. (7.5) is finite.

Consider now the case when the total spectral density $J_{\text{tot}}(\omega)$ is split in low and high frequency parts with crossover frequency ω_{c}, as given in Eq. (3.35). For the purpose of this section, details of the high-frequency part $J_{\text{hf}}(\omega)$ are irrelevant. For times $t \gg \omega_{\text{c}}^{-1}$, the only effect of the high frequency modes is mass renormalization,

$$M = M_0 + M_{\text{hf}} = M_0 + \frac{2}{\pi} \int_{0}^{\infty} d\omega \, \frac{J_{\text{hf}}(\omega)}{\omega^3} . \tag{7.6}$$

Now we are left with the problem of a Brownian particle of mass M coupled to a heat bath environment described by the spectral density $J(\omega)$ given in Eq. (3.38).

If the exponent s in Eq. (3.38) exceeds 2, the integral in Eq. (7.5) is indeed infrared-convergent. Then the environment in its entirety equips the particle with a polaron cloud which enhances inertia of the particle. The increment of the inertial mass due to the low-frequency spectral density (3.38) with exponential cutoff is [cf. Eq. (3.57)]

$$M_{\text{bath}} = M_{\text{lf}} \equiv \frac{2}{\pi} \int_{0}^{\infty} d\omega \, \frac{J(\omega)}{\omega^3} = \Gamma(s-2) \frac{2}{\pi} \frac{\gamma_s}{\omega_{\text{ph}}} \left(\frac{\omega_{\text{c}}}{\omega_{\text{ph}}} \right)^{s-2} M . \tag{7.7}$$

With Eq. (7.7), the spectral function $\tilde{\gamma}(\omega)$ behaves analogous to Eq. (3.37) as

$$\lim_{\omega \to 0} \tilde{\gamma}(\omega)/\omega = -i \, M_{\text{lf}}/M . \tag{7.8}$$

The dynamical susceptibility of the Brownian particle

$$\tilde{\chi}(\omega) = \frac{1}{M} \frac{1}{-\omega^2 - i\omega\tilde{\gamma}(\omega)} \tag{7.9}$$

thus has the low-frequency behavior

$$\lim_{\omega \to 0} \tilde{\chi}(\omega)\,\omega^2 \;=\; -\frac{1}{M + M_{\mathrm{lf}}}\,, \qquad\qquad s > 2\,. \tag{7.10}$$

From this we infer that, when $s > 2$, friction effectively vanishes at asymptotic time, and the Brownian particle behaves as a free particle with renormalized mass

$$M_{\mathrm{r}} \;=\; M_0 + M_{\mathrm{hf}} + M_{\mathrm{lf}} = M + M_{\mathrm{lf}}\,. \tag{7.11}$$

On the other hand, the integral in Eq. (7.7) is IR-divergent if $s \leqslant 2$. This indicates that the low-energy excitations of the reservoir produce real dynamical effects beyond simple mass renormalization which last until time infinity. The leading low-frequency behavior of the spectral damping function is [cf. Eq. (3.53)]

$$\tilde{\gamma}(\omega) \;=\; [\gamma_s/\sin(\pi s/2)]\,(-i\,\omega/\omega_{\mathrm{ph}})^{s-1}\,. \tag{7.12}$$

For $s < 2$, $\tilde{\gamma}(\omega) \propto \omega^{s-1}$ beats linear dependence on ω as $\omega \to 0$. Hence for $s < 2$ damping is effective at any time. In the sequel, we shall regard M as the mass which is already dressed by the reservoir's high frequency modes and focus on the effects induced by the spectral coupling function $J(\omega)$.

7.2 Displacement correlation and response function

If we take the limit $\omega_0 \to 0$ in the expression (6.127), the mean square of the position $\langle q^2 \rangle$ diverges because the position of a free particle is not bounded. In the displacement correlation function, the divergent initial value of $\langle q(t)\,q(0) \rangle$ is subtracted,

$$\begin{aligned}
D^+(t) &\equiv C^+(0) - C^+(t) = \langle\,[q(0) - q(t)]\,q(0)\,\rangle_\beta\,, \\
D^-(t) &\equiv C^-(0) - C^-(t) = \langle\,q(0)\,[q(0) - q(t)]\,\rangle_\beta\,,
\end{aligned} \tag{7.13}$$

The function $D^\pm(t)$ describes correlations between the displacement $q(0) - q(t)$ of the particle at time t and the initial position $q(0)$. In the split-up

$$D^\pm(t) \;=\; \tfrac{1}{2}\,D_{\mathrm{th}}(t) \mp i\,A(t)\,, \tag{7.14}$$

the real part for $t > 0$, $D_{\mathrm{th},+}(t) = \Theta(t)\,D_{\mathrm{th}}(t)$, is the mean square displacement. It is related to the symmetric position correlation function $S(t)$, defined in Eq. (6.13), as

$$D_{\mathrm{th},+}(t) \;\equiv\; \Theta(t)\langle\,[q(t) - q(0)]^2\,\rangle_\beta \;=\; \Theta(t)\,2[S(0) - S(t)]\,. \tag{7.15}$$

The imaginary part $A(t)$ is connected with the thermal expectation value of the commutator in the usual way,

$$A(t) \;=\; \langle\,[\,q(t),q(0)\,]\,\rangle_\beta/2\,i\,. \tag{7.16}$$

The response function is [cf. Eq. (6.15)]

$$\chi(t) \;=\; -\,(2/\hbar)\,\Theta(t)\,A(t)\,. \tag{7.17}$$

Since $\langle q(t) \rangle_\beta = 0$, the variance $\sigma^2(t)$ coincides with the mean square displacement,

$$\sigma^2(t) = D_{\text{th},+}(t) \qquad \text{for} \qquad t > 0 . \tag{7.18}$$

The Fourier integral representations of $D_{\text{th},+}(t)$ and of $\chi(t)$ are

$$D_{\text{th},+}(t) = \Theta(t) \frac{\hbar}{\pi} \int_{-\infty}^{+\infty} d\omega \, \tilde{\chi}''(\omega) \coth(\beta\hbar\omega/2) \left[1 - e^{-i\omega t} \right] , \tag{7.19}$$

$$\chi(t) = \Theta(t) \frac{i}{\pi} \int_{-\infty}^{+\infty} d\omega \, \tilde{\chi}''(\omega) \, e^{-i\omega t} . \tag{7.20}$$

Evaluation of the expressions (7.19) and (7.20) in the limit $t \to \infty$ shows that $\chi(t)$ is connected with the change of $D_{\text{th},+}(t)$ per unit time by the simple relation

$$\lim_{t\to\infty} \dot{D}_{\text{th},+}(t)/\chi(t) = 2k_{\text{B}}T . \tag{7.21}$$

The relation (7.21) is a version of the *Einstein relation*, as we shall see shortly. In the derivation of Eq. (7.21), no specific assumptions on $\tilde{\gamma}(\omega)$, and not even on $\tilde{\chi}(\omega)$ were made. Therefore, the relation (7.21) holds for any nonlinear system as well.

With the expression (7.19) one finds that the Laplace transform of the function $D_{\text{th},+}(t)$ is related to the dynamical susceptibility as

$$\hat{D}_{\text{th},+}(\lambda) = \frac{2\hbar}{\pi\lambda} \int_0^{+\infty} d\omega \, \frac{\omega^2}{\lambda^2 + \omega^2} \coth\left(\frac{\beta\hbar\omega}{2} \right) \tilde{\chi}''(\omega) . \tag{7.22}$$

This relation is a variant of the fluctuation-dissipation theorem and therefore holds for any linear or nonlinear quantum transport system. A by-product of this relation is

$$\lim_{\lambda\to 0} \lambda^2 \hat{D}_{\text{th},+}(\lambda) = 2k_{\text{B}}T \lim_{\omega\to 0} \omega \, \tilde{\chi}''(\omega) . \tag{7.23}$$

Consider finally the velocity response of the Brownian particle to a constant force switched on at time zero, $\delta F_{\text{ext}}(t) \to \Theta(t) F$, and to a δ-force $\delta F_{\text{ext}}(t) = Mv\,\delta(t)$ where F and v need not be infinitesimal. For these particular cases, the mean velocity of the particle at time t is found from Eq. (6.6) as

$$\langle \dot{q}(t) \rangle = \chi(t) \, F , \tag{7.24}$$

$$\langle \dot{q}(t) \rangle = \mathcal{R}(t) \, v , \qquad \text{with} \qquad \mathcal{R}(t) = M\dot{\chi}(t) . \tag{7.25}$$

7.3 Ohmic friction

We now study Brownian motion in the case of Ohmic friction in some detail.

7.3.1 Response function

It is practical knowledge from daily experience that the particle's drift velocity in a viscous medium under a constant force becomes asymptotically constant,

$$\lim_{t\to\infty} \langle \dot{q}(t) \rangle = \mu \, F . \tag{7.26}$$

The quantity μ is the dc mobility of the Brownian particle. The mobility μ is related to the response function $\chi(t)$ as

$$\mu = \lim_{t\to\infty} \chi(t) = \lim_{\omega\to 0} \omega \, \tilde{\chi}''(\omega = 0) \,. \qquad (7.27)$$

The first form results from Eq. (7.24), whereas the second form arises out of Eq. (7.20) with the representation $\delta(\omega) = \lim_{t\to\infty} \sin(\omega t)/(\pi\omega)$. Putting $\tilde{\gamma}(\omega) = \gamma$ in Eq. (7.9), the response function is found from the integral (7.20) in the familiar form

$$\chi(t) = \Theta(t) \frac{1}{M\gamma}\left(1 - e^{-\gamma t}\right) \,. \qquad (7.28)$$

Hence the mobility of a free Brownian particle in a viscous medium neither depends on temperature nor captures quantum effects,

$$\mu = \frac{1}{M\gamma} = \frac{1}{\eta} \,. \qquad (7.29)$$

Finally, the velocity response with respect to a δ-force is found from Eq. (7.25) as

$$\mathcal{R}(t) = M\dot{\chi}(t) = \Theta(t)\, e^{-\gamma t} \,. \qquad (7.30)$$

7.3.2 Mean square displacement

The time evolution of the mean square displacement can be found either by direct contour integration of the expression (7.19), or by observing Eq. (7.15) and taking the limit $\omega_0 \to 0$ in the expressions (6.65) and (6.66). Both routes yield

$$\sigma^2(t) = \frac{2}{M\beta\gamma}t - \frac{\hbar}{M\gamma}\cot(\beta\hbar\gamma/2)\left[1 - e^{-\gamma t}\right] + \frac{4\gamma}{M\beta}\sum_{n=1}^{\infty}\frac{1 - e^{-\nu_n t}}{\nu_n(\gamma^2 - \nu_n^2)} \,. \qquad (7.31)$$

In the high-temperature limit, this reduces to the classical mean square displacement

$$\sigma_{\mathrm{cl}}^2(t) = \frac{2}{M\beta\gamma^2}\left(\gamma t - 1 + e^{-\gamma t}\right) \,. \qquad (7.32)$$

At short time, there is $\sigma_{\mathrm{cl}}^2(t) = t^2/M\beta$, while at asymptotic time $\sigma_{\mathrm{cl}}^2(t)$ grows $\propto t$. Eq. (7.31) shows that quantum signatures in $\sigma^2(t)$ are relevant only for low temperature, $\beta\hbar\gamma/2\pi > 1$, and for short to intermediate time. With decreasing temperature, the regime where quantum fluctuations are relevant extends to longer time. However at any *finite* temperature, the dynamics at asymptotic time is dominated by the first term in Eq. (7.31), which is purely classical. We find for the diffusion coefficient

$$D \equiv \frac{1}{2}\lim_{t\to\infty}\frac{d}{dt}\sigma^2(t) = \frac{1}{M\beta\gamma} = \frac{k_{\mathrm{B}}T}{M\gamma} \,. \qquad (7.33)$$

The remaining task is to link the (linear) mobility μ to the diffusion coefficient. Upon inserting Eqs. (7.27) with (7.29) and (7.33) into (7.21), or by comparison of Eq. (7.29) with Eq. (7.33), we obtain the Einstein relation in the familiar form

$$D = k_{\mathrm{B}} T \mu .$$

(7.34)

Thus, the diffusion coefficient vanishes at zero temperature. This indicates that the increase of the mean square displacement slows down when T approaches zero.

As $T \to 0$, the Matsubara sum in Eq. (7.31) is converted into an integral,

$$\lim_{T \to 0} \frac{1}{\beta} \sum_{n=1}^{\infty} f(\nu_n) = \frac{\hbar}{2\pi} \int_0^{\infty} d\nu \, f(\nu) .$$

(7.35)

Since all other terms vanish, the time derivative of $\sigma_{T=0}^2(t)$ takes the form

$$\frac{d}{dt} \sigma_{T=0}^2(t) = \frac{2\hbar\gamma}{\pi M} \int_0^{\infty} d\nu \, \frac{e^{-\nu t}}{\gamma^2 - \nu^2} .$$

(7.36)

The integral can be expressed in terms of the exponential-integral function $\mathrm{Ei}(x)$,

$$\mathrm{Ei}(x) = \fint_{-\infty}^{x} du \, \frac{e^u}{u} .$$

(7.37)

The dashed integral symbol designates the principal value of the integral. Integration of Eq. (7.36) with initial value $\sigma_{T=0}^2(0) = 0$ yields the representation [88]

$$\sigma_{T=0}^2(t) = \frac{\hbar}{\pi M} \int_0^t ds \left(\mathrm{Ei}(\gamma s) \, e^{-\gamma s} - \mathrm{Ei}(-\gamma s) \, e^{\gamma s} \right) .$$

(7.38)

Evaluation of the integral for times $t \gg 1/\gamma$ yields sluggish logarithmic growth,

$$\sigma_{T=0}^2(t) = (2\hbar/\pi M\gamma) \ln(\gamma t) .$$

(7.39)

The logarithmic spread of the mean square displacemant at $T = 0$ actually holds true beyond free Brownian motion for any nonlinear quantum transport system with nonzero linear static mobility at $T = 0$,

$$\mu_0 = \mu(T = 0) = \lim_{\omega \to 0} \omega \, \tilde{\chi}''(\omega, T = 0) .$$

(7.40)

With this, there follows from Eq. (7.22) at $T = 0$, back in the time regime,

$$\lim_{t \to \infty} \hat{D}_{\mathrm{th},+}(t) = (2\hbar\mu_0/\pi) \ln(\gamma t) .$$

(7.41)

Hence, under condition (7.40) there is, in general, sluggish logarithmic growth. The logarithmic time dependence of $\sigma_{T=0}^2(t)$ at asymptotic time was found by Hakim and Ambegaokar [268]. It is obvious from the derivation that the logarithmic law (7.41) has its origin in the nonanalytic behavior of the expression (6.25) at $\omega = 0$ resulting from the *fluctuation-dissipation theorem*. In addition, the $\hbar$-dependence in Eq. (7.41) originates from Eq. (6.25) as well. Thus, the law (7.41) does not reflect intrinsic quantum mechanical features of the system.

7.3.3 Momentum spread

With the dynamical susceptibility (7.9), the integral expression (6.129) for the mean value of p^2 takes the form

$$\langle p^2 \rangle = \frac{\hbar M}{\pi} \int_0^\infty d\omega \, \frac{\omega \, \tilde{\gamma}'(\omega)}{[\omega - \tilde{\gamma}''(\omega)]^2 + \tilde{\gamma}'^2(\omega)} \coth(\beta \hbar \omega / 2) \,. \tag{7.42}$$

Alternatively, we may put $\omega_0 = 0$ in Eq. (6.131). The resulting series is

$$\langle p^2 \rangle = \frac{M}{\beta} \left(1 + 2 \sum_{n=1}^\infty \frac{\hat{\gamma}(\nu_n)}{\nu_n + \hat{\gamma}(\nu_n)} \right) \,. \tag{7.43}$$

In the strict Ohmic case, $\hat{\gamma}(z) = \gamma$, both the integral and the series expression are logarithmically divergent. Hence the obligatory high-frequency cutoff in the spectral coupling is actually a measurable quantity. With the Drude regularization (3.43), we obtain from (7.42) or from (7.43) the exact analytic expression

$$\langle p^2 \rangle = \frac{M}{\beta} + \frac{M\hbar}{\pi} \frac{\lambda_1 \lambda_2}{\lambda_1 - \lambda_2} \left[\psi\left(1 + \frac{\beta\hbar\lambda_1}{2\pi}\right) - \psi\left(1 + \frac{\beta\hbar\lambda_2}{2\pi}\right) \right] \,, \tag{7.44}$$

where

$$\lambda_{1,2} = \tfrac{1}{2}\omega_D \pm \tfrac{1}{2}\sqrt{\omega_D^2 - 4\gamma\omega_D} \,. \tag{7.45}$$

The first term in Eq. (7.44) is the leading one in the high-temperature regime. The residual terms become increasingly important as temperature is decreased. For zero temperature, we find in the range $\omega_D \gg 4\gamma$ logarithmic cutoff dependence

$$\langle p^2 \rangle_{T=0} = (M\hbar\gamma/\pi)\ln(\omega_D/\gamma) \,. \tag{7.46}$$

7.4 Frequency-dependent friction

For a power-law form of the damping function $\tilde{\gamma}(\omega)$, the response function can still be expressed in terms of known functions. To our knowledge, this is not possible for the mean square displacement, except in the long-time regime in which the dominant contribution to the Fourier integral of $\sigma^2(t)$ comes from frequencies near $\omega = 0$.

7.4.1 Response function and mobility

For memory-friction it is convenient to base the discussion of response functions on their Laplace integral representations. Putting $\omega_0 = 0$ in Eq. (6.33), the contour integral representations for the velocity response $\mathcal{R}(t)$ and position response $\chi(t)$ are

$$\mathcal{R}(t) = \frac{1}{2\pi i} \int_C dz \, \frac{e^{zt}}{z + \hat{\gamma}(z)} \,, \quad \text{and} \quad \chi(t) = \frac{1}{2\pi i} \int_C dz \, \frac{e^{zt}}{M z \, [\, z + \hat{\gamma}(z) \,]} \,. \tag{7.47}$$

In the regime $s < 2$, the first term in the series (3.53) for $\hat{\gamma}(z)$ is the leading one. With this, the position response function $\chi(t)$ takes the form

$$\chi(t) = \frac{1}{2\pi i} \int_C dz \, \frac{e^{zt}}{M z[\, z + \omega_s (z/\omega_s)^{s-1}\,]} = \Theta(t) \frac{t}{M} \frac{1}{2\pi i} \int_C dy \, \frac{e^y}{y^2} \frac{y^{2-s}}{y^{2-s} - x} \,, \qquad (7.48)$$

where $x = -(\omega_s t)^{2-s}$, and where

$$\omega_s = (\rho_s/\omega_{\rm ph})^{1/(2-s)} \omega_{\rm ph} \,, \qquad \text{with} \qquad \rho_s = \gamma_s/\sin(\pi s/2) \,. \qquad (7.49)$$

The integral is a generalized Mittag-Leffler function [269]

$$E_{\alpha,\beta}(x) = \frac{1}{2\pi i} \int_C dy \, \frac{y^{\alpha-\beta}}{y^\alpha - x} e^y = \sum_{k=0}^\infty \frac{x^k}{\Gamma(k\alpha + \beta)} \,, \qquad (7.50)$$

which has the asymptotic series

$$E_{\alpha,\beta}(x) = -\sum_{n=1}^\infty \frac{1}{\Gamma(\beta - n\alpha)} \frac{1}{x^n} \,. \qquad (7.51)$$

The specifications are $\alpha = 2 - s$, $\beta = 2$, and $x = -(\omega_s t)^{2-s}$. Thus we have

$$\chi(t) = \Theta(t) \frac{t}{M} E_{2-s,2}\left[-(\omega_s t)^{2-s}\right] \,. \qquad (7.52)$$

The integrand in Eq. (7.48) has three singularities:

(1) A branch point at $z = 0$ with a cut along the negative real axis.

(2) Two complex conjugate simple poles situated at $z_\pm = \omega_s \, e^{\pm i\pi/(2-s)}$.

In the sub-Ohmic regime $s < 1$, the singularities $z_\pm$ are located on the principal sheet at $z_\pm = -\Gamma_s \pm i\Omega_s$,

$$\Omega_s = \omega_s \cos\varphi_s \,, \qquad \Gamma_s = \omega_s \sin\varphi_s \,, \qquad \varphi_s = \frac{\pi}{2} \frac{s}{(2-s)} \,. \qquad (7.53)$$

The branch cut contributes an algebraic tail which dominates the behavior of $\chi(t)$ at long time. The resulting expression for $\chi(t)$ at asymptotic time is

$$\chi(t) = \frac{2}{(2-s)} \frac{\sin(\Omega_s t - \varphi_s) \, e^{-\Gamma_s t}}{M\omega_s} + \frac{t}{M} \sum_{n=1}^\infty \frac{(-1)^{n-1}}{\Gamma[2 - (2-s)n]} \frac{1}{(\omega_s t)^{(2-s)n}} \,. \qquad (7.54)$$

In the Ohmic case $s = 1$, the singularites of the integrand are simple poles at $z = 0$ and at $z = -\gamma$. The contributions of these poles yield the familiar expression (7.28).

In the super-Ohmic regime $1 < s < 2$, the simple poles $z_\pm$ poles are not on the principal sheet, and the cut contribution is a monotonically descending function of t. At long time, it is the asymptotic series given in Eq. (7.54) with the leading term

$$\chi(t) = \frac{\sin(\pi s/2)}{M\gamma_s\Gamma(s)} (\omega_{\rm ph} t)^{s-1} \,, \qquad (7.55)$$

which includes the Ohmic result $\chi(t \to \infty) = 1/M\gamma$ as special case.

The asymptotic form (7.55) suggests to define a generalized linear dc mobility

$$\mu_\ell^{(s)} \equiv \lim_{t\to\infty} \chi(t)/(\omega_{\mathrm{ph}}t)^{s-1} . \tag{7.56}$$

With this, the direct generalisation of the Ohmic mobility (7.29) emerges as

$$\mu_\ell^{(s)} = \frac{\sin(\pi s/2)}{M\gamma_s\Gamma(s)} , \qquad 0 < s < 2 . \tag{7.57}$$

Alternatively, we may consider the linear ac mobility

$$\tilde{\mu}(\omega) = -i\,\omega\tilde{\chi}(\omega) . \tag{7.58}$$

For $s < 2$, the ac mobility has the low-frequency behavior

$$\tilde{\mu}(\omega) = \frac{1}{M\tilde{\gamma}(\omega)} = \frac{\sin(\frac{1}{2}\pi s)}{M\gamma_s}\left(\frac{-i\,\omega}{\omega_{\mathrm{ph}}}\right)^{1-s} . \tag{7.59}$$

For $s = 2$, one finds with the expression (3.55) in the regime $\omega \ll \omega_{\mathrm{c}}$

$$\tilde{\mu}(\omega) = \frac{\pi}{2M\gamma_2}\frac{i\,\omega_{\mathrm{ph}}}{\omega\ln(\omega_{\mathrm{c}}/\omega)} . \tag{7.60}$$

Accordingly, the response function at asymptotic time $t \gg 1/\omega_{\mathrm{c}}$ exhibits logarithmic slowing of the free particle behavior,

$$\chi(t) = \frac{\pi}{2M\gamma_2}\frac{\omega_{\mathrm{ph}}t}{\ln(\omega_{\mathrm{c}}t)} . \tag{7.61}$$

Consider finally the regime $s > 2$, in which $\lim_{z\to 0}\hat{\gamma}(z) \propto z$ in leading order of z. Hence the asymptotic dynamics is primarily that of a free undamped particle with renormalized mass $M_{\mathrm{r}} = M + M_{\mathrm{bath}}$, where M_{bath} is defined in Eq. (7.7).

In the regime $2 < s < 4$, the leading correction in the series (3.53) is $\propto z^{s-1}$. The transient behavior of $\chi(t)$ is then described again in terms of a generalized Mittag-Leffler function $E_{\alpha,\beta}(x)$, but now $\alpha = s - 2$, $\beta = 2$, and $x = -(M_{\mathrm{r}}/M)(\omega_s t)^{s-2}$,

$$\chi(t) = \Theta(t)\frac{t}{M_{\mathrm{r}}}\left(1 - E_{s-2,2}\left[-(M_{\mathrm{r}}/M)(\omega_s t)^{s-2}\right]\right) . \tag{7.62}$$

With the leading asymptotic term of the Mittag-Leffler function added, one obtains

$$\chi(t) = \frac{t}{M_{\mathrm{r}}}\left(1 + \frac{M}{M_{\mathrm{r}}}\frac{1}{\Gamma(4-s)\sin[\pi(s/2-1)]}\frac{\gamma_s}{\omega_{\mathrm{ph}}}\frac{1}{(\omega_{\mathrm{ph}}t)^{s-2}}\right) . \tag{7.63}$$

The second term in the round bracket represents the leading deviation from behavior of a free undamped particle. For the special value $s = 3$, the expression (7.62) yields

$$\chi(t) = \Theta(t)\left(\frac{t}{M_{\mathrm{r}}} + \frac{1}{M_{\mathrm{r}}\Gamma_3}\left(1 - e^{-\Gamma_3 t}\right)\right) , \qquad \text{with} \qquad \Gamma_3 = \frac{M_{\mathrm{r}}}{M}\frac{\omega_{\mathrm{ph}}^2}{\gamma_3} . \tag{7.64}$$

The second term is the response of a free Ohmicly damped particle, Eq. (7.28).

The corresponding analysis in the super-Ohmic regimes $2n < s < 2n + 2$, where $n = 2, 3, \cdots$, is straightforward and left to the interested reader.

The velocity response $\mathcal{R}(t) = M\dot{\chi}(t)$ can be expressed in terms of generalized Mittag-Leffler functions as well. In the most relevant regime $0 < s < 4$, there is

$$
\begin{aligned}
\mathcal{R}(t) &= \Theta(t)\, E_{2-s,1}\left[-(\omega_s t)^{2-s}\right], & 0 < s < 2, \\
\mathcal{R}(t) &= \Theta(t)\,(M/M_{\mathrm{r}})\left(1 - E_{s-2,1}\left[-(M_{\mathrm{r}}/M)(\omega_s t)^{s-2}\right]\right), & 2 < s < 4.
\end{aligned}
\tag{7.65}
$$

7.4.2 Mean square displacement

The findings of the preceding subsection can be profitably deployed to determine the asymptotic time-dependence of the mean square displacement. At finite T, we can infer from the first integral of the relation (7.21) that there holds at asymptotic time

$$
\sigma^2(t) = \frac{2}{\beta}\int_0^t dt'\, \chi(t'), \qquad \text{for} \qquad t \gg \hbar\beta.
\tag{7.66}
$$

Above all, this relation holds for all times in the classical regime $k_{\mathrm{B}}T \gg \hbar\omega_s$, where again $\omega_s = (\rho_s/\omega_{\mathrm{ph}})^{1/(2-s)}\omega_{\mathrm{ph}}$. With the expressions (7.52) and (7.62) and with use of the series (7.50) one obtains from Eq. (7.66)

$$
\begin{aligned}
\sigma_{\mathrm{cl}}^2(t) &= \frac{2}{M\beta}\, t^2\, E_{2-s,3}[-(\omega_s t)^{2-s}], & 0 < s < 2, \\
\sigma_{\mathrm{cl}}^2(t) &= \frac{1}{M_{\mathrm{tot}}\beta}\, t^2\left\{1 - 2E_{s-2,3}[-(M_{\mathrm{r}}/M)(\omega_s t)^{s-2}]\right\}, & 2 < s < 4.
\end{aligned}
\tag{7.67}
$$

These expressions together with Eq. (7.61) yield the asymptotic time dependences

$$
\sigma^2(t) = \begin{cases}
\dfrac{2\sin(\pi s/2)}{M\beta\gamma_s\omega_{\mathrm{ph}}\Gamma(s+1)}(\omega_{\mathrm{ph}}t)^s, & s < 2, \\[2ex]
\dfrac{\pi}{2M\beta\gamma_2\omega_{\mathrm{ph}}}\dfrac{(\omega_{\mathrm{ph}}t)^2}{\ln(\omega_c t)}, & s = 2, \\[2ex]
\dfrac{1}{M_{\mathrm{r}}\beta}\, t^2, & s > 2.
\end{cases}
\tag{7.68}
$$

We may summarize the results as follows. For sub-Ohmic damping, $s < 1$, the ac mobility, the response function, and the mean square displacement in thermal equilibrium show sub-diffusive behaviors: the growth at asymptotic time is slower than in the diffusive regime. In contrast, in the super-Ohmic regime $1 < s < 2$, the growth of these quantities is faster than in the diffusive regime. For instance, the ac mobility $\tilde{\mu}(\omega)$ vanishes for $s < 1$ and diverges for $s > 1$ in the zero frequency limit. In the case $s = 2$, we find logarithmic modification of the free particle behavior. For $s > 2$, the asymptotic dynamics is that of a free particle with a renormalized mass. For a related discussion concerning ergodicity we refer to Section 3.1.10.

Finally, consider the position dispersion at $T = 0$, at which Eq. (7.19) reduces to

$$\sigma^2_{T=0}(t) \;=\; \frac{2\hbar}{\pi} \int_0^\infty d\omega \, \tilde{\chi}''(\omega) \left[1 - \cos(\omega t)\right]. \tag{7.69}$$

This function can be related to $\chi(t)$ by use of the integral representations (7.20) and

$$1 - \cos\omega t \;=\; \frac{2}{\pi} \mathop{\int\!\!\!\!\!-}_0^\infty dx \, \frac{\sin(\omega t x)}{x\,(1 - x^2)}. \tag{7.70}$$

Again, the dashed integral means the principal value. Thus we obtain

$$\sigma^2_{T=0}(t) \;=\; \frac{2\hbar}{\pi} \mathop{\int\!\!\!\!\!-}_0^\infty dx \, \frac{1}{x\,(1 - x^2)} \, \chi(xt). \tag{7.71}$$

Given the dynamics of $\chi(t)$ at long times, the asymptotic behavior of $\sigma^2_{T=0}(t)$ can now be found from the relation (7.71). With the forms for $\chi(t)$ given in Eqs. (7.55), (7.61) and (7.63), one obtains in the different damping regimes [88]

$$\sigma^2_{T=0}(t \to \infty) \;=\; \begin{cases} \sigma^2_\infty\,, & 0 < s < 1\,, \\[4pt] a_1 \ln(\gamma_1 t)\,, & s = 1\,, \\[4pt] a_s\,(\omega_{\mathrm{ph}}t)^{s-1}\,, & 1 < s < 2\,, \\[4pt] a_2\,\omega_{\mathrm{ph}}t/\ln^2(\omega_c t)\,, & s = 2\,, \\[4pt] a_s\,(\omega_{\mathrm{ph}}t)^{3-s}\,, & 2 < s < 3\,, \\[4pt] a_3 \ln(\Gamma_3 t)\,, & s = 3\,, \\[4pt] \overline{\sigma}^2_\infty\,, & s > 3\,. \end{cases} \tag{7.72}$$

In the regime $0 < s < 1$, $\chi(t)$ goes to zero as $t \to \infty$. Thus, with the substitution $x = u/t$ in Eq. (7.71) we find that $\sigma^2_{T=0}(t)$ approaches asymptotically the constant

$$\sigma^2_\infty \equiv \lim_{t\to\infty} \sigma^2(t) = \frac{2\hbar}{\pi} \int_0^\infty du \, \frac{\chi(u)}{u} = \frac{2\hbar}{\pi} \int_0^\infty dz \, \hat{\chi}(z) = \frac{2\hbar}{\pi M} \int_0^\infty dz \, \frac{1}{z^2 + z\hat{\gamma}(z)}. \tag{7.73}$$

With the leading terms of Eq. (3.53), $\hat{\gamma}(z) = \rho_s(z/\omega_{\mathrm{ph}})^{s-1} + \mathfrak{m}_1(s)z$, where $\mathfrak{m}_1(s) < 0$, the z-integration can be done in analytic form. The resulting expression for $s < 1$ is

$$\sigma^2_\infty \;=\; \frac{2/(2-s)}{\sin[\pi/(2-s)]} \left(\frac{[1 + \mathfrak{m}_1(s)]\,\omega_{\mathrm{ph}} \sin(\tfrac{1}{2}\pi s)}{\gamma_s} \right)^{\frac{1}{2-s}} \frac{\hbar}{[1 + \mathfrak{m}_1(s)]M\omega_{\mathrm{ph}}}. \tag{7.74}$$

We note that the singularity of σ^2_∞ occurring at $\mathfrak{m}_1(s) = -1$ in Eq. (7.74) is absent if we include in $\hat{\gamma}(z)$ the next-leading term $\propto z^3$ [cf. Eq. (3.53)].

In the regime $1 \leqslant s \leqslant 3$, the x-integral in Eq. (7.71) is calculated with the respective asymptotic forms for $\chi(xt)$. This yields the behaviors (7.72) with coefficients

$$
a_s = \begin{cases}
\dfrac{2}{\pi} \dfrac{\hbar}{M\gamma_1} \,, & \text{for} \quad s = 1 \,, \\[2ex]
\dfrac{\sin^2(\pi s/2)}{\cos[\pi(1 - s/2)]\Gamma(s)} \dfrac{\hbar}{M\gamma_s} \,, & \text{for} \quad 1 < s < 2 \,, \\[2ex]
\dfrac{\pi^2}{4} \dfrac{\hbar}{M\gamma_2} \,, & \text{for} \quad s = 2 \,, \\[2ex]
\dfrac{1}{\cos[\pi(1 - s/2)]\Gamma(4 - s)} \dfrac{\hbar M\gamma_s}{M_r^2\,\omega_{\text{ph}}^2} \,, & \text{for} \quad 2 < s < 3 \,, \\[2ex]
\dfrac{2}{\pi} \dfrac{\hbar}{M_r\Gamma_3} \,, & \text{for} \quad s = 3 \,.
\end{cases}
\tag{7.75}
$$

In the regimes $1 < s < 2$ and $2 < s < 3$, $\sigma_{T=0}^2(t)$ grows asymptotically as t^{s-1} and as t^{3-s}, respectively. For $s = 1$, the logarithmic behavior (7.39) is recovered. The same behavior is found for $s = 3$, because of the similarity of the expressions (7.64) and (7.28), and because the integral (7.71) with the extra term t/M_r in (7.64) is zero.

For the case $s = 2$, the logarithmic term of $\chi(xt)$ is handled as

$$
\frac{1}{\ln(\omega_c t x)} = \frac{1}{\ln(\omega_c t)} \frac{1}{1 + \ln(x)/\ln(\omega_c t)} \approx \frac{1}{\ln(\omega_c t)} - \frac{\ln x}{\ln(\omega_c t)^2} \,.
\tag{7.76}
$$

The contribution of the first term is zero, and the second term gives the above result.

For $s > 3$, the leading contribution to $\chi(t)$ grows linearly with t and therefore does not contribute to $\sigma_{T=0}^2(t)$. As the sub-leading term of $\chi(t)$ varies as t^{3-s}, the regime $s > 3$ is similar to the sub-Ohmic one, apart from subtraction of the free particle term. As a result, we find that $\sigma_{T=0}^2(t)$ approaches a constant $\overline{\sigma}_\infty^2$ as $t \to \infty$,

$$
\overline{\sigma}_\infty^2 = \frac{2\hbar}{\pi} \int_0^\infty dz \left(\hat{\chi}(z) - \frac{1}{M_r z^2} \right) .
\tag{7.77}
$$

In difference to the constant σ_∞^2 in the sub-Ohmic regime, the constant $\overline{\sigma}_\infty^2$ depends on the high-frequency properties of the spectral damping function.

7.5 Partition function and thermodynamic properties

7.5.1 Partition function

The partition function of a free particle is only defined if the particle is confined to a finite region. For a one-dimensional infinite square well of width L and constant inner potential $V_0 = 0$, the partition function is $Z_0 = \sum_{n=1}^\infty e^{-\beta E_g n^2/\hbar}$ with the ground-state energy $E_g = \hbar^2\pi^2/2ML^2$. In the temperature regime

$$
k_B T \geqslant E_0 = cE_g \,,
\tag{7.78}
$$

where c is a positive number sufficiently large so that the discreteness of the energy levels may be disregarded for $k_B T$ even at E_0. With the number of energy eigenstates

$g(E) = \sqrt{M/2E}\, L/\pi\hbar$ in the interval between E and $E + dE$, we then obtain the classical partition function of the free undamped particle as

$$Z_{0,cl}(\beta) = \int_0^\infty dE\, g(E)\, e^{-\beta E} = \frac{L}{\hbar}\sqrt{\frac{M}{2\pi\beta}} = \sqrt{\frac{\pi}{4\beta E_g}}\,. \tag{7.79}$$

For a damped particle, $Z_{0,cl}(\beta)$ is augmented by quantum fluctuations due to the bath coupling. The additional part can be derived under condition (7.78) from the infinite product in Eq. (6.80) by putting $\omega_0 = 0$. Accordingly, the reduced partition function of the free Brownian particle reads

$$Z(\beta) = Z_{0,cl}(\beta) \prod_{n=1}^\infty \frac{\nu_n}{\nu_n + \hat{\gamma}(\nu_n)}\,. \tag{7.80}$$

As in Eq. (6.80), the strict Ohmic case is excluded. For Drude damping (3.43) one gets

$$Z(\beta)/Z_{0,cl}(\beta) = \prod_{n=1}^\infty \frac{\nu_n(\nu_n + \omega_D)}{(\nu_n + \lambda_1)(\nu_n + \lambda_2)} = \frac{\Gamma(1 + \frac{\beta\hbar\lambda_1}{2\pi})\Gamma(1 + \frac{\beta\hbar\lambda_2}{2\pi})}{\Gamma(1 + \frac{\beta\hbar\omega_D}{2\pi})}\,. \tag{7.81}$$

Here $\lambda_{1,2}$ are given in Eq. (7.45). This reduces in the important regime $\omega_D \gg \gamma$, ν_1 to

$$Z(\beta) = Z_{0,cl}(\beta) \left(\frac{2\pi}{\beta\hbar\omega_D}\right)^{\beta\hbar\gamma/2\pi} \Gamma\left(1 + \frac{\beta\hbar\gamma}{2\pi}\right)\,. \tag{7.82}$$

The expression (7.81) for the reduced partition function holds for arbitrary ω_D, while Eq. (7.82) applies when ω_D is by far the largest frequency of the reduced system.

7.5.2 Internal and free energy for Drude damping

The internal energy $U = -\partial \ln Z/\partial \beta$ is found from the expression (7.80) as

$$U = \frac{1}{2\beta} + \frac{1}{\beta} \sum_{n=1}^\infty \frac{\hat{\gamma}(\nu_n) - \nu_n \partial\hat{\gamma}(\nu_n)/\partial\nu_n}{\nu_n + \hat{\gamma}(\nu_n)}\,. \tag{7.83}$$

A look at Eq. (7.43) reveals that U differs from the mean energy $\langle E \rangle = \langle p^2 \rangle/2M$ by the second term of the sum in (7.83). For Drude damping (3.43), the sum yields

$$U = \frac{1}{2\beta} + \frac{\hbar\omega_D}{2\pi}\psi\left(1 + \frac{\beta\hbar\omega_D}{2\pi}\right) - \frac{\hbar\lambda_1}{2\pi}\psi\left(1 + \frac{\beta\hbar\lambda_1}{2\pi}\right) - \frac{\hbar\lambda_2}{2\pi}\psi\left(1 + \frac{\beta\hbar\lambda_2}{2\pi}\right)\,, \tag{7.84}$$

where $\lambda_{1,2}$ are defined in Eq. (7.45). For $\omega_D \gg \gamma$, ν_1, this expression reduces to

$$U = \frac{1}{2\beta} + \frac{\hbar\gamma}{2\pi}\left[1 + \ln\left(\frac{\beta\hbar\omega_D}{2\pi}\right) - \psi\left(\frac{\beta\hbar\gamma}{2\pi}\right)\right]\,. \tag{7.85}$$

Next, consider the low-temperature series for the internal energy U and the free energy F. Writing in the latter case $Z = \kappa\, e^{-\beta F}$, where $\kappa = L\sqrt{M\gamma/2\pi\hbar}$, we obtain from the expression (7.81) the low temperature series[1]

---

[1]With the caveat that zero temperature means $T = E_0/k_B$ [cf. (7.78)], indicated by the label 0^+.

$$U = \hbar\epsilon_0 + b_1\frac{(k_\mathrm{B}T)^2}{\hbar\gamma} - b_2\frac{(k_\mathrm{B}T)^4}{(\hbar\gamma)^3} + \mathcal{O}(T^6) , \tag{7.86}$$

$$F = \hbar\epsilon_0 - b_1\frac{(k_\mathrm{B}T)^2}{\hbar\gamma} + \frac{b_2}{3}\frac{(k_\mathrm{B}T)^4}{(\hbar\gamma)^3} + \mathcal{O}(T^6) . \tag{7.87}$$

Here $\hbar\varepsilon_0 = \hbar[\lambda_1 \ln(\omega_\mathrm{D}/\lambda_1) + \lambda_2 \ln(\omega_\mathrm{D}/\lambda_2)]/2\pi$ is the ground-state energy, and

$$b_1 = \frac{\pi}{6}\left(1 - \frac{\gamma}{\omega_\mathrm{D}}\right) , \qquad \text{and} \qquad b_2 = \frac{\pi^3}{15}\left(1 - 3\frac{\gamma}{\omega_\mathrm{D}} - \frac{\gamma^3}{\omega_\mathrm{D}^3}\right) . \tag{7.88}$$

The disturbing point with the series (7.86) is that the coefficient b_1 is negative in the regime $\gamma > \omega_\mathrm{D}$ [270]. Prima facie, one might be worried with the thermodynamic stability of the open system in this regime. However, with the global system represented by the partition function $Z_\mathrm{tot} = Z\,Z_\mathrm{R}$ in mind, it is evident that the internal energies of the open system and the reservoir add up, $U_\mathrm{tot} = U + U_\mathrm{R}$. Now, since the total system is perfectly stable, the free energy U_tot is a monotonically increasing function of T. Apparently, the same property is not required by physical principles for a subsystem. The reason for the anomaly that the coefficient b_1 can be negative is discussed in the context of general frequency-dependent damping in the next subsection.

7.5.3 Anomalies in the specific heat

A free Brownian particle possesses a number of thermodynamic anomalies like a negative specific heat, dips in the specific heat as a signature of negative renormalized mass, or reentrant classicality. These surprising facts will be studied in the sequel.

The specific heat of the quantum Brownian particle results from the internal energy according to $C = \partial U/\partial T$. Based on the Matsubara representation (7.83), one finds

$$C/k_\mathrm{B} = \frac{1}{2} + \sum_{n=1}^{\infty}\left[\frac{[\hat{\gamma}(\nu_n) - \nu_n\,\partial\hat{\gamma}(\nu_n)/\partial\nu_n]^2}{[\nu_n + \hat{\gamma}(\nu_n)]^2} - \frac{\nu_n^2\,\partial^2\hat{\gamma}(\nu_n)/\partial\nu_n^2}{\nu_n + \hat{\gamma}(\nu_n)}\right] . \tag{7.89}$$

Alternatively, we may represent C as a spectral integral similar to Eq. (6.110),

$$C = C_\infty + \int_0^\infty d\omega\,\xi_\mathrm{fp}(\omega)[c_\mathrm{ho}(\omega,\beta) - k_\mathrm{B}] ,$$
$$= C_0 + \int_0^\infty d\omega\,\xi_\mathrm{fp}(\omega)c_\mathrm{ho}(\omega,\beta) . \tag{7.90}$$

Here, $c_\mathrm{ho}(\omega,\beta)$ is the specific heat of a single harmonic oscillator, defined in Eq. (6.109), and $\xi_\mathrm{fp}(\omega)$ is the density of states of the fictitious harmonic oscillators. The latter quantity can be determined from the spectral velocity response function $\tilde{\mathcal{R}}(\omega) = 1/[-i\,\omega + \tilde{\gamma}(\omega)]$ [cf. Eq. (3.123)] by the relation

$$\xi_\mathrm{fp}(\omega) = \frac{1}{\pi}\,\mathrm{Im}\,\frac{\partial\ln[\tilde{\mathcal{R}}(\omega)]}{\partial\omega} . \tag{7.91}$$

The value C_∞ is the specific heat in the classical limit, $C_\infty = \lim_{T\to\infty} C(T)$ and C_0 is the specific heat at the lowest temperature, at which the finite size of the particle's box is still negligible, $C_0 = \lim_{T\to E_0} C(T)$. Here E_0 is defined in Eq. (7.78). According to Eq. (7.90), the values C_∞ and C_0 are related by the sum rule [271]

$$C_\infty = C_0 + \Sigma_{fp}\, k_B \,, \qquad \text{with} \qquad \Sigma_{fp} = \int_0^\infty d\omega\, \xi_{fp}(\omega) \,. \tag{7.92}$$

Observing that the expression (7.91) is a partial derivative and $\mathrm{Re}\,\hat{\gamma}(-i\,\omega)$ is positive, Σ_{fp} is determined by the leading term of the low-frequency series (3.53) of $\hat{\gamma}(z)$, which is the first term for $s < 2$, and the second term for $s > 2$, yielding

$$\Sigma_{fp}(s) = \Theta(2-s)\left[1 - s/2\right] \,. \tag{7.93}$$

From the classical partition function $Z_{0,cl}(\beta) \propto \beta^{-1/2}$ there follows $C_\infty = \frac{1}{2}k_B$. Thus

$$C_0(s) = \begin{cases} \frac{1}{2}(s-1)k_B \,, & \text{for} \quad 0 < s < 2 \,. \\[2mm] \frac{1}{2}\, k_B \,, & \text{for} \quad s \geqslant 2 \,. \end{cases} \tag{7.94}$$

This is totally different from the harmonic oscillator, at which $\Sigma_{osc}(s) = 1$, $C_\infty = k_B$, and $C_0 = 0$ for all s (cf. Subsecs. 6.5.2 and 6.5.3).

Zero temperature limit

We see from Eq. (7.94) that $C(T = 0^+)$ [see footnote on page 201] is negative in the sub-Ohmic regime $0 < s < 1$ and positive in the super-Ohmic regime $s > 1$. Only for an Ohmic bath, $s = 1$, the specific heat is zero at $T = 0^+$, in agreement with the Third Law, regardless whether the box width L is finite or infinite. Reconciliation with the Third Law for $s \neq 1$ takes place in the ultra-narrow regime $0 < T < E_0/k_B$, in which the discreteness of the energy spectrum comes into effect [263].

Low temperature behavior

The leading correction to the specific heat at low T is obtained from the leading low-frequency term of $\xi_{fp}(\omega)$. With the series expression (3.53) we obtain from Eq. (7.91) in the regime $0 < s < 2$ the form

$$\begin{aligned} \xi_{fp}(\omega) &= c(s)(\omega/\omega_{ph})^{1-s}/\gamma_s \,, \\ c(s) &= \left[(2-s)/\pi\right]\sin(\pi s/2)^2\left[1 + \mathfrak{m}_1(s)\right] \,, \end{aligned} \qquad 0 < s < 2 \,. \tag{7.95}$$

Here, $\mathfrak{m}_1(s)$ is defined in Eq. (3.58). With the form (7.95) the leading behavior of the specific heat at low T is found from the expression (7.90) to read

$$C/k_B = \frac{(s-1)}{2} + b(s)\frac{\omega_{ph}}{\gamma_s}\left(\frac{k_B T}{\hbar\omega_{ph}}\right)^{2-s} \,, \qquad 0 < s < 2 \,, \tag{7.96}$$

where

$$b(s) = \Gamma(4-s)\zeta(3-s)\,c(s) \,, \tag{7.97}$$

and where $\zeta(x)$ is the Riemann zeta function. The prefactor $b(s)$ shows a striking anomaly. It changes sign when the dressed mass $M_r = \left[1 + \mathfrak{m}_1(s)\right] M$ goes through zero, i.e., when the damping strength γ_s goes through the critical value γ_s^* given in Eq. (3.59). Hence the decrease of the specific heat at low T with increasing temperature is a sign that the dressed mass is negative. At critical damping $\gamma_s = \gamma_s^*$, the term $\mu_3(s)z^3/\omega_c^2$ in Eq. (3.53) comes into play. With this term included in $\hat{\gamma}(z)$, one finds that the leading low temperature dependence of C at $\gamma = \gamma_s^*$ is T^{4-s} [271].

For the Drude cutoff in $J(\omega)$, Eq. (3.58) yields $\mathfrak{m}_1(s) = -(\rho_s/\omega_D)(\omega_D/\omega_{\mathrm{ph}})^{s-1}$. With this one finds in the Ohmic case $b(1) = (\pi/3)(1 - \gamma/\omega_D)$, which is in agreement with the T^2-term of the internal energy in Eq. (7.86).

In the regime $s > 2$, the term $\mathfrak{m}_1 z$ in the series (3.53) is the leading one, with $\mathfrak{m}_1(s)$ now given by the integral (3.57). This yields the low-frequency form

$$\xi_{\mathrm{fp}}(\omega) = c(s)(\gamma_s/\omega_{\mathrm{ph}}^2)(\omega/\omega_{\mathrm{ph}})^{s-3},$$

$$c(s) = -\frac{(s-2)}{\pi}\frac{1}{1+\mathfrak{m}_1(s)}, \qquad s > 2. \qquad (7.98)$$

Because the scaled renormalized mass $\mathfrak{m}_1(s)$ is consistently positive in the regime $s > 2$, the prefactor $c(s)$ is consistently negative. With Eq. (7.98) in Eq. (7.90), the specific heat at low T is found to read

$$C/k_{\mathrm{B}} = \frac{1}{2} - \frac{(s-2)}{\pi}\frac{\Gamma(s)\zeta(s-1)}{1+\mathfrak{m}_1(s)}\frac{\gamma_s}{\omega_{\mathrm{ph}}}(k_{\mathrm{B}}T/\hbar\omega_{\mathrm{ph}})^{s-2}, \qquad s > 2. \qquad (7.99)$$

Thus, the leading contribution to the specific heat at low T is consistently negative.

Quantum correction at high temperatures

According to Eq. (3.52), the leading behavior of $\hat{\gamma}(z)$ in the range $z \gg \omega_c$ is ρ/z. With this, the specific heat at high T is found from the expression (7.89) in the form

$$C/k_{\mathrm{B}} = \frac{1}{2} - 2\rho\sum_{n=1}^{\infty}\frac{1}{\nu_n^2}. \qquad (7.100)$$

Thus, the leading quantum correction to the specific heat at high T is negative. It universally behaves inversely proportional to T^2. For the sharp cutoff in the spectral density (3.52) one obtains with Eq. (3.52)

$$C/k_{\mathrm{B}} = \frac{1}{2} - \frac{1}{6\pi s}\frac{\gamma_s}{\omega_c}\left(\frac{\omega_c}{\omega_{\mathrm{ph}}}\right)^{s-1}\left(\frac{\hbar\omega_c}{k_{\mathrm{B}}T}\right)^2. \qquad (7.101)$$

This is the specific heat at high T holding for all s.

Reentrant classicality

For $s > 2$, the specific heat approaches its classical value $k_{\mathrm{B}}/2$ both in the limit $T \to \infty$ and in the limit $T \to 0^+$, and it goes through a minimum at intermediate temperature, as one can deduce from the limiting expressions (7.101) and (7.99). This remarkable behavior, known as *reentrant classicality* [271, 91], is owed to the small density of bath oscillators at low frequencies.

In the regime $0 < s < 2$, the specific heat starts out at temperature $T = 0^+$ from the value $(s - 1)/2\,k_B$ with a positive slope for $\gamma_s < \gamma_s^*$ and with a negative slope for $\gamma > \gamma_s^*$. In the former case, it monotonically increases with temperature towards the classical value $k_B/2$, while in the latter case it first goes through a minimum and then monotonically increases towards the classical value.

Drude damping

For Drude damping, $\hat{\gamma}(z) = \gamma/(1 + z/\omega_D)$, the density $\xi_{fp}(\omega)$ is found from (7.91) as

$$\xi_{fp}(\omega) = \frac{1}{\pi}\left(\frac{\lambda_1}{\omega^2 + \lambda_1^2} + \frac{\lambda_2}{\omega^2 + \lambda_2^2} - \frac{\omega_D}{\omega^2 + \omega_D^2}\right), \tag{7.102}$$

where λ_1 and λ_2 are given in Eq. (7.45). Thus, from the bath point of view, the main difference between the Drude-damped harmonic oscillator and Brownian free particle is an additional peak in the change of the bath density $\xi_{osc}(\omega)$, Eq. (6.93), compared with $\xi_{fp}(\omega)$. With Eq. (7.102) we obtain from (7.103), or directly from (7.89)

$$C/k_B = \frac{1}{2} + \frac{\hbar^2\beta^2}{4\pi^2}\left\{\sum_{i=1}^{2}\lambda_i^2\psi'\left(1 + \frac{\beta\hbar\lambda_i}{2\pi}\right) - \omega_D^2\psi'\left(1 + \frac{\beta\hbar\omega_D}{2\pi}\right)\right\}, \tag{7.103}$$

We have seen in Eq. (7.85) that the internal energy U diverges logarithmically, as $\omega_D \to \infty$. In contrast, the specific heat C becomes independent of ω_D in this limit,

$$C/k_B = \frac{1}{2} - \frac{\beta\hbar\gamma}{2\pi} + \frac{(\beta\hbar\gamma)^2}{4\pi^2}\psi'\left(1 + \frac{\beta\hbar\gamma}{2\pi}\right). \tag{7.104}$$

In the high-T regime $\beta\hbar\gamma \ll 1$, one finds depending on whether $\beta\hbar\omega_D \ll 1$ or $\beta\hbar\omega_D \gg 1$

$$C/k_B = \frac{1}{2} - \frac{1}{2\pi}\frac{\hbar\gamma}{k_BT} + \frac{1}{24}\left(\frac{\hbar\gamma}{k_BT}\right)^2, \qquad \beta\hbar\omega_D \ll 1, \tag{7.105}$$

$$C/k_B = \frac{1}{2} - \frac{1}{12}\frac{\hbar^2\gamma\omega_D}{(k_BT)^2}, \qquad \beta\hbar\omega_D \gg 1. \tag{7.106}$$

In the zero temperature limit, the specific heat goes to zero linearly with T,

$$C/k_B = \frac{\pi}{3}\left(1 - \frac{\gamma}{\omega_D}\right)\frac{k_BT}{\hbar\gamma} + \mathcal{O}(T^3). \tag{7.107}$$

The prefactor changes sign at $\gamma = \omega_D$, in line with the discussion below Eq. (7.97).

Concluding remarks

The thermodynamic anomalies found for the free Brownian particle are inherent to all systems for which the static susceptibility is zero. The anomalies fade away when a weak oscillator potential is turned on. The change of the thermodynamic properties in the transition from the free damped particle to the damped harmonic oscillator, including the entropy, is discussed in Ref. [263]. The role of the cutoff in the spectral density of the coupling $J(\omega)$, in particular algebraic versus sharp cutoff is analyzed in Ref. [91].

8 The thermodynamic variational approach

It is appealing to study quantum statistical expectation values with concepts familiar from classical statistical mechanics. The phase space representation of the reduced density matrix in terms of Wigner's function discussed in Subsection 4.4.2 is just one example. Another important concept which goes back to Feynman [272] is the *effective classical potential*.

In the effective classical potential method, all physical effects related to nonlinearity are exactly taken into account in the classical regime. In the quantum regime, the effective potential is optimized by use of a variational principle with a trial action which is quadratic in the dynamical variables. Feynman originally used a "free particle" trial action. Considerable improvement has been achieved by Giachetti and Tognetti [273], and independently by Feynman and Kleinert [274], by adding a harmonic potential term in the trial action, thereby introducing a second variational parameter. Further generalization is obvious: since linear dissipation is rendered by a "quadratic" action functional, it can be included in the variational approach with minor ancillary efforts.

In the following I present the main ideas. To be general, the discussion will be given for open systems with arbitrary linear dissipation.

8.1 Centroid and the effective classical potential

8.1.1 Centroid

A useful path partitioning concept has been developed by Feynman [3, 4]. The basic idea is to divide up the paths into equivalence classes and to rewrite the original path sum as a sum over paths belonging to the same equivalence class and an additional sum over the equivalence classes. In connection with the *effective classical potential*, it is convenient to consider as equivalence class all those paths which have the same mean position $q_c[q(\cdot)]$, usually referred to as the *centroid*,[1]

$$q_c[q(\cdot)] = \frac{1}{\hbar\beta} \int_0^{\hbar\beta} d\tau \, q(\tau) . \tag{8.1}$$

Within the centroid method, the path sum (4.60) for the reduced density matrix in thermal equilibrium is rearranged as a sum over all paths which are constrained to the same centroid and an additional ordinary integral over the centroid coordinate. We then have (in this chapter we use the normalization $\mathrm{tr}\,\rho_\beta = Z$)

---

[1]The meaning of the central dot in $q(\cdot)$ is explained in the footnote to Eq. (4.21).

$$\rho_\beta(q'', q') = \int_{-\infty}^{\infty} dq_c \, \rho_\beta(q'', q'; q_c) \,,$$

$$\rho_\beta(q'', q'; q_c) \equiv \int_{q(0)=q'}^{q(\hbar\beta)=q''} \mathcal{D}q(\cdot) \, \delta \left(q_c - q_c[q(\cdot)] \right) \exp\left(-S_{\text{eff}}[q(\cdot)]/\hbar \right) \,. \tag{8.2}$$

Correspondingly, we may also write the partition function $Z = \int dq' \, \rho_\beta(q', q')$ in the form of an integral over the centroid density,

$$Z = \int_{-\infty}^{\infty} dq_c \, \rho_{\beta,c}(q_c) \,, \tag{8.3}$$

$$\rho_{\beta,c}(q_c) \equiv \oint \mathcal{D}q(\cdot) \, \delta \left(q_c - q_c[q(\cdot)] \right) \exp\left(-S_{\text{eff}}[q(\cdot)]/\hbar \right) \,. \tag{8.4}$$

To proceed, we Fourier-expand the periodic path $q(\tau)$ with period $\hbar\beta$,

$$q(\tau) = q_c + \tilde{q}(\tau) = Q_0 + \sum_{\substack{n=-\infty \\ n \neq 0}}^{\infty} Q_n \, e^{i\nu_n\tau} \,. \tag{8.5}$$

The Fourier coefficient Q_0 represents the centroid q_c, as follows from definition (8.1). Next, we use the functional integral measure (4.231) and write the path sum in (8.2) as an infinite product of ordinary integrals. We then get the centroid density as

$$\rho_{\beta,c}(q_c) = \left(M/2\pi\hbar^2\beta \right)^{1/2} \exp\left[-\beta\mathcal{V}(q_c) \right] \,, \tag{8.6}$$

in which $\mathcal{V}(q_c)$ is the *effective classical potential*,

$$e^{-\beta\mathcal{V}(q_c)} = \oint \mathcal{D}\tilde{q} \, \exp\left(- M\beta \sum_{n=1}^{\infty} \left[\nu_n^2 + \nu_n\hat{\gamma}(\nu_n) \right] |Q_n|^2 \right.$$
$$\left. - \frac{1}{\hbar} \int_0^{\hbar\beta} d\tau \, V\left[q_c + \tilde{q}(\tau) \right] \right) \,. \tag{8.7}$$

The functional integral measure is limited to pure quantum fluctuations. It reads

$$\oint \mathcal{D}\tilde{q} \, \times \cdots \equiv \prod_{n=1}^{\infty} \int_{-\infty}^{\infty} \int_{-\infty}^{\infty} \frac{d\text{Re}\, Q_n \, d\text{Im}\, Q_n}{\pi/(M\beta\nu_n^2)} \, \times \cdots \,. \tag{8.8}$$

The expression (8.6) with (8.7) and (8.8) is still exact. In this decomposition, the quantum nature of the system is contained in the effective classical potential $\mathcal{V}(q_c)$.

8.1.2 The effective classical potential

Due to the kinetic action term $M\beta\nu_n^2|Q_n|^2$ in the exponent of (8.7), the Q_n are effectively of order $1/(\nu_n\sqrt{M\beta}) \propto 1/\sqrt{T}$. Therefore, the quantum fluctuations $\tilde{q}(\tau)$ in the potential term in Eq. (8.7) decrease when temperature is increased. In the classical limit, $V[q(\tau)]$ becomes independent of $\tilde{q}(\tau)$. Hence in this limit all the integrals in Eq. (8.7) are Gaussian. Since, in addition, $\hat{\gamma}(\nu_n)/\nu_n \to 0$ in the classical limit, the fluctuation integrals simply give a factor unity. Thus we have

$$\lim_{T\to\infty} \mathcal{V}(q_c) = V(q_c) , \tag{8.9}$$

and hence

$$\lim_{T\to\infty} Z \to Z^{(cl)} = \left(M/2\pi\hbar^2\beta\right)^{1/2} \int_{-\infty}^{\infty} dq_c \exp\left[-\beta V(q_c)\right] . \tag{8.10}$$

This consideration shows that the quantum statistical partition function of any particular smooth potential reduces to the classical partition function in the high temperature limit. Because of the similarity of Eq. (8.6) with (8.10), the function $\mathcal{V}(q_c)$ has been dubbed by Feynman and Kleinert [274] the *effective classical potential*.

To obtain first insights, consider the effective classical potential $\mathcal{V}(q_c)$ for the damped harmonic oscillator. Now, all the integrals contained in Eq. (8.7) with (8.8) are in strict Gaussian form at any temperature. We then get

$$\mathcal{V}_{osc}(q_c) = V_{osc}(q_c) - (1/\beta) \ln \mu_{osc}(\omega_0) , \tag{8.11}$$

where μ_{osc} is the pure quantum part of the partition function (6.80),

$$\mu_{osc}(\omega_0) = \prod_{n=1}^{\infty} \frac{\nu_n^2}{\nu_n^2 + \nu_n\hat{\gamma}(\nu_n) + \omega_0^2} = \beta\hbar\omega_0 Z_{osc}(\beta) . \tag{8.12}$$

For the Drude model (3.43), μ_{osc} can be expressed in terms of gamma functions,

$$\mu_{osc}(\omega_0) = \left(\frac{\beta\hbar\omega_0}{2\pi}\right)^2 \frac{\Gamma(\lambda_1/\nu_1)\,\Gamma(\lambda_2/\nu_1)\,\Gamma(\lambda_3/\nu_1)}{\Gamma(\omega_D/\nu_1)} , \tag{8.13}$$

where $\lambda_1, \lambda_2, \lambda_3$ are given in (6.45) with (6.46). In the absence of damping, we have

$$\mu_{osc}(\omega_0) = \frac{\hbar\omega_0/2k_B T}{\sinh(\hbar\omega_0/2k_B T)} . \tag{8.14}$$

Alternatively, we may express $\mathcal{V}_{osc}(q_c)$ in terms of the free energy $F_{osc} = -\ln Z_{osc}/\beta$,

$$\mathcal{V}_{osc}(q_c) = V_{osc}(q_c) - (1/\beta)\ln(\beta\hbar\omega_0) + F_{osc} . \tag{8.15}$$

Thus for the damped harmonic oscillator, $\mathcal{V}_{osc}(q_c)$ differs from $V_{osc}(q_c)$ only by a temperature-dependent energy shift. At high temperatures, the shift vanishes as $1/T$, $\mathcal{V}_{osc}(q_c) - V_{osc}(q_c) = \hbar^2[\omega_0^2 + \gamma\omega_D]/12k_B T + \mathcal{O}(1/T^3)$. In the opposite limit $T \to 0$, we have

$$\mathcal{V}_{osc}(q_c) = V_{osc}(q_c) + \hbar\varepsilon_0 , \tag{8.16}$$

where $\hbar\varepsilon_0$ is the ground state energy of the oscillator given in Eq. (6.98) or (6.99).

Utilization of the effective classical potential $\mathcal{V}(q_c)$ is physically very appealing. Use of $\mathcal{V}(q_c)$ for anharmonic potentials may seem of little aid, however, since it is impossible to carry out the integrals over all Q_n in Eq. (8.7) in analytic form. Fortunately, there is a variational method in which the path integral in question is compared with the path integral of a simpler problem solvable in analytic form. There is a minimum principle which leads to an upper bound for the effective classical potential. Upon choosing a suitable trial action, the partition function can be calculated quite accurately in the whole temperature range for many anharmonic systems.

8.2 Variational method

I now discuss a method which combines the path integral approach with a variational principle. The minimum principle is based on *Jensen's inequality* for convex functions $f(x)$, e.g. $f(x) = e^{-x}$, in probability theory.[2] For x being random, the average value of $f(x)$ always exceeds or equals the function of the average value of x, as long as x is real and the distribution of the underlying stochastic process is positive,

$$\langle f(x) \rangle \geqslant f(\langle x \rangle) . \tag{8.17}$$

This inequality for convex functions can be generalized to an exponential functional. Therefore it is very powerful in quantum statistical mechanics.

8.2.1 Variational method for the free energy

Suppose we have a trial action S_{tr} that is easier to work with. Then we may write

$$Z \equiv e^{-\beta F} = \oint \mathcal{D}q \, e^{-(S-S_{tr})/\hbar} \, e^{-S_{tr}/\hbar} = e^{-\beta F_{tr}} \left\langle e^{-(S-S_{tr})/\hbar} \right\rangle_{tr} , \tag{8.18}$$

where $Z_{tr} \equiv \exp(-\beta F_{tr}) = \oint \mathcal{D}q \exp(-S_{tr}/\hbar)$ is the partition function connected with S_{tr}, and $\langle O \rangle_{tr}$ is the average of the observable $O[q]$ with weight $\exp(-S_{tr}[q]/\hbar)$,

$$\langle O \rangle_{tr} \equiv \frac{\oint \mathcal{D}q(\cdot) \, O[q(\cdot)] \exp\left(-S_{tr}[q(\cdot)]/\hbar \right)}{\oint \mathcal{D}q(\cdot) \exp\left(-S_{tr}[q(\cdot)]/\hbar \right)} . \tag{8.19}$$

Using (8.17), we obtain from (8.18) the lower bound for the partition function

$$\exp(-\beta F) \geqslant \exp(-\beta F_{tr}) \exp\left(-\langle S - S_{tr} \rangle_{tr}/\hbar \right) . \tag{8.20}$$

This implies Feynman's upper bound for the free energy [3, 4],

$$F \leqslant F_{tr} + \langle S - S_{tr} \rangle_{tr}/\hbar\beta . \tag{8.21}$$

The minimum principle (8.21) has been used to determine the "best" free energy from a trial action $S_{tr}[q(\cdot)]$ for which the path integral can be worked out.

8.2.2 Variational method for the effective classical potential

Similar to the derivation of a global bound for the free energy, we can also deduce a local bound for the effective classical potential $\mathcal{V}(q_c)$, which is somewhat simpler to study. Since we wish to perform the variational principle in the quantum fluctuation part, we introduce a trial potential $V_{tr}(\tilde{q}; q_c)$ depending on $\tilde{q} = q - q_c$, and also the corresponding trial action $S_{tr}[\tilde{q}(\cdot); q_c]$. Here the extra variable q_c is to indicate

---

[2]For a discussion of bounds in quantum partition functions, cf. Ref. [275].

parametric dependence of the trial potential and of the trial action on the centroid q_c. By this ansatz, both the trial potential and the trial action become *nonlocal*.

For notational distinction from the full average $\langle O \rangle$, we introduce the double bracket symbol $\langle\!\langle O[q(\cdot)] \rangle\!\rangle_{q_c}^{(\mathrm{tr})}$ in order to denote the average of the observable $O[q(\cdot)]$ over the pure quantum fluctuations $\tilde{q}(\cdot)$ with weight $\exp\{-S_{\mathrm{tr}}[\tilde{q}(\cdot); q_c]/\hbar\}$,

$$\langle\!\langle O[q(\cdot)] \rangle\!\rangle_{q_c}^{(\mathrm{tr})} \equiv \frac{\oint \mathcal{D}\tilde{q}(\cdot)\, O[q_c + \tilde{q}(\cdot)] \exp\left(-S_{\mathrm{tr}}[\tilde{q}(\cdot); q_c]/\hbar\right)}{\oint \mathcal{D}\tilde{q}(\cdot)\, \exp\left(-S_{\mathrm{tr}}[\tilde{q}(\cdot); q_c]/\hbar\right)}. \tag{8.22}$$

With the definition (8.22), the effective classical potential may be written as

$$e^{-\beta \mathcal{V}(q_c)} = \mu_{\mathrm{tr}}(q_c) \left\langle\!\!\left\langle \exp\left\{-\int_0^{\hbar\beta} d\tau \left(V[q(\tau)] - V_{\mathrm{tr}}[\tilde{q}(\tau); q_c]\right)/\hbar\right\} \right\rangle\!\!\right\rangle_{q_c}^{(\mathrm{tr})}, \tag{8.23}$$

where $\mu_{\mathrm{tr}}(q_c)$ is the pure quantum contribution to the trial partition function,

$$\mu_{\mathrm{tr}}(q_c) \equiv e^{-\beta F_{\mathrm{tr}}(q_c)} = \oint \mathcal{D}\tilde{q}(\cdot)\, \exp\left(-S_{\mathrm{tr}}[\tilde{q}(\cdot); q_c]/\hbar\right). \tag{8.24}$$

Jensen's inequality (8.17) yields $\exp[-\beta \mathcal{V}(q_c)] \geq \exp[-\beta \mathcal{V}_{\mathrm{tr}}(q_c)]$, which implies for the effective classical potential the upper bound

$$\mathcal{V}(q_c) \leq \mathcal{V}_{\mathrm{tr}}(q_c). \tag{8.25}$$

The trial effective potential is given by

$$\mathcal{V}_{\mathrm{tr}}(q_c) = \left\langle\!\!\left\langle \int_0^{\hbar\beta} d\tau \left(V[q(\tau)] - V_{\mathrm{tr}}[\tilde{q}(\tau); q_c]\right)/\hbar\beta \right\rangle\!\!\right\rangle_{q_c}^{(\mathrm{tr})} - \frac{1}{\beta} \ln \mu_{\mathrm{tr}}(q_c). \tag{8.26}$$

Consider first an undamped system and use the trial action of a free particle,

$$S_{\mathrm{tr}}[\tilde{q}; q_c] \quad \rightarrow \quad S_0[q] = S_0[\tilde{q}] = M\hbar\beta \sum_{n=1}^{\infty} \nu_n^2 |Q_n|^2. \tag{8.27}$$

The simple trial action (8.27) does not depend on the centroid. Using the functional integration measure (8.8), we find from Eq. (8.24)

$$\mu_0 = 1. \tag{8.28}$$

Thus, according to (8.14), μ_0 is purely classical for a free particle trial action. The computation of the average in Eq. (8.26) is done by writing $V(q)$ as Fourier integral and by completing the square of the Q_n in the exponent. The resulting upper bound for the effective potential is a fuzzy copy of the original potential $V(q)$ [4],

$$\mathcal{V}_0(q_c) = \langle\!\langle V[q(\cdot)] \rangle\!\rangle_{q_c}^{(0)}. \tag{8.29}$$

Substituting the form (8.27) into Eq. (8.22), we obtain

$$\langle\!\langle V[q(\cdot)]\rangle\!\rangle_{q_c}^{(0)} \equiv \frac{1}{\sqrt{2\pi\langle\xi^2\rangle_0}} \int_{-\infty}^{\infty} dq\, V(q)\, \exp\left(-\frac{(q-q_c)^2}{2\langle\xi^2\rangle_0}\right). \qquad (8.30)$$

The convolution integral smears the original potential $V(q)$ out over a length scale $\sqrt{\langle\xi^2\rangle_0}$. The mean square spread of the Gaussian weight function in Eq. (8.30) is

$$\langle\xi^2\rangle_0 \equiv \langle\tilde{q}^2\rangle_0 = \frac{2}{M\beta}\sum_{n=1}^{\infty}\frac{1}{\nu_n^2} = \frac{\hbar^2\beta}{12M}. \qquad (8.31)$$

The Gaussian smearing of the original potential in the trial effective classical potential $\mathcal{V}_0(q_c)$ on the scale (8.31) is determined by the quantum fluctuations of a free particle.

A considerable improvement on Feynman's original variational method has been put forward by Giachetti and Tognetti [273], and independently by Feynman and Kleinert [274]. The essence of the refined variational ansatz is to include a harmonic potential in the trial action,

$$V_{\rm h}[q(\tau); q_c] = \frac{M}{2}\Omega^2(q_c)\,[q(\tau) - q_c]^2 = \frac{M}{2}\Omega^2(q_c)\,\tilde{q}^2(\tau), \qquad (8.32)$$

where $\Omega(q_c)$ is a local trial frequency to be optimally chosen. The trial potential $V_{\rm h}[q; q_c]$ is nonlocal because of the parametric dependence on the centroid q_c.

At this stage, we find it appropriate to generalize the discussion to include dissipation. For damped systems, it is natural to incorporate the dissipative action into the trial action. Thus we choose as trial action the general harmonic form

$$S_{\rm tr}[\tilde{q}; q_c]/\hbar \quad\rightarrow\quad S_{\rm h}[\tilde{q}; q_c]/\hbar = M\beta\sum_{n=1}^{\infty}\left[\nu_n^2 + \nu_n\hat{\gamma}(\nu_n) + \Omega^2(q_c)\right]|Q_n|^2. \qquad (8.33)$$

Substituting this form into Eq. (8.24), we find

$$\mu_{\rm h}[\Omega(q_c)] \equiv e^{-\beta F_{\rm h}[\Omega(q_c)]} = \prod_{n=1}^{\infty}\frac{\nu_n^2}{\nu_n^2 + \nu_n\hat{\gamma}(\nu_n) + \Omega^2(q_c)} = \beta\hbar\Omega(q_c)Z_{\rm h}(\beta). \qquad (8.34)$$

The quantity $F_{\rm h}[\Omega(q_c)]$ is the quantum part of the free energy, and $Z_{\rm h}(\beta)$ is the full partition function of the trial damped harmonic oscillator [cf. Eq.(6.80)]. Performing the average in Eq. (8.26) with the form (8.33) in the exponent of the weight function, we find the variational classical effective potential

$$\mathcal{V}_{\rm h}(q_c) = \langle\!\langle V[q(\cdot)]\rangle\!\rangle_{q_c}^{(\rm h)} + F_{\rm h}[\Omega(q_c)] - \frac{M}{2}\Omega^2(q_c)\langle\xi^2\rangle_{\rm h}. \qquad (8.35)$$

Here $\langle\!\langle V[q(\cdot)]\rangle\!\rangle_{q_c}^{(\rm h)}$ is the original potential $V(q)$ smeared with a Gaussian of half-width $\langle\xi^2\rangle_{\rm h}$ analogous to Eq. (8.30). The mean spread is given by

$$\langle\xi^2\rangle_{\rm h} \equiv \langle\tilde{q}^2\rangle_{\rm h} = \frac{2}{M\beta}\sum_{n=1}^{\infty}\frac{1}{\nu_n^2 + \nu_n\hat{\gamma}(\nu_n) + \Omega^2(q_c)}. \qquad (8.36)$$

Comparing the expression (8.36) with Eq. (6.128) we see that $\langle\xi^2\rangle_{\rm h}$ differs from the full coordinate dispersion by the missing $n = 0$ term, which is the classical dispersion

$\langle q^2 \rangle_\Omega^{(cl)} = 1/M\beta\Omega^2(q_c)$. While the missing term grows linearly with T, the fluctuation width $\langle \xi^2 \rangle_h$ shrinks with $1/T$ at large T, as can be seen from Eq. (6.136). The width $\langle \xi^2 \rangle_h$ is the pure quantum position dispersion of the damped harmonic oscillator, which is finite at all T,

$$\langle \xi^2 \rangle_h = \langle q^2 \rangle_\Omega - \langle q^2 \rangle_\Omega^{(cl)} . \tag{8.37}$$

The optimal trial function $\Omega^2(q_c)$ is determined by minimization of $\mathcal{V}_h(q_c)$,

$$\frac{d\mathcal{V}_h(q_c)}{d\Omega^2(q_c)} = \frac{\partial \mathcal{V}_h(q_c)}{\partial \Omega^2(q_c)} + \frac{\partial \mathcal{V}_h(q_c)}{\partial \langle \xi^2 \rangle_h} \frac{\partial \langle \xi^2 \rangle_h}{\partial \Omega^2(q_c)} = 0 . \tag{8.38}$$

Using the expressions (8.34) and (8.36), we obtain $\partial F_h/\partial \Omega^2 = \frac{1}{2}M\langle \xi^2 \rangle_h$. With this we find from the defining expression (8.35) that $\partial \mathcal{V}_h(q_c)/\partial \Omega^2(q_c)$ vanishes. Thus Eq. (8.38) reduces to $\partial \mathcal{V}_h(q_c)/\partial \langle \xi^2 \rangle_h = 0$. From this we see that the best choice is

$$\Omega^2(q_c) = \frac{2}{M} \frac{\partial \langle\!\langle V[q(\cdot)] \rangle\!\rangle_{q_c}^{(h)}}{\partial \langle \xi^2 \rangle_h} . \tag{8.39}$$

Combining Eq. (8.39) with (8.36), we get in general a transcendental equation for the function $\Omega^2(q_c)$, which has to be solved numerically. The expressions (8.36) and (8.39) represent the optimal choice for the functions $\langle \xi^2 \rangle_h$ and $\Omega^2(q_c)$ within the trial ansatz (8.33). With the optimal choice for $\langle \xi^2 \rangle_h$ and $\Omega(q_c)$, the function $\mathcal{V}_h(q_c)$ gives the best upper local bound for the true effective classical potential $\mathcal{V}(q_c)$, $\mathcal{V}(q_c) \leqslant \mathcal{V}_h(q_c)$. Accordingly, we find for the partition function the best lower bound

$$Z \geqslant \left(\frac{M}{2\pi\hbar^2\beta} \right)^{1/2} \int_{-\infty}^{\infty} dq_c \, \exp[\, -\beta\mathcal{V}_h(q_c) \,] . \tag{8.40}$$

For anharmonic systems at high temperature, the centroid integral has to be treated in most cases numerically. The second derivative of $\mathcal{V}_h(q_c)$ with respect to $\Omega^2(q_c)$ is non-negative, as can be easily demonstrated. This implies that the extremal effective potential $\mathcal{V}_h(q_c)$ is a local minimum. Clearly, for the particular case of the damped harmonic oscillator, $V(q) \propto q^2$, the variational method reproduces the exact partition function. On the other hand, in the limit $\Omega \to 0$, Feynman's original (non-optimal) choice is recovered. By construction, the variational result $\mathcal{V}_h(q_c)$ for $\mathcal{V}(q_c)$ becomes more accurate as the temperature is increased. Janke has shown [276] that the high temperature expansion of $\mathcal{V}_h(q_c)$ agrees with the Wigner expansion of $\mathcal{V}(q_c)$ up to terms of order β^3. In the classical limit, the mean spread $\langle \xi^2 \rangle_h$ becomes zero and $\mu_h(q_c) \to 1$, so that $\langle\!\langle V[q(\cdot)] \rangle\!\rangle_{q_c}^{(h)}$ is reduced to the original potential $V(q_c)$ and (8.40) maps on the classical partition function (8.10). In the opposite limit $T \to 0$, the variational method corresponds to a Gaussian trial ansatz for the ground state [274]. The Gaussian function is known to give extremely good ground-state energies for smooth single-well potentials. Thus, the effective classical potential turns out to be a good concept both in the high- and in the low-temperature limit, and in many cases at any temperature. Numerically, the approach gives quite reliable results for many smooth potentials at all temperatures, as emphasized in Refs. [6], [276] – [278]. The variational method works surprisingly well for hard-core potentials as well as for singular potentials except at very low temperatures where the cusp-like shape of the true wave functions becomes more important. The method has also been applied successfully to nonlinear field theories [279]. A friction term has been included in the variational method in Ref. [280].

In the path integral approach, imaginary-time position correlation functions are treated by adding the source term $\int_0^{\hbar\beta} d\tau' \, j(\tau')q(\tau')$ to the Euclidean action and performing functional differentiations with respect to the source $j(\tau)$. Using this method, we can also study position correlation functions within the variational method. The corresponding treatment is straightforward [6]. Analytic continuation to real time may be performed in the end.

A different view on the effective classical potential method has been put forward recently. It has been shown in Ref. [281] that, for standard Hamiltonians, the variational method is equivalent to a scheme in which the pure quantum fluctuations are treated self-consistently in a harmonic approximation (PQSCHA method). For nonstandard Hamiltonians like those describing magnetic systems, the Feynman-Jensen inequality is generally not applicable. The PQSCHA in phase space representation (see Subsection 8.2.4) is still meaningful for nonstandard Hamiltonians and allows to construct an effective classical potential. For a review and survey of applications of the PQSCHA, see Ref. [281, b].

8.2.3 Variational perturbation theory

A systematic improvement of the variational approach consists in performing a variational perturbation expansion [282, 6]. In the first step, the anharmonic action is expanded in powers of the fluctuations about the centroid, $\tilde{q}(\tau) = q(\tau) - q_c$,

$$S[q(\cdot)] \;=\; \hbar\beta V(q_c) + S_h[\tilde{q}(\cdot); q_c] + S_{int}[\tilde{q}(\cdot); q_c] \,, \qquad (8.41)$$

$$S_{int}[\tilde{q}(\cdot); q_c] \;=\; \int_0^{\hbar\beta} d\tau \, V_{int}[\tilde{q}(\tau); q_c] \,, \qquad (8.42)$$

where $V_{int}(\tilde{q}; q_c)$ is the deviation of the full potential from the trial harmonic potential,

$$V_{int}(\tilde{q}(\tau); q_c) \;\equiv\; \frac{1}{2!}\left[V''(q_c) - M\Omega^2(q_c)\right]\tilde{q}^2(\tau) + \frac{1}{3!}V'''(q_c)\,\tilde{q}^3(\tau) + \cdots \,, \qquad (8.43)$$

and where the prime denotes differentiation with respect to the coordinate. Substituting the expansion (8.43) into Eq. (8.23), we obtain the variational perturbation expansion for the effective classical potential in the form of a cumulant expansion,

$$\mathcal{V}(q_c) \;=\; \mathcal{V}_h(q_c) - \frac{1}{2!\,\hbar^2\beta}\left[\langle\!\langle S_{int}^2\rangle\!\rangle_{q_c}^{(h)} - \left(\langle\!\langle S_{int}\rangle\!\rangle_{q_c}^{(h)}\right)^2\right] \qquad (8.44)$$

$$+ \;\frac{1}{3!\,\hbar^3\beta}\left[\langle\!\langle S_{int}^3\rangle\!\rangle_{q_c}^{(h)} - 3\langle\!\langle S_{int}^2\rangle\!\rangle_{q_c}^{(h)}\langle\!\langle S_{int}\rangle\!\rangle_{q_c}^{(h)} + 2\left(\langle\!\langle S_{int}\rangle\!\rangle_{q_c}^{(h)}\right)^3\right] \pm \cdots \,.$$

The expression $\langle\!\langle S_{int}^m\rangle\!\rangle_{q_c}^{(h)}$ involves the average

$$\lang\!\langle V_{\text{int}}[\tilde{q}(\tau_1); q_c]\, V_{\text{int}}[\tilde{q}(\tau_2); q_c]\cdots V_{\text{int}}[\tilde{q}(\tau_m); q_c]\,\rangle\!\rangle_{q_c}^{(\text{h})} \tag{8.45}$$

over the quantum fluctuations $\tilde{q}(\cdot)$ with weight function $\exp\{-\mathcal{S}_{\text{h}}[\,\tilde{q}(\cdot); q_c\,]/\hbar\}$, where $\mathcal{S}_{\text{h}}[\,\tilde{q}(\cdot); q_c\,]$ is given in Eq. (8.33). The smearing formula can be found upon spatial Fourier expansion of $V_{\text{int}}[\tilde{q}(\tau); q_c]$. The average of the Fourier factors yields

$$\left\langle\!\!\left\langle \prod_{\ell=1}^{m} \exp[\,ik_\ell\tilde{q}(\tau_\ell)\,]\right\rangle\!\!\right\rangle_{q_c}^{(\text{h})} = \exp\left(-\frac{1}{2}\sum_{i=1}^{m}\sum_{j=1}^{m} k_i\xi_{\tau_i,\tau_j}^2 k_j \right), \tag{8.46}$$

where we have introduced the pure quantum coordinate autocorrelation function

$$\xi_{\tau,\tau'}^2 \equiv \langle\tilde{q}(\tau)\tilde{q}(\tau')\rangle_{\text{h}} = \frac{2}{M\beta}\sum_{n=1}^{\infty} \frac{\cos[\,\nu_n(\tau-\tau')\,]}{\nu_n^2 + \nu_n\hat{\gamma}(\nu_n) + \Omega^2(q_c)}. \tag{8.47}$$

The Fourier inversion of Eq. (8.46) gives again a Gaussian form, but with the inverse matrix. Thus we find in generalization of Eq. (8.30) the smearing formula [283]

$$\langle\!\langle F_1[q(\tau_1)]\, F_2[q(\tau_2)]\cdots F_m[q(\tau_m)]\rangle\!\rangle_{q_c}^{(\text{h})} = \prod_{k=1}^{m}\left(\int_{-\infty}^{\infty} dq_k\, F_k(q_k) \right) \tag{8.48}$$

$$\times \frac{1}{\sqrt{(2\pi)^m \det[\,\xi_{\tau,\tau'}^2\,]}} \exp\left[-\frac{1}{2}\sum_{i=1}^{m}\sum_{j=1}^{m}(q_i-q_c)\xi_{\tau_i,\tau_j}^{-2}(q_j-q_c) \right],$$

where $\xi_{\tau,\tau'}^{-2}$ is the inverse of the $m\times m$-matrix $\xi_{\tau,\tau'}^2$.

Truncation of the series (8.44) after the nth order defines the effective classical potential in order n, $\mathcal{V}^{(n)}(q_c)$. The first order corresponds to the variational effective classical potential, $\mathcal{V}^{(1)}(q_c) = \mathcal{V}_{\text{h}}(q_c)$. The infinite sum (8.44) is independent of the variational frequency $\Omega(q_c)$ by construction. For any finite order n, the optimal choice for $\Omega(q_c)$ is determined by the relation

$$\partial\mathcal{V}^{(n)}(q_c)/\partial\Omega(q_c) = 0. \tag{8.49}$$

The optimal frequency $\Omega^{(n)}(q_c)$ for a given order n has been dubbed *frequency of least dependence*. The variational perturbation expansion (8.44) is an alternating series, and the last cumulant in each order is positive for odd orders. This property leads to the following behavior. For odd n, $n = 2m + 1$, $\mathcal{V}^{(2m+1)}(q_c)$ is a minimum in Ω at the optimal frequency, whereas the second derivative for $n = 2m$ even, $\partial^2\mathcal{V}^{(2m)}(q_c)/(\partial\Omega)^2$, may change sign at the optimal frequency. With increasing odd or even n, the potential $\mathcal{V}^{(n)}(q_c)$ develops as a function of Ω a flat plateau at the optimal frequency. It has been demonstrated for the anharmonic oscillator and quartic double well that the third order effective classical potential $\mathcal{V}^{(3)}(q_c)$ is by far more accurate than the first order effective classical potential $\mathcal{V}^{(1)}(q_c)$.

A discussion of the smearing formula (8.48) and application of the variational perturbation method to the Coulomb potential for zero friction is given in Ref. [283].

For the anharmonic oscillator $V(q) = M\omega^2 q^2/2 + g\, q^4/4$, the energy eigenvalues $\{E_\nu(g)\}$ possess an essential singularity at $g = 0$. Thus, ordinary perturbation expansion in g has zero radius of convergence. The physical origin of the singular behavior is the possibility for tunneling when g is negative. Therefore, the form of the singularity may be analyzed using semiclassical methods. The important point now is that the variational perturbation expansion involves convergent infinite-order summations. This changes the ordinary expansion into a systematic expansion which has uniform convergence for arbitrary coupling strength, including strong-coupling [6].

8.2.4 Expectation values in coordinate and phase space

The results of Subsection 8.2.1 can be used directly to formulate quantum statistical averages of operators being functions of the position operator, $\hat{A}(\hat{q})$, or more generally of the position and momentum operator, $\hat{A}(\hat{p}, \hat{q})$. In the remainder of this subsection, we restrict the attention to Weyl ordering $\mathcal{Z}_0\{\hat{A}(\hat{p}, \hat{q})\}$, which is the case $\gamma = 0$ in Subsection 4.4.1. For the harmonic trial ansatz (8.33) or the PQSCHA, the optimally chosen restricted equilibrium density matrix $\rho(q'', q'; q_c)$ defined in Eq. (8.2) takes the particular form [284]

$$\rho(q'', q'; q_c) = \frac{\sqrt{M}\, e^{-\beta \mathcal{V}_h(q_c)}}{2\pi\hbar\sqrt{\beta\langle\xi^2\rangle_h}} \exp\left(-\frac{(q'' + q' - 2q_c)^2}{8\langle\xi^2\rangle_h} - \frac{\langle p^2\rangle_h}{2\hbar^2}(q'' - q')^2\right). \quad (8.50)$$

The effective classical potential $\mathcal{V}_h(q_c)$ and the mean square variation $\langle\xi^2\rangle_h$ are given in Eqs. (8.35) and (8.36). The momentum dispersion of the trial damped oscillator is

$$\langle p^2\rangle_h = \frac{M}{\beta} \sum_{n=-\infty}^{\infty} \frac{\Omega^2(q_c) + |\nu_n|\hat{\gamma}(|\nu_n|)}{\Omega^2(q_c) + \nu_n^2 + |\nu_n|\hat{\gamma}(|\nu_n|)}, \quad (8.51)$$

which is discussed in some detail in Section 6.6. The form (8.50) immediately follows from undoing the integration over the centroid in the expression (6.156) [cf. Eq. (8.2)].

The restricted Wigner function is a phase space density distribution which is connected with $\rho(q'', q'; q_c)$ as given by the correspondence (4.266). The restricted Wigner function resulting from Eq. (8.50) reads

$$f^{(\mathrm{W})}(p, q; q_c) = \frac{\exp\left[-\beta\mathcal{V}_h(q_c) - (q - q_c)^2/2\langle\xi^2\rangle_h - p^2/2\langle p^2\rangle_h\right]}{\sqrt{2\pi\hbar^2\beta/M}\,\sqrt{2\pi\langle\xi^2\rangle_h}\,\sqrt{2\pi\langle p^2\rangle_h}}. \quad (8.52)$$

In the self-consistent harmonic approximation (8.50) or (8.52), the quantum statistical average of an observable $A(q)$ takes the form of an integral over the centroid q_c,

$$\langle A\rangle = \frac{(M/2\pi\hbar^2\beta)^{1/2}}{Z_h} \int_{-\infty}^{\infty} dq_c \, \exp\left[-\beta\mathcal{V}_h(q_c)\right] \langle\!\langle A[q(\cdot)]\rangle\!\rangle_{q_c}^{(\mathrm{h})}, \quad (8.53)$$

in which $\langle\!\langle A[q(\cdot)]\rangle\!\rangle_{q_c}^{(\mathrm{h})}$ captures the average over the quantum fluctuations. The function $\langle\!\langle A[q(\cdot)]\rangle\!\rangle_{q_c}^{(\mathrm{h})}$ is a smeared portrayal of the original function $A(q)$ like in Eq. (8.30), but now the smearing width is $\sqrt{\langle\xi^2\rangle_h}$,

$$\langle\!\langle A[q(\cdot)]\rangle\!\rangle_{q_c}^{(\mathrm{h})} = \frac{1}{\sqrt{2\pi\langle\xi^2\rangle_h}} \int_{-\infty}^{\infty} dq \, A(q) \exp\left(-\frac{(q - q_c)^2}{2\langle\xi^2\rangle_h}\right). \quad (8.54)$$

Consider next the quantum statistical average of an observable $A(p, q)$ depending on position and momentum. The average of the operator in Weyl form $\mathcal{Z}_0\{\hat{A}(\hat{p}, \hat{q})\}$ over the pure quantum fluctuations in harmonic approximation turns out with the use of Eq. (8.52) as a double Gaussian average with half-widths $\langle \xi^2 \rangle_{\rm h}$ and $\langle p^2 \rangle_{\rm h}$,

$$\langle\!\langle A[q(\cdot), p(\cdot)]\rangle\!\rangle_{q_{\rm c}}^{\rm (h)} \equiv \frac{1}{2\pi} \iint_{-\infty}^{\infty} \frac{dq\, dp}{\sqrt{\langle \xi^2 \rangle_{\rm h} \langle p^2 \rangle_{\rm h}}}\, A(p, q)\, \exp\left(-\frac{(q - q_{\rm c})^2}{2\langle \xi^2 \rangle_{\rm h}} - \frac{p^2}{2\langle p^2 \rangle_{\rm h}} \right).$$

The final integral over the centroid is as given in Eq. (8.53). In the case of Ohmic dissipation, both the quantum part $\mu_{\rm h}$ of the partition function and the momentum dispersion $\langle p^2 \rangle_{\rm h}$ must be Drude-regularized since otherwise they diverge logarithmically (cf. Section 6.5 and 6.6). A different possibility for regularization of the effective classical potential would be to subtract the contribution of a free Brownian particle $(1/\beta)\ln(1/\mu_{\rm free})$ with $1/\mu_{\rm free} = \prod_{n=1}^{\infty}[1 + \hat{\gamma}(\nu_n)/\nu_n]$.

One remark is appropriate. Since the local density matrix $\rho(q'', q'; q_{\rm c})$ in Eq. (8.50) has a factor $\exp[-(q'' + q' - 2q_{\rm c})^2/8\langle \xi^2 \rangle_{\rm h}]$, the fluctuation width of the centroid is generally small at low and intermediate temperatures, and it becomes zero in the classical limit. Thus the special role of the centroid is less important for the density matrix than for the partition function. In numerical computations of the variational perturbation expansion, it turns out that one may treat the centroid in the same way as the quantum fluctuations. By this, the numerical efforts are reduced enormously.[3]

Historically, the variational method has been introduced by Feynman when he studied the polaron problem. For the action (4.128) with (4.130) there are two major reasons that a potential term in the trial action is not the appropriate representative to master the physical situation. First, the polaron is free to wander around in a crystal without giving preference to any particular place. Second, the polaron at position $q(\tau)$ interacts with itself at positions taken at any previous imaginary time τ'.

A trial action for the polaron which has these properties (except for a different spatial dependence) and which is analytically tractable is [3, 4]

$$S_0 = \frac{M}{2} \int_0^{\hbar\beta} d\tau\, \dot{\mathbf{q}}^2(\tau) + \frac{C}{2} \int_0^{\hbar\beta} d\tau \int_0^{\tau} d\tau'\, D_\omega(\tau - \tau')|\mathbf{q}(\tau) - \mathbf{q}(\tau')|^2 , \qquad (8.55)$$

where $D_\omega(\tau)$ is given in (3.83), and where C and ω are taken as adjustable parameters. According to the variational principle (8.21), we have for the ground-state energy

$$E \leqslant E_0 + \lim_{\beta \to \infty} \langle S - S_0 \rangle_0 / \hbar\beta , \qquad (8.56)$$

where E_0 is the ground-state energy of the model (8.55), and $\langle \cdots \rangle_0$ means average with weight $\exp(-S_0/\hbar)$. The lowest upper bound found by variation of the parameters ω and C gives for E values that are better than those obtained by all other methods, and in fact for all values of the polaron coupling constant α [4]. Unfortunately, Jensen's inequality does *not* hold in general for a polaron in a magnetic field (magneto-polaron) since the action is complex [285]. A slightly modified variational method tailored to the magneto-polaron case has been suggested in Ref. [286].

---

[3]I am indebted to A. Pelster for this communication.

9 Suppression of quantum coherence

There are two main lines of investigations on decoherence in quantum mechanics. On the one hand, one is interested in fundamental problems in the interpretation of quantum theory, in particular in questions connected with measurement and the quantum to classical transition (cf., e.g., Ref. [58]). The importance of decoherence in the appearance of a classical world in quantum theory is well known [287]. In cosmology, the decoherence which must have been occurred during the inflationary era of the Universe has been attributed to quantum fluctuations. On the other hand, environment-induced decoherence is omnipresent in the microscopic world. It is found, e.g., for an atom confined in a quantum optical trap, or for electron propagation in a mesoscopic device. Striking examples for coherence effects are the anomalous magnetoresistance in disordered systems due to weak localization [288], and Aharonov-Bohm type oscillations in the resistance of mesoscopic rings [289].

A system prepared in a non-equilibrium state and in contact with the environment relaxes to equilibrium. The decay of the off-diagonal states of the reduced density matrix (coherences) is denoted as decoherence or loosely speaking as dephasing [cf. the remark at the end of Section 9.1]. Evidently, the origin of decoherence is entanglement and interaction with a fluctuating environment. Therefore, decoherence depends on the spectral density of the environmental coupling, on temperature and on specific properties of the system. There follows from the formal structure of the influence functional that response and equilibrium correlation functions decay roughly on the same time scale as the coherences. We shall discuss this in some detail in Chapter 21.

There are experimental indications that dephasing of electrons in mesoscopic conductors persists down to zero temperature [290, 291, 292]. On the theoretical side, Altshuler and coworkers advocated that there is no dephasing at $T = 0$ [293], while Golubev and Zaikin found in their studies on diffusive conductors that dephasing of electrons levels off at low temperatures [294, 295]. There is still an ongoing debate about dephasing of electrons in diffusive conductors at zero temperature [296]–[299].

In a multitude of model systems, decoherence persists down to zero temperature, as we shall see in Section 9.3. This has been found also by others [171, 296, 300].

To determine decoherence quantitatively, it is natural to introduce a time scale τ_φ which characterizes the half-life of quantum interference. Basically, phase randomization of the system's wave function occurs through energy exchange processes with environmental modes, e.g., electron-electron or electron-phonon scattering.

Two complementary views about dephasing are possible [210]. Either one may treat the problem by studying the changes induced by the interfering particle in the states of the environment, or one may study the accumulation of the phase uncertainty due to quantum fluctuations of the reservoir coupling.

Below, the differences between a non-dynamical and a dynamical environment with regard to dephasing are discussed first. Then it is set out that there is no universal decoherence in models with delocalized bath modes. In the third section, a model with spatially localized bath modes is introduced, in which decoherence is independent of the actual geometry, and the half-life of quantum interference in such system is calculated semiclassically. Finally I dwell on electron dephasing in a diffusive conductor, and on the delicate question whether dephasing persists down to $T = 0$.

9.1 Nondynamical versus dynamical environment

An environment with no significant dynamics of its own acts upon the system as if it were a classically fluctuating potential. To study the loss of phase memory, consider a particle whose dynamics is governed by the Hamiltonian $H = H_0 + V_0(q) + \delta V(t)$. We assume that $\delta V(t)$ is a stochastic potential term with zero mean, $\langle \delta V(t) \rangle = 0$, and Gaussian statistics with covariance $\langle \delta V(t) \delta V(0) \rangle_{\mathrm{av}} = \langle \delta V^2 \rangle_{\mathrm{av}} v(t)$, where $v(0) = 1$. The stochastic impact $\delta V(t)$ gives rise to a fluctuating phase of the propagator,

$$K(q_2, q_1; t) = e^{-i\varphi(t)} K_0(q_2, q_1; t) \qquad \text{with} \qquad \varphi(t) = \frac{1}{\hbar} \int_0^t d\tau \, \delta V(\tau) . \qquad (9.1)$$

Taking the statistical average, $\langle K(q_2, q_1; t) \rangle_{\mathrm{av}} = \langle e^{-i\varphi(t)} \rangle_{\mathrm{av}} K_0(q_2, q_1; t)$, we have

$$\langle e^{-i\varphi(t)} \rangle_{\mathrm{av}} = e^{-\langle \varphi^2(t) \rangle_{\mathrm{av}}/2} \qquad\qquad\qquad (9.2)$$

$$\langle \varphi^2(t) \rangle_{\mathrm{av}} = \frac{2}{\hbar^2} \langle \delta V^2 \rangle_{\mathrm{av}} \int_0^t dt_2 \int_0^{t_2} dt_1 \, v(t_2 - t_1) . \qquad (9.3)$$

Now assume that the observation time t is large compared to the memory time of the correlation function $v(t)$. Then the phase uncertainty increases linearly with t,

$$\langle e^{-i\varphi(t)} \rangle_{\mathrm{av}} = e^{-\gamma_\varphi t} . \qquad\qquad\qquad (9.4)$$

It is natural to interpret γ_φ as the inverse time scale for dephasing. This is

$$\gamma_\varphi = \langle \varphi^2(t) \rangle / 2t = \frac{\langle \delta V^2 \rangle_{\mathrm{av}}}{\hbar^2} \int_0^\infty d\tau \, v(\tau) . \qquad (9.5)$$

In a fully quantized model with *dynamical* fluctuations, the situation is more intricate. Assume that the initial state of the global system (3.1) is factorized into a system state $|\psi_i\rangle$ and a bath state $|\chi\rangle$. Then the transition probability from state $|\psi_i\rangle$ to final state $|\psi_f\rangle$ is

$$P_{fi} = \sum_{\chi'} |\langle \psi_f | \langle \chi' | \exp[-i(H_{\mathrm{R}} + H_{\mathrm{S}} + H_{\mathrm{I}}) t/\hbar] |\chi\rangle |\psi_i\rangle|^2 . \qquad (9.6)$$

When the coupling is weak enough to leave the bath unchanged, we may put in place of the bath operators their expectation values. Assigning the expectation value $\langle H_{\mathrm{R}} \rangle_\chi$ to the phase of the bath state, we find for the transition amplitude of the system

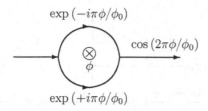

Figure 9.1: Schematic sketch of an Aharonov-Bohm device threaded by a magnetic flux ϕ.

$$A_{fi} = \langle \psi_f | \exp[-i(H_S + \langle H_I \rangle_\chi)\, t/\hbar] | \psi_i \rangle \,. \tag{9.7}$$

The amplitude is conveniently written as modulus times phase factor. Next assume that the initial bath state is given in the form of a mixture with weight c_χ for the state $|\chi\rangle$. Then, in analogy with Eq. (9.2), we take the average of the phase factor as

$$\langle e^{-i\varphi} \rangle = \sum_\chi c_\chi |A_{fi}(\chi)|\, e^{-i\varphi(\chi)} \Big/ \sum_\chi c_\chi |A_{fi}(\chi)| \,. \tag{9.8}$$

Evidently, this procedure breaks down for stronger coupling since the environmental states are not any more dwelling on their initial distribution. For a dynamically fluctuating potential it is therefore not possible to split the total phase into a system and a reservoir part. This demonstrates that in general it is impossible to assign to the subsystem a definite phase. We conclude from this that the whole concept of phase memory generally fails for an open quantum system. Therefore, one should speak of *decoherence* rather than of *dephasing*.

9.2 Suppression of transversal and longitudinal interferences

We recall to the reader's attention the Feynman-Vernon method for the reduced density matrix in which the motion of the particle in contact with the reservoir is described by the propagating function [cf. Eq. (5.12)]

$$J_{\mathrm{FV}}(q_f, q_f', t; q_i, q_i', 0) = \int \mathcal{D}q\, \mathcal{D}q' \exp\left\{ \frac{i}{\hbar}\left(S_S[q] - S_S[q'] \right) \right\} \mathcal{F}_{\mathrm{FV}}[q, q'] \,. \tag{9.9}$$

Let us now study whether decoherence depends on the properties of the reservoir coupling alone, or also on the particular geometry of the quantum interference system. Consider first an Aharonov-Bohm device in which a metallic ring is threaded by a magnetic flux ϕ as sketched in Fig. 9.1. The traversing paths are divided into two groups that pass the ring clockwise and anti-clockwise, thereby picking up a phase difference $\varphi = 2\pi\phi/\phi_0$, where ϕ_0 is the flux quantum. Since the paths are spatially separated in transversal direction, we shall refer to this type of interferences as *transversal*. For the global model (3.12), the decoherence factor is given by the noise action, Eq. (5.39) with (5.32). Assuming that the paths $q_1(t)$ and $q_2(t)$ are coupled to statistically independent environments, the cross terms in the noise action $\Phi^{(N)}$ are absent. Then we have

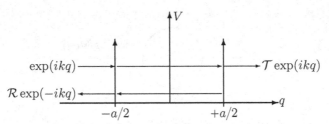

Figure 9.2: A resonant tunneling structure made up of two δ-potentials.

$$\left\langle e^{-i\varphi(t)} \right\rangle = \exp\left(-\sum_{j=1}^{2} \frac{1}{\hbar} \int_0^t dt_2 \int_0^{t_2} dt_1 \, q_j(t_2) L'(t_2 - t_1) q_j(t_1) \right). \qquad (9.10)$$

The same expression holds for a two-slit device. In the white noise limit, we recover for a slit distance q_0 the previous result (5.40).

The other device is a resonant tunneling structure with reflection and transmission coefficients $\mathcal{R}$ and $\mathcal{T}$, sketched in its simplest form in Fig. 9.2. In this device, the directly through-going path is interfering with a path which has made an additional roundtrip between the barriers. For δ-potentials, the two paths evolve in the scattering region as sketched in Fig. 9.3. Because they travel through the same spatial region and are only separated with respect to time, this interference type has been termed *longitudinal* [170]. Obviously, the interfering paths are influenced by the same environment. Putting $y(t) = q_2(t) - q_1(t)$, the resulting decoherence factor is

$$\left\langle e^{-i\varphi(t)} \right\rangle = \exp\left(-\frac{1}{\hbar} \int_0^t dt_2 \int_0^{t_2} dt_1 \, y(t_2) L'(t_2 - t_1) y(t_1) \right). \qquad (9.11)$$

The expressions (9.10) and (9.11) have been analyzed in Ref. [170]. One finds that they behave quite differently. For a ballistic path, e.g., the former exponent grows with the third power of t, whereas the latter increases quadratically in t. Observe also that Eq. (9.11) is space translational invariant while Eq. (9.10) is not. In models with delocalized bath modes, the suppression of interferences depends also strongly on spatial dimension, as emphasized in Refs. [210, 301]. In conclusion, in models with

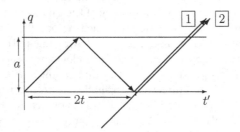

Figure 9.3: Sketch of two interfering path in the resonant tunneling device in Fig. 9.2.

delocalized bath modes, the decoherence process depends on the geometry of the interference device. Hence there is *no universal* decoherence behavior in these models.

9.3 Semiclassical decoherence

A particle moving in a real solvent or solid interacts with individual modes of the environment only within a spatially restricted region which we characterize hereafter by a length scale λ. When the particle is confined to a region of extent d where $d \ll \lambda$, as it may occur, e.g., in a two- or few-state system, then the actual coupling may be replaced by an effective coupling of the form (3.4) with (3.10). In this particular case, the simplified model with *delocalized* bath modes, Eq. (3.12), is appropriate. However, when the particle travels over distances large compared to the interaction range λ, one has to take into account the *localized* nature of the bath modes.

9.3.1 A model with localized bath modes

As in Sec. 3.1, we assume that the system-bath interaction is linear in the bath coordinates. Further, we suppose that the coupling is nonlinear in the system coordinate and confined to areas around distributed scattering centers. In one space dimension, we put

$$H_{\mathrm{I}} = -\left(\sum_{\alpha} c_{\alpha} x_{\alpha}\right) \lambda \sum_{q_0} u\left(\frac{q - q_0}{\lambda}\right) . \tag{9.12}$$

The function $u(x) = u(-x)$ has range unity and normalization $\int dx\, u(x) = 1$. The form function $u[(q - q_0)/\lambda$ limits the interaction of the particle with individual bath modes to a range of length λ about the scattering center q_0. To restore translational invariance of the coupling, the scatterers are assumed to be uniformly distributed in space. Then the interaction is characterized by the spatial autocorrelation function

$$\lambda^2 \int_{-\infty}^{\infty} dx\, u(x)u(q/\lambda - x) = \lambda^2 \mathcal{U}(q/\lambda) . \tag{9.13}$$

The elimination of the bath modes can be executed analogously to the case of delocalized modes discussed in Sec. 5.2. One then obtains the influence action

$$\Phi_{\mathrm{infl}}[q_1, q_2] = \lambda^2 \sum_{j=1}^{2} (-1)^{j-1} \int_0^t dt_2 \int_0^{t_2} dt_1 \Big\{ L(t_2 - t_1) \mathcal{U}\left(|q_j(t_2) - q_1(t_1)|/\lambda\right)$$

$$- L^*(t_2 - t_1) \mathcal{U}\left(|q_j(t_2) - q_2(t_1)|/\lambda\right) \Big\} . \tag{9.14}$$

The real part is the noise action, and the imaginary part is the friction action.

The model belongs to the general class of models describing state-dependent dissipation discussed in Subsections 3.1.2 and 4.2.2. It describes, e.g., the interaction with extended field modes. The electron-photon and the electron-phonon interaction, Eq. (3.175), are of the form (9.12). Another important example is the interaction of a charged particle with conduction electrons. The action resulting from the imaginary-time action (4.150) is in the form (9.14), in which $L(t)$ is the Ohmic heat bath kernel, and $\lambda^2 \mathcal{U}(\Delta q/\lambda) = (3/2k_F^2)\sin^2(k_F\Delta q)/(k_F\Delta q)^2$ with $\lambda = 1/k_F$.

The model with localized bath modes has been introduced in Ref. [170] in order to study suppression of interferences without dependence on geometry (see below). The same model has been used in Ref. [171] to study state-dependent quantum effects in Brownian motion. A number of new quantum effects have been found. For instance, the logarithmic law (7.39) for the spatial spreading in the Ohmic case at $T = 0$ holds for this model only on a scale $\sigma^2(t) \ll \lambda^2$. On a larger scale, there is a smooth crossover from a Gaussian to a frozen exponential profile. This genuine quantum effect can be understood from the interplay of the negative time correlations of the quantum noise with the spatial correlations described by the function $\mathcal{U}(q/\lambda)$.

In the remainder of this section, we base the discussion of the suppression of quantum interference on the action (9.14). At first, one observes that the self-terms are added up in the noise action $\Phi^{(N)} = \operatorname{Re} \Phi_{infl}$, whereas they appear with opposite sign in the friction action $\Phi^{(F)} = \operatorname{Im} \Phi_{infl}$. Next, suppose for simplicity that the self-terms of the two paths give the same contribution, and assume that the distance between the two interfering paths is large compared to the interaction range λ for most of the time, $\lambda \gg |q_1(t') - q_2(t'')|$. Then the self-terms in Eq. (9.14) are very large compared to the cross terms. But, since the self-terms cancel out in the friction action, this action depends only on the cross terms, and consequently

$$\Phi^{(F)}[q_1, q_2] \ll \Phi^{(N)}[q_1, q_2] \,. \tag{9.15}$$

Thus, in the present model friction plays only a secondary role in the decoherence process. Because of the estimate (9.15), we disregard the friction action in the remainder of this section.[1] There is a controversy whether Eq. (9.15) holds when fermionic many-body effects are relevant (see subsequent section). The noise action now reads

$$\Phi^{(N)}[q(\cdot)] = 2\lambda^2 \int_0^t dt_2 \int_0^{t_2} dt_1 \, L'(t_2 - t_1) \mathcal{U}\left[|q(t_2) - q(t_1)|/\lambda\right] \,. \tag{9.16}$$

While the action in Eq. (9.10) is proportional to the square of the slit distance q_0, the action (9.16) is proportional to λ^2. Since λ is a microscopic length, we usually have $q_0 \gg \lambda$. The actual geometry is irrelevant for the action (9.16). Decoherence by scattering events is as effective for small separation as for large separation of the interfering paths. Thus, the suppression of interferences is universal in this regime.

9.3.2 Dephasing rate formula

In a mesoscopic conductor, the electron's motion is strongly affected by dynamical fluctuations induced by the surroundings and by scattering at randomly distributed impurities. The dynamical fluctuations entail a dephasing factor of the form

---

[1] In the model with delocalized bath modes, Eq. (3.12), the situation is different. In this model, also the friction action $\Phi^{(F)}$ may become relevant for dephasing, as noticed in Ref. [211].

$$\left\langle e^{-i\varphi(t)} \right\rangle = \exp\left(-\Phi^{(N)}[\mathbf{q}(\cdot)]/\hbar\right) . \tag{9.17}$$

Taking in addition the average of the stochastic impurity scattering, we have

$$\left\langle\!\!\left\langle e^{-i\varphi(t)} \right\rangle\!\!\right\rangle_{\text{imp}} = \left\langle\exp\left(-\Phi^{(N)}[\mathbf{q}(\cdot)]/\hbar\right)\right\rangle_{\text{imp}} \approx \exp\left(-\left\langle\Phi^{(N)}[\mathbf{q}(\cdot)]/\hbar\right\rangle_{\text{imp}}\right) . \tag{9.18}$$

On condition that the observation time t is large compared to the memory time of the function $L'(\tau)$, we find, analogous to the proceeding in Eqs. (9.3) - (9.5), that $\Phi^{(N)}[\mathbf{q}(\cdot)]$ increases linearly with time. Observing the relation (9.4) we obtain

$$\gamma_\varphi = \lim_{t\to\infty} \frac{1}{t} \left\langle\Phi^{(N)}[\mathbf{q}(\cdot)]/\hbar\right\rangle_{\text{imp}} = \frac{\lambda^2}{\hbar} \int_{-\infty}^{\infty} d\tau\, L'(\tau) \left\langle \mathcal{U}\left[|\mathbf{q}(\tau) - \mathbf{q}(0)|/\lambda\right]\right\rangle_{\text{imp}} . \tag{9.19}$$

Chakravarty and Schmid [288] performed the impurity average using a quasiclassical approach. In this method, $\langle \cdots \rangle_{\text{imp}}$ means that the pair of paths $\mathbf{q}(t_2)$ and $\mathbf{q}(t_1)$ passing through the random impurity potential distribution is weighted with the probability distribution

$$P(\mathbf{r}, t_2 - t_1) = \left\langle\delta^n\left[\mathbf{r} - \mathbf{q}(t_2) + \mathbf{q}(t_1)\right]\right\rangle_{\text{imp}} , \tag{9.20}$$

where n is the number of space dimensions.

The averaged phase (9.19) can be written with the functions $P(\mathbf{r}, \tau)$ and

$$R(\mathbf{r}, \tau) = \lambda^2 L'(\tau)\mathcal{U}(\mathbf{r}/\lambda)/\hbar \tag{9.21}$$

as

$$\gamma_\varphi = \int_{-\infty}^{\infty} d\tau \int d^n\mathbf{r}\, R(\mathbf{r}, \tau)P(\mathbf{r}, \tau) . \tag{9.22}$$

The function $R(\mathbf{r}, \tau)$ includes the spatial and temporal correlations of the environment, while the function $P(\mathbf{r}, \tau)$ covers the weight of the two paths in the random impurity potential distribution. The rate expression (9.22) is written in $\mathbf{k}$-space as[2]

$$\gamma_\varphi = \int_{-\infty}^{\infty} d\tau \int \frac{d^n\mathbf{k}}{(2\pi)^n}\, \tilde{R}(\mathbf{k}, \tau)\tilde{P}(-\mathbf{k}, \tau) . \tag{9.23}$$

where

$$\begin{aligned} \tilde{P}(\mathbf{k}, \tau) &= \left\langle e^{-i\mathbf{k}\cdot[\mathbf{q}(\tau)-\mathbf{q}(0)]}\right\rangle_{\text{imp}} , \\ \tilde{R}(\mathbf{k}, \tau) &= \lambda^2 L'(\tau)\tilde{\mathcal{U}}(\lambda|\mathbf{k}|)/\hbar . \end{aligned} \tag{9.24}$$

Here $\tilde{\mathcal{U}}(\lambda|\mathbf{k}|) = \int d^n\mathbf{r}\, \mathcal{U}(|\mathbf{r}|/\lambda)\, e^{-i\mathbf{k}\cdot\mathbf{r}}$. Finally, the $\omega, \mathbf{k}$ representation of γ_φ is

$$\gamma_\varphi = \frac{1}{\pi} \int_0^{\infty} d\omega \int \frac{d^n\mathbf{k}}{(2\pi)^n}\, \tilde{R}(\mathbf{k}, \omega)\tilde{P}(-\mathbf{k}, -\omega) , \tag{9.25}$$

with

$$\tilde{R}(\mathbf{k}, \omega) = \lambda^2 J(\omega)\coth(\tfrac{1}{2}\beta\hbar\omega)\tilde{\mathcal{U}}(\lambda|\mathbf{k}|)/\hbar . \tag{9.26}$$

The form factor $\tilde{U}(\lambda|\mathbf{k}|)$ restricts the k-integration essentially to

$$|\mathbf{k}| \lesssim 1/\lambda . \tag{9.27}$$

The predominant contribution to the dephasing rate comes from environmental modes with wave numbers near $1/\lambda$. The remaining task is to calculate the weight funktion $\tilde{P}(\mathbf{k}, \omega)$ for ballistic and for diffusive motion, and then the integral (9.23).

---

[2]Similar dephasing rate formulas in the fermionic many-body context are discussed in Ref. [302]. The crucial physics of Fermi blocking is being controversially discussed (see Section 9.4).

9.3.3 Ballistic and diffusive motion

Statistical average of paths

The motion of a particle with velocity v in n space dimendions is ballistic on the scale of the interaction range λ when the mean free path ℓ is large compared to λ. With the diffusion coefficient $D = v\ell/n$ we then have as condition for ballistic motion

$$\kappa \equiv \lambda/\ell = \lambda v/nD \ll 1 \,. \tag{9.28}$$

In the ballistic regime (9.28) we have $\mathbf{r}(\tau) = \mathbf{r}(0) + \mathbf{v}\tau$. Since scattering processes are disregarded, the probability function (9.24) takes the simple form

$$\tilde{P}_{\mathrm{bal}}(\mathbf{k}, \tau) = e^{-i\mathbf{k}\cdot\mathbf{v}\tau} \,. \tag{9.29}$$

Hence the spectral probabilitiy function in the ballistic regime for a single particle is

$$\tilde{P}_{\mathrm{bal}}(\mathbf{k}, \omega) = 2\pi\,\delta(\omega - \mathbf{k}\cdot\mathbf{v}) \,. \tag{9.30}$$

In contrast, the motion of a particle is diffusive on the interaction range λ when

$$\kappa \gg 1 \,. \tag{9.31}$$

In the regime (9.31), the average over many collisions in Eq. (9.24) leads to diffusive spread of the probability distribution,

$$P_{\mathrm{diff}}(\mathbf{r}, \tau) = (4\pi D|\tau|)^{-n/2}\, e^{-|\mathbf{r}|^2/4D|\tau|} \tag{9.32}$$

$$\tilde{P}_{\mathrm{diff}}(\mathbf{k}, \tau) = e^{-D|\mathbf{k}|^2|\tau|} \,. \tag{9.33}$$

Hence the power spectrum of a diffusive particle has the familiar Lorentzian form

$$\tilde{P}_{\mathrm{diff}}(\mathbf{k}, \omega) = \frac{2D|\mathbf{k}|^2}{(D|\mathbf{k}|^2)^2 + \omega^2} \,. \tag{9.34}$$

We are now primed and ready to study the dephasing rate (9.25) in various limits. We assume in the remainder of this chapter that the cutoff in the spectral bath coupling is the largest frequency of the problem and thus plays no particular role.

Ballistic motion

In the ballistic case, the characteristic frequency scale ω_{b} and temperature scale T_{b} are determined by the inverse time scale for traversing the interaction region,

$$\omega_{\mathrm{b}} \equiv v/\lambda \,, \qquad \text{and} \qquad T_{\mathrm{b}} \equiv \hbar\omega_{\mathrm{b}}/k_{\mathrm{B}} \,. \tag{9.35}$$

In one dimension, the decoherence rate (9.25) with Eq. (9.30) takes the form [170]

$$\gamma_\varphi = \frac{\lambda^2}{\pi\hbar v} \int_0^\infty d\omega\, J(\omega)\coth(\tfrac{1}{2}\beta\hbar\omega)\,\tilde{\mathcal{U}}(\omega/\omega_{\mathrm{b}}) \,. \tag{9.36}$$

For $\omega_c \gg \omega_b$, the frequency integral is effectively cut off at ω near ω_b. If the environmental spectral density has a low-frequency cutoff ω_{min}, the decoherence rate drops to zero when the particle's velocity v falls below $\lambda\omega_{min}$.

The decoherence rate γ_φ can be calculated in analytic form for an algebraic spectral density $J(\omega) \propto \omega^s$ and a Lorentzian form of the spatial correlation function,

$$\mathcal{U}(r/\lambda) = 1/[1+r^2/\lambda^2] \qquad\qquad \tilde{\mathcal{U}}(\lambda k) = \lambda\pi e^{-\lambda|k|} . \qquad (9.37)$$

Employing for $J(\omega)$ the form (3.38) with exponential cutoff $e^{-\omega/\omega_b}$, the integral (9.36) yields for the decoherence rate the analytic expression

$$\gamma_\varphi = \left\{1 + 2(T/T_b)^{s+1}\,\zeta(s+1,1+T/T_b)\right\}\gamma_{\varphi,0} , \qquad (9.38)$$

where $\zeta(q,z)$ is a Riemann zeta function, and $\gamma_{\varphi,0}$ is the decoherence rate at zero temperature,

$$\gamma_{\varphi,0} = 2\pi\Gamma(1+s)K_s(\omega_b/\omega_{ph})^{s-1}\omega_b . \qquad (9.39)$$

The quantity $K_s \equiv \eta_s\lambda^2/2\pi\hbar$ is a dimensionless damping parameter analogous to the Kondo parameter K introduced in Eq. (4.159). Since the spectral density of the environmental coupling is gapless, there is decoherence even at zero temperature. The rate is roughly independent of T in the temperature range $0 \leqslant T \ll T_b$, and it varies with the velocity v of the particle as v^s. Well above T_b, the decoherence rate is determined by classical noise and thus depends linearly on T,

$$\gamma_\varphi = \frac{2T}{sT_b}\gamma_{\varphi,0} = 4\pi\Gamma(s)K_s\left(\frac{\omega_b}{\omega_{ph}}\right)^{s-1}\frac{k_BT}{\hbar} . \qquad (9.40)$$

The form (9.38) describes the crossover between the limiting cases (9.39) and (9.40).

In the Ohmic case, $s = 1$, the decoherence rate (9.38) takes the form

$$\gamma_\varphi = \left\{1 + 2(T/T_b)^2\psi'(1+T/T_b)\right\}2\pi K\omega_b , \qquad (9.41)$$

where $\psi'(z)$ is the trigamma function and $K = \eta\lambda^2/2\pi\hbar$. This reduces in the low and high temperature regimes to the forms

$$\gamma_\varphi = \eta\lambda v/\hbar = (\eta/M)\lambda/\lambda_B , \qquad\qquad 0 \leqslant T \ll T_b , \qquad (9.42)$$

$$\gamma_\varphi = 2\eta\lambda^2 k_BT/\hbar^2 = 2(\eta/M)\lambda^2/\lambda_{th}^2 , \qquad\qquad T \gg T_b . \qquad (9.43)$$

The second forms are written in terms of the de Broglie wave length $\lambda_B = \hbar/Mv$ and thermal wave length $\lambda_{th} = \hbar/(Mk_BT)^{1/2}$, respectively.

At zero temperature, the dephasing rate is proportional to the velocity of the particle. In the white-noise limit, Eq. (9.43), the rate is independent of the velocity and formally agrees with the result for the model with delocalized oscillators, Eq. (5.40). However, the respective length parameters q_0 and λ may differ drastically.

Diffusive motion

For diffusive motion of the particle in n space dimensions with isotropic spatial correlation function $\mathcal{U}(|\mathbf{r}|/\lambda)$, the decoherence rate (9.25) with (9.26) takes the form

$$\gamma_\varphi = \frac{\lambda^2}{\pi\hbar} \int_0^{1/\tau_c} d\omega \, J(\omega) \coth(\tfrac{1}{2}\hbar\beta\omega) \frac{\Omega_n}{(2\pi)^n} \int_0^\infty dk \, k^{n-1} \, \tilde{P}_{\text{diff}}(k,\omega) \tilde{\mathcal{U}}(\lambda k) , \qquad (9.44)$$

where $k = |\mathbf{k}|$, and Ω_n is the total solid angle, i.e., $\Omega_1 = 2$, $\Omega_2 = 2\pi$, and $\Omega_3 = 4\pi$. With Eqs. (9.28) and (9.35), the characteristic frequency and temperature scale for diffusive motion emerge as

$$\omega_d \equiv D/\lambda^2 = \omega_b/n\kappa , \qquad \text{and} \qquad T_d \equiv \hbar\omega_d/k_B = T_b/n\kappa . \qquad (9.45)$$

The ω-integral should be cut at the inverse collision time $1/\tau_c = v/\ell = v^2/nD$, since the assumption of diffusive transport is not valid anymore in the frequency range $\omega > 1/\tau_c$. We have $1/\tau_c = n\kappa^2\omega_d$ and hence in the diffusive regime $\omega_d\tau_c \ll 1$.

Consider now first the case of one space dimension, $n = 1$. In the classical regime, we have $\coth(\hbar\beta\omega/2) \to 2/\hbar\beta\omega$. With this, and with the spectral density $J(\omega) \propto \omega^s$, the frequency integral in Eq. (9.44) is ultraviolet-convergent in the regime $s < 2$ without cutoff. On the other hand, the high-frequency cutoff at $1/\tau_c$ is essential when $s \geqslant 2$. This shows that the classical regime corresponds the temperature regime $T \gg T_d$ for $s < 2$ and to the regime $T \gg \kappa^2 T_d$ for $s > 2$. With the space correlation function (9.37) we readily find

$$\gamma_\varphi = \begin{cases} \dfrac{\eta_s\lambda^2}{\pi\hbar} \dfrac{2\pi\Gamma(2s-1)}{\sin(\pi s/2)} \left(\dfrac{\omega_d}{\omega_{\text{ph}}}\right)^{s-1} \dfrac{k_B T}{\hbar} , & \text{for} \quad \tfrac{1}{2} < s < 2 , \\[3ex] \dfrac{\eta_2\lambda^2}{\pi\hbar} 16\ln(\kappa) \left(\dfrac{\omega_d}{\omega_{\text{ph}}}\right) \dfrac{k_B T}{\hbar} , & \text{for} \quad s = 2 , \\[3ex] \dfrac{\eta_s\lambda^2}{\pi\hbar} 8\Gamma(s-2) \left(\dfrac{\kappa^2\omega_d}{\omega_{\text{ph}}}\right)^{s-1} \dfrac{k_B T}{\hbar} & \text{for} \quad s > 2 . \end{cases} \qquad (9.46)$$

In the limit of Ohmic damping, $s \to 1$, the expression (9.46) becomes independent of the diffusion coefficient and coincides with the corresponding result for ballistic transport, Eq. (9.43).

In the limit $T \to 0$, the cutoff ω_c is relevant in the regime $s \geqslant 1$ and irrelevant when $s < 1$. We find from Eq. (9.44) the expressions

$$\gamma_\varphi = \begin{cases} \dfrac{\eta_s\lambda^2}{\pi\hbar} \dfrac{\pi\Gamma(1+2s)}{\cos(\pi s/2)} \left(\dfrac{\omega_{\text{ph}}}{\omega_d}\right)^{1-s} \omega_d , & \text{for} \quad s < 1 , \\[3ex] \dfrac{\eta_1\lambda^2}{\pi\hbar} 8\ln(\kappa) \, \omega_d , & \text{for} \quad s = 1 , \\[3ex] \dfrac{\eta_s\lambda^2}{\pi\hbar} 4\Gamma(s-1) \left(\dfrac{\kappa^2\omega_d}{\omega_{\text{ph}}}\right)^{s-1} \omega_d , & \text{for} \quad s > 1 . \end{cases} \qquad (9.47)$$

These expressions practically hold in the temperature regime $T \ll T_{\rm d}$ for $s < 1$ and $T \ll \kappa^2 T_{\rm d}$ for $s \geqslant 1$. Thus, this model with localized bath modes yields a finite decoherence rate in the diffusive regime at zero temperature.

The analysis can easily be generalized to n space dimensions. Assume that the correlation function in $\mathbf{k}$-space behaves as $\tilde{\mathcal{U}}(|\mathbf{k}| \to 0) \propto |\mathbf{k}|^{m-n}$. For spatial short-range interaction of Gaussian form, we have $m = n$. For spatial long-range power-law form, m is less than n. In the classical limit, $\coth(\beta\hbar\omega/2) \to 2/\beta\hbar\omega$, the ω-integral resulting from the k-integration in Eq. (9.44) is IR-convergent for $m > 2 - 2s$ [303]. If this condition is not met, the loophole is to place a time-dependent IR cutoff, $\omega_{\rm ir} = 1/t$, as explained on page 232.

An expression for γ_φ similar to Eq. (9.44) is discussed in Refs. [171, 300]. Saturation of τ_φ at $T = 0$ is also found in models which are free of form factors in the system-reservoir interaction [304]. Loss and Mullen presented a model for dephasing in which quantum fluctuations of the particle's path are coupled with a spatially localized environment [211], and they compared their findings with those of Ref. [210].

In the above semiclassical study of the model with localized bath modes, decoherence persists down to $T = 0$ when $J(\omega)$ is nonzero at frequencies below $\omega_{\rm b}$ for ballistic motion, and below $\omega_{\rm d}$ for diffusive motion. Decoherence endures because the reservoir can randomly absorb arbitrarily small amounts of energy down to $T = 0$ when the spectral density is gapless. Since $\omega_{\rm ph} \gg \kappa^2 \omega_{\rm d} \gg \omega_{\rm d}$, the dephasing rate at zero temperature is usually very small, and it becomes even smaller as the parameter s is raised from the sub-Ohmic sector $s < 1$ to the super-Ohmic regime $s > 1$. Decoherence decreases at low T with increasing s because of the much weaker super-Ohmic coupling at low frequencies compared to the Ohmic and sub-Ohmic case.

In Aharonov-Bohm ring devices with a quantum dot (QD) built in one arm of the ring, the phase shift of the electron passing through the quantum dot has been analyzed [305]. The experiments demonstrated that the electron can propagate coherently through a quantum dot. Such device offers the possibility to control dephasing rates by modifying the electromagnetic environment of the QD. An additional wire with a quantum point contact located close to the QD may act as sensitive measurement device to detect when the electron passes through the QD. This opens the possibility to build a "Which Path?" interferometer [306]. The influence of the measurement apparatus on dephasing is twofold. First, creation of real electron-hole pairs in the wire measures which path the electron took around the ring. Secondly, Ohmic damping resulting from the creation of virtual electron-hole pairs (cf. Subsection 4.2.8) leads to power-law suppression of the Aharonov-Bohm oscillations.

9.4 Decoherence of electrons

The conductivity of two-dimensional electron systems, e.g., thin disordered films or layers, exhibit reduction of the classical Drude conductivity. This phenomenon, generally called *weak localization*, is a significant quantum interference effect. The negative weak localization contribution to the conductivity can be proven to come mostly from quantum interference between self-crossing paths in which an electron propagates clockwise and counter-clockwise around a loop. Due to the coincident length of the two time-reversed paths, the quantum phases cancel each other exactly, and they vary sensitively when time-reversal invariance is broken by application of a magnetic field. Because in diagrammatic representation the selfcrossing path pairs are closely related to ladder diagrams in superconductivity, they are dubbed *Cooperon*.

Weak localization exists in one, two, and three dimensions. As it is much more likely to find self-crossing trajectories in low dimensions, the weak localization effect shows itself much stronger in low-dimensional systems (films and wires).

The study of decoherence in weak-localization systems is most easily described in a path integral approach with an influence functional accounting for the noisy quantum environment induced by the surrounding electrons. The influence functional method was originally devised for single-particle decoherence. The difficulty now is that the constituents of the reservoir and the dephasing electron are indistinguishable fermions and hence Fermi statistics has to be properly taken into account. For electrons in metals the dominant inherent interaction at low temperature is electron-electron scattering. As temperature is decreased, inelastic scattering processes get more and more rare, and the scattering time τ_{in} diverges as $1/T^2$. Because of this feature, it is largely thought that the electron decoherence time τ_φ should diverge as T approaches zero. About twenty years ago, Golubev and Zaikin (GZ) developed an influence functional approach for interacting fermions in a disordered conductor and studied the magnetoconductance in the weak-localization regime. They showed that, when focussing on single-particle response, the interaction with all the other fermions can be viewed as an effective environment. The Fermi statistics reveals itself in a complicated form of the resulting effective action which depends on the Fermi distribution function. With this they calculated the dephasing rate, and found it to be finite at zero temperature, $\gamma_\varphi^{\mathrm{GZ}}(T \to 0) = \gamma_{\varphi,0}^{\mathrm{GZ}}$ [294, 295]. However, this result is in contradiction to the widely accepted findings of Altshuler, Aronov, and Khmelnitskii (AAK) that $\gamma_\varphi^{\mathrm{AAK}}(T \to 0) \to 0$ [301]. There arose a fulminant controversy with impetuous critique from Altshuler and coworkers [293] and equally resolute rejection by GZ. An informative overview from GZ's point of view with most relevant references is given in Ref. [296]. Several years later, von Delft rederived the controversial influence functional of GZ for interacting electrons in disordered metals and confirmed that it is exact and takes proper account of the Pauli principle [297] (see also Ref. [307]). Since then the dispute is relocated to the question how one correctly evaluates the resulting path integral. Approximations utilized by GZ have been criticized [293, 297], in particular those concerning terms connected with Pauli blocking. While GZ elaborated the crucial terms in the time regime, von Delft analyzed them in the frequency regime. He then argued that this would allow for a somewhat more accurate treatment of recoil effects. By this, the GZ approach would readily reproduce the results of AAK for the dephasing rate. Heuristic arguments to reach the same conclusions as Ref. [297] were given in Ref. [298]. Subsequently, GZ objected that uncontrolled manipulations with diagrams were performed in Ref. [297], and that the resulting influence functional violates fundamental principles of quantum-statistical mechanics [299]. Let us now briefly embark on the subtle problem of decoherence in weak localization.

The density matrix of a quantum particle evolves according to the expression (5.11). For a single electron of charge $-e$, which moves in an impurity potential $V(\mathbf{r})$ and is subject to an external voltage $U_{\text{ext}}(\mathbf{r}, t)$, the propagating function is

$$J(\mathbf{r}_{1,\text{f}}, \mathbf{r}_{2,\text{f}}, t_f; \mathbf{r}_{1,\text{i}}, \mathbf{r}_{2,\text{i}}, t_i) = \int_{\mathbf{r}_1(t_i)=\mathbf{r}_{1,\text{i}}}^{\mathbf{r}_1(t_f)=\mathbf{r}_{1,\text{f}}} \mathcal{D}\mathbf{r}_1(\cdot) \int_{\mathbf{r}_2(t_i)=\mathbf{r}_{2,\text{i}}}^{\mathbf{r}_2(t_f)=\mathbf{r}_{2,\text{f}}} \mathcal{D}\mathbf{r}_2(\cdot) \tag{9.48}$$

$$\times \exp\left\{\frac{i}{\hbar}\left[S_0[\mathbf{r}_1(\cdot)] - S_0[\mathbf{r}_2(\cdot)] + \int_{t_i}^{t_f} dt' \left[eU_{\text{ext}}(\mathbf{r}_1(t')) - eU_{\text{ext}}(\mathbf{r}_2(t'))\right]\right]\right\},$$

where $S_0[\mathbf{r}] = \int_{t_i}^{t_f} dt' \left[\frac{1}{2}m\dot{\mathbf{r}}^2(t') - V(\mathbf{r}(t'))\right]$. The current density found from the density matrix is $\mathbf{j}(\mathbf{r}, t) = i(e/m)[\nabla_{\mathbf{r}_1}\rho(\mathbf{r}_1, \mathbf{r}_2, t) - \nabla_{\mathbf{r}_2}\rho(\mathbf{r}_1, \mathbf{r}_2, t)]|_{\mathbf{r}_1=\mathbf{r}_2=\mathbf{r}}$. Upon assuming a constant electrical field, $U_{\text{ext}} = -\mathbf{E} \cdot \mathbf{r}$, the linear conductivity emerges as

$$\sigma = \frac{e^2}{3m} \int_{-\infty}^{t} dt' \int d\mathbf{r}_1' \, d\mathbf{r}_2' \, [\nabla_{\mathbf{r}_{1,\text{f}}} - \nabla_{\mathbf{r}_{2,\text{f}}}] J_0(\mathbf{r}_{1,\text{f}}, \mathbf{r}_{2,\text{f}}, t; \mathbf{r}_1', \mathbf{r}_2', t')$$

$$\times [\mathbf{r}_1' - \mathbf{r}_2'] \rho_\beta(\mathbf{r}_1', \mathbf{r}_2') \Big|_{\mathbf{r}_{2,\text{f}}=\mathbf{r}_{1,\text{f}}}, \tag{9.49}$$

where J_0 is the propagator for $U_{\text{ext}} = 0$, and ρ_β is the thermal density matrix.

Drude conductivity

Consider now first classical charge transport in three space dimensions. When the impurity average $\langle \cdots \rangle_{\text{imp}}$ is performed for disordered systems with random potential $V(\mathbf{r})$, the contributions of different paths $\mathbf{r}_1(t') \neq \mathbf{r}_2(t')$ in the double path sum (9.48) cancel each other out. Instead, $\langle J_0 \rangle_{\text{imp}}$ is determined by the sum of probabilities resulting from pairs of identical classical paths $\mathbf{r}_1^{\text{cl}}(t') = \mathbf{r}_2^{\text{cl}}(t')$. In the quasiclassical approximation for the propagating function (see Subsec. 5.2.4), the expression (9.49) is equivalent to the standard linear response Kubo formula

$$\sigma = \frac{2e^2 N_0}{3} \int_0^\infty d\tau \, \langle \mathbf{v}(\tau)\mathbf{v}(0)\rangle_{\text{imp}} . \tag{9.50}$$

Here $N_0 = p_F^2/(2\pi^2\hbar^3 v_F)$ is the number of electrons per volume and per energy interval (and per spin polarization) at the Fermi surface. The conductivity σ has dimension frequency. In classical Boltzmann transport theory there is $\langle \mathbf{v}(\tau)\mathbf{v}(0)\rangle_{\text{imp}} = \mathbf{v}_F^2 e^{-\tau/\tau_c}$, where $\tau_c = \ell/|\mathbf{v}_F|$ is the mean collision time. With this, the expression (9.50) yields the Drude conductivity

$$\sigma = 2e^2 N_0 D , \tag{9.51}$$

where $D = \frac{1}{3}v_F^2\tau_c = \frac{1}{3}v_F\ell$ is the diffusion constant.

Weak localization

The leading quantum correction to the classical Drude conductivity, termed *weak localization*, arises from quantum interference between self-crossing paths, called Cooperon. Since the Cooperon paths traverse the same impurities, these interference terms are robust against disorder averaging [301, 288]. The Cooperon yield the negative conductivity contribution

$$\delta\sigma_{\text{WL}} = -\frac{2e^2 D}{\pi\hbar} \int_{\tau_c}^{\infty} dt \, P_{\text{diff},n}(0, t) \, e^{-f_n(t)} .$$ (9.52)

Here $P_{\text{diff},n}(0, t) = (4\pi Dt)^{-n/2}$ is the classical probability for a diffusive particle in n space dimensions to be found after time t at the initial point. The exponential factor $e^{-f_n(t)}$ is due to electron interactions and is expressed in terms of the noise and friction action [see Eq. (5.32)] as $e^{-f_n(t)} = \langle e^{-[\mathcal{S}^{(N)}+i\,\mathcal{S}^{(F)}]/\hbar} \rangle_{\text{imp}} = e^{-\langle[\mathcal{S}^{(N)}+i\,\mathcal{S}^{(F)}]/\hbar\rangle_{\text{imp}}}$. The effective action $\mathcal{S}^{(N)}+i\,\mathcal{S}^{(F)}$, which properly takes into account the Pauli principle, has been calculated by GZ [294] and has been rederived by von Delft [297]. It was argued by GZ in Ref. [294] that the friction action cancels out for a Cooperon when the impurity average is performed, $\langle\mathcal{S}^{(F)}(t)\rangle_{\text{imp}} = 0$, so that $f_n(t) = \langle\mathcal{S}^{(N)}(t)/\hbar\rangle_{\text{imp}}$.

The noise action $\mathcal{S}^{(N)}(t)$ can be analyzed in a simplified approach in which the electron interaction is emulated with a colored noise source[296]. The voltage noise $\delta U(t)$ obeys Gaussian statistics with correlator

$$\langle\delta U(\mathbf{r}, t)\delta U(0, 0)\rangle/\hbar \equiv I(\mathbf{r}, t) = \int \frac{d\omega\, d^3\mathbf{k}}{(2\pi)^4} \frac{\tilde{\mathcal{X}}_{\text{ee}}(\omega)}{|\mathbf{k}|^2} e^{i\mathbf{k}\cdot\mathbf{r}-i\omega t} .$$ (9.53)

Here $\tilde{\mathcal{X}}_{\text{ee}}(\omega)$ is essentially the power spectrum of the fluctuating voltage, analogous to Eq. (2.14),

$$\tilde{\mathcal{X}}_{\text{ee}}(\omega) = \text{Im}\left(\frac{-4\pi}{\epsilon_{\text{ee}}(\mathbf{k}, \omega)}\right) \coth(\tfrac{1}{2}\beta\hbar\omega) ,$$ (9.54)

where $\epsilon_{\text{ee}}(\mathbf{k}, \omega)$ is the electronic contribution to the dielectric susceptibility. With the noise correlator (9.53), the noise action reads

$$\mathcal{S}^{(N)}(t) = \frac{e^2}{2} \int_0^t dt' \int_0^t dt'' \left\{ I[\mathbf{r}_1(t') - \mathbf{r}_1(t''), t' - t''] + I[\mathbf{r}_2(t') - \mathbf{r}_2(t''), t' - t''] \right.$$

$$\left. - I[\mathbf{r}_1(t') - \mathbf{r}_2(t''), t' - t''] - I[\mathbf{r}_2(t') - \mathbf{r}_1(t''), t' - t''] \right\} .$$ (9.55)

Performing next the impurity average, the function $f_n(t)$ in Eq. (9.52) takes the form

$$f_n(t) = \frac{e^2}{\hbar} \int_0^t dt' \int_0^t dt'' \int d^n\mathbf{r}\, P_{\text{diff},n}(\mathbf{r}, |t' - t''|)[I(\mathbf{r}, t' - t'') - I(\mathbf{r}, t' + t'' - t)] .$$ (9.56)

The function $f_n(t)$ increases with time and controls the Cooperon decay via the expression (9.52). The analysis reveals that the first term in the square bracket is the leading one as $t \to \infty$. On condition $f_n(\tau_\varphi) \approx 1$, one readily obtains

$$f_{\mathrm{n}}(t) \equiv \frac{t}{\bar{\tau}_\varphi(\tau_{\mathrm{ir}})} = t \frac{2e^2}{\hbar} \int_{1/\tau_{\mathrm{ir}}}^{1/\tau_{\mathrm{c}}} \frac{d\omega}{2\pi} \, \tilde{\chi}_{\mathrm{ee}}(\omega) \frac{\Omega_n \, a^{n-3}}{(2\pi)^n} \int_0^\infty dk \, k^{n-1} \frac{P_{\mathrm{diff},n}(k,\omega)}{k^2} . \quad (9.57)$$

To avoid possible divergences of the ω-integral, there are placed IR- and UV-cutoffs in Eq. (9.57). The high frequency cutoff $1/\tau_{\mathrm{c}} = v_{\mathrm{F}}/\ell$ is put for the reason stated after Eq. (9.45). Suitable choices for τ_{ir}, if required, are $\tau_{\mathrm{ir}} = \tau_\varphi$ [294] and $\tau_{\mathrm{ir}} = t$ [297] (see below). In the first case, τ_φ is fixed selfconsistently. Furthermore, for $n = 1$, the quantity a^2 is the cross section of the wire, and for $n = 2$ the quantity a is the thickness of the film.

The former expression (9.44) is formally similar to the expression (9.57), if we identify $\tilde{\mathcal{U}}(\lambda k)$ with the Coulomb interaction $1/k^2$ in $\mathbf{k}$-space.

For a diffusive conductor with conductivity σ, the electronic contribution to the dielectric function is $\epsilon_{\mathrm{ee}}(\mathbf{k},\omega) = 1 + 4\pi\sigma/(-i\omega + D|\mathbf{k}|^2)$. As the term unity is small for typical metals, the power spectrum (9.54) becomes [295]

$$\tilde{\chi}_{\mathrm{GZ}}(\omega) = \frac{\omega}{\sigma} \coth(\tfrac{1}{2}\beta\hbar\omega) = \frac{\omega}{\sigma} \left(1 + 2\frac{e^{-\beta\hbar\omega}}{1 - e^{-\beta\hbar\omega}}\right) . \quad (9.58)$$

The first term in the round bracket leads to a finite decoherence time at zero temperature. It has been argued that this term is erroneous and originates from an erratic execution of the semiclassical averages [293]. Recently it was stated that the standard AAK result of divergent decoherence time at $T = 0$ can be reproduced using GZ's method by accounting for recoil effects when the electron with energy $\hbar\epsilon$ emits or absorbs a photon with energy $\hbar\epsilon$[293, 297, 298]. This would result in the substitution $\coth(\tfrac{1}{2}\beta\hbar\omega) \rightarrow \coth(\tfrac{1}{2}\beta\hbar\omega) + \tanh[\tfrac{1}{2}\beta\hbar(\epsilon - \omega)]$, where $\hbar\omega$ is the energy exchanged at an interaction vertex. The recoil term ensures that processes with $\omega > \epsilon$, $\omega \gg k_{\mathrm{B}}T/\hbar$ are suppressed, and thus the frequency integral in Eq. (9.57) would be UV-convergent. This would entail absence of decoherence at $T = 0$. In the influence functional approach, the recoil terms are supposed to arise from the friction action [298, 297].

Taking the thermal average of the electron's energy, Aleiner, Altshuler and Gershenson (AAG) found the modified power spectrum [293, 296]

$$\tilde{\chi}_{\mathrm{AAG}}(\omega) = \frac{\omega}{\sigma} \frac{\beta\hbar}{4} \int_{-\infty}^{\infty} d\epsilon \, \frac{\coth(\tfrac{1}{2}\beta\hbar\omega) + \tanh[\tfrac{1}{2}\beta\hbar(\epsilon - \omega)]}{\cosh^2(\tfrac{1}{2}\beta\hbar\epsilon)} = \frac{\omega}{\sigma} \frac{\tfrac{1}{2}\beta\hbar\omega}{\sinh^2(\tfrac{1}{2}\beta\hbar\omega)} . \quad (9.59)$$

To sum up, the AAG power spectrum drops to zero $\propto e^{-\beta\hbar\omega}$ as $\beta\hbar\omega \rightarrow \infty$, whereas the GZ power spectrum at zero temperature is

$$\tilde{\chi}_{\mathrm{GZ}}(\omega) = \omega/\sigma . \quad (9.60)$$

For this power spectrum, the ω-integral in Eq. (9.57) is IR-convergent. The inverse dephasing time or dephasing rate for $n = 1$, 2, and 3 space dimensions takes the form

$$\frac{1}{\tau_{\varphi,1}} = \frac{1}{\pi} \frac{e^2}{\hbar\sigma_1\tau_{\mathrm{c}}} \sqrt{2D\tau_{\mathrm{c}}} \,, \qquad\qquad n = 1 \,,$$

$$\frac{1}{\tau_{\varphi,2}} = \frac{1}{4\pi} \frac{e^2}{\hbar\sigma_2\tau_{\mathrm{c}}} \,, \qquad\qquad n = 2 \,, \qquad (9.61)$$

$$\frac{1}{\tau_{\varphi,3}} = \frac{1}{3\pi^2} \frac{e^2}{\hbar\sigma\tau_{\mathrm{c}}} \frac{1}{\sqrt{2D\tau_{\mathrm{c}}}} \,, \qquad\qquad n = 3 \,.$$

Here $\sigma_1 = a^2\sigma$ is the conductivity of a wire with cross section a^2, and $\sigma_2 = a\,\sigma$ is the conductivity of a film with thickness a. The first form is the disputed GZ dephasing rate leading to decoherence at $T = 0$. In conclusion, the power spectrum (9.60) for $T = 0$ leads to dephasing at zero temperature for any space dimension.

Interestingly enough, the power spectra $\tilde{\chi}_{\mathrm{GZ}}(\omega)$ and $\tilde{\chi}_{\mathrm{AAG}}(\omega)$ have the same high temperature form

$$\tilde{\chi}(\omega) = 2/\beta\hbar\sigma \,, \qquad\qquad (9.62)$$

while subleading terms differ. With this, an IR cutoff $1/\tau_{\mathrm{ir}}$ in Eq. (9.57) is required in the case of one space dimension, whereas this is not required for $n = 2$ and 3. Putting in the 1D-case $\tau_{\mathrm{ir}} = \tau_{\varphi,1}$, and then determine $\tau_{\varphi,1}$ from Eq (9.57) [301], one finds

$$\frac{1}{\tau_{\varphi,1}} = \left(\frac{2}{\pi} \frac{e^2}{\hbar\sigma_1} \sqrt{2D} \frac{k_{\mathrm{B}}T}{\hbar}\right)^{2/3} \,, \qquad\qquad n = 1 \,,$$

$$\frac{1}{\tau_{\varphi,2}} = \frac{1}{2\pi} \frac{e^2}{\hbar\sigma_2} \ln(k_{\mathrm{B}}T\tau_{\mathrm{c}}/\hbar) \frac{k_{\mathrm{B}}T}{\hbar} \,, \qquad\qquad n = 2 \,, \qquad (9.63)$$

$$\frac{1}{\tau_{\varphi,3}} = \frac{2}{\pi^2} \frac{e^2}{\hbar\sigma} \frac{1}{\sqrt{2D}} \left(\frac{k_{\mathrm{B}}T}{\hbar}\right)^{3/2} \,, \qquad\qquad n = 3 \,.$$

The expression (9.52) with (9.57) and with the rates (9.63) describes Cooperon decay $e^{-t/\tau_{\varphi,n}}$. However, the Cooperon in one space dimension is known to decay as

$$e^{-f_1(t)} = e^{-(t/\tau_{\varphi,1})^{3/2}} \,, \qquad\qquad (9.64)$$

i.e., with power $\frac{3}{2}$ in the exponent [301]. This particular time dependence of the Cooperon decay in 1D is exactly found from Eq. (9.57) if we put $\tau_{\mathrm{ir}} = t$. The resulting expression for $\tau_{\varphi,1}$ in Eq. (9.64) coincides with the expression given in Eq. (9.63).

PART III:

QUANTUM STATISTICAL DECAY

In the second part, I considered the exactly solvable case of linear dissipative quantum systems, the thermodynamic variational approach useful for nonlinear quantum systems, and issues of quantum decoherence. Now I turn to a study of the frequent situation in which a metastable state is separated from the outside region by a free energy barrier. The decay of a metastable state plays a central role in many scientific areas including low-temperature physics, nuclear physics, chemical kinetics and transport in biomolecules. At high temperatures, the system escapes from the metastable well predominantly over the barrier by thermal activation. At zero temperature on the other hand, the system is in the localized ground state of the metastable well and can escape only by quantum mechanical tunneling through the barrier. I shall focus the discussion on the influence of friction on the decay process. The treatment will be based on an effective method which allows to study this problem in a unified manner for temperatures ranging from $T = 0$ up to the classical regime.

10 Introduction

There are many processes in physics, chemistry, and biology in which a system makes transitions between different states by traversing a barrier. The theory of rate coefficients for barrier crossing has a long history extending back to the work by Arrhenius in 1889 [308]. H. A. Kramers' article of 1940 [266] represents a cornerstone in the quantitative analysis of thermally activated rate processes. This work provided a thorough theoretical description in the classical regime both for very *weak* and for *moderate-to-strong damping*. It includes important limiting cases such as the transition state theory or the Smoluchowski model of diffusion controlled processes.

The investigation of quantum mechanics for macroscopic variables has been stimulated considerably by Leggett's discussion concerning the validity of quantum mechanics at the macroscopic level [309, 310]. Strong impulse to the field was given further by the work of Caldeira and Leggett [78], who studied *quantum tunneling in the presence of dissipation* at zero temperature. In subsequent years, quantum reaction theory has become a very active and challenging area in chemical physics and theoretical chemistry, both from the analytical and numerical point of view. Parallel to these developments, the functional integral method has provided a unified description of the thermal quantum decay in the entire temperature range [311]. A comprehensive review of many of the theoretical concepts and ideas in reaction rate theory extending from classical rate theory to more recent quantum versions has been given by Hänggi, Talkner and Borkovec [312]. I also refer to a recent account of the semiclassical approach to quantum tunneling in complex systems by Ankerhold [313].

Assume that the metastable system in question is characterized by a generalized coordinate q. It may be visualized as a particle of mass M moving in an external potential $V(q)$ which has a single metastable minimum, chosen to lie at the origin of both the coordinate and potential axes, $V(0) = 0$. I assume that the potential $V(q)$ is fairly smooth and has the general form depicted in Fig. 10.1. In particular, $V(q)$ is taken to be negative in the region $q > q_{ex}$, where q_{ex} is the nonzero value of q for which $V(q) = 0$. The point q_{ex} is called the "exit point" from the barrier region. Once the particle has left the metastable well, it will not return in finite time. With this assumption, we can disregard quantum coherence effects. In physical systems, the coordinate q is the tunneling degree of freedom, and the thermal initial state is characterized by the partition function Z_0 of the well region. In chemical reactions, one thinks of q as the reaction coordinate and Z_0 denotes the partition function for reactants. Among the chemical applications are dissociation and recombination reactions and transfer processes of atoms and electrons.

The concept of metastability is useful when the barrier is large enough that the decay time of the metastable state is very long compared with all other characteristic

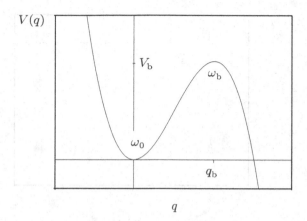

Figure 10.1: A metastable "quadratic-plus-cubic" potential well.

time scales of the system dynamics. There exist many time scales, e.g., the correlation time τ_c of the noise, the time of relaxation in the locally stable well τ_r, the thermal time $\hbar\beta$, and the time scales $\tau_0 = \omega_0^{-1}$ and $\tau_b = \omega_b^{-1}$ which are related to the curvature of the potential at the metastable minimum and at the barrier top. Here, ω_0 is the angular frequency of small oscillations around the metastable minimum

$$\omega_0 = [V''(0)/M]^{1/2} , \tag{10.1}$$

and the so-called barrier frequency

$$\omega_b = [-V''(q_b)/M]^{1/2} \tag{10.2}$$

characterizes the width of the parabolic top of the barrier hindering the decay process. It represents the angular frequency of small oscillations around the minimum of the upside-down potential $-V(q)$ (cf. Fig. 10.2). All these various time scales will become important in a precise description of the escape rate. Weak metastability implies that the activation energy V_b is by far the largest energy of the problem, in particular

$$V_b \gg k_B T , \qquad \text{and} \qquad V_b \gg \hbar\omega_0 . \tag{10.3}$$

In the sequel, I shall not put restrictions on the numerical value of the ratio $\hbar\omega_0/k_B T$. It is convenient to parameterize the rate of escape from the metastable well in the form

$$k = A\,e^{-B} . \tag{10.4}$$

The quantity B is a dimensionless measure of the immenseness of the barrier the particle has to overcome. The prefactor A is a kind of an attempt frequency of the particle in the well towards the barrier. We shall be concerned with the question on which scale the parameters A and B are influenced by dissipation. The findings are briefly summarized as follows. In the classical regime, the exponent B is independent

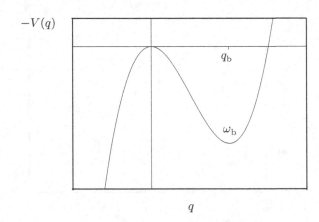

Figure 10.2: The upside-down potential $-V(q)$.

of damping, whereas the prefactor A is reduced by environmental coupling. In the quantum regime, both the exponent and the prefactor of the rate significantly depend on the strength and on the spectral properties of the environmental coupling. As damping is increased, the classical regime extends to lower temperatures.

In the last decade, the problem of quantum tunneling of macroscopic variables has attracted a lot of interest. The research was considerably stimulated by experiments on macroscopic quantum tunneling in Josephson systems (cf., e.g., Ref. [314], and references therein), and by theoretical work put forward by Caldeira and Leggett [78]. As often in science, the problem of tunneling in the presence of coupling to (infinitely) many degrees of freedom, e.g., to phonons, quasiparticles, and electrons, has several precursors. Among them are the work on deep-inelastic collisions of heavy ions by Brink *et al.* [315], and by Möhring and Smilansky [316]. On the other hand, quantum tunneling in the presence of phonon modes has a long history in solid state physics. The study of multi-phonon effects in the polaron problem was initiated by Holstein [131] and by Emin and Holstein [317]. Further development includes the study of polaron effects on quantum tunneling which received enormous impact through the work by Flynn and Stoneham [134]. A critical review of the early work which relied mainly on the *Condon approximation* has been given by Sethna [173].

Before embarking on a closer look at multidimensional thermal quantum decay, the classical regime is roamed in Chapter 11. The following Chapter 12 covers the theoretical methods in the various regimes of thermal quantum decay.

11 Classical rate theory: a brief overview

11.1 Classical transition state theory

Let us begin the discussion with the simplified description of thermally activated decay by transition state theory (TST). In its simplest form, which is the case of a single degree of freedom in the absence of a reservoir, one makes two *ad hoc* assumptions:

(i) Thermal equilibrium prevails in the well, e.g., through the action of Maxwell's demon, so that the metastable state is represented by a canonical distribution.

(ii) There is no return once the barrier top has been crossed even infinitesimally.

Upon identifying the rate with the probability current across the barrier, and taking the Boltzmann weight of the normalized current density at the barrier top, we obtain

$$k_{cl}^{(TST)} = \frac{\frac{1}{2\pi\hbar} \int dp\, dq\, e^{-\beta[p^2/2M + V(q)]} \delta(q - q_b)\, \Theta(p)\, p/M}{\frac{1}{2\pi\hbar} \int_{q < q_b} dp\, dq\, e^{-\beta[p^2/2M + V(q)]}}. \tag{11.1}$$

Here I have chosen the barrier to the right of the well with the maximum at $q = q_b$, $V_b = V(q_b)$. The expression (11.1) yields the well-known transition state formula

$$k_{cl}^{(TST)} = f_{cl}^{(TST)}\, e^{-V_b/k_B T} \quad \text{with} \quad f_{cl}^{(TST)} = \frac{1}{2\pi\hbar}\frac{k_B T}{Z_{0,cl}} = \frac{\omega_0}{2\pi}, \tag{11.2}$$

where $Z_{0,cl} = k_B T/\hbar\omega_0$ is the classical partition function of the well region. The Arrhenius law for the escape rate $k_{cl} \propto e^{-V_b/k_B T}$ reflects the exponentially small tail of the canonical initial state at the threshold energy $E = V_b$. Note that the classical transition state formula (11.2) does not depend on the width of the barrier. The attempt frequency $f_{cl}^{(TST)} = \omega_0/2\pi$ is just the frequency of small oscillations in the well. The classical rate vanishes as temperature is lowered to absolute zero.

11.2 Moderate-to-strong-damping regime

In a pioneering paper published in 1940, Kramers reported a groundbreaking study of the classical escape of a particle from a metastable well in various limits [266]. In his treatment of the reaction rate, the starting point is nonlinear Brownian motion in phase space, which is analogous to the Markovian Langevin equation in coordinate space, Eq. (2.3) with (2.1) and (2.2). Following Kramers, the stochastic dynamics for the reaction coordinate q and the velocity $v = \dot{q}$ (or momentum) is conveniently described in terms of a probability density $p(q, v; t)$. The dynamical equation found for $p(q, v; t)$ is of the Fokker-Planck type, nowadays termed Klein-Kramers equation,

$$\frac{\partial p(q, v; t)}{\partial t} = \left(-\frac{\partial}{\partial q} v + \frac{\partial}{\partial v}\left(\frac{1}{M}\frac{\partial V(q)}{\partial q} + \gamma v\right) + \gamma \frac{k_\mathrm{B}T}{M}\frac{\partial^2}{\partial v^2}\right) p(q, v; t) . \quad (11.3)$$

A comprehensive review of the Fokker-Planck equation and of the common techniques for tracing solutions is given in the book by Risken [318]. In Sec. 6.8, I have sketched the derivation of the Kramers equation (11.3) for the damped linear oscillator. By taking in the generalized Fokker-Planck equation (6.187) the classical limit, $\langle q^2 \rangle = 1/M\beta\omega_0^2$, $\langle p^2 \rangle = M/\beta$, $\gamma_q = \omega_0^2$ and $\gamma_p = \gamma$, one arrives at the Klein-Kramers equation (11.3) for the potential $V(q) = \frac{1}{2}M\omega_0^2 q^2$. Equation (11.3) gives a complete description of the classical stochastic process described by the Langevin equation (2.1)–(2.3).

The Klein-Kramers equation (11.3) has two different time-independent solutions. One solution satisfying $\partial p(q, v; t)/\partial t = 0$ is the canonical equilibrium state

$$p_\mathrm{eq}(q, v) = N \exp\left\{-\left[\tfrac{1}{2}Mv^2 + V(q)\right]/k_\mathrm{B}T\right\} , \quad (11.4)$$

which describes the thermal distribution of q and v in the well (N is a normalization constant). The population of the well is found in Gaussian approximation as

$$n \equiv \int_{-\infty}^{q_\mathrm{b}} dq \int_{-\infty}^{\infty} dv\, p_\mathrm{eq}(q, v) = \frac{2\pi k_\mathrm{B}T}{M\omega_0} N . \quad (11.5)$$

The other stationary solution of Eq. (11.3) represents a steady state with a probability flow across the barrier. The steady state describes the following situation. Particles are continuously injected at the bottom of the well. Afterwards they stay sufficiently long in the well region so that they thermalize. Finally, those particles which have gained the barrier energy leave the barrier region. Outside the barrier region, they are absorbed by a particle sink. To obtain the steady-flow state for a potential of the form sketched in Fig. 10.1, I follow Kramers [266] and make the ansatz

$$p_\mathrm{flow}(q, v) = G[u(q, v)]\, p_\mathrm{eq}(q, v) , \quad (11.6)$$

where $G[u(q = 0, v)]$ captures the flow property. I impose the boundary conditions

$$G[u(q = 0, v)] = 1 , \quad \text{and} \quad G[u(q = q_+, v)] \approx 0 . \quad (11.7)$$

Here, q_+ is a point far beyond the barrier region ($q_+ \gg q_\mathrm{b}$). The first condition determines that the steady state (11.6) matches on the equilibrium state (11.4) in the well region. The second condition arranges that the particles are removed on the other side of the barrier. The constraints (11.7) imply that the function $G[u(q, v)]$ depends on q and v only through a linear combination of q and v. With the choice

$$u(q, v) = q - q_\mathrm{b} - \varrho v/\omega_\mathrm{b} \quad (11.8)$$

in the ansatz (11.6), the Klein-Kramers equation (11.3) is transformed into an ordinary differential equation for $G(u)$, if we choose the parameter ϱ as a solution of the quadratic equation

$$\varrho^2 + (\gamma/\omega_b)\varrho - 1 = 0 . \tag{11.9}$$

The resulting differential equation for $G(u)$ reads

$$\kappa u G'(u) + G''(u) = 0 \tag{11.10}$$

with

$$\kappa = M\omega_b^3/(\varrho\gamma k_B T) . \tag{11.11}$$

The solution of Eq. (11.10) with the boundary conditions (11.7) is

$$G(u) = \sqrt{\frac{\kappa}{2\pi}} \int_u^\infty du_1 \exp\left(-\kappa\frac{u_1^2}{2}\right) . \tag{11.12}$$

The quantity κ must be positive in order that $G(u)$ vanishes for q near to q_+. This entails that we must choose the positive root of Eq. (11.9) which is

$$\varrho = \sqrt{1 + \left(\frac{\gamma}{2\omega_b}\right)^2} - \frac{\gamma}{2\omega_b} . \tag{11.13}$$

Consider now the outgoing flow $\mathcal{F}$ of the steady state $p_{\text{flow}}(q, v)$ at the barrier top,

$$\mathcal{F}(q_b) \equiv \int_{-\infty}^\infty dv\, v p_{\text{flow}}(q_b, v) . \tag{11.14}$$

Insertion of the form (11.6) with (11.12) yields an integral expression which can be transformed upon integration by parts into a standard Gauss integral. We readily get

$$\mathcal{F}(q_b) = \varrho\frac{k_B T}{M} N \exp\left(-V_b/k_B T\right) . \tag{11.15}$$

The rate of escape over the barrier equals outgoing probability flow at the barrier top divided by the population of the well. Using the expressions (11.5) and (11.14), we get

$$k_{\text{cl}} \equiv \frac{\mathcal{F}(q_b)}{n} = f_{\text{cl}} \exp\left(-V_b/k_B T\right) , \tag{11.16}$$

where the classical attempt frequency f_{cl} is given by

$$f_{\text{cl}} = \varrho f_{\text{cl}}^{(\text{TST})} = \varrho\frac{\omega_0}{2\pi} . \tag{11.17}$$

The form (11.16) with (11.17) is Kramers' widely known classical escape rate holding for moderate-to-strong damping. Strictly speaking, the expression (11.17) applies when damping is strong enough that the particle relaxes to the thermal equilibrium state in the well before it escapes over the barrier. The attempt frequency (11.17) differs from the transition state value $f_{\text{cl}}^{(\text{TST})}$ by the transmission factor ϱ, which captures all effects of damping in the classical regime.

In the light of the derivation given here in the framework of the Klein-Kramers equation, the transmission factor is reduced, $\rho < 1$, because of multiple re-crossing runs at the barrier which the particle's noisy trajectory makes.

Alternative derivations of the Kramers rate formula (11.16) with (11.17) are given below. In Section 15.1, I employ the Im F method, and in Section 15.4 I consider multi-dimensional transition state theory of the system-plus-environment complex. This will yield a physical interpretation of the transmission factor $\rho < 1$ which is different from the conception of recrossing processes at the barrier top (cf. concluding remarks in Section 15.4).

11.3 Strong damping regime

For strong damping or large viscosity, the inertia term in Eq. (2.3) is small compared to the friction term. On this condition, an arbitrary initial distribution $p(q, v; t)$ will relax within short time for all q to a Maxwell velocity distribution. After this initial slip regime, slow diffusion of the density distribution in coordinate space will take place. The corresponding time evolution of the reduced probability density

$$p(q, t) \equiv \int_{-\infty}^{\infty} dv \, p(q, v; t) \tag{11.18}$$

is described in the classical regime by the Smoluchowski diffusion equation [312, 318]

$$\frac{\partial p(q, t)}{\partial t} = \frac{1}{M\gamma} \frac{\partial}{\partial q} \hat{L}_{\rm cl}(q) \, p(q, t) \tag{11.19}$$

with the Smoluchowski operator

$$\hat{L}_{\rm cl}(q) \equiv \frac{\partial V(q)}{\partial q} + k_{\rm B} T \frac{\partial}{\partial q} . \tag{11.20}$$

For position-dependent diffusion coefficient $D(q)$, the Smoluchowski operator reads

$$\hat{L}_{\rm cl}(q) \equiv \frac{\partial V(q)}{\partial q} + \frac{\partial}{\partial q} D(q) . \tag{11.21}$$

The particle current or flow connected with the distribution $p(q, t)$ is given by

$$\mathcal{F}(q, t) = -\frac{1}{M\gamma} \left(V'(q) + \frac{\partial}{\partial q} D(q) \right) p(q, t) . \tag{11.22}$$

The Langevin equation which corresponds to the diffusion equation (11.19) with the Smoluchowski operator (11.21) is found to read in Ito representation [318]

$$M\gamma \dot{q}(t) = -V'(q) + \sqrt{2M\gamma D(q)} \, \zeta(t) , \tag{11.23}$$

where $\zeta(t)$ is Gaussian white noise with zero mean and correlation $\langle \zeta(t)\zeta(0) \rangle = \delta(t)$.

One stationary solution of the Smoluchowski equation (11.19) with (11.21) is the equilibrium state in the potential well

$$p_{\rm eq}(q) = N \, e^{-\phi(q)} , \tag{11.24}$$

where N is a normalization constant, and the "effective potential"

$$\phi(q) = \ln D(q) + \psi(q) \qquad \text{with} \qquad \psi(q) = \int_0^q dq_1 \frac{V'(q_1)}{D(q_1)} . \tag{11.25}$$

The other stationary solution of the Smoluchowski equation is a steady-flow state with constant current or flow $\mathcal{F}$,

$$p_{\text{flow}}(q) = M\gamma\mathcal{F} \, e^{-\phi(q)} \int_q^{q_+} dq_1 \, e^{\psi(q_1)} . \tag{11.26}$$

The steady state $p_{\text{flow}}(q)$ is exponentially small at the other side of the potential barrier. In the steady-flow state, the population of the well is

$$n \equiv \int_{-\infty}^{q_b} dq \, p_{\text{flow}}(q) . \tag{11.27}$$

Next, we use that the integrations over the well and the barrier regions decouple and that they can be done in Gaussian approximation. Observing further that the escape rate k_{cl} is the current divided by the population of the well, we obtain the expression

$$k_{\text{cl}} = \frac{\mathcal{F}}{n} = \frac{D(0)}{2\pi M\gamma} \sqrt{\phi''(0)|\psi''(q_b)|} \, e^{\psi(0)-\psi(q_b)} , \tag{11.28}$$

where we have assumed $\phi'(0) = \psi'(q_b) = 0$.

In the case of state-independent diffusion $D = k_B T$, this reduces to the expression

$$k_{\text{cl}} = \frac{\omega_0 \omega_b}{2\pi\gamma} \exp\left(-V_b/k_B T\right) . \tag{11.29}$$

Thus we find that the escape rate in the Smoluchowski limit is again in the form (11.16) with (11.17), in which the transmission factor ϱ is inversely proportional to the friction coefficient,

$$\varrho = \omega_b/\gamma . \tag{11.30}$$

This form coincides with the strong-friction limit of Kramers' expression (11.13).

Finally I remark that we take up the expression (11.29) in Section 15.3. There we discuss quantum corrections of the escape rate in the Smoluchowski limit.

11.4 Weak-damping regime

For very weak damping, the treatment given in the previous two sections fails since damping is not strong enough to maintain thermal equilibrium in the well region. In the steady-flux state, the escape over the barrier is limited by the short supply of energy which is required to raise the particle to the barrier top. The escape over the barrier results in a loss of population compared with the canonical population in an energy band of width $k_B T$ just below the barrier top. For weak friction, the motion of the particle in the well is oscillatory, and the effect of damping is a gradual change of the distribution in energy space. The mean energy loss ΔE during one round trip in the well is

$$\Delta E = \gamma \mathcal{I}(E) , \tag{11.31}$$

where $\mathcal{I}(E)$ is the abbreviated (Minkowskian) action[1] for one round trip at total energy E in the well with turning points q_1 and q_2. There holds

$$\mathcal{I}(E) = \oint dq \, p(q) = M \oint dt \, \dot{q}^2(t) = 2 \int_{q_1}^{q_2} dq \, \sqrt{2M[E - V(q)]} . \tag{11.32}$$

For weak damping, the energy E is a slowly varying with time. The weak time-dependence of E can well be described by a probability density function $p(E,t)$ describing a diffusion process in energy space. The diffusion equation for the probability density $p(E,t)$ reads [312]

$$\dot{p}(E,t) = \gamma \frac{\partial}{\partial E} \mathcal{I}(E) \left(1 + k_B T \frac{\partial}{\partial E} \right) \frac{\omega(E)}{2\pi} p(E,t) , \tag{11.33}$$

where $\omega[\mathcal{I}(E)] \equiv \omega(E)$ is the angular frequency at abbreviated action $\mathcal{I}(E)$.

When the particle has built up energy as large as the threshold energy V_b, it escapes from the well with probability 1. The rate is therefore given by the particle flow in energy space through the energy point V_b, divided by the population n of the well. The current $\mathcal{F}$ associated with the steady-flow state $p_{\text{flow}}(E)$ of Eq. (11.33) is

$$\mathcal{F} = -\gamma \mathcal{I}(E) \left(1 + k_B T \frac{\partial}{\partial E} \right) \frac{\omega(E)}{2\pi} p_{\text{flow}}(E) . \tag{11.34}$$

With use of the relation

$$p(E) = (\partial \mathcal{I}/\partial E) \, \tilde{p}(\mathcal{I}) = [2\pi/\omega(E)] \, \tilde{p}(\mathcal{I}) , \tag{11.35}$$

the current $\mathcal{F}$ may be expressed in terms of the steady-flow state $\tilde{p}_{\text{flow}}(\mathcal{I})$,

$$\mathcal{F} = -\gamma \mathcal{I} \left(1 + \frac{2\pi k_B T}{\omega(E)} \frac{\partial}{\partial \mathcal{I}} \right) \tilde{p}_{\text{flow}}(\mathcal{I}) . \tag{11.36}$$

Removing the particle at $\mathcal{I} = \mathcal{I}(E = V_b) \equiv \mathcal{I}_b$ from the well region means putting $\tilde{p}_{\text{flow}}(\mathcal{I} = \mathcal{I}_b) = 0$. With this boundary condition, the solution of Eq. (11.36) reads

$$\tilde{p}_{\text{flow}}(\mathcal{I}) = \frac{\mathcal{F}}{2\pi\gamma k_B T} e^{-E(\mathcal{I})/k_B T} \int_{\mathcal{I}}^{\mathcal{I}_b} d\mathcal{I}_1 \frac{\omega(\mathcal{I}_1)}{\mathcal{I}_1} e^{E(\mathcal{I}_1)/k_B T} . \tag{11.37}$$

The escape rate is again flow over population, this time in abbreviated-action space,

$$k_{\text{cl}} = \mathcal{F} \Big/ \int_0^{\mathcal{I}_b} d\mathcal{I} \, \tilde{p}_{\text{flow}}(\mathcal{I}) . \tag{11.38}$$

Next we insert the expression (11.37) and calculate the integrals in Gaussian approximation. We then obtain the rate k_{cl} again in the form (11.16) with Eq. (11.17), in which the transmission factor ϱ is given by

$$\varrho = \gamma \mathcal{I}(V_b)/k_B T . \tag{11.39}$$

---

[1]The Minkowskian [Euclidean] abbreviated action is denoted by $\mathcal{I}(E)$ [$\mathbb{W}(E)$].

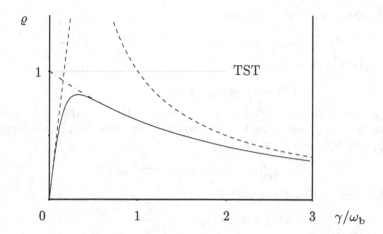

Figure 11.1: Turnover of the transmission factor ϱ from energy-diffusion-limited behavior [Eq. (11.39), straight dashed line] to spatial-diffusion-controlled behavior [Eq. (11.13), dashed-dotted curve]. The Smoluchowski limit $\varrho = \omega_b/\gamma$ is approached for large damping.

Here $\mathcal{I}(V_b)$ is the abbreviated action for one round trip in the well at energy $E = V_b$.

In summary, the transmission across the barrier is reduced compared to the transition state value $\varrho = 1$ by two physically different effects. For moderate-to-strong damping, ϱ is diminished because the trajectory of the particle is stochastic and thus may cross the position q_b of the barrier top several times before the particle eventually escapes. The corresponding expression is given in Eq. (11.13). In the opposite weak-damping limit, $\gamma < k_B T/\mathcal{I}(V_b)$, the stationary distribution at energies near to the barrier top, $p(E \approx V_b)$, is depleted compared with the canonical distribution. As a result, the transmission factor decreases linearly with γ, as in Eq. (11.39).

As already noted by Kramers, the limiting behaviors (11.13) and (11.39) imply a maximum of the rate at a damping value intermediate in strength between the above two limits. The actual transmission factor ϱ undergoes a *turnover* near $\gamma = k_B T/\mathcal{I}(V_b)$ in the form of a bell-shaped curve as sketched in Fig. 11.1. The turnover theory bridges between the spatial-diffusion-controlled and the energy-diffusion-limited formulas for the activated escape. It will be discussed in some detail in Section 14.2. There, light is shed also on the multi-dimensional aspects of the classical decay process.

The simplest analytic form of a metastable potential is the "quadratic-plus-cubic" potential sketched in Fig. 10.1 with

$$V(q) \;=\; \frac{1}{2}M\omega_0^2 q^2 \left(1 - \frac{2q}{3q_b}\right) \;=\; \frac{1}{2}M\omega_0^2 q^2 \left(1 - \frac{q}{q_{\mathrm{ex}}}\right). \tag{11.40}$$

Here, q_b is the position of the barrier top, and q_{ex} is the exit point from the barrier for $E = 0$. Hence $q_{\mathrm{ex}} = 3q_b/2$ is the tunneling distance from the bottom of the well.

243

Barrier height and barrier frequency are

$$V_b = M\omega_0^2 q_b^2/6 = 2M\omega_0^2 q_{ex}^2/27 , \qquad \text{and} \qquad \omega_b = \omega_0 . \tag{11.41}$$

The potential has the symmetry

$$V(q_b - q) = V_b - V(q) . \tag{11.42}$$

The round trip path in the potential $V(q)$ at energy $E = V_b$, which leaves the point $q = q_b$ at time $-\infty$, bounces back at $q = -q_b/2$ at time zero, and then returns to the starting point at time $+\infty$, reads

$$q_{E=V_b}(t) = q_b - (3q_b/2) \operatorname{sech}^2(\omega_0 t/2) . \tag{11.43}$$

The abbreviated action of this path is

$$\mathcal{I}(V_b) = 36V_b/5\omega_0 . \tag{11.44}$$

We find with use of Eq. (11.44) that Kramers' formula (11.16) with (11.17) and (11.13) is correct in the classical regime when

$$\gamma/\omega_b \gtrsim 5k_B T/36V_b . \tag{11.45}$$

We see from the conditions (10.3) and (11.45) that the moderate-to-strong damping transmission factor (11.13) is valid down to considerably weak damping.

The classical escape rate (11.16) decreases exponentially fast, as T is lowered. Thus, at very low T, the metastable state is mainly depopulated via quantum tunneling.

12 Quantum rate theory: basic methods

With the advent of quantum mechanics, the first who introduced quantum tunneling was Friedrich Hund in 1927, when he described the intramolecular rearrangement in ammonia molecules [319]. Shortly later, the tunneling effect was popularized by Oppenheimer [320], who used it to explain the ionization of atoms in strong electric fields, and by Gamow [321], and Gurney and Condon [322], who explained the radioactive decay of nuclei. Perhaps the oldest guess of a quantum transition state theory was made by Wigner, when he proposed as a quantum mechanical generalization of the classical TST expression (11.1) the formula [323]

$$k_{\rm W} \;=\; \frac{1}{Z_0}\frac{1}{2\pi\hbar}\int dp\,dq\,\frac{p}{M}\Theta(p)\delta(q-q_{\rm b})f^{({\rm W})}(p,q)\,, \tag{12.1}$$

where Z_0 is the partition function of the well region, and $f^{({\rm W})}(p,q)$ is the quantum Wigner distribution (4.266). The expression (12.1) gives the correct first order quantum correction to the classical transition state rate expression and has been used by Miller [324] to derive a semiclassical transition state theory that involves a periodic classical trajectory in the upside-down potential surface (see Sections 12.3 and 15.4). In this chapter we consider the quantum rate theory in one space dimension.

12.1 Formal rate expressions in terms of flux operators

Based on quantum scattering theory, Miller and coworkers put forward a rate expression in the form of a Boltzmann average of reactive quantum fluxes [325, 326],

$$k \;=\; {\rm Re}\left\{{\rm tr}\left(e^{-\beta\hat{H}}\hat{F}\hat{\mathcal{P}}\right)\right\}/Z_0 \;=\; {\rm tr}\left(e^{-\beta\hat{H}/2}\hat{F}_s\,e^{-\beta\hat{H}/2}\hat{\mathcal{P}}\right)/Z_0\,. \tag{12.2}$$

Here, $\hat{H}$ is the Hamiltonian, Z_0 is the quantum partition function of the reactant, i.e. of the well region, and tr denotes the trace operation. The operator $\hat{F}_s$ is the symmetrized *flux through dividing surface* operator. For the metastable potential of the form sketched in Fig. 10.1, the flux operator $\hat{F}_s$ is given by

$$\hat{F}_s \;=\; \tfrac{1}{2}(\hat{F}+\hat{F}^\dagger)\,, \qquad \hat{F} \;=\; \delta\left(\hat{q}-q_{\rm b}\right)\hat{p}/M\,, \tag{12.3}$$

where p is the momentum conjugate to q. Finally, $\hat{\mathcal{P}}$ is a projection operator projecting onto outgoing states, i.e., states with positive momentum in the infinite future,

$$\hat{\mathcal{P}} \;=\; \lim_{t\to\infty}e^{i\hat{H}t/\hbar}\,\Theta(\hat{p})\,e^{-i\hat{H}t/\hbar} \;=\; \lim_{t\to\infty}e^{i\hat{H}t/\hbar}\,\Theta(\hat{q}-q_{\rm b})\,e^{-i\hat{H}t/\hbar}\,. \tag{12.4}$$

The equivalence of both forms was shown in Ref. [325]. A major advantage of the expression (12.2) is that the exact thermal quantum rate expression can be directly given without necessity of first calculating energy-dependent rates and then perform a thermal averaging. The expression (12.2) for the quantum crossing rate involves the Hamiltonian in two ways: first, in the Boltzmann operator $e^{-\beta \hat{H}}$, and secondly through the projection operator $\hat{\mathcal{P}}$ in Eq. (12.4). The infinite time limit of this projection can be handled correctly only by a quantum scattering calculation.

The expression for the crossing rate given in Eq. (12.2) is dynamically exact, with the only assumption that the total system is initially in thermal equilibrium. Thus, the formula covers both the spatial-diffusion-limited regime for moderate-to-strong damping and the energy-diffusion-limited regime for weak damping.

Different formally exact expressions for the thermal escape rate were derived from Eq. (12.2) in Ref. [325]. An equivalent form of the rate, which is similar to a result obtained by Yamamoto using Kubo's linear response formalism, is [327]

$$k = \frac{1}{Z_0} \int_0^\infty dt\, C(t) , \tag{12.5}$$

where $C(t)$ is the flux-flux autocorrelation function

$$C(t) = \text{tr}\left[e^{i\hat{H}t/\hbar} e^{-\beta \hat{H}/2} \hat{F}_s e^{-\beta \hat{H}/2} e^{-i\hat{H}t/\hbar} \hat{F}_s \right] . \tag{12.6}$$

To evaluate the rate expression (12.5) with (12.6), one has to deal with the quantum scattering problem. This amounts to the evaluation of the matrix element of the complex-time evolution operator $< q'|\, e^{-i\hat{H}(t-i\hbar\beta/2)/\hbar}\, |q'' >$. The propagation of this quantity is the source of difficulty in the numerically exact calculation of reaction rates. In a considerable simplification, the time propagation with the full Hamiltonian is replaced by the propagation with that of a parabolic barrier [328]. With the simpler projection operator

$$\hat{\mathcal{P}} = \Theta(\hat{q} - q_b + \hat{p}/M\omega_b) , \tag{12.7}$$

which unlike Eq. (12.4) is free of real-time propagation, the thermal rate may be written as

$$k = \text{tr}\left(e^{-\beta \hat{H}/2} \hat{F}_s e^{-\beta \hat{H}/2} \Theta(\hat{q} - q_b + \hat{p}/M\omega_b) \right) \Big/ Z_0 . \tag{12.8}$$

Importantly, this expression depends only on matrix elements of the thermal density operator $e^{-\beta \hat{H}}$.

The flux operator $\hat{F}_s$ in Eq. (12.6) is of low rank. In one dimension, there are only two nonzero eigenvalues, one positive and one negative, corresponding to flux in forward and backward direction. The eigenvalues of the thermal flux operator $e^{-\beta \hat{H}/2} \hat{F}_s e^{-\beta \hat{H}/2}$ can be calculated efficiently using the iterative Lanczos scheme [329]. This allows to compute only those eigenvalues which are nonzero and contribute to the rate. The calculated basis functions are then propagated without numerical difficulties. The low rank of $\hat{F}_s$ implies a similar low rank for the full operator in Eq. (12.6). The trace can then be computed in this much smaller basis. Numerical stability of the real-time propagation can be considerably improved by introducing a complex absorbing potential. Following these lines, the numerically exact solution of Eq. (12.5) with (12.6) is feasible for systems with several degrees of freedom [330].

12.2 Quantum transition state theory

The quantum transition state approximation is based on the assumption that there are only outgoing particles at the barrier top, just as in classical transition state theory, Eq. (11.1). This constraint is provided by the projection operator [326, 331]

$$\hat{P} = \Theta(\hat{p}) \,. \tag{12.9}$$

As the expression (12.9) solely depends on the momentum, it is easier than Eq. (12.7). With this limitation, the flux operator at the barrier top in Eq. (12.2) is given by

$$\hat{F}\hat{P} \xrightarrow{\text{TST}} \delta\,(\hat{q}-q_{\rm b})\,\frac{\hat{p}}{M}\,\Theta(\hat{p}) = \frac{1}{2M}\,\delta\,(\hat{q}-q_{\rm b})\,[\,|\hat{p}| + \hat{p}\,] \,. \tag{12.10}$$

The second term in the square bracket of Eq. (12.10) does not contribute in thermal equilibrium because of detailed balance. The choice (12.9) eliminates the troublesome projection operator (12.4). The expression (12.10) is free of real-time propagation and therefore circumvents the need to know the full scattering dynamics. The Hamiltonian $\hat{H}$ enters only via the canonical operator $e^{-\beta \hat{H}}$. With the identity

$$e^{-\beta \hat{H}} = \frac{1}{\pi\hbar}\lim_{\epsilon\to0^+}\text{Im}\int_0^\infty dE\,e^{-\beta E}\int_0^\infty d\tau\,e^{(E+i\epsilon-\hat{H})\tau/\hbar} \,, \tag{12.11}$$

we may rewrite the formal expression (12.2) in the form

$$k^{(\text{TST})} = \frac{1}{Z_0}\frac{1}{2\pi\hbar}\int_0^\infty dE\,p(E)\,e^{-\beta E} \,, \tag{12.12}$$

where $p(E)$ is the dimensionless distribution function

$$\begin{aligned}
p(E) &= \lim_{\epsilon\to0^+}\text{Im}\int_0^\infty d\tau\,e^{(E+i\epsilon)\tau/\hbar} \\
&\quad \times \int dq\,\delta(q-q_{\rm b})\,|\dot{q}\,|_{q=q_{\rm b}} < q\,|\,e^{-\hat{H}\tau/\hbar}\,|\,q > \,.
\end{aligned} \tag{12.13}$$

The expression (12.12) with (12.13) is the quantum mechanical rate expression for a one-dimendional system in the transition state approximation. The function $p(E)$ describes the transmission probability at energy E. From now on we write k for $k^{(\text{TST})}$.

12.3 Semiclassical limit

In the semiclassical quantum transition state theory [324], the imaginary-time propagator $< q''|\,e^{-\hat{H}\tau/\hbar}\,|q' >$ in Eq. (12.13) is taken into account in *semiclassical* approximation.[1] The semiclassical form of $< q''|\,e^{-\hat{H}\tau/\hbar}\,|q' >$ is found either from the

---

[1] For diverse applications of semiclassical propagators and periodic orbit theory to quantum chaos, I refer to the monographs by M. Gutzwiller [233] and by F. Haake [332].

semiclassical limit of the path integral (4.18) or from the real-time Van Vleck propagator (5.104) with (5.105) by analytic continuation from real time t to imaginary time τ, $t \to -i\tau$. By virtue of the relation[2]

$$-\frac{\partial^2 S_{cl}(q'', \tau; q', 0)}{\partial q'' \, \partial q'} = \frac{1}{\dot{q}(\tau)\,\dot{q}(0)} \frac{\partial^2 S_{cl}(q'', \tau; q', 0)}{\partial \tau^2} \,, \tag{12.14}$$

the semiclassical propagator resulting from a single classical path may be written as

$$< q'' \,|\, e^{-\hat{H}\tau/\hbar} \,|\, q' > \;=\; \frac{e^{-in\pi/2}}{\sqrt{2\pi\hbar}} \sqrt{\left|\frac{\partial^2 S_{cl}/d\tau^2}{\dot{q}(\tau)\,\dot{q}(0)}\right|}\; e^{-S_{cl}(\tau)/\hbar} \,, \tag{12.15}$$

where

$$S_{cl}(\tau) \;=\; \int_0^\tau d\tau' \left(\tfrac{1}{2}M\dot{q}^2(\tau') + V[q(\tau')]\right) \tag{12.16}$$

is the Euclidean action for the path obeying $M\ddot{q}(\tau) = \partial V/\partial q$ with boundaries $q(0) = q'$ and $q(\tau) = q''$, and n is the number of conjugate points along this path. The propagator (12.15) satisfies a fixed-point property analogous to Eq. (5.107). As initial- and endpoint coincide for the propagator in (12.13), the classical path is a periodic orbit[3] of period τ in the inverted potential $-V(q)$, and $S_{cl}(\tau)$ is the associated action. With Eq. (12.15) the transmission probability (12.13) for one cycle takes the form

$$P_1(E) \;=\; \frac{i\,\mathcal{J}_1}{\sqrt{2\pi\hbar}} \int_C d\tau \, \sqrt{|\partial^2 S_{cl}/\partial\tau^2|} \; e^{-[S_{cl}(\tau) - E\tau]/\hbar} \,. \tag{12.17}$$

Here $\mathcal{J}_1$ is the phase factor arising from the conjugate points. For one space dimension, the conjugate points are the turning points of the periodic orbit. Since each turning point yields a factor $e^{-i\pi/2}$, the phase factor for m cycles is $\mathcal{J}_m = e^{-im\pi}$.

The second step of the semiclassical approximation consists in the calculation of the time integral in Eq. (12.17) with the method of steepest descent. The exponential factor $\exp\{-[S_{cl}(\tau) - E\tau]/\hbar\}$ is stationary for Euclidean time $\tau = \bar{\tau}(E)$,

$$\left.\frac{\partial S_{cl}(\tau)}{\partial \tau}\right|_{\tau = \bar{\tau}(E)} \;=\; E \,. \tag{12.18}$$

The condition (12.18) fixes the period $\bar{\tau}$ of one cycle of the periodic orbit at energy E, $\bar{\tau} = \bar{\tau}(E)$. Expanding $S(\tau) - E\tau$ about $\tau = \bar{\tau}(E)$, one gets

$$S_{cl}(\tau) - E\tau \;=\; W(E) + \frac{1}{2}\left.\frac{\partial^2 S_{cl}}{\partial \tau^2}\right|_{\tau = \bar{\tau}} (\tau - \bar{\tau})^2 + \cdots \,, \tag{12.19}$$

where $W(E)$ is the abbreviated *Euclidean* action

$$W(E) \;=\; M \oint d\tau \, \dot{q}^2(\tau) \;=\; 2 \int_{q_1}^{q_2} dq \, \sqrt{2M[V(q) - E]} \,. \tag{12.20}$$

---

[2]The Euclidean action is denoted by S. The Euclidean and Minkowskian abbreviated actions are termed $W(E)$ and $\mathcal{I}(E)$ [cf. Eq. (11.32)].

[3]The periodic path is deliberately called *orbit* in anticipation of the multi-dimensional case.

The boundaries q_1 and q_2 are the zeros of the integrand and thus represent the turning points of the classical motion in the upside-down potential $-V(q)$ at conserved energy

$$E = V[q(\tau)] - \tfrac{1}{2}M\dot{q}^2(\tau) . \tag{12.21}$$

Since the second variation of the action $S_{cl}(\tau)$ emerges as negative,

$$\partial^2 S_{cl}/\partial \tau^2|_{\tau=\bar{\tau}} = 1/(\partial \tau/\partial E)|_{\tau=\bar{\tau}} < 0 , \tag{12.22}$$

the periodic orbit is unstable against small perturbations. The saddle point is located on the positive-real axis of the complex-τ plane at $\bar{\tau}(E) = -\partial W(E)/\partial E$, and the direction of steepest descent is perpendicular to the imaginary-time axis in the complex τ-plane. The transmission probability at energy E resulting from one cycle then is

$$p_1(E) = -\mathcal{J}_1\, e^{-W(E)/\hbar} . \tag{12.23}$$

The motion of the particle in the well of the upside-down potential is bounded and periodic in the regime $0 \leqslant E < V_b$. For $E \ll V_b$, the particle spends most of the time in the vicinity of the inner turning point q_1 and it bounces back from the outer turning point q_2. For this reason, the periodic path satisfying Eq. (12.21) is called *bounce* [109]. We shall denote this path by $q_B(\tau)$. It is convenient to choose the phase of the bounce such that $q_B(\tau = 0) = q_2$. We then have $q_B(\tau) = q_B(-\tau)$, and $q_B(\pm\bar{\tau}/2) = q_1$. In the limit $E \to 0$, we have $\bar{\tau} \to \infty$, $q_2 \to q_{ex}$ and $q_1 \to 0$.

A periodic path with m bounces or cycles at energy E takes time $\tau_m = m\bar{\tau}(E)$. It has abbreviated action $m\,W(E)$, and the related phase factor is $\mathcal{J}_m = e^{-i\pi m}$.

Taking into account the single-bounce and all multi-bounce contributions, we get

$$p(E) = \sum_{m=1}^{\infty}(-1)^{m-1} e^{-mW(E)/\hbar} = \frac{1}{1 + e^{W(E)/\hbar}} . \tag{12.24}$$

In the semiclassical limit $\hbar \to 0$ with $T/T_0 < 1$ held fixed, where

$$k_B T_0 = \frac{1}{\beta_0} \equiv \frac{\hbar\omega_b}{2\pi} , \tag{12.25}$$

the Boltzmann average in Eq. (12.12) is dominated by the periodic orbit with one cycle of period $\bar{\tau} \equiv -\partial W/\partial E = \hbar\beta$,

$$k = \frac{1}{Z_0}\frac{1}{2\pi\hbar}\int_0^{V_b} dE\, p_1(E)\, e^{-\beta E} = \frac{1}{Z_0}\frac{1}{2\pi\hbar}\int_0^{V_b} dE\, e^{-\beta E - W(E)/\hbar} . \tag{12.26}$$

Calculation of the integral by steepest descent yields

$$k = \frac{1}{Z_0}\frac{1}{\sqrt{2\pi\hbar|\tau_B'|}}\, e^{-S_B/\hbar} , \qquad\qquad T < T_0 , \tag{12.27}$$

$$\tau_B' \equiv \frac{\partial \tau}{\partial E}\bigg|_{E=E_{\hbar\beta}} = -\frac{\partial^2 W(E)}{\partial E^2}\bigg|_{E=E_{\hbar\beta}} = \frac{1}{\partial^2 S/\partial \tau^2}\bigg|_{\tau=\hbar\beta} . \tag{12.28}$$

For a barrier which is wider than a parabolic one, we have $\tau'_B < 0$.

The quantity S_B is the action (12.16) of the bounce with period $\hbar\beta$ and associated total energy $E = E_{\hbar\beta}$. Using Eqs. (12.16) and (12.21), we obtain

$$S_B \equiv S_{cl}(\hbar\beta) = W(E_{\hbar\beta}) + E_{\hbar\beta}\hbar\beta . \tag{12.29}$$

The expression (12.27) with Eqs. (12.28) and (12.29) represents the semiclassical quantum rate formula in the transition state approach [333]. This form is generally valid in the temperature regime $0 \leqslant T < T_0$. The semiclassical expression (12.27) is rounded off with the semiclassical partition function of the well region

$$Z_0 = \frac{1}{2\sinh(\beta\hbar\omega_0/2)} , \tag{12.30}$$

which is the exact quantum mechanical partition function for a harmonic well.

In the regime $T > T_0$, the integral in Eq. (12.12) is dominated by energies $E \gtrsim V_b$. Then the multi-bounce expression (12.24) applies. Taking for $W(E)$ the abbreviated action for a parabolic barrier matched with the actual barrier, we then have

$$p(E) = \frac{1}{1 + e^{2\pi(V_b - E)/\hbar\omega_b}} . \tag{12.31}$$

Upon using the forms (12.30) and (12.31), and extending the lower integration bound in Eq. (12.12) to $E = -\infty$, the thermal rate is found as

$$k = \frac{\omega_b}{2\pi} \frac{\sinh(\frac{1}{2}\beta\hbar\omega_0)}{\sin(\frac{1}{2}\beta\hbar\omega_b)} e^{-\beta V_b} , \qquad T > T_0 . \tag{12.32}$$

As T is raised to the classical regime, $T \gg \hbar\omega_0/k_B, \hbar\omega_b/k_B$, this expression reduces to the classical transition state formula (11.2).

Observing that $\sin(\frac{1}{2}\beta\hbar\omega_b) = \sin[\frac{1}{2}(\beta_0 - \beta)\hbar\omega_b]$, we see that the rate expression (12.32) diverges $\propto 1/(\beta_0 - \beta)$ or $\propto 1/(T - T_0)$. At T slightly above T_0, we have

$$k = \frac{1}{Z_0} \frac{1}{2\pi\hbar(\beta_0 - \beta)} e^{-\beta V_b} , \qquad T - T_0 \ll T_0 . \tag{12.33}$$

The singularity is removed by taking into account that the actual barrier is wider than the parabolic one. With the leading correction included, the abbreviated action is

$$W(E) = \frac{2\pi}{\omega_b}(V_b - E) + \frac{|\tau'_b|}{2}(V_b - E)^2 , \tag{12.34}$$

where $\tau'_b = \partial\tau/\partial E|_{E=V_b} < 0$. For the "quadratic-plus-cubic" potential (11.40), there holds $|\tau'_b| = 5\pi/(18\omega_0 V_b)$, as follows from the "multi-dimensional" expression (16.33) in the limit $\hat{\gamma}(z) \to 0$. Upon inserting the relation (12.34) into the integral expression

$$k = \frac{1}{Z_0} \frac{1}{2\pi\hbar} \int_{-\infty}^{V_b} dE\, e^{-\beta E - W(E)/\hbar} , \tag{12.35}$$

and completing the square in the exponent, we readily get

$$k = \frac{1}{2Z_0} \frac{1}{\sqrt{2\pi\hbar|\tau_b'|}} \, \mathrm{erfc}\left(\frac{\hbar(\beta_0 - \beta)}{\sqrt{2\hbar|\tau_b'|}} \right) \exp\left(-\beta V_b + \frac{\hbar(\beta - \beta_0)^2}{2|\tau_b'|} \right), \qquad (12.36)$$

where $\mathrm{erfc}(z)$ is the complementary error function [90]

$$\mathrm{erfc}(z) \equiv \frac{2}{\sqrt{\pi}} \int_z^\infty dt \; e^{-t^2} . \qquad (12.37)$$

The expression (12.36) smoothly interpolates between the rate expression (12.27) valid for temperatures below T_0 and the expression (12.33) which holds for temperatures slightly above T_0. The bridging expression (12.36) is required in a narrow temperature regime about T_0 which is $|T - T_0|/T_0 \lesssim k_B T_0 \sqrt{2|\tau_b'|/\hbar}$. We shall consider this crossover regime in some more detail in Section 16.3.

12.4 Quantum tunneling regime

Consider now the decay rate in the low-temperature regime $T \ll T_0$ more closely. In the regime $E \ll V_b$, the abbreviated Euclidean action (12.20) can be expanded as

$$\mathbb{W}(E) = \mathbb{W}_h(E) + \mathbb{W}(0) - \mathbb{W}_h(0) + E\left[\partial\mathbb{W}/\partial E - \partial\mathbb{W}_h/\partial E \right]_{|E=0} + \cdots, \qquad (12.38)$$

where $\mathbb{W}_h(E)$ is the abbreviated action for the harmonic potential $V_h(q) = \frac{1}{2}M\omega_0^2 q^2$. The action of the bounce with period $\hbar\beta = \infty$ and associated energy $E = 0$ is

$$\mathbb{S}_0 \equiv \mathbb{S}_{cl}(\tau \to \infty) = \mathbb{W}(0) = 2\int_0^{q_{ex}} dq \sqrt{2MV(q)} , \qquad (12.39)$$

where the boundaries $q = 0$ and $q = q_{ex}$ are the zeros of the potential $V(q)$. The bounce path $q_B(\tau)$ at $E = 0$ is the solution of $\dot{q} = \sqrt{2V(q)/M}$. With the symmetric choice $q_B(\tau) = q_B(-\tau)$, it hits the bounce point q_{ex} at $\tau = 0$ and approaches the origin $q = 0$ with exponential slow-down

$$\lim_{\tau \to \pm\infty} q_B(\tau) = C_0 \sqrt{\frac{\mathbb{S}_0}{2M\omega_0}} \, e^{-\omega_0|\tau|} . \qquad (12.40)$$

Here we have parametrized the pre-exponential factor conveniently. The constant C_0 is a numerical factor which depends on the shape of the barrier. With Eqs. (12.39) and (12.40) we obtain from Eq. (12.38) the low-energy expansion

$$\mathbb{W}(E) = \mathbb{S}_0 - \frac{E}{\omega_0} - \frac{E}{\omega_0} \ln\left(\frac{C_0^2 \mathbb{S}_0 \omega_0}{E} \right) + \mathcal{O}\left[(E/V_b)^2 \right] . \qquad (12.41)$$

We shall use this expression shortly.

Suppose that the particle is initially prepared in the ground state of the metastable well. The probability per unit time that it escapes from this state by quantum tunneling through the barrier is given by the standard WKB formula [109, 110]

$$\gamma_0 = \omega_0 C_0 \sqrt{\mathbb{S}_0/2\pi\hbar} \; e^{-\mathbb{S}_0/\hbar} . \qquad (12.42)$$

If the particle is initially placed in the nth excited state of the well with energy $E_n = \hbar\omega_0(n + \frac{1}{2})$, the rate of escape from the well in the semiclassical limit is [110]

$$\gamma_n = \frac{1}{n!}\left(C_0^2 \frac{S_0}{\hbar}\right)^n \gamma_0 . \tag{12.43}$$

For the cubic metastable potential (11.40) with the symmetry relation (11.42), we have $S_0 = \mathcal{I}(V_b) = 36V_b/5\omega_0$, and the bounce at $E = 0$ has the bell-shaped form $q_B(\tau) = \frac{3}{2}q_b \operatorname{sech}^2(\frac{1}{2}\omega_0\tau)$ [cf. Eqs. (11.44) and (11.40)], yielding $C_0 = \sqrt{60}$. With these expressions, the rate (12.42) takes the compact form

$$\gamma_0 = \frac{6\omega_0}{\sqrt{\pi}}\sqrt{\frac{6V_b}{5\hbar\omega_0}} \exp\left(-\frac{36}{5}\frac{V_b}{\hbar\omega_0}\right) . \tag{12.44}$$

We now define a thermal rate by averaging the decay rates for the individual energy levels in the well with the canonical distribution,

$$k \equiv \sum_n \gamma_n \, e^{-\beta E_n} \Big/ \sum_n e^{-\beta E_n} . \tag{12.45}$$

With the semiclassical expression (12.43) for γ_n, the sum in Eq. (12.45) can be carried out explicitly, yielding

$$k = \frac{e^{-\hbar\omega_0/2k_B T}}{Z_0} \exp\left(C_0^2 \frac{S_0}{\hbar} e^{-\hbar\omega_0/k_B T}\right)\gamma_0 . \tag{12.46}$$

The expression (12.46) represents the leading thermal enhancement of the quantum statistical rate due to thermal occupation of excited states in the well. It applies when the thermal energy is very small compared to $\hbar\omega_0$. At higher temperature, the sum of partial rates in Eq. (12.45) is replaced by the integral expression

$$k = \frac{1}{Z_0} \int_0^\infty dE \, \rho(E)\gamma(E) \, e^{-\beta E} . \tag{12.47}$$

Here the density of states is constant, $\rho(E) = 1/\hbar\omega_0$, and $\gamma(E)$ is the WKB expression of the transmission probability

$$\gamma(E) = \frac{\omega_0}{2\pi} e^{-W(E)/\hbar} , \tag{12.48}$$

which applies for elevated energy $E \gg \hbar\omega_0$.

The rate expression (12.47) with Eq. (12.48) coincides with the expression (12.26). Hence the steepest descent approximation for the energy integral yields the semiclassical formula (12.27), which we rewrite for convenience,

$$k = \frac{1}{Z_0} \frac{1}{\sqrt{2\pi\hbar|\tau_B'|}} e^{-S_B/\hbar} , \qquad T < T_0 . \tag{12.49}$$

We now show that the expression (12.49) is valid down to zero temperature. To this, we first calculate the energy of the bounce with asymptotically large period $\hbar\beta$. Inversion of the relation $|\partial W/\partial E|_{E=E_{\hbar\beta}} = \hbar\beta$ with $W(E)$ given in Eq. (12.41) yields

$$E_{\hbar\beta} = C_0^2 S_0 \, \omega_0 \, e^{-\beta\hbar\omega_0} . \tag{12.50}$$

With this and with (12.41) the bounce action (12.29) for period $\hbar\beta \gg 1/\omega_0$ emerges as

$$S_B = S_{cl}(\hbar\beta) , \qquad \text{with} \qquad S_{cl}(\tau) = \left(1 - C_0^2 \, e^{-\omega_0\tau}\right) S_0 . \tag{12.51}$$

Upon employing the relations (12.28) and (12.51), we then obtain

$$|\tau_B'|^{-1} = |\partial^2 S_{cl}(\tau)/\partial\tau^2|_{\tau=\hbar\beta} = C_0^2 S_0 \, \omega_0^2 \, e^{-\beta\hbar\omega_0} . \tag{12.52}$$

Now all constituents of the expression (12.49) are on hand. With the expressions (12.51) and (12.52) we directly obtain the previous result (12.46). This verifies that the expression (12.49) is valid down to zero temperature [110, 333].

The agreement of Eq. (12.49) with the expression (12.46) at low T is surprising for two reasons. First, in the initial expression (12.47) the discreteness of the low-lying states is disregarded. Secondly, the steepest-descent transmission formula (12.48) is not correct for low-lying states [110].[4] Apparently, these two flaws cancel out each other when we proceed via steepest descent approximation to the formula (12.49).

12.5 Free energy method

An independent and powerful line of reasoning is to calculate the quantum rate with a pure thermodynamic equilibrium method. The thermodynamic method was pioneered by Langer [334] in 1967 and reviewed by himself in 1980 in Ref. [335]. In the original treatment, Langer calculated the classical nucleation rate governing the early stage of a first-order phase transition. In this approach the quantity of interest is the free energy of the metastable system. Because of states of lower energy on the other side of the barrier, the partition function of the metastable well is defined by means of an analytical continuation from a stable potential to the metastable one depicted in Fig. 10.1. The corresponding deformation of the integration contour leads to a unique imaginary part of the free energy associated with the metastable state. This quantity is then related to the decay probability of the system, analogous to the interpretation of imaginary energies of resonances in quantum field theory. Interestingly and importantly enough, this method is not restricted to the classical regime. As we shall see, all the results of Section 12.3 are reproduced by the free energy (Im F) method. In a sense, the Im F method is easier than the method presented in Section 12.3 since the energy and time integrals in Eqs. (12.12) and (12.13) are obviated.

---

[4]The formula (12.48) for $\gamma(E_n)$ follows from the expression (12.43) by substituting Stirling's asymptotic formula for the factorial $n!$. This approximation is only justified when n is large.

In this section, we briefly sketch the Im F method for the onedimensional case. The Im F method turns out to be particularly appropriate in the case of a system-plus-environment complex where the number N of bath degrees of freedom is very large or even infinity. The respective treatment is given in Chapters 15 – 17.

The general argument is as follows [110, 333]. For a metastable system, the partition function may be written as

$$Z = \sum_n e^{-\beta z_n} = \sum_n e^{-\beta(E_n - i\hbar\gamma_n/2)} , \qquad (12.53)$$

where the $z_n = E_n - i\hbar\gamma_n/2$ are the complex energies of the individual states, and $\sum_n$ denotes summation over all states. For weak metastability, $V_b/\hbar\omega_0 \gg 1$, we have $\hbar\gamma_n \ll E_n$ for all relevant n. Thus we may write

$$Z = Z' + iZ'' \approx \sum_n e^{-\beta E_n} + i\frac{\hbar\beta}{2}\sum_n \gamma_n e^{-\beta E_n} . \qquad (12.54)$$

The imaginary part Z'' of the partition function is generally small compared to the real part Z'. The real part Z' is (apart from exponentially small corrections) determined by properties of the well, $Z' = Z_0$, and the imaginary part Z'' is determined by properties of the barrier, $Z'' = \text{Im}\, Z_b$. Using the defining relation of the free energy

$$F = -\ln Z/\beta , \qquad (12.55)$$

we obtain with the expression (12.54)

$$\text{Im}\, F = -\frac{1}{\beta}\frac{\text{Im}\, Z_b}{Z_0} = -\frac{\hbar}{2}\frac{\sum_n \gamma_n e^{-\beta E_n}}{\sum_n e^{-\beta E_n}} . \qquad (12.56)$$

Thus the quantum-statistical rate (12.45) is found in the Im F-method as

$$k = -\frac{2}{\hbar}\text{Im}\, F = \frac{2}{\hbar\beta}\frac{\text{Im}\, Z_b}{Z_0} . \qquad (12.57)$$

The formula (12.57) is the generalization of the ground state relation $\gamma_0 = -2\,\text{Im}\,E_0/\hbar$ to finite temperature. Actually, the relation (12.57) is only valid in the regime $T < T_0$, as we shall see below in this section.

Let us study Im F in the semiclassical limit. The dominant stationary point in function space for the barrier contribution Z_b to the partition function in the regime $T < T_0$ is the periodic bounce with period $\hbar\beta$ and associated action S_B. To second order in the fluctuations about the extremal path, $q(\tau) = q_B(\tau) + \xi(\tau)$, the action takes the form

$$S[q(\cdot)] = S_B + \frac{M}{2}\int_0^{\hbar\beta} d\tau\, \xi(\tau)\,\Lambda[q_B(\tau)]\,\xi(\tau) \qquad (12.58)$$

with the fluctuation operator $\Lambda[q_B(\tau)] = -\partial^2/\partial\tau^2 + V''[q_B(\tau)]/M$ acting in the space of $\hbar\beta$-periodic functions. Expanding $\xi(\tau)$ into the eigenmodes $\{\chi_n(\tau)\}$ of $\Lambda[q_B(\tau)]$ with orthonormality condition $\int_0^{\hbar\beta} d\tau\, \chi_n(\tau)\chi_m(\tau)/\hbar\beta = \delta_{nm}$,

$$\xi(\tau) = \sum_n c_n \chi_n(\tau) , \tag{12.59}$$

the action becomes

$$\mathbb{S}[q(\cdot)] = \mathbb{S}_{\mathrm{B}} + \frac{M\hbar\beta}{2} \sum_n \lambda_n[q_{\mathrm{B}}(\cdot)] c_n^2 . \tag{12.60}$$

The $\lambda_n[q_{\mathrm{B}}]$ are the eigenvalues of $\mathbf{\Lambda}[q_{\mathrm{B}}(\tau)]$ for periodic boundary conditions. According to the analysis in Subsection 4.3.2, the ratio $Z^{(\mathrm{b})}/Z_0$ would then be

$$\frac{Z^{(\mathrm{b})}}{Z_0} = \sqrt{\frac{\det\{-\partial^2/\partial\tau^2 + \omega_0^2\}}{\det\{-\partial^2/\partial\tau^2 + V''[q_{\mathrm{B}}(\tau)]/M\}}}\, e^{-\mathbb{S}_{\mathrm{B}}/\hbar} = \prod_n \sqrt{\frac{\lambda_n^{(0)}}{\lambda_n[q_{\mathrm{B}}(\cdot)]}}\, e^{-\mathbb{S}_{\mathrm{B}}/\hbar} , \tag{12.61}$$

with

$$\lambda_n^{(0)} = \nu_n^2 + \omega_0^2 , \qquad \lambda_n[q_{\mathrm{B}}(\cdot)] = \nu_n^2 + V''[q_{\mathrm{B}}(\cdot)]/M . \tag{12.62}$$

However, this expression is not quite correct for two reasons.

(1) Differentiation of the equation of motion $\ddot{q}_{\mathrm{B}}(\tau) - V'[q_{\mathrm{B}}(\tau)]/M = 0$ with respect to time τ shows that $\dot{q}_{\mathrm{B}}(\tau)$ is an eigenmode of the fluctuation operator $\mathbf{\Lambda}[q_{\mathrm{B}}(\tau)]$ with eigenvalue zero. The normalized zero mode is

$$\chi_1(\tau) = \sqrt{M\hbar\beta/\mathbb{W}_{\mathrm{B}}}\, \dot{q}_{\mathrm{B}}(\tau) , \tag{12.63}$$

where

$$\mathbb{W}_{\mathrm{B}} \equiv M \int_0^{\hbar\beta} d\tau\, \dot{q}_{\mathrm{B}}^2(\tau) . \tag{12.64}$$

The zero mode $\chi_1(\tau)$ induces an infinitesimal time translation of the bounce,

$$q_{\mathrm{B}}(\tau + \delta\tau_1) = q_{\mathrm{B}}(\tau) + \sqrt{\mathbb{W}_{\mathrm{B}}/M\hbar\beta}\, \chi_1(\tau)\, \delta\tau_1 . \tag{12.65}$$

The eigenvalue of the mode $\chi_1(\tau)$ is zero because the bounce action is invariant under changes of the phase. As a result, the integral over the zero mode amplitude c_1 diverges for the second order action (12.60). With Eq. (12.65) however, the flawed fluctuation integral can be transcribed into a good-natured integral over the range of the arbitrary phase time τ_1 of the bounce $q_{\mathrm{B}}(\tau + \tau_1)$ [324, 336, 337]. Reverse transaction of the divergent zero mode contribution, and use of the Jacobian $dc_1/d\tau_1 = \sqrt{\mathbb{W}_{\mathrm{B}}/M\hbar\beta}$, leads to the substitution rule

$$\frac{1}{\sqrt{\lambda_1[q_{\mathrm{B}}(\cdot)]}} \rightarrow \sqrt{\frac{M\hbar\beta}{2\pi\hbar}} \int_{-\infty}^{\infty} dc_1 \exp\left(-M\hbar\beta\lambda_1[q_{\mathrm{B}}(\cdot)]c_1^2/2\hbar\right)$$

$$\rightarrow \sqrt{\frac{\mathbb{W}_{\mathrm{B}}}{2\pi\hbar}} \int_0^{\hbar\beta} d\tau_1 = \sqrt{\frac{\mathbb{W}_{\mathrm{B}}}{2\pi\hbar}}\, \hbar\beta . \tag{12.66}$$

(2) While the bounce $q_B(\tau)$ is free of nodes, the zero mode (12.63) has one node. Thus, by the node-counting theorem, there exists a nodeless fluctuation mode $c_0\chi_0(\tau)$ with a negative eigenvalue $-|\lambda_0[q_B(\cdot)]|$. The negative eigenvalue manifests that the bounce is a saddle point of the action, and not a minimum. This feature is owed to the fact that we are dealing with the partition function of a metastable well. The mode χ_0 must be dealt with by distortion of the c_0-contour into the half-plane $\mathrm{Im}\, c_0 > 0$. Thereby the contribution of this mode to the barrier partition function $Z^{(b)}$ gathers up an imaginary part. This term is given by

$$\sqrt{\frac{M\hbar\beta}{2\pi\hbar}}\int_0^{i\infty} dc_0 \exp\left(M\hbar\beta|\lambda_0[q_B(\cdot)]|c_0^2/2\hbar\right) = i\,\frac{1}{2}\,\frac{1}{\sqrt{|\lambda_0[q_B(\cdot)]|}}. \qquad (12.67)$$

The factor i originates from the analytic continuation, which is required if one deforms an absolute stable potential to a metastable one [338]. The additional factor $\frac{1}{2}$ appears because the integration in Eq. (12.67) covers only one wing of the Gaussian function.

With items (1) and (2) taken into account, the flawed expression (12.61) is converted into the correct semiclassical result

$$\frac{\mathrm{Im}\,Z_b}{Z_0} = \frac{\hbar\beta}{2}\sqrt{\frac{W_B}{2\pi\hbar}}\left|\frac{\det\{-\partial^2/\partial\tau^2 + \omega_0^2\}}{\det'\{-\partial^2/\partial\tau^2 + V''[q_B(\tau)]/M\}}\right|^{1/2} e^{-S_B/\hbar}. \qquad (12.68)$$

The determinants are calculated for eigenfunctions obeying periodic boundary conditions, as discussed in Subsection 4.3.2. The prime in $\det'[\cdots]$ indicates that the zero eigenvalue is omitted.

Clearly, it would be inconsistent to keep a real contribution to $Z^{(b)}/Z^{(0)}$ resulting from $Z^{(b)}$, since it would be exponentially small by the factor $e^{-\beta V_b}$, while non-Gaussian corrections to $Z^{(0)}$ are disregarded anyhow. On the other hand, it is consistent to keep the exponentially small imaginary part (12.68) as it is the leading one.

Alternatively, we may calculate the bounce contribution to the partition function by starting from the semiclassical imaginary-time propagator (12.15). Putting $\tau = \hbar\beta$, and taking into account that the distortion of the integration contour at the saddle point into the direction of steepest descent yields an factor $i/2$ compared with Eq. (12.15), we obtain

$$\mathrm{Im} <q|e^{-\beta\hat{H}}|q>_{|\text{barrier}} = \frac{1}{2}\frac{1}{\sqrt{2\pi\hbar}}\frac{1}{|\dot{q}_B(q)|}\sqrt{\left|\frac{\partial^2 S_{cl}(q,\tau;q,0)}{\partial\tau^2}\right|_{\tau=\hbar\beta}}\; e^{-S_B/\hbar}. \qquad (12.69)$$

Writing the trace integral $\oint dq \cdots$ as $\int_0^{\hbar\beta} d\tau\, \dot{q}(\tau)\cdots$, and observing the relations (12.28), we obtain

$$\mathrm{Im}\,Z_b = \frac{1}{2}\frac{\hbar\beta}{\sqrt{2\pi\hbar|\tau_B''|}}\, e^{-S_B/\hbar}. \qquad (12.70)$$

Upon equating Eq. (12.68) with Eq. (12.70), we can express the ratio of the determinants in Eq. (12.68) directly in terms of the classical mechanics of the bounce,

$$Z_0^2 \left| \frac{\det\{-\partial^2/\partial\tau^2 + \omega_0^2\}}{\det'\{-\partial^2/\partial\tau^2 + V''[q_B(\tau)]/M\}} \right| = \frac{1}{W_B|\tau_B'|} . \tag{12.71}$$

This relation agrees with the result of a direct computation [339].

By inserting Eq. (12.70) into Eq. (12.57) we see that the resulting rate expression coincides with the form (12.27). This consistency affirms validity of the relation (12.57) in the temperature regime $T \leqslant T_0$.

For $T > T_0$, there is no periodic orbit. The stationary point determining the barrier partition function is the constant path $q(\tau) = q_b$. As a result, the zero mode is missing, but there is again a mode with negative eigenvalue. The obligatory deformation of the respective integration contour yields an imaginary barrier contribution $\operatorname{Im} Z_b$,

$$\operatorname{Im} F = -\frac{1}{\beta}\frac{\operatorname{Im} Z_b}{Z_0} \quad \text{with} \quad \operatorname{Im} Z_b = \frac{1}{4\sin(\beta\hbar\omega_b/2)} e^{-\beta V_b} . \tag{12.72}$$

The expressions (12.72) and (12.32) are in accordance if there holds the relation

$$k = -\frac{2}{\hbar}\frac{\beta}{\beta_0}\operatorname{Im} F = \frac{2}{\hbar\beta_0}\frac{\operatorname{Im} Z_b}{Z_0} \quad \text{for} \quad T > T_0 . \tag{12.73}$$

These findings, which go back to Affleck [333], are summarized as follows. The relation (12.57) applies in the regime $T \leqslant T_0$, in which $\operatorname{Im} Z_b$ is determined by the periodic bounce $q_B(\tau)$. Because the phase of the bounce $q_B(\tau)$ is arbitrary, there is a zero mode in the deviations about $q_B(\tau)$. For this reason, $\operatorname{Im} Z_b$ is proportional to $\hbar\beta$. The factor β cancels out in the expression (12.70) for $\operatorname{Im} F$. Above T_0, there is no zero mode because the saddle point trajectory is the constant path $q(\tau) = q_b$. Because of the absence of the zero mode above T_0, the rate for $T > T_0$ is given by the modified formula (12.73). Clearly, this is merely a plausible argument and certainly not a proof. The expressions (12.57) and (12.73) can be written in the combined form

$$k = -2\sigma \operatorname{Im} F/\hbar , \tag{12.74}$$

where σ is a piecewise constant numerical factor given by

$$\begin{aligned} \sigma &= 1 &\text{for} &\quad \beta\hbar\omega_b \geqslant 2\pi , \\ \sigma &= \beta\hbar\omega_b/2\pi &\text{for} &\quad \beta\hbar\omega_b < 2\pi . \end{aligned} \tag{12.75}$$

We anticipate that in multidimensional tunneling systems an effective barrier frequency ω_R takes the place of the bare barrier frequency ω_b (see Section 14.3).

A general remark is in order. We have seen that quantum tunneling opens a new channel for barrier crossing as the temperature is lowered below the classical regime. Therefore, the escape rate is enhanced compared with the classical rate expression. At intermediate temperature, a rough estimate of the rate is obtained naively by adding to the classical rate a quantum mechanical counterpart, $k = k_{cl} + k_{qm}$ [340].

257

In a first guess, k_{qm} may be identified with the ground-state tunneling rate (12.42). With similar reasoning, a plausible criterion for the crossover temperature T_0, at which roughly the transition from thermally activated to quantum mechanical decay occurs, was given by Goldanskii already in 1959 [341]. By equating the Arrhenius exponent with the Gamow factor for the transmission of a parabolic barrier with height V_b and barrier frequency ω_b,

$$\beta_0 V_b = 2\pi V_b / \hbar \omega_b , \qquad (12.76)$$

the crossover temperature of an undamped system is found just in the form (12.25).

The above considerations shed light on the significance of the crossover temperature T_0. For $T < T_0$, there is a periodic bounce trajectory with a period $\hbar/k_B T$ in the upside-down potential $-V(q)$. At the crossover temperature, $T = T_0$, the bounce takes the shortest possible period $2\pi/\omega_b$. At this temperature, the bounce path represents a harmonic oscillation about q_b in the inverted potential $-V(q)$, and the respective actions are the same. Interestingly, these actions coincide also with the action of the constant path $q = q_b$ at $T = T_0$, as we see from the relation (12.76).

The crossover temperature T_0 can be quite large for very light particles, e.g., above room temperature for electrons. On the other hand, for macroscopic quantum tunneling phenomena in Josephson systems, the effective mass is typically very high so that the crossover temperature can well be in the milli-Kelvin region [314].

Since the Im F method is basically a thermodynamic method, it is employable when thermal equilibrium prevails in the well region. This implies that the time scale for relaxation in the well is short compared to the average time for escape from the well. The Im F method can also be applied in certain limits, when the barrier is time-dependent. First, it is applicable in the adiabatic limit of slow-frequency driving [342]. In this case, the escaping particle is subject to the current barrier potential, and the total rate is found from the average of the relevant potential configurations. Secondly, the Im F method is also applicable in the case of fast driving, in which the escape occurs in an effective potential resulting from an average over one period of driving. Clearly, the Im F method is not applicable when the driving frequency is of the order of the bare or dressed well frequency.

Let us conclude this section with a general remark on a recent controversy. Tunneling has been investigated by considering classical real-time trajectories with energies higher than the barrier top [343]. In this approach, the smallness of the tunneling rate results from destructive interference of the various path contributions. Subsequently, it has been shown that the numerical real-time Fourier integral method which employs input only from the energy region above the barrier top is inadequate in the deep tunneling regime [344]. In the semiclassical quantum transition state theory discussed above and in subsequent chapters, tunneling is regarded as an imaginary-time process which can be quantitatively described in terms of a classical trajectory on the upside-down potential surface and of the fluctuations about this path.

12.6 Centroid method

Dynamical simulations of barrier crossing processes are impaired by the fact that the tunneling paths are rare events. The efficiency of the computation is enormously enhanced by filtering out irrelevant paths that do not cross the surface dividing the reactant from the reaction product and therefore do not contribute to the crossing rate. An efficient numerical strategy consists in a direct sampling of the centroid density $\rho_c(q_c)$ discussed in Section 8.1 by employing imaginary-time path integral Monte Carlo techniques for the constrained path integral in Eq. (8.4). Upon tuning the centroid variable q_c, the effective potential $\mathcal{V}(q_c)$ can be computed. Generally, the effective potential $\mathcal{V}(q_c)$ may not necessarily have the same shape as the original potential $V(q)$. As noticed by Gillan [345] and subsequently treated more rigorously by Voth *et al.* [346], the reaction rate is directly connected with the change of the effective classical potential $\mathcal{V}(q_c)$ when the centroid is tuned from the reactant state to the barrier top (dividing surface).

Assume that the effective potential has a local minimum at $q_c = q_{min}$ with well frequency ω_{min}, and a maximum at $q_c = q_{max}$ with barrier frequency ω_{max}. Unless $V(q)$ is symmetric, q_{max} is different from the position q_b of the barrier top of $V(q)$. Following the analysis given in Section 12.5, the point $q_c = q_{min}$ is a stable stationary point, and integration over q_c in the well region [cf. Eq. (8.3) with Eq. (8.6)] gives the partition function of reactants in the Gaussian approximation, $Z_0 = (1/\hbar\beta\omega_{min}) \exp[-\beta\mathcal{V}(q_{min})]$. In contrast, the point q_{max} is a saddle point which requires distortion of the integration contour into the complex q_c-plane. We then find in steepest descent the barrier contribution

$$\text{Im}\,Z_b = \frac{1}{2\beta\hbar\omega_{max}}\, e^{-\beta\mathcal{V}(q_{max})}\,. \tag{12.77}$$

Using Eq. (12.74), we obtain for the rate the expression

$$k = \frac{\sigma}{\hbar\beta}\,\frac{\omega_{min}}{\omega_{max}}\, e^{-\beta[\mathcal{V}(q_{max})-\mathcal{V}(q_{min})]}\,, \tag{12.78}$$

which for $\sigma = 1$ is Gillan's centroid formula [345]. For temperatures $k_B T \geqslant \hbar\omega_{max}/2\pi$, Affleck's factor σ in Eq. (12.78) is $\beta\hbar\omega_{max}/2\pi$ [cf. Eq. (12.75)], which is the centroid result obtained in Refs. [346, 347].

Instead of computing the effective classical potential numerically using quantum Monte Carlo techniques, one may estimate the effective potential by employing the variational methods discussed in Subsection 8.2.2, or one may apply semiclassical methods [348]. In the semiclassical limit, we have [cf. Eq. (8.11) with Eq. (8.14), and corresponding expressions for the parabolic barrier]

$$\begin{aligned}
\mathcal{V}(q_{min}) &= V(q_{min}) + \ln[\sinh(\beta\hbar\omega_{min}/2)/(\beta\hbar\omega_{min}/2)]/\beta\,,\\
\mathcal{V}(q_{max}) &= V(q_{max}) + \ln[\sin(\beta\hbar\omega_{max}/2)/(\beta\hbar\omega_{max}/2)]/\beta\,.
\end{aligned} \tag{12.79}$$

It is straightforward to see that Eq. (12.78) with Eq. (12.79) agrees for $T > T_0$ with the previous result in Eq. (12.32).

We remark again that the effective potential includes all quantum effects of the partition function. The result (12.78) involves only two Gaussian approximations and therefore seems "less approximate" than the semiclassical expression in which all fluctuation modes are treated in the Gaussian approximation. Generally one finds from numerical studies that the centroid approximation gives reasonable results for a symmetric or almost symmetric barrier and not too low temperatures [348]. It is seemingly superior to semiclassical methods for low barriers. Being nearly exact about T_0, the accuracy of the centroid method deteriorates as temperature is decreased. For very low T, the semiclassical bounce or periodic orbit method appears to be more accurate. With increasing asymmetry, the centroid approximation becomes less reliable. For a metastable potential that is not bounded from below, the centroid density diverges because of the unlimited potential drop beyond the barrier leading to infinite values for the rate. The generalization of the centroid method to the multidimensional case along the lines given in Section 8.1 and in Chapters 14–16 is straightforward.

13 Multidimensional quantum rate theory

13.1 The global metastable potential

We now generalize the treatment of quantum-statistical decay to a system with $N+1$ degrees of freedom and eventually take the limit $N \to \infty$. The composite system is assumed to be described by the system-plus-reservoir model (3.12) in which $V(q)$ is metastable as sketched in Fig. 10.1. We shall base the discussion on the thermodynamic $\mathrm{Im}\, F$ method. Some of the results thus obtained will be confirmed by the multidimensional generalization of the semiclassical method discussed in Sec. 12.3.

The thermodynamic approach starts out from the functional integral representation of the partition function of the damped system. The exact formal path integral expression for the reduced partition function Z is written in Eq. (4.214), and the effective action for the motion in the upside-down potential is given in Eq. (4.61). The multi-dimensional upside-down potential landscape $-V(q, \mathbf{x})$, where $V(q, \mathbf{x})$ is defined in Eq. (3.14) and replicated in Eq. (13.1), is sketched in Fig. 13.1.

For $N+1$ degrees of freedom, the potential $-V(q, \mathbf{x})$ for $q > 0$ is concave (positive curvature) in one direction, giving rise to periodic motion, but convex (negative curvature) in N directions, and thus unstable with respect to small perturbations in these directions. The total energy

$$
\begin{aligned}
E &= V[q(\tau), \mathbf{x}(\tau)] - \frac{M}{2}\dot{q}^2(\tau) - \frac{1}{2}\sum_{\alpha=1}^{N} m_\alpha \dot{x}_\alpha^2(\tau)\,, \\
V(q, \mathbf{x}) &= V(q) + \frac{1}{2}\sum_{\alpha=1}^{N} m_\alpha \omega_\alpha^2 \Big(x_\alpha - \frac{c_\alpha}{m_\alpha \omega_\alpha^2}q\Big)^2\,.
\end{aligned}
\tag{13.1}
$$

is usually the only constant of motion. The periodic orbit of the imaginary-time motion is then characterized by N stability angular frequencies [349].

13.2 Periodic orbit and bounce

13.2.1 The case $T = 0$

The tunneling contribution to the free energy is dominated by the periodic orbit in the $(q, \mathbf{x})$-plane. At $T = 0$, the total energy E is zero and the period of the orbit is infinity. The condition $V(q, \mathbf{x}) = 0$ defines the point P_{m} situated at the maximum of the parabolic hill of the upside-down potential ($q_{\mathrm{m}} = 0$, $\mathbf{x}_{\mathrm{m}} = 0$), as well as the outer N-dimensional hyperbolic surface $\Sigma_{\mathrm{ex}}(q, \mathbf{x})$. The point P_{m} is the generalization of the turning point $q_1 = 0$ in Eq. (12.20) for $E = 0$, whereas the surface Σ_{ex} is the generalization of the exit point q_{ex} for N bath degrees of freedom. The surface Σ_{ex} separates the well and barrier region from the classically accessible outside region. On the periodic orbit, the particle stays infinitely long in the infinite past at the point P_m and then gradually starts moving towards the well of the upside-down potential. Eventually, it rushes through the well (which actually is a saddle point, concave in one and convex in N dimensions), and then bounces back from the surface Σ_{ex} at time $\tau = 0$. The particle hits the surface Σ_{ex} perpendicularly at a particular point denoted by P_{ex}. On the way back it rushes again through the well on the same path, and finally comes to rest at time infinity at the starting point P_m. The orbit is parabolic near the surface Σ_{ex} and is therefore very sensitive to a small perturbation. This reflects the fact that the periodic orbit is unstable. The envelope surface of the parabolic paths is the caustic which touches the surface Σ_{ex} at the point P_{ex}. If the particle would not approach the surface Σ_{ex} perpendicularly, it could not return on the same path, but would run down for ever the slope of the potential $-V(q, \mathbf{x})$ [cf. Fig. 13.1, left]. The particular role of the unstable periodic orbit in multi-dimensional quantum decay has been emphasized by Banks *et al.* already in 1973 [350]. The periodic orbit is often referred to as the *most probable escape path* (MPEP).

13.2.2 The case $0 < T < T_0$

As T is raised slightly above zero, the energy E of the periodic orbit is raised too, and the period becomes $\hbar\beta$. Further, the point P_m dissolves into a N-dimensional elliptic surface Σ_m around $q = 0$, $\mathbf{x} = 0$. The inner elliptic surface Σ_m as well as the outer parabolic surface Σ_{ex} are defined by the condition $E - V(q, \mathbf{x}) = 0$. A particle on the periodic orbit leaves the surface Σ_m vertically at a particular point (denoted again by P_m) at a time, chosen as $-\hbar\beta/2$, then rushes through the well of the upside-down potential and bounces off the outer surface Σ_{ex} vertically at time $\tau = 0$ at a particular point P_{ex}. Then it returns to the starting point P_m on Σ_m at time $\hbar\beta/2$. The periodic orbit meets the same points in the $\{q, \mathbf{x}\}$-space on the way forth and back.

When the energy E is slightly below the barrier top V_b, the periodic orbit is degenerated to a harmonic oscillation around the saddle point along the stable direction with renormalized frequency ω_R. The effective barrier frequency ω_R is found from the normal mode analysis at the barrier top of $V(q, \mathbf{x})$, which is given below in Sec. 14.1. Since $2\pi/\omega_R$ is the shortest period of a periodic orbit, the temperature $T_0 \equiv \hbar\omega_R/2\pi k_B$ is a crossover temperature above which a periodic orbit no longer exists.

In Fig. 13.1, we have sketched the potential landscape (left) and the contour lines of equal potential height (right). We have also drawn the periodic orbit for a particular value of E with turning points on the surfaces Σ_m and Σ_{ex}, respectively. The periodic orbit or MPEP is a curved path in the $\{q, \mathbf{x}\}$-space with conserved total energy (13.1). The q-component of the periodic orbit is usually referred to as bounce $q_B(\tau)$. This path starts out from the position q_m at imaginary time $-\hbar\beta/2$, bounces back at time $\tau = 0$ from the point q_{ex} and returns to the starting point at time $\hbar\beta/2$.

The projection of the periodic orbit in $\{q, \mathbf{x}\}$-space onto the q-axis is accomplished upon elimination of the bath variables $\mathbf{x}(\tau)$, as explained in Section 4.2. The bounce

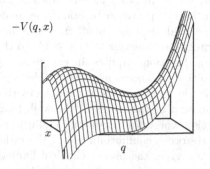 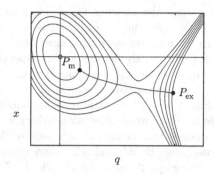

Figure 13.1: Sketch of the potential landscape of the upside-down potential $-V(q, \mathbf{x})$ for a single bath degree of freedom (left). The contour lines of equal potential height are shown for the potential $-V(q, \mathbf{x})$ (right). The contour lines around the top $q = 0$, $x = 0$ (symbol $\circ$) are elliptic, while those in the outside region $q > q_{\mathrm{b}}$ are hyperbolic. The curved periodic orbit with $E > 0$ has the inner turning point P_{m} on the elliptic contour Σ_{m} and the outer turning point P_{ex} on the hyperbolic contour Σ_{ex}.

path is a stationary point in function space of the action $S_{\mathrm{eff}}[q(\cdot)]$ defined in Eq. (4.61), $\delta S_{\mathrm{eff}}[q(\cdot)]/\delta q(\cdot)|_{q(\cdot)=q_{\mathrm{B}}(\cdot)} = 0$. It has period $\hbar\beta$ and satisfies, as a result of the projection, the time–non-local equation of motion

$$- M\,\ddot{q}_{\mathrm{B}}(\tau) + V'[q_{\mathrm{B}}(\tau)] + \int_{-\hbar\beta/2}^{\hbar\beta/2} d\tau'\, k(\tau - \tau') q_{\mathrm{B}}(\tau') = 0 . \qquad (13.2)$$

There is no conserved energy associated with this path.

The related bounce action is $S_{\mathrm{B}} = S_{\mathrm{eff}}[q_{\mathrm{B}}(\cdot)]$, written out in full

$$S_{\mathrm{B}} = \int_{-\hbar\beta/2}^{\hbar\beta/2} d\tau \left[\tfrac{1}{2} M\dot{q}_{\mathrm{B}}(\tau)^2 + V[q_{\mathrm{B}}(\tau)] \right] + \int_{-\hbar\beta/2}^{\hbar\beta/2} d\tau \int_{-\hbar\beta/2}^{\tau} d\tau'\, k(\tau - \tau')\, q_{\mathrm{B}}(\tau) q_{\mathrm{B}}(\tau') . \qquad (13.3)$$

As in the one-dimensional case discussed above in Section 12.5, there is again a crossover temperature T_0 at which the bounce action coincides with the action of the trivial constant path $q = q_{\mathrm{b}}$, $x_\alpha = (c_\alpha/m_\alpha\omega_\alpha^2)q_{\mathrm{b}}$ ($\alpha = 1, \cdots, N$). The crossover at $T = T_0$ from thermal to quantum decay will be discussed in Chapter 14. Special treatment of certain fluctuation modes shall be required in a narrow temperature range around T_0. The corresponding analysis is given in Chapter 16.

For a globally stable potential, the stationary points are minima of the action. In contrast, the bounce path occurring in a metastable potential is a saddle point, i.e., there is a direction in function space along which the action gets smaller. This phenomenon is because the potential has another local minimum beyond the barrier. As a result, there appears a fluctuation mode about the bounce which has a negative eigenvalue: the nodeless amplitude fluctuation mode. In order to get a meaningful expression for the bounce contribution to the barrier partition function, the mode with the negative eigenvalue must be treated by analytical continuation, as specified in Eq. (12.67). By deforming the respective integration contour from a stable to the actual unstable situation, the barrier partition function acquires an exponentially small imaginary part which can be assigned to the free energy as in Eq. (12.56).

Since the $\operatorname{Im} F$ method relies on equilibrium thermodynamics, it cannot describe effects which are related to nonthermal occupation of the states in the well. Such deviations may occur for extremely weak friction where the condition (11.45) is violated and the internal relaxation times for restoring equilibrium are of the order of the escape times or larger. The $\operatorname{Im} F$ method gives the proper classical limit of the escape in the damping regime (11.45). In the quantum regime, nonequilibrium effects are weaker. Hence the $\operatorname{Im} F$ method is valid down to considerably weaker damping than specified by the constraint (11.45). As we shall see, the rate formula for a damped system obtained by the $\operatorname{Im} F$ method approaches at high temperatures Kramers' expression (11.16) with Eq. (11.17) and Eq. (11.13). Moreover, the effects of frequency-dependent damping [351] as well as quantum corrections to thermally activated decay calculated by dynamical methods [352] are reproduced *quantitatively* by the thermodynamic method.

Among the first researchers who studied the effects of dissipation on the quantum mechanical decay of a metastable ground state were Caldeira and Leggett in 1983 [78]. They used the "bounce"-technique within the functional integral formulation [109]. Shortly later, the bounce method was found to be also an effective scheme for the calculation of dissipative quantum tunneling at finite temperature [353] – [356]. A unified description of quantum statistical decay of a metastable state in the entire temperature range was given in Ref. [311].

It is worth noting that a sporadic debate in the literature whether dissipation increases or decreases the tunneling probability is related to the absence or presence of the potential counter term $\Delta V(q)$, Eq. (3.6) with (3.10), in the system-reservoir coupling term defined in Eq. (3.4). As long as one wishes to study solely the effects of friction and decoherence, and to disregard the potential renormalization induced by the coupling, one has to choose $\Delta V(q)$ as given in Eq. (3.6). Then one finds, as we shall see, that damping *always* decreases the decay probability. This is because the environmental coupling in the Hamiltonian (3.12) does not change the barrier height, but the tunneling distance along the curved periodic orbit gets larger as the coupling is increased (see Fig. 13.1). On the other hand, if we choose $\Delta V(q) = 0$, the barrier height is ultimately reduced by the coupling. In fact, enhancement of tunneling by weakening the barrier overcompensates counteractive reduction of tunneling by the curvature of the periodic orbit. Overall, if $\Delta V(q) = 0$, the decay of a metastable state is facilitated by the coupling.

The functional integral method provides a unified approach to weak metastability of a dissipative system in large regions of the parameter space, as we shall see.

14 Crossover from thermal to quantum decay

Under the condition of weak metastability, Eq. (10.3), the functional integral (4.214) for the partition function is dominated by the stationary points of the action. Besides the periodic bounce $q_B(\tau)$, there are two trivial constant solutions of Eq. (4.215). On the one hand, we have $\bar{q}(\tau) = q_b$, in which the particle sits in the minimum of the upside-down potential $-V(q)$, and accordingly on the barrier top of the original potential. On the other hand, there is the constant path $\bar{q}(\tau) = 0$, in which the particle sits in the minimum of the original potential, and accordingly at the local maximum of the upside-down potential (see Figs. 10.1 and 10.2).

For T below the crossover temperature T_0, the bounce with period $\hbar/k_B T$ exists. In the absence of damping, the shortest period is $\hbar\beta_0 = 2\pi/\omega_b$, at which the bounce is a harmonic oscillation about q_b. The respective temperature $T_0 = \hbar\omega_b/2\pi k_B$ is in agreement with Goldanskii's conjecture, Eq. (12.76). At temperatures below T_0, the action S_B of the bounce $q_B(\tau)$ is smaller than the action $\hbar\beta V_b$ of the constant path $\bar{q}(\tau) = q_b$. As a result, the barrier contribution to the partition function is dominated by the bounce $q_B(\tau)$, instead by the trivial constant path $\bar{q}(\tau) = q_b$. From this we infer that T_0 is the temperature where roughly the transition between thermally activated decay and quantum tunneling decay occurs.

We now turn to the discussion of a damped particle. Again, the crossover temperature is defined as the temperature at which the action of the bounce trajectory and the action of the constant path $q = q_b$ coincide. For a study of this regime it is expedient to perform a normal mode analysis of the global system at the barrier top.

14.1 Normal mode analysis at the barrier top

The dynamics near the barrier top can be solved in analytic form with a normal mode analysis of the system-plus-reservoir Hamiltonian for a parabolic barrier [357]. The mixed quadratic form of the Hamiltonian is

$$H^{(b)} = \frac{1}{2}M\dot{q}^2 + \frac{1}{2}\sum_{\alpha=1}^{N} m_\alpha \dot{x}_\alpha^2 + V^{(b)}(q, \mathbf{x}) ,$$

$$V^{(b)}(q, \mathbf{x}) = V_b - \frac{1}{2}M\omega_b^2(q - q_b)^2 + \frac{1}{2}\sum_{\alpha=1}^{N} m_\alpha \omega_\alpha^2 \left(x_\alpha - \frac{c_\alpha}{m_\alpha \omega_\alpha^2} q \right)^2 . \tag{14.1}$$

The barrier top is at $q = q_b$, $x_\alpha = (c_\alpha/m_\alpha\omega_\alpha^2)q_b$. Choosing for the deviations from the barrier top the mass weighted coordinates (4.233), one obtains

$$H^{(b)} = V_b + \frac{1}{2}\sum_{\alpha=0}^{N} \dot{u}_\alpha^2 + \frac{1}{2}\sum_{\alpha=0}^{N}\sum_{\alpha'=0}^{N} U_{\alpha\alpha'}^{(b)} u_\alpha u_{\alpha'} . \tag{14.2}$$

As the force constant matrix at the barrier top $\mathbf{U}^{(b)}$ covers a repelling spring force for the central mode, $U^{(b)}_{00} = -\omega^2_b$, it differs from the matrix $\mathbf{U}^{(0)}$ of the stable harmonic system given in Eq. (4.235) in the element $U^{(b)}_{00}$,

$$U^{(b)}_{\alpha\alpha'} = U^{(0)}_{\alpha\alpha'} - \delta_{\alpha,0}\delta_{\alpha',0}\left(\omega^2_0 + \omega^2_b\right). \tag{14.3}$$

On the other hand, the normal mode representation of $H^{(b)}$ is

$$H^{(b)} = V_b + \frac{1}{2}\sum_{\alpha=0}^{N}\left(\dot{y}^2_\alpha + \mu^{(b)\,2}_\alpha y^2_\alpha\right). \tag{14.4}$$

Here, $y_0(\tau)$ is the unstable mode with squared frequency $\mu^{(b)\,2}_0 \equiv -\omega^2_R$, while the normal modes $y_\alpha(\tau)$, $\alpha \neq 0$, are stable bath modes with frequency $\mu^{(b)}_\alpha$.

Similar to the relation (4.245), we may express the determinant of $\mathbf{U}^{(b)}$ both with the original frequencies of the system and with the eigenfrequencies,

$$\det \mathbf{U}^{(b)} = \prod_{\alpha=0}^{N}\mu^{(b)\,2}_\alpha = -\omega^2_b\prod_{\alpha=1}^{N}\omega^2_\alpha. \tag{14.5}$$

Solution of the secular equation $\det\left(\mathbf{U}^{(b)} - \lambda^2\,\mathbf{1}\right) = 0$ yields for the eigenvalues of the force constant matrix $\mathbf{U}^{(b)}$ the implicit relations [357]

$$\omega^2_R = \omega^2_b \bigg/ \left(1 + \frac{1}{M}\sum_{\alpha'=1}^{N}\frac{c^2_{\alpha'}}{[m_{\alpha'}\omega^2_{\alpha'}(\omega^2_{\alpha'} + \omega^2_R)]}\right), \tag{14.6}$$

$$\mu^{(b)\,2}_\alpha = \omega^2_\alpha \bigg/ \left(1 + \frac{1}{M}\sum_{\alpha'=1}^{N}\frac{c^2_{\alpha'}}{[m_{\alpha'}\omega^2_{\alpha'}(\omega^2_{\alpha'} - \mu^{(b)\,2}_\alpha)]}\right), \qquad \alpha \neq 0. \tag{14.7}$$

The original system coordinate q is a linear combination of the mass weighted normal coordinates $\{y_\alpha\}$ according to the orthogonal transformation

$$q = q_b + \frac{1}{\sqrt{M}}u_{00}\,y_0 + \frac{1}{\sqrt{M}}\sum_{\alpha=1}^{N}u_{\alpha 0}\,y_\alpha, \tag{14.8}$$

with the matrix elements

$$u_{00} = \left(1 + \frac{1}{M}\sum_{\alpha'=1}^{N}\frac{c^2_{\alpha'}}{m_{\alpha'}(\omega^2_{\alpha'} + \omega^2_R)^2}\right)^{-1/2}. \tag{14.9}$$

$$u_{\alpha 0} = \left(1 + \frac{1}{M}\sum_{\alpha'=1}^{N}\frac{c^2_{\alpha'}}{m_{\alpha'}(\omega^2_{\alpha'} - \mu^{(b)\,2}_\alpha)^2}\right)^{-1/2}. \tag{14.10}$$

The frequency ω_R of the unstable mode $y_0(\tau)$ may equivalently be expressed in terms of the spectral density $J(\omega)$ or in terms of the friction kernel $\hat{\gamma}(z)$. There follows from Eq. (14.6) with use of the definitions (3.25) and (3.29)

$$\omega_R^2 = \omega_b^2 \left(1 + \frac{2}{\pi M}\int_0^\infty d\omega \, \frac{J(\omega)}{\omega}\frac{1}{(\omega^2 + \omega_R^2)}\right)^{-1} = \frac{\omega_b^2}{1 + \hat{\gamma}(\omega_R)/\omega_R}. \qquad (14.11)$$

The latter equality may equivalently be written as the relation

$$\omega_R^2 + \omega_R \hat{\gamma}(\omega_R) = \omega_b^2. \qquad (14.12)$$

Since $\hat{\gamma}(\omega_R) > 0$, the effective barrier frequency ω_R is smaller than the bare barrier frequency ω_b. As visible in Fig. 13.1, the curvature of the MPEP at the saddle point becomes smaller when the coupling to the bath modes is turned on, because the latter modes yield negative contributions to the curvature.

Equation (14.12) has a unique real positive solution for ω_R. The renormalized barrier frequency ω_R is the familiar Kramers-Grote-Hynes frequency appearing in the theory of non-Markovian rate processes [351].

14.2 Turnover theory for activated rate processes

The development of a unified theory for the classical attempt frequency bridging between the spatial-diffusion limited transmission factor $\varrho = \omega_R/\omega_b$ [cf. Eq. (11.13) in the Ohmic case] and the energy-diffusion limited expression (11.39) is known as the *Kramers turnover problem*. Near to the barrier top, the total energy E of the global system becomes the sum of the normal mode energies E_α, $E = \sum_{\alpha=0}^N E_\alpha$. Since the normal modes are not coupled, the probability to cross the barrier depends entirely on the energy E_0 left in the unstable mode y_0. When $E_0 < V_b$, the particle moving along the escape coordinate y_0 encounters a turning point and thereafter it returns to the well. When $E_0 > V_b$, the particle leaves the well in each attempt with probability one. The particular role of the unstable mode has been utilized in Refs. [358, 359]. By calculating the energy left in the unstable mode y_0, rather than in the physical coordinate q, inconsistencies in an earlier theory [360] could be resolved. Following Kramers [266], imagine that particles are constantly injected near to the bottom of the well. Then the system will approach a steady flux state. When the probability in the well is normalized as to describe a single particle, the flux equals the classical escape rate k_{cl}. Let $p(E_0)dE_0$ be the probability per unit time to find the particle in the unstable mode y_0 near the barrier top in the energy interval between E_0 and $E_0 + dE_0$. Since particles with $E_0 > V_b$ escape, the rate is given by the expression

$$k_{cl} = \int_{V_b}^\infty dE_0 \, p(E_0). \qquad (14.13)$$

In the general case, $p(E_0)$ is a nonequilibrium probability distribution.

Now assume as a first crude guess a thermal distribution in the well, in fact also in the vicinity of the barrier top. Upon writing the canonical distribution in the normal mode representation (14.4) at the barrier top, and the normalization factor in the normal mode representation (4.237) of the well region, the equilibrium distribution of the unstable mode takes the form

$$W_{eq}(\dot{y}_0, y_0) = \frac{\prod_{\alpha=1}^{N} \int d\dot{y}_\alpha \, dy_\alpha \exp\left(-\frac{1}{2}\beta\left[\dot{y}_\alpha^2 + \mu_\alpha^{(b)\,2} y_\alpha^2\right]\right)}{\prod_{\alpha=0}^{N} \int d\dot{\rho}_\alpha \, d\rho_\alpha \exp\left(-\frac{1}{2}\beta\left[\dot{\rho}_\alpha^2 + \mu_\alpha^{(0)\,2} \rho_\alpha^2\right]\right)} \, e^{-\beta E_0} , \tag{14.14}$$

where $E_0 = \frac{1}{2}\dot{y}_0^2 + V_b - \frac{1}{2}\omega_R^2 y_0^2$ is the energy accumulated in the unstable mode y_0. With the relations (4.245) and (14.5), and with $\mu_0^{(b)\,2} = -\omega_R^2$, we readily obtain

$$p_{eq}(E_0) \equiv W_{eq}(\dot{y}_0, y_0) = \frac{\beta}{2\pi}\sqrt{\frac{\det \mathbf{U}^{(0)} \mu_0^{(b)\,2}}{\det \mathbf{U}^{(b)}}} \, e^{-\beta E_0} = \frac{\beta \omega_0}{2\pi} \frac{\omega_R}{\omega_b} \, e^{-\beta E_0} . \tag{14.15}$$

With the equilibrium probability (14.15) the rate (14.13) takes the form (11.16),

$$k_{cl} = f_{cl} \, e^{-\beta V_b} , \tag{14.16}$$

with the classical attempt frequency

$$f_{cl} = \varrho \, \frac{\omega_0}{2\pi} \qquad \text{with} \qquad \varrho = \frac{\omega_R}{\omega_b} . \tag{14.17}$$

This epression is the generalization of the Ohmic form to general frequency-dependent friction. We see that memory-friction manifests itself through the Grote-Hynes frequency ω_R, which is implicitly given by the relation (14.12) [351].

Consider now the probability distribution $p(E_0)$ more closely. Because of the outgoing flux, the actual distribution $p(E_0)$ may deviate from the thermal distribution $p_{eq}(E_0)$ (14.15). In general, $p(E_0)$ obeys the steady-state condition

$$p(E_0) = \int_0^{V_b} dE_0' \, P(E_0|E_0') \, p(E_0') . \tag{14.18}$$

Here $P(E_0|E_0')$ is the conditional probability that the particle leaving the barrier region with energy E_0' in the unstable mode y_0 returns to the barrier with an energy E_0. The conditional probability satisfies detailed balance,

$$P(E_0|E_0') \, e^{-\beta E_0'} = P(E_0'|E_0) \, e^{-\beta E_0} . \tag{14.19}$$

As a result of the relation (14.19), the distribution $p(E_0)$ matches with $p_{eq}(E_0)$ at energy E_0 considerably below V_b.

The key quantity now is the energy transferred from the unstable mode to the bath as the particle moves from the barrier to the well and then back to the barrier. The equation of motion for the stable mode α governed by the full Hamiltonian is

$$\ddot{y}_\alpha(t) + \mu_\alpha^{(b)\,2} y_\alpha(t) = g_\alpha F(t) , \qquad \alpha \neq 0 , \tag{14.20}$$

where

$$g_\alpha = u_{\alpha 0}/u_{00} . \tag{14.21}$$

The force pulse $F(t)$ comes from the anharmonic part of the potential

$$V^{(\mathrm{nl})}(q) = V(q) - V_{\mathrm{b}} + \tfrac{1}{2}M\omega_{\mathrm{b}}^2(q - q_{\mathrm{b}})^2 \ . \tag{14.22}$$

The coupling of the unstable mode to the reservoir is weak on the condition

$$\varepsilon \equiv \sum_{\alpha=1}^{N} g_\alpha^2 = \frac{1}{u_{00}^2} - 1 \ll 1 \ . \tag{14.23}$$

In the regime $\varepsilon \ll 1$ we may disregard the back-action of the stable modes on the force pulse in Eq. (14.20). Then the nonlinear force $F(t)$ depends on the time dependence of the unstable mode according to the relation

$$F(t) = -u_{00}\frac{1}{M}\frac{\partial V^{(\mathrm{nl})}(q)}{\partial q}\Bigg|_{q = q_{\mathrm{b}} + u_{00}y_0(t)/\sqrt{M}} \ . \tag{14.24}$$

Equation (14.20) describes a forced harmonic oscillator. It is easily solved for arbitrary initial values of y_α and $\dot{y}_\alpha$ at time $t = 0$. The energy transferred on average per cycle from the unstable mode y_0 to the stable mode y_α is

$$\begin{aligned}
\Delta E_{\mathrm{cycle}} &= \frac{1}{2}g_\alpha^2 \int_{-\infty}^{\infty} dt \int_{-\infty}^{\infty} dt' \cos\left[\mu_\alpha^{(\mathrm{b})}(t - t')\right] F(t)F(t') \\
&\quad + g_\alpha \int_{-\infty}^{\infty} dt \left[\dot{y}_\alpha(0)\cos\left(\mu_\alpha^{(\mathrm{b})}t\right) - y_\alpha\mu_\alpha^{(\mathrm{b})}\sin\left(\mu_\alpha^{(\mathrm{b})}t\right)\right] F(t) \ .
\end{aligned} \tag{14.25}$$

The time integrations cover the history of the unstable trajectory $y_0(t)$ which starts at time $t = -\infty$ at the barrier, then moves through the well, and finally returns to the barrier at time $t = +\infty$. To leading order, the equation of motion of the unstable mode is decoupled from the bath modes and takes the form

$$\ddot{y}_0(t) - \omega_{\mathrm{R}}^2 y_0 + u_{00}\frac{1}{M}\frac{\partial V^{(\mathrm{nl})}(q)}{\partial q}\Bigg|_{q = q_{\mathrm{b}} + u_{00}y_0(t)/\sqrt{M}} = 0 \ . \tag{14.26}$$

Next, we choose thermal initial conditions according to

$$\begin{aligned}
\mu_\alpha^{(\mathrm{b})\,2}\langle y_\alpha^2(0)\rangle &= \langle \dot{y}_\alpha^2(0)\rangle = k_{\mathrm{B}}T \ , \\
\langle y_\alpha(0)\rangle &= \langle \dot{y}_\alpha(0)\rangle = \langle y_\alpha(0)\dot{y}_\alpha(0)\rangle = 0 \ .
\end{aligned} \tag{14.27}$$

Then the energy transferred on average from the particle to the reservoir per cycle is

$$\Delta E \equiv \langle \Delta E_{\mathrm{cycle}}\rangle = \frac{1}{2}\int_{-\infty}^{\infty} dt \int_{-\infty}^{\infty} dt' \,\Phi(t - t')F(t)F(t') \tag{14.28}$$

with the dissipation kernel

$$\Phi(t - t') = \sum_{\alpha=1}^{N} g_\alpha^2 \cos\left[\mu_\alpha^{(\mathrm{b})}(t - t')\right] \ . \tag{14.29}$$

The Laplace transform of the kernel $\Phi(t - t')$ may be written as

269

$$\hat{\Phi}(z) = \sum_{\alpha=1}^{N} g_\alpha^2 \frac{z}{z^2 + \mu_\alpha^{(b)2}} = \frac{1}{u_{00}} \frac{z}{z^2 + z\hat{\gamma}(z) - \omega_b^2} - \frac{z}{z^2 - \omega_R^2} . \qquad (14.30)$$

The first expression is the Laplace transform of (14.29). The second form expresses $\hat{\Phi}(z)$ in terms of the spectral damping function $\hat{\gamma}(z)$. The derivation of this form is less direct and involves the inversion of the force constant matrix $\mathbf{U}^{(b)}$ [359]. Finally, the uncertainty of the mean square energy loss for the initial state is

$$\delta E^2 \equiv \langle \Delta E_{\text{cycle}}^2 \rangle - \langle \Delta E_{\text{cycle}} \rangle^2 = 2k_BT\Delta E . \qquad (14.31)$$

Observing the relations (14.28) and (14.31), the conditional probability $P(E_0|E_0')$ is well described by the Gaussian distribution

$$P(E_0|E_0') = \left(4\pi k_BT\Delta E\right)^{-1/2} \exp\left[-\left(E_0 - E_0' + \Delta E\right)^2 / 4k_BT\Delta E\right] . \qquad (14.32)$$

Considering that the solution of Eq. (14.18) matches the thermal distribution $p_{\text{eq}}(E_0)$ in the regime $V_b - E_0 \gg k_BT$, it is convenient to put

$$p(E_0) = p_{\text{eq}}(E_0) \, e^{\beta(E_0-V_b)/2} \, \phi[\beta(E_0 - V_b)] . \qquad (14.33)$$

The ansatz (14.33) transforms Eq. (14.18) into a Wiener-Hopf equation for the function $\phi(x)$ with a symmetric kernel which can be solved by standard techniques [360]. With the resulting expression for $p(E)$ the rate expression (14.13) is found in the form (11.16) with the classical attempt frequency (11.17), $f_{\text{cl}} = \varrho\omega_0/2\pi$, but now the transmission factor is

$$\varrho = \frac{\omega_R}{\omega_b} \exp\left\{\frac{1}{\pi}\int_{-\infty}^{\infty} \frac{dx}{1+x^2} \ln\left[1 - \exp\left(-\frac{1+x^2}{4}\frac{\Delta E}{k_BT}\right)\right]\right\} . \qquad (14.34)$$

For $\Delta E \gg k_BT$, the transmission factor is close to the Grote-Hynes value ω_R/ω_b, $\varrho = (\omega_R/\omega_b)[1 - \exp(-\Delta E/4k_BT)]$. Nonequilibrium effects in $p(E_0)$ are only important when ΔE is of the order of k_BT or smaller. When the energy loss is very small, $\Delta E \ll k_BT$, the expression (14.34) yields linear dependence on ΔE,

$$\varrho = \frac{\omega_R}{\omega_b}\frac{\Delta E}{k_BT} , \qquad \text{for} \qquad \Delta E \ll k_BT . \qquad (14.35)$$

Importantly, ΔE coincides for weak damping with the energy loss given in Eq. (11.31), and thus the transmission factor (14.35) matches Kramers' finding (11.39).

A deeper analytical and numerical study of the energy loss formula (14.28) with (14.29) is given in Ref. [359]. A multidimensional generalization of the turnover theory for activated rate processes is reported in Ref. [361]. This concludes our discussion of the Kramers turnover problem.

14.3 The crossover temperature

At $T = 0$, the period of the bounce is infinity, and the amplitude, i.e., the distance taken across the barrier, is maximal. With increasing temperature, both the period and the amplitude of the bounce decrease. Eventually, the bounce has contracted to a small-amplitude oscillation around the barrier point q_b. In the $N + 1$-dimensional configuration space, there is left a harmonic oscillation of the mode $y_0(\tau)$ in imaginary time with frequency ω_R. Importantly, there is no periodic orbit with a frequency smaller than the Grote-Hynes ω_R. The crossover temperature [362, 354]

$$T_0 \equiv \hbar\omega_R/2\pi k_B , \qquad (14.36)$$

is the temperature at which the action S_B of the bounce $q_B(\tau)$ coincides with the action of the harmonic oscillation of the mode $y_0(\tau)$, and also with the action of the trivial path $\bar{q}(\tau) = q_b$. This action is $S_B(T=T_0) = \hbar\beta_0 V_b = 2\pi V_b/\omega_R$. Roughly speaking, at temperature $T = T_0$ there is the crossover between thermal hopping and quantum tunneling. The narrow temperature window around T_0 will be considered more closely in Chapter 16.

The relation (14.36) with (14.12) is valid for any linear dissipation mechanism. Since $\hat{\gamma}(\omega_R)$ in Eq. (14.12) is positive, the crossover temperature T_0 is always lowered with onset of damping. On the other hand, T_0 increases monotonously towards the value $\hbar\omega_b/2\pi k_B$ of the undamped case as the memory-friction relaxation time is increased while $\hat{\gamma}(\omega = 0)$ is held fixed. In the particular case of Ohmic damping, $\hat{\gamma}(z) = \gamma$, and a cubic potential of the form (11.40), for which $\omega_b = \omega_0$, we obtain from Eq. (14.12)

$$\omega_R = \omega_b \left(\sqrt{1 + \alpha^2} - \alpha\right) . \qquad (14.37)$$

Here $\alpha = \gamma/2\omega_0$ is the usual dimensionless damping parameter. With this we get

$$k_B T_0 = (\hbar\omega_b/2\pi) \left(\sqrt{1 + \alpha^2} - \alpha\right) . \qquad (14.38)$$

In the strong damping limit, Eq. (14.38) simplifies to the form

$$k_B T_0 = \frac{\hbar}{2\pi} \frac{\omega_b^2}{\gamma} = \frac{\hbar\omega_0}{4\pi\alpha} . \qquad (14.39)$$

The latter form holds if $\omega_b = \omega_0$. The monotonous decrease of the crossover temperature with increasing damping strength is sketched in Fig. 14.1.

With the effective barrier frequency ω_R, the general relation (12.74) between the rate and the imaginary part of the free energy takes the form

$$k = -2\sigma \operatorname{Im} F/\hbar , \qquad (14.40)$$

where

$$\begin{aligned} \sigma &= 1 & &\text{for} & \beta\hbar\omega_R &\geqslant 2\pi , \\ \sigma &= \beta\hbar\omega_R/2\pi & &\text{for} & \beta\hbar\omega_R &< 2\pi . \end{aligned} \qquad (14.41)$$

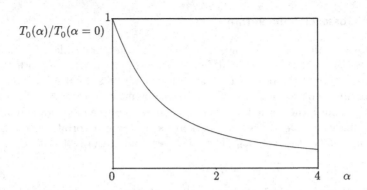

Figure 14.1: The normalized crossover temperature $T_0(\alpha)/T_0(\alpha = 0)$ is shown for Ohmic damping as a function of the parameter $\alpha = \gamma/2\omega_{\mathrm{b}}$.

Additional reasons for the validity of the expressions (14.40) with (14.41) are presented in Section 15.4. The unified formula (14.40) with (14.41) has been derived alternatively in a study of the reactive flux through a parabolic barrier [347].

The various temperature regions in quantum statistical decay are shown in Fig. 14.2. Below T_0, the functional integral is dominated by the bounce trajectory $q_{\mathrm{B}}(\tau)$, and the predominant escape mechanism is quantum mechanical tunneling. Above T_0, the functional integral is dominated by the constant trajectory $\bar{q}(\tau) = q_{\mathrm{b}}$, which leads to the Arrhenius factor $e^{-\beta V_{\mathrm{b}}}$, and the relevant decay process is thermally activated escape over the barrier. The fluctuation modes about $\bar{q}(\tau) = q_{\mathrm{b}}$ yield quantum corrections in the prefactor of the rate formula thereby enhancing the rate above the classical expression. They become increasingly important as the temperature is lowered. Special care is required in the evaluation of the functional integral in a narrow crossover region about T_0 (cf. Chapter 16). We have now prepared the ground to work out the rate formulas in the various temperature regimes.

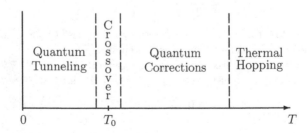

Figure 14.2: Sketch of the dominant escape mechanism as a function of temperature.

15 Thermally activated decay

15.1 Rate formula above the crossover regime

Consider the decay of the metastable state at temperatures fairly above T_0 so that thermal activation prevails. The rate may be written in the form

$$k = f_{\text{cl}} c_{\text{qm}} \, e^{-V_b/k_B T} , \tag{15.1}$$

where f_{cl} is the classical attempt frequency, and c_{qm} is a dimensionless factor covering enhancement of the classical rate by quantum fluctuations. We now compute f_{cl} and c_{qm} for arbitrary frequency-dependent damping.

We proceed by splitting the partition function (4.214) of the damped system into the well contribution $Z^{(0)}$ and the barrier contribution $Z^{(b)}$, which result from the Gaussian fluctuations about the constant paths $\bar{q}(\tau) = 0$ and $\bar{q}(\tau) = q_b$, respectively,

$$Z = Z^{(0)} + Z^{(b)} = Z^{(0)} \left(1 + Z^{(b)}/Z^{(0)} \right) . \tag{15.2}$$

Next, we recall the analysis of the fluctuation modes for the partition function of a damped harmonic system given in Section 4.3. A path with $\hbar\beta$-periodic fluctuations about the constant path $\bar{q}(\tau) = 0$ is written as

$$q(\tau) \longrightarrow q^{(0)}(\tau) = \sum_{n=-\infty}^{\infty} Y_n \, e^{i\nu_n \tau} , \tag{15.3}$$

where $\nu_n = 2\pi n/\hbar\beta$. Upon inserting the Fourier series (15.3) into the action $S_{\text{eff}}[q]$ given in Eq. (4.61), and disregarding terms of third and higher order of Y_n, we find[1]

$$S_2[q^{(0)}(\cdot)] = \frac{1}{2} M \hbar\beta \sum_{n=-\infty}^{\infty} \Lambda_n^{(0)} Y_n Y_{-n} . \tag{15.4}$$

The $\{\Lambda_n^{(0)}\}$ are the eigenvalues of the fluctuation operator $\Lambda^{(0)}$ defined in Eq. (4.223). With the findings in Subsection 4.3.3 [see Eq. (4.224)] we have

$$\Lambda_n^{(0)} = \nu_n^2 + \omega_0^2 + |\nu_n| \hat{\gamma}(|\nu_n|) . \tag{15.5}$$

Similarly, the path $q^{(b)}(\tau)$ with $\hbar\beta$-periodic fluctuations about q_b

$$q(\tau) \longrightarrow q^{(b)}(\tau) = q_b + \sum_{n=-\infty}^{\infty} X_n \, e^{i\nu_n \tau} \tag{15.6}$$

---

[1] In the sequel, we use for $S_{\text{eff}}[q]$ the simpler notation $S[q]$.

has the second order action

$$S_2[q^{(b)}(\cdot)] = \hbar\beta V_b + \frac{1}{2}M\hbar\beta \sum_{n=-\infty}^{\infty} \Lambda_n^{(b)} X_n X_{-n} \tag{15.7}$$

with

$$\Lambda_n^{(b)} = \nu_n^2 - \omega_b^2 + |\nu_n|\,\hat{\gamma}(|\nu_n|) . \tag{15.8}$$

The modes $\{Y_n\}$ and $\{X_n\}$, apart from the mode X_0, have positive eigenvalues. The contributions of these modes to the partition function is found in the usual way by carrying out the Gaussian integrals over the amplitude sets $\{Y_n\}$ and $\{X_n\}$ with use of the functional measure (4.231). The mode X_0 needs special treatment. Since the corresponding eigenvalue $\Lambda_0^{(b)} = -\omega_b^2$ is negative, the integral over X_0 is divergent. The divergence is related to the fact that the action for the constant path $\bar{q}(\tau) = q_b$ is a saddle point in function space with the *unstable* direction along X_0. In fact, this should not be a surprise since we are trying to evaluate the free energy of an unstable system. Langer [334] was the first who showed that the functional integral can still be defined by deforming the integration contour of the variable X_0 into the upper half of the complex plane along the direction of steepest descent. This leads to a positive imaginary part of the barrier contribution $Z^{(b)}$ with the familiar factor $\frac{1}{2}$, as we have explained in item (2) in Section 12.5. The resulting imaginary part of the free energy

$$\mathrm{Im}\, F = -\,\mathrm{Im}\, Z^{(b)}/(\beta\, Z^{(0)}) \tag{15.9}$$

is expressed again in terms of a ratio of determinants,

$$\mathrm{Im}\, F = -\frac{1}{2\beta}\sqrt{D^{(0)}/|D^{(b)}|}\; e^{-\beta V_b} . \tag{15.10}$$

Here, $D^{(0)}$ and $D^{(b)}$ are the determinants related to the second order actions (15.4) and (15.7) for the well and barrier regions. In diagonal representation, we have

$$D^{(0)} = \omega_0^2 \prod_{n=1}^{\infty} \left(\Lambda_n^{(0)}\right)^2 , \qquad D^{(b)} = -\omega_b^2 \prod_{n=1}^{\infty} \left(\Lambda_n^{(b)}\right)^2 , \tag{15.11}$$

where $\Lambda_n^{(0)}$ and $\Lambda_n^{(b)}$ are given in Eqs. (15.5) and (15.8), respectively.

With use of Eqs. (12.73), (15.10) and (14.36) the rate above T_0 is found to read

$$k = \frac{1}{\hbar\beta_0}\sqrt{\frac{D^{(0)}}{|D^{(b)}|}}\; e^{-\beta V_b} = \frac{1}{\hbar\beta_0}\frac{\omega_0}{\omega_b}\prod_{n=1}^{\infty}\frac{\Lambda_n^{(0)}}{\Lambda_n^{(b)}}\; e^{-\beta V_b}$$

$$= \frac{\omega_0}{2\pi}\frac{\omega_R}{\omega_b}\prod_{n=1}^{\infty}\frac{\nu_n^2 + \omega_0^2 + \nu_n\hat{\gamma}(\nu_n)}{\nu_n^2 - \omega_b^2 + \nu_n\hat{\gamma}(\nu_n)}\; e^{-\beta V_b} . \tag{15.12}$$

Upon comparing Eq. (15.12) with Eq. (15.1), we find for the classical attempt frequency the previous form (14.17), and for the quantum mechanical enhancement factor the infinite product form [354]

$$c_{\mathrm{qm}} = \prod_{n=1}^{\infty} \frac{\Lambda_n^{(0)}}{\Lambda_n^{(b)}} = \prod_{n=1}^{\infty} \frac{\nu_n^2 + \omega_0^2 + \nu_n \hat{\gamma}(\nu_n)}{\nu_n^2 - \omega_b^2 + \nu_n \hat{\gamma}(\nu_n)} . \tag{15.13}$$

The result (15.13) is corroborated by an independent dynamical approach [352]. In the classical limit $T/T_0 \to \infty$, the factor c_{qm} approaches unity. Hence the rate expression (15.12) reduces to the proper classical form [cf. Eq. (14.16) with (14.17)]

$$k_{\mathrm{cl}} = \frac{\omega_0}{2\pi} \frac{\omega_{\mathrm{R}}}{\omega_{\mathrm{b}}} e^{-\beta V_{\mathrm{b}}} . \tag{15.14}$$

Compared with the TST formula (11.2), the classical rate is reduced by the transmission factor $\varrho = \omega_{\mathrm{R}}/\omega_{\mathrm{b}}$, which describes the effect of *recrossings* of the barrier top by the particle in the moderate to large damping region. The result agrees with the rate expression obtained from the extension of Kramers' approach to the case of memory friction [351].[2] The expression (15.14) also agrees with the classical rate expression obtained below in Section 15.4 from multidimensional transition state theory. There, however, the interpretation of the reduction factor ϱ will be different.

The effect of memory friction on the rate is reflected by the renormalization of the barrier frequency. Interestingly, the renormalized frequency ω_{R} is the same which enters the definition of the crossover temperature T_0. For frequency-independent damping, $\hat{\gamma}(z) = \gamma$, the attempt frequency (14.17) simplifies to Kramers' celebrated result, Eq. (11.17) with Eq. (11.13).

In summary, the key assumptions underlying the generalized Kramers formula (15.14) are as follows [312]:

(1) The coupling to the environment is so strong that the rate is controlled by the diffusive dynamics near to the barrier top.

(2) The effective potential is harmonic near the bottom of the well and has a parabolic shape around the barrier top.

(3) The stochastic dynamics is represented by the *linear* Langevin equation (2.4) with a Gaussian random force with classical power spectrum (2.9).

15.2 Quantum corrections in the pre-exponential factor

At temperatures well below the classical regime but still above the crossover temperature, the rate is enhanced by quantum corrections,

$$k = c_{\mathrm{qm}} k_{\mathrm{cl}} , \tag{15.15}$$

where c_{qm} is given in Eq. (15.13).

---

[2]In the literature of chemical physics, the factor $\varrho = \omega_{\mathrm{R}}/\omega_{\mathrm{b}}$ is known as the Grote-Hynes correction to classical TST.

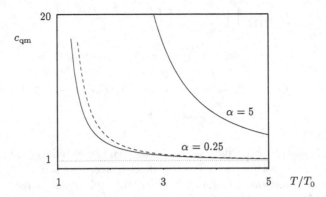

Figure 15.1: The quantum correction factor c_{qm} is shown as a function of the scaled temperature T/T_0 for a system with $\omega_0 = \omega_{\mathrm{b}}$ and frequency-independent damping for $\alpha = 0.25$ and $\alpha = 5$. The dashed curve is a plot of the formula (15.16), where $T_0 = T_0(\alpha = 0.25)$.

The leading quantum correction at high T is found upon rewriting the expression (15.13) as an exponential of a sum of logarithms and expanding each logarithm in powers of $1/\nu_n$. One then finds in leading order of $1/\nu_n$ a surprisingly simple expression which is independent of damping for arbitrary spectral coupling [362],

$$c_{\mathrm{qm}} \; = \; \exp\left((\omega_0^2 + \omega_{\mathrm{b}}^2) \sum_{n=1}^{\infty} \frac{1}{\nu_n^2} \right) \; = \; \exp\left(\frac{\hbar^2}{24} \frac{(\omega_0^2 + \omega_{\mathrm{b}}^2)}{(k_{\mathrm{B}}T)^2} \right) . \qquad (15.16)$$

The escape is enhanced by two different effects. First, quantum fluctuations increase the mean energy in the well. Second, when a particle is thermally excited almost to the barrier top, quantum fluctuations allow for tunneling through the remaining small barrier. Both effects add up and lead to an effective reduction of the barrier.

For frequency-independent damping, the infinite product in Eq. (15.13) can be expressed in terms of gamma functions [352]

$$c_{\mathrm{qm}} \; = \; \frac{\Gamma(1 + \lambda_{\mathrm{b}}^+/\nu)\,\Gamma(1 + \lambda_{\mathrm{b}}^-/\nu)}{\Gamma(1 + \lambda_0^+/\nu)\,\Gamma(1 + \lambda_0^-/\nu)} , \qquad (15.17)$$

where $\nu = \nu_1 = 2\pi k_{\mathrm{B}}T/\hbar$ is the smallest Matsubara frequency, and where

$$\lambda_{\mathrm{b}}^\pm \; = \; \frac{\gamma}{2} \pm \left(\frac{\gamma^2}{4} + \omega_{\mathrm{b}}^2 \right)^{1/2} , \qquad \lambda_0^\pm \; = \; \frac{\gamma}{2} \pm \left(\frac{\gamma^2}{4} - \omega_0^2 \right)^{1/2} . \qquad (15.18)$$

Plots of the enhancement factor c_{qm} versus temperature are shown in Fig. 15.1.

At strong friction $\gamma \gg \omega_0$, ω_{b} and $T \gg T_0$, we get from the expression (15.17)

$$c_{\mathrm{qm}} \; = \; \exp\left\{ \frac{\hbar\beta(\omega_{\mathrm{b}}^2 + \omega_0^2)}{2\pi\gamma} \left[\psi\left(1 + \frac{\hbar\beta\gamma}{2\pi}\right) - \psi(1) \right] \right\} , \qquad (15.19)$$

where $\psi(z)$ is the digamma function. This form reduces in the high temperature regime $T \gg \hbar\gamma/2\pi k_B$ to the leading quantum correction given above in Eq. (15.16).

In the temperature regime

$$T_0 \ll T \ll \hbar\gamma/2\pi k_B , \qquad (15.20)$$

where T_0 is given in Eq. (14.39), the expression (15.17) takes the form

$$c_{qm} = \exp\left\{ \frac{\hbar\beta(\omega_b^2 + \omega_0^2)}{2\pi\gamma} \ln\left(\frac{\hbar\beta\gamma}{2\pi}\right) \right\} = \left(\frac{\gamma^2 T_0}{\omega_b^2 T}\right)^{(1+\omega_0^2/\omega_b^2)T_0/T} . \qquad (15.21)$$

This factor can enhance the rate quite significantly, even well above the crossover temperature. For instance, for $T = 4T_0$ and $\omega_0 = \omega_b$, one gets $c_{qm} = \gamma/2\omega_b \gg 1$.

In the limit $T \to T_0$ from above, the eigenvalue $\Lambda_1^{(b)}$ in Eq. (15.13) drops to zero. As a result, the quantum correction factor c_{qm} becomes singular, as visible in Fig. 15.1. The singularity points to the fact that treatment of the fluctuation modes $X_{\pm 1}$, introduced in Eq. (15.6), in Gaussian approximation breaks down as T approaches T_0. The proper handling of these modes in the crossover regime is given in Chapter 16.

15.3 The quantum Smoluchowski equation approach

At strong friction $\gamma \gg \omega_0$, the evolution of the probability density in coordinate space $p(q,t)$ of a classical Brownian particle is described by the Smoluchowski diffusion equation (11.19). In the temperature regime (15.20), the dynamics of the Brownian particle can still be described by a diffusion equation of the form (11.19), but the diffusion coefficient D is above the classical value $D_{cl} = k_B T$ because of quantum fluctuations. When the drift potential is harmonic, $V(q) = \frac{1}{2}M\omega_0^2 q^2$, it is quite natural to assume that the diffusion coefficient is increased by quantum fluctuations in accordance with the enhancement of the position spread,

$$D_{cl} = k_B T = M\omega_0^2 \langle q^2 \rangle_{cl} \longrightarrow \tilde{D}_{qm} = M\omega_0^2 [\langle q^2 \rangle_{cl} + \lambda] , \qquad (15.22)$$

where $\lambda \equiv \langle q^2 \rangle_{qm} = \langle q^2 \rangle - \langle q^2 \rangle_{cl}$ is the quantum mechanical part of position spread at strong friction. Explicit expressions for λ are given in Eqs. (6.139) and (6.140).

For a nonlinear potential $V(q)$, it is natural to generalize the diffusion coefficient (15.22) to a position-dependent diffusion coefficient $\tilde{D}_{qm}(q) = k_B T + \lambda V''(q)$. We then arrive at the Smoluchowski diffusion equation analogous to Eq. (11.19) with (11.21),

$$\frac{\partial p(q,t)}{\partial t} = \frac{1}{M\gamma} \frac{\partial}{\partial q} \hat{L}_{qm}(q) p(q,t) \qquad (15.23)$$

with the position-dependent quantum Smoluchowski flux operator

$$\hat{L}_{qm}(q) = \frac{\partial V(q)}{\partial q} + \frac{\partial}{\partial q} \tilde{D}_{qm}(q) . \qquad (15.24)$$

Here the quantum Smoluchowski equation (QSE) has been deduced in a heuristic manner. The quantum correction in the diffusion coefficient $\tilde{D}_{qm}(q)$ has been verified by a path integral evaluation of the propagating function J introduced in Section 5.3.3 [363], and by different reasoning in Ref. [364].

The QSE with the diffusion coefficient $\tilde{D}_{qm}(q)$ has a weakness which shows up in order λ^2 and in higher orders of λ. Namely, for a periodic potential $V(q) = V(q+nL)$ the diffusion coefficient $\tilde{D}_{qm}(q)$ invalidates corresponding periodicity of the equilibrium potential $\psi(q) = \int_0^q dx\, V'(x)/\tilde{D}_{qm}(x)$ [cf. Eq. (11.25)] in order λ^2 and in higher orders of λ. As a result, for a periodic drift potential $V(q)$, there would be a nonzero stationary equilibrium current, which would violate the second law of thermodynamics in order λ^2. The appropriate modification of the diffusion coefficient, which coincides with $\tilde{D}_{qm}(q)$ in first order of λ, is [365]

$$D_{qm}(q) = k_B T/[1 - \lambda V''(q)/k_B T] . \tag{15.25}$$

With this form, $\psi(q)$ has the periodicity of $V(q)$,

$$\psi(q) = \beta\{V(q) - \tfrac{1}{2}\lambda\beta[V'(q)]^2\} . \tag{15.26}$$

Hence the QSE is free of a stationary current and thus fully reconciled with the second law of thermodynamics. The resulting equilibrium position distribution is

$$p_{eq}(q) = \frac{N_0}{D_{qm}(q)} e^{-\psi(q)} . \tag{15.27}$$

For a harmonic well, the equilibrium potential $\psi(q)$ agrees in order λ with the exponent $\tfrac{1}{2}q^2/\langle q^2\rangle$ of the equilibrium position distribution $< q|\hat{\rho}_\beta|q >$ given in Eq. (6.156). Since $\langle q^2\rangle$ gets smaller as γ is increased, the equilibrium distribution for a particle in a well is squeezed by friction quite substantially.

We may now study the escape of the quantum Brownian particle from a metastable well within the QSE. With use of the expressions (15.26) and (15.25), we immediately obtain from Eq. (11.28) the expression

$$k = \frac{\sqrt{V''(0)|V''(q_b)|}}{2\pi M\gamma} e^{-\beta V_b}\, \tilde{c}_{qm} , \qquad \text{with} \qquad \tilde{c}_{qm} = e^{\beta\lambda\{V''(0)+|V''(q_b)|\}/2} . \tag{15.28}$$

This result differs from the expression (11.29) by the factor $\tilde{c}_{qm}$. This term describes enhancement of the classical rate in the overdamped Smoluchowski regime by quantum fluctuations. With the explicit forms (6.136), (6.139) or (6.140) applicable in the respective temperature regimes, the enhancement factor $\tilde{c}_{qm}$ coincides in the respective limits with the earlier expressions (15.16), (15.19) and (15.21), which have been calculated within the path integral approach from the fluctuation determinants.

Finally, we mention that, in difference to the original work [363, 365], the QSE (15.23) with the flux operator (15.24) and the diffusion coefficient (15.25) has no quantum correction in the drift potential. If we would add the term $\Delta V_{qm} = \tfrac{1}{2}\lambda V''(q)$ to the drift potential $V(q)$, as done in Refs. [363, 365], the quantum mechanical enhancement in Eq. (15.28) would be doubled. The absence of quantum fluctuations in the drift potential has been noticed also in Refs. [313] and [364].

15.4 Multidimensional quantum transition state theory

With the above considerations, the escape problem for temperatures above T_0 is basically solved. As the occurrence of the factor σ in the formula (14.40) with (14.41) has been questioned in the early literature [366, 310], it seems useful to confirm the results of the Im F method by the independent periodic orbit approach.

An independent unique approach to quantum statistical decay is based on the multidimensional quantum transition state theory (MQTST) put forward by Miller [324]. In the semiclassical limit, the Euclidean propagator is dominated by *periodic orbits* in the upside-down potential landscape $-V(q, \mathbf{x})$ of the $(N + 1)$-dimensional configuration space. The periodic orbit is a solution of the equations of motion (4.35) and (4.36) analytically continued to imaginary time. It satisfies conservation of total energy according to Eq. (13.1). The qualitative behavior of the periodic orbit has been discussed in Chapter 13.

Following Gutzwiller [349], we now introduce a coordinate z_0 which measures distance along the curved periodic trajectory and N coordinates $\{z_j\}$, $j = 1, \cdots, N$, being displacements locally orthogonal to z_0. We shall call z_0 the *escape coordinate*, and $z_0^{(PO)}(\tau)$ the periodic orbit trajectory, which progresses along the escape coordinate. It is convenient to choose the phase of the periodic orbit such that

$$z_0^{(PO)}(\pm\hbar\beta/2) = z_0^{(m)}, \quad \text{and} \quad z_0^{(PO)}(0) = z_0^{(ex)}, \tag{15.29}$$

where $z_0^{(m)}$ and $z_0^{(ex)}$ are the turning points on the surfaces Σ_m and Σ_{ex} introduced in Chapter 13, respectively.

Next, we put a flat dividing surface Σ_b which delimits the domain of attraction about the metastable minimum from the exterior region. The escape path crosses the dividing plane Σ_b perpendicularly at $z_0^{(b)}$, where $z_0^{(b)}$ is the point on the escape coordinate at which the full potential $V(q, \mathbf{x})$ has a maximum. The transition state approximation implies that the escaping particle crosses the dividing surface Σ_b only once. With this choice of coordinates, we then have similar to Eq. (12.10)

$$\hat{F}\hat{P} \xrightarrow{\text{TST}} \delta\left(\hat{z}_0 - z_0^{(b)}\right) \frac{\hat{p}_0}{M}\Theta(\hat{p}_0) = \frac{1}{2M}\delta\left(\hat{z}_0 - z_0^{(b)}\right)\left[|\hat{p}_0| + \hat{p}_0\right]. \tag{15.30}$$

Again, the second term in the square bracket does not contribute because of detailed balance. The multi-dimensional generalization of Eq. (12.12) with (12.13) is

$$k = \frac{1}{Z_{\text{tot}}^{(0)}} \frac{1}{2\pi\hbar} \int_0^\infty dE \, p(E) \, e^{-\beta E}, \tag{15.31}$$

where $Z_{\text{tot}}^{(0)}$ is the partition function of the total system, and

$$p(E) = \lim_{\epsilon \to 0^+} \text{Im} \int_0^\infty d\tau \, e^{(E+i\epsilon)\tau/\hbar}$$
$$\times \int d\mathbf{z} \, \delta\left(z_0 - z_0^{(b)}\right) |\dot{z}_0|_{z_0 = z_0^{(b)}} < \mathbf{z} \, | \, e^{-\hat{H}\tau/\hbar} \, | \, \mathbf{z} > . \tag{15.32}$$

The function $p(E)$ represents the inclusive transmission probability at the total energy E. As the quantum statistical decay rate (15.31) is the Boltzmann-averaged transmission probability, we should expect that $p(E)$ includes all possible partitions of the total energy E into the energy left in the escape coordinate and the individual energies in the transverse degrees of freedom. This is the case, indeed, as we shall see shortly [cf. Eq. (15.40) below].

According to Gutzwiller's semiclassical analysis [349], a single periodic orbit in the multi-dimensional upside-down potential $-V(q, \mathbf{x})$ [see Eq. (13.1)] yields in generalization of Eq. (12.17) the contribution

$$p_1(E) = \frac{i\,\mathcal{J}_1}{\sqrt{2\pi\hbar}} \int_C d\tau \left| \frac{\partial^2 S_{cl}}{\partial \tau^2} \right|^{1/2} \left(\prod_{\alpha=1}^N \frac{1}{2\sinh[\tau\mu_\alpha(z)/2]} \right) e^{-[S_{cl}(\tau) - E\tau]/\hbar} . \tag{15.33}$$

The Euclidean action for a single periodic orbit $z_0^{(PO)}(\tau)$ with period $\bar{\tau}$ is

$$S_{cl}(\bar{\tau}) = W(E) + E\bar{\tau} , \tag{15.34}$$

where E is the conserved total energy (13.1) of this particular path, and

$$W(E) \equiv \int_0^{\bar{\tau}} d\tau' \left(M\dot{q}^2(\tau') + \sum_{\alpha=1}^N m_\alpha \dot{x}_\alpha^2(\tau') \right) \tag{15.35}$$

is the abbreviated action. The factor $\mathcal{J}_1 = -1$ keeps track of the phase factors $e^{-i\pi/2}$ picked up by the periodic orbit at each of the two conjugate points. Finally, the N-fold product in Eq. (15.33) results from the integrations of the Gaussian fluctuations perpendicular to the reaction coordinate. The set $\{\mu_\alpha(z)\}$ ($\alpha = 1, 2, \cdots, N$) represents the dynamical stability (angular) frequencies of the periodic orbit. Computation of the integral over τ in Eq. (15.33) by steepest descent, as in Section 12.3, yields in generalization of Eq. (12.23)

$$p_1(E) = -\mathcal{J}_1 \left(\prod_{\alpha=1}^N \frac{1}{2\sinh[\bar{\tau}(E)\mu_\alpha(E)/2]} \right) e^{-W(E)/\hbar} . \tag{15.36}$$

Consider next the contribution of n cycles at total energy E. Evidently, the abbreviated action is $nW(E)$, the total time spent is $\tau_n = n\bar{\tau}$, and the overall phase factor is $\mathcal{J}_n = (-1)^n$. Summation of the contributions from all numbers of cycles yields

$$p(E) = \sum_{n=1}^\infty (-1)^{n-1} e^{-nW(E)/\hbar} \prod_{\alpha=1}^N \frac{1}{2\sinh[n\bar{\tau}(E)\mu_\alpha(E)/2]} . \tag{15.37}$$

In the further processing, we expand each of the $\sinh^{-1}(\cdots)$-functions into a geometrical series of exponentials, and put $\bar{\tau} = -W'(E)$. With this we can perform the sum over n. The resulting expression is

$$p(E) = \sum_{n_1=0}^{\infty} \sum_{n_2=0}^{\infty} \cdots \sum_{n_N=0}^{\infty}$$

$$\times \left\{ 1 + \exp\left[\frac{1}{\hbar} \left(\mathbb{W}(E) - \mathbb{W}'(E) \sum_{\alpha=1}^{N} (n_\alpha + \tfrac{1}{2}) \hbar \mu_\alpha(E) \right) \right] \right\}^{-1} . \tag{15.38}$$

For a given set of quantum numbers $\{n_\alpha\}$ in the transverse degrees of freedom, the energy left in the escape coordinate z_0 is

$$E_{\rm esc} = E - E_\perp , \qquad \text{with} \qquad E_\perp = \sum_{\alpha=1}^{N} (n_\alpha + \tfrac{1}{2}) \hbar \mu_\alpha(E) . \tag{15.39}$$

With the observation that the argument of the exponential function in Eq. (15.38) represents the leading terms of a Taylor expansion of $W(E_{\rm esc})$ around $W(E)$, the micro-canonical cumulative transmission probability can be written as [367, 312]

$$p(E) = \sum_{n_1=0}^{\infty} \sum_{n_2=0}^{\infty} \cdots \sum_{n_N=0}^{\infty} \frac{1}{e^{\mathbb{W}(E_{\rm esc})/\hbar} + 1} . \tag{15.40}$$

For $T > T_0$, the main contribution to the integral in Eq. (15.31) comes from the energy range $E_{\rm esc} \gtrsim V_{\rm b}$. In this regime, the barrier is parabolic and the global system is conveniently given in the normal mode representation (14.4). Near the barrier top the escape coordinate z_0 coincides with the normal mode y_0, and the stability angular frequencies μ_α of the periodic orbit correspond to the eigenfrequencies $\mu_\alpha^{(b)}$, $(\alpha = 1, 2, \cdots, N)$. In the parabolic barrier approximation, the periodic orbit z_0 is oscillating along y_0 with period $2\pi/\omega_{\rm R}$, and the abbreviated action simply is

$$\mathbb{W}(E_{\rm esc}) \approx \mathbb{W}_{\rm harm}(E_{\rm esc}) = 2\pi(V_{\rm b} - E_{\rm esc})/\omega_{\rm R} . \tag{15.41}$$

Upon interchanging in Eq. (15.31) the integration over E with the summations in the expression (15.40) for $p(E)$, one obtains [367]

$$k = \frac{1}{Z_{\rm tot}^{(0)}} \frac{1}{2\pi\hbar} \sum_{n_1=0}^{\infty} \sum_{n_2=0}^{\infty} \cdots \sum_{n_N=0}^{\infty} e^{-\beta(E_\perp + V_{\rm b})} \int_0^{\infty} dE \, \frac{e^{-\beta(E - E_\perp - V_{\rm b})}}{1 + e^{-\beta_0(E - E_\perp - V_{\rm b})}} , \tag{15.42}$$

where $\beta_0 = 2\pi/\hbar\omega_{\rm R}$, and where $E_\perp$ is the energy in the transverse modes.

The partition function $Z_{\rm tot}^{(0)}$ of the well regime plus environment is conveniently expressed in terms of the eigenfrequencies $\mu_\alpha^{(0)}$, as discussed in Subsec. 4.3.5 and specified in Eq. (4.238). Extending in Eq. (15.42) the lower integration limit to $E = -\infty$, the integral yields the function $\pi/[\beta_0 \sin(\pi\beta/\beta_0)]$. Corrections due to the actually finite lower bound of the integral in Eq. (15.42) are exponentially small in the range $T_0 < T \ll V_{\rm b}/k_{\rm B}$. It is now straightforward to perform the summations, yielding

$$k = \frac{\omega_{\rm R}}{2\pi} \frac{\sinh(\beta\hbar\mu_0^{(0)}/2)}{\sin(\beta\hbar\omega_{\rm R}/2)} \prod_{\alpha=1}^{N} \frac{\sinh(\beta\hbar\mu_\alpha^{(0)}/2)}{\sinh(\beta\hbar\mu_\alpha^{(b)}/2)} e^{-\beta V_{\rm b}} . \tag{15.43}$$

Next, the product of the sinh factors can be transformed to read

$$
\prod_{\alpha=0}^{N} \frac{\sinh(\beta\hbar\mu_\alpha^{(0)}/2)}{|\sinh(\beta\hbar\mu_\alpha^{(b)}/2)|} = \prod_{\alpha=0}^{N} \left(\frac{\mu_\alpha^{(0)}}{|\mu_\alpha^{(b)}|} \prod_{n=1}^{\infty} \frac{\mu_\alpha^{(0)\,2} + \nu_n^2}{\mu_\alpha^{(b)\,2} + \nu_n^2} \right)
$$

$$
= \frac{\omega_0}{\omega_b} \prod_{n=1}^{\infty} \frac{\nu_n^2 + \omega_0^2 + \nu_n\hat{\gamma}(\nu_n)}{\nu_n^2 - \omega_b^2 + \nu_n\hat{\gamma}(\nu_n)} .
$$

(15.44)

In the first equality we have written the sinh function as infinite product representation (4.238). The second equality follows with Eqs. (4.241), (4.243) and (4.245), and with the corresponding relations in which $-\omega_b^2$ is substituted for ω_0^2. Thus we find

$$
k = \frac{\omega_0}{2\pi} \frac{\omega_R}{\omega_b} \prod_{n=1}^{\infty} \frac{\nu_n^2 + \omega_0^2 + \nu_n\hat{\gamma}(\nu_n)}{\nu_n^2 - \omega_b^2 + \nu_n\hat{\gamma}(\nu_n)} \, e^{-\beta V_b} .
$$

(15.45)

The expression (15.45) is in agreement with the earlier result (15.12) found with the Im F method. Thus we have verified not only the rate expression (15.12) by an independent method, but also the occurrence of the factor $\sigma = \beta\hbar\omega_R/2\pi$ for $T > T_0$ in Eq. (14.40).

The periodic orbit approach applies also to the quantum tunneling regime $T < T_0$. In this case, the tunneling rate is again given by Eq. (15.31) with Eq. (15.40). However, the abbreviated action $W(E_{esc})$ must take into account the actual barrier shape. The regime $T < T_0$ will be discussed in Section 17.1.

In the MQTST ansatz (15.30), recrossings of the dividing surface Σ_b are excluded. In this method, the reduction factor $\varrho = \omega_R/\omega_b$ appears because the barrier frequency ω_R of the potential $V(q, \mathbf{x})$ along the escape coordinate y_0 is smaller than the barrier frequency of the bare potential $V(q)$. In contrast to MQTST, in the standard treatment of the decay by Langevin or Fokker-Planck equation methods, the reduction factor $\varrho = \omega_R/\omega_b$ originates from diffusive recrossings of the barrier top along the particle's coordinate q [312]. Thus, although the two physical pictures are different, the results are in correspondence.

The form (15.43) displays the multi-dimensional character of the barrier crossing process. This feature is somewhat hidden in the formula (15.45). The latter form is especially convenient for systems in which one eventually takes the limit $N \to \infty$.

16 The crossover region

We have seen in Section 15.2 that the quantum correction factor c_{qm} increases with decreasing temperature, and that it diverges as $T \to T_0$. The divergence is, because the eigenvalue $\Lambda_1^{(b)} = \Lambda_{-1}^{(b)} = \nu_1^2 - \omega_b^2 + |\nu_1|\hat{\gamma}(|\nu_1|)$ in Eq. (15.7) vanishes at $T = T_0$. Near $T = T_0$, the eigenvalue $\Lambda_1^{(b)}$ may be expanded in powers of the parameter

$$\varepsilon \equiv (T_0 - T)/T_0 , \tag{16.1}$$

which, for laster purpose, is chosen negative for $T > T_0$. In leading order of ε there is

$$\Lambda_1^{(b)} = -\varepsilon\,\Omega^2 , \qquad \text{with} \qquad \Omega^2 = \omega_b^2 + \omega_R^2[1 + \partial\hat{\gamma}(\omega_R)/\partial\omega_R] . \tag{16.2}$$

With Eq. (16.2), the rate (15.12) can be written for T slightly above T_0 as

$$k = \frac{\omega_R}{2\pi}\frac{\Omega^2}{\Lambda_1^{(b)}} A\, e^{-\beta V_b} = \frac{\omega_R}{2\pi}\frac{A}{-\varepsilon}\, e^{-\beta V_b} . \tag{16.3}$$

The dimensionless factor A depends weakly on temperature near $T = T_0$. It is given by (we put again $\nu = \nu_1 \equiv 2\pi/\hbar\beta$)

$$A \equiv \frac{\omega_0}{\omega_b}\frac{\nu^2 + \omega_0^2 + \nu\hat{\gamma}(\nu)}{\Omega^2}\prod_{n=2}^{\infty}\frac{\nu_n^2 + \omega_0^2 + \nu_n\hat{\gamma}(\nu_n)}{\nu_n^2 - \omega_b^2 + \nu_n\hat{\gamma}(\nu_n)} . \tag{16.4}$$

The factor A can equivalently be expressed in terms of the partition function of the global system $Z_{\text{tot}}^{(0)}$ and the eigenfrequencies $\mu_\alpha^{(b)}$ for $\alpha \geqslant 1$. Extracting the factor $1/\sin(\beta\hbar\omega_R/2) \approx -1/\pi\varepsilon$, and using the form (4.238), we find from Eq. (15.43)

$$A = \frac{1}{2\pi}\frac{1}{Z_{\text{tot}}^{(0)}}\prod_{\alpha=1}^{N}\frac{1}{2\sinh(\beta\hbar\mu_\alpha^{(b)}/2)} . \tag{16.5}$$

In the Ohmic case, the infinite product in Eq. (16.4) can be expressed again in terms of gamma functions. We find with Eqs. (15.17) and (15.18)

$$A = \frac{\omega_0}{\omega_b}\frac{\nu^2}{\Omega^2}\frac{\Gamma(2 + \lambda_b^+/\nu)\Gamma(2 + \lambda_b^-/\nu)}{\Gamma(1 + \lambda_0^+/\nu)\Gamma(1 + \lambda_0^-/\nu)} . \tag{16.6}$$

Eq. (16.3) with (16.5) is the multi-dimensional extension of the expression (12.33).

The zero of the eigenvalue $\Lambda_1^{(b)}$ at $T = T_0$, Eq. (16.2), points to the fact that at this temperature the action of the constant path $q(\tau) = q_b$ is degenerate with the action of the periodic bounce path $q(\tau) = q_B(\tau)$. The singularity in Eq. (16.3) at $T = T_0$ is due to the Gaussian treatment of the fluctuation modes $X_{\pm 1}$ in Eq. (15.7). The singularity is absent, if one treats these modes as non-Gaussian modes. Integration of these modes for T slightly above and slightly below T_0 are slightly different, as we shall see. The cases $T > T_0$ and $T < T_0$ were discussed in Refs. [356] and [354], respectively. The reader is also referred to the discussion in Ref. [311].

16.1 Beyond steepest descent above T_0

To regularize the annoying fluctuation integrals, the second-order action (15.7) must be supplemented by terms of higher order in the amplitudes X_1 and X_{-1}. Expanding the potential $V(q)$ about the barrier top beyond the parabolic barrier form,

$$V(q) = V_b - \frac{1}{2} M\omega_b^2 (q - q_b)^2 + \sum_{k=3}^{\infty} \frac{1}{k} M c_k (q - q_b)^k , \qquad (16.7)$$

the extension of the second order action $S_2[q^{(b)}(\cdot)]$ given in Eq. (15.7) is found as

$$S_4[q^{(b)}(\cdot)] = S_2[q^{(b)}(\cdot)] + \frac{M\hbar\beta}{2} \left[2c_3 \left(X_{-2}X_1^2 + X_2 X_{-1}^2 + 2X_0 X_1 X_{-1} \right) + 3c_4 X_1^2 X_{-1}^2 \right] ,$$

where we have kept terms up to the fourth order in $X_{\pm 1}$. After integrating out the amplitudes X_0 and $X_{\pm n}$ ($n \geqslant 2$), the action of the residual modes $X_{\pm 1}$ becomes

$$\Delta S_1^{(b)}[X_{\pm 1}] = \frac{1}{2} M\hbar\beta \left(2\Lambda_1^{(b)} X_1 X_{-1} + B_4 X_1^2 X_{-1}^2 \right) . \qquad (16.8)$$

The coefficient B_4 measures the strength of the leading anharmonic contribution,

$$B_4 = 4c_3^2/\omega_b^2 - 2c_3^2/\Lambda_2^{(b)} + 3c_4 . \qquad (16.9)$$

It is positive for the usual case of a barrier which is wider than a parabolic barrier.[1]

We now withdraw the Gaussian approximation for the fluctuation modes X_1 and X_{-1} which has lead us to the spurious form (16.1) near $T = T_0$. Instead we substitute [cf. the functional measure (4.231)]

$$\frac{1}{\Lambda_1^{(b)}} \longrightarrow \frac{1}{\tilde{\Lambda}_1^{(b)}} \equiv i \frac{M\beta}{2\pi} \int_{-\infty}^{\infty} dX_1 \int_{-\infty}^{\infty} dX_{-1} \exp\left[-\Delta S_1^{(b)}[X_{\pm 1}]/\hbar \right] . \qquad (16.10)$$

Upon introducing polar coordinates (r, φ) defined by $X_{\pm 1} \equiv (r/\sqrt{2}) e^{\pm i\varphi}$, we obtain

$$\frac{1}{\tilde{\Lambda}_1^{(b)}} = M\beta \int_0^{\infty} dr\, r \exp\left[-\tfrac{1}{2} M\beta \Lambda_1^{(b)} r^2 - \tfrac{1}{8} M\beta B_4 r^4 \right] , \qquad (16.11)$$

which may be written in terms of the function erfc(z) given in Eq. (12.37) as

$$\frac{1}{\tilde{\Lambda}_1^{(b)}} = \left(\frac{\pi M\beta}{2B_4} \right)^{1/2} \mathrm{erfc}\left(\Lambda_1^{(b)} \left(\frac{M\beta}{2B_4} \right)^{1/2} \right) \exp\left(\Lambda_1^{(b)\, 2} \frac{M\beta}{2B_4} \right) . \qquad (16.12)$$

With the form (16.2) for $\Lambda_1^{(b)}$, we then get the compact expression

$$\Omega^2/\tilde{\Lambda}_1^{(b)} = \sqrt{\pi}\, \kappa\, \mathrm{erfc}(-\kappa\varepsilon)\, e^{\kappa^2 \varepsilon^2} , \qquad (16.13)$$

in which the dimensionless parameter κ is given by

---

[1] A brief discussion of the role of the coefficient B_4 is given at the end of Section 16.3.

$$\kappa = \sqrt{\frac{M\Omega^4 \beta_0}{2B_4}} . \tag{16.14}$$

Finally, upon substituting $\Omega^2/\tilde{\Lambda}_1^{(b)}$ for $\Omega^2/\Lambda_1^{(b)}$ in Eq. (16.3), the thermal rate becomes

$$k = \frac{\omega_R}{2\pi} A \sqrt{\pi} \kappa \operatorname{erfc}(-\kappa\varepsilon) e^{\kappa^2\varepsilon^2 - \beta V_b} . \tag{16.15}$$

To summarize, the problematic term $1/\Lambda_1^{(b)}$ has been regularized by taking into account non-Gaussian fluctuations of the corresponding mode. The improved rate formula (16.15) is valid in the range extending from very high T, where pure thermal activation prevails, down to $T = T_0$. Before taking up discussion of the result achieved, let us consider the crossing rate for T slightly below T_0.

16.2 Beyond steepest descent below T_0

Below T_0, the bounce trajectory $q_B(\tau)$ exists as a third extremal action path besides the constant paths $q(\tau) = 0$ and $q(\tau) = q_b$. Since the bounce action $S_B \equiv S_{\text{eff}}^{(E)}[q_B]$ is smaller than the action $\hbar\beta V_b$ of the trivial saddle point $q(\tau) = q_b$, the bounce gives the leading contribution to $\operatorname{Im} Z^{(b)}$, and thus to $\operatorname{Im} F$. Because the bounce is a periodic path with period $\hbar\beta$, it may be written as Fourier series,

$$q_B(\tau) = q_b + \sum_{n=-\infty}^{\infty} Q_n e^{i\nu_n\tau} . \tag{16.16}$$

The bounce action is invariant under change of the bounce phase, $q_B(\tau) \to q_B(\tau + \tau_0)$. The time translation invariance of the bounce action entails a *zero mode*, i.e. a mode with zero eigenvalue, in the quantum fluctuations about the bounce. It is convenient to choose $q_B(\tau) = q_B(-\tau)$ which implies $Q_n = Q_{-n}$. For T slightly below T_0, the amplitudes Q_n are small and they can be calculated from the equation of motion (13.2) as a power series in $\sqrt{\varepsilon}$, where again $\varepsilon \equiv (T_0 - T)/T_0$. The leading terms are

$$Q_1 = \left(\frac{\varepsilon\Omega^2}{B_4}\right)^{1/2} , \qquad Q_0 = -\frac{2c_3}{\omega_b^2} Q_1^2 , \qquad Q_2 = \frac{c_3}{\Lambda_2^{(b)}} Q_1^2 . \tag{16.17}$$

With these forms and with Eq. (16.14) the bounce action can be written as

$$S_B = \hbar\beta V_b - \hbar\kappa^2\varepsilon^2 + \mathcal{O}(\varepsilon^3) . \tag{16.18}$$

In the next step, we equip the bounce trajectory with quantum fluctuations,

$$q(\tau) = q_B(\tau) + \sum_{n=-\infty}^{\infty} F_n e^{i\nu_n\tau} , \tag{16.19}$$

and calculate the change of the action. We purposefully write

$$S_4[q] = S_B + S^{(2)} + S^{(4)} , \tag{16.20}$$

where $\mathbb{S}^{(2)}$ is the second order action, and the relevant anharmonic contributions are in $\mathbb{S}^{(4)}$. With use of the expressions (16.17) the action $\mathbb{S}^{(2)}$ may be written as

$$\mathbb{S}^{(2)} = \frac{M\hbar\beta}{2} \left(-\omega_b^2 G_0^2 + 2\Lambda_2^{(b)} G_2 G_{-2} + \varepsilon\Omega^2 (F_1 + F_{-1})^2 + 2\sum_{n=3}^{\infty} \Lambda_n^{(b)} F_n F_{-n} \right) . \quad (16.21)$$

Here we have kept only the terms of leading order in ε, and we have introduced

$$\begin{aligned} G_0 &\equiv F_0 - (2c_3/\omega_b^2)(\varepsilon\Omega^2/B_4)^{1/2}(F_1 + F_{-1}) , \\ G_{\pm 2} &\equiv F_{\pm 2} + (2c_3/\Lambda_2^{(b)})(\varepsilon\Omega^2/B_4)^{1/2} F_{\pm 1} . \end{aligned} \quad (16.22)$$

The fluctuation mode G_0 reduces the action $\mathbb{S}^{(2)}$. This verifies that the bounce is a saddle point. The integration of the amplitude G_0 must again be carried out by deformation of the contour as described in Section 12.5. This leads to an imaginary contribution to the partition function from the barrier region with the already familiar extra factor $1/2$ [cf. Eq. (12.70)].

Next, observe that the eigenvalues $\Lambda_1^{(b)}$ and $\Lambda_{-1}^{(b)}$, which are degenerate above T_0 [see Eq. (15.8)], split into the eigenvalues $\Lambda_{1,-}^{(B)}$ and $\Lambda_{1,+}^{(B)}$ of the second order action (16.21) as T passes through T_0 from above. The degeneracy is removed because the constant path bifurcates into the bounce path at the crossover temperature. With substitution of $F_{\pm 1} \equiv (F_{1,+} \pm F_{1,-})/\sqrt{2}$ into Eq. (16.21), we find that the eigenvalue $\Lambda_{1,-}^{(B)}$ of the mode $F_{1,-}$ remains zero,[2] while the eigenvalue of the mode $F_{1,+}$ is $\Lambda_{1,+}^{(B)} = 2\varepsilon\Omega^2$. The eigenvalues with index $n \geqslant 2$ are independent of ε in leading order. In summary, the fluctuation determinant D_B' connected with the second-order action functional (16.21) is found as

$$|D_B'| = 2\varepsilon \Omega^2 |D_B''| \qquad \text{with} \qquad |D_B''| = |D^{(b)''}| \equiv \omega_b^2 \left(\prod_{n=2}^{\infty} \Lambda_n^{(b)} \right)^2 , \quad (16.23)$$

where terms of order ε^2 are disregarded. The single prime in D_B' is to indicate that the zero eigenvalue is omitted.

Because of the smallness of the eigenvalues $\Lambda_{1,\pm}^{(B)}$, the Gaussian approximation for the modes $F_{1,\pm}$ breaks down. Hence the action (16.21) must be supplemented by terms of third and fourth order in the amplitudes F_1 and F_{-1}. These higher-order terms include the nonlinear couplings between the amplitudes F_1 and F_{-1}, and the couplings with the amplitudes F_0 and $F_{\pm 2}$. The relevant contribution has the form

$$\begin{aligned} \mathbb{S}^{(4)} = \; &- M\hbar\beta \left[c_3 \left(F_2 F_{-1}^2 + F_{-2} F_1^2 + 2 F_0 F_1 F_{-1} \right) \right. \\ &\left. + 3c_4 Q_1 \left(F_1^2 F_{-1} + F_1 F_{-1}^2 \right) + 3c_4 F_1^2 F_{-1}^2/2 \right] . \end{aligned} \quad (16.24)$$

The Gaussian integrations over the amplitudes F_0 and $F_{\pm 2}$ are performed by completing the square. Then we are left with the integral over the amplitudes F_1 and F_{-1} weighted with the action factor $\exp(-\Delta\mathbb{S}_1^{(B)}[F_{\pm 1}]/\hbar)$, where

---

[2]The zero eigenvalue is due to the time-translation symmetry of the bounce mentioned before.

$$\Delta S_1^{(B)}[F_{\pm 1}] = \frac{MB_4\hbar\beta}{4}\left[Q_1^2(F_1 + F_{-1})^2 + 2Q_1(F_1 + F_{-1})F_1F_{-1} + F_1^2F_{-1}^2\right]. \quad (16.25)$$

Here the terms of third and fourth order stem from the action $S^{(4)}$. Upon introducing polar cordinates (ρ, φ) with origin at $-Q_1$,

$$F_{\pm 1} = \left(\rho/\sqrt{2}\right)e^{\pm i\varphi} - Q_1, \quad (16.26)$$

the action $\Delta S^{(B)}[F_{\pm 1}]$ turns out to be independent of φ. Hence the φ mode is the zero mode inducing changes of the phase of the bounce, while the ρ mode describes amplitude fluctuations. These fluctuations can be as large as the bounce amplitude Q_1. Therefore, we have to keep all terms up to the order ρ^4.

Reverse transaction of the respective Gaussians approximation amounts to the substitution

$$\frac{1}{\sqrt{\Lambda_{1,+}^{(B)}\Lambda_{1,-}^{(B)}}} \longrightarrow \frac{1}{\Lambda_1^{(B)}} \equiv i\frac{M\beta}{2\pi}\int_{-\infty}^{\infty}dF_1\int_{-\infty}^{\infty}dF_{-1}\exp\left[-\Delta S_1^{(B)}[F_{\pm 1}]/\hbar\right]$$

$$= M\beta\int_0^{\infty}d\rho\,\rho\exp\left[-\tfrac{1}{8}\beta MB_4\left(\rho^2 - 2Q_1^2\right)^2\right]. \quad (16.27)$$

Thus we arrive again at an error function,

$$1/\Lambda_1^{(B)} = \left(\sqrt{\pi}\kappa/\Omega^2\right)\mathrm{erfc}\left(-\kappa\varepsilon\right). \quad (16.28)$$

As in the case $T > T_0$, the dimensionless quantities ε and κ are defined in Eqs. (16.1) and (16.14). With the assignment (16.27), the expression (15.10) is converted into

$$\mathrm{Im}\,F = -\frac{1}{2\beta\Lambda_1^{(B)}}\left(\frac{D^{(0)}}{|D_B''|}\right)^{1/2}e^{-S_B/\hbar}, \quad (16.29)$$

where the determinant $|D_B''|$ is given in Eq. (16.23). With use of the relation (12.74) and the expressions (15.11), (16.4) and (16.18), the decay rate is readily found as

$$k = \frac{1}{\hbar\beta}A\sqrt{\pi}\,\kappa\,\mathrm{erfc}\left(-\kappa\varepsilon\right)e^{\kappa^2\varepsilon^2 - \beta V_b}. \quad (16.30)$$

Now observe that at $T = T_0$, which is equivalent to $1/\hbar\beta = \omega_R/2\pi$, the formula (16.30) exactly coincides with the previous result (16.15) found for temperatures slightly above T_0. From this we infer that the expression (16.30) is valid in the entire crossover region above and below T_0.

For T slightly below T_0, the coefficient B_4, defined in Eq. (16.9), can be related to the change of the bounce period per unit energy. This is shown as follows. First, we expand the bounce action about the period $\hbar\beta_0$ of the bounce

$$S(\hbar\beta) = \hbar\beta V_b + \frac{1}{2}S''(\hbar\beta_0)\left(\hbar\beta - \hbar\beta_0\right)^2 + \mathcal{O}\left((\hbar\beta - \hbar\beta_0)^3\right). \quad (16.31)$$

Next, we equate the expressions (16.31) and (16.18) and use the relations[3]

$$\mathsf{S}''(\tau) = -\frac{1}{\mathsf{W}''(E)} = -\frac{1}{|\tau'(E)|} , \qquad (16.32)$$

where $\mathsf{W}(E)$ is the abbreviated action for one periodic orbit at total energy E and where $\tau(E)$ is the associated period [cf. Eq.(15.34)]. We then find for the change of the bounce period per unit energy

$$|\tau_b'| \equiv \left| \frac{\partial \tau}{\partial E} \right|_{E=V_b} = \frac{B_4}{M\Omega^4} \hbar\beta_0 = \frac{\hbar}{2} \frac{\beta_0^2}{\kappa^2} . \qquad (16.33)$$

Solving Eq. (16.33) for κ and substituting this into Eq. (16.30), we obtain [367]

$$k = \frac{\pi A}{\sqrt{2\pi\hbar|\tau_b'|}} \operatorname{erfc}\left(\frac{\hbar(\beta_0 - \beta)}{\sqrt{2\hbar|\tau_b'|}} \right) \exp\left(-\beta V_b + \frac{\hbar(\beta - \beta_0)^2}{2|\tau_b'|} \right) . \qquad (16.34)$$

This expression holds again for T around T_0.

16.3 The scaling region

As the energy $M\Omega^4/B_4$ is usually of the order of the barrier height, the parameter κ is of the order of $(V_b/\hbar\omega_R)^{1/2}$. Thus we have $\kappa \gg 1$ in the semiclassical limit. The sophisticated formula (16.15) [or (16.30)] is appropriate in the region $|\kappa\varepsilon| \lesssim 1$, or

$$|T - T_0| \lesssim T_0/\kappa , \qquad (16.35)$$

in which the argument of the erfc function is of order one or smaller. For $\kappa \gg 1$, the crossover region is narrow on the temperature scale T_0.

For T well above this regime, which is $\kappa\varepsilon < -1$, we may use the asymptotic form

$$\sqrt{\pi}\,\kappa\, e^{\kappa^2\varepsilon^2} \operatorname{erfc}(-\kappa\varepsilon) \approx -1/\varepsilon . \qquad (16.36)$$

whereby Eq. (16.15) reduces to the expression (16.3).

Consider next the temperature region slightly below the crossover regime. We then have $\kappa\varepsilon > 1$ and $\operatorname{erfc}(-\kappa\varepsilon) \approx 2$. The factor A may be written with use of Eq. (16.23) as $A = \sqrt{(2\varepsilon/\Omega)\,D^{(0)}/|D_B'|}$. Next we relate the parameter ε to the normalization factor W_B of the zero mode [see Eq. (12.64)]

$$\mathsf{W}_B \equiv M \int_0^{\hbar\beta} d\tau\, \dot{q}_B^2(\tau) . \qquad (16.37)$$

Observe that W_B differs from the (multidimensional) abbreviated action (15.35) by the missing kinetic contribution of the reservoir degrees of freedom. Using for the periodic bounce path the form (16.16) with (16.17), we obtain in leading order in ε

---

[3]The prime (double prime) denotes the first (second) derivative with respect to the argument.

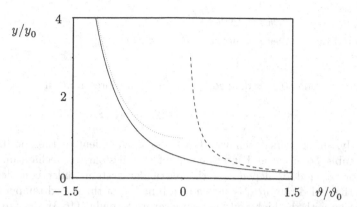

Figure 16.1: The scaled rate y/y_0 is shown as a function of the scaled temperature ϑ/ϑ_0. The high temperature formula (16.3) is represented by a dashed line and the low temperature formula (16.39) as a dotted line. The crossover function (16.44) smoothly matches onto these functions below and above the crossover region.

$$\mathbb{W}_B = 2M\hbar\beta\nu_1^2 Q_1^2 = 8\pi^2 \frac{M\Omega^2}{B_4\hbar\beta_0}\varepsilon \overset{(16.14)}{=} 16\pi^2 \left(\frac{\kappa}{\hbar\beta_0\Omega}\right)^2 \hbar\varepsilon . \tag{16.38}$$

Solving this expression for ε, we readily may write the rate expression (16.30) as

$$k = \sqrt{\frac{\mathbb{W}_B}{2\pi\hbar}\frac{D^{(0)}}{|D'_B|}}\, e^{-S_B/\hbar} , \tag{16.39}$$

where $|D'_B|$ and S_B are given in Eqs. (16.23) and (16.18), respectively.

Another useful expression for the rate results from Eq. (16.34) with substitution of the form (16.5) for the quantity A. The result is

$$k = \frac{1}{Z_{\text{tot}}^{(0)}}\left(\prod_{\alpha=1}^{N}\frac{1}{2\sinh(\hbar\beta\mu_\alpha^{(b)}/2)}\right)\frac{1}{\sqrt{2\pi\hbar|\tau'_b|}}\, e^{-S_B/\hbar} . \tag{16.40}$$

The expressions (16.39) and (16.40) hold in the regime $\kappa^{-1} < \varepsilon \ll 1$. They smoothly match well below T_0 on the semiclassical quantum tunneling rate formula which is discussed subsequently in Section 17.1. Finally, we note that the expression (16.40) represents the multidimensional generalization of the earlier result (12.27).

The quantum-statistical decay rate possesses in the crossover region a universal scaling behavior [311]. Defining a temperature scale ϑ_0 and a frequency scale y_0,

$$\vartheta_0 \equiv T_0/\kappa , \qquad y_0 = (\omega_R/2\pi)\sqrt{\pi\kappa A} , \tag{16.41}$$

there follows from Eq. (16.30) that the quantity

$$y/y_0 \equiv (k/y_0)\, e^{\beta V_{\rm b}} \tag{16.42}$$

is represented by a *universal* function of $\vartheta/\vartheta_0 \equiv \kappa(T - T_0)/T_0$,

$$y/y_0 = U(\vartheta/\vartheta_0)\,. \tag{16.43}$$

The scaling function $U(z)$ is defined by the integral representation

$$U(z) = \frac{2}{\sqrt{\pi}} \int_0^\infty dt\, e^{-t^2 - 2zt} = \mathrm{erfc}(z)\, e^{z^2}\,. \tag{16.44}$$

Interestingly, the universal behavior (16.43) is independent of the specific form of the metastable potential and independent of the dissipative mechanism. Only the scale factors y_0 and ϑ_0 depend on the particular system under consideration. In Fig. 16.1, the scaled rate y/y_0 is shown as a function of the scaled temperature ϑ/ϑ_0 in comparison with the high and low temperature formulas (16.3) and (16.39).

Let us finally briefly discuss the significance of the anharmonic contribution in the action (16.8). When the coefficient B_4 is positive, the bounce amplitude and the bounce action increase monotonously with the bounce period. Hence the transition from the constant path to the periodic path at $T = T_0$ is a second order phase transition [355, 356]. In the limit $B_4 \to 0$, the parameter κ diverges, as we see from Eq. (16.14). Thus, the scaling function (16.44) becomes meaningless in this limit. To regularize the interpolating function in the case $B_4 = 0$, we have to include in the action $\Delta S_1^{(\rm b)}$ in Eq. (16.8) the term of sixth order,

$$\Delta S_1^{(\rm b)} = \frac{1}{2} M \hbar \beta \left(2\Lambda_1^{(\rm b)} X_1 X_{-1} + B_6 X_1^3 X_{-1}^3 \right)\,. \tag{16.45}$$

For the potential (16.7), the coefficient B_6 is

$$B_6 = 20 c_6/3 - 2 c_4^2/\Lambda_3^{(\rm b)}\,. \tag{16.46}$$

The significant eigenvalue $\tilde{\Lambda}_1^{(\rm b)}$ is given by the expression (16.10), but now with the action (16.45). Eventually, we find again the universal scaling form (16.43) for the thermal-to-quantum crossover. However, the scaling function $U(z)$ and the scale factor κ are different. The dimensionless parameter κ is

$$\kappa = \Omega^2 \left(M^2 \beta^2/4 B_6 \right)^{1/3}\,, \tag{16.47}$$

and the scaling function $U(z)$ has the integral representation

$$U(z) = \frac{2}{\sqrt{\pi}} \int_0^\infty dt\, e^{-t^3 - 2zt}\,. \tag{16.48}$$

Last of all a short additional view. When B_4 is negative, the barrier is narrower than the harmonic one. Then the bounce action is not any more a monotonous function of temperature. Rather, the classical-to-quantum transition has the characteristics of a phase transition of first order. Physically, this is because tunneling near to the barrier top is less favorable than at lower level. The case $B_4 < 0$ is exotic and of minor physical importance.

17 Dissipative quantum tunneling

We are now skilled and ready to study the decay of a metastable state at temperature T well below the crossover regime at which the escape from the well goes off by dissipative quantum tunneling through the barrier.

17.1 The quantum rate formula

At temperature T below T_0, the relevant stationary point of the action accounting for the barrier region is the single-bounce trajectory $q_B(\tau)$. It obeys the equation of motion (13.2) with the dissipative kernel (4.47). To second order in the fluctuations $\xi(\tau)$ about the extremal path, $q(\tau) = q_B(\tau) + \xi(\tau)$, the action takes the form

$$\mathbb{S}[q(\cdot)] = \mathbb{S}_B + \frac{M}{2} \int_0^{\hbar\beta} d\tau\, \xi(\tau)\, \Lambda[q_B(\tau)]\, \xi(\tau) , \qquad (17.1)$$

where $\mathbb{S}_B$ is the bounce action (13.3), and the fluctuation operator $\Lambda[q_B(\tau)]$ is defined in Eq. (4.218). Expanding $\xi(\tau)$ into the eigenmodes $\{\chi_n(\tau)\}$ of the operator $\Lambda[q_B(\tau)]$,

$$\xi(\tau) = \sum_n c_n \chi_n(\tau) , \qquad (17.2)$$

where $\chi_n(\tau)$ is normalized on the interval $(0, \hbar\beta)$, the action becomes

$$\mathbb{S}[q(\cdot)] = \mathbb{S}_B + \frac{M\hbar\beta}{2} \sum_n \Lambda_n[q_B(\cdot)]\, c_n^2 . \qquad (17.3)$$

The $\Lambda_n[q_B(\cdot)]$ are the eigenvalues of $\Lambda[q_B(\cdot)]$ for periodic boundary conditions,

$$\Lambda_n[q_B(\cdot)] = \nu_n^2 + |\nu_n|\hat{\gamma}(|\nu_n|) + V''[q_B(\cdot)]/M . \qquad (17.4)$$

Thus the determinant of the operator $\Lambda[q_B(\cdot)]$ in diagonal representation is

$$D[q_B(\cdot)] \equiv \det\left(\Lambda[q_B(\cdot)]\right) = \prod_n \Lambda_n[q_B(\cdot)] . \qquad (17.5)$$

Since the bounce is time-periodic, there is again a zero mode due to the time-translational invariance of the action, and a mode with a negative eigenvalue accounting for metastability. These modes must be precisely dealt with as explained in items (1) and (2) in Section 12.5.

Following the lines drawn in Section 12.5, the barrier contribution to the partition function becomes imaginary. With the generalization of the expression (12.68) to the multidimensional case, the single-bounce contribution is found to read

$$\frac{\operatorname{Im} Z_{\text{tot}}^{(B)}}{Z_{\text{tot}}^{(0)}} = \frac{\hbar\beta}{2} \sqrt{\frac{\mathbb{W}}{2\pi\hbar}} \sqrt{\frac{D^{(0)}}{|D'[q_B(\cdot)]|}} \, e^{-S_B/\hbar} . \tag{17.6}$$

Here, $\mathbb{W}_B$ is twice the kinetic part of the bounce action, Eq. (16.37), $D^{(0)}$ is the fluctuation determinant of the well regime given in Eq. (15.11), and $D'[q_B(\cdot)]$ is the determinant of the fluctuations about the bounce path, Eq. (17.5), and the prime indicates that the zero eigenvalue is omitted.

With use of the relation $k = (2/\hbar\beta) \operatorname{Im} Z_{\text{tot}}^{(B)}/Z_{\text{tot}}^{(0)}$, the quantum decay rate at temperature T is concisely written as

$$k = f_{\text{qm}} \, e^{-S_B/\hbar} . \tag{17.7}$$

The exponent is determined by the bounce action (13.3) of a single bounce. The prefactor or quantum mechanical attempt frequency is

$$f_{\text{qm}} = \sqrt{\frac{W_B}{2\pi\hbar}} \sqrt{\frac{D^{(0)}}{|D'[q_B(\cdot)]|}} . \tag{17.8}$$

The rate expression (17.7) with (17.8) holds down to zero temperature. It matches slightly below T_0 the previous expression (16.39) with the actions S_B and W_B given in Eq. (16.18) and Eq. (16.38), respectively. In addition, slightly below T_0 the determinant $|D'[q_B(\cdot)]|$ matches the determinant of the fluctuations near the barrier top $|D'_B|$ given in Eq. (16.23).

As in the case $T > T_0$ discussed in Section 15.4, it is possible to rewrite Eq. (17.7) with Eq. (17.8) in a form which directly displays the multi-dimensional character of the tunneling process. In the "many-dimensional WKB" approach, the system-plus-reservoir complex is visualized as tunneling entity which moves along the (curved) most probable escape path (periodic orbit) $z_0^{(PO)}(\tau)$ in the $(N+1)$-dimensional configuration space. In this picture, the bounce trajectory $q_B(\tau)$ is the projection of the periodic orbit[1] onto the reaction coordinate q. In the multi-dimensional representation, the exponent in the tunneling formula is given by the expression [cf. Eqs. (15.34) and (15.35)]

$$S_B/\hbar = \frac{2}{\hbar} \int_{z_0^{(m)}}^{z_0^{(\text{ex})}} dz_0 \sqrt{2M[V(q,\mathbf{x}) - E]} + E\beta , \tag{17.9}$$

where dz_0 is the infinitesimal path element along the curved orbit with period $\hbar\beta$, and where E is the conserved energy of this path. The turning points are as in Eq. (15.29). At zero temperature we have $E = 0$ and $\hbar\beta = \infty$.

Consider next the pre-exponential factor of the tunneling rate. It is convenient to split the determinant of the fluctuations about the periodic orbit into the contribution from *longitudinal* fluctuations along the escape coordinate

$$\xi_{\parallel}(\tau) \equiv y_0(\tau) - y_0^{(PO)}(\tau) , \tag{17.10}$$

---

[1] As a result of the projection, the energy of the path $q_B(\tau)$ is not conserved, and the equation of motion becomes nonlocal in time.

and into the *transverse* determinant describing fluctuations in the coordinates $y_n(\tau)$ $(n = 1, 2, \cdots, N)$ which are locally perpendicular to the escape coordinate. The latter determinant is conveniently expressed in terms of the dynamical stability frequencies $\mu_\alpha^{(B)}$ $(\alpha = 1, 2, \cdots, N)$ of the periodic orbit. We then have [312, 367]

$$k = \frac{1}{Z_{\text{tot}}^{(0)}} \left(\prod_{\alpha=1}^{N} \frac{1}{2 \sinh(\hbar\beta\mu_\alpha^{(B)}/2)} \right) \frac{1}{\sqrt{2\pi\hbar|\tau'(E_{\hbar\beta})|}} \, e^{-S_B/\hbar} \,, \qquad (17.11)$$

where $E_{\hbar\beta}$ is the total energy of the periodic orbit with period $\hbar\beta$. This form of the quantum statistical tunneling rate is valid down to zero temperature. It smoothly matches on the previous formula (16.40) at temperatures slightly below T_0.

An alternative derivation of the formula (17.7) with (17.8) has been given for zero temperature by Schmid [368] using a many-dimensional WKB approach for the quasi-stationary ground-state wave function. The generalization to finite temperatures is presented in Ref. [369].

For weak Ohmic dissipation, the bounce action and the attempt frequency may be calculated perturbatively in the Ohmic coupling $\alpha = \gamma/2\omega_0$. For the cubic metastable potential (11.41), one obtains as extension of the WKB action $S_0 = 36V_b/5\omega_0$ in Eq. (12.44) at $T = 0$ the bounce action

$$S_B(T = 0) = \frac{36}{5} \frac{V_b}{\omega_0} \left(1 + \frac{45\,\zeta(3)}{\pi^3} \alpha + \mathcal{O}(\alpha^2) \right) \,, \qquad (17.12)$$

where $\zeta(3)$ is a Riemann number, and the quantum mechanical prefactor [371]

$$f_{\text{qm}}(T = 0) = 12\sqrt{6\pi} \, \frac{\omega_0}{2\pi} \sqrt{\frac{V_b}{\hbar\omega_0}} \left(1 + 2.86\,\alpha + \mathcal{O}(\alpha^2) \right) \,. \qquad (17.13)$$

Dissipative quantum tunneling from the ground state in a metastable well was studied first by Caldeira and Leggett [78]. The qualitative conclusion of their study was that damping suppresses quantum tunneling by an exponential factor, which depends linearly on α for weak damping, as given in Eq. (17.12) .

17.2 Thermal enhancement of macroscopic quantum tunneling

With increasing temperature, the probability, that the particle tunnels through the barrier from an excited state in the well, is growing. Hence the quantum-statistical decay probability is growing as well. In the semiclassical limit, the leading thermal enhancement at low T originates from the temperature dependence of the bounce action. It is convenient to write

$$k(T) = k(0)\, e^{\mathcal{A}(T)} \quad \text{with} \quad \mathcal{A}(T) = [\,S_B(0) - S_B(T)\,]/\hbar \,. \qquad (17.14)$$

The asymptotic expansion of the bounce action for T slightly above zero is discussed in some detail in Ref. [372] for a general metastable potential and arbitrary frequency-dependent damping. By use of the asymptotic Euler-Maclaurin expansion [90], the

Fourier coefficients Q_n of the bounce at low T may be related to the Fourier representation $Q(\omega)$ of the zero temperature bounce. The leading contribution to $\mathcal{A}(T)$ is due to the temperature dependence of the dissipative kernel in the action $S_B = S_{\text{eff}}[q_B(\cdot)]$ given in Eq. (13.3). Upon introducing the width of the zero temperature bounce

$$\tau_{\text{w}} \equiv \frac{1}{q_b} \int_{-\infty}^{\infty} d\tau \, q_B^{(0)}(\tau) , \qquad (17.15)$$

and using the relation (4.50), we obtain

$$\mathcal{A}(T) = \frac{q_b^2 \tau_{\text{w}}^2}{2\hbar} \left[K_T(0) - K_{T=0}(0) \right] = \frac{q_b^2 \tau_{\text{w}}^2}{2\pi\hbar} \int_0^{\infty} d\omega \, J(\omega) \left(\coth(\tfrac{1}{2}\beta\hbar\omega) - 1 \right) , \qquad (17.16)$$

In the second form, we have used for the kernel the spectral representation (4.54).

When $J(\omega)$ behaves as ω^s, the kernel $k(\tau)$ in the equation of motion decays asymptotically for $T = 0$ as $|\tau|^{-(1+s)}$. Therefore, we have $q_B(|\tau| \rightarrow \infty) \propto |\tau|^{-(1+s)}$. This ensures that the bounce width (17.15) is finite for all $s > 0$.

For the spectral density $J(\omega) = M\gamma_s \omega_{\text{ph}}^{1-s} \omega^s$ [cf. Eqs (3.38) and (3.39)], the enhancement function (17.16) takes the form

$$\mathcal{A}(T) = 2\Gamma(1+s)\zeta(1+s)\frac{M\gamma_s q_b^2}{2\pi\hbar}\omega_{\text{ph}}^2 \tau_{\text{w}}^2 \left(\frac{T}{T_{\text{ph}}}\right)^{1+s} , \qquad (17.17)$$

where $\Gamma(z)$ is Euler's gamma function and $\zeta(z)$ is Riemann's zeta function. Interestingly, the thermal enhancement is sensitive to the spectral form of the dissipative mechanism, and it qualitatively differs from the undamped case. For an undamped system, the enhancement is exponentially weak, $\mathcal{A}(T) \propto e^{-\hbar\omega_0/k_B T}$, as we can see from the expression (12.46). In contrast, for a damped system, we have algebraic enhancement, $\mathcal{A}(T) \propto T^{1+s}$. The exponent of the algebraic law is independent of the particular form of the metastable potential and is therefore a distinctive feature of the predominant spectral damping [353, 372]. Measurement of the power of the algebraic enhancement of the tunneling probability at low T gives direct information about the spectral features of the environmental coupling. The properties of the barrier and additional dependence on friction enter only through the square of the factor $\tau_{\text{w}} q_b$.

For the important case of Ohmic damping, $J(\omega) = M\gamma\omega$, we have

$$\mathcal{A}(T) = \frac{\pi^2}{3} \frac{M\gamma q_b^2}{2\pi\hbar} \left(\frac{\tau_{\text{w}} k_B T}{\hbar}\right)^2 , \qquad (17.18)$$

and the next-to-leading order term varies with temperature as T^4. Consider next the bounce width at $T = 0$. In the limit $\gamma \rightarrow 0$, the width of the bounce in the cubic potential, $q_B^{(0)}(\tau) = \tfrac{3}{2} q_b \, \text{sech}^2(\tfrac{1}{2}\omega_0\tau)$ [cf. Section 12.4] is found from Eq. (17.15) as $\tau_{\text{w}} = 6/\omega_0$. For strong damping, $\alpha \equiv \gamma/2\omega_0 \gg 1$, we obtain $\tau_{\text{w}} = 4\pi\alpha/\omega_0$, as follows from the analytic form of the bounce given below in Eq. (17.26). From these limiting cases we infer that the bounce width τ_{w} is growing with increasing damping. This is again a signature of the fact that the tunneling distance along the curved orbit in the $\{q, \mathbf{x}\}$ configuration space gets larger with increasing system-reservoir coupling, as we have discussed already in Chapter 13.

The universal enhancement (17.17) and (17.18) is due to thermally excited states of the environment, and *not* to thermally excited states in the well. A discussion of the formula (17.18) from the view of thermal quantum noise-theory is given in Ref. [373]. Finally, we remark that a similar *universal* low temperature behavior $\propto T^{s+1}$ occurs in the free energy of damped quantum systems [224] (cf. Sections 6.5 and 22.9), and, e.g., in transport properties [374] (see Subsection 27.1).

17.3 Quantum decay in a cubic potential for Ohmic friction

In this section, we consider the quantum-statistical decay of the metastable state in the cubic potential (11.40) for Ohmic friction characterized by the damping constant

$$\alpha \equiv \gamma/2\omega_0 \,. \tag{17.19}$$

For general α, the quantum mechanical rate expression (17.7) with (17.8) must be computed numerically.

In the regime $\alpha \gg 1$, the quantum statistical tunneling rate can be calculated in analytic form for the cubic metastable potential and for the tilted cosine potential. This is possible, since for strong damping $\alpha \gg 1$, the inertia term in the equation of motion (13.2) is small compared to the friction term, and therefore may be disregarded.

We now turn to the discussion of metastability in the quadratic-plus-cubic potential. The tilted washboard potential will be discussed in the subsequent section.

17.3.1 Bounce action and quantum mechanical prefactor

For the cubic metastable potential (11.40), the bounce obeys the equation of motion

$$-\ddot{q}_{\rm B}(\tau) + \omega_0^2\, q_{\rm B}(\tau) - \frac{\omega_0^2}{q_{\rm b}}q_{\rm B}^2(\tau) + \frac{1}{M}\int_{-\hbar\beta/2}^{\hbar\beta/2} d\tau'\, k(\tau-\tau')q_{\rm B}(\tau') \;=\; 0\,, \tag{17.20}$$

or equivalently in Fourier space upon writing $q_{\rm B}(\tau) = \sum_\ell Q_\ell\, e^{i\nu_\ell \tau}$ with $\nu_\ell = 2\pi\ell/\hbar\beta$,

$$\left(\nu_\ell^2 + \gamma|\nu_\ell| + \omega_0^2\right)Q_\ell \;=\; \frac{\omega_0^2}{q_{\rm b}}\sum_m Q_{\ell-m}Q_m\,. \tag{17.21}$$

With use of the equation of motion (17.21), the bounce action (13.3) can be written as

$$S_{\rm B} \;=\; \frac{M\omega_0^2}{6q_{\rm b}}\,\hbar\beta\sum_{\ell,m} Q_\ell Q_m Q_{-\ell-m}\,, \tag{17.22}$$

and the zero mode normalization factor (16.37) is $W_{\rm B} = M\,\hbar\beta\sum_m \nu_m^2 Q_m Q_{-m}$. The nonlinear equation (17.21) may be solved numerically by successive iteration, starting for instance with the zero order ansatz $Q_m = Q_{-m} \propto q_{\rm b}\, e^{-|m|}$.

Consider next the fluctuations $y(\tau)$ about the bounce, $q(\tau) = q_B(\tau) + y(\tau)$. Writing $y(\tau) = \sum_\ell Y_\ell e^{i\nu_\ell \tau}$, the eigenvalue equation for the fluctuation operator $\Lambda[q_B(\tau)]$ reads

$$[\nu_\ell^2 + \gamma|\nu_\ell| + \omega_0^2]Y_\ell^{(n)} - \frac{2\omega_0^2}{q_b} \sum_m Q_{\ell-m}Y_m^{(n)} = \Lambda_n^{(B)} Y_\ell^{(n)} . \qquad (17.23)$$

The translational mode is $y_{1,-}(\tau) \propto \dot{q}_B(\tau)$, where $\dot{q}_B(\tau) = i\sum_\ell \nu_\ell Q_\ell e^{i\nu_\ell \tau}$. This mode is odd and has one node, and the eigenvalue $\Lambda_{1,-}^{(B)}$ is zero.

Since the Fourier coefficient Q_m is small for large m, the eigenvalue $\Lambda_n(B)$ for $n \gg 1$ can be calculated by iteration with the starting value $\Lambda_n^{(B)} = \Lambda_n^{(b)}$. For small n on the other hand, the eigenvalue $\Lambda_n^{(B)}$ can be calculated numerically by diagonalization of the truncated dynamical matrix of Eq. (17.23). In the numerical implementation, low eigenvalues require large rank N of the dynamical matrix with N up to $N = 250$.

A numerical study of tunneling rates at $T = 0$ was given first by Chang and Chakravarty [378]. Accurate numerical calculations at finite temperatures in the range $0.1\,T_0 \leqslant T \leqslant T_0$ and $0.1 \leqslant \alpha \leqslant 1$ are presented in Ref. [311]. This paper shows tables both for the exponent and the prefactor of the quantum rate formula (17.7). The numerical data are in remarkable agreement with the measured lifetime of the metastable zero-voltage state in the current-biased Josephson junction [314, 379].

The characteristic features of the quantum decay rate as a function of friction strength and temperature are summarized below in Section 17.5.

17.3.2 Analytic results for strong Ohmic dissipation

For strong Ohmic damping, $\alpha \gg 1$, the inertia term $\nu_\ell^2 Q_\ell$ in Eq. (17.21) can be disregarded. In the absence of the inertia term, the equation of motion for the Fourier coefficients Q_ℓ is solved in analytic form with the ansatz

$$Q_\ell = a\, e^{-b|\ell|} . \qquad (17.24)$$

The coefficients are found to read

$$a = q_b T/T_0 , \qquad \text{and} \qquad b = \operatorname{artanh}(T/T_0) , \qquad (17.25)$$

where $k_B T_0 \equiv \hbar\omega_0/4\pi\alpha = \hbar\omega_0^2/2\pi\gamma$ is the crossover temperature in the strong damping limit, Eq. (14.39). With the expressions (17.24) and (17.25), both the bounce and the bounce action are obtained in analytic form [356]. In the regime $T \leqslant T_0$, there is

$$q_B(\tau) = q_b \frac{T}{T_0} \sum_\ell e^{-b|\ell|} e^{i\nu_\ell \tau} = q_b \frac{(T/T_0)^2}{1 - \sqrt{1 - (T/T_0)^2} \cos(2\pi k_B T\tau/\hbar)} , \qquad (17.26)$$

$$\mathsf{S}_B(T) = 6\pi\alpha \frac{V_b}{\omega_0}\left[1 - \frac{1}{3}\left(\frac{T}{T_0}\right)^2\right] = \frac{2\pi}{9}\gamma M q_{ex}^2\left[1 - \frac{1}{3}\left(\frac{T}{T_0}\right)^2\right] . \qquad (17.27)$$

The periodic bounce (17.26) shrinks to the constant path $q_B(\tau) = q_b = 2q_{ex}/3$, and $\mathsf{S}_B(T_0)/\hbar$ merges into the Arrhenius exponent $V_b/k_B T_0$, as $T \to T_0$,

In the limit $T \to 0$, the bounce (17.26) takes the Lorentzian form [355]

$$q_{\text{B}}^{(T=0)}(\tau) = \frac{2\alpha}{\omega_0} q_{\text{b}} \int_{-\infty}^{+\infty} d\nu \, e^{-2\alpha|\nu|/\omega_0} e^{i\nu\tau} = \frac{2q_{\text{b}}}{1 + (\omega_0\tau/2\alpha)^2} . \qquad (17.28)$$

For this path, the bounce point or turning point is $q_{\text{turn}} = 2q_{\text{b}}$, whereas $q_{\text{turn}} = 3q_{\text{b}}/2$ for $\alpha = 0$. The turning point q_{turn} increases monotonously with α in the position range $3q_{\text{b}}/2 \leqslant q_{\text{turn}} \leqslant 2q_{\text{b}}$. Note that $q_{\text{turn}} > 3q_{\text{b}}/2$ is possible because the bounce does not conserve energy when α is nonzero.[2]

The action at $T = 0$, with the leading correction from the inertia term, is [311]

$$S_{\text{B}}(T = 0) = 6\pi\alpha\frac{V_{\text{b}}}{\omega_0}\left[1 + \frac{1}{4\alpha^2} + \mathcal{O}(\alpha^{-4})\right]. \qquad (17.29)$$

With the width $\tau_{\text{w}} = 4\pi\alpha/\omega_0$ of the bounce (17.28), the thermal enhancement exponent (17.18) is in agreement with the T^2 contribution in the action (17.27). Interestingly enough, we see from the expression (17.27) that the T^2 law strictly holds in the range $0 \leqslant T \leqslant T_0$, and there is no other temperature dependence in the exponential factor of the rate for strong damping $\alpha \gg 1$.

Consider next the fluctuations $y(\tau)$ about the bounce. For the low-energy modes we may disregard the inertia term $\nu_\ell^2 Y_\ell$ in Eq. (17.23). The truncated eigenvalue equation for the even modes with zero and with two nodes is solved for all ℓ with the ansatz $Y = (1 + c|\ell|) e^{-b|\ell|}$. The resulting constraints for c are

$$\begin{aligned} c\tanh(b) &= -\Lambda/\omega_0^2 , \\ c(1 + \Lambda/\omega_0^2) &= -\tanh(b)/\sinh^2(b) . \end{aligned} \qquad (17.30)$$

Elimination of c yields a mixed quadratic equation for Λ. The resulting eigenvalues for the modes with zero and with two nodes are

$$\begin{aligned} \Lambda_0^{(\text{B})} &= -\omega_0^2\left(\sqrt{1 + 4[1 - (T/T_0)^2]} + 1\right)/2 , \\ \Lambda_{1,+}^{(\text{B})} &= \omega_0^2\left(\sqrt{1 + 4[1 - (T/T_0)^2]} - 1\right)/2 . \end{aligned} \qquad (17.31)$$

Owing to metastability, the lowest eigenvalue $\Lambda_0^{(\text{B})}$ is negative. Further, the odd mode with one node is the time-translational mode $y_0(\tau) \propto \dot{q}_{\text{B}}(\tau)$ with eigenvalue $\Lambda_{1,-}^{(\text{B})} = 0$.

Next we turn to the higher eigenvalues $\Lambda_n^{(\text{B})}$ with $|n| \geqslant 2$. First, we observe that at the crossover temperature $T = T_0$, at which the periodic orbit is degenerated to a point, $q_{\text{B}}(\tau) = q_{\text{b}}$, the eigenvalues $\Lambda_n^{(\text{b})}$ ($|n| \geqslant 2$) are given by the expression (15.8), where $\nu_n = 2\pi n/\hbar\beta_0$. Obsering that for the cubic metastable potential (11.40) there is $\omega_{\text{b}}^2 = \omega_0^2$, we may write

$$\Lambda_n^{(\text{b})} = \nu_n^2 + \gamma|\nu_n| + \omega_0^2 - 2\omega_{\text{b}}^2 = \nu_n^2 + \gamma|\nu_n| + \omega_0^2 - 4\pi\gamma/\hbar\beta_0 , \qquad (17.32)$$

---

[2]The periodic orbit in the full $\{q, \mathbf{x}\}$ space has conserved total energy. The point q_{turn} is the projection of the exit point P_{ex} of the periodic orbit onto the q-axis (cf. Fig. 13.1 in Chapter 13).

where $\beta_0 = 1/k_{\rm B}T_0$. It is tempting to replace in the second form the inverse crossover temperature β_0 by the actual inverse temperature β, and to assume that the resulting expression for $\Lambda_n^{(\rm B)}$ holds in the entire regime $\beta > \beta_0$. This asssumption yields

$$\Lambda_n^{(\rm B)} = \nu_n^2 + \gamma|\nu_n| + \omega_0^2 - 2\gamma\nu , \qquad (|n| \geqslant 2) , \qquad (17.33)$$

where $\nu = \nu_1 = 2\pi/\hbar\beta$. The expression (17.33) found by this intuitive argument is indeed correct in the regime $0 < T \leqslant T_0$, as was shown in Ref. [355]. We would like to add here that the kinetic term ν_n^2 in the expression (17.33) is requisite for large $|n|$, while it is irrelevant for small $|n|$. Finally, the action $\mathbb{W}_{\rm B}$, which fixes the normalizationof the zero mode is calculated with the expression (17.26) as

$$\mathbb{W}_{\rm B} = M \int_{-\hbar\beta/2}^{\hbar\beta/2} d\tau \, \dot{q}_{\rm B}^2(\tau) = 6\pi\frac{V_{\rm b}}{\gamma}\left(1 - \frac{T^2}{T_0^2}\right) . \qquad (17.34)$$

To obtain the pre-exponential factor (17.8), we gather up the expression (15.11) for the determinant D_0 and the expressions (17.31), (17.33) and (17.34). The resulting quantum mechanical attempt frequency of the rate expression is

$$f_{\rm qm} = A_1 A_2 \omega_0 . \qquad (17.35)$$

Here A_1 includes the zero-mode normalization factor and the eigenvalues of the two lowest even modes, and A_2 covers all higher eigenvalues,

$$A_1 = \sqrt{\frac{\mathbb{W}_{\rm B}}{2\pi\hbar}} \frac{\omega_0^2}{\sqrt{|\Lambda_0^{(\rm B)}|\Lambda_{1,+}^{(\rm B)}}} , \qquad A_2 = \frac{\prod_{n=1}^{\infty}[\nu_n^2 + \gamma\nu_n + \omega_0^2]}{\omega_0^2 \prod_{n=2}^{\infty} \Lambda_n^{(\rm B)}} . \qquad (17.36)$$

With Eqs. (17.31) and (17.34) the temperature dependence cancels out in A_1, yielding

$$A_1 = \sqrt{3V_{\rm b}/\hbar\gamma} . \qquad (17.37)$$

The infinite products in the expression for A_2 can be expressed again in terms of gamma functions as in Eq. (15.17),

$$A_2 = \frac{\omega_0^2 + \nu^2 - \gamma\nu}{\omega_0^2} \frac{\Gamma(1 + \lambda_{\rm B}^+/\nu)\,\Gamma(1 + \lambda_{\rm B}^-/\nu)}{\Gamma(1 + \lambda_0^+/\nu)\,\Gamma(1 + \lambda_0^-/\nu)} . \qquad (17.38)$$

For $\gamma \gg 2\omega_0$ we have $\lambda_0^+ = \gamma$, $\lambda_0^- = \omega_0^2/\gamma$ and $\lambda_{\rm B}^{\pm} = \lambda_0^{\pm} \pm 2\nu$. With these expressions, the partial prefactor A_2 is readily evaluated in leading order in γ/ω_0 as

$$A_2 = (\gamma/\omega_0)^4 = 16\,\alpha^4 . \qquad (17.39)$$

Thus, the pre-exponential factor of the rate or attempt frequency turns out to be temperature-independent for large α. Upon collecting the factors A_1 and A_2, and the bounce action (17.27), the thermal rate(17.7) is found in the regime $0 \leqslant T \leqslant T_0$ in the analytic form

$$k = 8\sqrt{6}\,\alpha^3\omega_0\sqrt{\alpha\frac{V_b}{\hbar\omega_0}}\,\exp\left[-6\pi\frac{\alpha V_b}{\hbar\omega_0}\left(1 - \frac{T^2}{3T_0^2}\right)\right]. \tag{17.40}$$

Subleading contributions to the rate expression are given in Ref. [311]. This concludes our discussion of the quantum-statistical escape of an overdamped particle from the metastable well of a cubic potential in the semiclassical limit.

17.4 Quantum decay in a tilted cosine potential

The phenomenon of incoherent tunneling of a macroscopic variable has become clearly visible in superconducting quantum interference devices. In Josephson junctions, the phase jump of the Landau-Ginsburg order parameter across the junction plays the rôle of the tunneling coordinate, as discussed in Subsec. 3.2.2. The measured lifetimes of the zero voltage state in current-biased Josephson systems were found to be in excellent agreement [375] with the theoretical predictions.

The standard phenomenological model for a Josephson junction is the Resistively and Capacitively Shunted Junction (RCSJ) model [111] introduced in Subsec. 3.2.2. In the RCSJ model, the Josephson junction is equipped with a shunt capacitance C and an effective shunt resistance R. The corresponding deterministic equation of motion for the phase jump ψ across the junction is given in Eq. (3.150). It reads[3]

$$C\left(\frac{\phi_0}{2\pi}\right)^2\frac{d^2\psi}{dt^2} + \frac{1}{R}\left(\frac{\phi_0}{2\pi}\right)^2\frac{d\psi}{dt} + E_J\sin\psi - I_{\text{ext}}\frac{\phi_0}{2\pi} = 0, \tag{17.41}$$

where $\phi_0 = h/2e$ is the flux quantum, I_{ext} is the externally applied bias current, and $E_J = I_c\phi_0/2\pi$ is the Josephson coupling energy. The current I_c is the maximum supercurrent which the junction can sustain. Equation (17.41) describes classical damped motion of the phase in the potential (4.200).

The dynamical equation (17.41) is formally equivalent to the equation of motion of a Brownian particle of mass M and position X in the absence of fluctuations,

$$M\ddot{X}(t) + M\gamma\dot{X}(t) + \partial V(X)/\partial X = 0. \tag{17.42}$$

The potential $V(X)$ is of trigonometric form with a sloping background resulting in a potential drop $2\pi V_{\text{tilt}}$ per length X_0 as sketched in Fig. 17.1,

$$V(X) = -V_0\cos\left(2\pi X/X_0\right) - V_{\text{tilt}}\,2\pi X/X_0. \tag{17.43}$$

In the mapping of the two models, the equivalence relations are

$$\begin{aligned}
X/X_0 &\;\hat{=}\; \psi/2\pi, & M X_0^2 &\;\hat{=}\; C\phi_0^2, & \gamma &\;\hat{=}\; 1/RC, \\
V_0 &\;\hat{=}\; E_J, & V_{\text{tilt}} &\;\hat{=}\; I_{\text{ext}}\,\phi_0/2\pi.
\end{aligned} \tag{17.44}$$

The model (17.42) with (17.43) is archetypal for many transport systems in condensed matter physics.

---

[3]The RCSJ model is Markovian and corresponds to the substitution $Y^*(\omega) = 1/R$ in Eq. (3.220).

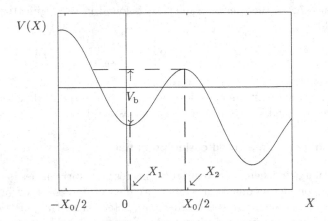

Figure 17.1: Sketch of the tilted washboard potential.

The two macroscopically distinguishable states in the current-biased Josephson junction are the zero voltage state and the voltage state. According to Josephson's second relation $\dot\psi = 2eU/\hbar$, the first state corresponds to the particle being trapped in a well, and the second to the particle sliding down the cascade of wells. The particle in the well oscillates with the "plasma" frequency

$$\omega_0 \; = \; (2\pi/X_0)\sqrt{V_0/M} \qquad \text{provided that} \qquad V_0 \equiv \sqrt{V_0^2 - V_{\text{tilt}}^2} \; > \; 0 \,. \quad (17.45)$$

For the potential $V(X)$ given in Eq. (17.43), the barrier frequency ω_{b} coincides again with the well frequency ω_0. The minima and maxima of the tilted washboard potential are located in the principal interval $0 \leqslant X < X_0$ at

$$X_1 \; = \; (X_0/2\pi)\arctan\left(V_{\text{tilt}}/V_0\right), \qquad \text{and} \qquad X_2 \; = \; X_0/2 - X_1 \,, \quad (17.46)$$

and the barrier height is given by

$$V_{\text{b}} \; = \; V(X_2) - V(X_1) \; = \; 2[\,V_0 - V_{\text{tilt}}\,\text{arccot}(V_{\text{tilt}}/V_0)\,]\,. \quad (17.47)$$

As $V_{\text{tilt}} \to V_0$, or $V_0 \to 0$, the cascade of barriers ceases to exist. In the regime $V_{\text{tilt}} > V_0$, the slope of the potential is negative everywhere, so that the particle slides all the way down and cannot be trapped even in the absence of inertia.

In the regime $V_{\text{tilt}} < V_0$, the Ohmic damping strength is characterized by the dimensionless viscosity

$$K \; \equiv \; \frac{\eta X_0^2}{2\pi\hbar} \; = \; 2\pi\alpha\,\frac{2V_0}{\hbar\omega_0}\,. \quad (17.48)$$

The first form corresponds to the Kondo parameter K introduced in Eq. (4.159). The second form combines K with the usual dimensionless damping parameter

$$\alpha \; \equiv \; \gamma/2\omega_0 \,. \quad (17.49)$$

For a potential near degeneracy, $V_{\text{tilt}} \ll V_0$, we have $K \approx 2\pi\alpha V_b/\hbar\omega_0$.

Consider next the crossing rate from the higher to the lower well, which we denote by k^+ henceforth. To that end it is useful to introduce the thermal energies

$$k_B T_0 \equiv V_0/K = \hbar\omega_b/4\pi\alpha , \quad \text{and} \quad k_B T_1 \equiv V_{\text{tilt}}/K . \qquad (17.50)$$

Here, T_0 is the crossover temperature between quantum tunneling and thermal hopping in the strong damping limit, Eq. (14.39), while the thermal energy $k_B T_1$ characterizes the effective strength of the bias. In the tunneling regime $T < T_0$, the crossing rate is dominated by the one-bounce contribution. Upon partial integration of the time–non-local term in Eq. (13.2) and use of the expression (4.57), the bounce trajectory in imaginary time obeys the equation of motion

$$- M\ddot{X}_B(\tau) + V_0(2\pi/X_0)\sin\left[2\pi X_B(\tau)/X_0\right] - 2\pi V_{\text{tilt}}/X_0$$
$$+ 2\alpha M(\omega_0/\hbar\beta) \int_{-\hbar\beta/2}^{\hbar\beta/2} d\tau' \cot[\pi(\tau - \tau')/\hbar\beta]\dot{X}_B(\tau') = 0 . \qquad (17.51)$$

In the regime of high friction, $\alpha \gg 1$, the inertia term $M\ddot{X}_B(\tau)$ in Eq. (17.51) is negligibly small compared to the friction term and can therefore be dropped. The truncated equation of motion can be solved in analytic form, as discovered by Korshunov [376]. In the range $T < T_0$ and $\alpha \gg 1$, the bounce with period $\hbar/k_B T$ is

$$X_B(\tau) = X_1 + \frac{X_0}{\pi} \arctan\left(\frac{T^2/T_0 T_1}{1 - \sqrt{[1 - T^2/T_0^2][1 + T^2/T_1^2]}\cos(2\pi k_B T\tau/\hbar)}\right) . \qquad (17.52)$$

The exponent $\mathbb{B}$ in the rate expression $k^+ = f_{\text{qm}}\, e^{-\mathbb{B}}$ is given by the bounce action (13.3) reduced by the action of the temporally constant trajectory X_1 dwelling in the minimum of the well, $\mathbb{B} = \mathcal{S}_B(T)/\hbar - V(X_1)/k_B T$. Upon disregarding the inertia term in $\mathcal{S}_B(T)$ one readily finds

$$\mathbb{B} = K \ln\left(\frac{V_0^2}{V_{\text{tilt}}^2 + K^2(k_B T)^2}\right) + 2K\left[1 - \frac{T_1}{T}\arctan\left(\frac{T}{T_1}\right)\right] . \qquad (17.53)$$

In the limit $T \to 0$, the bounce (17.52) and the action (17.53) take the forms

$$X_{B,T=0}(\tau) = X_1 + \frac{X_0}{\pi}\arctan\left(\frac{2T_0/T_1}{1 + (\omega_0\tau/2\alpha)^2}\right) , \qquad (17.54)$$

$$\mathbb{B}_{T=0} = 2K \ln(V_0/V_{\text{tilt}}) . \qquad (17.55)$$

As $T \to T_0$, the bounce (17.52) reduces to the constant path $X_B(\tau) = X_2$, and the exponent $\mathbb{B}$ matches with the Arrhenius exponent $V_b/k_B T_0$.

Consider next the determinant of the fluctuation operator. The eigenvalues of the symmetric modes with zero and with two nodes $\Lambda_0^{(B)}$ and $\Lambda_{1,+}^{(B)}$ are found as

$$\Lambda_0^{(B)} = -\omega_0^2\left\{\sqrt{\frac{1}{4} + \frac{V_{\text{tilt}}^4}{V_0^4}\left(1 + \frac{T^2}{T_1^2}\right)\left(1 - \frac{T^2}{T_0^2}\right)} - \frac{1}{2} + \frac{V_{\text{tilt}}^2}{V_0^2}\left(1 + \frac{T^2}{T_1^2}\right)\right\} ,$$

$$\Lambda_{1,+}^{(B)} = \omega_0^2\left\{\sqrt{\frac{1}{4} + \frac{V_{\text{tilt}}^4}{V_0^4}\left(1 + \frac{T^2}{T_1^2}\right)\left(1 - \frac{T^2}{T_0^2}\right)} + \frac{1}{2} - \frac{V_{\text{tilt}}^2}{V_0^2}\left(1 + \frac{T^2}{T_1^2}\right)\right\} . \qquad (17.56)$$

The lowest eigenvalue $\Lambda_0^{(B)}$ is again negative because of the underlying metastability. We now have

$$|\Lambda_0^{(B)}| \, \Lambda_{1,+}^{(B)} = \omega_0^4 \frac{V_{\text{tilt}}^2}{V_0^2} \left(1 + \frac{T^2}{T_1^2}\right)\left(1 - \frac{T^2}{T_0^2}\right). \tag{17.57}$$

The eigenvalues above $\Lambda_{1,+}^{(B)}$ are in the form (17.33) with ω_0 given in Eq. (17.45). The action which determines the normalization of the zero mode is

$$\mathbb{W}_B \equiv M \int_0^{\hbar\beta} d\tau \, \dot{X}_B^2(\tau) = 4\pi \frac{(V_0^2 - V_{\text{tilt}}^2)^{3/2}}{\gamma V_0^2}\left(1 - \frac{T^2}{T_0^2}\right). \tag{17.58}$$

Writing the pre-exponential factor of the rate as in Eq. (17.35) with (17.36), we find

$$A_1 \equiv \sqrt{\frac{\mathbb{W}_B}{2\pi\hbar}} \frac{\omega_0^2}{\sqrt{|\Lambda_0^{(B)}|\Lambda_{1,+}^{(B)}}} = \frac{1}{\sqrt{\alpha\hbar\omega_0}} \frac{M^{3/2}\omega_0^3}{V_{\text{tilt}}} \frac{(X_0/2\pi)^3}{\sqrt{1 + (T/T_1)^2}}, \tag{17.59}$$

$$A_2 \equiv \frac{\prod_{n=1}^{\infty}[\nu_n^2 + \gamma\nu_n + \omega_0^2]}{\omega_0^2 \prod_{n=2}^{\infty}\Lambda_n^{(B)}} = 16\,\alpha^4.$$

After all, the quantum mechanical attempt frequency is found to read

$$f_{\text{qm}} = A_1 A_2 \omega_0 = \frac{2\,\alpha^{7/2}}{\pi^3} \frac{(M\omega_0^2 X_0^2)^{3/2}}{\sqrt{\hbar\omega_0}V_{\text{tilt}}} \frac{1}{\sqrt{1 + (T/T_1)^2}}\,\omega_0 \tag{17.60}$$

The expressions (17.53) and (17.60) are the required components which determine the semiclassical tunneling rate to the next lower well (cf. Ref. [376]),

$$k^+ = f_{\text{qm}}\, e^{-B}. \tag{17.61}$$

Interestingly, the problem of the overdamped particle in the tilted washboard potential can be transformed into that of the overdamped particle in the quadratic-plus-cubic potential system if we put

$$V_{\text{tilt}} = 12\,V_{\text{b}}(X_0/2\pi q_{\text{b}})^3, \tag{17.62}$$

$$V_0 = V_{\text{tilt}}\sqrt{1/4 + (X_0/2\pi q_{\text{b}})^2},$$

and then perform the limit $X_0 \to \infty$. Under these conditions, the tilted washboard potential

$$V_{\text{per}}(q) = V(X_1 + q) - V(X_1) \tag{17.63}$$

is transformed into to the metastable potential,

$$\lim_{X_0\to\infty} V_{\text{per}}(q) \to V(q) = 3V_{\text{b}}\,(1 - 2q/3q_{\text{b}})\,q^2/q_{\text{b}}^2, \tag{17.64}$$

treated in Section 17.3. In the same limit, the rate expression (17.61) with the prefactor (17.60) and the exponent (17.53) turns in fact into the rate expression (17.40) for the cubic metastable potential.

The opposite scenario is when the bias energy V_{tilt} is small on the scale of the corrugation energy V_0. This regime is discussed next.

17.4.1 The case of weak bias

For a weak slope,

$$V_{tilt} \ll V_b = 2V_0 , \qquad (17.65)$$

neighboring wells are nearly degenerate and separated by a barrier of height V_b. Then the bounce consists of a weakly bound and widely spaced instanton–anti-instanton pair. Under condition (17.65), the distance τ between the instanton and the anti-instanton is large compared to the width α/ω_b of the instanton, and the action of the bounce changes only weakly when the distance τ is moved away from the stationary point $\tau = \tau_{st}$. To study the weak-bias case in some detail, we write the associated action as a function of the temporal spacing τ,

$$S(\tau) = 2S_{inst} + \hbar W(\tau) - \hbar\epsilon\tau . \qquad (17.66)$$

The first term is twice the inherent action of an instanton. The second term $\hbar W(\tau)$ represents the interaction between the instantons at distance τ, and the last term is the bias action for potential drop $\hbar\epsilon$ per spatial distance X_0. The bias frequency is

$$\epsilon = 2\pi V_{tilt}/\hbar . \qquad (17.67)$$

Now we anticipate that the interaction $W(\tau)$ has the spectral representation (4.77), and takes for the Ohmic spectral density $G(\omega) = 2K\omega$ the form (18.55),

$$W(\tau) = 2K \ln\left[(\hbar\beta\gamma/\pi)\sin(\pi\tau/\hbar\beta)\right] . \qquad (17.68)$$

Here we have equated for convenience the occurring reference frequency with the damping frequency γ. The action (17.66) is extremal at $\tau = \tau_{st}$, where

$$\tau_{st} = (\hbar\beta/\pi)\arctan(2\pi K/\hbar\beta\epsilon) . \qquad (17.69)$$

The instanton action S_{inst} is determined by requiring that the extremal action $S(\tau_{st})/\hbar$ coincides with the bounce action (17.53), $\mathbb{B} = S(\tau_{st})/\hbar$. This condition yields

$$S_{inst}/\hbar = K\left[1 + \ln(\pi V_0/K\hbar\gamma)\right] . \qquad (17.70)$$

Observe that the reference frequency γ drops out in the combined expression (17.66). Near to degeneracy, $V_{tilt} \ll V_0$, the absolute value of the negative eigenvalue is found from the expression (17.56) to read

$$|\Lambda_0^{(B)}| = (2\alpha/K)^2[\epsilon^2 + (2\pi K/\hbar\beta)^2] . \qquad (17.71)$$

With this form the rate expression (17.61) with (17.53) and (17.60) can be written as

$$k^+ = (2\alpha)^3\sqrt{\frac{K}{\pi}}\,\omega_0^2\,e^{-2S_{inst}/\hbar}\,\frac{e^{\epsilon\tau_{st}-W(\tau_{st})}}{|\Lambda_0^{(B)}|^{1/2}} , \qquad (17.72)$$

Since $(2\alpha/K)^2 = (\hbar\omega_0/\pi V_b)^2 << 1$, there is $|\Lambda_0^{(B)}| \ll \omega_0^2$. This indicates that the Gaussian approximation for this mode may be inappropriate and thus the expression (17.72) may be not correct in the regime of weak bias and low temperature.

To eliminate this flaw, we take one step back before the breathing mode is integrated out. The dubious Gaussian integration is reverted by the replacement.

$$\frac{e^{\epsilon\tau_{st}-\mathcal{W}(\tau_{st})}}{|\Lambda_0^{(B)}|^{1/2}} \rightarrow \frac{1}{2\alpha}\sqrt{\frac{K}{\pi}}\,\mathrm{Im}\int_{\mathcal{C}} d\tau\, e^{\epsilon\tau-\mathcal{W}(\tau)}\,. \tag{17.73}$$

Here, $\mathcal{C}$ is a contour being deformed to pass through the saddle point of the action at $\tau = \tau_{st}$ in the direction of steepest descent, which is perpendicular to the real τ-axis. With this substitution, the rate (17.72) may be written in the form

$$k^+ = \frac{\overline{\Delta}^2}{4}\,2\,\mathrm{Im}\int_{\tau_s}^{\tau_s+i\infty} d\tau\, e^{\epsilon\tau-\mathcal{W}(\tau)}\,. \tag{17.74}$$

The tunneling amplitude $\overline{\Delta}/2$ is determined by intrinsic properties of the instanton,

$$\overline{\Delta}/2 = f_{\mathrm{inst}}\,e^{-S_{\mathrm{inst}}/\hbar}\,, \qquad \text{with} \qquad f_{\mathrm{inst}} = \sqrt{K/2\pi}\,\gamma\,. \tag{17.75}$$

The analytically continued integral captures the breathing of the bounce.

With the instanton action (17.70) the tunneling amplitude is found to read

$$\overline{\Delta} = 2\sqrt{K/2\pi}\,\gamma\,e^{-K}(K\hbar\gamma/\pi V_0)^K\,, \qquad K \gg 1\,. \tag{17.76}$$

Eq. (17.76) expresses the transfer amplitude $\overline{\Delta}$ with the parameters of the original potential system. Interestingly, the relation (17.76) is generalized to hold for all K with the subsitution $\overline{\Delta} \rightarrow \Delta\,e^{C_E}$, where

$$\Delta = \Gamma(1+K)\,[\Gamma(1+1/K)]^K\,(\gamma/\pi)^{1+K}\,(V_0/\hbar)^{-K}\,. \tag{17.77}$$

With the asympotic representation of $\Gamma(z)$ for large values of $|z|$, the expression $\Delta\,e^{C_E}$ matches the form (17.76) in the limit $K \rightarrow \infty$. The relation (17.77), viz. Eq. (28.122) with $\omega_c = \gamma$, is derived in Sec. 28.5 based on a self-duality property of this model.

Incidentally, we may express the attempt frequency f_{inst} with the eigenvalues of the fluctuation operator about the instanton– anti-instanton pair,

$$f_{\mathrm{inst}} = \omega_0\sqrt{\widetilde{A}_1 A_2}\,, \tag{17.78}$$

where A_2 is given in Eq. (17.59). The term

$$\widetilde{A}_1 = \sqrt{\frac{K}{\pi}\frac{\omega_0}{2\gamma}}\sqrt{\frac{W_B}{2\pi\hbar}}\sqrt{\omega_0^2/\Lambda_{1,+}^{(B)}}\,, \tag{17.79}$$

differs from the factor A_1 given in Eq. (17.36) by the Jacobian $\sqrt{K/\pi}\,\omega_0/2\gamma$, and by absence of the eigenvalue of the breathing mode $|\Lambda_0^{(B)}|$. With the expressions (17.56), (17.58), and (17.59), one arrives in turn at the expression $f_{\mathrm{inst}} = \sqrt{K/2\pi}\,\gamma$.

If we would evaluate the integral (17.74) by steepest descent, we would recover the previous result (17.72). Importantly, however, the integration can be done exactly.

Observing that the analytically continued function $Q(z) = \mathcal{W}(\tau = iz)$ is an analytic function of z in the strip $0 > \mathrm{Im}\, z > -\hbar\beta$ of the complex z-plane, we may write

$$k^+ = \frac{\Delta^2 e^{2C_{\mathrm{E}}}}{4} \int_{-\infty-ic}^{\infty-ic} dz\ e^{i\epsilon z - Q(z)}, \tag{17.80}$$

where the constant c is chosen in the interval $0 < c < \hbar\beta$ such that the integration contour lies to the right of the singularities of the integrand. For $K < \frac{1}{2}$, we may put $c = 0^+$. We then obtain, using the expression (17.68),

$$k^+ = \frac{\Delta^2 e^{2C_{\mathrm{E}}}}{2} \left(\frac{\pi}{\hbar\beta\omega_c}\right)^{2K} \int_0^\infty dt\ \frac{\cos(\epsilon t - \pi K)}{\sinh^{2K}(\pi t/\hbar\beta)}. \tag{17.81}$$

yielding [90]

$$k^+(T,\epsilon) = \Gamma(1 + \tfrac{1}{K})^{2K}\, \Gamma(1 + K)^2\, \frac{2^{2K} e^{2C_{\mathrm{E}}}}{8\pi^3}\, \hbar\beta\gamma^2\, \frac{|\Gamma(K + i\hbar\beta\epsilon/2\pi)|^2}{(\beta V_0)^{2K}\, \Gamma(2K)}\, e^{\hbar\beta\epsilon/2}. \tag{17.82}$$

As the integral in Eq. (17.80) is a contour integral which encircles the singularity at $z = 0$, the analytic expression (17.82) holds not only in the regime $K < \frac{1}{2}$ but also in the regime $K \geqslant \frac{1}{2}$.[4] Further discussion of this point is given below Eq. (20.74).

In the absence of the bias, $\epsilon = 0$, the crossing rate takes the form

$$k^+(T,0) = \frac{\Gamma(1 + \tfrac{1}{K})^{2K}\, \Gamma(1 + K)^3\, e^{2C_{\mathrm{E}}}}{4\pi^{3/2}\, K\, \Gamma(\tfrac{1}{2} + K)}\, \frac{\gamma}{\pi}\, \frac{\hbar\gamma}{V_0} \left(\frac{k_{\mathrm{B}}T}{V_0}\right)^{2K-1}. \tag{17.83}$$

In the opposite limit $T = 0$ and nonzero bias, we find

$$k^+(0,\epsilon) = \frac{\Gamma(1 + \tfrac{1}{K})^{2K}\, \Gamma(1 + K)^2\, e^{2C_{\mathrm{E}}}}{2\pi\, \Gamma(2K)}\, \frac{\gamma}{\pi}\, \frac{\hbar\gamma}{V_0} \left(\frac{\hbar\epsilon}{\pi V_0}\right)^{2K-1}. \tag{17.84}$$

The temperature $T_1 = \hbar|\epsilon|/2\pi K k_{\mathrm{B}}$, is a kind of a crossover temperature. In the low temperature regime $T \ll T_1$, the thermal energy is negibly small compared with the bias energy, so that the expression (17.84) is appropriate. For $T \gg T_1$, the bias energy may be disregarded, and thus the form (17.83) is adequate.

Up to now, the crossing rate from the higher to the lower well, $k = k^+$, has been studied. The backward rate from the lower to the higher well obeys detailed balance,

$$k^-(T,\epsilon) = e^{-\hbar\epsilon/k_{\mathrm{B}}T}\, k^+(T,\epsilon). \tag{17.85}$$

Within steepest descent for the breathing mode, the backward rate k^- is determined by a different branch of the arctan function in the action (17.53). A general discussion of the detailed balance property is given below in Subsection 20.2.1.

It is straightforward to see that the leading thermal enhancement calculated from the expression (17.82) is in agreement with the expression (17.18).

---

[4]The contour integral (17.80) with the singular integrand $e^{-W(iz)}$, where $W(\tau)$ is given in Eq. (17.68), is analogous to Hankel's contour integral of the reciprocal gamma function [377].

These results can directly be applied to macroscopic quantum tunneling in a voltage-biased Josephson junction. The corresponding model is introduced in Subsection 3.4.3 [see also Eq. (17.41)]. In the mapping of the Brownian particle model to the Josephson junction model, we have the parameter identifications $K = 1/\rho$, where $\rho = R/R_Q$ ($R_Q = 2\pi\hbar/4e^2$ and $E_c = 2e^2/C$ are the resistance quantum and charging energy for Cooper pairs), $V_0 = E_J$, $\hbar e = 2eU/\rho$, and $\gamma = E_c/\pi\hbar\rho$. The overdamped limit corresponds to $1/4\alpha^2 = 2\pi^2\rho^2 E_J/E_c \ll 1$, as follows with the correspondence relations (17.44). The dc current through the Ohmic impedance R in the circuit sketched in Fig. 3.3 (see Subsection 3.4.3) is $I = [U - \hbar\langle\dot\psi\rangle/2e]/R$. For $\rho \ll 1$ and small V_x, the phase slip is determined by the single-bounce contribution (17.61), $\langle\dot\psi\rangle = 2\pi k^+$. Thus we have

$$I(U) = [U - (\hbar\pi/e)k^+(1 - e^{-\beta 2eU/\rho})]/R . \tag{17.86}$$

For $T = 0$, the rate k^+ in the overdamped limit is given in Eq. (17.84). With this we readily obtain

$$I(U) = \frac{U}{R}\left(1 - \frac{\pi^{-3/2-2/\rho}\,e^{2C_E}}{2\rho^3}\frac{\Gamma(1/\rho)\Gamma(\rho)^{2/\rho}}{\Gamma(1/2 + 1/\rho)}\left(\frac{eU}{E_J}\right)^{2/\rho}\left(\frac{E_c}{eU}\right)^2\right), \tag{17.87}$$

where terms of order $(eU/E_J)^{4/\rho^2}$ are disregarded.

When the junction is in the zero voltage state, the voltage drop occurs at the resistor, yielding Ohm's law. Macroscopic quantum tunneling into a finite voltage state yields the *nonlinear* second term in the $I(U)$ characteristics. Further discussion of the voltage-biased Josephson junction is given in Subsec. 20.3.4 for weak Josephson coupling, in Sec. 28.3 for the special cases $\rho = 1/2$ and $\rho = 2$, and in Subsec. 28.5.2, where we examine the selfduality map of the Josephson current for the cases ρ and $1/\rho$.

17.5 Concluding remarks

The imaginary-time functional integral approach has provided an almost complete description of the quantum statistical decay of a metastable state extending from thermally activated decay at high temperatures down to very low temperatures where the system tunnels out of the ground state. The main features of the decay are summarized in the Arrhenius plot shown in Fig. 17.2. In this diagram, the classical rate is represented by a falling straight line. The rate flattens out towards a finite value at $T = 0$ due to quantum tunneling. For an undamped system ($\alpha = 0$), the transition between the classical and quantum regime is rather sharp. In the presence of damping, the classical rate is reduced because the attempt frequency is dressed by the factor ω_R/ω_b. The corresponding constant shift is not visible in Fig. 17.2. In contrast, the tunneling rate at zero temperature is exponentially reduced. In addition, the crossover temperature is lowered and the transition between thermally activated decay and tunneling becomes more gradual when damping is increased. For strong Ohmic dissipation, there is a large region in which thermal and quantum fluctuations interplay and thermal enhancement is governed by the power law given in Eq. (17.18).

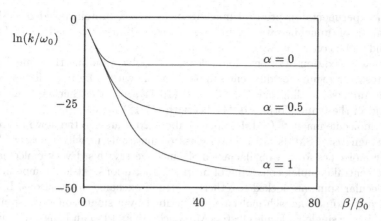

Figure 17.2: Arrhenius plot of the decay rate of a metastable system with cubic potential (barrier height $V_b = 5\hbar\omega_0$) and Ohmic damping ($\alpha = \gamma/2\omega_0$) for various values of α.

The qualitative features of the tunneling rate can be seen rather directly in the "many-dimensional WKB" approach. The minimum of the global potential $V(q, \mathbf{x})$ of the system-environment complex on the hyperplane of constant q is located away from the q-axis at $x_\alpha = (c_\alpha/m_\alpha\omega_\alpha^2)\,q$, and the value at the minimum is $V(q)$ and hence independent of the bath coupling, as we see from Eq. (3.14). The reduction of the tunneling rate by the environmental coupling can therefore be viewed as being due to the fact that the path under the barrier of the global potential $V(q, \mathbf{x})$, i.e. the path between the turning points in Eq. (17.9), is longer, both for zero and finite temperature, compared to the under-barrier-distance or tunneling length of the undamped system in the potential $V(q)$.

The phenomenon of macroscopic quantum tunneling has been observed experimentally in a large number of physical and chemical systems [312]. An especially attractive physical system is the current-biased Josephson junction or rf superconducting quantum interference device (SQUID) system. We have shown in Part I that for the case of a general linear impedance mechanism knowledge of the classical dissipative equation of motion of the flux uniquely determines its quantum-mechanical behavior. Since in Josephson devices all relevant parameters can be measured independently, these systems provided accurate tests of the MQT theory without adjustable parameters.

On a $\log k$ versus T^2 plot, the experimentally observed slope can be compared with the predicted behavior (17.18). Indeed, experiments have confirmed most of the theoretical predictions very accurately [375, 314]. Reviews of experimental results in MQT are given in Refs. [379, 380].

Evidence both for non-Ohmic and Ohmic dissipation has been discovered experimentally in diverse systems. Thermal enhancement with $s = 5$ in Eq. (17.17) was found in experiments on proton tunneling in hydrated protein powders [381]. The data analysis of tunneling rates for Li^+ impurities in diluted perovskite $K_{1-x}Li_xTaO_3$ was found to be consistent with $s = 3$ [382].

Dielectric relaxation of orientational defects in polycrystalline H_2O and D_2O ices in the crossover regime for different samples was shown tos be in perfect agreement with the universal scaling law (16.43) with (16.44), and the thermal enhancement was found in the Ohmic form, Eq. (17.14) with (17.18) [383].

Single-molecule magnets (SMMs) opened about 20 years ago the new research field of nanomagnetism. SMMs are a novel class of materials in which nearly identical magnetic molecules form a regular assembly in large crystals. Every molecular cluster has a central complex consisting of magnetic metal ions. At low temperature the intramolecular spins are locked together by strong exchange interactions. In the giant spin approximation each molecule is described by a single collective spin S. For instance, in the single-molecule clusters Mn_{12} acetate and Fe_8 an Ising-type magneto-crystalline anisotropy energetically favors and stabilizes giant spin states with magnetic quantum number $m = \pm 10$ and generates an energy barrier for the reversal of the magnetization of about 70 K for $Mn_{12}ac$ and 25 K for Fe_8.

SMMs have attracted much interest in recent years because of their quantum interference properties which appear due to the alignment of the magnetic molecules on a macroscopic scale and yield information on single-molecule properties [384]. Experiments with SMMs provided clear evidence of quantum tunneling of the magnetization through the magnetic anisotropy barrier [159, 161, 164] and revealed potential use of SMMs in quantum computing and data storage [385].

Other candidates for MQT physics are magnetic nanoparticles consisting of 10^3 to 10^6 magnetic moments. Magnetic systems have the advantage that several parameters which can tune the tunneling probability can be varied in a controlled manner.

Towards a quantitative picture of tunneling rates in MQT and of decoherence rates in MQC it is an indispensable prerequisite to understand the interaction of the giant spin with the environment. In particular, interactions with nuclear spins and with the lattice vibrations of the crystal are sources of energy dissipation and decoherence and therefore are of particular interest. Microwave radiation induces transitions between different spin states and can therefore provide information on the spin dynamics of the molecules. Given an effective Hamiltonian for the giant-spin-plus-reservoir complex, one may then derive, with the methods presented here, MQT rates for the diverse parameter regimes. Altogether, macroscopic quantum tunneling of magnetic particles is an effective field to check experimentally the theoretical predictions assembled in this part.

PART IV:

THE DISSIPATIVE TWO-STATE SYSTEM

One of the most intriguing aspects of quantum theory is the phenomenon of constructive and destructive interference. The phase coherence between different quantum mechanical states becomes apparent in oscillatory behaviors of observables. The simplest model involving quantum coherence is a two-state system. The problem of a quantum system whose state is effectively confined to a two-dimensional Hilbert space is often encountered in physics and chemistry. For instance, imagine a quantum mechanical particle tunneling clockwise forth and back between two different localized states. In reality, such system is strongly influenced by the surroundings. The coupling can lead to qualitative changes in the behaviors: the environment-induced fluctuations can destroy quantum coherence and can even lead to a "phase transition" to a state in which quantum tunneling is quenched. A manipulable two-state system is the quantum version of the classical binary bit, usually called qubit or quantum bit. In quantum computing, a qubit is the basic unit of quantum information and quantum computing devices. Since decoherence is the biggest enemy of qubit implementation, understanding, control and strategies for reduction of environmental influences are indispensable.

18 Introduction

Anderson *et al.* [121] and, independently, Phillips [122] postulated in 1972 the existence of two-level systems in glasses to explain low temperature anomalies of the specific heat in these amorphous materials. Although the true microscopic nature of the tunneling entities in glasses is unclear, they can be visualized by a particle tunneling in a double well along an (unknown) reaction coordinate. It turned out in ultrasonic experiments as an important difference between dielectric and metallic glasses that the lifetime of the tunneling eigenstates is drastically shorter in the case of metals. Golding *et al.* [386] explained this unexpected behavior by a nonadiabatic coupling of the tunneling entity to the conduction electrons in metals. Shortly later, Black and Fulde [387] extended this idea to describe relaxation processes in a superconducting environment by following the lines sketched above in Subsections 3.3.3 and 4.2.8. The theoretical predictions were confirmed by G. Weiss *et al.* [388] in ultrasonic experiments on superconducting amorphous metals. They showed that the lifetime is reduced when the environment is switched from superconducting to normalconducting, thus demonstrating the significance of the electronic coupling. A survey of the early developments and the perturbative treatment of this coupling is given in Ref. [123]. For a nonperturbative treatment see Ref. [389]. The nonlinear acoustic response of amorphous metals has been studied by Stockburger *et al.* [390], and the results for the dynamical susceptibility were found in good agreement with experiments [391]. A comprehensive review of the physics of tunneling systems in amorphous and crystalline solids with emphasis on the thermodynamic, acoustic, dielectric and optical properties has been given in Ref. [392].

The tunneling of light particles like small polarons, hydrogen isotopes or muons in solids has been thoroughly studied since several decades. While earlier work was mainly concerned with the significance of polaron effects [130]–[136], more recent attention has focused on the singular transient response of the fermionic environment in metals at low temperatures. The nonadiabatic influence of conduction electrons on the motion of interstitials in metals was proposed by Kondo [393] to explain the anomalous temperature dependence of muon diffusion in host metals at low temperature. This mechanism has been confirmed experimentally for incoherent tunneling of μ^+ and p in Cu, Al, and Sc [102], for defects in mesoscopic wires [106, 105], and for tunneling of H and D in metals and superconductors. Comprehensive reviews of experiment [394] and theory [395] are available. The particle tunnels either randomly or coherently between particular interstitial sites. Experimentally, one passes from the incoherent to the coherent dynamics by lowering temperature. The transition becomes apparent, e.g., in a change from quasi-elastic to inelastic neutron scattering.

Electron transfer reactions are ubiquitous in chemical and biological systems. In its simplest form, an electron localized at a donor site is tunneling to the acceptor site. Marcus' theory [107, 108] provides the appropriate scenario to describe such processes. Often the interaction between the charge and the polarization cloud of the environment is so strong that tunneling is possible only with the assistance of favorable equilibrium fluctuations in the environmental modes. Because of the small mass, electron transfer is strongly influenced by quantum effects even at room temperature.

Examples of quantum coherence in an effective two-state system are the inversion resonance of the NH_3 molecule, strangeness oscillations of a neutral K-meson beam, and coherent tunneling of light interstitials in tunneling centers. Another important, but apparently quite different example of a two-state system is a rf SQUID ring threaded by an external flux near half a flux quantum (cf. Subsection 3.2.2). Such a system might be the appropriate vehicle for the observation of "macroscopic quantum coherence" (MQC) on a true macroscopic scale [310, 396].

The behavior of the two-state system (TSS) is strongly influenced by the dissipative coupling to the heat bath's dynamical degrees of freedom. We shall consider linear couplings to the heat bath that are sensitive to the value of σ_z. For instance, a dipole-local-field coupling provides a simple physical model for this type of coupling. To be definite, we choose $H_I = -q \sum_\alpha c_\alpha x_\alpha$, where $q = \sigma_z q_0/2$ with q_0 being the spatial distance of the two localized states. It should be adequate in many cases of interest that the response of the environment to a perturbation can be considered as linear. Then, a bath which is represented by a set of harmonic oscillators with a coupling linear in the coordinates captures the essential physics we wish to describe. Since the TSS is like a spin, the corresponding model has become known in the literature as the "spin-boson" model. We have introduced this model already in Section 3.2. The relevant Hamiltonian is given in Eq. (3.145) or in Eq. (3.147).

Despite its apparent simplicity, the spin-boson model cannot be solved exactly by any known method (apart from some limited regimes of the parameter space). Not only is the spin-boson model nontrivial mathematically, it is also nontrivial physically.

The environment acts on the TSS by a fluctuating force $\xi(t) = \sum_\alpha c_\alpha x_\alpha(t)$. For a linearly responding bath, the modes $x_\alpha(t)$ obey Gaussian statistics. Therefore the dynamics of the bath is fully characterized by the force autocorrelation function in thermal equilibrium $\langle \xi(t)\xi(0) \rangle_\beta$, which is simply a superposition of harmonic oscillator correlation functions [cf. Eq. (5.33)]. In the formal path integral expression for the reduced density matrix, the environment reveals itself through an influence functional $\mathcal{F}$. In Chapters 4 and 5 we have given for $\mathcal{F}$ several useful forms applicable to thermodynamics and dynamics, respectively.

18.1 Truncation of the double-well to the two-state system

18.1.1 Shifted bath oscillators and orthogonality catastrophe

In Section 3.2, I have already briefly addressed the reduction of the double well system to the two-state system. Before I resume the related discussion for the dissipative TSS, consider now first the adiabatic limit, in which the environmental modes adapt themselves instantaneously to the particle's position. The oscillator part of the Hamiltonian (3.145) for the two positions $\sigma_z = \pm 1$ of the particle reads

$$H_\pm = \frac{1}{2} \sum_\alpha \left(\frac{p_\alpha^2}{m_\alpha} + m_\alpha \omega_\alpha^2 x_\alpha^2 \mp q_0 c_\alpha x_\alpha \right) = \sum_\alpha \left[\hbar \omega_\alpha b_\alpha^\dagger b_\alpha \mp \tfrac{1}{2} \hbar \lambda_\alpha \left(b_\alpha + b_\alpha^\dagger \right) \right] . \quad (18.1)$$

In the second line, we have introduced phonon creation and annihilation operators, and we put $c_\alpha = \sqrt{2\hbar m_\alpha \omega_\alpha} \, \lambda_\alpha / q_0$. Upon introducing the shifted operators

$$b_{\pm,\alpha} = b_\alpha \mp \tfrac{1}{2} \lambda_\alpha / \omega_\alpha , \quad (18.2)$$

the terms linear in b_α, $b_\alpha^\dagger$ cancel out, and we get the Hamiltonian $H_\pm$ in normal form

$$H_\pm = \sum_\alpha H_{\pm,\alpha} \quad \text{with} \quad H_{\pm,\alpha} = \hbar \left(\omega_\alpha \, b_{\pm,\alpha}^\dagger b_{\pm,\alpha} - \frac{\lambda_\alpha^2}{4\omega_\alpha} \right) . \quad (18.3)$$

Since the shifted operators obey the same commutation relations as the original ones, the two different vacuum states for the boson α are defined by $b_{\pm,\alpha} |0_{\pm,\alpha}\rangle = 0$. The normalized many-boson states $\{ |n_{\pm,\alpha}\rangle \}$ can be created from the vacuum state $|0_{\pm,\alpha}\rangle$ in the usual way, $|n_{\pm,\alpha}\rangle = (b_{\pm,\alpha}^\dagger)^{n_{\pm,\alpha}} |0_{\pm,\alpha}\rangle / \sqrt{n_{\pm,\alpha}!}$. Application of the one annihilation operator on the vacuum state of the other yields

$$b_{\mp,\alpha} |0_{\pm,\alpha}\rangle = \pm \frac{\lambda_\alpha}{\omega_\alpha} |0_{\pm,\alpha}\rangle . \quad (18.4)$$

The Hamiltonians $H_{\pm,\alpha}$ describe two harmonic oscillators with eigenfrequency ω_α, and with centers at spatial distance

$$s_\alpha = \frac{c_\alpha}{m_\alpha \omega_\alpha^2} q_0 = \sqrt{\frac{2\hbar}{m_\alpha \omega_\alpha}} \frac{\lambda_\alpha}{\omega_\alpha} . \quad (18.5)$$

The vacuum or ground states of $H_{-,\alpha}$ and $H_{+,\alpha}$ can mutually be transformed into each other with the displacement operation

$$|0_{\pm,\alpha}\rangle = e^{i\Omega_{\mp,\alpha}} |0_{\mp,\alpha}\rangle \quad \text{with} \quad \Omega_{\mp,\alpha} = \pm \frac{s_\alpha p_\alpha}{\hbar} = \pm i \frac{\lambda_\alpha}{\omega_\alpha} \left[b_{\mp,\alpha}^\dagger - b_{\mp,\alpha} \right] . \quad (18.6)$$

With use of the commutation relation of $b_{\pm,\alpha}$ with $b_{\pm,\alpha}^\dagger$ it appears that the vacuum state of the one oscillator can be expressed in terms of a coherent state of the displaced other oscillator,

$$|0_{\pm,\alpha}\rangle = \exp \left(- \frac{\lambda_\alpha^2}{2\omega_\alpha^2} \right) \sum_{n=0}^\infty \frac{(\pm 1)^n}{\sqrt{n!}} \left(\frac{\lambda_\alpha}{\omega_\alpha} \right)^n |n_{\mp,\alpha}\rangle . \quad (18.7)$$

The probability to excite n bosons with energy $\hbar\omega_\alpha$ in a sudden transition from the state $\sigma_z = -1$ to the state $\sigma_z = 1$ is Poissonian distributed,

$$p_{n,\alpha} = |\langle 0_{-,\alpha} | n_{+,\alpha}\rangle|^2 = \exp \left(- \frac{\lambda_\alpha^2}{\omega_\alpha^2} \right) \frac{1}{n!} \left(\frac{\lambda_\alpha}{\omega_\alpha} \right)^{2n} , \quad (18.8)$$

and the mean particle number is $\bar{n}_\alpha \equiv \sum_n n \, p_{n,\alpha} = \lambda_\alpha^2 / \omega_\alpha^2$.

In the absence of the tunneling coupling, the two lowest energy eigenstates of the *symmetric* global system are the symmetric and anti-symmetric degenerate states

$$|\psi_\pm\rangle = \frac{1}{\sqrt{2}}\left(|+\rangle \prod_{\alpha \in hf} |0_{+,\alpha}\rangle \pm |-\rangle \prod_{\alpha \in hf} |0_{-,\alpha}\rangle\right). \tag{18.9}$$

Here, $|\pm\rangle$ are the localized states with eigenvalues $\sigma_z = \pm 1$ of the TSS, and $|0_{\pm,\alpha}\rangle$ are the related shifted ground states of the α'th oscillator.

At zero temperature, the phonon overlap in the tunneling amplitude Δ is made up of phonons residing in the shifted ground states. Thus we have

$$\Delta \equiv \langle -|+\rangle \prod_\alpha \langle 0_{-,\alpha}|0_{+,\alpha}\rangle = \Delta_0 \exp\left(-\frac{1}{2}\sum_\alpha \frac{\lambda_\alpha^2}{\omega_\alpha^2}\right). \tag{18.10}$$

Here, the bare tunneling or transition amplitude Δ_0 is the bare overlap amplitude between the two localized states, $\Delta_0 = \langle -|+\rangle$, and $\langle 0_{-,\alpha}|0_{+,\alpha}\rangle$ is the overlap amplitude between the oscillators in the shifted ground states. Thus, the dressed amplitude is always smaller than the bare amplitude. Metaphorically speaking, the particle is dragging around a phonon cloud which hampers the transition because of the displacement of the cloud in the transition. The exponential dressing factor is known as Franck-Condon factor. In the continuum limit, the spectral coupling $G(\omega) = \sum_\alpha \lambda_\alpha^2 \delta(\omega - \omega_\alpha)$, introduced in Eq. (3.148), and related to $J(\omega)$ in Eq. (3.148), is a smooth function of ω. Then we have

$$\Delta = \Delta_0 \exp\left(-\frac{1}{2}\int_0^\infty d\omega \frac{G(\omega)}{\omega^2}\right). \tag{18.11}$$

When the integral is infrared-divergent, the two displaced boson clouds become orthogonal, i.e., they represent two different "worlds" which do not mutually interfere. This situation is dubbed *orthogonality catastrophe*. Ultraviolet-divergence of the integral does not occur, since there is always a cutoff in the spectral function $G(\omega)$ for physical reasons, as we have emphasized in Subsection 3.1.3.

In the super-Ohmic case $G(\omega) \propto \omega^s$, where $s > 1$, the integral is infrared-convergent. In the absence of the orthogonality catastrophe, elastic tunneling processes without dynamical involvement of the reservoir are possible.

When the spectral density is Ohmic or sub-Ohmic, the integral is infrared-divergent, and hence tunneling is quenched in the adiabatic limit. This is a clear indication that the adiabatic approximation breaks down for slow modes, and even leads to qualitatively wrong behavior in the Ohmic and sub-Ohmic regime. Proper treatment of these modes in the nonadiabatic regime is discussed in Sec. 20.2.

At finite T, the Franck-Condon factor is extended by the possibility for coherent multi-phonon emission and absorption in the transition, $\Delta = \Delta_0 \exp[-\frac{1}{2}(\rho_e + \rho_a)]$. The socalled Huang-Rhys factors [397] represent the mean number of emitted (e) and absorbed (a) phonons, summed over the entire spectrum (see also Subsec. 20.2.2),

$$\rho_e = \int_0^\infty d\omega \frac{G(\omega)}{\omega^2} \frac{1}{e^{\beta\hbar\omega} - 1}, \qquad \rho_a = \int_0^\infty d\omega \frac{G(\omega)}{\omega^2} \frac{1}{1 - e^{-\beta\hbar\omega}}. \tag{18.12}$$

18.1.2 Adiabatic renormalization

Consider a quantum particle moving in a double-well potential $V(q)$ as described in Sect. 3.2. Assume that the particle interacts with an environment as given in Eq. (3.12). We now wish to see under which conditions the continuous system can be reduced to a discrete two-state system. Roughly speaking, the reduction is possible if there is a wide separation of the relevant energy scales.

In the first stage of the truncation procedure, the system is coupled only to bath modes with frequency $\omega > \omega_c$. This leads (i) to adiabatic dressing of the mass analogous to Eq. (3.37), $M = M_0 + \Delta M_{hf}$, and hence reduction of the well frequency, $\omega_{0,r} = \sqrt{M_0/M}\,\omega_0$, and (ii) to adiabatic dressing of the tunnel splitting $\Delta_0 \to \Delta$,

$$\Delta = \Delta_0 \exp\left(-\frac{1}{2} \sum_{\alpha \in hf} \frac{\lambda_\alpha^2}{\omega_\alpha^2} \right) = \Delta_0 \exp\left(-\frac{1}{2} \int_0^\infty d\omega \, \frac{G_{hf}(\omega)}{\omega^2} \right). \tag{18.13}$$

Next, ω_c is chosen to satisfy the conditions

$$\hbar|\epsilon|, \, k_B T \ll \hbar\omega_c \ll \hbar\omega_{0,r}, \quad \text{and} \quad \Delta \ll \omega_c. \tag{18.14}$$

Under conditions (18.14), excited states in either well will not be significantly populated. Therefore, they can be ignored. As a result, we may replace the Hamiltonian of the continuous double-well system by an effective two-state Hamiltonian with an adiabatically dressed tunnel splitting. In the second stage of the reduction scheme, the remaining low-frequency bosons with $\omega < \omega_c$ are dynamically coupled to the discrete system. Thus we arrive at the dissipative two-state or spin-boson Hamiltonian [cf. Eq. (3.147) with (3.148)]

$$\hat{H}_{SB} = -\frac{\hbar\Delta}{2}\sigma_x - \frac{\hbar\epsilon}{2}\sigma_z - \frac{1}{2}\sigma_z \sum_{\alpha \in lf} \hbar\lambda_\alpha \left(b_\alpha + b_\alpha^\dagger \right) + \sum_{\alpha \in lf} \hbar\omega_\alpha b_\alpha^\dagger b_\alpha. \tag{18.15}$$

The associated spectral density $G(\omega)$ covers the coupling to modes with $\omega \lesssim \omega_c$,

$$G(\omega) = \sum_{\alpha \in lf} \lambda_\alpha^2 \, \delta(\omega - \omega_\alpha). \tag{18.16}$$

The spectral density $G(\omega)$ includes the coupling to all bath modes that are slow on the time scale Δ^{-1}. These modes will severely influence the transitions between the two low-lying states, as we shall see, The coupling will cause pivotal influences and distinct long-ranged memory effects in the parameter regimes of interest.

Under condition $k_B T \ll \hbar\omega_c$, the Huang-Rhys factors are taken at $T = 0$, and thus the Franck-Condon factor (18.13) applies. It is interesting to note that the form (18.13) does not not distinguish whether the dressed amplitude Δ originates from a single bath mode, or whether there is in reality a continuum of bath states in the regime $\omega > \omega_c$ [398].

The reduction procedure relying on the expression (18.13) has been studied by Sethna [173] in the context of tunneling centers in solids which are described by a spectral density $G(\omega) \propto \omega^3$ (cf. Subsection 4.2.4). In this case, also low-frequency

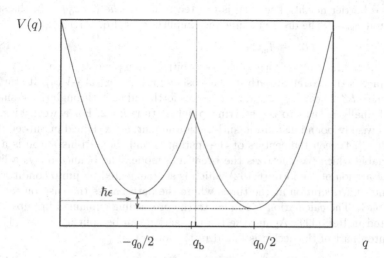

Figure 18.1: The asymmetric double well with a cusp as described by Eq. (18.18).

modes may be included in the Franck-Condon factor, as the integral is IR-convergent. The case of Ohmic dissipation is more subtle due to an inherent infrared divergence, as addressed already in the preceding subsection (see also Subsection 18.1.4).

In the sequel, we shall regard Δ as the tunneling matrix element which is already dressed by the high-frequency modes of the environment as given in Eq. (18.13).

18.1.3 Tunneling amplitude in a double well

Let us now turn to a particle in a weakly biased double well subjected to Ohmic bath coupling [cf. Eq. (4.158)] with cutoff at $\omega = \omega_c$,

$$G(\omega) = 2K\omega\,\Theta(\omega_c - \omega)\,. \tag{18.17}$$

Double well with a cusp-shaped barrier

For the slightly asymmetric double well potential sketched in Fig. 18.1,

$$V(q) = \begin{cases} \frac{1}{2}M\omega_0^2(q + \frac{1}{2}q_0)^2 + \frac{1}{2}\hbar\epsilon\,, & q < q_b\,, \\ \frac{1}{2}M\omega_0^2(q - \frac{1}{2}q_0)^2 - \frac{1}{2}\hbar\epsilon\,, & q > q_b\,, \end{cases} \tag{18.18}$$

the Kondo coupling parameter [cf. Eqs. (17.19) and (17.48)] is ($\alpha = \gamma/2\omega_0$)

$$K \equiv \frac{\eta q_0^2}{2\pi\hbar} = \frac{M\gamma q_0^2}{2\pi\hbar} = \frac{8}{\pi}\,\alpha\,v_b\,, \tag{18.19}$$

$$v_b \equiv \frac{V_b}{\hbar\omega_0} = \frac{1}{8}\frac{M\omega_0^2 q_0^2}{\hbar\omega_0} \tag{18.20}$$

is the scaled barrier height. The cusp is located at $q_{\rm b} = -\hbar\epsilon/(M\omega_0^2 q_0)$. In the semi-classical limit $v_{\rm b} \gg 1$, the dressed tunneling amplitude [cf. Eq. (17.75)]

$$\Delta/2 = f_{\rm inst}\, e^{-S_{\rm inst}/\hbar} \tag{18.21}$$

can be calculated in analytic form [399] for arbitrary damping strength.

The bounce is the extremal path in the upside-down potential $-V(q)$. It starts at thermal time $-\hbar\beta/2$ near $-q_0/2$ and then moves forth and back through the V-shaped valley, and finally returns to the starting point at time $\hbar\beta/2$. For a weak bias, the bounce is a weakly-bound instanton–anti-instanton pair, as explained in Subsec. 17.4. The length τ betweeen the centers of the instanton and the anti-instanton is a collective variable which parametrizes the breathing mode. The bounce obeys a linear equation of motion of the form (6.162) which is supplemented by jump conditions in velocity and in acceleration at the times where the path passes the cusp on the way forth and back. The calculation is similar to the proceeding explained in Subsec. 6.7 and reported in Ref. [399]. Again, the bounce action can be split as in Eq. (17.66). The instanton part of the action is found in the form

$$S_{\rm inst}/\hbar = 2v_{\rm b}(1 - 2\alpha^2)\, g(\alpha) + K\left[\, C_{\rm E} + \ln(\omega_0/\omega_{\rm c})\,\right], \tag{18.22}$$

where $g(\alpha)$ is given in Eq. (6.102), $C_{\rm E} = 0.5772\ldots$ is Euler's constant, and $\omega_{\rm c}$ is a reference frequency as in Eq. (17.70).

In the weak damping limit, $\alpha \ll 1$, we obtain from Eq. (18.22)

$$S_{\rm inst}/\hbar = 2v_{\rm b} + K\left[\,\ln(\omega_0/\omega_{\rm c}) - c_0\,\right] + \mathcal{O}(K^2), \tag{18.23}$$

with the numerical constant $c_0 = 1/2 - C_{\rm E} \approx -0.0772$.

In the opposite strong-damping limit $\alpha \gg 1$, in which $g(\alpha) = (2/\pi\alpha)\ln(2\alpha)$, we find from Eq. (18.22) with Eq. (18.19), and with $c_\infty = \ln(\pi/4) - C_{\rm E} \approx -0.82$

$$S_{\rm inst}/\hbar = K\left[\,\ln(\omega_0/\omega_{\rm c}) + \ln(v_{\rm b}) - \ln K - c_\infty\,\right]. \tag{18.24}$$

In the zero-damping limit, the prefactor $f_{\rm inst}$ in Eq. (18.21) is given by

$$f_{\rm inst} = \omega_0\sqrt{2v_{\rm b}/\pi}\,, \tag{18.25}$$

whereas for strong damping $\alpha \gg 1$, the prefactor (17.78) takes the form [399]

$$f_{\rm inst} = \omega_0\sqrt{\tilde{A}_1 A_2} = \omega_0\sqrt{K/4\pi\alpha^2}\,\sqrt{2.042\ln\alpha}\,. \tag{18.26}$$

Smooth double well

For a symmetric double well with a smooth barrier, the instanton width $\tau_{\rm inst}$ is the inverse of the effective barrier frequency $\omega_{\rm R}$. In the regime $\alpha \gg 1$, there is $\omega_{\rm R} = \omega_{\rm b}/2\alpha$, as follows from Eq. (14.37). For given barrier height $V_{\rm b}$, bare well frequency ω_0, and spacing q_0 between the wells, the Kondo parameter $K = M\gamma q_0^2/2\pi\hbar$ is related to the damping parameter $\alpha = \gamma/2\omega_0$ by $K/\alpha = \lambda v_{\rm b}$, where $v_{\rm b} = V_{\rm b}/\hbar\omega_0$, and λ is a numerical constant depending on the particular form of the double well. There is $\lambda = 8/\pi$ for the cusp potential, $\lambda = 2\pi$ for the cosine potential, and $\lambda = 32/\pi$ for the quartic potential (3.131). The frequency $\omega_{\rm R} = \lambda \omega_{\rm b} v_{\rm b}/2K$ provides an effective high-frequency cutoff for the instanton action. As a result, the expression (18.24) for the instanton action generally holds for any symmetric double well in the strong damping limit. Only the numerical factor c_∞ depends on the particular shape of the barrier of the double well potential.

The instanton action for the quartic double-well potential, Eq. (3.131), has been estimated in Ref. [401]. The numerical constant for weak damping was found, using a variational method, to be given by $c_0 \approx -0.2392$. In the strong damping limit, it was found the form (18.24) with $c_\infty = \ln(3\pi/4) - 5/2 - C_{\rm E} \approx -2.2202$.

The instanton action (17.70) for the cosine potential (17.43) at large α can be written in the form (18.24) with the constant $c_\infty = -\ln(2/\pi) \approx 0.4516$. The prefactor for this potential at large α, Eq. (17.75), can be written in the form

$$f_{\rm inst} = \omega_0 \sqrt{K/8\pi\alpha^2} \, 4\alpha^2 \,. \tag{18.27}$$

The striking dfference between the prefactors (18.26) and (18.27) originates from the factor $\sqrt{A_2}$, where A_2 is defined in Eq. (17.36). The factor $\sqrt{A_2}$ covers the higher eigenvalues of the fluctuation modes, and thus sounds out the barrier region. It is $\propto \alpha^2$ for the cosine potential (smooth barrier) and $\propto \sqrt{\ln \alpha}$ for the cusp potential (singular barrier frequency). It is plausible to assume that $\tilde{A}_1 \propto K/\alpha^2$ for any double well with smooths wells and $A_2 \propto \alpha^4$ for any double well with smooth wells and smooth barrier. It remains to demonstrate that the unphysical cutoff frequency $\omega_{\rm c}$ in the truncated model (18.15) with (18.16) for Ohmic coupling cancels out in the calculation of physical quantities. To this, we first recall that $\Delta \propto \exp[-K \ln(\omega_0/\omega_{\rm c})]$, as follows from Eq. (18.21) with (18.22). Second, Δ and the instanton pair interaction $\mathcal{W}(\tau)$ are combined in the form $\Delta^2 \exp[-\mathcal{W}(\tau)]$ [cf. the rate expression (17.74)]. Thus, with the scaling form (17.68) for $\mathcal{W}(\tau)$, holding in the regime $\omega_{\rm c}\tau \gg 1$, the cutoff frequency $\omega_{\rm c}$ actually cancels out. In the core regime $\omega_{\rm c}\tau \lesssim 1$, however, the cancellation of $\omega_{\rm c}$ in $\Delta^2 \exp[-\mathcal{W}(\tau)]$ is only partial. From this we infer that the cancellation is only in leading order of $\epsilon/\omega_{\rm c}$, $\omega_{\rm c}/\omega_{\rm b}$, $\omega_{\rm c}/\omega_0$, and $1/\hbar\beta\omega_{\rm c}$ [401].

18.1.4 Renormalized tunneling matrix element

We now resume the adiabatic renormalization scheme in which high frequencies are integrated out successively. In order to extend the adiabatic renormalization (18.13) to lower frequencies. we define the dressed tunneling amplitude

$$\Delta_{\rm d}(\omega_\ell) \equiv \Delta \exp\left(-\frac{1}{2}\int_{\omega_\ell}^\infty d\omega \, \frac{G(\omega)}{\omega^2}\right). \tag{18.28}$$

Since $\Delta_{\rm d} < \Delta$, we may iteratively integrate out oscillators with frequencies in the range $\omega_\ell(\Delta) > \omega > \omega_\ell(\Delta_{\rm d}) \equiv p\,\Delta_{\rm d}$. Here p is unspecified, but generally large, $p \gg 1$. The procedure converges towards a dressed tunneling amplitude

$$\Delta_r = \rho \Delta \,, \tag{18.29}$$

where ρ is self-consistently determined by the relation [86]

$$\rho = \exp\left(-\frac{1}{2} \int_{p\Delta_r}^{\infty} d\omega \, \frac{G(\omega)}{\omega^2} \right) . \tag{18.30}$$

The treatment of the slow modes can be improved by employing the flow equation formalism by Wegner [403] in which the Hamiltonian is diagonalized by continuous unitary transformations. This method leads to the self-consistent relation[1] [405, 406]

$$\Delta_r = \Delta \exp\left(-\frac{1}{2} \int_{0}^{\infty} d\omega \, \frac{G(\omega)}{\omega^2 - \Delta_r^2} \right) , \tag{18.31}$$

where the dash denotes the principal value of the improper integral. Eq. (18.30) disregards frequencies below $p\Delta_r$. In contrast, slow modes $\omega < \Delta_r$ in the relation (18.31) lead to an increase of the renormalized tunneling amplitude. The difference between Eq. (18.31) and (18.30) is negligible in the Ohmic and super-Ohmic case.

Consider next a spectral density of power-law form with exponential cutoff,

$$G(\omega) = 2\delta_s \omega_{\mathrm{ph}}^{1-s} \omega^s \, e^{-\omega/\omega_c} \,. \tag{18.32}$$

Here, δ_s is a dimensionless parameter.[2] In the super-Ohmic case, $s > 1$, the integral in Eq.(18.28) remains regular in the limit $\omega_\ell \to 0$. It is then natural to introduce a Franck-Condon factor which includes *all* modes of the reservoir. Thus we define

$$\begin{aligned} \Delta_{\mathrm{eff}} &\equiv \Delta_{\mathrm{d}}(\omega_\ell = 0) = \Delta \, e^{-B_s} \,, \\ B_s &= \delta_s \Gamma(s-1)(\omega_{\mathrm{D}}/\omega_{\mathrm{ph}})^{s-1} \,. \end{aligned} \qquad s > 1 \,, \tag{18.33}$$

Here we have used an exponential cutoff with Debye frequency ω_{D}, as common in the super-Ohmic regime.

For Ohmic damping, we have $d\ln[\Delta_{\mathrm{d}}(\omega_\ell)]/d\ln(\omega_\ell) = K$. Hence the effects of the low-frequency modes at $T = 0$ crucially depend on the value of K. For $K > 1$, the renormalized tunneling amplitude Δ_r is iterated to zero, $\Delta_r = \Delta_{\mathrm{d}}(\omega_\ell = 0) = 0$. This is the localization phenomenon found by Chakravarty [144] and by Bray and Moore [145] using renormalization group methods. For $K < 1$, we find from Eq. (18.30) or from Eq. (18.31), disregarding an unspecified numerical constant,

$$\Delta_r = (\Delta/\omega_c)^{K/(1-K)} \Delta \qquad \text{for} \qquad K < 1 \,. \tag{18.34}$$

This form was also found with a variational treatment of the sluggish modes [402].

For subsequent convenience, we also introduce a slightly modified frequency scale in which Δ_r is multiplied by the factor $[\Gamma(1 - 2K)\cos(\pi K)]^{1/2(1-K)}$,

---

[1]A similar integral is also found for the finite-T modification, Eq. (19.28) with (19.27).

[2]Here we distinguish the frequency scale ω_{ph} from the cutoff ω_c. In the super-Ohmic case, we shall sometimes identify ω_c with the Debye frequency ω_{D}. For $s = 1$, the parameter δ_1 coincides with Kondo's parameter K and with the parameter α introduced by Leggett *et al.* [86].

$$\Delta_{\text{eff}} = [\,\Gamma(1 - 2K)\cos(\pi K)\,]^{1/2(1-K)}\Delta_{\text{r}}\,. \tag{18.35}$$

The additional factor is of order one for all $K \lesssim \frac{1}{2}$ and is singular at $K = 1$. The effective amplitude Δ_{eff} is equal to Δ for $K = 0$ and equal to $\pi\Delta^2/2\omega_{\text{c}}$ for $K = \frac{1}{2}$. We shall see in Subsec. 22.1.1 that Δ_{eff} is the proper inverse time scale for $0 \leqslant K < 1$.

In the sub-Ohmic case, $s < 1$, we may put $\omega_{\text{c}} \to \infty$ since the adiabatic dressing integral is UV-convergent. In the adiabatic scheme, Eq. (18.30), Δ_{r} is iterated to zero with an essential singularity. Thus we would expect that tunneling is quenched at zero temperature, i.e., the particle is effectively trapped in one of the two localized states. In contrast, the improved relation (18.31) predicts different behavior for very weak sub-Ohmic coupling [407] (c is a numerical constant of order unity)

$$\delta_s^{1/(1-s)}\omega_{\text{ph}} < c\,\Delta\,. \tag{18.36}$$

In this regime, the relation (18.31) gives a nonzero renormalized tunneling matrix element. The findings are in accord with results from a study of the related Ising model [cf. Section 19.5] using a variational method [398].

At finite temperature the iteration procedure stops at $\omega_\ell = 1/\hbar\beta$. We then obtain $\Delta_{\text{r}} = \Delta\exp[-\delta_s(\hbar\beta\omega_{\text{ph}})^{1-s}/(1 - s)]$, which is qualitatively correct [86], as we shall see in Subsection 20.2.7.

In connection with quantum state engineering the coherent sub-Ohmic regime (18.36) is of particular interest. The low-frequency modes $\omega \lesssim \Delta$ or $\omega \lesssim 1/\hbar\beta$ must be treated nonadiabatically in this regime, as we have seen above. One finds that these modes entail pure dephasing. Discussion of this issue is given in Section 22.4 and Subsection 22.5.3.

18.1.5 Polaron transformation

With regard to the spin-boson Hamiltonian (18.15) it is useful to introduce a basis of dressed states in which the oscillator α is shifted by a displacement $-\frac{1}{2}s_\alpha\sigma_z$, where $s_\alpha = q_0 c_\alpha/m_\alpha\omega_\alpha^2$, as discussed in Subsection 18.1.1. The unitary operator $\mathcal{U}$ transforms to the basis of the displaced harmonic oscillator states,

$$\mathcal{U} = \exp[-\tfrac{1}{2}i\sigma_z\Omega] \quad \text{with} \quad \Omega = \sum_\alpha \frac{s_\alpha p_\alpha}{\hbar} = i\sum_\alpha \frac{\lambda_\alpha}{\omega_\alpha}[\,b_\alpha^\dagger - b_\alpha\,]\,. \tag{18.37}$$

The so-called polaron transformation diagonalizes the last three terms in the Hamiltonian (18.15). The transformed Hamiltonian $\tilde{H} = \mathcal{U}^{-1}H\mathcal{U}$ takes the exact form

$$\tilde{H} = -\frac{\hbar\Delta}{2}\Big(|R\rangle\langle L|\,e^{i\Omega} + |L\rangle\langle R|\,e^{-i\Omega}\Big) - \frac{\hbar\epsilon}{2}\sigma_z + \sum_\alpha \hbar\omega_\alpha b_\alpha^\dagger b_\alpha\,. \tag{18.38}$$

Here we have dropped an irrelevant constant.

Alternatively, instead of transforming the Hamiltonian, we may transform the tunneling operator, $\tilde{\sigma}_x = \mathcal{U}\sigma_x\mathcal{U}^{-1}$. The polaron-transformed tunneling operator $\tilde{\sigma}_x$ acts in the full system-plus-reservoir space and takes the form

$$\tilde{\sigma}_x = |R\rangle\langle L| \, e^{-i\Omega} + \text{h.c.} = |R\rangle\langle L| \prod_\alpha \int dx_\alpha \, |x_\alpha\rangle\langle x_\alpha - s_\alpha| + \text{h.c.} . \tag{18.39}$$

The bare tunneling operator σ_x acts in the space of the TSS alone, whereas the polaron-dressed operator $\tilde{\sigma}_x$ acts in the full space of TSS and bath. The application of $\tilde{\sigma}_x$ transfers the particle from one localized state to the other and simultaneously shifts the set of bath oscillators $\{\alpha\}$ by the set of displacements $\{s_\alpha\}$. In the dressed basis, pictorially, the particle drags behind it a polaronic cloud. The equilibrium autocorrelation function of the dressed tunneling operator evolving under the full untransformed Hamiltonian, $\tilde{\sigma}_x(t) = e^{iHt/\hbar}\tilde{\sigma}_x(0)\,e^{-iHt/\hbar}$, differs from the autocorrelation function of the bare σ_x. This will be discussed in Subsec. 22.6.5.

18.2 Pair interaction in the charge representation

18.2.1 Analytic expression for spectral density with any power s

We have introduced in Subsec. 4.2.3 a charge representation for the Euclidean influence functional of the spin-boson model in which the charges are lined up with alternating sign. The imaginary-time charge interaction $\mathcal{W}(\tau)$, introduced in Eq. (4.77), is the second integral of the kernel $\mathcal{K}(\tau)$,

$$\mathcal{W}(\tau) = \int_0^\infty d\omega \, \frac{G(\omega)}{\omega^2} \, \frac{\cosh(\omega\hbar\beta/2) - \cosh[\omega(\hbar\beta/2 - \tau)]}{\sinh(\omega\hbar\beta/2)} . \tag{18.40}$$

The function $\mathcal{W}(\tau)$ is real for real τ and satisfies the reflection property

$$\mathcal{W}(\tau) = \mathcal{W}(\hbar\beta - \tau) \qquad \text{for} \qquad 0 \leqslant \tau < \hbar\beta . \tag{18.41}$$

Advancement from thermodynamics to dynamics involves analytic continuation of the function $\mathcal{W}(\tau)$ to real time, $Q(t) \equiv \mathcal{W}(\tau = it)$. There holds

$$Q(t) = \int_0^\infty d\omega \, \frac{G(\omega)}{\omega^2} \left\{ \coth\left(\frac{\beta\hbar\omega}{2}\right)\left(1 - \cos(\omega t)\right) + i\sin(\omega t) \right\} . \tag{18.42}$$

For complex time $z = t - i\tau$, the pair interaction $Q(z)$ has the symmetry

$$Q(-z - i\hbar\beta) = Q(z) . \tag{18.43}$$

The function $Q(z)$ is analytic in the strip $0 \geqslant \operatorname{Im} z > -\hbar\beta$.

The integral (18.42) can be executed in analytic form for the spectral density

$$G(\omega) = 2\delta_s \omega_{\rm ph}^{1-s} \omega^s \, e^{-\omega/\omega_{\rm c}} . \tag{18.44}$$

The resulting expression of the charge pair interaction for general s is [408]

$$Q(t) = 2\delta_s \Gamma(s-1) \Big\{ (\omega_{\rm c}/\omega_{\rm ph})^{s-1} \left[1 - (1 + i\,\omega_{\rm c} t)^{1-s} \right]$$
$$+ 2(\beta\hbar\omega_{\rm ph})^{1-s} \left[\zeta(s-1, 1+\kappa) - \operatorname{Re}\zeta(s-1, 1+\kappa + i\,t/\hbar\beta) \right] \Big\} . \tag{18.45}$$

Here, $\zeta(z, q)$ is the generalized zeta function [89], $\Gamma(z)$ is the gamma function, and

$$\kappa = 1/\beta\hbar\omega_c . \tag{18.46}$$

The imaginary-time correlator $\mathcal{W}(\tau)$ is found from Eq. (18.45) by analytic continuation, $t = -i\tau$. For later use, we also introduce the function $X(t) \equiv Q(t - i\hbar\beta/2)$,

$$X(t) \equiv X_1 - X_2(t) = \int_0^\infty d\omega \frac{G(\omega)}{\omega^2} \left(\coth(\tfrac{1}{2}\beta\hbar\omega) - \frac{\cos(\omega t)}{\sinh(\tfrac{1}{2}\beta\hbar\omega)} \right) . \tag{18.47}$$

The representation (18.47) follows from Eq. (18.42). For $s > 2$, the functions X_1 and $X_2(t)$ are individually IR-convergent. For $s \leqslant 2$ only the sum $X(t)$ is IR-convergent. The function $X(t)$ is found either with direct integration or from the expression (18.45) with the relation $\zeta[x, q + 1] = \zeta[x, q] - q^{-x}$. Here, X_1 is the Huang-Rhys exponent (18.12), and $2B_s$ is the T-independent part,

$$X_1 = \rho_a + \rho_e = 2B_s + 2D_s(T) ,$$

$$B_s = \delta_s \Gamma[s - 1](\omega_c/\omega_{ph})^{s-1} , \tag{18.48}$$

$$D_s(T) = d_s(\kappa)(T/T'_{ph})^{s-1} ,$$

with the temperature scale T'_{ph} and coefficient $d_s(\kappa)$ given by

$$T'_{ph} = T_{ph}/(2\delta_s\pi^2)^{1/(s-1)} , \qquad d_s(\kappa) = \Gamma(s-1)\zeta[s-1, 1+\kappa]/\pi^2 . \tag{18.49}$$

For $s = 3$, T'_{ph} coincides with the temperature scale used in the literature [444, 445]. The time-dependent part $X_2(t)$ is

$$X_2(t) = (2/\pi^2)\Gamma(s-1)(T/T'_{ph})^{s-1} \operatorname{Re}\zeta(s-1, \tfrac{1}{2} + \kappa + it/\hbar\beta) . \tag{18.50}$$

The expressions (18.48) and (18.50) hold for general s, any T and any ω_c.

We see from (18.42) that for a bosonic bath $Q'(t) = \operatorname{Re}Q(t)$ is temperature-dependent, whereas $Q''(t) = \operatorname{Im}Q(t)$ is not. A spin-$\frac{1}{2}$ bath described by the spectral density Eq. (3.257) shows opposite behavior, i.e., $Q'(t)$ is temperature-independent, and the temperature dependence is captured by $Q''(t)$. The explicit calculation of the spin-bath correlation function for the various spectral densities and the study of the implications with respect to thermodynamics and dynamics is left to the reader.

18.2.2 Ohmic dissipation and universality limit

Upon taking in Eqs. (18.45) and (18.50) the limit $s \to 1$, we obtain the pair interaction for the Ohmic spectral density $G(\omega) = 2K\omega e^{-\omega/\omega_c}$ as (we put $\delta_1 = K$)

$$Q(t) = 2K \ln \left(\frac{\Gamma[\kappa]\Gamma(1+\kappa)}{\Gamma(\kappa + it/\hbar\beta)\Gamma(1+\kappa - it/\hbar\beta)} \right) , \tag{18.51}$$

$$X(t) = 2K \ln \left(\frac{\Gamma(\kappa)\Gamma(1+\kappa)}{\Gamma(\tfrac{1}{2} + \kappa + it/\hbar\beta)\Gamma(\tfrac{1}{2} + \kappa - it/\hbar\beta)} \right) . \tag{18.52}$$

We shall refer to the case in which $\hbar\omega_c$ is the largest energy scale of the open system as *strictly* Ohmic. Formally, this case corresponds to the limit $\kappa \to 0$ in Eqs. (18.51) and (18.52). We then arrive at the so-called scaling form[3]

$$Q_{sc}(t) \equiv Q'_{sc}(t) + i\,Q''_{sc}(t) = 2K\ln\left(\frac{\beta\hbar\omega_c}{\pi}\sinh\frac{\pi|t|}{\hbar\beta}\right) + i\,\pi K\,\mathrm{sgn}(t)\,, \quad (18.53)$$

$$X_{sc}(t) = 2K\ln\left(\frac{\beta\hbar\omega_c}{\pi}\cosh\frac{\pi t}{\hbar\beta}\right). \quad (18.54)$$

Correspondingly, the charge interaction in imaginary time has the scaling form

$$\mathcal{W}_{sc}(\tau) = 2K\ln\left[(\beta\hbar\omega_c/\pi)\sin(\pi\tau/\hbar\beta)\right]\,, \quad 0 < \tau < \hbar\beta\,. \quad (18.55)$$

In the Coulomb gas representation for the partition function, the charge interactions can be divided into dipole self-interactions and interdipole interactions. Each factor Δ^2 is associated with a dipole self-interaction, which for dipole length τ is $e^{-\mathcal{W}_{sc}(\tau)} \propto \omega_c^{-2K}$. Correspondingly, in real-time propagation from a diagonal to a diagonal state of the RDM, each factor Δ^2 comes with a factor $e^{-Q'_{sc}(t)}$, as can be seen from the exact formal expressions given below in Subsection 21.2.2. Thus, with the forms (18.53) and (18.55), Δ and ω_c occur only in the combination $\Delta^2/\omega_c^{2K} = \Delta_r^{2-2K}$, where Δ_r is defined in Eq. (18.34). We call an observable which is a function of Δ_r without other dependence on ω_c as universal. Any extra dependence on the cutoff ω_c is called non-universal. The universality or scaling limit is

$$\omega_c \to \infty \quad \text{with} \quad \Delta_r \quad \text{held fixed}\,. \quad (18.56)$$

In the scaling limit, the Ohmic TSS is equivalent to the anisotropic Kondo model, and the energy scale $\hbar\Delta_r$ corresponds to the Kondo energy, up to a numerical factor which is of order 1 for $0 < K \lesssim \frac{1}{2}$. The correspondence is discussed in Section 19.4.

For $K \geq \frac{1}{2}$, the interaction factor $e^{-\mathcal{W}(\tau)}$ with the form (18.55) may require a short-time regularisation of the breathing mode integral of the dipole. In such case, it is usually convenient to put a hard-sphere repulsion at a distance $\tau_c = 1/\omega_c$ between the charges [cf. Eqs. (19.90) and (19.99) given below]. As a result, the partition function explicitly depends on the cutoff for $K \geq \frac{1}{2}$. In contrast, the specific heat is found to be universal for $K < \frac{3}{2}$ and to depend on the cutoff for $K \geq \frac{3}{2}$ (cf. the discussion in Subsection 19.2.1).

---

[3]In the scaling limit, effects of the core regime of the pair interaction, $\tau \lesssim 1/\omega_c$, are disregarded.

19 Thermodynamics

In this chapter, we consider the environmental influences on thermodynamic properties of selected quantum systems. First, we investigate the equilibrium properties of the open two-state system. Then we discuss, based on exact formal expressions for the partition function, the relationship of the Ohmic spin-boson model with variants of the Kondo model and with the $1/r^2$ Ising model. Finally, we calculate quantum-statistical tunneling rates with the method of analytic continuation of the free energy. Emphasis is put on electron transfer in a solvent and on interstitial tunneling in solids.

19.1 Partition function and specific heat

We now study the partition function of the dissipative TSS based on the power series in Δ^2. We discuss various limits and introduce adequate approximations. Diverse applications will be given in subsequent sections.

19.1.1 Static susceptibility and specific heat

The key quantity of the thermodynamics of the spin boson model is the partition function $Z(\beta)$, defined as the trace of the system states of the reduced canonical density operator $\hat{\rho}_{\mathrm{eq}} \equiv \mathrm{tr}_{\mathrm{bath}}\, e^{-\beta \hat{H}_{\mathrm{SB}}}$, where $\hat{H}_{\mathrm{SB}}$ is the spin-boson Hamiltonian (18.15),

$$Z \equiv \mathrm{tr}_{\mathrm{S}}\, \hat{\rho}_{\mathrm{eq}} = <R|\hat{\rho}_{\mathrm{eq}}|R> + <L|\hat{\rho}_{\mathrm{eq}}|L> = Z_{\mathrm{R}} + Z_{\mathrm{L}} . \tag{19.1}$$

The expectation of σ_z in the canonical state yields the difference of the occupation probabilities $P_{\mathrm{R,eq}}$ and $P_{\mathrm{L,eq}}$ of the two localized states in thermal equilibrium,

$$\langle \sigma_z \rangle_{\mathrm{eq}} \equiv \mathrm{tr}\left\{\sigma_z \hat{\rho}_{\mathrm{eq}}\right\}/Z = Z_{\mathrm{R}}/Z - Z_{\mathrm{L}}/Z = P_{\mathrm{R,eq}} - P_{\mathrm{L,eq}} . \tag{19.2}$$

With the bias term $\propto \epsilon\, \hat{\sigma}_z$ in $\hat{H}_{\mathrm{SB}}$, we may relate $\langle \sigma_z \rangle_{\mathrm{eq}}$ to the free energy F,

$$\langle \sigma_z \rangle_{\mathrm{eq}} = \frac{2}{\hbar\beta} \frac{1}{Z} \frac{\partial Z}{\partial \epsilon} = \frac{2}{\hbar\beta} \frac{\partial \ln Z}{\partial \epsilon} = -\frac{2}{\hbar} \frac{\partial F}{\partial \epsilon} . \tag{19.3}$$

The response to an external bias is described by the nonlinear susceptibility[1]

---

[1] We mark the static susceptibility by a bar, and the dynamical susceptibility by a tilde. Throughout part IV, we choose a normalization which differs by a factor of 4 compared to the 1993 edition in order to have agreement with the standard form (22.165) of the fluctuation-dissipation theorem for the σ_z autocorrelation function. There is also a factor 4 difference with respect to the literature on the s-d model, since there the corresponding susceptibility χ_{sd} is normalized according to the correlation function for the spin $S_z = \sigma_z/2$. The equivalence relation is $\overline{\chi}_z = 4\chi_{sd}/(g\mu_{\mathrm{B}})^2$.

$$\overline{\chi}_z(T, \epsilon) \equiv \frac{2}{\hbar} \frac{\partial \langle \sigma_z \rangle_{\text{eq}}}{\partial \epsilon} = \frac{4}{\beta \hbar^2} \frac{\partial^2 \ln Z}{\partial \epsilon^2} = -\frac{4}{\hbar^2} \frac{\partial^2 F}{\partial \epsilon^2} . \tag{19.4}$$

In connection with the Kondo problem, $\overline{\chi}_z$ describes the response of the impurity to a magnetic field perturbation $-g\mu_B \hbar \sigma_z / 2$. In the two-state system, the static susceptibility $\overline{\chi}_z$ is the response to a static strain field or bias. We shall see in Subsec. 21.2.4 that the static susceptibility can also be calculated in a dynamical approach.

Finally, the specific heat is found from Z via the thermodynamic relation

$$c/k_B = \beta^2 \partial^2 \ln Z / \partial \beta^2 . \tag{19.5}$$

The path sum approach will give the partition function as a power series in Δ^2. There follows from the relations (19.2)–(19.5) that one has to perform a resummation of the resulting series in order to obtain the thermodynamic response functions in the form of power series representations in Δ^2. If one investigates the thermodynamic response functions within a *dynamical* approach, exact formal series expressions in Δ^2 can be derived directly both for $\langle \sigma_z \rangle_{\text{eq}}$ and $\ln Z$. As an advantage over the series for Z, these give upon differentiation directly the respective series expansions for the thermodynamic response functions. The latter route will be taken in Chapter 21.

19.1.2 Exact formal expression for the partition function

In the semiclassical limit, the partition function is determined by the classical paths in the upside-down potential. In the upside-down double well, the only classical paths available at low temperatures [apart from the trivial paths $q(\tau) = \pm q_0/2$] are sequences of well-separated instantons or kinks. In the tight-binding two-state limit, the *spin*-path $\sigma(\tau) = 2q(\tau)/q_0$ performs sudden flips at random times s_j between the eigenvalues ± 1 of the Pauli matrix σ_z, as sketched in Fig. 4.1. Since each kink (forward move) is followed by an anti-kink (backward move), it is useful to group a given path with $2m$ transitions into m kink–anti-kink pairs (bounces). We shall denote the lengths of the bounces by $\{\tau_j\}$ and the intervals between bounces by $\{\rho_j\}$,

$$\tau_j = s_{2j} - s_{2j-1} ; \qquad \rho_j = s_{2j+1} - s_{2j} , \qquad (j = 1, \ldots, m) , \tag{19.6}$$

where $s_0 = 0$ and $s_{2m+1} = \hbar\beta$. In the equivalent charge picture, the bounces are superseded by dipoles, and the intervals τ_j and ρ_j are intra- and inter-dipole lengths. Each kink–anti-kink pair or dipole contributes to the partition function a term $\Delta^2/4$ times a factor depending on the bias. The effects of the environment are in the interactions between the charges. With this background, it is straightforward to write the partition function as a power series in Δ^2. The contribution from the right/left well, which is the lower/higher well for $\epsilon > 0$, is

$$Z_{R/L} = \sum_{m=0}^{\infty} \left(\frac{\Delta}{2}\right)^{2m} \int_0^{\hbar\beta} ds_{2m} \int_0^{s_{2m}} ds_{2m-1} \cdots \int_0^{s_2} ds_1 \, \mathbb{B}_{R/L,m}(\{s_j\}) \mathbb{F}_m(\{s_j\}) . \tag{19.7}$$

The kink centers or charge positions s_j are the collective coordinates of the problem. The function $\mathbb{B}_{R/L,m}(\{s_j\})$ includes the dependence on the bias,

$$\mathbb{B}_{R/L,m} = \exp\left(\pm \tfrac{1}{2}\hbar\beta\epsilon \mp \epsilon \sum_{j=1}^{2m}(-1)^j s_j\right) = \exp\left(\pm \tfrac{1}{2}\hbar\beta\epsilon \mp \epsilon \sum_{j=1}^{m}\tau_j\right). \quad (19.8)$$

The first form gives the bias factor in terms of the flip times $\{s_j\}$, whereas the second form is expressed in terms of the periods $\{\tau_j\}$ spent in the left/right well.

The factor $\mathbb{F}_m$ carries the environmental influences in the form of kink or charge interactions. The influence function $\mathbb{F}_m$ has been given in Eq. (4.77) and is

$$\mathbb{F}_m = \exp\left\{\sum_{j=2}^{2m}\sum_{i=1}^{j-1}(-1)^{i+j}\,\mathcal{W}(s_j - s_i)\right\}. \quad (19.9)$$

The expression (19.7) with Eqs. (19.8) and (19.9) is the "Coulomb gas" representation of the partition function. It is in the form of a grand-canonical sum of unit charges ± 1 stringed with alternating sign. All the influences of the environment are conveyed by the charge interaction $\mathcal{W}(\tau)$ defined in Eq. (18.40).

Alternatively, the sequence of alternating charges may be viewed as a string of ordered dipoles. Accordingly, we may rewrite Eq. (19.9) as

$$\mathbb{F}_m = \exp\left(-\sum_{j=1}^{m}\mathcal{W}(\tau_j)\right)\exp\left(-\sum_{n=1}^{m-1}\sum_{j=1}^{m-n}\Lambda_{j+n,j}\right). \quad (19.10)$$

Here the consecutively numbered dipoles are arranged in chronological order. The first exponential factor represents the self-interactions of the m dipoles, and the second one describes the interactions between these units. The terms $\Lambda_{j+1,j}$ are the nearest-neighbor terms, the terms $\Lambda_{j+2,j}$ are the next-nearest-neighbor terms, etc. The interaction of dipole j with a chronologically later dipole k is given by

$$\Lambda_{k,j} = \mathcal{W}(s_{2k} - s_{2j-1}) + \mathcal{W}(s_{2k-1} - s_{2j}) - \mathcal{W}(s_{2k} - s_{2j}) - \mathcal{W}(s_{2k-1} - s_{2j-1}). \quad (19.11)$$

We recall that Δ is already adiabatically dressed by the high-frequency modes, as specified in Eq. (18.13). Therefore, if the two-state system was actually obtained from an extended system as described in Section 18.1, the physically relevant quantity Δ_{eff} becomes independent of ω_c to leading order. Just as well, the spin-boson model may be considered as an independent model in which Δ is a free parameter which does not depend on ω_c at all. The latter view is taken in Sections 19.4 and 19.5 in which the spin-boson model is linked to the Kondo model and to the Ising model, respectively.

It is convenient to write the series (19.7) in Laplace representation,

$$Z_{R/L} = \frac{1}{2\pi i}\int_{\mathcal{C}} d\lambda\, e^{\lambda\hbar\beta}\, z_{R/L}(\lambda). \quad (19.12)$$

Here $\mathcal{C}$ is the standard Bromwich contour, i. e., any contour from $-i\infty$ to $+i\infty$ lying entirely to the right of all singularities of $z_{R/L}(\lambda)$. We then obtain

$$z_{R/L}(\lambda) = \frac{1}{\lambda \mp \epsilon/2} \sum_{m=0}^{\infty} \left(\frac{\Delta}{2}\right)^{2m}$$

$$\times \prod_{j=1}^{m} \left(\int_0^{\infty} d\tau_j\, e^{-(\lambda \pm \epsilon/2)\tau_j} \int_0^{\infty} d\rho_j\, e^{-(\lambda \mp \epsilon/2)\rho_j} \right) \mathbb{F}_m. \tag{19.13}$$

The partition function of the two-state system is

$$Z = Z_R(\epsilon) + Z_L(\epsilon) = Z_R(\epsilon) + Z_R(-\epsilon) = Z_L(\epsilon) + Z_L(-\epsilon). \tag{19.14}$$

The expressions (19.12)–(19.14) represent the exact formal solution for the partition function of the spin-boson model. They hold for linear friction with arbitrary frequency dependence.

19.1.3 The self-energy method

In the absence of dissipation, we have $\mathcal{F}_m^{(E)} = 1$. Then the series (19.13) is a geometrical series which is added up to the form (for later use, we write Δ instead of Δ_0)

$$z_{R/L}^{(0)}(\lambda) = \frac{1}{(\lambda \mp \epsilon/2)} \frac{1}{\left(1 - \frac{\Delta^2}{4} \frac{1}{(\lambda^2 - \epsilon^2/4)}\right)} = \frac{\lambda \pm \epsilon/2}{\lambda^2 - (\Delta^2 + \epsilon^2)/4}. \tag{19.15}$$

The expression (19.15) has simple poles at $\lambda = \pm\Delta_b/2$, where

$$\Delta_b \equiv (\Delta^2 + \epsilon^2)^{1/2}. \tag{19.16}$$

The transform $z_{R/L}^{(0)}(\lambda)$ is easily inverted to obtain $Z_{R/L}^{(0)}(\beta)$,

$$Z_{R/L}^{(0)}(\beta) = \cosh(\beta\hbar\Delta_b/2) \pm (\epsilon/\Delta_b)\sinh(\beta\hbar\Delta_b/2). \tag{19.17}$$

As a side note, the same form (19.17) is found by writing in the defining expression $Z_{R/L}^{(0)}(\beta) = <R/L|\, e^{-\beta H_{\mathrm{TSS}}}\, |R/L>$ the localized states $|R/L>$ as linear combinations of the eigenstates $|g/e>$ of the Hamiltonian H_{TSS}, as given in Eq. (3.138) with (3.139).

The specific heat resulting from $Z^{(0)}(\beta)$ exhibits the familiar Schottky anomaly

$$c/k_B = (\beta\hbar\Delta_b/2)^2 \operatorname{sech}^2(\beta\hbar\Delta_b/2). \tag{19.18}$$

This includes that the specific heat of the undamped system is exponentially small at low temperature $k_B T \ll \hbar\Delta_b$, $c/k_B = (\hbar\Delta_b/k_B T)^2 \exp(-\hbar\Delta_b/k_B T)$, while it depends algebraically on temperature, $c/k_B = (\hbar\Delta_b/2k_B T)^2$, in the regime $k_B T \gg \hbar\Delta_b$.

As regards the partition function of the open system, we may write

$$z_{R/L}(\lambda) = \frac{1}{[\, z_{R/L}^{(0)}(\lambda)\,]^{-1} - \Sigma_{R/L}(\lambda)}. \tag{19.19}$$

Following the Green function terminology, we shall refer to the function $\hbar\Sigma_{R/L}(\lambda)$ as the self-energy. By definition, the self-energy $\Sigma_{R/L}(\lambda)$ captures all spin-environment correlations. Formally, the self-energy may be regarded as an isobaric ensemble of interacting kink–anti-kink pairs or dipoles. In graphical terms, the self-energy includes all *irreducible* diagrams. These are those diagrams which can not be divided into disconnected sub-diagrams without cutting inter-dipole interaction lines. For this purpose, we now introduce a modified influence functional which is irreducible by construction. There holds, of course, $\tilde{\mathbb{F}}_1 = e^{-\mathcal{W}(\tau)} - 1$. Subtraction of all reducible components of the influence functional for $n > 1$ is achieved in the series

$$\tilde{\mathbb{F}}_n \equiv \mathbb{F}_n - \sum_{j=2}^{n} (-1)^j \sum_{m_1,\cdots,m_j \geq 1} \mathbb{F}_{m_1} \mathbb{F}_{m_2} \cdots \mathbb{F}_{m_j} \delta_{m_1+\cdots+m_j,n} \cdot \tag{19.20}$$

Again we put the convention that in each term the dipoles are consecutively numbered and arranged in chronological order. In the subtracted terms, the factors like $\mathbb{F}_{m_j}$ and $\mathbb{F}_{m_i}$ are not correlated with each other. To illustrate the general expression, we explicitly give the case $n = 3$,

$$\begin{aligned} \tilde{\mathbb{F}}_3 &\equiv \mathbb{F}_3 - \mathbb{F}_2 \mathbb{F}_1 - \mathbb{F}_1 \mathbb{F}_2 + \mathbb{F}_1 \mathbb{F}_1 \mathbb{F}_1 \\ &= e^{-\mathcal{W}(\tau_1)-\mathcal{W}(\tau_2)-\mathcal{W}(\tau_3)} \\ &\quad \times \left\{ \left(e^{-\Lambda_{3,1}} - 1\right) e^{-\Lambda_{3,2}-\Lambda_{2,1}} + \left(e^{-\Lambda_{3,2}} - 1\right) \left(e^{-\Lambda_{2,1}} - 1\right) \right\}. \end{aligned} \tag{19.21}$$

Linearization of $\tilde{\mathbb{F}}_3$ in the interaction $\mathcal{W}(\tau)$ yields $\tilde{\mathbb{F}}_3 = -\Lambda_{3,1}$, which directly evidences irreducibility.

With the irreducible influence functional (19.20), the series in Δ^2 for $\Sigma_{R/L}(\lambda)$ reads

$$\Sigma_{R/L}(\lambda) = \sum_{m=1}^{\infty} \Sigma_{R/L}^{(m)}(\lambda), \tag{19.22}$$

where

$$\Sigma_{R/L}^{(1)}(\lambda) = \frac{\Delta^2}{4} \int_0^{\infty} d\tau \, e^{-(\lambda \pm \epsilon/2)\tau} \left(e^{-\mathcal{W}(\tau)} - 1\right), \tag{19.23}$$

$$\Sigma_{R/L}^{(m)}(\lambda) = \left(\frac{\Delta^2}{4}\right)^m \prod_{j=1}^{m} \left(\int_0^{\infty} d\tau_j \, e^{-(\lambda \pm \epsilon/2)\tau_j}\right) \prod_{k=1}^{m-1} \left(\int_0^{\infty} d\rho_k \, e^{-(\lambda \mp \epsilon/2)\rho_k}\right) \tilde{\mathbb{F}}_m,$$

The expression (19.19) represents a formally exact rearrangement of the original series (19.13). Upon expanding Eq. (19.19) in powers of $\Sigma_{R/L}(\lambda)$ and substituting the series expansion (19.22) with (19.23), the original series (19.13) is actually recovered.

In practical computations, one may truncate the series (19.22) for the self-energy at a given order, and then invert the Laplace transform (19.19) to obtain $Z_{R/L}(\beta)$. In cases where explicit inversion is difficult, much can nevertheless be learned from a study of the singularities of $z_{R/L}(\lambda)$.

19.1.4 Weak-damping limit

In the weak-damping limit, the coupling to the environment can be treated as a perturbation. Obviously, a perturbative expansion of the partition function in Δ is not very meaningful. Rather we aim at a calculation of the self-energy in linear order of the correlation $\mathcal{W}(\tau)$, usually called one-phonon approximation There are two contributions. The first accounts for the internal correlation of a single dipole, and the second for the correlations between two dipoles. Importantly, in the interval ρ between these correlated dipoles, the system may make any even number of uncorrelated transitions. The grand-canonical ensemble of the corresponding noninteracting dipoles can be summed yielding the imaginary-time propagator [cf. Eq. (19.17)],

$$Z_{R/L}^{(0)}(\rho) = \cosh(\rho\Delta_b/2) \pm (\epsilon/\Delta_b)\sinh(\rho\Delta_b/2) . \tag{19.24}$$

With this insertion between the two correlated dipoles, one readily obtains in the weak-coupling (wc) regime

$$\begin{aligned}
\Sigma_{R/L}^{(wc)}(\lambda) = {}& -\frac{\Delta_{eff}^2}{4}\int_0^\infty d\tau\, \mathcal{W}(\tau)\, e^{-(\lambda \pm \epsilon/2)\tau} \\
& -\frac{\Delta_{eff}^4}{16}\int_0^\infty d\tau_1\, d\rho\, d\tau_2\, e^{-(\lambda \pm \epsilon/2)(\tau_1+\tau_2)}\, e^{-\lambda\rho}\, Z_{R/L}^{(0)}(\rho) \\
& \times \left[\mathcal{W}(\tau_1 + \rho + \tau_2) + \mathcal{W}(\rho) - \mathcal{W}(\tau_1 + \rho) - \mathcal{W}(\rho + \tau_2)\right] .
\end{aligned} \tag{19.25}$$

This expression is irreducible in the sense specified in the previous subsection. The second term describes one-phonon thermal processes in which the system makes any number of tunneling transitions in the interval between emission and absorption of the phonon. This term is disregarded in the noninteracting–kink-pair approximation.

The expression (19.25) is easily converted into the spectral representation of the self-energy upon using for $\mathcal{W}(\tau)$ the representation (18.40). The ensuing triple-integration over the time intervals is trivial. The self-energy shifts the poles of $z_{R/L}(\lambda)$ in Eq. (19.19) from $\lambda = \pm\Delta_b/2$ to $\lambda = \pm\Omega/2$. The change of the squared level splitting due to single-phonon exchange is found from Eq. (19.19) with (19.15) as

$$\Omega^2 = \Delta_b^2 + 4(\lambda \pm \epsilon/2)\Sigma_{R/L}^{(wc)}(\lambda)|_{\lambda=\pm\Delta_b/2} = \Delta_b^2 - \Delta^2 v(\Delta_b) . \tag{19.26}$$

The function $v(z)$ is the spectral principle value integral

$$v(z) = v_0(z) + v_1(z) = \int_0^\infty d\omega\, \frac{G(\omega)}{\omega^2 - z^2}\left(1 + \frac{2}{e^{\hbar\beta\omega} - 1}\right) . \tag{19.27}$$

The term $v_0(\Delta_b)$ is the one-phonon contribution to the dressed tunneling amplitude (18.33) and (18.35), respectively. With exponentiation of this term, we readily obtain

$$\Omega^2 = \Delta_{eff}^2 + \epsilon^2 - \Delta_{eff}^2\, v_1\left(\sqrt{\Delta_{eff}^2 + \epsilon^2}\right) . \tag{19.28}$$

Picking up the poles at $\lambda = \pm\Omega/2$ by the contour integral (19.12), the partition function $Z_{R/L}$ is exactly found in the form (19.17) in which Δ_b is replaced by Ω,

$$Z_{R/L}(\beta) = \cosh(\beta\hbar\Omega/2) \pm (\epsilon/\Omega)\sinh(\beta\hbar\Omega/2) . \qquad (19.29)$$

To summarize, the functional form of the partition function is left unchanged for weak damping, but the level splitting becomes renormalized and depends on the spectral density of the coupling to the reservoir and on temperature.

19.1.5 Noninteracting-dipole approximation (NIDA)

With increasing damping strength, the dipole self-interactions in the interaction factor (19.10) cause that the average dipole length $\langle\tau\rangle$ gets small compared to the average interval $\langle\rho\rangle$ between the dipoles. The picture we then have is that the dipoles form a *dilute gas* in the *fixed volume* $\hbar\beta$, and we may argue that the interdipole interactions $\Lambda_{j,k}$ may safely be ignored relative to the intra-dipole or self-interactions. Within the noninteracting–dipole or –kink-pair approximation, we have $\mathsf{F}_{2m} = \exp[-\sum_{j=1}^{m}\mathcal{W}(\tau_j)]$. As a result, the self-energy is reduced to the single-dipole contribution $\Sigma_{R/L}^{(1)}(\lambda)$ given in Eq. (19.23). Thus the expression (19.19) turns into

$$z_{R/L}^{(1)}(\lambda) = \frac{1}{[\,z_{R/L}^{(0)}(\lambda)\,]^{-1} - \Sigma_{R/L}^{(1)}(\lambda)} = \frac{1}{[\,\lambda \mp \epsilon/2 - K_{R/L}(\lambda)\,]} . \qquad (19.30)$$

The kernel $K_{R/L}(\lambda)$ represens the breathing mode of the dipole,

$$K_{R/L}(\lambda) = \frac{\Delta^2}{4} \int_0^\infty d\tau\, e^{-(\lambda\pm\epsilon/2)\tau}\, e^{-\mathcal{W}(\tau)} . \qquad (19.31)$$

When the attractive self-interaction of the dipole is sufficiently strong at short distances, the kernel $K_{R/L}(\lambda)$ can be expanded about $\lambda = \pm\epsilon/2$,

$$K_{R/L}(\lambda) = K_{R/L}(\pm\epsilon/2) + (\lambda \mp \epsilon/2)\, K'_{R/L}(\pm\epsilon/2) + \cdots . \qquad (19.32)$$

The average dipole length $\langle\tau\rangle$ may be be estimated as $|K'_{R/L}(\pm\epsilon/2)|/K_{R/L}(\pm\epsilon/2)$, whereas the average interval $\langle\rho\rangle$ roughly is $1/K_{R/L}(\pm\epsilon/2)$. Thus the dipoles form a dilute gas and the noninteracting–dipole approximation is justified provided that

$$|K'_{R/L}(\pm\epsilon/2)| \equiv \frac{\Delta^2}{4} \int_0^\infty d\tau\, \tau\, e^{\mp\epsilon\tau - \mathcal{W}(\tau)} \ll 1 . \qquad (19.33)$$

When this condition is met, $z_{R/L}(\lambda)$ has a simple pole located at $\lambda = \lambda_{R/L}^{(0)}(\beta)$, and thus, with a hard-sphere boundary added,

$$Z_{R/L}^{(1)}(\beta) = e^{\hbar\beta\lambda_{R/L}^{(0)}(\beta)} , \qquad \text{with} \qquad \lambda_{R/L}^{(0)}(\beta) = \pm\frac{\epsilon}{2} + \frac{\Delta^2}{4} \int_{1/\omega_c}^\infty d\tau\, e^{\mp\epsilon\tau}\, e^{-\mathcal{W}(\tau)} . \qquad (19.34)$$

The free energy associated with the expression (19.34) is

$$F_{R/L}^{(1)}(\beta) = \mp\frac{\hbar\epsilon}{2} - \frac{\hbar\Delta^2}{4}\int_{1/\omega_c}^{\infty} d\tau\, e^{\mp\epsilon\tau}\, e^{-\mathcal{W}(\tau)}\,. \tag{19.35}$$

This expression applies, when the integral is dominated by the hard-sphere boundary. When the integral is dominated by the bulk region, the noninteracting–kink-pair approximation fails.[2] The approximation is inconsistent in particular regimes since it does not properly take into account the reflection property (18.41) of the interactions. Moreover, in the weak-coupling limit, the interactions $\Lambda_{j,k}$ render contributions to the first order in δ_s even for zero bias, as we can see from the expression (19.7) with (19.10). Thus the noninteracting-dipole approximation is inconsistent in the regime of weak-to-moderate damping and/or low-to-moderate temperature.

19.1.6 The limit of high temperatures

in the regime $T \gg \hbar\Delta_{\text{eff}}/k_B$, the series (19.7) may be truncated at the $m = 1$ term,

$$Z_{R/L} = e^{\pm\beta\hbar\epsilon/2}\left(1 + \frac{\Delta^2}{4}\int_0^{\hbar\beta} d\tau\,[\hbar\beta - \tau]\, e^{\mp\epsilon\tau - \mathcal{W}(\tau)} + \mathcal{O}(\Delta^4)\right)\,. \tag{19.36}$$

With the reflection symmetry $\mathcal{W}(\tau) = \mathcal{W}(\hbar\beta - \tau)$, the interval $\hbar\beta/2 \leqslant \tau \leqslant \hbar\beta$ can be mapped onto the interval $0 \leqslant \tau \leqslant \hbar\beta/2$. With this, $Z = Z_R + Z_L$ can be written as

$$Z(\beta) = 2\cosh(\tfrac{1}{2}\beta\hbar\epsilon) + \frac{\hbar\beta\Delta^2}{2}\int_0^{\hbar\beta/2} d\tau\,\cosh[\epsilon(\tfrac{1}{2}\hbar\beta - \tau)]\, e^{-\mathcal{W}(\tau)}\,. \tag{19.37}$$

In summary, at high T the partition function and derivatives of it are dominated by the one-dipole contribution.

19.2 Ohmic dissipation

So far, the study of the partition function has been general, as the spectral coupling has not been specified. Consider now the Ohmic case in the scaling regime $k_B T \ll \hbar\omega_c$, in which the kink pair interaction is given by the expression (18.55). First, we work out results in various limits, and then turn to the analytically solvable case $K = 1/2$.

19.2.1 Specific heat and Wilson ratio

The weak-damping regime $K \ll 1$

For weak-damping and general ϵ, the level splitting $\hbar\Omega$ has been calculated in Subsection 19.1.4. With the Ohmic spectral density $G(\omega) = 2K\omega$, the integral (19.27)

---

[2]Compared to the noninteracting-blip approximation (NIBA) which is useful in the calculation of dynamical quantities [cf. Section 21.3], this approximation is valid in a smaller region of the parameter space. In the NIBA, there are partial cancellations among interblip correlations. Corresponding cancellations are absent in the present case.

can be calculated in analytic form. There arise contributions which can be assigned to the dressed tunneling amplitude Δ_{eff} defined in Eq. (18.35). In addition, it is appropriate to replace Δ consistently by Δ_{eff}. With this one finds for all T the effective temperature dependent tunnel splitting frequency as

$$\Omega = \sqrt{\Delta_{\text{eff}}^2 + \epsilon^2} \left\{ 1 + K\varrho^2 \left[\text{Re}\, \psi\left(\frac{i}{2\pi\varrho\,\theta}\right) + \ln(2\pi\,\theta) \right] \right\} . \qquad (19.38)$$

Here $\psi(z)$ is the digamma function, and θ is a scaled temperature,

$$\theta = k_{\text{B}}T/\hbar\Delta_{\text{eff}} \qquad \varrho = \Delta_{\text{eff}}/\sqrt{\Delta_{\text{eff}}^2 + \epsilon^2} . \qquad (19.39)$$

Inserting Eq. (19.38) in Eq. (19.29), the asymptotic low temperature series of the specific heat at weak damping $K \ll 1$ is found as[3] (B_m is a Bernoulli number)

$$c/k_{\text{B}} = K\pi\varrho^2 \sum_{n=0}^{\infty} (2n+1)|B_{2n+2}|(2\pi\varrho\,\theta)^{2n+1} . \qquad (19.40)$$

In leading order, the specific heat varies linearly with T, which indicates that the equivalent Kondo problem (see Section 19.4) shows Fermi liquid behavior at low T.

In this context, a meaningful quantity is the so-called Wilson ratio

$$R_1 \equiv \lim_{T \to 0} \frac{4\,c(T,\epsilon)/k_{\text{B}}}{\overline{\chi}_z(T,\epsilon)k_{\text{B}}T} , \qquad (19.41)$$

where $\overline{\chi}_z(T,\epsilon)$ is the static nonlinear susceptibility defined in (19.4). The factor of 4 is chosen for convenience [see footnote to Eq. (19.4)]. For weak damping, Eq. (19.29) yields the form $\overline{\chi}_z(0,\epsilon) = 2\varrho^3/(\hbar\Delta_{\text{eff}})$. Combining this expression with the result (19.40), we obtain the Wilson ratio for weak damping $K \ll 1$ and general bias as

$$R_1 = 2K\pi^2/3 . \qquad (19.42)$$

Most interestingly, the value (19.42) of the Wilson ratio holds in the extended regime $0 < K < 1$ for any bias [224]. The corresponding general proof is given within a real-time approach in Section 22.9.

The regime $0 < K < 1$

Consider first for zero bias the high temperature regime $\theta \gg 1$ in which the expression (19.37) for the partition function applies. The integral (19.37) with (18.55) is regular in the regime $0 < K < 1/2$. The resulting expression for the specific heat (19.5) is

$$c/k_{\text{B}} = 2\pi(1-K)\frac{\Gamma[\frac{1}{2}+K]\Gamma[\frac{3}{2}-K]}{\Gamma[1-K]^2}(2\pi\,\theta)^{2K-2} , \qquad \theta \gg 1 . \qquad (19.43)$$

---

[3]The noninteracting–kink-pair approximation discussed in Subsection 19.1.5 is inappropriate in the low temperature regime. It would render spurious terms with even powers of θ.

In fact, this expression for the specific heat applies also in the extended damping regime $1/2 \leqslant K < 1$. In the regime $0 < K < 1$ and $\theta \gg 1$, the specific heat decreases with increasing temperature. Because of the linear increase with T in the regime $\theta \ll 1$, the specific heat shows a Schottky-like peak near $\theta = 1/2\pi$, but with algebraic slopes to the left and right of the peak [408].

The regime $K > 1$

The case $K > 1$ corresponds to the ferromagnetic sector of the Kondo problem. Upon using standard functional relations of the gamma functions, the expression (19.43) can be analytically continued to the regime $1 < K < 3/2$, yielding

$$c/k_B = (K-1)^2 \frac{\pi^{3/2}}{2} \frac{\Gamma(3/2 - K)}{\Gamma(2 - K)} \frac{\Delta^2}{\omega_c^2} \left(\frac{\pi k_B T}{\hbar \omega_c} \right)^{2K-2} . \tag{19.44}$$

For zero bias, the high-temperature expression (19.44) holds down to zero temperature. The Schottky peak is missing, and the specific heat is monotonically increasing $\propto T^{2K-2}$ in the entire regime $0 < k_B T \ll \hbar \omega_c$ (cf. also page 447).

In the regime $K > 3/2$, the integral in Eq. (19.34) is determined by the hard-sphere boundary. With the short-time expansion

$$e^{-W_{sc}(\tau)} = \frac{1}{(\omega_c \tau)^{2K}} \left[1 + K(\pi/\beta)^2 \tau^2/3 \right], \tag{19.45}$$

one finds from Eq. (19.34) for the specific heat the expression

$$c/k_B = \frac{\pi^2}{6} \frac{K}{(2K-3)} \frac{\Delta^2}{\omega_c^2} \frac{k_B T}{\hbar \omega_c}, \qquad K > 3/2. \tag{19.46}$$

Thus the system shows Fermi liquid behavior for all T below $\hbar\omega_c/k_B$ [408].

In the limit $K \to \frac{3}{2}$, the expressions (19.44) and (19.46) resulting from the bulk and from the boundary of the τ-integral, respectively, diverge. However, as the singularities have opposite signs, the sum of both contributions is regular, yielding

$$c/k_B = \frac{\pi^2}{4} \frac{\Delta^2}{\omega_c^2} \frac{k_B T}{\hbar \omega_c} \left[\ln\left(\frac{2}{\pi} \frac{\hbar\omega_c}{k_B T} \right) - \frac{5}{3} \right], \qquad K = 3/2. \tag{19.47}$$

19.2.2 The special case $K = \frac{1}{2}$

For the special value $K = \frac{1}{2}$, the influence interaction factor $\mathbb{F}_m$, Eq. (19.9) with the pair interaction (18.55), can be decomposed into a sum of $m(m-1)$ contributions, which represent the combinatorical possibilities of assigning m dipoles to the m positive and m negative charges. In each assignment, the given sequence of charges is left unchanged and only the related intra-dipole interactions are taken into account. Consider, for example, the influence function of four alternating charges with intervals $\tau_1 = t_2 - t_1$, $\tau_2 = t_4 - t_3$, and $\rho = t_3 - t_2$, and interaction (18.55),

$$\mathbb{F}_2 \equiv e^{-W_{sc}(\tau_1) - W_{sc}(\tau_2) - W_{sc}(\rho) - W_{sc}(\tau_1 + \rho + \tau_2) + W_{sc}(\tau_1 + \rho) + W_{sc}(\rho + \tau_2)} . \tag{19.48}$$

This factor is disassembled exactly into two contributions,

$$\mathbb{F}_2 = e^{-\mathcal{W}_{sc}(\tau_1) - \mathcal{W}_{sc}(\tau_2)} + e^{-\mathcal{W}_{sc}(\rho) - \mathcal{W}_{sc}(\tau_1 + \rho + \tau_2)} . \tag{19.49}$$

The equivalence of Eq. (19.49) with Eq. (19.48), and corresponding relationships for arbitrary integer m, only hold for the Ohmic scaling form (18.55) at $K = \frac{1}{2}$. With the respective decomposition of $\mathcal{F}_m^{(E)}$ in the series (19.7), the integrals over the intervals $\{\tau_j\}$ and $\{\rho_i\}$ can be calculated in analytic form [408]. The coefficients of the Taylor series in Δ are expressed in terms of the integral

$$K_m = \int_{x_0}^{\infty} dx \, \frac{x^m}{\sinh x} , \tag{19.50}$$

where $x_0 = \pi/\hbar\beta\omega_c$ is a hard-sphere cutoff at short distance. In reality, the cutoff is required only for K_0. The partition function for the unbiased system is found as

$$Z = 2\exp\left(\sum_{m=1}^{\infty} \frac{(-1)^{m-1}}{m!} \frac{K_{m-1}}{(2\pi\theta)^m} \right) = 2\exp\left(\int_{x_0}^{\infty} dx \, \frac{1 - e^{-x/2\pi\theta}}{x \sinh x} \right), \tag{19.51}$$

where θ is given in Eq. (19.39). For the current value $K = \frac{1}{2}$ there holds

$$\theta = \frac{k_B T}{\hbar\gamma} , \qquad \text{where} \qquad \gamma \equiv \Delta_{\text{eff}}(K = \frac{1}{2}) = \frac{\pi}{2} \frac{\Delta^2}{\omega_c} . \tag{19.52}$$

Use of the expression (19.51) in Eq. (19.5) yields for the specific heat

$$c/k_B = \frac{1}{2\pi\theta} - \frac{1}{2(2\pi\theta)^2} \int_0^{\infty} dx \, \frac{x \, e^{-x/2\pi\theta}}{\sinh x} , \tag{19.53}$$

The integral is a trigamma function $\psi'(z)$ [89], and thus

$$c/k_B = \frac{1}{2\pi\theta} - \frac{1}{2(2\pi\theta)^2} \psi'\left(\frac{1}{2} + \frac{1}{4\pi\theta} \right) . \tag{19.54}$$

At this point, we anticipate that the case $K = \frac{1}{2}$ corresponds to the "Toulouse limit" of the anisotropic Kondo model [143], which in turn is equivalent to the resonance level model [410, 412] with zero Coulomb interaction.[4] The respective model describes a d-level near the Fermi surface which is coupled to a bath of spinless fermions with a band width $2\hbar\omega_c$ around the Fermi level. The coupling to the fermions is assumed constant within the band. We put $\hbar\omega_c \ll E_F$. Then the dispersion relation can be linearized about the Fermi wave vector. Measuring the momentum from the reference value $p_F = mv_F$, we have

$$E(k) - E_F \equiv \hbar\epsilon_k = \hbar v_F k . \tag{19.55}$$

The Toulouse Hamiltonian reads (L is a normalization length)

---

[4]The equivalence of the Ohmic two-state model to these models is discussed in Section 19.4.

$$H_{\text{Toul}} = \sum_k \hbar\epsilon_k c_k^\dagger c_k + \hbar\epsilon_d d^\dagger d + \frac{V}{\sqrt{L}} \sum_k \left(c_k^\dagger d + d^\dagger c_k \right) . \tag{19.56}$$

To uncover the equivalence, consider first the partition function of this model and place for the moment the d-level at the Fermi energy, $\epsilon_d = 0$. In the interaction picture, the series in V^2 can be written as

$$Z_{\text{Toul}} = 2 \sum_{m=0}^{\infty} \left(\frac{V^2}{L\hbar^2} \right)^m \int_0^{\hbar\beta - \tau_c} ds_{2m} \int_0^{s_{2m} - \tau_c} ds_{2m-1} \cdots \int_0^{s_2 - \tau_c} ds_1 \, \mathbb{G}_m(\{s_j\}) . \tag{19.57}$$

$$\mathbb{G}_m(\{s_j\}) \equiv \sum_{k_1, \cdots, k_{2m}} \exp\left(\sum_{j=1}^{m} \left(\epsilon_{k_{2j-1}} s_{2j-1} - \epsilon_{k_{2j}} s_{2j} \right) \right) \langle c_{k_{2m}} c_{k_{2m-1}}^\dagger \cdots c_{k_2} c_{k_1}^\dagger \rangle_\beta ,$$
$$\tag{19.58}$$

Here $\langle \cdots \rangle_\beta$ means thermal average of products of fermionic reservoir operators, and the time-dependent exponential factors are already extracted from the average.

Using Wick's theorem, the thermal average in Eq. (19.58) can be decomposed into a series which represents all possibilities of building products of thermal averages of electron-hole contractions. These are [cf. Eq. (4.135)]

$$\langle c_k^\dagger c_{k'} \rangle_\beta = \delta_{kk'} f(\epsilon_k) , \quad \text{and} \quad \langle c_k c_{k'}^\dagger \rangle_\beta = \delta_{kk'} [1 - f(\epsilon_k)] , \tag{19.59}$$

where $f(\epsilon)$ is the Fermi function. The thermal fermion propagator is given by

$$D(\tau) \equiv \sum_k \langle c_k^\dagger c_k \rangle_\beta e^{\epsilon_k \tau} = \sum_k \langle c_k c_k^\dagger \rangle_\beta e^{-\epsilon_k \tau} . \tag{19.60}$$

Taking the continuum limit and observing that the density of states per unit length

$$\rho \equiv \frac{1}{2\pi} \frac{dk}{d\epsilon_k} = \frac{1}{2\pi v_{\text{F}}} \tag{19.61}$$

is constant, we have

$$D(\tau) = \rho L \int_{-\omega_c}^{\omega_c} d\epsilon \, \frac{e^{-\epsilon\tau}}{1 + e^{-\hbar\beta\epsilon}} . \tag{19.62}$$

The integral in Eq. (19.62) is well-defined for infinite limits. Hence we may remove the UV cutoffs, and, instead of that, introduce a hard-sphere repulsion at short distance $\tau = \tau_c \equiv 1/\omega_c$. Thus we find

$$D(\tau) = \Theta(\tau - \tau_c) \frac{\pi \rho L}{\hbar\beta} \frac{1}{\sin(\pi\tau/\hbar\beta)} = \Theta(\tau - \tau_c) \rho L \omega_c \, e^{-W_{sc}(\tau)} . \tag{19.63}$$

In the second form, we have expressed the thermal fermionic propagator in terms of the intra-dipole interaction factor $e^{-W_{sc}(\tau)}$, where $W_{sc}(\tau)$ is given in Eq. (18.55).

Using the correspondence relation (19.63), one immediately realizes that the Wick decomposition of $\mathbb{G}_m(\{s_j\})$ is equivalent to the decomposition of the influence function $\mathbb{F}_m(\{s_j\})$ discussed before Eq. (19.48). In detail, there holds

$$\mathbb{G}_m(\{s_j\}) = (\rho L \omega_c)^m \, \mathbb{F}_m(\{s_j\}) . \tag{19.64}$$

This relation provides the correspondence of the resonance level model (19.56) with the Ohmic spin-boson model for the special damping parameter $K = \frac{1}{2}$. With the relation (19.64) the partition function (19.57) takes the form

$$Z_{\text{Toul}} = 2 \sum_{m=0}^{\infty} \left(\frac{\rho V^2 \omega_c}{\hbar^2} \right)^m \int_0^{\hbar\beta - \tau_c} ds_{2m} \int_0^{s_{2m} - \tau_c} ds_{2m-1} \cdots \int_0^{s_2 - \tau_c} ds_1 \, \mathbb{F}_m(\{s_j\}) . \quad (19.65)$$

Apparently, this series is identical in form with the corresponding series for the partition function of the Ohmic two-state system at $K = \frac{1}{2}$. The mapping is completed by the parameter identifications (in the sequel a nonzero bias is allowed)

$$\gamma \equiv \pi \Delta^2 / 2\omega_c = 2\pi \rho V^2 / \hbar^2 , \qquad \text{and} \qquad \epsilon = \epsilon_{\text{d}} . \quad (19.66)$$

The latter relation connects the bias of the TSS with the energy of the d-level.

The thermal average in Eq. (19.58) can be decomposed with Wick's theorem into a series which represents all possibilities of building products of electron-hole contractions. The equations of motion for the single-particle imaginary-time Green functions can be solved in Fourier space. The resulting fermionic Matsubara representations are

$$\mathcal{G}(\tau) \equiv \langle T_\tau d^\dagger(\tau) d(0) \rangle_\beta = \frac{1}{\hbar\beta} \sum_{n\,\text{odd}} \frac{e^{-i\nu_n \tau}}{[-i\nu_n - \epsilon - i\gamma\,\text{sgn}(\nu_n)/2]} , \quad (19.67)$$

$$g_k(\tau) \equiv \langle T_\tau c_k^\dagger(\tau) d(0) \rangle_\beta = \frac{1}{\hbar\beta} \sum_{n\,\text{odd}} \frac{V/\hbar\sqrt{L}}{(-i\nu_n - \epsilon_k)} \frac{e^{-i\nu_n \tau}}{[-i\nu_n - \epsilon - i\gamma\,\text{sgn}(\nu_n)/2]} .$$

We are now prepared to calculate the partition function. Putting $H(\lambda) = H_0 + \lambda H_1$, where H_1 is the interaction in the Hamiltonian (19.56), the change of the free energy by the interaction is given by [172]

$$\Delta F = \int_0^1 d\lambda \langle \hat{H}_1 \rangle_\lambda = \frac{V}{\sqrt{L}} \int_0^1 d\lambda \sum_k \langle c_k^\dagger d + d^\dagger c_k \rangle_\lambda , \quad (19.68)$$

where $\langle \cdots \rangle_\lambda$ denotes the *thermal average* $\langle H_1 \rangle_\lambda = \text{tr}\{ e^{-\beta H(\lambda)} H_1 \} / \text{tr}\, e^{-\beta H(\lambda)}$. With use of the expression (19.67) for $g_k(0)$, the thermal expectation value and finally the free energy (19.68) are easily calculated. Following this road, one arrives at the expression [the hard sphere radius τ_c in (19.65) is replaced by a high-frequency cutoff]

$$Z_{\text{Toul}} = 2\cosh\left(\frac{\beta\hbar\epsilon}{2} \right) \exp\left\{ \frac{\beta\hbar}{2\pi} \int_{-\omega_c}^{\omega_c} d\omega \tanh\left(\frac{\beta\hbar\omega}{2} \right) \arctan\left(\frac{\gamma}{2(\omega - \epsilon)} \right) \right\} . \quad (19.69)$$

Here the first factor is due to the d-level term $\hbar\epsilon\, d^\dagger d$ of the Hamiltonian, and the second factor results from the coupling term. To transform the "fermionic" expression (19.69) into the bosonic form (19.51), we employ the integral representations

$$\tanh\left(\frac{\beta\hbar\omega}{2} \right) = \frac{2}{\pi} \int_0^\infty dx \, \frac{\sin(\beta\hbar\omega x/\pi)}{\sinh x} ,$$

$$1 - e^{-x/2\pi\theta} = \frac{\hbar\beta}{\pi^2} \int_{-\infty}^\infty d\omega \, \sin\left(\frac{\beta\hbar\omega x}{\pi} \right) \arctan\left(\frac{\gamma}{2\omega} \right) . \quad (19.70)$$

With these, the expression (19.69) is converted into

$$Z_{\text{Toul}} = 2\cosh\left(\frac{\beta\hbar\epsilon}{2}\right)\exp\left\{\int_{x_0}^{\infty}dx\,\frac{\left[1 - e^{-x/2\pi\theta}\right]\cos(x\,\epsilon/\pi\gamma\theta)}{x\sinh x}\right\}, \tag{19.71}$$

where $\epsilon = \epsilon_{\text{d}}$. Finally, by virtue of the integral representation

$$\cosh\left(\frac{\beta\hbar\epsilon}{2}\right) = \lim_{x_0\to 0}\exp\left\{\int_{x_0}^{\infty}dx\,\frac{1 - \cos(x\,\epsilon/\pi\gamma\theta)}{x\sinh x}\right\}, \tag{19.72}$$

the expression (19.71) takes the form

$$Z_{\text{Toul}} = 2\exp\left\{\int_{x_0}^{\infty}dx\,\frac{1 - e^{-x/2\pi\theta}\cos(x\,\epsilon/\pi\gamma\theta)}{x\sinh x}\right\}. \tag{19.73}$$

The expression (19.73) for the partition function is the extension of the former result (19.51) to nonzero bias. It is straightforward to calculate the specific heat from Eq. (19.73). The resulting generalization of the expression (19.54) is

$$c(\theta,\epsilon)/k_{\text{B}} = \frac{1}{2\pi\theta} - \frac{1}{8\pi^2\theta^2}\text{Re}\left\{(1 + i\,2\epsilon/\gamma)^2\,\psi'\left(\frac{1}{2} + \frac{1 + i\,2\epsilon/\gamma}{4\pi\theta}\right)\right\}, \tag{19.74}$$

where $\psi'(z) = d\psi(z)/dz$ is the trigamma function and $\theta = k_{\text{B}}T/\hbar\gamma$. The leading low-temperature behavior of the specific heat found from Eq. (19.74) is

$$c_{\text{as}}(T,\epsilon)/k_{\text{B}} = \frac{2\pi}{3}\frac{\gamma^2}{\gamma^2 + 4\epsilon^2}\frac{k_{\text{B}}T}{\hbar\gamma}. \tag{19.75}$$

Next, the static *nonlinear* susceptibility is found from the relation (19.4) with the expression (19.73) in analytic form,

$$\overline{\chi}_z(\theta,\epsilon) = \frac{2}{\pi^2}\frac{1}{\hbar\gamma\theta}\text{Re}\,\psi'\left(\frac{1}{2} + \frac{1 + i\,2\epsilon/\gamma}{4\pi\theta}\right). \tag{19.76}$$

This simplifies in the zero temperature limit to the expression

$$\overline{\chi}_z(0,\epsilon) = \frac{8}{\pi\hbar\gamma}\frac{\gamma^2}{\gamma^2 + 4\epsilon^2}. \tag{19.77}$$

Finally, with the expressions (19.75) and (19.77), the Wilson ratio emerges as

$$R_1 \equiv 4\frac{c_{\text{as}}(T,\epsilon)/k_{\text{B}}}{k_{\text{B}}T\overline{\chi}_z(0,\epsilon)} = \frac{\pi^2}{3}. \tag{19.78}$$

This is in agreement with the previous result (19.42).

We shall resume the case $K = \frac{1}{2}$, but with emphasis on dynamics, in Section 22.6.

19.3 Non-Ohmic spectral densities

In this section, the specific heat for a non-Ohmic spectral density of the form (18.44) is studied. The imaginary-time pair correlation function at $T = 0$ for $s \neq 1$ can be found from the expression (18.45) by analytic continuation. We get

$$
\begin{aligned}
\mathcal{W}_{T=0}(\tau) &= a(s) \left[(\omega_c \tau)^{1-s} - 1 \right], \\
a(s) &= 2\delta_s \left[\Gamma(s)/(1-s) \right] (\omega_{\mathrm{ph}}/\omega_c)^{1-s},
\end{aligned}
\qquad \omega_c \tau \gg 1 . \qquad (19.79)
$$

The function $a(s)$ is positive for $0 < s < 1$ and negative for $s > 1$. An important difference between the sub-Ohmic and the super-Ohmic case is that the pair correlation function at $T = 0$ increases monotonously with τ in the former case, $0 < s < 1$, while it approaches a finite constant for super-Ohmic damping. As a result, the integral in Eq. (19.37) is dominated by the hard-sphere boundary at $\tau = 1/\omega_c$ for sub-Ohmic, and by the bulk for super-Ohmic damping.

19.3.1 The sub-Ohmic case

Sub-Ohmic damping is usually large down to zero temperature, and under condition (19.33), which is $\rho_s \equiv \delta_s^{2/(1-s)}(\omega_{\mathrm{ph}}/\Delta)^2 \gg 1$, the NIDA is justified. If, in addition, there holds $\delta_s^{(1+s)/(1-s)}(T_{\mathrm{ph}}/T)^{1+s} \gg 1$, temperature dependence is determined, analogous to Eq. (19.45), by the short-time dependence of the one-phonon contribution,

$$
e^{-\mathcal{W}(\tau)} = e^{-\mathcal{W}_{T=0}(\tau)} \left\{ 1 + 2\delta_s \Gamma(1+s)\zeta(1+s)(T/T_{\mathrm{ph}})^{s-1} (\tau/\hbar\beta)^2 \right\} . \qquad (19.80)
$$

With this, the τ-integral in Eq. (19.34) is an incomplete gamma function, and the resulting specific heat is

$$
\begin{aligned}
c/k_{\mathrm{B}} &= b(s)\, e^{a(s)}\, \Gamma[\, 3/(1-s),\, a(s)\,] \, (T/T_{\mathrm{ph}})^s , \\
b(s) &= s^2(s+1)\zeta(1+s)[2\delta_s\Gamma(s)/(1-s)]^{-(s+2)/(1-s)}\Delta^2/4\omega_{\mathrm{ph}}^2 .
\end{aligned}
\qquad (19.81)
$$

Taking in Eq. (19.81) the limit $\omega_c \to \infty$, the specific heat is simplified the form

$$
c/k_{\mathrm{B}} = b(s)\,\Gamma[3/(1-s)]\,(T/T_{\mathrm{ph}})^s . \qquad (19.82)
$$

In the regime $\rho_s \gg 1$, the expression (19.82) holds in the range $0 \leqslant T < T_{\max}$, where $T_{\max}$ can be much larger than $\hbar\Delta/k_{\mathrm{B}}$. Hence the Schottky peak is absent and the specific heat is steadily increasing with T like in the Ohmic strong-damping regime.

If one is interested in the limit $s \to 1$, one must keep ω_c finite. As $s \to 1$, both arguments of the incomplete gamma function diverge, but the ratio stays constant. One can directly deduce from the Euler integral of the incomplete gamma function

$$
\lim_{x \to \infty} e^x\, \Gamma(\lambda x,\, x) = x^{\lambda x - 1}/(1-\lambda) \qquad (19.83)
$$

With this asymptotic formula, the expression (19.81) directly reduces to the Ohmic specific heat expression (19.46) holding for strong damping $K > 3/2$.

19.3.2 The super-Ohmic case

The density of low energy excitations is low for super-Ohmic damping. Thus, at low T, one usually is in the weak-coupling regime, and one may expand the self-energy in a power series in the coupling δ_s. With a dimensional analysis one sees that one-phonon exchange is the dominant process in the regime $\delta_s(\omega_c/\omega_{\rm ph})^{s-1} \ll 1$. The one-phonon contribution to the specific heat is calculated via insertion of the expression (19.28) with (19.27) in (19.29). One readily obtains the asymptotic low temperature series

$$c/k_{\rm B} = \delta_s\, \varrho^{3-s}\left(\frac{\Delta_{\rm eff}}{\omega_{\rm ph}}\right)^{s-1}\sum_{n=0}^{\infty}(2n+s)\Gamma(2n+2+s)\zeta(2n+1+s)\,(\varrho\,\theta)^{2n+s}. \tag{19.84}$$

Here $\zeta(z)$ is the Riemann zeta function, and $\theta = k_{\rm B}T/\hbar\Delta_{\rm eff}$ with $\Delta_{\rm eff}$ given in Eq. (18.33). This is the extension of the series (19.40) to super-Ohmic damping. The specific heat varies as T^s in the low temperature regime $\theta \ll 1$. In the opposite high temperature limit $\theta \gg 1$, one may put $G(\omega)/(\omega^2 - z^2) \to -G(\omega)/z^2$ in Eq. (19.28). With this one readily finds

$$c/k_{\rm B} = \frac{1}{4\theta^2}\left\{1 - 2\delta_s(3-s)(2-s)\Gamma(s-1)\zeta(s-1)\left(\Delta_{\rm eff}/\omega_{\rm ph}\right)^{s-1}\theta^{s-1}\right\}. \tag{19.85}$$

For a discussion of other parameter regimes, we refer to Ref. [408].

19.4 Relation with the Kondo model

In this section we consider the connection of the Ohmic spin-boson model with the Kondo model. In its simplest form, the Kondo problem shows up when a single magnetic impurity of spin 1/2 interacts via an exchange scattering potential with a band of conduction electrons. In the adiabatic scheme, the constant density of electron-hole excitations near the Fermi surface gives rise to Anderson's orthogonality catastrophe [413], and to logarithmic infrared divergences in a variety of physical quantities, e. g., in the static magnetic susceptibility, or in the soft X-ray absorption and emission of metals [190, 414, 143]. These phenomena have been popularized by Kondo as Fermi surface effects [138]. The relationship between the spin-boson and the Kondo model is due to the fact that the low-energy electron-hole excitations have bosonic character and can be interpreted in terms of density fluctuations [191].

For a constant density of these excitations, there is an exact mapping between fermions and bosons [415]. Here we restrict our attention to the anisotropic Kondo model and to the resonance level model (see Ref. [143]). Generalizations to multi-channel Kondo impurity models which show non-Fermi liquid behavior, and the possible experimental relevance are reviewed in Ref. [416]. In Subsection 19.2.2 we have already discussed a particular case of the correspondence. There we have shown that the Ohmic spin-boson model with the Kondo parameter $K = \frac{1}{2}$ maps on the Toulouse model which is a special case of the Kondo model.

19.4.1 Anisotropic Kondo model

The Kondo model is, at first glance, a simple model in which a magnetic impurity interacts with the conduction electrons via a pointlike exchange interaction, so that only s-wave scattering occurs. Further, it is assumed that, out of the many different bands which may exist in a solid, only a narrow band of half-width $\hbar\omega_c = \hbar/\tau_c$, chosen symmetric about the Fermi energy E_F, interacts with the impurity. Linearizing the dispersion relation as in Eq. (19.55), the Kondo Hamiltonian takes the form

$$H_K = \hbar v_F \sum_{k,\sigma} k c_{k,\sigma}^\dagger c_{k,\sigma} + J\, \mathbf{S}\cdot \mathbf{s}(0)/\hbar \,, \tag{19.86}$$

which is also referred to as the s-d Hamiltonian. The operator $c_{k,\sigma}^\dagger$ creates a conduction electron with momentum $\hbar k$ and spin polarization $\sigma = \pm 1$. Further, $S_i = \frac{1}{2}\hbar\tau^i$ is the spin operator of the impurity, where τ^i ($i = 1, 2, 3$) are the Pauli matrices (to make a distinction from the TSS operators, we choose τ^i). The effective spin operator due to the conduction electrons at the impurity site $\mathbf{r} = 0$ is conveniently expressed in terms of Wannier operators for electrons localized at the origin with spin polarization σ,

$$c_\sigma^\dagger(0) = L^{-1/2} \sum_k c_{k,\sigma}^\dagger \,, \tag{19.87}$$

where $k = 2\pi n/L$ ($n \in \{0, \pm 1, \ldots\}$), and L is a normalization length. We then have

$$s_i(0) = \frac{\hbar}{2} \sum_{\sigma,\sigma'} c_\sigma^\dagger(0)\tau_{\sigma,\sigma'}^i c_{\sigma'}(0) \,. \tag{19.88}$$

In the original Kondo problem, the exchange interaction is rotational invariant, $J_\parallel = J_\perp$. In order to relate the Kondo problem to the spin-boson problem, it is essential to generalize the isotropic coupling J to the case where the exchange constants $J_\parallel$ for the term $S_z s_z$ and $J_\perp$ for the term $S_x s_x + S_y s_y$ are independent parameters and arbitrarily large. The corresponding anisotropic Kondo Hamiltonian is

$$H_K = \hbar v_F \sum_{k,\sigma} k c_{k,\sigma}^\dagger c_{k,\sigma} + \tfrac{1}{4}\hbar J_\parallel \tau_z \sum_\sigma \sigma c_\sigma^\dagger c_\sigma + \tfrac{1}{2}\hbar J_\perp \left(\tau_+ c_\downarrow^\dagger c_\uparrow + \tau_- c_\uparrow^\dagger c_\downarrow \right) \tag{19.89}$$

with $\tau_\pm = (\tau^x \pm i\tau^y)/2$. The $J_\parallel$-term describes scattering of the fermions at the impurity in which the spin polarization is conserved while the $J_\perp$-term describes spin-flip scattering. The parameters $J_\parallel$ and $J_\perp$ have dimension velocity. The dimensionless coupling constants are $\rho J_\parallel$ and $\rho J_\perp$, and the constant density of states is $\rho = (2\pi v_F)^{-1}$.

To show the equivalence between the anisotropic Kondo model, Eq. (19.89), and the spin-boson model, we follow the steps of Yuval and Anderson [417], who have cast the partition function for the impurity into the "Coulomb gas" representation

$$Z_K = \sum_{m=0}^\infty \left(\frac{\rho J_\perp \cos^2 \delta_K}{2\tau_c} \right)^{2m} \int_0^{\hbar\beta - \tau_c} ds_{2m} \int_0^{s_{2m}-\tau_c} ds_{2m-1} \cdots \int_0^{s_2 - \tau_c} ds_1 \tag{19.90}$$

$$\times \exp\left\{ 2\left(1 - \frac{2}{\pi}\delta_K\right)^2 \sum_{j>k=1}^{2m} (-1)^{j+k} \ln\left[\frac{\hbar\beta}{\pi\tau_c} \sin\left(\frac{\pi(s_j - s_k)}{\hbar\beta} \right) \right] \right\} \,.$$

339

Here, $\tau_c = 1/\omega_c$ is the hard-sphere distance of neighbouring charges, and δ_K is the scattering phase shift at the singular non-spin-flip potential $J_\parallel \, \delta(x)/4$. The series (19.90) is in correspondence with the Coulomb gas representation (19.7) with (19.9), and with the Ohmic form (18.55) for $\mathcal{W}(\tau)$, if we equate

$$\Delta = \rho J_\perp \cos^2(\delta_K)/\tau_c \, ; \qquad K = (1 - 2\delta_K/\pi)^2 \, . \tag{19.91}$$

The scattering phase shift $\delta_K(J_\parallel)$ depends on the regularization prescription for the singular scattering potential [143]. If we choose a separable form, we obtain

$$\delta_K(J_\parallel) = \arctan(\pi\rho J_\parallel/4) \, , \tag{19.92}$$

while the smoothing prescription $\delta(x) \to (2\pi)^{-1} \int_{-k_c}^{k_c} dk \, e^{ikx}$, where k_c is large, gives $\delta_K = \pi\rho J_\parallel/4$. In Eq. (19.92), only the contribution which is linear in $J_\parallel$ is universal. This implies that the correspondence between the anisotropic Kondo model and the Ohmic TSS is universal in the regime $\rho J_\perp = \Delta/\omega_c \ll 1$ and $\rho|J_\parallel| = |1 - K| \ll 1$.

A local magnetic field h_m in z-direction which couples to the impurity spin, and not to the conduction electrons, gives to H_K the additional contribution $-\frac{1}{2}g\mu_B h_m \tau^z$. Thus, the bias energy $\hbar\epsilon$ in the TSS corresponds to $g\mu_B h_m$. Introducing the static susceptibility χ_{sd} with zero Landé factor for the electron band, the Wilson ratio is found as [cf. Eq. (57) in Ref. [410], and footnote to Eq. (19.4)]

$$R_1 = \lim_{T \to 0} \frac{(g\mu_B)^2 \, c(T)/k_B}{T\chi_{sd}} = \frac{2\pi^2}{3} \left(1 - \frac{2\delta_K}{\pi}\right)^2 , \qquad \frac{\chi_{sd}}{(g\mu_B)^2} = \frac{\overline{\chi}_z}{4} \, . \tag{19.93}$$

This result is in agreement with the Wilson ratio for the Ohmic TSS (see the discussion in Section 22.9). We see from Eq. (19.92) that the regime $-\infty < \rho J_\parallel < \infty$ maps on the regime $4 > K > 0$. The critical coupling $K = 1$ separates the ferromagnetic Kondo regime $\rho J_\parallel < 0$, which corresponds to the regime $K > 1$, from the more intriguing antiferromagnetic regime $\rho J_\parallel > 0$, i. e., $K < 1$. The special choice $\delta_K(J_\parallel) = (2 - \sqrt{2})\pi/4$, which corresponds to the case $K = \frac{1}{2}$, is known in the literature as the Toulouse limit of the anisotropic Kondo model. This limiting case is exactly solvable in closed form [143].

In the antiferromagnetic regime, the Kondo temperature $T_K \propto \Delta_r$ separates the perturbative regime $T > T_K$, in which physical quantities, e.g., the magnetic susceptibility, can be expanded in powers of $J_\perp$, from the nonperturbative regime. In the regime $\rho J_\perp \ll \rho J_\parallel \ll 1$, which is equivalent to $\Delta/\omega_c \ll 1 - K \ll 1$, the Kondo temperature is given by [143]

$$k_B T_K = \frac{2}{\pi} \left(\frac{J_\perp}{2J_\parallel}\right)^{1/\rho J_\parallel} \hbar\omega_c = \frac{2}{\pi} \left(\frac{1}{2(1 - K)}\right)^{1/(1-K)} \hbar\Delta_r \, . \tag{19.94}$$

The second equality follows with the equivalence relations (19.91), and with (18.34). For $T \ll T_K$, the properties in the antiferromagnetic sector can be described phenomenologically in the spirit of Landau's theory of a Fermi liquid [411].

19.4.2 Resonance level model

The other model of interest is the *resonance level model*. This model has been introduced as a generalization of the Toulouse model (19.56) by Schlottmann [412], and independenly by Vigmann and Finkels'teĭn [410] (for a review of the resonance level model, see Ref. [143]). The corresponding Hamiltonian describes a d-level near the Fermi surface interacting with a band of spinless fermions. In generalization of the model (19.56), the fermions interact with each other through a repulsive contact potential which represents the (screened) Coulomb interaction. Linearizing again the dispersion relation about the Fermi momentum, the Hamiltonian takes the form

$$
\begin{aligned}
H_{\mathrm{RL}} \;=\; & \hbar v_{\mathrm{F}} \sum_k k c_k^\dagger c_k + \hbar \epsilon_d \, d^\dagger d + (V/\sqrt{L}) \sum_k \left(c_k^\dagger d + d^\dagger c_k \right) \\
& + (U/L) \sum_{k,k'} \left(c_k^\dagger c_{k'} - c_{k'} c_k^\dagger \right) \left(d^\dagger d - \tfrac{1}{2} \right) .
\end{aligned}
\tag{19.95}
$$

The partition function of this model is equivalent to Eq. (19.7) if we identify

$$
\Delta^2 \;=\; 4\rho\,\omega_c (V/\hbar)^2 \cos^2(\delta_{\mathrm{RL}}) ; \quad K \;=\; (1 - 2\delta_{\mathrm{RL}}/\pi)^2 / 2 ; \quad \epsilon \;=\; \epsilon_d , \tag{19.96}
$$

where $\delta_{\mathrm{RL}} = \pi \rho U/\hbar + \mathcal{O}(U^2)$ is the scattering phase for the contact potential $U\delta(x)$, and where again $\rho = (2\pi v_{\mathrm{F}})^{-1}$. In the absence of the repulsive interaction, $U = 0$, we have $K = \tfrac{1}{2}$. This case corresponds to the Toulouse limit of the antiferromagnetic Kondo Hamiltonian. Perturbation expansion in ρU in the resonant level model corresponds to perturbation expansion about $K = \tfrac{1}{2}$ in the two-state model.

Here we have discussed the similarity between the Ohmic two-state system and the Kondo problem on the level of the partition function. The comparison of these models on the basis of their Hamiltonians is set out in Ref. [86]. We should like to remark that the mapping of the Ohmic two-state system neither to the anisotropic Kondo model nor to the resonant level model is exact. However, since the low-energy excitations of these models are very similar, the physical phenomena of these models at low temperature are closely related.

19.5 Relation with the $1/r^2$ Ising model

The inverse square ferromagnetic Ising model is another specific realization of the Coulomb gas or Kondo model. The Ising spins represent the imaginary time history of the single impurity spin of the Kondo model. The hard-sphere regularization at $\tau = \tau_c \equiv 1/\omega_c$ is provided by the lattice, $N\tau_c = \hbar\beta$. It has been shown by Anderson and Yuval [418] that the Ising spin model with ferromagnetic $1/r^2$ spin-pair interactions can be mapped on the Kondo model. With the correspondence of the Kondo model to the TSS explained in the previous section, the inverse-square Ising model is also equivalent with the Ohmic TSS. Direct correspondence of these models has been shown by Spohn and Dümcke by means of functional integral techniques [398].

Consider the one-dimensional N-site Ising model for spins $\{S_j = \pm 1\}$ with ferromagnetic long-range interactions, a magnetic field term, and periodic boundary conditions, described by the Hamiltonian

$$H_{\text{Ising}} = (\hbar/N\tau_{\text{c}})\left(-\sum_{j>i} V(j-i)\, S_j S_i + \frac{h_{\text{m}}}{2}\sum_j S_j\right),$$

$$V(n) = (K/2)\,(\pi/N)^2/\sin^2(\pi n/N)\,, \qquad n \geqslant 2\,, \tag{19.97}$$

where $N = \hbar\beta/\tau_{\text{c}}$. In the state representation, the partition function reads

$$Z_{\text{Ising}} \equiv \operatorname{tr} e^{-\beta H_{\text{Ising}}} = \sum_{S_1,\cdots,S_N} \exp\left\{\sum_{j>i} V(j-i)\, S_j S_i - \frac{h_{\text{m}}}{2}\sum_j S_j\right\}. \tag{19.98}$$

The correspondence with the spin-boson model is most easily seen in the charge representation, usually referred to as Coulomb gas representation [419],

$$Z_{\text{Ising}} = \sum_{n=0}^{\infty} y^{2n} \int_0^{\hbar\beta-\tau_{\text{c}}} \frac{ds_{2n}}{\tau_{\text{c}}} \int_0^{s_{2n}-\tau_{\text{c}}} \frac{ds_{2n-1}}{\tau_{\text{c}}} \cdots \int_0^{s_2-\tau_{\text{c}}} \frac{ds_1}{\tau_{\text{c}}} \tag{19.99}$$

$$\times \exp\left(4\sum_{j=2}^{2n}\sum_{i=1}^{j-1} (-1)^{j+i} U[(s_j-s_i)/\tau_{\text{c}}] + h_{\text{m}}\sum_{j=1}^{2n} (-1)^j s_j/\tau_{\text{c}}\right),$$

where $y = \exp[2U(0)]$ is the fugacity of the Coulomb gas. The spin interaction $-V(n)$ is the second derivative of the charge interaction $U(n)$ in discretized form [cf. Eqs. (4.76) and (4.77)], $V(n) = 2U(n) - U(n+1) - U(n-1)$, and thus

$$U(n) = (K/2)\ln\left[(N/\pi)\sin(\pi n/N)\right], \qquad n \geqslant 1\,. \tag{19.100}$$

With this one finds $V(1) = -U(0) - KC/2$.[5] Accord with the charge representation (14.2) of the spin boson model is achieved by putting $y = \Delta/2\omega_{\text{c}}$ and $h_{\text{m}} = -\epsilon\tau_{\text{c}}$.

In standard form, the Hamiltonian of the Ising model reads

$$H_{\text{Ising}} = -\frac{\hbar}{2N\tau_{\text{c}}}\left(J_{\text{NN}}\sum_j S_{j+1} S_j + J_{\text{LR}}\sum_{j>i} \frac{(\pi/N)^2 S_j S_i}{\sin^2[\pi(j-i)/N]} - h_{\text{m}}\sum_j S_j\right), \tag{19.101}$$

where $J_{\text{LR}} = K$, and $J_{\text{NN}} + J_{\text{LR}} = 2V(1)$. Correspondence with the Ohmic TSS holds for $\Delta/2\omega_{\text{c}} = \exp[-J_{\text{NN}} - (1+C)J_{\text{LR}}]$ and $K = J_{\text{LR}}$.

The equivalence of the Ohmic TSS with the $1/r^2$ Ising model has been utilized in Refs. [420, 421]. In these studies, the imaginary-time correlation function is calculated by Monte Carlo simulations on the Ising system and then continued to real-time by employing a Padé approximation.

The correspondence of the TSS with a long-range ferromagnetic Ising model can also be established for non-Ohmic spectral coupling. For $G(\omega) \propto \omega^s$, the spin interaction $V(j-i)$ in the Ising model at zero temperature falls off as $(j-i)^{-(1+s)}$. Upon using the correspondence, a number of properties proven rigorously for the Ising model [398, 422] can directly be transferred to the dissipative two-state system.

---

[5]The constant C depends on the particular hard-sphere regularization chosen. Here $C = \ln 2$.

20 Electron transfer and incoherent tunneling

Before we turn in Chapter 21 to the real-time dynamics of the TSS , we first study quantum-statistical tunneling with the thermodynamic method discussed above in Part III. The results shall be confirmed in Chapter 21 by a dynamical approach.

20.1 Electron transfer

Electron transfer plays an important role in many processes in chemistry and biology. A striking example is the ultrafast primary electron transfer step found in photosynthetic reaction centers which is responsible for the high efficiency of the photosynthetic mechanism. Significant progress with the understanding of the essence of electron transfer (ET) processes in condensed matter has been made by Marcus [107, 108]. He was the first who discovered the importance of the solvent environment. To describe the effects of the solvent coupled to the electronic degree of freedom, he employed linear response theory. In the usual ET reaction, there is a free energy barrier separating reactants and products. The barrier is because the donor and acceptor states are strongly solvated, and the transfer of the electron then requires a reorganization of the environment. At normal temperature, the electron has to tunnel through the barrier from the reactant to the product state. The tunneling is only effective if suitable bath fluctuations bring reactant and product energy levels into resonance. In the classical limit, the transfer rate is determined by two factors. First, a Boltzmann factor with the activation energy required for bath fluctuations. Second, an attempt frequency prefactor describing the probability for tunneling once the levels are in resonance. The latter factor is mainly determined by the overlap between the electronic wave functions localized on different redox sites. For large electronic coupling the reaction is adiabatic, whereas the reaction is nonadiabatic for weak electronic coupling,

It is implied by the success of the classical Marcus theory that the electron can be described in terms of a discrete variable at all temperatures of interest. Here we restrict the attention to the important case of a two-state system or a spin-$1/2$ system with localized states at $q = \frac{1}{2}q_0\sigma$: a donor state D with eigenvalue $\sigma = -1$ of σ_z, and an acceptor state A with eigenvalue $\sigma = +1$, and a linearly responding heat bath represents the solvent. Then the proper Hamiltonian is the spin-boson Hamiltonian, as given in Eq. (3.145) or (3.147), or in Eq. (18.15),

$$H = -\tfrac{1}{2}\hbar\Delta\,\sigma_x - \tfrac{1}{2}\hbar\epsilon\,\sigma_z - \tfrac{1}{2}\mathfrak{E}(t)\,\sigma_z + H_{\mathrm{R}} . \tag{20.1}$$

Here, $H_{\mathrm{TSS}} = -\frac{1}{2}\hbar\Delta\,\sigma_x - \frac{1}{2}\hbar\epsilon\,\sigma_z$ describes the tunneling of the electron between the donor and the acceptor state, and H_{R} represents the solvent as a Gaussian reservoir. The collective bath mode $\mathfrak{E}(t) = -\mu\mathcal{E}(t) = q_0\sum_\alpha c_\alpha x_\alpha(t)$ is coupled to the spin operator σ_z. It can be thought of as a fluctuating dynamical polarization energy, whereby $\mathcal{E}(t)$ is the polarization field, and μ is the difference of the dipole moments of the two electronic states. The polarization field $\mathcal{E}(t)$ vanishes on average.

All effects of the solvent on the electron transfer are contained in the autocorrelation function of the polarization energy [cf. Eq. (4.75)] for complex time $z = t - i\tau$,

$$\frac{\langle \mathcal{E}(z)\mathcal{E}(0) \rangle_\beta}{\hbar^2} = \frac{\mu^2}{\hbar^2}\langle \mathcal{E}(z)\mathcal{E}(0) \rangle_\beta = \ddot{Q}(z) = \int_0^\infty d\omega\, G(\omega)\, \frac{\cosh[\omega(\hbar\beta/2 - iz)]}{\sinh[\omega\hbar\beta/2]} , \quad (20.2)$$

where $Q(z)$ is the pair interaction (18.42).

20.1.1 Adiabatic bath

An adiabatic bath is a sluggish bath. It is characterized by only zero frequency fluctuations and is therefore not dynamical. In practice, electron transfer (ET) systems are adiabatic when $\omega_c \ll \Delta$ and $\omega_c \ll k_B T/\hbar$, where ω_c is a typical bath frequency, and when frequencies significantly higher than ω_c are absent in the bath. The case of an adiabatic bath can be solved in analytical form [423].

In the adiabatic regime, Eq. (20.2) reduces to

$$\mu^2\langle \mathcal{E}(z)\mathcal{E}(0) \rangle_\beta \longrightarrow \mu^2\langle |\mathcal{E}|^2 \rangle_\beta = 2\Lambda_{\rm cl}/\beta , \quad (20.3)$$

where

$$\Lambda_{\rm cl} = \hbar \int_0^\infty d\omega\, \frac{G(\omega)}{\omega} \quad (20.4)$$

is the classical *reorganization energy*, or *solvation energy*, or *coincidence energy*.

For $G(\omega) = 2\delta_s \omega_{\rm ph}^{1-s}\omega^s e^{-\omega/\omega_c}$, the reorganization energy is given by

$$\Lambda_{\rm cl} = 2\delta_s \Gamma(s)\, (\omega_c/\omega_{\rm ph})^{s-1}\hbar\omega_c . \quad (20.5)$$

In the particular case of Ohmic coupling, $s = 1$ and $\delta_1 = K$, there holds

$$\Lambda_{\rm cl} = 2K\hbar\omega_c . \quad (20.6)$$

For a linearly responding bath, the polarization field $\mathcal{E}$ with mean zero and mean square (20.3) has the normalized Gaussian distribution

$$\rho(\mathcal{E}) = \frac{\mu}{2}\sqrt{\frac{\beta}{\pi\Lambda_{\rm cl}}} \exp\left(-\beta\mu^2\mathcal{E}^2/4\Lambda_{\rm cl}\right) . \quad (20.7)$$

In the adiabatic limit, the polarization is slow, and the Born-Oppenheimer method gives for the Hamiltonian $H_{\rm TSS} + \frac{1}{2}\mu\mathcal{E}\sigma_z$ two electronic eigenstates with energies

$$E_\pm(\mathcal{E}) = \pm\tfrac{1}{2}\hbar\Omega(\mathcal{E}) , \qquad \Omega(\mathcal{E}) \equiv \sqrt{\Delta^2 + (\epsilon - \mu\mathcal{E}/\hbar)^2} . \quad (20.8)$$

The adiabatic partition function is an average of Boltzmann factors $e^{-\beta E_\pm(\mathcal{E})}$ with the Gaussian probability distribution (20.7),

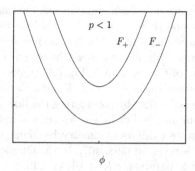 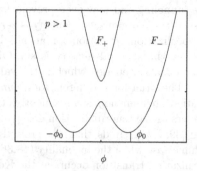

Figure 20.1: The adiabatic Born-Oppenheimer surfaces $F_+(\phi)$ and $F_-(\phi)$ for the symmetric spin-boson model, Eq.(20.13). The parameter p is chosen as $p = 1/2$ (left) and $p = 8$ (right). The distance between the surfaces at $\phi = 0$ is $\hbar\Delta$.

$$Z_{\text{ad}} = \int_{-\infty}^{\infty} d\mathcal{E}\, \rho(\mathcal{E}) \left(e^{-\beta E_+(\mathcal{E})} + e^{-\beta E_-(\mathcal{E})}\right) = 2\int_{-\infty}^{\infty} d\mathcal{E}\, \rho(\mathcal{E}) \cosh\left[\tfrac{1}{2}\beta\hbar\Omega(\mathcal{E})\right] . \quad (20.9)$$

The expression (20.9) coincides with the partition function of a 1D Ising model with *infinite range* interactions. To discuss the qualitative features, we rewrite Eq. (20.9) as

$$Z_{\text{ad}} = \frac{\mu}{2}\sqrt{\frac{\beta}{\pi\Lambda_{\text{cl}}}} \int_{-\infty}^{\infty} d\mathcal{E} \left(e^{-\beta F_+(\mathcal{E})} + e^{-\beta F_-(\mathcal{E})}\right) . \quad (20.10)$$

Here

$$F_\pm(\mathcal{E}) = \frac{\mu^2\mathcal{E}^2}{4\Lambda_{\text{cl}}} \pm \frac{\hbar}{2}\sqrt{\Delta^2 + (\epsilon - \mu\mathcal{E}/\hbar)^2} \quad (20.11)$$

are the potential energies for the polarization energy $\mu\mathcal{E}$ in the adiabatic limit.

Consider first the symmetric case $\epsilon = 0$, where the two redox states are solvated with the same energy. Depending on the numerical value of the parameter

$$p = \Lambda_{\text{cl}}/\hbar\Delta , \quad (20.12)$$

the adiabatic surfaces are qualitatively different. Measuring the polarization energy in units of $\hbar\Delta$, $\phi \equiv \mu\mathcal{E}/\hbar\Delta$, the expression (20.11) is written as

$$F_\pm(\phi) = \frac{\hbar\Delta}{2}\left(\frac{\phi^2}{2p} \pm \sqrt{1+\phi^2}\right) . \quad (20.13)$$

The potential $F_+(\phi)$ is monostable with the minimum at $\phi = 0$ for all p, while $F_-(\phi)$ is monostable with minimum at $\phi = 0$ for $p \leqslant 1$ only. If $p > 1$, the potential $F_-(\phi)$ is bistable with minima $F_-^{(\text{min})} = -[p + 1/p]\hbar\Delta/4$ at $\phi = \pm\phi_0$, where $\phi_0 = \sqrt{p^2-1}$. This behavior is illustrated in Fig. 20.1. Thus, if $p > 1$, the symmetry of the ground state of the spin is spontaneously broken, $\langle\sigma_z\rangle \neq 0$. This is a type of fluctuation-induced *self-trapping* in the *adiabatic* spin-boson model. Whether trapping for an adiabatic bath occurs is decided by the competition between the reorganization energy Λ_{cl} and the energy for resonant tunneling $\hbar\Delta$. Slowly fluctuating fields may impede tunneling and may even prevent transitions. When spontaneous fluctuations are absent, the reorganization energy is zero, and then the particle tunnels clockwise.

The localization transition for an adiabatic bath, $\omega_c \ll k_B T/\hbar$, is not sensitive to whether the spectral density is sub-Ohmic, Ohmic or super-Ohmic, since only the reorganization energy, which is the value of the integral in Eq. (20.4), decisively matters. The situation is different for a *dynamical* bath, $\omega_c \gg k_B T/\hbar$ and $\omega_c \gg \Delta$. The localization phenomenon is conveniently studied either by considering the flow of the parameters in the renormalization group equations [86] or by employing a variational method [398]. One finds that the ground state exhibits symmetry breaking in the sub-Ohmic case when the inequality (18.36) is reversed [398, 407]. In the Ohmic case, the localization transition occurs at the Kondo parameter $K = 1$ [144, 145, 86]. Here again slow modes are responsible for the symmetry breaking. The phase diagram as a function of Δ/ω_c for $s \leqslant 1$ is discussed in Ref. [398]. It is shown by using an energy-entropy argument that there is *no* phase transition to localization when

$$\int_0^\infty d\omega \, G(\omega)/\omega^2 < \infty . \tag{20.14}$$

Thus, the symmetry of the ground state is not spontaneously broken for super-Ohmic dissipation of any strength.

When the bath is adiabatic, the dynamics can be solved in analytic form in two steps. Consider the thermal equilibrium correlation function $C^+(t) \equiv \langle \sigma_z(t)\sigma_z(0) \rangle_\beta$. In the first step, the dynamics of the TSS is calculated for a constant polarization energy $\mu\mathcal{E}$. This yields[1]

$$C_0^+(t;\mathcal{E}) = \frac{(\epsilon - \mu\mathcal{E}/\hbar)^2}{\Omega^2} + \frac{\Delta^2}{\Omega^2}\cos(\Omega t) - i\tanh\left(\tfrac{1}{2}\beta\hbar\Omega\right)\frac{\Delta^2}{\Omega^2}\sin(\Omega t) . \tag{20.15}$$

The tunneling frequency $\Omega = \Omega(\mathcal{E})$ is defined in Eq. (20.8). In the second step, the correlation function (20.15) is averaged with the adiabatic weight functional (20.9),

$$C_{ad}^+(t) = \frac{2}{Z_{ad}} \int_{-\infty}^\infty d\mathcal{E}\, \rho(\mathcal{E}) \cosh\left[\tfrac{1}{2}\beta\hbar\Omega(\mathcal{E})\right] C_0^+(t;\mathcal{E}) . \tag{20.16}$$

Thus, the final step towards the system dynamics is simple quadrature. The two-state correlation function of the free system is broadened by adiabatic bath fluctuations. Due to the time-independent part in Eq. (20.15), the "adiabatic" thermal correlation function (20.16) does not drop to zero at asymptotic time. Correct decay of the correlation $C^+(t)$ is provided by dynamical fluctuations (cf. Chapter 21).

20.1.2 Marcus theory for electron transfer

In the absence of nuclear (solvent) tunneling, the electron transfer dynamics is well understood in two different limits. In the *adiabatic limit*, the electronic coupling Δ

---

[1]We assume that the two eigenstates with energies $E_\pm(\mathcal{E})$ are thermally occupied. The real part of Eq. (20.15) is the result of a standard quantum mechanical calculation for a biased TSS, and the imaginary part follows with the fluctuation-dissipation theorem [cf. Eqs. (6.58) and (6.59)].

is so large that the parameter p is of order one or smaller. When p is above one, but not extremely large, the rate is controlled by the motion of the solvent on the lower electronic surface [cf. Fig. 20.1 (right)] and well described by the classical activated rate theory [108]. For $\epsilon = 0$, the rate depends on the electronic coupling Δ like

$$k \propto \exp[-\beta(F_-^* - \hbar\Delta/2)] \,, \tag{20.17}$$

where $F_-^* = \Lambda_{\rm cl}/4$ is the free energy barrier for $\Delta = 0$ [cf. Fig. 20.2 (a)].

In the opposite *nonadiabatic limit*, the electronic coupling is weak, $p \gg 1$. To lowest order in the coupling Δ, the transfer rate is given by the Golden Rule formula

$$k \propto \Delta^2 \exp(-\beta F_-^*) \,. \tag{20.18}$$

The two opposite limits are distinguished by different time scales. The adiabatic limit is characterized by slow fluctuations of the solvent (nuclear degree of freedom) so that the electron transfer may be treated in Born-Oppenheimer approximation. In the nonadiabatic limit, the solvent fluctuations are assumed to be fast. Then the ET is determined by thermal excitations of the solvent to the vicinity of the crossing region which are followed by a fast transversal of this region.

The nonadiabatic regime is conveniently discussed with the adiabatic potentials (20.11). For $\Delta = 0$, these free energy curves become two intersecting parabola,

$$F_\pm(\mathcal{E}) = \pm \frac{\hbar\epsilon}{2} + \frac{1}{4\Lambda_{\rm cl}} (\mu\mathcal{E} \mp \Lambda_{\rm cl})^2 - \frac{\Lambda_{\rm cl}}{4} \,. \tag{20.19}$$

which are known in the literature as the *Marcus parabola*. The parabola are the adiabatic energy functions for *diabatic* states ("diabatic" states are the eigenstates for $\Delta = 0$). In Fig. 20.2 the Marcus parabola are displayed schematically for three different values of the bias parameter ϵ. Here we choose ϵ positive. The higher local minimum in Figs. 20.2 (b) and (c) is the donor (D) state. This state is located at $\mathcal{E} = \Lambda_{\rm cl}/\mu$. The lower local minimum is the acceptor (A) state. This state is situated at $\mathcal{E} = -\Lambda_{\rm cl}/\mu$. Pictorially, the reorganization energy is the difference of the free energy $F_-(\mathcal{E})$ between the donor state and the acceptor state,

$$\Lambda_{\rm cl} = F_-(\Lambda_{\rm cl}/\mu) - F_-(-\Lambda_{\rm cl}/\mu) \,. \tag{20.20}$$

Since the nuclear degree of freedom (polarization) is slow compared to the electronic degree of freedom, the transition is determined by the *Franck-Condon principle*. This principle states that transitions between electronic states occur vertically in Fig. 20.2; that is, the nuclear reaction coordinate $\mathcal{E}$ does not change during the electronic transition. The vertical transition, however, is energetically unfavorable unless the polarization field $\mathcal{E}$ is close to $\mathcal{E}^* \equiv \hbar\epsilon/\mu$, which is the *Landau-Zener* regime. The point $\mathcal{E} = \mathcal{E}^*$ is the point of intersection of the two curves. The activation (free) energies for this particular bath polarization relative to the D and A state are

$$\begin{aligned}
F_D^* &\equiv F_+(\mathcal{E}^*) - F_+(\Lambda_{\rm cl}/\mu) = (\hbar\epsilon - \Lambda_{\rm cl})^2/4\Lambda_{\rm cl} \,, \\
F_A^* &\equiv F_-(\mathcal{E}^*) - F_-(-\Lambda_{\rm cl}/\mu) = (\hbar\epsilon + \Lambda_{\rm cl})^2/4\Lambda_{\rm cl} \,.
\end{aligned} \tag{20.21}$$

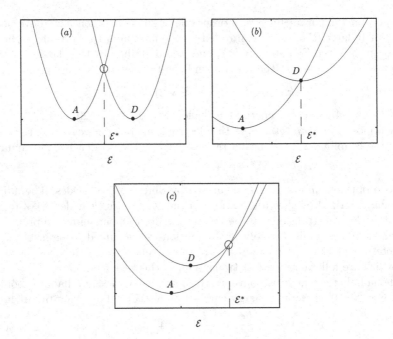

Figure 20.2: Free energy of the diabatic ($\Delta = 0$) electronic states as a function of the nuclear reaction coordinate $\mathcal{E}$ ("Marcus parabola"). The figures show the schematic dependence on the bias ϵ. (a) Symmetric case $\epsilon = 0$. (b) Activationless case $\epsilon = \Lambda_{\mathrm{cl}}/\hbar$. (c) Inverted regime $\epsilon > \Lambda_{\mathrm{cl}}/\hbar$. The intersection point of the two parabola is at $\mathcal{E}^* = \hbar\epsilon/\mu$.

In the symmetric case $\epsilon = 0$, we have $F_D^* = F_A^* = \Lambda_{\mathrm{cl}}/4$. With increasing bias the activation energy related to the donor state becomes smaller, and therefore the $D \to A$ rate becomes larger. When the condition $\hbar\epsilon = \Lambda_{\mathrm{cl}}$ is reached, the free energy barrier has disappeared. Since at this point $F_D^* = 0$, there is activationless electron transfer from D to A, which depends only weakly on temperature. As the bias is increased further, the activation energy grows again. For this reason, the regime $\hbar\epsilon > \Lambda_{\mathrm{cl}}$ is called the *inverted regime*. Altogether, the Marcus theory predicts that the transfer rate is a non-monotonous function of the bias. The sweeping success of Marcus' theory is founded on the discovery of this characteristic behavior in ET measurements.

In the low temperature regime, nuclear tunneling becomes relevant [108]. This collective tunneling phenomenon is generally thought to prevail in the inverted regime. However, one finds that nuclear tunneling at low T may dominate the transfer dynamics even for a symmetric system. The quantum effects depend on the low-frequency behavior of the spectral density, the coupling strength, and the cutoff frequency. These features will be discussed in some detail for the nonadiabatic regime in Sec. 20.2.

In a *classical* description of activated barrier crossing, the rate formula consists of the Boltzmann factor $\exp(-\beta F_{D/A}^*)$ and of the pre-exponential factor, viz. the attempt frequency. The Boltzmann factor represents the probability for a fluctuation so that $\mathcal{E} = \mathcal{E}^*$ (see Fig. 20.2). In the nonadiabatic limit, the transfer rate is made up by a single tunneling event, ie., it is of lowest order in the electronic coupling Δ. The corresponding golden rule classical rate or Marcus rate is [cf. Eq. (20.54]

$$k_{\text{cl,GR}}^{\pm}(T,\epsilon) = \frac{\pi\hbar\Delta^2}{2} \frac{1}{\sqrt{4\pi\Lambda_{\text{cl}}k_{\text{B}}T}} \exp\left(-\frac{(\hbar\epsilon \mp \Lambda_{\text{cl}})^2}{4\Lambda_{\text{cl}}k_{\text{B}}T}\right). \qquad (20.22)$$

Attempts were made to include higher orders of Δ in the attempt frequency by putting a Landau-Zener pre-exponential factor aimed to describe the probability for tunneling on resonance $\mathcal{E} = \mathcal{E}^*$. This factor depends on the details of the dynamics in the Landau-Zener transition region [108, 424, 425].

When the collective bath motion is slow compared with the motion of the TSS, all orders of the electronic coupling can be taken into account as a geometrical series in the prefactor of the rate [426, 427]. This analysis is performed in a real-time path sum approach. The resulting classical rate k_{cl}^+ from the D to the A state, and the respective backward rate k_{cl}^-, take the Zusman-type form

$$k_{\text{cl}}^{\pm}(T,\epsilon) = \frac{\pi\hbar\Delta^2}{2} \frac{1}{1 + \pi\hbar\Delta^2/2\omega_{\text{R}}\Lambda_{\text{cl}}} \frac{1}{\sqrt{4\pi\Lambda_{\text{cl}}k_{\text{B}}T}} \exp\left(-\frac{(\hbar\epsilon \mp \Lambda_{\text{cl}})^2}{4\Lambda_{\text{cl}}k_{\text{B}}T}\right), \qquad (20.23)$$

The prefactor describes dynamical recrossings of the dividing surface. The rates $k_{\text{cl}}^{\pm}$ are related by detailed balance,

$$k_{\text{cl}}^-(T,\epsilon) \equiv k_{\text{cl}}^+(T,-\epsilon) = k_{\text{cl}}^+(T,\epsilon)\exp(-\beta\hbar\epsilon). \qquad (20.24)$$

The frequency scale ω_{R} in Eq. (20.23) depends on details of the solvent model [426]. It may be identified, e.g., with the renormalized frequency defined in Eq. (14.12). The classical forward rate has a maximum at $\hbar\epsilon = \Lambda_{\text{cl}}$, and it is *symmetric* around this maximum. At very high temperatures, the Arrhenius factor is saturated, and the rate shows a universal $T^{-1/2}$ temperature dependence. In the adiabatic ("large" Δ) limit, the Marcus parabola are split up and the transfer is described by Transition State Theory (TST) on the lower energy surface. In this limit, the rate expression (20.23) becomes independent of the electronic coupling,

$$k_{\text{cl,TST}}^+(T,\epsilon) = \frac{\omega_{\text{R}}}{2\pi} \left(\frac{\pi\Lambda_{\text{cl}}}{k_{\text{B}}T}\right)^{1/2} \exp\left(-\frac{(\hbar\epsilon - \Lambda_{\text{cl}})^2}{4\Lambda_{\text{cl}}k_{\text{B}}T}\right). \qquad (20.25)$$

In the opposite nonadiabatic limit, (20.23) reduces to the *Golden Rule* rate (20.22). Whether formula (20.25) or (20.22) applies depends on the *adiabaticity parameter*

$$g = \pi\hbar\Delta^2/2\omega_{\text{R}}\Lambda_{\text{cl}}. \qquad (20.26)$$

For $g \ll 1$, we have the nonadiabatic rate expression (20.22), while for $g \gg 1$, we obtain the Δ independent adiabatic rate (20.25). The rate in the adiabatic limit is generally much larger than in the nonadiabatic limit.

Attempts for a unified treatment of the crossover region between the adiabatic and nonadiabatic limits, in which the assumed separation of time scales is violated, have been largely founded on an imaginary-time formulation and on an assumed relationship between the ET rate and an analytic continuation of the free energy (cf. Section 12.5). Among these approaches are treatments which are based on the centroid free energy method (cf. Section 12.6) put forward, e.g., by Gehlen *et al.* [428] and extended later on to complex-valued centroid coordinates by Stuchebrukhov and Song [429], and on numerical instanton methods [430].

A profound theoretical treatment of the crossover region which is free of questionable assumptions is based on the real-time path integral formulation. The respective discussion is given in Section 22.8.

20.2 Incoherent tunneling in the nonadiabatic regime

When the reduced system is ergodic (cf. Section 3.1.10), it approaches thermal equilibrium as time passes by. For sufficiently high temperature and/or strong coupling, the system exhibits overdamped exponential relaxation towards the equilibrium state. This regime is mainly studied with focus on the evolution of occupation probabilities. The corresponding real-time study is postponed to the next chapter. In this section, our aim is to calculate the transition rates of the spin-boson model using the free energy method, which is an imaginary-time approach. Later on, we shall confirm that the results obtained agree with those of the real-time approach.

In Section 12.1, we have given formally exact quantum mechanical rate expressions which form the basis of numerical computation. Here we confine ourselves to the limit of nonadiabatic electron transfer. This is the limit in which the tunneling amplitude Δ is small compared to characteristic vibrational frequencies and to the reorganization frequency $\Lambda_{cl}/\hbar$. Then perturbation theory in the tunneling amplitude Δ is appropriate. The "Golden Rule" approach to the spin-boson tunneling rate, widely used in chemical physics and in condensed matter physics, treats the system-environment coupling to all orders and the tunneling amplitude in second order. In the terminology of the extended system, the nonadiabatic rate corresponds to the single-bounce rate expression. The nonadiabatic regime is relevant, e.g., for electron transfer and for defect tunneling in solids (cf. the recent review in Ref. [392]). Systematic corrections to the nonadiabatic rate are discussed within a dynamical approach in Subsec. 22.8.

20.2.1 General expressions for the nonadiabatic rate

In the nonadiabatic regime, the second term in the expression (19.35), which is the breathing mode contribution of the instanton–anti-instanton pair, is the leading tunneling contribution to the free energy associated with the right/left state. We recall that the stationary point τ_{st} of the integrand $e^{\pm\epsilon\tau - \mathcal{W}(\tau)}$, obeying $\pm\epsilon - \mathcal{W}'(\tau_{st}) = 0$, is a saddle point with the direction of steepest descent along real time [cf. Subsec. 17.4.1].

By deformation of the integration path along steepest descent, the free energy (19.35) acquires an imaginary part $\mathrm{Im}F_{L/R}$, which is related to the forward/backward tunneling rate $k^{\pm}$ from the energetically higher/lower state $\sigma = \mp 1$ to the lower/higher state $\sigma = \pm 1$ by [cf. Eq (14.40) with $\kappa = 1$]

$$k^{\pm} = -(2/\hbar)\,\mathrm{Im}\,F_{L/R}\,. \tag{20.27}$$

It is appropriate, as will become clear shortly, to write the nonadiabatic rate as

$$k^{\pm}(\epsilon) = \tfrac{1}{2}\pi\hbar\Delta^2\,P(\pm\hbar\epsilon)\,. \tag{20.28}$$

The function $P(E)$ is found from Eq. (20.27) to be given by the integral

$$P(E) = \frac{1}{i\,2\pi\hbar}\int_{\tau_{\mathrm{st}}-i\infty}^{\tau_{\mathrm{st}}+i\infty} d\tau\, e^{\,E\tau/\hbar - \mathcal{W}(\tau)}\,. \tag{20.29}$$

Since $\mathcal{W}(\tau)$ is analytic in the strip $0 \leqslant \mathrm{Re}\,\tau < \hbar\beta$, one may use τ_{st} as a free parameter in this regime. Favorable values are $\tau_{\mathrm{st}} = 0$ and $\tau_{\mathrm{st}} = \hbar\beta/2$. In the first case, we obtain

$$P(E) = \frac{1}{2\pi\hbar}\int_{-\infty}^{\infty} dt\, e^{\,i\,Et/\hbar - Q(t)}\,, \tag{20.30}$$

where $Q(t) = \mathcal{W}(i\,t)$ is the correlation function (18.42). In the second case, we get

$$P(E) = \frac{e^{\beta E/2}}{2\pi\hbar}\int_{-\infty}^{\infty} dt\, e^{\,iEt/\hbar - X(t)}\,, \tag{20.31}$$

where $X(t) = Q(t - i\,\hbar\beta/2)$ has the integral representation (18.47).

The nonadiabatic rate is nonperturbative in the bath coupling and perturbative (golden rule) in the electronic coupling. Equivalent forms have been obtained already in the sixties [397, 131], and were rederived later on repeatedly. The nonadiabatic rate obtained via the $\mathrm{Im}\,F$ method matches the rate obtained in the noninteracting-blip approximation (NIBA) of the dynamical approach [cf. Eq. (21.82)].

In electron transfer, the function $P(E)$ represents the nuclear contribution to the rate. In a sense, $P(E)$ represents the thermally averaged nuclear Franck-Condon overlap integral [424, 108]. In single-charge tunneling, $P(E)$ includes the influence of the electromagnetic environment [431], as discussed below in Section 20.3.

20.2.2 General properties of the $P(E)$ function

For a biased system, the energy of the particle before and after the transition differs by $E = \hbar\epsilon$. Exactly this amount of energy must be absorbed into or emitted from the bath ($E > 0$ or $E < 0$) in order that the transfer can take place. The supply of the necessary energy exchanged with the environment is regulated by the function $P(E)$.

The function $P(E)$ is the Fourier transform of the correlation function $e^{-Q(t)}$. It has a number of properties holding for general spectral coupling $G(\omega)$. First, $P(E)$

is a real positive function of E. This is, because $X(t)$ is a real positive function of t, as follows from Eq. (18.47). Secondly, it is normalized as

$$\int_{-\infty}^{\infty} dE\, P(E) = e^{-Q(0)} = 1 , \tag{20.32}$$

and thirdly, as $X(t) = X(-t)$, it obeys the *principle of detailed balance*,

$$P(-E) = e^{-\beta E}\, P(E) . \tag{20.33}$$

Therefore, the backward rate $k^-(\epsilon) \equiv k^+(-\epsilon)$ is related to the forward rate $k^+(\epsilon)$ by

$$k^-(\epsilon) = e^{-\beta\hbar\epsilon}\, k^+(\epsilon) . \tag{20.34}$$

Formally, detailed balance is an outcome of the analytic properties of the function $Q(z)$ in the strip $0 \geqslant \operatorname{Im} z > -\hbar\beta$, and the reflection property (18.43). These features show that $P(E)|_{E>0}$ is the probability that the system transfers the amount of energy E to the reservoir during the tunneling event, and $P(-E)|_{E>0}$ is the probability that the reservoir transfers the energy E from the reservoir to the system [432].

Because of the inherent property (20.33), $P(E)$ can generally be written as [433]

$$P(\hbar\omega) = \frac{D(|\omega|)}{\hbar\omega_c}\, f(-\omega) . \tag{20.35}$$

Here $f(\omega)$ is the Fermi function, and $D(\omega) = D(-\omega)$ can be regarded as a kind of a spectral density of fermionic quasiparticles being symmetric about the Fermi frequency $\epsilon_F = 0$ (cf. Sec. 20.3). Further, ω_c is the cutoff in the spectral bath coupling.

Probability for emission and absorption of energy

To substantiate the meaning of $P(E)$, we rely on an old technique originally introduced by Lax [397] in connection with optical line shapes, and employed in the ET theory by Jortner [434] (see also Refs. [424, 431, 157]).

Consider first the case where the thermal reservoir consists of a single oscillator with eigenfrequency ω_0. The respective spectral density is

$$G(\omega) = \lambda_0^2\, \delta(\omega - \omega_0) . \tag{20.36}$$

Assume that the oscillator is in the ground state. Then there is $Q(z) = \rho[\,1 - e^{-i\omega_0 z}\,]$, where $\rho = (\lambda_0/\omega_0)^2$. Upon expanding $e^{-Q(z)}$ in powers of $e^{-i\omega_0 z}$, then the integral expression (20.30) for $P(E)$ takes the form of an infinite series of delta functions,

$$P(E) = \sum_{m=0}^{\infty} p_m\, \delta(E - m\hbar\omega_0) , \qquad p_m = \frac{\rho^m}{m!}\, e^{-\rho} . \tag{20.37}$$

The quantities $p_1, p_2, \cdots, p_m$ are the probabilities that the oscillator absorbs $1, 2, \cdots, m$ quanta of energy $\hbar\omega_0$. The probabilities $\{p_m\}$ form a normalized Poisson distribution with mean number ρ and mean energy $\langle E \rangle = \rho\hbar\omega_0$ of absorbed quanta.

At finite T, the bath oscillator can not only absorb but can also emit quanta. The mean numbers of absorbed and emitted bosons are $\rho_a = (1 + n)\rho$, and $\rho_e = n\rho$, where $n = 1/[e^{\beta\hbar\omega_0} - 1]$ is the Bose distribution function. The quantities ρ_a and ρ_e are usually referred to as Huang-Rhys factors [397]. With the Poissonian probabilites $p_{m,a} = \rho_a^m e^{-\rho_a}/m!$ for absorption of m quanta and $p_{\ell,e} = \rho_e^\ell e^{-\rho_e}/\ell!$ for emission of ℓ quanta, the probability for exchange of energy E of the quantum oscillator in thermal equilibrium with the TSS takes in generalization of Eq. (20.37) the form

$$P(E) = e^{-(\rho_e + \rho_a)} \sum_{\ell,m} \frac{\rho_e^\ell \rho_a^m}{\ell!\, m!}\, \delta\left[E - (m - \ell)\hbar\omega_0\right] . \tag{20.38}$$

With $k = m - \ell$ as summation variable, the sum over ℓ turns into the ascending series of the modified Bessel functions $I_k(z)$ [90]. Putting $\mu = \hbar\beta\omega_0/2$, one readily obtains

$$P(E) = e^{-\rho\coth\mu} \sum_{k=-\infty}^{\infty} I_k(\rho/\sinh\mu)\, e^{k\mu}\, \delta(E - k\hbar\omega_0) . \tag{20.39}$$

The very same expression for $P(E)$ is found, when we start out from the representation (20.31). First, the expression (18.47) with the (20.36) for $G(\omega)$ yields

$$e^{-X(t)} = e^{-\rho\coth\mu}\, e^{\rho\cos(\omega_0 t)/\sinh\mu} = e^{-\rho\coth\mu} \sum_{k=-\infty}^{\infty} I_k(\rho/\sinh\mu)\, e^{-i k\omega_0 t} . \tag{20.40}$$

Here we have utilized that the second exponential is the generating function of the modified Bessel functions [90]. The subsequent time integration provides the δ-constraint for the energy and thus the expression (20.39) in total. With the distribution (20.38), the mean energy takes the balance $\langle E \rangle = (\rho_a - \rho_e)\hbar\omega_0 = \rho\,\hbar\omega_0$.

It is straightforward to generalize the discussion to two and three, and finally to infinitely many reservoir oscillators. This confirms that $P(E)$ is the probability for exchange of energy with the environment resulting from uncorrelated exchange of bosons between the reservoir and the TSS in compliance with detailed balance.

Moments, cumulants and generating functions

Most interestingly, the generating function for the moments of E with distribution $P(E)$ is just the function $\mathcal{Z}(t) = e^{-Q(t)}$. This can be seen as follows. First, we use the relation $E^n e^{i Et/\hbar} = (-i\hbar)^n d^n e^{i Et/\hbar}/dt^n$ and subsequently transfer the differentiations in the integral (20.30) upon partial integration to the function $e^{-Q(t)}$. The remaining E-integral is just $2\pi\hbar\,\delta(t)$, and thus

$$\langle E^n \rangle = (i\hbar)^n d^n e^{-Q(t)}/dt^n\big|_{t=0} . \tag{20.41}$$

Next, $\ln\mathcal{Z}(t) = -Q(t)$ is the generating function of the cumulants of E,

$$\langle E^n \rangle_c = -(i\hbar)^n d^n Q(t)/dt^n\big|_{t=0} . \tag{20.42}$$

Eq. (20.42) covers the whole statistics of the energy transfer in the nonadiabatic limit. The balance for the mean energy absorbed and emitted by the reservoir is

$$\langle E \rangle \equiv \int_{-\infty}^{\infty} dE \, E P(E) \; = \; \langle E \rangle_{\text{abs}} - \langle E \rangle_{\text{em}} \, . \tag{20.43}$$

With the spectral representations (18.42) and (20.4) of $Q(z)$ and of Λ_{cl} one finds that the net energy absorbed by the bath is the classical reorganization energy,

$$\langle E \rangle \; = \; -i\hbar \, dQ(t)/dt|_{t=0} \; = \; \Lambda_{\text{cl}} \, . \tag{20.44}$$

In addition, the mean spread of the net absorbed energy is

$$\langle \delta E^2 \rangle \equiv \langle (E - \langle E \rangle)^2 \rangle \; = \; \hbar^2 \, d^2 Q(t)/dt^2|_{t=0} \, . \tag{20.45}$$

This expression is generally valid. In the classical regime, it turns into

$$\langle \delta E^2 \rangle = 2k_{\text{B}} T \, \Lambda_{\text{cl}} \, . \tag{20.46}$$

Other features

Often in nature, a system is influenced by a number of different dissipative mechanisms connected with different bath correlation functions $Q_1(t), \cdots, Q_n(t)$. Then the correlator $e^{-Q(z)}$ is factorized into contributions $e^{-Q_j(z)}$ associated with partial probabilities $P_j(E)$, $j = 1, \cdots, n$. The function $P(E)$ gathering up the entirety of energy exchange processes is a convolution of the partial probabilities $P_j(E)$ [434],

$$P(E) \; = \; \int_{-\infty}^{\infty} dE_1 \cdots dE_{n-1} \, P_1(E - E_1) P_2(E_1 - E_2) \cdots P_n(E_{n-1}) \, . \tag{20.47}$$

This is written in the representation (20.35) for the allocation $Q(t) = Q_1(t) + Q_2(t)$ as

$$P(\hbar\omega) = \frac{D(|\omega|)}{\hbar\omega_{\text{c}}} f(-\omega) = \int_{-\infty}^{\infty} d\omega' \frac{D_1(|\omega - \omega'|) D_2(|\omega'|)}{(\hbar\omega_{\text{c}})^2} f(\omega' - \omega) f(-\omega') \, , \tag{20.48}$$

This relation will be used in connection with single charge tunneling in Sec. 20.3.

In numerical computation of $P(E)$, the calculation of $P(E)$ for $T = 0$ and E near zero is difficult, as $Q_0(t) \equiv Q_{T=0}(t)$ depends algebraically on t for $s \neq 1$, and logarithmically on t for $s = 1$. The computation is facilitated with an integral equation for $P_0(E)$. Following Ref. [435], we start out from the relation

$$\frac{d}{dt} e^{-Q_0(t)} \; = \; -i \int_0^{\infty} d\omega \, \frac{G(\omega)}{\omega} e^{-i\omega t} e^{-Q_0(t)} \, . \tag{20.49}$$

Taking the Fourier transform and integrating by parts the resulting integral on the l.h.s., we obtain for $P_0(E)$ the integral equation [424, b] (see also Refs. [436, 157]),

$$E \, P_0(E) \; - \; \int_0^{E} dE' \, \frac{G[(E - E')/\hbar]}{(E - E')/\hbar} \, P_0(E') \, . \tag{20.50}$$

Numerical integration of Eq. (20.50) yields $P_0(E)$ up to a multiplicative constant, the starting value $P_0(0)$. Subsequently, the constant can be fixed by the normalization condition (20.32). This route avoids calculation of $P_0(E)$ via Eq. (20.30).

At finite temperature, we write $Q_1(t) = Q(t) - Q_0(t)$. The direct integration of Eq. (20.30) for $P_1(E)$ is without difficulties, since $Q_1(t)$ is exponentially cut off. Finally, $P(E)$ is found as a convolution of $P_0(E)$ and $P_1(E)$ [cf. Eq. (20.47)].

Steepest descent

Consider next the integral (20.30) in steepest descent approximation (SDA). Writing

$$R(t) = Q(t) - i\,Et/\hbar\,, \qquad (20.51)$$

the SDA formula with the saddle point at $t^* = -i\,\tau_{st}$, where $0 < \tau_{st} \leqslant \hbar\beta/2$, is

$$P(E) = \frac{1}{\hbar\sqrt{2\pi R''(t^*)}}\,e^{-\beta F_{qm}} \qquad \text{with} \qquad F_{qm} = R(t^*)/\beta\,. \qquad (20.52)$$

Here, F_{qm} and $R''(t^*)$ are positive real, F_{qm} is the "quantum activation free energy" [437] of the donor state, and $R''(t^*)$ accounts for quantum effects in the prefactor. The expression (20.52) is formal. In general, it is not possible to invert the condition $R'(t^*) = 0$ to find t^* in analytic form. Various cases in which the formula (20.52) is justified and explicit expressions are available, are discussed in the sequel.

Classical regime

In the classical regime $\beta\hbar\omega_c \ll 1$, the short-time expansion of $Q(t)$ is primarily determined by the classical reorganization energy Λ_{cl} defined in Eq. (20.4),

$$Q_{cl}(t) = i\,\Lambda_{cl}t/\hbar + (\Lambda_{cl}/\beta)\,t^2/\hbar^2 + \mathcal{O}(t^3)\,. \qquad (20.53)$$

With this, the $P(E)$-function resulting from the SDA formula (20.52) emerges as

$$P_{cl}(E) = \sqrt{\beta/4\pi\Lambda_{cl}}\,e^{-\beta F_D^*} \qquad \text{with} \qquad F_D^* = (E - \Lambda_{cl})^2/4\Lambda_{cl}\,. \qquad (20.54)$$

This expression is in correspondence with the classical rate of Marcus, Eq. (20.22).

20.2.3 The spectral probability for absorption at $T = 0$

In the limit $T \to 0$, the pair interaction (18.45) takes the analytic form

$$\begin{aligned}
Q_{T=0}(z) &= 2\delta_s\,[\Gamma(s)/(s-1)]\,(\omega_c/\omega_{ph})^{s-1}[1 - (1 + i\,\omega_c z)^{1-s}]\,,\\
Q_{T=0}(z) &= 2K\ln(1 + i\,\omega_c z)\,, \qquad \text{for} \qquad s = 1\,.
\end{aligned} \qquad (20.55)$$

Because $Q_{T=0}(z)$ is analytic in the halfplane $\text{Im}(z) < 0$, the integral (20.30) directly yields $P(-|E|) = 0$. This approves that any bath at $T = 0$ can not release energy.

With the form (20.55), the saddle point condition $R'(t^*) = 0$ yields for $E > 0$ $t^* = -i\,\tau_{st}$ with $\tau_{st} = [\,(\Lambda_{cl}/E)^{1/s} - 1\,]/\omega_c$. Here Λ_{cl} is the solvation energy (20.5). The length τ_{st} is the width of the bounce (kink–anti-kink pair) in imaginary time. With this, the SDA formula (20.52) takes for $E > 0$ the form

$$P_{T=0}(E) = \frac{(\Lambda_{\text{cl}}/E)^{(1+s)/2s}}{\sqrt{2\pi s \Lambda_{\text{cl}} \hbar \omega_c}} \exp \left\{ -\frac{E}{\hbar \omega_c} + \frac{1}{s-1} \frac{\Lambda_{\text{cl}}}{\hbar \omega_c} \left[s \left(\frac{E}{\Lambda_{\text{cl}}} \right)^{\frac{s-1}{s}} - 1 \right] \right\} . \tag{20.56}$$

This expression is asymptotically exact in the parameter range

$$[2\delta_s \Gamma(s)]^{1/s} (E/\hbar \omega_{\text{ph}})^{(s-1)/s} = (E/\Lambda_{\text{cl}})^{(s-1)/s} \Lambda_{\text{cl}}/\hbar \omega_c \gg (s+2)(s+1)/8s . \tag{20.57}$$

which precisely is the regime of multi-phonon absorption into the bath. It roughly corresponds to the regime $E \lesssim \hbar \omega_{\text{ph}} \delta_s^{1/(1-s)}$ in the sub-Ohmic case, $E \gtrsim \hbar \omega_{\text{ph}}/\delta_s^{1/(s-1)}$ in the super-Ohmic case, and to $\delta_1 \equiv K \gtrsim 1$ in the Ohmic case.

The probability for transfer of energy to the reservoir is maximal at $E = E_{\text{max}}$, where E_{max} obeys the transcendental equation

$$[1 - (\Lambda_{\text{cl}}/E_{\text{max}})^{1/s}] E_{\text{max}} + [(s+1)/2s] \hbar \omega_c = 0 . \tag{20.58}$$

The expressions (20.56) – (20.58) hold for general s. The steepness of the ascending flank for E well below E_{max} depends on the particular value of s.

For $s < 1$, the expression (20.56) is valid down to $E = 0$. In the limit $\omega_c \to \infty$, E_{max} becomes independent of ω_c and turns with the expression (20.5) for Λ_{cl} into

$$E_{\text{max}} = [2s/(1+s)]^{s/(1-s)} [2\delta_s \Gamma(s)]^{1/(1-s)} \hbar \omega_{\text{ph}} . \tag{20.59}$$

Because of the high density of $G(\omega)$ at low ω, elastic tunneling is absent, and the probability of the TSS to lose only a small amount of energy is exponentially small,

$$P_{T=0}(E \to 0) \propto E^{-\frac{1+s}{2s}} \exp \left(-\frac{s}{(1-s)} \frac{\Lambda_{\text{cl}}}{\hbar \omega_c} \left(\frac{\Lambda_{\text{cl}}}{E} \right)^{(1-s)/s} \right) . \tag{20.60}$$

Therefore, barrier crossing of a symmetric system at zero temperature is fully suppressed [with the exception of the very weak damping regime (18.36)].

The descending flank of $P_{T=0}(E)$ for E above E_{max} is due to multi-phonon emission. These processes gradually die out as E is increased further, so that the steepest descent formula (20.56) ceases to be valid. For $E \gg E_{\text{max}}$, the influence of the environment is only weak and therefore the probability for energy absorption into the reservoir is determined by one-phonon exchange. Putting $e^{-Q(t)} \approx 1 - Q(t)$ and using the representation (18.42) for $Q(t)$, we obtain[2] for $E \gg E_{\text{max}}$ and $s < 1$

$$P_{T=0}(E) = \frac{1}{E} \frac{G(E/\hbar)}{E/\hbar} = \frac{1}{\Gamma(s)} \frac{\Lambda_{\text{cl}}}{(\hbar \omega_c)^2} \left(\frac{\hbar \omega_c}{E} \right)^{2-s} e^{-E/\hbar \omega_c} . \tag{20.61}$$

Hence, apart from a factor $\hbar/E^2$, the one-phonon contribution to $P(E)$ directly reproduces the spectral density of the system-reservoir coupling.

The *super*-Ohmic regime $s > 2$ is opposite. Now the low-energy tail of $P_{T=0}(E)$ is shaped by the one-phonon process, Eq. (20.61). With increasing E, multi-phonon processes gradually join in and form the ascending flank of $P_{T=0}(E)$. For $E \gtrsim E_{\text{max}}$,

---

[2] At finite T, the one-phonon absorption process has an additional factor $1/[1 - \exp(-\beta E)]$.

the asymptotic multi-phonon expression (20.56) applies. The maximum of $P_{T=0}(E)$ is situated at $E_{max} = \Lambda_{cl} - \frac{1}{2}(1 + s)\hbar\omega_c$. In the limit $E \to \infty$, $P_{T=0}(E)$ drops to zero faster than algebraically due to the exponential cutoff in the spectral density $G(\omega)$.

Consider next the Ohmic case. Taking in Eq. (20.56) the limit $s \to 1$, we obtain

$$P_{T=0}(E) = \Theta(E) \sqrt{\frac{K}{\pi}} \frac{e^{2K}}{(2K)^{2K}} \frac{e^{-E/\hbar\omega_c}}{E} \left(\frac{E}{\hbar\omega_c}\right)^{2K}. \tag{20.62}$$

For $K > \frac{1}{2}$, the probability for absorption of energy by the reservoir shows a maximum at $E_{max} = (2K - 1)\hbar\omega_c = \Lambda_{cl} - \hbar\omega_c$. Well below the maximum, the function $P_{T=0}(E)$ has power law form $\propto E^{2K-1}$.

Interestingly, for the Ohmic correlation function $Q_{T=0}(t) = 2K \ln(1 + i\omega_c t)$ [cf. Eq. (18.51) for $T = 0$], the integral (20.30) can be done in analytic form, yielding

$$P_{T=0}(E) = \frac{1}{\Gamma(2K)} \frac{e^{-E/\hbar\omega_c}}{E} \left(\frac{E}{\hbar\omega_c}\right)^{2K}, \qquad P_{T=0}(-E) = 0. \tag{20.63}$$

This expression reduces to the steepest descent expression (20.62) if we replace $\Gamma(2K)$ by Stirling's asymptotic formula, which requires large values of K.

20.2.4 Crossover from quantum to classical behavior

At finite T and E near zero, the saddle point in the integral (20.29) is near $\tau = \hbar\beta/2$. Hence the integral representation (20.31), in which the saddle point is near $t = 0$, is appropriate. The expansion of $X(t)$ about $t = 0$ to second order in t yields

$$X(t) = \Lambda_1 \beta/4 + (\Lambda_2/\beta) t^2/\hbar^2, \tag{20.64}$$

where Λ_1 and Λ_2 are quantum mechanical reorganization energies,

$$\Lambda_1 = \frac{4}{\beta} \int_0^\infty d\omega \frac{G(\omega)}{\omega^2} \frac{\cosh(\frac{1}{2}\beta\hbar\omega) - 1}{\sinh(\frac{1}{2}\beta\hbar\omega)}, \qquad \Lambda_2 = \frac{\hbar^2\beta}{2} \int_0^\infty d\omega \frac{G(\omega)}{\sinh(\frac{1}{2}\beta\hbar\omega)}. \tag{20.65}$$

For the algebraic spectral density $G(\omega) = 2\delta_s \omega_{ph}^{1-s} \omega^s e^{-\omega/\omega_c}$ one obtains

$$\Lambda_1 = 4\left[2B_s - \lambda_1(s,\kappa)\left(\frac{T}{T'_{ph}}\right)^{s-1}\right] k_B T, \qquad \Lambda_2 = \lambda_2(s,\kappa)\left(\frac{T}{T'_{ph}}\right)^{s-1} \frac{k_B T}{4}, \tag{20.66}$$

where B_s and T'_{ph} are given in Eqs. (18.48) and (18.49), and where

$$\begin{aligned}
\lambda_1(s,\kappa) &= 2\Gamma(s-1)\left[\zeta(s-1, \tfrac{1}{2}+\kappa) - \zeta(s-1, 1+\kappa)\right]/\pi^2, \\
\lambda_2(s,\kappa) &= 4\Gamma(s+1)\zeta(s+1, \tfrac{1}{2}+\kappa)/\pi^2.
\end{aligned} \tag{20.67}$$

According to the representations (20.65), there holds $\Lambda_1 \geqslant \Lambda_2$, and both Λ_1 and Λ_2 turn into the classical reorganization energy Λ_{cl}, Eq. (20.4), in the classical limit $\kappa \equiv 1/\hbar\beta\omega_c \to \infty$. In the opposite quantum limit $\kappa \to 0$, the coefficients read

$$\lambda_1(s,0) = \Gamma(s-1)\left[2^s - 4\right]\zeta(s-1)/\pi^2 ,$$
$$\lambda_2(s,0) = 4\Gamma(s+1)\left[2^{s+1} - 1\right]\zeta(s+1)/\pi^2 . \tag{20.68}$$

With the expression (20.64) in Eq. (20.31), the probability for exchange of zero energy is found as (cf. Eq. (4.4) in Ref. [438])

$$P(0) = \sqrt{\beta/4\pi\Lambda_2}\, e^{-\beta\Lambda_1/4} . \tag{20.69}$$

In the bounce picture discussed in Subsec. 17.4.1, the formula (20.69) has the following interpretation. The bounce consists of a weakly bound instanton–anti-instanton pair with imaginary-time distance $\hbar\beta/2$, which is just half of the bounce period. The exponent $\beta\Lambda_1/4$ is the pair interaction

$$\beta\Lambda_1/4 = \mathcal{W}(\hbar\beta/2) = Q(-i\hbar\beta/2) = X(0) , \tag{20.70}$$

and Λ_2/β is a measure for the stiffness of the bounce width, and thus is directly connected with the eigenvalue of the breathing mode.

The leading correction to the formula (20.69) arises from the order t^4 of $X(t)$. This contribution is small when the condition

$$32\delta_s \left(\frac{T}{T_{\mathrm{ph}}}\right)^{s-1} \frac{\left[\Gamma(s+1)\,\zeta(s+1, \tfrac{1}{2}+\kappa)\right]^2}{\Gamma(s+3)\,\zeta(s+3, \tfrac{1}{2}+\kappa)} \gg 1 \tag{20.71}$$

is satisfied. Physically, in this regime multi-phonon processes govern the energy exchange. In the Ohmic case, the condition (20.71) is reduced to $K \gg 1$. In the super-Ohmic case, it is met for temperatures well above T_{ph}. For a sub-Ohmic bath, the condition (20.71) holds at temperatures T fairly below T_{ph}.

When the energy E exchanged with the environment is in the range

$$E \ll (1 + 2\kappa)\Lambda_2 , \tag{20.72}$$

the saddle point τ_{st} is still near to $\hbar\beta/2$, and the expression (20.64) is still valid. With this, the integral (20.31) yields as a generalisation of the expression (20.69)

$$P(E) = \sqrt{\beta/4\pi\Lambda_2}\, e^{-\beta F_{\mathrm{qm}}} \qquad \text{with} \qquad F_{\mathrm{qm}}(E) = \Lambda_1/4 - E/2 + E^2/4\Lambda_2 . \tag{20.73}$$

This expression is the extension of the SDA formula (20.54) beyond the classical regime. The quantities Λ_1 and Λ_2 are quantum mechanical reorganization energies. Corrections to the formula (20.73) are small if the conditions (20.71) and (20.72) are met. The energy range (20.72) sensitively depends on temperature.

Nuclear tunneling is disclosed by the specific temperature dependencies of the quantum activation energies Λ_1 and Λ_2.

20.2.5 The Ohmic case

Consider first the scaling regime $k_{\mathrm{B}}T$, $\hbar|\epsilon| \ll \hbar\omega_{\mathrm{c}}$, in which the pair interactions are $Q_{\mathrm{sc}}(t)$ and $X_{\mathrm{sc}}(t)$ given in Eqs. (18.53) and (18.54), respectively. With these functions, both the integrals (20.30) and (20.31) can be calculated in analytic form. The resulting analytic expression holding for general K, T, and ϵ is [440, 441, 442]

$$k^+(T,\epsilon) = \frac{1}{4}\frac{\Delta^2}{\omega_c}\left(\frac{\hbar\omega_c}{2\pi k_B T}\right)^{1-2K}\frac{|\Gamma(K+i\,\hbar\epsilon/2\pi k_B T)|^2}{\Gamma(2K)}\,e^{\hbar\epsilon/2k_B T}\,. \qquad (20.74)$$

For $K < 1$, one may combine Δ and ω_c in the effective amplitude Δ_{eff} defined in Eq. (18.35). In the universality limit (18.56), there is no other dependence on ω_c,

$$k^+(T,\epsilon) = \Delta_{\text{eff}}\frac{\sin(\pi K)}{2\pi}\left(\frac{\hbar\Delta_{\text{eff}}}{2\pi k_B T}\right)^{1-2K}|\Gamma(K+i\,\hbar\epsilon/2\pi k_B T)|^2\,e^{\hbar\epsilon/2k_B T}\,. \qquad (20.75)$$

With the relation

$$\Gamma(x+i\,y)\,\Gamma(x-i\,y) \approx \frac{1}{x^2+y^2}\frac{\pi y}{\sinh(\pi y)}\,, \qquad \text{for} \quad |x|\lesssim 0.1\,,\; y\lesssim 3\,, \qquad (20.76)$$

the rate (20.75) takes for $K\lesssim 0.1$ and moderate to high T, $\hbar\epsilon/2\pi k_B T)\lesssim 3$, the form

$$k^+(T,\epsilon) = \left(\frac{2\pi k_B T}{\hbar\Delta_{\text{eff}}}\right)^{2K}\frac{\pi K\Delta_{\text{eff}}^2}{\left[(2\pi K k_B T/\hbar)^2+\epsilon^2\right]}\frac{\epsilon}{\left[1-\exp(-\hbar\epsilon/k_B T)\right]}\,. \qquad (20.77)$$

For $T = 0$, one finds for general K, ω_c, and ϵ, upon drawing on the result (20.63),

$$k^+(0,\epsilon) = \frac{\pi}{2\Gamma(2K)}\frac{\Delta^2}{\omega_c}\left(\frac{\epsilon}{\omega_c}\right)^{2K-1}e^{-\epsilon/\omega_c}\,, \qquad k^-(0,\epsilon) = 0\,. \qquad (20.78)$$

Observe the change of sign in the power of ϵ^{2K-1} at $K = 1/2$.

Consider next the leading thermal enhancement at low T for large damping $K \gg 1$. With the leading thermal contribution $\Delta Q_{\text{sc}}(t) = (K/3)(\pi/\hbar\beta)^2 t^2$ to $Q_{\text{sc}}(t)$, the rate can formally be written as $k^+(T,\epsilon) = \exp[\,K(\pi^2/3)\partial^2/(\partial\hbar\beta\epsilon)^2\,]\,k^+(0,\epsilon)$. Keeping in each differentation the term with leading order in K, one obtains

$$k^+(T,\epsilon) = k^+(0,\epsilon)\exp\left\{\frac{\pi^2}{3}K\left(2K\frac{k_B T}{\hbar\epsilon}\right)^2\right\}. \qquad (20.79)$$

The curly bracket corresponds to the enhancement function $\mathcal{A}(T)$ for a metastable potential, Eq. (17.14) with (17.18). Upon equating $q_b\tau_w$ with $q_0\tau_s$, where $\tau_s = 2K/\epsilon$ is the width of the instanton–anti-instanton pair or dipole length [cf. (17.69)], they match exactly. Again the T^2 law in Eq. (20.79) is a signature of Ohmic dissipation.

In the zero bias limit $\epsilon \to 0$, the expressions (20.74) and (20.75) turn into

$$k^\pm(T,0) = \frac{\sqrt{\pi}\,\Gamma(K)}{4\Gamma(K+\frac{1}{2})}\frac{\Delta^2}{\omega_c}\left(\frac{\pi k_B T}{\hbar\omega_c}\right)^{2K-1} = \frac{\Delta_{\text{eff}}}{2}\frac{\Gamma(K)}{\Gamma(1-K)}\left(\frac{\hbar\Delta_{\text{eff}}}{2\pi k_B T}\right)^{1-2K}\,. \qquad (20.80)$$

Most interestingly, the rate increases as temperature is decreased if $K < \frac{1}{2}$, This striking behavior was originally discovered by Kondo. Later on, it was confirmed in experiments on diffusion of charged interstitials in metals (see Ref. [393, 138], and references therein). The increase of the rate with decreasing temperature is a signature that the dynamics becomes coherent at lower temperature. The incoherent-coherent transition is discussed in Subsection 22.2.2, and the relation with the diffusion coefficient is discussed in Subsection 27.1.

In the regime $0 < K < 1$, the above nonadiabatic or Golden Rule expressions are valid if either the temperature or the bias exceeds a certain value (cf. Subsec. 22.1.1). For $K < \frac{1}{2}$, low temperature and weak bias, the dynamics may actually be coherent. Therefore, it cannot be simply described in terms of a single rate. For $\frac{1}{2} < K < 1$, the Golden Rule approximation may break down at low T. In contrast, in the range $K > 1$ the rate expression (20.74) holds true down to zero temperature.

Next, consider the crossover from quantum to classical in the bias range (20.72). The quantum activation energies Λ_1 and Λ_2 are found from Eq. (20.65) or from (20.66) to read ($\kappa = 1/\hbar\beta\omega_c$)

$$
\begin{aligned}
\Lambda_1 &= 8K\,k_B T \ln[\hbar\beta\omega_c \Gamma^2(1+\kappa)/\Gamma^2(\tfrac{1}{2}+\kappa)]\,, \\
\Lambda_2 &= 2K\,k_B T\,\psi'(\tfrac{1}{2}+\kappa)\,,
\end{aligned}
\tag{20.81}
$$

With these, the rate formula (20.28) with the expression (20.73) turns into

$$
k^+(T,\epsilon) = \frac{\hbar\beta\Delta^2}{4(\beta\hbar\omega_c)^{2K}} \sqrt{\frac{\pi\,e^{\beta\hbar\epsilon}}{2K\psi'(\tfrac{1}{2}+\kappa)}} \frac{\Gamma^{4K}(\tfrac{1}{2}+\kappa)}{\Gamma^{4K}(1+\kappa)} \exp\left\{ -\frac{(\beta\hbar\epsilon)^2}{8K\psi'(\tfrac{1}{2}+\kappa)} \right\}.
\tag{20.82}
$$

That simplifies in the quantum regime $k_B T \ll \hbar\omega_c$ ($\kappa \ll 1$) to the form

$$
k^+(T,\epsilon) = \frac{\sqrt{\pi}}{4\sqrt{K}} \frac{\Delta^2}{\omega_c} \left(\frac{\pi}{\beta\hbar\omega_c}\right)^{2K-1} \exp\left\{ \frac{\beta\hbar\epsilon}{2} - \frac{1}{K}\left(\frac{\beta\hbar\epsilon}{2\pi}\right)^2 \right\},
\tag{20.83}
$$

which coincides for $\epsilon \ll 2\pi^2 K k_B T/\hbar$ and $K \gg 1$, with the general expression (20.74). In the classical regime, $k_B T \gg \hbar\omega_c$, Eq. (20.82) reduces to the Marcus rate (20.22).

Thus, the rate expression (20.82) covers the crossover between quantum and classical behavior. The above rate expressions are of direct interest for a large variety of tunneling problems involving an Ohmic bath coupling, e.g., interstitials in metals and charges in resistive electromagnetic environments. Finally, we remark that the theoretical interest in Ohmic dissipation is based to a large extent on the power-law behaviors of the transfer rate discussed in this subsection.

20.2.6 Exact nonadiabatic rates for $K = \frac{1}{2}$ and $K = 1$

For the special value $K = \frac{1}{2}$, the Golden Rule rate (20.28) with (20.29) can be calculated in analytic form for arbitrary $\kappa = 1/\beta\hbar\omega_c$, any bias, and any temperature. Inserting for $Q(t)$ the form (18.51) and closing the contour in the half-plane Im $t > 0$, the simple poles of $\Gamma(\kappa + it/\hbar\beta)$ located at $t = i(n+\kappa)\hbar\beta$, where $n = 0, 1, 2, \cdots$, are encircled. The residua of these poles are proportional to $\Gamma(n+2\kappa+1)$. By writing this function as an Euler integral [89], all the residua can summed. Lastly, the remaining integral can be done in analytic form. The resulting expression is [443]

$$k^+(T,\epsilon; K = \tfrac{1}{2}) = \frac{\Gamma(\tfrac{1}{2} + \kappa)}{\sqrt{\pi}\Gamma(1+\kappa)} \frac{e^{\beta\hbar\epsilon/2}}{[\cosh(\beta\hbar\epsilon/2)]^{1+2\kappa}} \frac{\gamma}{2}, \qquad (20.84)$$

where $\gamma \equiv \Delta_{\text{eff}}(K = \tfrac{1}{2}) = \pi\Delta^2/2\omega_c$. One directly sees that the expression (20.84) satisfies the detailed balance condition (20.34).

For the symmetric system, the rate (20.84) simplifies to the form

$$k^+(T,0; K = \tfrac{1}{2}) = k^-(T,0; K = \tfrac{1}{2}) = \frac{\Gamma(\tfrac{1}{2} + \kappa)}{\sqrt{\pi}\Gamma(1+\kappa)} \frac{\gamma}{2}. \qquad (20.85)$$

In the quantum regime at low T, $\kappa \ll 1$, the rate (20.85) is temperature-independent, $k_+ = k_- = \gamma/2$. This form agrees with the expression (20.80) at $K = \tfrac{1}{2}$.

In the opposite classical regime, $\kappa \gg 1$, there holds

$$k^+(T,0; K = \tfrac{1}{2}) = \frac{\Delta^2}{4\,\omega_c} \left(\frac{\pi\hbar\omega_c}{k_B T}\right)^{1/2} \propto \frac{1}{\sqrt{T}}. \qquad (20.86)$$

As the solvation energy is $\Lambda_{\text{cl}} = \hbar\omega_c$, this expression matches the prefactor of the Marcus rate (20.22).

For fixed T, the rate (20.84) is maximal at the bias $\epsilon = \epsilon_{\text{max}}$, where

$$\hbar\epsilon_{\text{max}} = k_B T \ln(1 + \hbar\omega_c/k_B T). \qquad (20.87)$$

In the classical limit, ϵ_{max} approaches the solvation energy ω_c, and the rate simplifies to the Marcus form (20.22). As T is lowered, the bias ϵ_{max} is lowered too. As T approaches zero, ϵ_{max} drops to zero, and the forward rate becomes $k^+(0,\epsilon) = e^{-\epsilon/\omega_c}\gamma$.

Having calculated the rate for $K = 1/2$, one can also calculate the rate for $K = 1$ in analytic form by use of the general relation resulting from the property (20.28),

$$k^+(\epsilon; 2K) = \frac{2}{\pi\Delta^2} \int_{-\infty}^{\infty} d\epsilon' \, k^+(\epsilon'; K)k^+(\epsilon - \epsilon'; K), \qquad (20.88)$$

Inserting the form (20.84) in Eq. (20.88), one finds

$$k^+(T,\epsilon; K = 1) = \frac{2^{2\kappa}}{\hbar\beta} \frac{\Delta^2}{2\,\omega_c^2} \frac{\Gamma^2(\tfrac{1}{2} + \kappa)}{\Gamma^2(1+\kappa)} \frac{\text{Re}\, \mathcal{Q}_{2\kappa}[\coth(\beta\hbar\epsilon/2)]}{[\sinh(\beta\hbar\epsilon/2)]^{1+2\kappa}} e^{\beta\hbar\epsilon/2}, \qquad (20.89)$$

where $\mathcal{Q}_\nu(z)$ is the Legendre function of the second kind [89],

$$\mathcal{Q}_\nu(z) = \frac{\Gamma(\nu+1)\Gamma(\tfrac{1}{2})}{2^{\nu+1}\Gamma(\nu+\tfrac{3}{2})} z^{-1-\nu} \,_2F_1\left[\tfrac{1}{2}(\nu + 2), \tfrac{1}{2}(\nu + 1); \tfrac{1}{2}(2\nu + 3); 1/z^2\right]. \qquad (20.90)$$

The rate (20.89) satisfies detailed balance, $k_-(\epsilon) = e^{-\beta\hbar\epsilon}k_+(\epsilon)$, as follows with the relation (20.90). For zero bias, there results from the expression (20.89)

$$k^+(T,0; K = 1) = \frac{2^{2\kappa}}{\hbar\beta} \frac{\Delta^2}{4\,\omega_c^2} \frac{\Gamma^3(\tfrac{1}{2} + \kappa)}{\Gamma(1+\kappa)\Gamma(\tfrac{3}{2} + 2\kappa)}. \qquad (20.91)$$

In the limit $\kappa \to 0$ we then have $k^+(T, 0; K = 1) = (\Delta^2/4\omega_c^2)\, 2\pi k_B T/\hbar$, which matches the expression (20.80) for $K = 1$. In the opposite high-temperature limit, the familiar $T^{-1/2}$ Marcus law is recovered. At high temperatures, the rate expression (20.89) is symmetric in the bias around the maximum. This is at $\epsilon = \epsilon_{\text{max}}$, where $\epsilon_{\text{max}} = \Lambda_{\text{cl}}/\hbar = 2\omega_c$. With decreasing temperature, ϵ_{max} is gradually shifted from $\Lambda_{\text{cl}}/\hbar$ to $\Lambda_{\text{cl}}/2\hbar$, and the rate becomes asymmetric around this value.

20.2.7 The sub-Ohmic case $(0 < s < 1)$

As the sub-Ohmic case is distinguished by the high spectral density of the coupling at low frequencies, the dissipative effects become most noticeable at low T and weak bias. The forward rate for a biased system at zero temperature is given by

$$k^+(0, \epsilon) = (\pi\hbar/2)\,\Delta^2 P_{T=0}(\hbar\epsilon)\,. \tag{20.92}$$

The behavior of $P_{T=0}(\hbar\epsilon)$ has been discussed in Subsection 20.2.3.

Consider next the low-temperature corrections in the regime $k_B T \ll \hbar\epsilon$. When the temperature is raised above zero, the particle can also absorb energy. Thus we should expect that the ascending shoulder of $P(E)$ is shifted to lower energies. This phenomenon can easily be studied asymptotically. Since the form (20.55) cuts off the Fourier integral (20.30) for $P(E)$ at $|t| \gtrsim 1/(\omega_{ph}\delta_s^{1/(1-s)})$, the temperature-dependent term of $Q(t)$ in Eq. (18.45) may be expanded in powers of t^2,

$$\Delta Q(t) = 2\delta_s\Gamma(s+1)\zeta(s+1)\,(\hbar\beta\omega_{ph})^{-(s+1)}(\omega_{ph}t)^2 + \mathcal{O}[(\omega_{ph}t)^4]\,. \tag{20.93}$$

The leading temperature correction for $k_B T \ll E_{max}$ is given by the factor $\exp[-\Delta Q(t^*)]$ where $t^* = -i\tau_{st}(\epsilon) = -i\left[(\Lambda_{cl}/\hbar\epsilon)^{1/s} - 1\right]/\omega_c$ is the stationary point for $T = 0$. Since $\Delta Q(t^*)$ is negative, this term actually leads to an enhancement of the probability function and hence of the crossing rate. We readily find

$$k^+(T, \epsilon) = k^+(0, \epsilon)\exp\left\{2\delta_s\Gamma(s+1)\zeta(s+1)[\omega_{ph}\tau_{st}(\epsilon)]^2\left(\frac{T}{T_{ph}}\right)^{1+s}\right\}, \tag{20.94}$$

where we have put $\hbar\omega_{ph} = k_B T_{ph}$. Upon equating τ_{st} with the bounce width τ_W, one finds that the exponent in Eq. (20.94) coincides with the enhancement function (17.17) obtained for a general extended metastable potential. This function varies as T^{1+s}, which is in agreement with the conclusions reported in Section 17.2.

When s approaches one from below, the condition (20.57) reduces to the condition $K \gg 1$. In this limit and under condition $\hbar\epsilon \ll \Lambda_{cl}$, the expression (20.94) with (20.56) matches with the Ohmic result (20.79) with (20.78).

Finally, we turn to the discussion of the rate for a symmetric system. In the low-temperature regime determined by the condition (20.71) we may use the rate expression (20.69) with the activation energies (20.66). Thus we find

$$k^+(T, 0) = \frac{\Delta^2\,e^{2|B_s|}}{4\omega_{ph}}\sqrt{\frac{\pi}{a_s(\kappa)}}\left(\frac{T_{ph}}{T}\right)^{\frac{1+s}{2}}\exp\left\{-\frac{b_s(\kappa)}{4}\left(\frac{T_{ph}}{T}\right)^{1-s}\right\}, \tag{20.95}$$

where $a_s(\kappa)$ and $b_s(\kappa)$ are defined in Eq. (20.67). As T approaches zero, the rate vanishes with an essential singularity, as elastic tunneling dies out in this limit.

20.2.8 The super-Ohmic case $(s > 1)$

In the regime $s > 1$, we rely on $X(t) = X_1 - X_2(t)$ given in X_1 is given in Eqs. (18.48) and (18.50). In addition, we identify the cutoff ω_c with the Debye frequency ω_D.

Consider first the regime well below the Debye temperature, $T \ll T_D \equiv \hbar\omega_D/k_B$, or equivalently $\kappa \ll 1$. The adiabatic Huang-Rhys exponent X_1 results in a temperature-dependent renormalized tunneling amplitude,

$$\tilde{\Delta}(T) = \Delta\, e^{-X_1(T)/2} = \Delta_{\text{eff}}\, e^{-D_s(T)}\,, \quad \text{with} \quad \Delta_{\text{eff}} = \Delta\, e^{-B_s}\,. \tag{20.96}$$

Here, Δ_{eff} is the polaron-dressed tunneling amplitude at $T = 0$ given in Eq. (18.33). The Franck-Condon or Huang-Rhys factor $e^{-D_s(T)}$ with $D_s(T)$ given in (18.48) with (18.49) describes the reduction of the amplitude by thermal excitations of the polaron cloud. Here we restrict the attention to the most relevant regime

$$\hbar\Delta_{\text{eff}}/k_B \ll T'_{\text{ph}} \ll T_D\,. \tag{20.97}$$

For $T \ll T'_{\text{ph}}$, the dressed matrix element $\tilde{\Delta}$ is practically temperature-independent. With $\kappa = 0$ in Eq. (18.50), there results

$$X_2(t) = 2[\,\Gamma(s-1)/\pi^2\,]\,(T/T'_{\text{ph}})^{s-1}\,\mathrm{Re}\,\zeta(s-1,\tfrac{1}{2}+it/\hbar\beta)\,. \tag{20.98}$$

With these preliminaries, the forward tunneling rate is given by

$$k^+(T,\epsilon) = \frac{\tilde{\Delta}^2}{4}\, e^{\beta\hbar\epsilon/2} \int_{-\infty}^{\infty} dt\, e^{i\epsilon t}\, e^{X_2(t)}\,. \tag{20.99}$$

Expansion of the integrand in powers of $X_2(t)$ gives the multi-phonon series for the rate. The term linear in $X_2(t)$ is the one-phonon contribution,

$$k^+(T,\epsilon) = \frac{\pi}{2}\frac{\tilde{\Delta}^2}{\epsilon^2}\frac{G(\epsilon)}{1-\exp(-\beta\hbar\epsilon)} = \frac{1}{2\pi}\frac{\tilde{\Delta}^2}{\epsilon}\frac{(\hbar\epsilon/k_B T'_{\text{ph}})^{s-1}}{1-\exp(-\beta\hbar\epsilon)}\,. \tag{20.100}$$

With the backward rate $k^-(T,\epsilon) = e^{-\beta\hbar\epsilon}k^+(T,\epsilon)$ added, the one-phonon contribution to the relaxation rate reads

$$\gamma_r \equiv k^+ + k^- = (\pi\tilde{\Delta}^2/2\epsilon^2)\,G(\epsilon)\coth(\beta\hbar\epsilon/2)\,. \tag{20.101}$$

Thus, for nonzero bias and T above $\hbar\epsilon/k_B$, the one-phonon rate is proportional to T.

For $s < 3$, the one-phonon rate diverges in the limit $\epsilon \to 0$, while it remains finite for $s = 3$ in this limit,

$$k^+(T,0) = k^-(T,0) = (\hbar\tilde{\Delta}^2/2\pi k_B T'_{\text{ph}})\,T/T'_{\text{ph}}\,. \tag{20.102}$$

These rates vary in the regime $T \ll T'_{\text{ph}}$, in which $\tilde{\Delta} \approx \Delta_{\text{eff}}$, *linearly* with T [136].

For $s > 3$, the one-phonon exchange rate vanishes as $\epsilon \to 0$. Thus for a symmetric system, the leading contribution for $T \ll T'_{\text{ph}}$ is "two-phonon assisted" exchange,

$$k^+(T,0) = \frac{\pi\tilde{\Delta}^2}{8}\int_0^{\infty}\frac{d\omega}{\omega^4}\frac{G^2(\omega)}{\sinh^2(\tfrac{1}{2}\beta\hbar\omega)} = \frac{\Gamma(2s-3)\zeta(2s-4)}{2\pi^3}\frac{\hbar\tilde{\Delta}^2}{k_B T}\left(\frac{T}{T'_{\text{ph}}}\right)^{2s-2}\,. \tag{20.103}$$

In the range $T \ll T'_{\text{ph}}$, the rate varies as T^{2s-3}. This gives for the diffusion constant[3] in the small-polaron model [131] at $s = 5$ the familiar T^7 law [134] .

At this point, we should like to remark that the present treatment relies on the assumption that the dynamics is incoherent. However, for $T \ll T'_{\text{ph}}$, the dynamics is actually coherent and therefore the Golden Rule approach is inadequate. The relevant discussion within a dynamical approach is given in Section 22.3. There it turns out that the rate (20.101) is the rate describing relaxation of the incoherent part of $\langle \sigma_z \rangle_t$ towards the equilibrium value $\langle \sigma_z \rangle_\infty$ for $\epsilon \gg \tilde{\Delta}$ [cf. Eqs. (22.93), (22.92), and (22.97)]. Similar conclusions hold for the two- and multi-phonon rate. The discussion of the dynamics beyond the one-phonon process is given in Subsection 22.1.2.

For T of the order of T'_{ph} or larger, phonon processes of higher order contribute, and for $T \gg T'_{\text{ph}}$ the full multi-phonon series has to be taken into account. In this regime, the SDA formula (20.73) applies. With the expressions (20.66) with (20.67) for Λ_1 and Λ_2, one obtains

$$
k^+(T, \epsilon) = \frac{\hbar \Delta^2 e^{-2B_s}}{2 k_B T'_{\text{ph}}} \sqrt{\frac{\pi}{\lambda_2(s, \kappa)}} \left(\frac{T'_{\text{ph}}}{T} \right)^{(s+1)/2} \tag{20.104}
$$
$$
\times \ \exp \left\{ \frac{\hbar \epsilon}{2 k_B T} + \lambda_1(s, \kappa) \left(\frac{T}{T'_{\text{ph}}} \right)^{s-1} - \frac{1}{\lambda_2(s, \kappa)} \left(\frac{\hbar \epsilon}{k_B T'_{\text{ph}}} \right)^2 \left(\frac{T'_{\text{ph}}}{T} \right)^{s+1} \right\} .
$$

This expression virtually holds in the entire regime $T \gtrsim T'_{\text{ph}}$.[4] In the temperature regime $T'_{\text{ph}} \lesssim T \ll T_D$, in which $\kappa \equiv 1/\beta\hbar\omega_D \approx 0$, the coefficients $\lambda_1(s, 0)$ and $\lambda_2(s, 0)$, given in Eq. (20.68) are temperature independent. For $\epsilon = 0$, the rate varies as $k^+ \propto T^{-(s+1)/2} \exp[\lambda_1(s, 0)(T/T'_{\text{ph}})^{s-1}]$. Since $\lambda_1(s, 0) > 0$ for $s > 1$, the thermal polaronic effects lead to an exponential enhancement of the rate. The function $\lambda_1(s, 0)$ is regular at $s = 2$, whereas it is singular at $s = 1$. However, as the term B_s in Eq. (20.66) has the same singularity with opposite sign, the quantum activation energy Λ_1 is regular at $s = 1$.

At higher temperature, the dependence on κ becomes relevant. In the classical regime $\kappa \gg 1$, the expression (20.104) matches the classical rate (20.22). In this limit, the adiabatic dressing factor e^{-2B_s} is fully compensated by thermal increase of the rate. Thus, dressing of the tunneling amplitude has faded away. The regime $T \gg T_D$ has been studied for the small-polaron problem in Refs. [131, b] and [446].

In conclusion, the rate formula (20.104) covers the entire parameter regime in which multi-phonon exchange dominates the transition rate.

20.2.9 Incoherent defect tunneling in metals

The tunneling of a defect or interstitial in a metal at low temperature is predominantly influenced by the interaction with conduction electrons. At higher temperature, the

---

[3]In the incoherent tunneling regime, the diffusion constant is proportional to the tunneling rate [cf. Subsection 25.2.1].

[4]The conditions for the validity of the expression (20.104) are given in Eqs. (20.71) and (20.72).

coupling to acoustic phonons join in. The coupling to electrons leads to a decrease of the rate with increasing temperature for $K < 1/2$ and $\epsilon = 0$, as we may see from Eq. (20.80). The latter coupling causes phonon-assisted exponential enhancement of the rate at higher temperature, as described by Eq. (20.104). Therefore, the joint influences lead to a minimum of the rate as a function of temperature.

In order to study incoherent tunneling near the minimum of the rate we take as a basis the combined spectral density

$$G_{\mathrm{lf}}(\omega) = 2K\omega\, e^{-\omega/\omega_c} + 2\delta_3\omega_{\mathrm{ph}}^{-2}\,\omega^3\, e^{-\omega/\omega_D} \,. \tag{20.105}$$

Interestingly, for the spectral density (20.105), the nonadiabatic tunneling rate can be found in closed analytical form in the entire temperature regime.

Consider first the regime $T \ll T_{\mathrm{D}}$, $\hbar\omega_c/k_{\mathrm{B}}$. For $\kappa_{\mathrm{el}} = \kappa_{\mathrm{ph}} \approx 0$, we obtain from the general expression (18.50) for the pair interaction $X(t)$ the form

$$
\begin{aligned}
X(t) &= X_{\mathrm{el}}(t) + X_{\mathrm{ph}}(t) \,, \\
X_{\mathrm{el}}(t) &= 2K \ln\left[(\beta\hbar\omega_c/\pi) \cosh(\pi t/\hbar\beta)\right] \,, \\
X_{\mathrm{ph}}(t) &= 2B_3 + \phi/3 - \phi/\cosh^2(\pi t/\hbar\beta) \,,
\end{aligned}
\tag{20.106}
$$

where

$$\phi \equiv (T/T_{\mathrm{ph}}')^2 = 2\pi^2\delta_3(k_{\mathrm{B}}T/\hbar\omega_{\mathrm{ph}})^2 \,. \tag{20.107}$$

With use of the expressions (20.106) the tunneling rate may be written as

$$k^+(T,\epsilon;K) = k_{\mathrm{el}}^+(T,\epsilon;K)\,\mathcal{A}_{\mathrm{ph}}(T,\epsilon) \,, \tag{20.108}$$

where k_{el}^+ is the forward rate in the presence of the conduction electrons alone, Eq. (20.74) or Eq. (20.75), and where $\mathcal{A}_{\mathrm{ph}}(T,\epsilon)$ represents the activation factor due to the contact with the phonon bath in the presence of the fermionic excitations. Expanding the factor $\exp[\phi/\cosh^2(\pi t/\hbar\beta)]$ in the integrand of Eq. (20.31) in a power series in $\phi/\cosh^2(\pi t/\hbar\beta)$, and integrating each term, the resulting series turns out as a generalized hypergeometric series $_2F_2(a,b;c,d;z)$ [90]. The phonon activation factor takes the form [444]

$$\mathcal{A}_{\mathrm{ph}}(T,\epsilon) = e^{-2B_3}\, e^{-\phi/3}\, {}_2F_2\left(K + i\hbar\beta\epsilon/2\pi,\ K - i\hbar\beta\epsilon/2\pi;\ K,\ K + \tfrac{1}{2};\ \phi\right) \,. \tag{20.109}$$

Eq. (20.108) with (20.109) and (20.74) or (20.75) constitutes the exact nonadiabatic rate expression in the regime $T \ll T_{\mathrm{D}}$, $\hbar\omega_c/k_{\mathrm{B}}$. Implementation of the hypergeometric series in Eq. (20.108) yields the expansion of the tunneling rate in terms of multiphonon exchange processes in presence of coupling to conduction electrons.

It is instructive to consider this result more closely in various limits. For zero bias, $k_{\mathrm{el}}(T,0)$ is given in Eq. (20.80), and the phonon factor (20.109) reduces to a degenerate hypergeometric function (Kummer function) [90]. We then have

$$k^\pm(T,0;K) = \frac{\sqrt{\pi}\,\Gamma(K)}{4\Gamma(K+\tfrac{1}{2})}\frac{\Delta^2}{\omega_c}\left(\frac{\hbar\omega_c}{\pi k_{\mathrm{B}}T}\right)^{1-2K} e^{-(2B_3+\phi/3)}\, {}_1F_1\left(K;\ K+\tfrac{1}{2};\ \phi\right) \,. \tag{20.110}$$

In the regime $T \ll T'_{\rm ph}$, we may expand the Kummer function in a power series in ϕ. This corresponds to categorization of the rate in terms of the one-phonon process, two-phonon process, etc., but with the Ohmic contribution fully taken into account [447].

In the opposite regime $T \gg T'_{\rm ph}$ ($\phi \gg 1$) (but still well below the Debye temperature), we may substitute the asymptotic expression of the Kummer function . Then the rate expression (20.110) takes the form

$$k^+(T,0;K) = \frac{1}{4\sqrt{\pi}} \frac{\hbar\Delta^2 e^{-2B_3}}{k_{\rm B}T'_{\rm ph}} \left(\frac{\pi k_{\rm B}T}{\hbar\omega_{\rm c}}\right)^{2K} \left(\frac{T'_{\rm ph}}{T}\right)^2 \exp\left(\frac{2T^2}{3T'^2_{\rm ph}}\right). \qquad (20.111)$$

This expression is nonperturbative in the phonon coupling. For $K = 0$, Eq. (20.111) coincides with an early result by Holstein [131, b] and by Pirc and Gosar [448].

On the other hand, for $K = 0$ and $\epsilon \neq 0$, the expression (20.108) with (20.109) and with $\tilde{\Delta}^2 = \Delta^2 e^{-(2B_3 + \phi/3)}$ turns into

$$k^+(T,\epsilon;0) = \frac{1}{2\pi} \frac{\epsilon}{1 - e^{-\hbar\beta\epsilon}} \left(\frac{\hbar\tilde{\Delta}}{k_{\rm B}T'_{\rm ph}}\right)^2 {}_2{\rm F}_2\left(1 + i\frac{\hbar\beta\epsilon}{2\pi}, 1 - i\frac{\hbar\beta\epsilon}{2\pi}; 2, \frac{3}{2}; \phi\right). \qquad (20.112)$$

Next, consider a symmetric system in the absence of the coupling to conduction electrons, $K = 0$. We obtain from Eq. (20.112) the multi-phonon series expansion

$$k^+(T,0;0) = \frac{\tilde{\Delta}}{2\pi} \frac{\hbar\tilde{\Delta}}{k_{\rm B}T'_{\rm ph}} \sum_{n=1}^{\infty} \frac{\Gamma(\frac{3}{2})}{n\Gamma(n + \frac{1}{2})} \left(\frac{T}{T'_{\rm ph}}\right)^{2n-1}. \qquad (20.113)$$

The terms $n = 1$ and $n = 2$ correspond to the one-phonon process, Eq. (20.102), and two-phonon process, Eq. (20.103). Alternatively, we can derive the form (20.113) from Eq. (20.110). However, to circumvent a divergence in the limit $K \to 0$, the diagonal (zero-phonon) process has to be subtracted [131, 86]. The asymptotic representation of the series (20.113) in the limit $T \gg T'_{\rm ph}$ is given in Eq. (20.111) for $K = 0$.

Consider finally the generalization of the rate formula (20.111) to the regime $T \gtrsim T_{\rm D}$, but further on $T \ll T_{\rm c,el}$. With the solvation energies (20.65), $\Lambda_1 = \Lambda_{1,\rm el} + \Lambda_{1,\rm ph}$ and $\Lambda_2 \approx \Lambda_{2,\rm ph}$, as $\Lambda_{2,\rm el} \ll \Lambda_{2,\rm ph}$, the SDA formula (20.73) for $\epsilon = 0$ yields

$$k^+(T,0;K) = \frac{\hbar\Delta^2 e^{-2B_3}}{2k_{\rm B}T'_{\rm ph}} \sqrt{\frac{\pi}{\lambda_2(3,\kappa_{\rm ph})}} \left(\frac{\pi k_{\rm B}T}{\hbar\omega_{\rm c}}\right)^{2K} \left(\frac{T'_{\rm ph}}{T}\right)^2 e^{\lambda_1(3,\kappa_{\rm ph})(T/T'_{\rm ph})^2}, \qquad (20.114)$$

where $\kappa_{\rm ph} = T/T_{\rm D}$. The coefficients $\lambda_{1,2}(3,\kappa)$ are found from Eq. (20.67) as

$$\lambda_1(3,\kappa) = 2[\zeta\left(2, \tfrac{1}{2} + \kappa\right) - \zeta\left(2, 1 + \kappa\right)]/\pi^2, \quad \lambda_2(3,\kappa) = 24\zeta\left(4, \tfrac{1}{2} + \kappa\right)/\pi^2. \qquad (20.115)$$

The expression (20.114) describes the crossover from the multi-phonon quantum rate (20.111) holding for $\kappa_{\rm ph} = 0$ to the generalized Marcus rate (we assume $\omega_{\rm c} \gg \omega_{\rm D}$)

$$k^+(T,0;K) = \frac{\hbar\Delta^2}{4} \left(\frac{\pi k_{\rm B}T}{\hbar\omega_{\rm c}}\right)^{2K} \left(\frac{\pi}{\Lambda_{\rm ph,cl} k_{\rm B}T}\right)^{1/2} \exp\left(-\frac{\Lambda_{\rm ph,cl}}{4k_{\rm B}T}\right), \qquad (20.116)$$

in which the phonon bath is considered as classical bath, $\Lambda_{\mathrm{ph,cl}} = 2\delta_3(\omega_\mathrm{D}/\omega_{\mathrm{ph}})^2\hbar\omega_\mathrm{D}$, and the electron bath as a quantum mechanical one, $\Lambda_{2,\mathrm{el}} = 2K\ln(\pi k_\mathrm{B}T/\hbar\omega_\mathrm{c})$.

The theory of tunneling and diffusion of light interstitials in metals has been reviewed in Ref. [395]. The cooperation of the electron bath with the phonon bath leads to a minimum of the tunneling rate as a function of temperature for $K < \frac{1}{2}$. The minimum has been observed for muon diffusion in Al and in Cu [102], for incoherent hydrogen tunneling in Niobium [101, 103], and for defect tunneling in mesoscopic Bi wires [106, 105, 445]. Recently, the data for jump rates of defects in mesoscopic Bi wires [105] have been analyzed by using a roughly guessed formula interpolating between the one-phonon rate and the asymptotic expression (20.111) [449].

For tunneling of hydrogen in Niobium, the minimum of the crossing rate is found for T near T_D. The corresponding rate has been studied numerically in this regime [450]. It would now be interesting to apply the formula (20.114) to this problem. One final remark is appropriate. Because of disorder in real systems, the above rate expressions must be averaged with a distribution of bias energies if a quantitative comparison of theory with experiment is attempted.

20.3 Single charge tunneling

A tunnel junction embedded in an electrical circuit forms a quantum system violating Ohm's law. The tunneling transitions of electrons through the barrier are liable to inelastic scattering processes with creation of electromagnetic modes. Thereby the current is reduced at low voltage, an effect called dynamical Coulomb blockade (DCB). The Coulomb blockade regime is most developed in the weak-tunneling limit, $R_\mathrm{T} \gg R_\mathrm{K}$, where R_T is the tunneling resistance, and $R_\mathrm{K} = 2\pi\hbar/e^2$ is the resistance quantum. Control of single electron tunneling processes by tuning of gate voltages may be achieved in the DCB regime. The common principle of single electron devices, such as turnstiles, pumps, and transistors, is to transfer electrons one by one in systems of small tunnel junctions (cf. the contributions by D. Esteve, and by D. V. Averin and K. H. Likharev in Ref. [156]). As an indispensable ground work, it is necessary to understand the DCB effect in charge tunneling through a single junction.

20.3.1 Weak-tunneling regime

For weak tunneling of electrons through a normal junction, the wave function of an excess electron is localized near to the barrier. Then the discrete charge representation (3.239) applies and tunneling through the barrier can be treated in Golden Rule approximation. This corresponds formally to incoherent tunneling in a two-state system in the nonadiabatic limit. The effects of the electromagnetic environment are included in the phase-phase equilibrium correlation function $Q_\varphi(t)$ given in Eq. (3.230). We rely on the global model (3.238) introduced in Subsec. 3.4.1, and we assume a constant tunneling amplitude, $T_{\mathbf{k},\mathbf{k}'} = T_\mathrm{T}$, and a constant density of states in the electron band around the Fermi energy E_F. It is convenient to combine the relevant junction properties in the tunneling resistance R_T,

$$\frac{1}{R_T} = \frac{4\pi e^2}{\hbar} N_R(0) N_L(0) \Omega_R \Omega_L |T_T|^2 . \tag{20.117}$$

Here $N_{R/L}(0)$ and $\Omega_{R/L}$ are the density of states and the volume of the right/left electrode, respectively. The tunneling curent is (we put $v = eU/\hbar$)

$$I(U) = e\left[k^+(v) - k^-(v) \right] = e\left[1 - e^{-\beta\hbar v} \right] k^+(v) . \tag{20.118}$$

Here $k^+(v)$ is the (forward) rate for tunneling from the left electrode to the right electrode, and $k^-(v)$ is the backward rate. In the second form, we have assumed thermal equilibrium in the electrodes.

When the tunneling resistance R_T is large compared to the resistance quantum $R_K = 2\pi\hbar/e^2$, the tunneling rate $k^+(v)$ is well described in golden rule approximation. There results in frequency representation

$$k^+(v) = \frac{\hbar}{2\pi} \frac{R_K}{R_T} \int_{-\omega_c}^{\omega_c} d\omega' \int_{-\omega_c}^{\omega_c} d\omega'' \, f(\omega') \, f(-\omega'') \, P_{em}(\hbar\omega' + \hbar v - \hbar\omega'') . \tag{20.119}$$

The energy $\hbar v$ represents the mismatch of the quasiparticle energies of the left and right electrodes in the quasiparticle Hamiltonian (3.237) due to the applied voltage eU. The cutoff ω_c in the electron band is in general the largest frequency of the problem. Therefore, we eventually may take the limit $\omega_c \to \infty$.

The formula (20.119) has a neat interpretation. The integrand is proportional to the spectral probability density for an occupied state at energy $\hbar\omega'$ in the left electrode and an empty state at energy $\hbar\omega''$ in the right electrode times the probability density $P_{em}(\hbar\omega' + \hbar v - \hbar\omega'')$ that the electromagnetic environment absorbs the energy $\hbar(\omega' + v - \omega'')$. The bias energy $\hbar v = eU$ expresses the difference of the Fermi energies in the two electrodes due to the ideal voltage source U.

With the substitution $\omega'' = \omega + \omega'$, the ω'-integration in Eq. (20.119) can be done,

$$\int_{-\infty}^{\infty} d\omega' \, f(\omega') \, f(-\omega' - \omega) = \int_{-\infty}^{\infty} d\omega' \frac{f(\omega') - f(\omega' + \omega)}{1 - e^{-\hbar\beta\omega}} = \frac{\omega}{1 - e^{-\hbar\beta\omega}} . \tag{20.120}$$

With this relation, the rate expression (20.119) can be written as

$$k^+(v) = \hbar \int_{-\infty}^{\infty} d\omega \, k_0^+(\omega) P_{em}(\hbar v - \hbar\omega) ,$$

$$k_0^+(v) = \frac{1}{2\pi} \frac{R_K}{R_T} \int_{-\infty}^{\infty} d\omega' \frac{f(\omega') - f(\omega' + v)}{1 - e^{-\hbar\beta v}} = \frac{1}{2\pi} \frac{R_K}{R_T} \frac{v}{1 - e^{-\hbar\beta v}} . \tag{20.121}$$

For "elastic" tunneling there is $P_{em}(\hbar v - \hbar\omega) = \delta(\hbar v - \hbar\omega)$, and thus $k^+(v) = k_0^+(v)$. The rate $k_0^+(v)$ describes tunneling of the fermionic quasiparticle through the barrier in the absence of coupling with the electromagnetic surroundings.

The probability that the surroundings absorb the energy E is

$$P_{\text{em}}(E) = \frac{1}{2\pi\hbar} \int_{-\infty}^{\infty} dt\, e^{iEt/\hbar - Q_{\text{em}}(t)} . \tag{20.122}$$

For the electromagnetic environment described by the Hamiltonian (3.217), the function $Q_{\text{em}}(t)$ corresponds to the phase correlation function $Q_{\varphi}(t)$ which describes the time correlations of the fluctuations of the phase jump $\varphi(t)$ at the junction due to the electromagnetic environment, $Q_{\text{em}}(t) = Q_{\varphi}(t)$,

$$e^{-Q_{\text{em}}(t)} \equiv \left\langle e^{i\varphi(t)} e^{-i\varphi(0)} \right\rangle_{\beta} = e^{-\langle [\varphi(0)-\varphi(t)]\varphi(0)\rangle_{\beta}} . \tag{20.123}$$

To obtain the second form, we have employed the Gaussian statistical properties of the fluctuating phase $\varphi(t)$. The phase correlation function $Q_{\text{em}}(t) \equiv \langle [\varphi(0)-\varphi(t)]\varphi(0)\rangle_{\beta}$ can be written as [cf. Eqs. (3.230) and (3.231)]

$$Q_{\text{em}}(t) = \int_{0}^{\infty} d\omega\, \frac{G_{\text{em}}(\omega)}{\omega^2} \left\{ \coth\left(\frac{\hbar\beta\omega}{2}\right)\left(1 - \cos(\omega t)\right) + i\sin(\omega t) \right\} \tag{20.124}$$

with

$$G_{\text{em}}(\omega) = (e^2/\pi\hbar)\,\omega^2 \tilde{\chi}''(\omega) = 2\omega \operatorname{Re} Z_t^*(\omega)/R_{\text{K}} . \tag{20.125}$$

An Ohmic resistor R with the junction capacitance C in parallel is represented by the spectral density [cf. Eq. (3.241)]

$$G_{\text{em}}(\omega) = \frac{2\alpha\,\omega}{1 + (\omega/\omega_{\text{R}})^2} , \tag{20.126}$$

where $\alpha = R/R_{\text{K}}$, $\omega_{\text{R}} = 1/RC = E_{\text{c}}/\pi\hbar\alpha$, and where $E_{\text{c}} = e^2/2C$ is the charging energy. For such environment, the reorganization energy Λ_{cl} or absorbed net energy (20.44) coincides with the charging energy,

$$\Lambda_{\text{cl}} = E_{\text{c}} . \tag{20.127}$$

Thus, for a high-impedance environment, the junction behaves classically down to fairly low temperature. In this regime, $P_{\text{em}}(E)$ has the Gaussian form (20.22),

$$P_{\text{em}}(E) = \frac{1}{(4\pi E_{\text{c}} k_{\text{B}} T)^{1/2}} \exp\left(-\frac{(E - E_{\text{c}})^2}{4E_{\text{c}} k_{\text{B}} T}\right) . \tag{20.128}$$

With increasing quality factor Q_{qual} in the spectral density (3.244), the probability function $P_{\text{em}}(E)$ changes from the smooth form (20.128) to the resonant characteristics given in Eq. (20.39), where ω_0 corresponds to ω_{L}.

It is instructive to see that in general the tunneling rate for a fermion can be transformed into a tunneling rate for a boson, and vice versa. To this, we initially write the $P(E)$ function as in Eq. (20.35), which ensures detailed balance for $P(E)$. For the scaling form $Q_{\text{sc}}(t)$ given in Eq. (18.53) one finds from Eq. (20.74) [433]

$$D_{\text{sc}}(\omega, K) = \frac{1}{\Gamma(2K)} \left(\frac{\hbar\beta\omega_{\text{c}}}{2\pi}\right)^{1-2K} \frac{|\Gamma(K + i\hbar\beta\omega/2\pi)|^2}{|\Gamma(1/2 + i\hbar\beta\omega/2\pi)|^2} \Theta(\omega_{\text{c}} - |\omega|) . \tag{20.129}$$

The function $D_{sc}(\omega, K)$ is an effective density of states of fermionic quasiparticles in a band of width $2\omega_c$. For the particular values $K = \frac{1}{2}$ and $K = 1$, there is

$$
\begin{aligned}
D_{sc}(\omega,\, K = \tfrac{1}{2}) &= 1 \,, \\
D_{sc}(\omega,\, K = 1) &= \omega \coth(\tfrac{1}{2}\hbar\beta\omega)/\omega_c \,.
\end{aligned} \qquad |\omega| < \omega_c \,. \qquad (20.130)
$$

As a result of (20.35) with (20.130), the Fermi function can be written as the integral

$$
f(\omega) = \frac{\omega_c}{2\pi} \int_{-\infty}^{\infty} dt\, e^{-i\omega t}\, e^{-Q_{sc}(t,\, K=1/2)} \,. \qquad (20.131)
$$

Next, we insert the Fourier representation (20.131) both for $f(\omega')$ and $f(-\omega'')$ and the Fourier representation (20.122) for $P_{em}(t)$ into the rate expression (20.119). Then one directly obtains with $2Q_{sc}(t, \frac{1}{2}) = Q_{sc}(t, 1)$,

$$
k^+(v) = \left(\frac{\omega_c}{2\pi}\right)^2 \frac{R_K}{R_T} \int_{-\infty}^{\infty} dt\, e^{ivt}\, e^{-Q_{sc}(t;K=1)-Q_{em}(t)} \,. \qquad (20.132)
$$

Thus we have found a surprising result: tunneling of a fermion through a barrier embedded in a conductor is like tunneling of a boson subjected to Ohmic dissipation with Kondo parameter $K = 1$. The correlation function $Q_{em}(t)$ representing the electromagnetic environment is simply added to the Ohmic correlation function.

For an Ohmic impedance, the spectral density $G_{em}(\omega)$ is given by the Drude form (20.126). The correlation function $Q_{em}(t; \alpha)$ for the algebraic cutoff at frequency ω_R differs in essence from the Ohmic scaling form (18.53 by an adiabatic correction,

$$
\begin{aligned}
Q_{em}(t;\alpha) &= Q_{sc}(t;\alpha) + \Delta Q_{em}(\alpha) \,, \\
\Delta Q_{em}(\alpha) &= 2\alpha \ln(\omega_R/\omega_c) + 2\alpha\, \zeta_D \,, \\
\zeta_D &= \psi\left(1 + \frac{\hbar\beta\omega_R}{2\pi}\right) - \psi(1) - \ln\left(\frac{\hbar\beta\omega_R}{2\pi}\right) - \frac{\pi}{\hbar\beta\omega_R} \overset{\hbar\beta\omega_R \gg 1}{\Longrightarrow} C_E \,,
\end{aligned} \qquad (20.133)
$$

where C_E is Euler's constant. The relation $\zeta_D = C_E$ holds in the regime $\hbar\beta\omega_R \gg 1$. With the first relation of (20.133), the forward tunneling rate (20.132) takes the form

$$
k^+(v) = \left(\frac{\omega_c}{2\pi}\right)^2 \frac{R_K}{R_T}\, e^{-\Delta Q_{em}(\alpha)} \int_{-\infty}^{\infty} dt\, e^{ivt}\, e^{-Q_{sc}(t;K=1+\alpha)} \,. \qquad (20.134)
$$

Thus, apart from the adiabatic dressing $e^{-\Delta Q_{em}(\alpha)}$ weak tunneling of electrons coupled to a resistive electromagnetic environment behaves exactly like weak tunneling of bosons embedded in an Ohmic environment with Ohmic damping parameter

$$
K = 1 + \alpha \,. \qquad (20.135)
$$

One remark is in order. The expression (20.134) is the leading contribution to the rate in the weak-tunneling limit. Generalization to the full weak-tunneling power series in R_K/R_T, and to the related strong-tunneling series will be given in Chapter 31.

Alternatively, the electromagnetic correlations can be transferred into an effective frequency- and temperature-dependent tunneling density. For this purpose, we write as a generalization of Eq. (20.131) and in accordance with the relation (20.35),

$$\frac{1}{2\pi\hbar} \int_{-\infty}^{\infty} dt\, e^{-i\omega t}\, e^{-Q_{sc}(t;K=1/2)-Q_{em}(t)/2} = \frac{D_{\text{eff}}(|\omega|)}{\hbar\omega_c} f(\omega) . \tag{20.136}$$

Following the relation (20.48), the forward rate (20.132) is readily transformed into

$$k^+(v) = \frac{1}{2\pi} \frac{R_K}{R_T} \int_{-\omega_c}^{\omega_c} d\omega\, D_{\text{eff}}(|\omega|) D_{\text{eff}}(|v-\omega|) f(-\omega)\, f(\omega - v) . \tag{20.137}$$

In this formulation, the effects of the environment are included in an effective frequency-dependent tunneling density of states of the fermionic quasiparticle. In the absence of the environmental coupling, $D_{\text{eff}}(\omega)$ is unity. Then the rate $k^+(v)$ matches with the rate $k_0^+(v)$ given in Eq. (20.121).

In conclusion, the expressions (20.132) and (20.137) are equivalent. They are bosonic and fermionic representations of charge transfer through a strong barrier.

20.3.2 The current-voltage characteristics

With use of the expression (20.118) with (20.121), the current-voltage characteristics of a single junction embedded in an electromagnetic environment can be written as

$$I(U) = \int_{-\infty}^{\infty} dE\, \frac{1 - e^{-\beta eU}}{1 - e^{-\beta E}}\, P_{em}(eU - E)\, I_0(E/e) , \tag{20.138}$$

where

$$I_0(U) = U/R_T \tag{20.139}$$

is the Ohmic tunneling current. The expression (20.138) satisfies the obvious relation $I(-U) = -I(U)$, as follows with use of the detailed balance relation (20.33).

When the impedance of the environment at zero frequency $Z(\omega = 0)$ is small compared to the resistance quantum R_K, the charge transfer through the junction is mainly elastic, $P_{em}(E) \approx \delta(E)$. With this we obtain the form (20.121) for the forward tunneling rate, and finally with Eq. (20.118) or directly from Eq. (20.138) the Ohmic law $I(U) = U/R_T$. This warrants the interpretation of R_T as tunneling resistance.

In the opposite limit of a high-impedance setting, the energy absorbed at low T equals the charging energy $E_c = e^2/2C$, $P_{em}(E) = \delta(E - E_c)$. The resulting current has the Coulomb gap $eU > E_c$ in correspondence with the energy balance (3.211),

$$I(U) = \Theta(U - E_c/e)\, [U - E_c/e]/R_T . \tag{20.140}$$

For $T = 0$, the bath cannot supply energy. Then the expression (20.138) turns into

$$I(U) = \Theta(U)\, \frac{1}{eR_T} \int_0^{eU} dE\, (eU - E)\, P_{em}(E) . \tag{20.141}$$

Direct information about the distribution of energy absorbed by the environment is gained from the second derivative of the current with respect to the applied voltage,

$$\frac{d^2 I(U)}{dU^2} = \frac{e}{R_T} P_{em}(eU) . \tag{20.142}$$

In the regime $eU \gg k_B T$ and $P_{em}(|eU|) \ll P_{em}(0)$ there results from Eq. (20.138)

$$I(U) = \frac{1}{eR_T} \int_{-eU}^{eU} dE \, (eU - E) \, P_{em}(E) . \tag{20.143}$$

Using the sum rules (20.32) and (20.44), and supposing a resistive environment for which Λ_{cl} equals the charging energy, Eq. (20.127), we find the linear current-voltage characteristics in the regime $k_B T \ll e^2/2C < eU$ as

$$I(U) = [U - e/2C]/R_T . \tag{20.144}$$

The voltage shift $\Delta U = e/2C$ in the linear $I(U)$ characteristics is the manifestation of the Coulomb blockade in the classical regime.

The current-voltage characteristics in the quantum regime $k_B T \ll eU \ll e^2/2C$ is considered next. One sees from the nonadiabatic rate formula (20.134) that the analytic rate expressions (20.74)–(20.80) directly apply to single-electron tunneling in a resistive environment described by the spectral density (20.126) if we assign

$$K \to 1 + \alpha , \qquad \text{and} \qquad \frac{\Delta^2}{4} \to \left(\frac{\omega_c}{2\pi}\right)^2 \frac{R_K}{R_T} e^{-\Delta Q_{em}(\alpha)} . \tag{20.145}$$

With use of these correspondences and of the rate expression (20.78), the current-voltage characteristics in the regime $k_B T \ll eU \ll \hbar \omega_R$ is found to read

$$I(U) = \frac{e^{-2\alpha C_E}}{\Gamma(2 + 2\alpha)} \frac{U}{R_T} \left(\frac{\pi \alpha \, eU}{E_c}\right)^{2\alpha} , \qquad \alpha = R/R_K . \tag{20.146}$$

This expression describes the dynamical Coulomb blockade, in which the sharp Coulomb gap of the classical regime, Eq. (20.140) is rounded by quantum fluctuations. The dynamical Coulomb blockade is characterized by a super-linear behavior $I(U) \propto U^{1+2\alpha}$ in the regime $k_B T \ll eU$, or equivalently, by the zero-bias anomaly $dI(U)/dU \propto U^{2\alpha}$, instead of the voltage-independent conductance in the absence of the environment. Because of the substitution $K \to 1 + \alpha$ in the correspondence, the current shows the characteristic power laws already encountered in Subsection 20.2.5. The tunneling density of states associated with the Ohmic impedance is obtained from Eq. (20.129) with (20.145) at $T = 0$ for $\omega \ll \omega_R$ as

$$D_{em}(\omega) = e^{-\alpha C_E} (|\omega|/\omega_R)^\alpha / \Gamma(1 + \alpha) . \tag{20.147}$$

The tunneling density of states is nonanalytic at the Fermi energy and it is thinned down around it for $\alpha > 0$ compared with the constant tunneling density of states for $\alpha = 0$. In the expression (20.137), the zero-bias anomaly originates from the low density of states near $\omega = 0$. The power law in Eq. (20.146) directly reflects the nonanalytic behavior $D_{em}(\omega \to 0) \propto |\omega|^{\alpha}$.

We should like to remark that we have the same power-law form as in Eq. (20.146) for any other environment with a finite impedance at $\omega = 0$, where $\alpha = Z(0)/R_K$. Only the prefactor depends on the spectral properties of the impedance.

Consider next the linear conductance at finite temperature. Using the correspondences (20.145), we obtain from Eq. (20.118) with Eq. (20.80) the expression

$$\frac{dI}{dU}\bigg|_{U=0} = \frac{e^{-2\alpha C_E}}{R_T} \frac{\sqrt{\pi}\,\Gamma(1+\alpha)}{2\Gamma(\frac{3}{2}+\alpha)} \left(\frac{\pi^2 \alpha\, k_B T}{E_c}\right)^{2\alpha}. \tag{20.148}$$

In the high voltage regime $eU \gg E_c$, we may use in Eq. (20.141) for $P_{em}(E)$ the form (20.61). Upon substituting for the spectral density $G_{em}(\omega)$ the expression (20.126), we obtain the leading correction to the strict Coulomb gap form (20.140),

$$I(U) = \frac{1}{R_T}\left(U - \frac{e}{2C} + \frac{\alpha}{\pi^2}\frac{e^2}{4C^2}\frac{1}{U}\right), \qquad \text{for} \qquad eU \gg E_c. \tag{20.149}$$

Thus, the actual offset is smaller than the offset in Eq. (20.140).

Let us finally consider a LC transmission line described by the sub-Ohmic spectral density (3.246). Upon putting $s = \frac{1}{2}$ in the expression (20.56), we find for the probability function $P_{em}(E)$ in the limit $\omega_c \to \infty$ the expression[5]

$$P_{em}(E) = \sqrt{\frac{eV_c}{4\pi E^3}} \exp\left(-\frac{eV_c}{4E}\right) \qquad \text{with} \qquad V_c = \frac{4eR_0}{C_0 R_K}. \tag{20.150}$$

The function $P_{em}(E)$ has a maximum at $E = eV_c/6$. Insertion of this form into the current (20.141) yields exponential reduction of the current $\propto \exp(-V_c/4V_a)$ for $V_a \ll V_c$ instead of the power law reduction in the Ohmic case, Eq. (20.146). The strong weakening of the charge transfer is due to the much higher density of low-frequency excitations for the LC transmission line. At higher voltage, the current characteristics is found to be similar to the case of an Ohmic resistive environment.

20.3.3 Weak tunneling of 1D interacting electrons

In 1D quantum wires, the electron-electron interaction invalidates the Fermi liquid model. The low-energy excitations of the correlated system are not anymore fermionic quasiparticles, but collective density fluctuations of a harmonic fluid. The appropriate model is the Tomonaga-Luttinger liquid model. This model is described in boson representation in Chapter 31.

The transport of charge across a barrier or impurity in a 1D quantum wire is treated below in Sec. 31.1. Nevertheless, it is quite appropriate to set up already now a formal

---

[5]The expression for V_c is misprinted in Ref. [157].

similarity between correlated electron-tunneling through a strong barrier and weak single-electron tunneling in the presence of a resistive electromagnetic environment. To this aim, we only need to anticipate that the equilibrium correlation function of the phase jump $\bar{\phi}(t)$ across a strong barrier of the fermion field (31.2) in the Luttinger liquid state matches the Ohmic correlation function $Q_{sc}(t; K)$ given in Eq. (18.53), $4\pi \langle \, [\,\bar{\phi}(0) - \bar{\phi}(t)\,]\,\bar{\phi}(0)\,\rangle_\beta = Q_{sc}(t; 1/g)$. Here, g is a dimensionless contact interaction constant in the Luttinger model. The regime $g < 1$ corresponds to repulsive electron interaction, and $g = 1$ is the Fermi liquid point. The phase correlation function $Q_{sc}(t; 1/g)$ coincides with the Ohmic correlation function (18.53), if we put $g = 1/K$,

$$Q_{sc}(t; 1/g) \equiv \frac{2}{g} \ln \left[\frac{\beta \hbar \omega_c}{\pi} \sinh \left(\frac{\pi |t|}{\hbar \beta} \right) \right] + i \frac{\pi}{g} \, \text{sgn}(t) \,. \tag{20.151}$$

In analogy with the boson representation (20.134), the tunneling rate of a charge e across an impurity with applied voltage $U = \hbar v/e$ may be written as

$$k^+(v) = \left(\frac{\omega_c}{2\pi} \right)^2 \frac{R_K}{R_T} \int_{-\infty}^{\infty} dt \, e^{ivt} \, e^{-Q_{sc}(t; 1/g)} \,. \tag{20.152}$$

This form reveals a direct formal correspondence between weak tunneling of 1D interacting electrons and weak tunneling of electrons coupled to an electrical circuit with Ohmic impedance. In the correspondence we have

$$1 + \alpha = 1/g \,. \tag{20.153}$$

We shall discuss below in Chapter 31, in particular in Subsection 31.1.4, that the correspondence also holds for joint tunneling of two, three and many charges, and that it is also valid in the weak-barrier, or equivalent strong-tunneling regime.

It is natural to resume the previous reasoning by introducing an effective tunneling density of states of the fermionic entity, $D_{sc}(\omega; 1/g)$, as in Eq. (20.129). Then one obtains analogous to Eq. (20.48) the expressions [433]

$$\begin{aligned}
k^+(v) &= \frac{1}{2\pi} \frac{R_K}{R_T} \int_{-\omega_c}^{\omega_c} d\omega \, D_{sc}(|\omega|, \tfrac{1}{2g}) D_{sc}(|v - \omega|, \tfrac{1}{2g}) f(-\omega) f(\omega - v) \,, \\
&= \frac{\omega_c}{2\pi} \frac{R_K}{R_T} D_{sc}(|v|, \tfrac{1}{g}) f(-v) \,.
\end{aligned} \tag{20.154}$$

With the second form, the current is found to be given by

$$I(U) = \frac{\hbar \omega_c}{e R_T} \left(1 - e^{-\beta eU} \right) f(-eU/\hbar) \, D_{sc}(e|U|/\hbar, 1/g) \,. \tag{20.155}$$

One can directly verify with use of the expression (20.130) that the current-voltage relation (20.155) reduces at the Fermi liquid point $g = 1$ to Ohm's law $I(U) = U/R_T$.

In a double junction system, besides sequential tunneling, higher order tunneling processes may occur in which the Coulomb barrier is bypassed by virtual occupation of the island. A theoretical description of resonant tunneling for correlated electrons using the above concept of effective tunneling densities is reported in Ref. [451].

A nonperturbative formalism in the presence of strong Coulomb interactions has been developed and applied to resonant tunneling in Ref. [452]. A review of mesoscopic electron transport is given in Ref. [453]. Electrical conduction in single-molecule circuits is discussed in a monograph by Cuevas and Scheer [454].

20.3.4 Tunneling of Cooper pairs

Tunneling of Cooper pairs through a Josephson junction is affected by the electromagnetic environment as well. For weak Josephson coupling energy, $E_J \ll E_c = 2e^2/C$, we may calculate the crossing rate for a Cooper pair in Golden Rule approximation with respect to the Josephson coupling energy. Since the tunneling entities are bosons, the rate can be written in a form analogous to Eq. (20.30) [cf. Fig. 3.3 on page 61, and Subsec. 3.4.3]. With the externally applied voltage V_a, the rate expression is

$$k^+ = \frac{E_J^2}{4\hbar^2} \int_{-\infty}^{\infty} dt \, e^{i\,2eUt/\hbar} \, e^{-Q_\psi(t)} \, . \tag{20.156}$$

The bias energy is $2eU$ because the charge transferred by a Cooper pair across the junction is $2e$, and the phase correlation function for Cooper pairs $Q_\psi(t)$ is given in Eq. (3.249). With the correspondences

$$\hbar\Delta \,\hat{=}\, E_J \,=\, (\hbar/2e)I_c \, , \qquad \hbar\epsilon \,\hat{=}\, 2eU \, , \qquad K \,\hat{=}\, \rho \,=\, R/R_Q \, , \tag{20.157}$$

where $R_Q = 2\pi\hbar/4e^2$, and with

$$Q_\psi(t) \equiv Q_{sc}(t; K = \rho) + 2\rho \ln(\omega_R/\omega_c) + 2\rho\zeta \, , \tag{20.158}$$

we can directly apply the results for incoherent tunneling of a boson with Ohmic friction presented in Section 20.2 to Cooper pair tunneling. The logarithmic term in Eq. (20.158) takes into account that the cut-off frequency is

$$\omega_R = 1/[Z(0)C] = E_c/(\pi\rho\hbar) \, , \tag{20.159}$$

instead of ω_c. The quantity ζ accounts for the particular high-frequency dependence of the total impedance $Z_t^*(\omega)$ in adiabatic approximation,

$$\zeta = \zeta_D + \int_0^\infty \frac{d\omega}{\omega} \left(\frac{\mathrm{Re}\, Z_t^*(\omega)}{\rho R_Q} - \frac{1}{1 + (\pi\rho\hbar\omega/E_c)^2} \right) . \tag{20.160}$$

The term ζ_D, given in Eq. (20.133), arises from the Drude form (20.126), and the integral accounts for deviation of the actual $Z_t^*(\omega)$ from this behavior.

Introducing the function $P_\psi(E)$ as the Fourier transform of $e^{-Q_\psi(t)}$ [see Subsection 20.2.2 for the physical meaning and the properties of $P_\psi(E)$], the current-voltage characteristics takes the form

$$I(U) = 2e(k^+ - k^-) = \frac{\pi e E_J^2}{\hbar} \left(1 - e^{-2\beta eU} \right) P_\psi(2eU) \, . \tag{20.161}$$

Using the rate expression (20.74) and the above correspondences, the current-voltage characteristics is found to read [455]

$$I(U) = \frac{\pi e \rho}{\hbar} \frac{(E_{\mathrm{J}} e^{-\rho \zeta})^2}{E_{\mathrm{c}}} \left(\frac{\beta E_{\mathrm{c}}}{2\pi^2 \rho}\right)^{1-2\rho} \frac{|\Gamma(\rho + i\,\beta eU/\pi)|^2}{\Gamma(2\rho)} \sinh(\beta eU). \qquad (20.162)$$

This expression holds for a wide range of temperatures in the weak tunneling regime $E_{\mathrm{J}} \ll E_{\mathrm{c}}$. For small ρ, the supercurrent-voltage characteristics (20.162) shows a peak at voltage $U = \pi \rho / e\beta$ which becomes increasingly distinct as temperature is lowered.

The current at zero temperature is found from (20.162) as

$$I(U) = \frac{\pi^{5/2} \rho}{2\,\Gamma(\rho)\Gamma(\rho + \frac{1}{2})} \left(\frac{E_{\mathrm{J}} e^{-\rho \zeta}}{eU}\right)^2 \left(\frac{\pi \rho eU}{E_{\mathrm{c}}}\right)^{2\rho} \frac{U}{R}. \qquad (20.163)$$

Hence the supercurrent exhibits the zero bias anomaly $I(U) \propto U^{2\rho-1}$ [436]. It describes suppression of the current by the Coulomb blockade effect for $\rho > 1$. Also the zero bias conductance at finite temperature exhibits power law behavior,

$$\frac{dI}{dU}\bigg|_{U=0} = \frac{1}{R_{\mathrm{Q}}} \frac{\sqrt{\pi}}{2} \frac{\Gamma(\rho)}{\Gamma(\rho + \frac{1}{2})} \left(\frac{E_{\mathrm{J}} e^{-\rho \zeta}}{E_{\mathrm{c}}/\rho \pi^2}\right)^2 \left(\frac{\beta E_{\mathrm{c}}}{\rho \pi^2}\right)^{2-2\rho}. \qquad (20.164)$$

For low impedance impedance $\rho \lesssim 0.1$, and moderate to high temperature $\beta eU/\pi \lesssim 3$, we may simplify the expression (20.162) with use of the relation (20.76). This yields with use of $E_{\mathrm{J}} = (\hbar/2e)\,I_{\mathrm{c}}$

$$I(U) = \frac{I_{\mathrm{c}}^2 e^{-2\rho \zeta}}{2} \left(\frac{2\pi^2 \rho}{\beta E_{\mathrm{c}}}\right)^{2\rho} \frac{RU}{U^2 + (\pi\rho/\beta e)^2}. \qquad (20.165)$$

In the usual case of low impedance, $\rho < 1$, the conductance for $T = 0$ found from Eq. (20.163) diverges in the zero bias limit. Also the zero-bias conductance (20.164) diverges in the limit $T \to 0$ as $T^{2\rho-2}$. This indicates onset of strong tunneling in the low-energy regime. The singularity in the weak-tunneling conductance is smoothed out by taking into account terms of higher order in E_{J}^2 (see Chapter 28). At this point, a preliminary final remark is appropriate. The case of weak Josephson coupling [Eq. (20.163)] is related to the case of large Josephson coupling [Eq. (17.87)] by a duality symmetry. This issue is discussed in Section 28.3.

20.3.5 Tunneling of quasiparticles

Quasiparticle tunneling in a superconducting junction is similar to that in a normal junction. Observing detailed balance for the tunneling rate, Eq. (20.34), we may write the current again in the form (20.118). In modification of Eq. (20.119), the forward rate for quasiparticle tunneling is

$$k^+(v) = \frac{\hbar}{2\pi} \frac{R_{\mathrm{K}}}{R_{\mathrm{T}}} \int_{-\omega_{\mathrm{c}}}^{\omega_{\mathrm{c}}} d\omega' \int_{-\omega_{\mathrm{c}}}^{\omega_{\mathrm{c}}} d\omega'' \, \mathcal{N}_{\mathrm{qp}}(\omega') \mathcal{N}_{\mathrm{qp}}(\omega'') \, f(\omega') f(-\omega'') \, P_{\mathrm{em}}[\hbar(\omega' + v - \omega'')].$$

The essential difference from the expression (20.119) is that the density of states of the BCS quasiparticles $\mathcal{N}_{\mathrm{qp}}(\omega)$ strongly depends on frequency, as given in Eq. (4.171), and repeated here

$$\mathcal{N}_{\mathrm{qp}}(\omega) = \Theta\left(|\omega| - \Delta_{\mathrm{g}}\right) |\omega|/\sqrt{\omega^2 - \Delta_{\mathrm{g}}^2} \, . \tag{20.166}$$

Other forms analogous to those given for quasiparticle tunneling in a normal junction are easily found. For instance, in analogy with Eq. (20.138) the quasiparticle current through the junction in the presence of the environment, $I_{\mathrm{qp,em}}$, is related to the quasiparticle current in the absence of it, $I_{\mathrm{qp,0}}$, by the integral expression [436]

$$I_{\mathrm{qp,em}}(U,T) = \int_{-\infty}^{\infty} dE \, \frac{1 - e^{-\beta eU}}{1 - e^{-\beta E}} \, P_{\mathrm{em}}(eU - E) \, I_{\mathrm{qp,0}}(E/e, T) \, , \tag{20.167}$$

$$I_{\mathrm{qp,0}}(U,T) = \frac{\hbar}{eR_{\mathrm{T}}} \int_{-\infty}^{\infty} d\omega \, \mathcal{N}_{\mathrm{qp}}(\omega)\mathcal{N}_{\mathrm{qp}}(\omega + eU/\hbar) \left[f(\omega) - f(\omega + eU/\hbar)\right] .$$

The ω-integral can be done in analytic form for zero temperature. The resulting expression is a linear combination of two hypergeometric functions $_2F_1(z)$,

$$I_{\mathrm{qp,0}}(U,0) = \Theta(U - U_{\mathrm{g}}) \, (U/R_{\mathrm{T}}) \, U/[2(U + U_{\mathrm{g}})] \tag{20.168}$$

$$\times \left[B(\tfrac{1}{2},\tfrac{1}{2}) \, _2F_1(\tfrac{1}{2},\tfrac{1}{2};1;z) - (1 - U_{\mathrm{g}}/U)^2 \, B(\tfrac{3}{2},\tfrac{1}{2}) \, _2F_1(\tfrac{1}{2},\tfrac{3}{2};2;z) \right] ,$$

with the gap voltage $U_{\mathrm{g}} = 2\hbar\Delta_{\mathrm{g}}/e$ and the variable $z = (U - U_{\mathrm{g}})^2/(U + U_{\mathrm{g}})^2$. This expression can alternatively be written with the complete elliptic integrals of the first and second kind, $K(y)$ and $E(y)$, respectively, as [157]

$$I_{\mathrm{qp,0}}(U,0) = \Theta(U - U_{\mathrm{g}}) \left[E(y) - (U_{\mathrm{g}}^2/2U^2) \, K(y) \right] U/R_{\mathrm{T}} \, , \tag{20.169}$$

where $y = 1 - (U_{\mathrm{g}}/U)^2$.

The quasiparticle current $I_{\mathrm{qp,0}}(U,0)$ is zero in the voltage regime $U < U_{\mathrm{g}}$. At the gap voltage, $U = U_{\mathrm{g}}$, the quasiparticle current jumps from zero to the value

$$I_{\mathrm{qp,0}}(U_{\mathrm{g}},0) = (\pi/4) \, U_{\mathrm{g}}/R_{\mathrm{T}} \, . \tag{20.170}$$

As U is tuned to the large voltage regime, the quasiparticle current becomes Ohmic,

$$I_{\mathrm{qp,0}}(U \gg U_{\mathrm{g}}, 0) = U/R_{\mathrm{T}} \, . \tag{20.171}$$

Consider finally the implications of an electromagnetic environment. For the Ohmic form $G_{\mathrm{em}}(\omega) = 2\alpha\omega$, where $\alpha = R/R_{\mathrm{K}}$, one obtains from Eq. (20.167), using findings from Subsec. 20.2.5, in particular Eq. (20.78), the anomalous threshold behavior

$$I_{\mathrm{qp,em}}(U,0) \propto \Theta(U - U_{\mathrm{g}}) \, (U - U_{\mathrm{g}})^{2\alpha} \, . \tag{20.172}$$

Thus the jump of height (20.170) at the threshold is smeared by quantum fluctuations of the Ohmic resistance to a power law form. At voltage $U \gg U_{\mathrm{g}}$, the quasiparticle current varies as

$$I_{\mathrm{qp,em}}(U,0) \propto U^{2\alpha+1} \, . \tag{20.173}$$

The deviation of $I_{\mathrm{qp,em}}(U,0)$ from the Ohmic law $I(U) \propto U$ is again a signature of dynamical Coulomb blockade by an Ohmic electromagnetic environment.

21 Two-state dynamics: basics and methods

So far, we have mainly discussed quantum statistical properties of the open two-state system. As far as dynamics is concerned, we have been limited to the study of nonadiabatic tunneling rates in the incoherent regime. We now turn towards real-time dynamics based on the methods given in Chapter 5. The approach will cover the full dynamics in a unified manner for different kinds of the initial preparation and for arbitrary linear dissipation, both in the incoherent and oscillatory regime. We shall provide explicit expressions in most regions of the parameter space. Emphasis is put on the regime in which the system shows quantum coherent oscillations.

21.1 Initial preparation, expectation values, and correlations

We have discussed in Sections 5.2 and 5.3 initial conditions, preparation functions and propagating functions for a general global system. We now deal with the specification for the open two-state system or spin boson model with the emphasis on issues of experimental relevance. Our considerations are built upon the basics of the spin-boson model introduced in Subsec. 3.2.1.

21.1.1 Expectation values

The dynamics of the expectation values of the two-state system are captured by the reduce density matrix (RDM) $\rho(t)$. The diagonal elements $\rho_{R,R}$ and $\rho_{L,L}$ are the population probabilities of the right and left state, and the off-diagonal elements $\rho_{L,R}$ and $\rho_{R,L}$ are the coherences. Upon writing the Pauli matrices in the localized state representation (3.137), the expectation values $\langle \sigma_j(t) \rangle = \mathrm{tr}\{\sigma_j \rho(t)\}$ are found as

$$
\begin{aligned}
\langle \sigma_z \rangle_t &= \rho_{R,R}(t) - \rho_{L,L}(t) \,, \\
\langle \sigma_x \rangle_t &= \rho_{L,R}(t) + \rho_{R,L}(t) \,, \\
\langle \sigma_y \rangle_t &= i\,\rho_{L,R}(t) - i\,\rho_{R,L}(t) \,.
\end{aligned}
\tag{21.1}
$$

The quantity $\langle \sigma_z \rangle_t$ describes the difference of the populations of the two localized states It gives immediate information about the tunneling dynamics and is most directly relevant in studies of "macroscopic quantum coherence" (MQC). The understanding of the TSS dynamics is completed by the knowledge of the coherences $\langle \sigma_x \rangle_t$ and $\langle \sigma_y \rangle_t$ [cf. Section 4.1]. We obtain from Eq. (4.5) the bound

$$
\langle \sigma_x \rangle_t^2 + \langle \sigma_y \rangle_t^2 + \langle \sigma_z \rangle_t^2 \;\leqslant\; 1 \,.
\tag{21.2}
$$

Equality holds in the absence of damping. With the Heisenberg representation

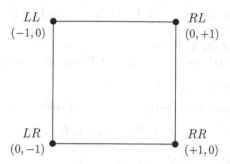

Figure 21.1: Graphical sketch of the four states $\zeta = (\eta, \xi)$ of the reduced density matrix. The diagonal states are $(\pm 1, 0)$, and the off-diagonal states $(0, \pm 1)$.

$$\sigma_j(t) = e^{iHt/\hbar}\,\sigma_j\,e^{-iHt/\hbar}\,, \qquad\qquad j = x,\,y,\,z\,, \qquad (21.3)$$

and the commutation relation $(H, \sigma_z) = i\,\hbar\Delta\,\sigma_y$, one finds the relation

$$\Delta\langle\sigma_y\rangle_t = -d\langle\sigma_z\rangle_t/dt\,. \qquad (21.4)$$

Hence the coherence $\langle\sigma_y\rangle_t$ is proportional to the tunneling current.

For the isolated TSS, the expectation values (21.1) are conveniently obtained by means of the associated transition amplitudes. With the rotation (3.138) from the localized basis (3.136) to the eigen basis (3.142) of $H_{\rm TSS}$, one finds

$$\begin{aligned} A_{\rm R,R} &\equiv\; <R|\,e^{-iH_{\rm TSS}\,t/\hbar}\,|R> =\; \cos(\tfrac{1}{2}\Delta_{\rm b}t) + i\,(\epsilon/\Delta_{\rm b})\sin(\tfrac{1}{2}\Delta_{\rm b}t)\,, \\ A_{\rm L,R} &\equiv\; <L|\,e^{-iH_{\rm TSS}\,t/\hbar}\,|R> =\; i\,(\Delta/\Delta_{\rm b})\sin(\tfrac{1}{2}\Delta_{\rm b}t)\,. \end{aligned} \qquad (21.5)$$

With these, and with the initial condition $\rho_{\rm R,R}(t=0) = 1$, one finally gets

$$\begin{aligned} \langle\sigma_z\rangle_t^{(0)} &\equiv\; A_{\rm R,R}^*A_{\rm R,R} - A_{\rm L,R}^*A_{\rm L,R} =\; \epsilon^2/\Delta_{\rm b}^2 + (\Delta^2/\Delta_{\rm b}^2)\cos(\Delta_{\rm b}t)\,, \\ \langle\sigma_x\rangle_t^{(0)} &\equiv\; A_{\rm L,R}^*A_{\rm R,R} + A_{\rm R,R}^*A_{\rm L,R} =\; (\epsilon\Delta/\Delta_{\rm b}^2)[\,1 - \cos(\Delta_{\rm b}t)\,]\,, \\ \langle\sigma_y\rangle_t^{(0)} &\equiv\; i(A_{\rm L,R}^*A_{\rm R,R} - A_{\rm R,R}^*A_{\rm L,R}) =\; (\Delta/\Delta_{\rm b})\sin(\Delta_{\rm b}t)\,. \end{aligned} \qquad (21.6)$$

These expressions saturate the bound in Eq. (21.2). The phase-coherent dynamics shows up in the RDM oscillations with frequency $\Delta_{\rm b} = \sqrt{\Delta^2 + \epsilon^2}$. The expression for $\langle\sigma_z\rangle_t^{(0)}$ is sometimes called Rabi's formula.[1] For $\epsilon = 0$, we have $(H, \sigma_x) = 0$, and hence $\sigma_x(t) = \sigma_x(0)$. Therefore, $\langle\sigma_x\rangle_t^{(0)}$ is frozen up at the initial value, which is zero.

The expectation values $\langle\sigma_j\rangle_t$ $(j = x,\,y,\,z)$ of the *dissipative* two-state system are conveniently expressed in terms of the conditional propagating function $J(\zeta, t; \zeta_0, t_0)$. Here, ζ denotes one of the four states of the RDM, as sketched in Fig. 21.1. The

---

[1]For a discussion of the bare TSS and the fictitious spin $\tfrac{1}{2}$ system, we refer the reader to Ref. [458].

diagonal states (populations) of the RDM are labelled by $\zeta = (\eta, 0) = \eta$, and the off-diagonal states (coherences), are labelled by $\zeta = (0, \xi) = \xi$.

We now assume that the TSS starts out at time zero from the diagonal state $\eta_0 = 1$. Propagation of the RDM under the full Hamiltonian is then given by

$$\rho_\zeta(t) = J(\zeta, t; \eta_0 = 1, 0) , \qquad (21.7)$$

and the expectation values of the populations and coherences take the form

$$\langle \sigma_z \rangle_t = \sum_{\eta = \pm 1} \eta \, J(\eta, t; \eta_0 = 1, 0) ,$$

$$\langle \sigma_x \rangle_t = \sum_{\xi = \pm 1} J(\xi, t; \eta_0 = 1, 0) , \qquad (21.8)$$

$$\langle \sigma_y \rangle_t = i \sum_{\xi = \pm 1} \xi \, J(\xi, t; \eta_0 = 1, 0) .$$

We shall take up these expressions below in Subsection 21.2.3.

According to Chapter 5, the propagating function J may be written as a double path sum over the spin paths $\sigma(s) = (2/q_0) q(s)$ and $\sigma'(s) = (2/q_0) q'(s)$

$$J = \int \mathcal{D}\sigma(\cdot) \int \mathcal{D}\sigma'(\cdot) \, \mathcal{A}[\sigma(\cdot)] \mathcal{A}^*[\sigma'(\cdot)] \mathcal{F}[\sigma(\cdot), \sigma'(\cdot)] \qquad (21.9)$$

with appropriate boundary conditions. The functional $\mathcal{A}[\sigma(\cdot)]$ is the probability amplitude for the isolated TSS [analogous to Eq. (21.5)] to follow the path $\sigma(s)$, and $\mathcal{F}[\sigma(\cdot), \sigma'(\cdot)]$ is the real-time influence functional discussed in Chapter 5. For the two-state system, the paths $\sigma(s)$ and $\sigma'(s)$ jump between the two discrete values $+1$ and -1, as depicted in Fig. 4.1 on page 84. Finally, it is expedient, following Eq. (5.30), to introduce antisymmetric and symmetric spin paths $\xi(s)$ and $\eta(s)$,

$$\xi(s) \equiv \tfrac{1}{2} \left[\sigma(s) - \sigma'(s) \right] ; \qquad \eta(s) \equiv \tfrac{1}{2} \left[\sigma(s) + \sigma'(s) \right] , \qquad 0 \leqslant s \leqslant t . \qquad (21.10)$$

Both paths jump back and forth between the values ± 1 under condition $\xi(s) \eta(s) = 0$.

21.1.2 Initial state with shifted and unshifted thermal reservoir

Consider now the case in which the spin or TSS is initially in the diagonal state η_i and the reservoir in an unshifted ($\kappa = 0$) or shifted ($\kappa = 1$) thermal state,

$$\mathbb{W}_{\text{th}}(t = 0, \eta_i; \kappa) = |\eta_i \rangle < \eta_i| \otimes e^{-\beta[\hat{H}_R + \kappa \hat{H}_I[q_i = \eta_i q_0/2]]} / Z_R . \qquad (21.11)$$

When for the product initial state $\mathbb{W}_{\text{th}}(t = 0, \eta_i; \kappa = 0)$ the system-bath coupling is switched on at time zero, this scenario is referred to as preparation class A in the sequel. In class A, the system starts propagation before the bath has relaxed to the related shifted canonical state. Class A applies, e.g., in electron transfer reactions where a particular electronic donor state is suddenly prepared by photoinjection.[2]

---

[2]The observability of electronic coherence in ET reactions for preparation A is studied in Refs. [456, 457], and references therein.

The other important choice is $\kappa = 1$ in Eq. (21.11), referred to as preparation class *B* in the sequel. This initial state may be prepared by holding the TSS for some large time in the state η_i , e.g, by applying a strong negative bias $-\eta_i \hbar \epsilon_0 \Theta(-t)$ with $\epsilon_0 \gg \Delta$, so that the environment could have come into equilibrium with it. At time zero the constraint is released. Such initial state may be designed, e.g., in a rf SQUID device with a suitable choice of the applied magnetic field (cf. Subsection 3.2.2).

Next it is convenient to introduce $\mathfrak{L}(t) = q_0^2 \, L(t)/\hbar$ and the function $Q(t)$ obeying $\ddot{Q}(t) = \mathfrak{L}(t)$. With these, the influence functional tailored to the thermal initial state (21.11) is found from (5.123) with the relations (5.32) and (5.29) as

$$\mathcal{F}_{\mathrm{FV},\kappa}[\eta(\cdot), \xi(\cdot)]/\hbar = e^{-S_{\mathrm{FV},\kappa}[\eta(\cdot), \xi(\cdot)]/\hbar} \,,$$

$$S_{\mathrm{FV},\kappa}[\eta(\cdot), \xi(\cdot)]/\hbar = \int_0^t ds_2 \, \xi(s_2) \int_0^{s_2} ds_1 \, [\mathfrak{L}'(s_2 - s_1)\xi(s_1) + i\,\dot{Q}''(s_2 - s_1)\eta(s_1)]$$

$$+ \, i\,(1 - \kappa)\eta_i \int_0^t ds \, \dot{Q}''(s)\xi(s) \,. \tag{21.12}$$

The function $Q(t)$ has the spectral representation (18.42). In the Ohmic scaling limit, $\dot{Q}''(t) \propto \delta(t)$, the last term in Eq. (21.12) is zero, irrespective of the choice for κ, because $\xi(0) = 0$. In the general case, the slippage term vanishes with increasing t on the time scale of order $1/\omega_c$. Therefore, the effects of different preparation of the initial state are practically relevant only in the adiabatic limit, in which ω_c is of the order of Δ or smaller.

21.1.3 Thermal initial state

In most cases of interest, the system under consideration, e.g. a tunneling system in a solid, can not be prepared in a particular pure state. In the usual experimentally reproducible situation, the system has relaxed to the thermal equilibrium state, which is entangled with the environment, before a measurement takes place. Then, a certain observable is measured at time zero and again at time t. Repeated measurement cycles establish the equilibrium autocorrelation function of this observable.

The equilibrium autocorrelation functions of the TSS are ($j = x,\, y,\, z$)

$$C_j^{\pm}(t) \equiv \langle \sigma_j(\pm t)\sigma_j(0)\rangle_\beta - \langle \sigma_j \rangle_{\mathrm{eq}}^2 = \mathrm{tr}\left\{ e^{-\beta H}\, \sigma_j(\pm t)\sigma_j(0)\right\}/Z - \langle \sigma_j \rangle_{\mathrm{eq}}^2 \,, \tag{21.13}$$

where $\sigma_j(t)$ is given in Eq. (21.3). The equilibrium average $\langle \sigma_j \rangle_{\mathrm{eq}}^2$ is subtracted for $j = x,\, z$ in order that the Fourier transforms are well-defined. Actually, $C_y^{\pm}(t)$ can be found from $C_z^{\pm}(t)$ by differentiation, $C_y^{\pm}(t - t') = (\partial^2/\partial t\, \partial t')\, C_z^{\pm}(t - t')/\Delta^2$, as follows with use of the commutation relation $(H, \sigma_z) = i\hbar\Delta\,\sigma_y$. The real part of $C_j^{\pm}(t)$ is the symmetric equilibrium autocorrelation function of the observable σ_j, while the imaginary part of $C_j^{\pm}(t)$ is connected with the linear response $\chi_j(t)$ to an external force coupled to σ_j. We have

$$S_j(t) \equiv \mathrm{Re}\, C_j^{\pm}(t) = \tfrac{1}{2}\langle\, [\sigma_j(t)\sigma_j(0) + \sigma_j(0)\sigma_j(t)\,]\rangle_\beta - \langle \sigma_j \rangle_{\mathrm{eq}}^2$$

$$\hbar\chi_j(t) \equiv -2\,\Theta(t)\, \mathrm{Im}\, C_j^+(t) = i\,\Theta(t)\langle\, [\sigma_j(t)\sigma_j(0) - \sigma_j(0)\sigma_j(t)\,]\rangle_\beta \,. \tag{21.14}$$

If, firstly, we disregard the bath correlations, we get with the representations (3.137)

$$
\begin{aligned}
C_z^+(t) &= \frac{1}{Z}\Big\{ <R|\,e^{-\beta H}|R> <R|\sigma_z(t)|R> \; - \; <L|\,e^{-\beta H}|L> <L|\sigma_z(t)|L> \\
&\quad + \; <R|\,e^{-\beta H}|L> <L|\sigma_z(t)|R> \; - \; <L|\,e^{-\beta H}|R> <R|\sigma_z(t)|L> \Big\} - \langle\sigma_z\rangle_{\text{eq}}^2 \,,
\end{aligned}
$$

$$
\begin{aligned}
C_x^+(t) &= \frac{1}{Z}\Big\{ <R|\,e^{-\beta H}|R> <R|\sigma_x(t)|L> \; + \; <L|\,e^{-\beta H}|L> <L|\sigma_x(t)|R> \\
&\quad + \; <R|\,e^{-\beta H}|L> <L|\sigma_x(t)|L> \; + \; <L|\,e^{-\beta H}|R> <R|\sigma_x(t)|R> \Big\} - \langle\sigma_x\rangle_{\text{eq}}^2 \,.
\end{aligned}
$$

Next, we take into account that the TSS is actually entangled with the environment when at time zero the first measurement takes place. To deal with this case, we proceed as follows. Suppose that at time $t_{\text{p}} < 0$ the system is released from the RDM state ζ_p, and then propagates under the full Hamiltonian. At time $t_0 = 0$, the system is measured to be in the state ζ_0. At later time t, a second measurement is performed in which it is found in the state ζ. The corresponding three-time conditional propagating function for this sequence of events is denoted by $J(\zeta, t;\, \zeta_0, 0;\, \zeta_p, t_{\text{p}})$. For an ergodic system, the correlation of a dynamical variable at time t with the same (or a different) dynamical variable at time zero does not depend on the particular initial state chosen at time t_{p}, if t_{p} is shifted backwards into the infinite past. Thus, for convenience, we may choose at time t_{p} a product initial state with the reservoir in thermal equilibrium and the system in a diagonal state, say in the state $\eta_{\text{p}} = +1$.[3] Then, before the first measurement of a TSS observable at time zero is performed, the TSS has already equilibrated with the environment and is in an entangled canonical state of the system-plus-reservoir complex. With these preliminary thoughts we see that we may study equilibrium correlation functions within the standard real-time influence functional approach for propagating functions introduced in Subsec. 5.3.3.

Consider first the correlation function $C_z^+(t)$, in which the system finally ends up in a diagonal state of the RDM. Regarding the symmetrized part, the system dwells at time zero in a diagonal state, whereas for the anti-symmetrized part it is at this time in an off-diagonal state. Taking into account the weight factors of these states, we get for $S_z(t) = \operatorname{Re} C_z^\pm(t)$ and $\hbar\chi_z(t) = -2\Theta(t)\operatorname{Im} C_z^+(t)$ the representations

$$
S_z(t) = \lim_{t_{\text{p}}\to-\infty} \sum_{\eta=\pm 1}\sum_{\eta_0=\pm 1} \eta\,\eta_0\, J(\eta, |t|;\, \eta_0, 0;\, \eta_{\text{p}}, t_{\text{p}}) - \langle\sigma_z\rangle_{\text{eq}}^2 \,, \tag{21.15}
$$

$$
\hbar\chi_z(t) = \lim_{t_{\text{p}}\to-\infty} 2i\,\Theta(t) \sum_{\eta=\pm 1}\sum_{\xi_0=\pm 1} \eta\,\xi_0\, J(\eta, t;\, \xi_0, 0;\, \eta_{\text{p}}, t_{\text{p}}) \,. \tag{21.16}
$$

The function $S_z(t)$ is relevant, e.g., for inelastic neutron scattering at interstitials or defects in a lattice. The differential cross section of this process is given by [214]

$$
\frac{\partial^2 \Sigma_{\text{inel}}}{\partial\Omega\,\partial\omega} = \frac{\Sigma_{\text{inc}}}{4\pi}\frac{k_{\text{f}}}{k_{\text{i}}}\, S_{\text{inel}}(\mathbf{k}, \omega) \,, \tag{21.17}
$$

---

[3]The product state may be prepared, e.g., according to the preparation class A, or according to the preparation class B, which have been introduced in Subsection 21.1.2.

where $\hbar\mathbf{k}$ and $\hbar\omega$ are momentum and energy transfer of the neutrons, and Σ_{inc} is the incoherent cross section of the scatterer. The scattering function, which is the dynamic structure factor for a scatterer with position $\mathbf{q}(t)$ at time t, is given by

$$S_{\mathrm{inel}}(\mathbf{k}, \omega) = \frac{1}{2\pi} \int_{-\infty}^{\infty} dt\, e^{i\omega t} \left(\langle e^{-i\mathbf{k}\cdot\mathbf{q}(0)} e^{i\mathbf{k}\cdot\mathbf{q}(t)} \rangle_\beta - \left| \langle e^{i\mathbf{k}\cdot\mathbf{q}} \rangle_\beta \right|^2 \right) . \tag{21.18}$$

For a defect tunneling between positions $\pm\frac{1}{2}\mathbf{q}_0$, we have $\mathbf{q}(t) = \frac{1}{2}\mathbf{q}_0\,\sigma_z(t)$. This yields

$$S_{\mathrm{inel}}(\mathbf{k}, \omega) = \sin^2\left(\tfrac{1}{2}\mathbf{k}\cdot\mathbf{q}_0\right) \tilde{C}_z^-(\omega)/2\pi . \tag{21.19}$$

Here, $\tilde{C}_z^-(\omega)$ is the Fourier transform of the pseudo-spin correlation function $C_z^-(t)$,

$$\tilde{C}_z^-(\omega) = \int_{-\infty}^{\infty} dt\, e^{i\omega t} \left(\langle \sigma_z(0)\sigma_z(t) \rangle_\beta - \langle \sigma_z \rangle_{\mathrm{eq}}^2 \right) . \tag{21.20}$$

With the definition $\tilde{S}_z(\omega) = [\tilde{C}_z^+(\omega) + \tilde{C}_z^-(\omega)]/2$ and the relation (6.21) one obtains

$$\tilde{C}_z^-(\omega) = \frac{2}{1 + e^{\beta\hbar\omega}}\, \tilde{S}_z(\omega) . \tag{21.21}$$

With this, the differential cross section for *inelastic* neutron scattering takes the form

$$\frac{\partial^2 \Sigma_{\mathrm{inel}}}{\partial\Omega\,\partial\omega} = \frac{\Sigma_{\mathrm{inc}}}{4\pi^2} \frac{k_{\mathrm{f}}}{k_{\mathrm{i}}} \sin^2\left(\tfrac{1}{2}\mathbf{k}\cdot\mathbf{q}_0\right) \frac{\tilde{S}_z(\omega)}{1 + e^{\beta\hbar\omega}} . \tag{21.22}$$

Hence, inelastic neutron scattering yields data directly for the spectral function $\tilde{S}_z(\omega)$.

The other important equilibrium correlation function is that of the tunneling or coherence operator σ_x. Since the polaron unitary operator (18.37) does not commute with σ_x, the correlation function of the bare coherence operator,

$$C_{x,\mathrm{bare}}^+(t) = \mathrm{tr}\left\{ e^{-\beta H} e^{iHt/\hbar}\, \sigma_x\, e^{-iHt/\hbar}\, \sigma_x \right\}/Z - \left[\mathrm{tr}\left\{ e^{-\beta H}\, \sigma_x \right\}/Z \right]^2 , \tag{21.23}$$

differs from that of the polaron-dressed coherence operator (18.39),

$$\begin{aligned} C_{x,\mathrm{pd}}^+(t) &= \mathrm{tr}\left\{ e^{-\beta H} e^{iHt/\hbar}\, \tilde{\sigma}_x\, e^{-iHt/\hbar}\, \tilde{\sigma}_x \right\}/Z - \left[\mathrm{tr}\left\{ e^{-\beta H}\, \tilde{\sigma}_x \right\}/Z \right]^2 , \\ &= \mathrm{tr}\left\{ e^{-\beta\tilde{H}} e^{i\tilde{H}t/\hbar}\, \sigma_x\, e^{-i\tilde{H}t/\hbar}\, \sigma_x \right\}/Z - \left[\mathrm{tr}\left\{ e^{-\beta\tilde{H}}\, \sigma_x \right\}/Z \right]^2 . \end{aligned} \tag{21.24}$$

Consider next both for the bare and pd correlation functions, the symmetric part $S_x(t) = \mathrm{Re}\, C_x^+(t)$ and antisymmetric part $\hbar\chi_x(t) = -2\Theta(t)\,\mathrm{Im}\, C_x^+(t)$. In path representation, these functions are described in terms of a four-time propagating function $J(\zeta, t; \zeta_+, 0_+; \zeta_-, 0_-, \zeta_{\mathrm{p}}, t_{\mathrm{p}})$. Due to the operation of σ_x at time zero, the TSS makes at this time a transition from an off-diagonal to a diagonal state, referred to as group A, or a transition from a diagonal to an off-diagonal state, referred to as group B. Finally, the system ends up off-diagonal. Assuming again preparation of the TSS in a diagonal state at time t_{p}, we get $S_x(t) = S_x^{(\mathrm{A})}(t) + S_x^{(\mathrm{B})}(t) - \langle \sigma_x \rangle_{\mathrm{eq}}^2$, where

$$S_x^{(A)}(t) = \lim_{t_p \to -\infty} \frac{1}{2} \sum_{\{\xi, \xi_+, \eta_- = \pm 1\}} J(\xi, |t|; \xi_+, 0_+; \eta_-, 0_-; \eta_p, t_p) \,,$$

$$S_x^{(B)}(t) = \lim_{t_p \to -\infty} \frac{1}{2} \sum_{\{\xi, \eta_+, \xi_- = \pm 1\}} J(\xi, |t|; \eta_+, 0_+; \xi_-, 0_-; \eta_p, t_p) \,, \tag{21.25}$$

and for the linear response function $\chi_x(t) = \chi_x^{(A)}(t) + \chi_x^{(B)}(t)$ with

$$\hbar \chi_x^{(A)}(t) = \lim_{t_p \to -\infty} i\,\Theta(t) \sum_{\{\xi, \xi_+, \eta_- = \pm 1\}} \xi_+ \eta_-\, J(\xi, t; \xi_+, 0_+; \eta_-, 0_-; \eta_p, t_p) \,,$$

$$\hbar \chi_x^{(B)}(t) = \lim_{t_p \to -\infty} i\,\Theta(t) \sum_{\{\xi, \eta_+, \xi_- = \pm 1\}} \eta_+ \xi_-\, J(\xi, t; \eta_+, 0_+; \xi_-, 0_-; \eta_p, t_p) \,. \tag{21.26}$$

The bare and polaron-dressed cases of the expressions (21.25) and (21.26) differ in the bath correlations, as will be discussed in Subsec. 21.2.5.

21.2 Exact formal expressions for the system dynamics

21.2.1 Sojourns and blips

Since the spin paths $\xi(s)$ and $\eta(s)$ are piecewise constant with occasional sudden jumps in between, it is expedient in consideration of the relation

$$\mathfrak{L}(t_2 - t_1) = -\frac{\partial^2}{\partial t_2\, \partial t_1} Q(t_2 - t_1) \,, \tag{21.27}$$

to perform integrations by parts in the expression (21.12). Then the influence action (21.12) takes the velocity representation

$$\mathcal{S}_{\mathrm{FV},\kappa}[\eta(\cdot), \xi(\cdot)]/\hbar = -\int_{t_0}^{t} ds_2\, \dot{\xi}(s_2) \int_{t_0}^{s_2} ds_1 [\, Q'(s_2 - s_1)\dot{\xi}(s_1) + i\, Q''(s_2 - s_1)\dot{\eta}(s_1) \,]$$

$$- i\,(1 - \kappa)\eta_i \int_0^t ds\, Q''(s)\dot{\xi}(s) \,. \tag{21.28}$$

The double path sum (21.9) can be visualized as a path sum for a single path that visits the four states of the reduced density. The combinatorial problem facing us then is the sum of paths a walker can go along the edges of the square sketched in Fig. 21.1 on 379. A period the path spends in a diagonal state has been dubbed *sojourn* by Leggett *et al.* [86], and a period the path dwells in an off-diagonal state has been termed *blip*. During a sojourn, the function $\xi(\tau)$ is zero, whereas during a blip interval the function $\eta(\tau)$ is zero. There are two sojourn states, labelled by $\eta_j = +1$ [state (RR)] and $\eta_j = -1$ [state (LL)]. Similarly, there are two blip states, labelled by $\xi_j = +1$ [state (RL)] and $\xi_j = -1$ [state (LR)].

A general sojourn-to-sojourn path along the edges of the square in Fig. 21.1 making $2n$ transitions at intermediate times t_j in chronological order $0 \leqslant t_1 \leqslant t_2 \cdots \leqslant t_{2n} \leqslant t$, and initial time $t_0 = 0$, is parametrized by

$$\eta^{(n)}(s) = \sum_{j=0}^{n} \eta_k [\Theta(s - t_{2k}) - \Theta(s - t_{2k+1})] ,$$

$$\xi^{(n)}(s) = \sum_{j=1}^{n} \xi_j [\Theta(s - t_{2j-1}) - \Theta(s - t_{2j})] . \qquad (21.29)$$

The intervals spent in a blip state and in a sojourn state are labelled τ_j and s_j,

$$\tau_j = t_{2j} - t_{2j-1} , \qquad s_j = t_{2j+1} - t_{2j} . \qquad (21.30)$$

In a pictorial representation, the $\{\xi_j\}$ are blip charges, and the $\{\eta_k\}$ are sojourn charges. Because of the opposite signs of the two terms in the square brackets in Eq. (21.29), a blip with label ξ_j represents a blip dipole with successive charges ξ_j and $-\xi_j$, and sojourn with label η_k represents a sojourn dipole with successive charges η_k and $-\eta_k$. The influence action (21.28) with the paths (21.29) may be regarded as entirety of correlations among the chronological sequence of sojourn and blip dipoles. The real part of the action covers the intra- and inter-blip correlations, and the imaginary part the correlations between the sojourns and the subsequent blips. There are no sojourn-sojourn correlations. For $n + 1$ sojourns and n interjacent blips, and with use of the notation $Q_{j,k} = Q(t_j - t_k)$, the resulting influence function is

$$\mathcal{F}_n = G_n H_n ,$$

$$G_n = \exp\left[-\sum_{j=1}^{n} Q'_{2j,2j-1} \right] \exp\left[-\sum_{j=2}^{n} \sum_{k=1}^{j-1} \xi_j \Lambda_{j,k} \xi_k \right] , \qquad (21.31)$$

$$H_n = \exp\left\{ i \sum_{k=0}^{n-1} \eta_k \phi_{k,n} \right\} , \qquad \text{with} \qquad \phi_{k,n} = \sum_{j=k+1}^{n} \xi_j X_{j,k} .$$

The first exponential in G_n represents the self-interactions of the n blips, and the second exponential contains the interactions between the blips.[4] The exponent of H_n includes the blip-sojourn interactions. The phase $\phi_{k,n}$ describes the correlations between sojourn k and all subsequent blips. The last sojourn η_n spanning the interval $t - t_{2n}$ is free of bath correlations. The blip-blip correlations $\Lambda_{j,k}$ and sojourn-blip correlations $X_{j,k}$ consist in each case of four charge interactions,

$$\Lambda_{j,k} = Q'_{2j,2k-1} - Q'_{2j-1,2k-1} + Q'_{2j-1,2k} - Q'_{2j,2k} ,$$

$$X_{j,k} = Q''_{2j,2k+1} - Q''_{2j-1,2k+1} + (1 - \kappa \delta_{k,0})[Q''_{2j-1,2k} - Q''_{2j,2k}] . \qquad (21.32)$$

The contribution with prefactor κ in $X_{j,0}$ ensures that the correlations of the sojourn charge η_0 located at initial time zero with the later blip charges are absent in case of the shifted initial bath state, which is the case $\kappa = 1$ in Eq. (21.11).

---

[4]The factor G_n may be regarded as a charge interaction factor for $2n$ charges $\xi_j = \pm 1$. For a transition at time t_j from ζ-state $(\pm 1, 0)$ to $(0, +1)$ and from $(0, -1)$ to $(\pm 1, 0)$, the charge is $\xi_j = +1$, and for a transition in reverse direction the charge is ξ_j is -1.

The function G_n acts as a filtering function which suppresses long blips. Therefore, it prolongs dwells of the system in sojourn states on average. Physically, this is because the environment is continuously measuring σ_z and therefore reduces occupation of blip states and, as a consequence, quantum interference between the eigenstates of σ_z.

It is straightforward to write down the various weight factors resulting from the amplitude product $\mathcal{A}[\sigma]\,\mathcal{A}^*[\sigma']$ for the TSS Hamiltonian (3.136). The weight to switch per unit time from a diagonal state η to an off-diagonal state ξ, or vice versa, is

$$-i\,\xi\eta\,\Delta/2\,. \tag{21.33}$$

Thus we have the weight factor $-i\Delta/2$ for transitions $(RL) \Longleftrightarrow (RR)$ and $(LL) \Longleftrightarrow (LR)$, and $i\Delta/2$ for transitions $(RL) \Longleftrightarrow (LL)$ and $(RR) \Longleftrightarrow (LR)$.

The weight factor to stay in a sojourn is unity, while the weight factor to stay in the jth blip with label ξ_j and length $\tau_j = t_{2j} - t_{2j-1}$ is $\exp(i\,\epsilon\,\xi_j\tau_j)$. The impact of the bias in a n-blip sequence is accumulated in the bias weight factor

$$B_n \;=\; \exp\left(i\,\epsilon \sum_{j=1}^{n} \xi_j\tau_j \right). \tag{21.34}$$

The sum over all paths along the edges of the square sketched in Fig. 21.1 now means (i) to arrange the paths according the number n of transitions, (ii) to integrate out the n time-ordered flip times in the interval from zero to t, (iii) to sum over all possible intermediate sojourn and blip states the path may take in the given order, and (iv) to sum up the contributions of all orders of transitions

$$\sum_{\text{all paths}} \cdots \;\to\; \sum_{n=0}^{\infty} \sum_{\{\eta_j=\pm 1\}} \sum_{\{\xi_j=\pm 1\}} \int_{t\geqslant t_n\geqslant t_{n-1}\cdots\geqslant t_1\geqslant 0} dt_n\,dt_{n-1}\cdots dt_1 \cdots\,. \tag{21.35}$$

Correlated system-reservoir initial states at time zero, e.g. in equilibrium correlation functions, are taking into account by extending the evolution of the system since the infine past as discussed in Subsec. 5.3.3 (cf. Fig. 5.4) and Subsec. 21.1.3. To this end, we divide the integrations over the time-ordered flip times $\{t_j\}$ into a negative and a positive time branch. We shall use the compact notation

$$\int_{t_p}^{t} \mathcal{D}_{k,\ell}\{t_j\} \;\times\; \cdots \;\equiv\; \int_{0}^{t} dt_{\ell+k} \cdots \int_{0}^{t_{\ell+2}} dt_{\ell+1} \int_{t_p}^{0} dt_\ell \cdots \int_{t_p}^{t_2} dt_1\,\Delta^{k+\ell} \;\times\; \cdots\,. \tag{21.36}$$

Here, ℓ is the number of flips in the negative time branch $t_p \leqslant t' \leqslant 0$ and k is the number of flips in the positive time branch, $0 \leqslant t' \leqslant t$. The integration symbol includes, for convenience, the tunneling amplitude factor. Eventually, the preparation time t_p is sent to the infinite past. We are now well equipped to deal with the various conditional propagating functions for the boundary conditions of interest.

21.2.2 Conditional propagating functions

We have seen in Section 21.1 that the relevant expectation values and equilibrium autocorrelation functions can be expressed in terms of two-time, three-time, and four-time conditional propagating functions.

If the expectation values $\langle \sigma_j \rangle_t$ ($j = x, y, z$) imply a factorizing initial state of the global system at time zero, they can be expressed in terms of the two-time conditional propagating function $J(\zeta, t; \eta_0, 0)$. Here the initial state of the system is the sojourn state η_0, and the preparation of the reservoir may be done either according to class A or according to class B. When the final state is again a sojourn, the number of transitions the system makes is even, while it is odd, when the final state is a blip.

Upon collecting the various weight factors discussed in the previous subsection we obtain for the two-time sojourn-sojourn correlation function the series expression

$$J(\eta, t; \eta_0, 0) = \delta_{\eta,\eta_0} + \eta\eta_0 \sum_{m=1}^{\infty} \frac{(-1)^m}{2^{2m}} \int_0^t \mathcal{D}_{2m,0}\{t_j\} \sum_{\{\xi_j=\pm1\}} G_m B_m \sum_{\{\eta_j=\pm1\}'} H_m . \quad (21.37)$$

In contrast, the series for the sojourn-blip correlation function reads

$$J(\xi, t; \eta_0, 0) = -i\xi\eta_0 \sum_{m=1}^{\infty} \frac{(-1)^{m-1}}{2^{2m-1}} \int_0^t \mathcal{D}_{2m-1,0}\{t_j\} \sum_{\{\xi_j=\pm1\}'} G_m B_m \sum_{\{\eta_j=\pm1\}'} H_m . \quad (21.38)$$

The prime in $\{\eta_j = \pm1\}'$ and $\{\xi_j = \pm1\}'$ indicates that the boundary charges are those indicated in the argument of the propagating functions.

The three-time conditional propagating function for being in sojourn states at the preparation time t_p, at time zero, and again at time t reads

$$J(\eta, t; \eta_0, 0; \eta_p, t_p) = \delta_{\eta,\eta_0}\delta_{\eta_0,\eta_p} \quad (21.39)$$

$$+ \delta_{\eta,\eta_0}\eta_0\eta_p \sum_{n=1}^{\infty} \frac{(-1)^n}{2^{2n}} \int_{t_p}^t \mathcal{D}_{0,2n}\{t_j\} \sum_{\{\xi_j=\pm1\}} G_n B_n \sum_{\{\eta_j=\pm1\}'} H_n$$

$$+ \delta_{\eta_0,\eta_p}\eta\eta_0 \sum_{m=1}^{\infty} \frac{(-1)^m}{2^{2m}} \int_{t_p}^t \mathcal{D}_{2m,0}\{t_j\} \sum_{\{\xi_j=\pm1\}} G_m B_m \sum_{\{\eta_j=\pm1\}'} H_m$$

$$+ \eta\eta_p \sum_{m=1}^{\infty}\sum_{n=1}^{\infty} \frac{(-1)^{n+m}}{2^{2(n+m)}} \int_{t_p}^t \mathcal{D}_{2m,2n}\{t_j\} \sum_{\{\xi_j=\pm1\}} G_{m+n} B_{m+n} \sum_{\{\eta_j=\pm1\}'} H_{m+n} .$$

In each order of Δ the charge sum is again over all possible arrangements of the intermediate charges, and the boundary charges are those indicated in the argument of the propagating function.

The analogous series for the case in which a blip state is occupied at time zero is

$$J(\eta, t; \xi_0, 0; \eta_p, t_p) = \eta\eta_p \sum_{m=1}^{\infty}\sum_{n=1}^{\infty} \frac{(-1)^{n+m-1}}{2^{2(n+m-1)}} \int_{t_p}^t \mathcal{D}_{2m-1,2n-1}\{t_j\}$$

$$\times \sum_{\{\xi_j=\pm1\}'} G_{m+n-1} B_{m+n-1} \sum_{\{\eta_j=\pm1\}'} H_{m+n-1} . \quad (21.40)$$

387

Here the sum is again over all intermediate blip and sojourn states.

21.2.3 The expectation values $\langle \sigma_j \rangle_t$ $(j = x, y, z)$

Consider first the population $\langle \sigma_z \rangle_t$. Inserting the series (21.37) for the two-time conditional propagating function into Eq. (21.8), and performing the summation over the intermediate sojourns $\{\eta_j = \pm 1\}'$, we obtain the exact formal series expression [86]

$$\langle \sigma_z \rangle_t = 1 + \sum_{m=1}^{\infty} (-1)^m \int_0^t \mathcal{D}_{2m,0}\{t_j\} \frac{1}{2^m} \sum_{\{\xi_j = \pm 1\}} \left(F_m^{(+)} B_m^{(s)} - F_m^{(-)} B_m^{(a)} \right). \tag{21.41}$$

Here the bias dependence is divided up into the even and odd contributions

$$B_m^{(s)} = \cos\left(\epsilon \sum_{j=1}^{m} \xi_j \tau_j \right); \qquad B_m^{(a)} = \sin\left(\epsilon \sum_{j=1}^{m} \xi_j \tau_j \right), \tag{21.42}$$

and the effects of the environment are in the influence functions

$$F_m^{(+)} = G_m \prod_{k=0}^{m-1} \cos\left(\phi_{k,m} \right), \qquad F_m^{(-)} = G_m \sin\left(\phi_{0,m} \right) \prod_{k=1}^{m-1} \cos\left(\phi_{k,m} \right). \tag{21.43}$$

The term G_m describes the intra- and inter-blip correlations, and the product of phase factors represents the entirety of sojourn-blip correlations (21.31). Again, the sum over the labels $\{\xi_j\}$ captures all possible intermediate blip states, $\{\xi_j = \pm 1\}$.

For later convenience, we introduce the notation

$$\langle \sigma_z \rangle_t \equiv P(t) = P_R(t) - P_L(t) = 2P_R(t) - 1, \tag{21.44}$$

where $P_{R/L}(t)$ is the occupation probability of the right/left well, respectively. There holds $P(0) = 1$. The function $P(t)$ may be split into the components which are symmetric (s) and antisymmetric (a) under inversion of the bias $(\epsilon \to -\epsilon)$,

$$P(t) = P_s(t) + P_a(t). \tag{21.45}$$

When the damping of the system persists until time infinity, the system is ergodic and relaxes to the canonical state.[5] The even part $P_s(t)$ goes to zero as $t \to \infty$, and the odd part $P_a(t)$ approaches the equilibrium value $P_\infty \equiv \langle \sigma_z \rangle_{eq}$,

$$\langle \sigma_z \rangle_{eq} \equiv P_\infty = \lim_{t \to \infty} \sum_{m=1}^{\infty} (-1)^{m-1} \int_0^t \mathcal{D}_{2m,0}\{t_j\} \frac{1}{2^m} \sum_{\{\xi_j = \pm 1\}} F_m^{(-)} B_m^{(a)}. \tag{21.46}$$

This expression is the outcome of the dynamical nonequilibrium approach. On the other hand, one may specify $\langle \sigma_z \rangle_{eq}$ in the framework of thermodynamics, as given in Eqs. (19.2) and (19.3). For an ergodic system, the resulting expressions are in correspondence with those found from Eq. (21.46) (see also the discussion in Sec. 22.9).

---

[5] The criteria for the spectral density $G(\omega)$ of an ergodic open system are specified in Sec. 3.1.10.

With the series (21.38) for the two-time correlation function $J(\xi, t; \eta_0, 0)$ and the relations (21.8), the coherences are found to be given by [459]

$$\langle \sigma_x \rangle_t = \sum_{m=1}^{\infty} (-1)^{m-1} \int_0^t \mathcal{D}_{2m-1,0}\{t_j\} \frac{1}{2^m} \sum_{\{\xi_j = \pm 1\}} \xi_m \left(F_m^{(+)} B_m^{(a)} + F_m^{(-)} B_m^{(s)} \right), \quad (21.47)$$

$$\langle \sigma_y \rangle_t = \sum_{m=1}^{\infty} (-1)^{m-1} \int_0^t \mathcal{D}_{2m-1,0}\{t_j\} \frac{1}{2^m} \sum_{\{\xi_j = \pm 1\}} \left(F_m^{(+)} B_m^{(s)} - F_m^{(-)} B_m^{(a)} \right). \quad (21.48)$$

The series expressions (21.41), (21.47) and (21.48) are exact formal expressions for the time evolution of the reduced density matrix. The expressions (21.41) and (21.48) are consistent with the relation (21.4).

Finally, we touch upon the different initial conditions discussed in Subsec. 21.1.2. In class A, the initial sojourn starts at time zero, whereas in class B it starts in the infinite past. We then have for $j = 2, 3, \cdots$

$$\begin{aligned}
X_{j,0}^{(A)} &= Q_{2j,1}'' + Q_{2j-1,0}'' - Q_{2j,0}'' - Q_{2j-1,1}'' , \\
X_{j,0}^{(B)} &= Q_{2j,1}'' - Q_{2j-1,1}'' ,
\end{aligned} \qquad (21.49)$$

and for $j = 1$

$$\begin{aligned}
X_{1,0}^{(A)} &= Q_{2,1}'' + Q_{1,0}'' - Q_{2,0}'' , \\
X_{1,0}^{(B)} &= Q_{2,1}'' .
\end{aligned} \qquad (21.50)$$

In the Ohmic scaling limit (18.53), the correlation function $Q''(t)$ is instantaneous, $Q''(t) = \pi K \operatorname{sgn}(t)$. Then the two preparation classes coincide, $X_{j,0}^{(A)} = X_{j,0}^{(B)}$, and thus $X_{j,k} = \delta_{j,k+1} X_{k+1,k}$. This yields $\phi_{k,m} = \xi_{k+1} X_{k+1,k} = \xi_{k+1} \pi K$. As a result, the influence functions (21.43) are simplified to the concise expressions

$$F_m^{(+)} = [\cos(\pi K)]^m G_m ; \qquad F_m^{(-)} = \xi_1 \tan(\pi K) F_m^{(+)} . \qquad (21.51)$$

We shall use these forms below in Section 22.6.

21.2.4 Correlation and response function of the populations

With insertion of the series (21.39) into Eq. (21.15) and summation over the intermediate sojourns, the symmetric correlation function takes the form [223]

$$\begin{aligned}
S_z(t) &= S_{z,\text{unc}}(t) + R(t) , \\
S_{z,\text{unc}}(t) &= P_s(|t|) + P_\infty [P_a(|t|) - P_\infty] .
\end{aligned} \qquad (21.52)$$

Here, $P_{s,a}(t)$ are the symmetric and antisymmetric parts of $\langle \sigma_z \rangle_t$ in Eq. (21.41), and $P_\infty \equiv P_a(t \to \infty) = \langle \sigma_z \rangle_{\text{eq}}$ is the equilibrium value of σ_z. The term $S_{z,\text{unc}}(t)$ is the contribution to $S_z(t)$ in which the entanglement of the TSS with the bath at time zero is disregarded. This formally corresponds to the neglect of the bath correlations $Q(\tau)$ between the negative-time and the positive-time branch of the full spin path. If we had chosen a product initial state for the system-plus-reservoir complex, just before the first measurement takes place, then $S_{z,\,\mathrm{unc}}(t)$ would be the result for the symmetric σ_z autocorrelation function. The residual term $R(t)$ in Eq. (21.52) describes the dynamical effects at time t due to the entangled system-bath initial state,

$$
\begin{aligned}
R(t) \;=\;& \lim_{t_p \to -\infty} \sum_{m=1}^{\infty} \sum_{n=1}^{\infty} (-1)^{m+n-1} \int_{t_p}^{|t|} \mathcal{D}_{2m,2n}\{t_j\} \\
&\times \frac{1}{2^{m+n}} \sum_{\{\xi_j = \pm 1\}} \Big[F_{m,n} B_m^{(s)} B_n^{(s)} - \big(F_{m,n} - F_m^{(-)} F_n^{(-)} \big) B_m^{(a)} B_n^{(a)} \Big],
\end{aligned}
\tag{21.53}
$$

$$
F_{m,n} \;=\; G_{m+n} \sin\big(\phi_{n,m+n}\big) \sin\big(\phi_{0,m+n}\big) \prod_{k=1\,(k \neq n)}^{m+n-1} \cos\big(\phi_{k,m+n}\big). \tag{21.54}
$$

With the scaling form (18.53), this expression simplifies to

$$
F_{m,n} \;=\; \xi_1 \xi_{n+1} \big[\sin(\pi K) \big]^2 \big[\cos(\pi K) \big]^{m+n-2} G_{m+n}. \tag{21.55}
$$

The term $F_m^{(-)} F_n^{(-)}$ in the round bracket in Eq. (21.53) is a subtraction term in which cross correlations are absent. This term is just the counter term of the contribution $P_\infty P_a(|t|)$ in Eq.(21.52). The piece $R(t)$ is the major contribution to $S_z(t)$ at low temperature and long time, as we shall see below. Since $\lim_{t\to\infty} P(t) = P_\infty$ and $\lim_{t\to\infty} R(t) = 0$, there holds $\lim_{t\to\infty} S_z(t) = 0$, due to the subtraction in (21.13).

Similarly, the response function (21.16) is found with use of Eq. (21.40) as [223]

$$
\begin{aligned}
\chi_z(t) \;=\;& \Theta(t) \lim_{t_p \to -\infty} \frac{4}{\hbar} \sum_{m=1}^{\infty} \sum_{n=1}^{\infty} \frac{(-1)^{m+n}}{2^{m+n}} \\
&\times \int_{t_p}^{t} \mathcal{D}_{2m-1,2n-1}\{t_j\} \sum_{\{\xi_j = \pm 1\}} \xi_n F_{m+n-1}^{(-)} B_{m+n-1}^{(s)}.
\end{aligned}
\tag{21.56}
$$

An essential difference between the functions $S_z(t)$ and $\chi_z(t)$ is that in the former the TSS is in a sojourn at time zero, while in the latter it is in a blip at this time.

The static nonlinear susceptibility is defined by the relations

$$
\overline{\chi}_z \;\equiv\; \tilde{\chi}_z(\omega = 0) \;=\; \int_0^\infty dt\, \chi_z(t). \tag{21.57}
$$

With insertion of the series (21.56), displacement of the time axis, and rearrangement of the double series, we readily obtain the path sum expression

$$
\overline{\chi}_z \;=\; \lim_{t \to \infty} \frac{2}{\hbar} \sum_{m=1}^{\infty} (-1)^{m-1} \int_0^t \mathcal{D}_{2m,0}\{t_j\} \frac{1}{2^m} \sum_{\{\xi_j \pm 1\}} F_m^{(-)} B_m^{(s)} \Big(\sum_{n=1}^{m} \xi_n \tau_n \Big). \tag{21.58}
$$

In the derivation, the integrations over the subintervals $\tau_n^- = -t_{2n-1}$ and $\tau_n^+ = t_{2n}$ of the blip length $\tau_n = t_{2n} - t_{2n-1}$ have been handled as

$$\int_0^\infty d\tau_n^- \, d\tau_n^+ \, f(\tau_n^- + \tau_n^+) \;=\; \int_0^\infty d\tau_n \, \tau_n f(\tau_n) \, . \tag{21.59}$$

The series (21.58) represents the static susceptibility as long-time limit of a dynamical quantity. Upon comparing the series (21.58) for $\overline{\chi}_z$ with the corresponding series (21.46) for $\langle\sigma_z\rangle_{\rm eq}$, we see that both are related by

$$\overline{\chi}_z \;=\; \frac{2}{\hbar} \frac{\partial \langle\sigma_z\rangle_{\rm eq}}{\partial\epsilon} \, . \tag{21.60}$$

Thus, the dynamical approach confirms the thermodynamic relation (19.4).

21.2.5 Correlation and response function of the coherences

Coherence correlations are more complex than population correlations for two reasons. First, the system makes an additional flip at time zero. The flip is enforced by the operation of σ_x at this time. Secondly, the system ends up in an off-diagonal state at time t. The final off-diagonal state gives rise to a boundary term in the influence function which exactly appears as if there would be an extra flip at time t back to a diagonal state. The additional flips at times zero and t are not dynamical.

For the mentioned two reasons, the influence function for the bare coherence correlation function introduces additional bath correlations which result in an overall factor ρ^2 in the bare tunneling correlation $C_{x,{\rm bare}}^+(t)$ and an overall factor ρ in the coherences $\langle\sigma_x\rangle_t$ and $\langle\sigma_y\rangle_t$. Here $\rho \equiv \Delta_{\rm r}/\Delta$ is the adiabatic dressing factor discussed in Subsec. 18.1.4. Using for $\Delta_{\rm r}$ the Ohmic form (18.34), where $0 < K < 1$, one gets

$$\rho \;=\; (\Delta_{\rm r}/\omega_{\rm c})^K \, , \tag{21.61}$$

which depends explicitly on the cutoff $\omega_{\rm c}$. Thus the adiabatic dressing factor ρ drops to zero in the Ohmic scaling limit $\omega_{\rm c} \to \infty$ with $\Delta_{\rm r}$ held fixed. Hence, the bare correlation function $C_{x,{\rm bare}}(t)$ as well as the coherences $\langle\sigma_x\rangle_t$ and $\langle\sigma_y\rangle_t$ are non-universal, as observed first by Guinea [460], and they drop to zero in the scaling limit.

Universality is restored, however, by taking into account that, when the particle tunnels from the one localized state to the other, it drags behind it a phonon cloud, as described in Subsec. 18.1.5. The phonon drag is taken into account in the correlation function $C_{x,{\rm pd}}^+(t)$ of the *polaron-dressed* tunneling operator, Eq. (21.24). Since the dressed operator $\tilde\sigma_x$ acts in the full system-plus-reservoir space and the correlated initial state involves a particular preparation of the reservoir, the elimination of the bath modes has to be reconsidered. The related analysis has been given in Ref. [461]. In the end one finds a modified influence functional and restrictions on the contributing paths. The constraints in the expressions (21.25) and (21.26) are

$$\xi = \xi_+ = -\eta_- \quad \text{(group A)} \, , \qquad \xi = -\xi_- = \eta_+ \quad \text{(group B)} \, . \tag{21.62}$$

Paths which violate the constraints (21.62) would yield vanishing contributions in the scaling limit. In the modified influence functional, the system's flips at time zero and at time t are not correlated with the dynamical jumps induced by the Hamiltonian. In the charge picture, the boundary charges are removed by the polaron transformation.

Because of the additional flip the system makes at time zero without a factor Δ, it is convenient to use instead of Eq. (21.36) the modified integration symbol

$$\int_{t_p}^{t} \tilde{\mathcal{D}}_{k,\ell}\{t_j\} \times \cdots \equiv \int_0^t dt_{\ell+k+1} \cdots \int_0^{t_{\ell+3}} dt_{\ell+2} \int_{t_p}^0 dt_\ell \cdots \int_{t_p}^{t_2} dt_1 \, \Delta^{k+\ell} \times \cdots , \quad (21.63)$$

and the extra flip is at time zero, $t_{\ell+1} = 0$. For the groups A and B specified below Eq. (21.26) the system is finally in a blip state. After execution of the η-sum, the terms (21.64) of the symmetric correlation function $S_{x,\mathrm{pd}}(t)$ are found to read [461]

$$S_x^{\mathrm{A}}(t) = \frac{1}{2} \sum_{m=1}^{\infty} [-\cos(\pi K)/2]^{m-1} \int_{-\infty}^{|t|} \tilde{\mathcal{D}}_{2m-2,0}\{t_j\} \sum_{\{\xi_j=\pm 1\}_{\mathrm{A}}} G_m^{\mathrm{A}} B_m^{(\mathrm{s})} , \quad (21.64)$$

$$S_x^{\mathrm{B}}(t) = -\sum_{m=2}^{\infty} \sum_{n=1}^{\infty} [-\cos(\pi K)/2]^{n+m-1} \sin^2(\pi K) \quad (21.65)$$

$$\times \int_{-\infty}^{|t|} \tilde{\mathcal{D}}_{2m-1,2n-1}\{t_j\} \sum_{\{\xi_j=\pm 1\}_{\mathrm{B}}} \xi_{m+n} \, \xi_1 \, G_{m+n}^{\mathrm{B}} B_{m+n}^{(\mathrm{s})} .$$

Correspondingly, Eq. (21.26) yields for $\chi_{x,\mathrm{pd}}(t)$ the contributions [461]

$$\chi_x^{\mathrm{A}}(t) = \frac{\Theta(t)}{\hbar} \sum_{m=1}^{\infty} \sum_{n=1}^{\infty} [-\cos(\pi K)/2]^{m+n-1} \tan(\pi K) \int_{-\infty}^{t} \tilde{\mathcal{D}}_{2m-2,2n}\{t_j\} \quad (21.66)$$

$$\times \sum_{\{\xi_j=\pm 1\}_{\mathrm{A}}} \xi_{m+n} \, \xi_1 \, G_{m+n}^{\mathrm{A}} B_{m+n}^{(\mathrm{s})} ,$$

$$\chi_x^{\mathrm{B}}(t) = \frac{\Theta(t)}{\hbar} \sum_{m=1}^{\infty} \sum_{n=1}^{\infty} [-\cos(\pi K)/2]^{m+n-1} \tan(\pi K) \int_{-\infty}^{t} \tilde{\mathcal{D}}_{2m-1,2n-1}\{t_j\} \quad (21.67)$$

$$\times \sum_{\{\xi_j=\pm 1\}_{\mathrm{B}}} [\sin^2(\pi K)\, \xi_{n+1} + \cos^2(\pi K)\, \xi_{n+m}] \, \xi_1 \, G_{m+n}^{\mathrm{B}} B_{m+n}^{(\mathrm{s})} .$$

The subscripts $\{\ldots\}_{\mathrm{A}}$ and $\{\ldots\}_{\mathrm{B}}$ indicate that the blip labels are subject to the constraints $\xi_{m+n} = \xi_{n+1}$ for group A, and $\xi_{m+n} = -\xi_n$ for group B. The blip correlation factor factors G_{m+n}^{A} and G_{m+n}^{B} differ from the form (21.31) for G_{m+n} by the absence of the blip charges situated at times zero and t. With this rule, and with the scaling form (18.53) for $Q(t)$, Δ and ω_{c} are universally combined in the renormalized tunneling amplitude (18.34), and there is no extra dependence on ω_{c}.

Equations (21.64)–(21.67) are exact formal expressions for the polaron-dressed coherence correlation function in the scaling limit. The series expressions for the expectation and correlation functions presented above in Subsecs. 21.2.3–21.2.5 are summed in analytic form for the Toulouse model in Sec. 22.6.

21.2.6 Generalized exact master equation and integral relations

We see from the exact formal expressions for the expectation values $\langle \sigma_j \rangle_t$ that the system's transitions are correlated with each other through the influence functions $F_m^{(\pm)}(\{t_j\})$. Notwithstanding this formidable complexity, one can describe the dynamics of the conditional populations $P(i, t; j, 0)$ [i and j label diagonal states] for any N-state system in terms of a set of exact generalized master equations (GME)

$$\dot{P}(i, t; j, 0) = -\sum_{k=1}^{N} \int_0^t dt' \, K(i, t; k, t') P(k, t'; j, 0) , \qquad t > 0 . \qquad (21.68)$$

Conservation of probability provides the sum rule $\sum_i K(i, t; j, t') = 0$.

Consider next the case of two states, which we label with the eigenvalues ± 1 of σ_z. Now, the sum rule yields $K(\mp 1, t; \pm 1, t') = -K(\pm 1, t; \pm 1, t')$. With the kernels $K_z^{(s,a)}(t, t') = K(-1, t; -1, t') \pm K(1, t; 1, t')$, which are even and odd under bias inversion, the GME for the population difference $\langle \sigma_z \rangle_t = P(1, t; 1, 0) - P(-1, t; 1, 0)$ of the two diagonal states with product initial state at time zero is found to read [459, 462]

$$\frac{d \langle \sigma_z \rangle_t}{dt} = \int_0^t dt' \left[K_z^{(a)}(t, t') - K_z^{(s)}(t, t') \langle \sigma_z \rangle_{t'} \right] . \qquad (21.69)$$

Irreducible kernels and self-energies

The kernels are, by definition, the irreducible components in the exact formal series (21.41). Irreducibility of a kernel means that it cannot be cut into two uncorrelated pieces at an intermediate sojourn without removing bath correlations across this sojourn. Accordingly, it is expedient to define *irreducible* influence clusters $\tilde{F}_n^{(\pm)}$ by subtraction of the reducible components in $F_n^{(\pm)}$, which, by definition, are in product form. Since the associated bias factor factorizes also in the subtractions, as we can see from the cluster-wise $\{\xi\}$-summations, we consider the product $F_n^{(\pm)} B_n^{(s,a)}$. For a path with n blips and time growing from right to left in each term, we find

$$\tilde{F}_n^{(\pm)} B_n^{(s,a)} \equiv F_n^{(\pm)} B_n^{(s,a)} \qquad (21.70)$$

$$- \sum_{j=2}^{n} (-1)^j \sum_{m_1, \cdots, m_j} F_{m_1}^{(+)} B_{m_1}^{(s)} F_{m_2}^{(+)} B_{m_2}^{(s)} \cdots F_{m_j}^{(\pm)} B_{m_j}^{(s,a)} \delta_{m_1 + \cdots + m_j, n} .$$

The inner sum is over positive integers m_j. The subtractions represent all possibilities of factorizing the influence functions into clusters. For instance, the $n = 3$ term reads

$$\tilde{F}_3^{(\pm)} B_3^{(s,a)} = F_3^{(\pm)} B_3^{(s,a)} - F_2^{(+)} B_2^{(s)} F_1^{(\pm)} B_1^{(s,a)} \qquad (21.71)$$

$$- F_1^{(+)} B_1^{(s)} F_2^{(\pm)} B_2^{(s,a)} + F_1^{(+)} B_1^{(s)} F_1^{(+)} B_1^{(s)} F_1^{(\pm)} B_1^{(s,a)} .$$

In the subtractions, the bath correlations are only inside the individual cluster factors $F_{m_j}^{(\pm)}$, and there are no bath correlations between the clusters.

The kernels are determined by matching the iterative solution of Eq. (21.69) with the exact formal series expression (21.41). Eventually, one finds

$$K_z^{(s,a)}(t,t') = \Delta^2 F_1^{(\pm)}(t,t') B_1^{(s,a)}(t,t')$$

$$+ \sum_{n=2}^{\infty} (-1)^{n-1} \frac{\Delta^{2n}}{2^n} \int_{t'}^t dt_{2n-1} \cdots \int_{t'}^{t_3} dt_2 \sum_{\{\xi_j = \pm 1\}} \tilde{F}_n^{(\pm)} B_n^{(s,a)} . \tag{21.72}$$

The function $\tilde{F}_n^{(\pm)} B_n^{(s,a)}$ depends on $2n$ flip times. The first flip just occurs at $t_1 = t'$, and the last one at $t_{2n} = t$. The intermediate flip times are time-ordered. The GME (21.69) with the kernels (21.72) is exact for general linear dissipation.

The coherence $\langle \sigma_x \rangle_t$ is connected with $\langle \sigma_z \rangle_t$ by the exact integral relation [459]

$$\langle \sigma_x \rangle_t = \int_0^t dt' [K_x^{(s)}(t,t') + K_x^{(a)}(t,t') \langle \sigma_z \rangle_{t'}] . \tag{21.73}$$

The kernel $K_x^{(s,a)}(t,t')$ is found by matching (21.73) with the formal series (21.47),

$$K_x^{(s,a)}(t,t') = \Delta F_1^{(\mp)}(t,t') B_1^{(s,a)}(t,t')$$

$$+ \sum_{n=2}^{\infty} (-1)^{n-1} \frac{\Delta^{2n-1}}{2^n} \int_{t'}^t dt_{2n-1} \cdots \int_{t'}^{t_3} dt_2 \sum_{\{\xi_j = \pm 1\}} \xi_n \tilde{F}_n^{(\mp)} B_n^{(s,a)} . \tag{21.74}$$

The GME (21.69) and the integral relation (21.73) also hold when the bias and the tunneling amplitude are time-dependent. The corresponding generalization of these quantities is given below in Eq. (23.3) with (23.2), and in Eq. (23.5).

In the absence of external driving, the kernels $K_z^{(s,a)}(t,t')$ and $K_x^{(s,a)}(t,t')$ depend only on the relative time $t - t'$. Then, the integral expressions (21.69) and (21.73) are convolutions which can be solved in Laplace space.[6] In this way, one obtains

$$\langle \sigma_z(\lambda) \rangle = \frac{1 + \hat{K}_z^{(a)}(\lambda)/\lambda}{\lambda + \hat{K}_z^{(s)}(\lambda)} = \frac{1 + \Sigma_z^{(a)}(\lambda)/\lambda}{\lambda + \lambda \Delta^2/(\lambda^2 + \epsilon^2) + \Sigma_z^{(s)}(\lambda)} , \tag{21.75}$$

$$\langle \sigma_x(\lambda) \rangle = \frac{1}{\lambda} \Sigma_x^{(s)}(\lambda) + \left[\epsilon \Delta/(\lambda^2 + \epsilon^2) + \Sigma_x^{(a)}(\lambda) \right] \langle \sigma_z(\lambda) \rangle . \tag{21.76}$$

Here we have introduced the self-energies $\hbar \Sigma_z^{(s,a)}(\lambda)$ and $\hbar \Delta \Sigma_x^{(s,a)}(\lambda)$. The self-energies are, apart from trivial factors $\hbar$ and $\hbar \Delta$, the Laplace transforms of the kernels (21.72) and (21.74) in which the respective kernels of the bare TSS are subtracted,[7]

$$\Sigma_z^{(s)}(\lambda) \equiv \hat{K}_z^{(s)}(\lambda) - \lambda \Delta^2/(\lambda^2 + \epsilon^2) , \qquad \Sigma_z^{(a)}(\lambda) \equiv \hat{K}_z^{(a)}(\lambda) ,$$

$$\Sigma_x^{(a)}(\lambda) \equiv \hat{K}_x^{(a)}(\lambda) - \epsilon \Delta/(\lambda^2 + \epsilon^2) , \qquad \Sigma_x^{(s)}(\lambda) \equiv \hat{K}_x^{(s)}(\lambda) . \tag{21.77}$$

---

[6] We use for the Laplace transform the convention as given in Eqs. (6.33) and (6.34).
[7] The quantity $\Sigma_z^{(s,a)}(\lambda)$ has dimension frequency, whereas $\Sigma_x^{(s,a)}(\lambda)$ is dimensionless.

By definition, the self-energies $\hbar\Sigma_z^{(s,a)}(\lambda)$ and $\hbar\Delta\Sigma_x^{(s,a)}(\lambda)$ are zero for vanishing coupling to the reservoir. Once the kernels or self-energies are known, the expressions (21.75) and (21.76) can be inverted by contour integration to obtain $\langle\sigma_z\rangle_t$ and $\langle\sigma_x\rangle_t$.

When the reduced system is ergodic, the expectation values relax to the respective thermal equilibrium state. The residua of the pole at $\lambda = 0$ of $\langle\sigma_z(\lambda)\rangle$ and $\langle\sigma_x(\lambda)\rangle$ give us directly $\langle\sigma_z\rangle_{t\to\infty} = \langle\sigma_z\rangle_{eq}$ and $\langle\sigma_x\rangle_{t\to\infty} = \langle\sigma_x\rangle_{eq}$, Since all other singularities have a negative real part, their contributions fade away in the course of time. The equilibrium values are given by the exact formal expressions

$$\langle\sigma_z\rangle_{eq} = \lim_{\lambda\to 0} \frac{\Sigma_z^{(a)}(\lambda)}{\Sigma_z^{(s)}(\lambda)}, \quad \langle\sigma_x\rangle_{eq} = \lim_{\lambda\to 0} \left[\Sigma_x^{(s)}(\lambda) + \left(\Delta/\epsilon + \Sigma_x^{(a)}(\lambda)\right)\langle\sigma_z\rangle_{eq}\right]. \quad (21.78)$$

When explicit inversion of the transforms (21.75) and (21.76) is not possible, much about the dynamics can nevertheless be learned from a study of the singularities.

Despite the progress achieved by the exact results (21.75) and (21.76), the formal expressions for the kernels as well as the series expression for $C_z^{(\pm)}(t)$ obtained in the preceding section, while exact, are extremely cumbersome. Nevertheless, they can be evaluated in certain limits by analytical methods. This is discussed next. Some exact results in the regime of incoherent exponential relaxation at $T = 0$ are discussed in Subsection 22.8.1.

21.3 The noninteracting-blip approximation (NIBA)

21.3.1 Assumptions

In the preceding section, exact formal expressions for the system dynamics have been presented. I now turn to the discussion of the dynamics within the noninteracting-blip approximation (NIBA). This approach will turn out to be exact in various limits [86].

The simple assumption underlying the NIBA is that the average time $\langle s\rangle$ spent by the system in a diagonal or sojourn state is very large compared to the average time $\langle\tau\rangle$ spent in an off-diagonal or blip state. On this condition we may take the view that the blip or dipole gas is dilute and noninteracting. In detail, the NIBA assumption $\langle s\rangle \gg \langle\tau\rangle$ leads to two simple prescriptions regarding the sojourn-blip correlations $X_{j,k}$ and interblip correlations $\Lambda_{j,k}$ given in Eq. (21.32).

(1) Set the sojourn-blip correlations $X_{j,k}$ equal to zero when $j \neq k + 1$, and put $X_{k+1,k} = Q''(t_{2k+2} - t_{2k+1})$. Thus, the correlations between a sojourn k and the subsequent blips reduce to the intrablip phase correlation $\phi_{k,n} = \xi_{k+1}Q''(\tau_{k+1})$.

(2) Set all interblip interactions $\Lambda_{j,k}$ in the factor G_n in Eq. (21.31) equal to zero.

With these specifications, the influence functional (21.31) reduces to a factorized form of intra-blip bath correlations in which the sign of the individual blip phase depends on the label of the respective preceding sojourn,

$$\mathcal{F}_{\text{NIBA}}^{(n)} = \prod_{j=1}^{n} \exp\left\{ -Q'(\tau_j) + i\xi_j \eta_{j-1} Q''(\tau_j) \right\}. \tag{21.79}$$

In the Ohmic scaling limit (18.53), the assumption (1) is exact. Thus in this important case, the NIBA consists in the neglect of the interblip interactions $\Lambda_{j,k}$. Generally, the NIBA is a valid approximation in at least three different regimes [86]:

(a) Weak-coupling and zero bias: the interblip correlations $\Lambda_{j,k}$ cancel out in linear order of δ_s on the $\{\xi_j\}$-summations in Eq. (21.72), while the intrablip correlations contribute in this order. Hence the NIBA for $\langle \sigma_z \rangle_t$ is appropriate for zero bias in the weak damping (one-phonon exchange) limit. In contrast, the interblip correlations of first order cancel out only partially in the kernel $K_z^{(s,a)}(\tau)$ for nonzero bias, and in the kernel $K_x^{(s,a)}(\tau)$ for zero and nonzero bias. Therefore in these cases, the neglect of interblip correlations may result in flaws. I will touch upon these at the end of this subsection.

(b) For $s > 1$ and $T = 0$, as well as for $s > 2$ and $T > 0$, the values of $Q'(t)$ for $t \approx \Delta^{-1}$ and asymptotic time $t \to \infty$ differ only slightly. As a result, the interblip interactions $\Lambda_{j,k}$ are small compared to the intrablip interactions, when the average length of a sojourn is of the order of Δ^{-1} or larger.

(c) Long blips are suppressed when the function $Q'(t)$ increases with t at long times. This occurs at $T = 0$ for sub-Ohmic damping, $s < 1$, and at finite temperatures for $s < 2$.[8] Therefore, the NIBA is justified (1) in the sub-Ohmic case for all T, (2) in the Ohmic case for large damping and/or high temperature, and (3) in the super-Ohmic case $1 < s < 2$ for high temperature. Long blips are generally suppressed when the bias is very large, $\epsilon \gg \Delta$.

Thus, the NIBA gives consistent results in three physically quite different regimes.

The picture we now have is that of a noninteracting gas of blips confined in an interval of length t. The entire impact of the environment is contained in the intrablip correlation factor (21.79).[9] In the absence of interblip correlations, the irreducible influence functions $\tilde{F}_n(\pm)$ in Eq. (21.70) are zero for $n = 2, 3, \cdots$. Thus, in the NIBA, the kernels are determined by the one-blip contribution, which are the respective first terms in the series expressions (21.72) and (21.74),[10]

---

[8]For $T = 0$, there is $Q'(t \to \infty) \propto t^{1-s}$, whereas for $T > 0$ there is $Q'(t \to \infty) \propto t^{2-s}$.

[9]A simple derivation of the NIBA in the Heisenberg picture is given in Ref. [463].

[10]The generalization for time-dependent driving is studied in Section 23.1.

$$\hat{K}_z^{(s)}(\lambda) = \Delta^2 \int_0^\infty d\tau\, e^{-\lambda\tau} \cos(\epsilon\tau)\, e^{-Q'(\tau)} \cos[Q''(\tau)] \,,$$

$$\hat{K}_z^{(a)}(\lambda) = \Delta^2 \int_0^\infty d\tau\, e^{-\lambda\tau} \sin(\epsilon\tau)\, e^{-Q'(\tau)} \sin[Q''(\tau)] \,,$$

$$\hat{K}_x^{(s)}(\lambda) = \Delta \int_0^\infty d\tau\, e^{-\lambda\tau} \cos(\epsilon\tau)\, e^{-Q'(\tau)} \sin[Q''(\tau)] \,,$$

$$\hat{K}_x^{(a)}(\lambda) = \Delta \int_0^\infty d\tau\, e^{-\lambda\tau} \sin(\epsilon\tau)\, e^{-Q'(\tau)} \cos[Q''(\tau)] \,.$$

(21.80)

The related selfenergies are $\hbar\Sigma_z^{(s/a)}(\lambda)$ and $\hbar\Delta\Sigma_x^{(s/a)}(\lambda)$, where analogous to (21.77)

$$\Sigma_z^{(s)}(\lambda) \equiv \hat{K}_z^{(s)}(\lambda) - \lambda\Delta^2/(\lambda^2 + \epsilon^2) \,, \qquad \Sigma_z^{(a)}(\lambda) \equiv \hat{K}_z^{(a)}(\lambda) \,,$$

$$\Sigma_x^{(a)}(\lambda) \equiv \hat{K}_x^{(a)}(\lambda) - \epsilon\Delta/(\lambda^2 + \epsilon^2) \,, \qquad \Sigma_x^{(s)}(\lambda) \equiv \hat{K}_x^{(s)}(\lambda) \,.$$

(21.81)

Laplace inversion of the expressions (21.75) and (21.76) with (21.81) gives the evolution of the damped TSS in the NIBA. One finds that the NIBA reproduces the exact dynamics qualitatively (and often quantitatively) in the most interesting time regime of several Δ_r^{-1}. However, the NIBA does not correctly describe at low T the long-time behavior of $\langle\sigma_j\rangle_t$ and of $C_j^\pm(t)$ $(j = x, y, z)$, not to mention the algebraic long-time tails at zero temperature.

21.3.2 Limitations

Before we embark on the discussion of $\langle\sigma_j\rangle_t$ and $C_z^\pm(t)$ for particular spectral densities, we point out two flaws of the NIBA which arise for low temperature at asymptotic time. First, we observe that the function $R(t)$ in Eq. (21.52), which is due to the entangled initial state, is zero in the NIBA.[11] Hence the function $S_z(t)$ reduces to $S_z^{(unc)}(t)$, in which the components $P_{s,a}(t)$ and $P_\infty = \langle\sigma_z\rangle_{eq}$ are the respective NIBA expressions. Evidently, the algebraic long-time tails of $S_z(t)$ at $T = 0$ carried by the function $R(t)$ (cf. Subsec. 6.4.1 and Sec. 22.7) are disregarded in the NIBA. Secondly, the equilibrium values $\langle\sigma_z\rangle_{eq}$ for nonzero bias and $\langle\sigma_x\rangle_{eq}$ for zero or nonzero bias are qualitatively incorrect at low temperature. In the NIBA we have

$$\Sigma_z^{(s)}(0) = k^+ + k^- \,, \qquad \Sigma_z^{(a)}(0) = k^+ - k^- \,,$$

(21.82)

where $k^\pm$ are the nonadiabatic rate expressions discussed in Subsec. 20.2.1. With the detailed balance relation (20.34), the expressions (21.78) turn into the NIBA forms

$$\langle\sigma_z\rangle_{eq} = \tanh(\hbar\beta\epsilon/2) \,,$$

$$\langle\sigma_x\rangle_{eq} = (\Delta/\epsilon)\tanh(\hbar\beta\epsilon/2) \,.$$

(21.83)

Thus the NIBA predicts symmetry breaking and strict localization in the lower well at $T = 0$, even when the bias is infinitesimal, $\langle\sigma_z\rangle_{eq\,|T=0} = \mathrm{sgn}(\epsilon)$. Furthermore, we get $\langle\sigma_x\rangle_{eq\,|T=0} = \Delta/|\epsilon|$, which violates in the regime $|\epsilon| < \Delta$ the bound $|\langle\sigma_x\rangle_t| \leqslant 1$.

---

[11]The term $R(t)$ is the analogue of the Matsubara contribution $S_2(t)$ in the expression (6.64).

To find the correct weak-damping forms for $\langle\sigma_z\rangle_{eq}$ and $\langle\sigma_x\rangle_{eq}$ we follow a simple thought. First, assuming that the eigenstates of H_{TSS} are thermally occupied, we get

$$
\begin{aligned}
\langle\sigma_z\rangle_{eq} &= \left(<R|e^{-\beta H_{TSS}}|R> - <L|e^{-\beta H_{TSS}}|L>\right)/Z, \\
\langle\sigma_x\rangle_{eq} &= \left(<R|e^{-\beta H_{TSS}}|L> + <L|e^{-\beta H_{TSS}}|R>\right)/Z.
\end{aligned}
\tag{21.84}
$$

Next, we express the localized states $|R>$ and $|L>$ with the eigenstates $|g>$ and $|e>$ of the Hamiltonian H_{TSS}. With use of the relations (3.138) – (3.140), we then obtain

$$
\begin{aligned}
\langle\sigma_z\rangle_{eq} &= (\epsilon/\Delta_b)\tanh(\hbar\beta\Delta_b/2) = \cos\varphi\,\langle\tau_z\rangle_{eq}, \\
\langle\sigma_x\rangle_{eq} &= (\Delta/\Delta_b)\tanh(\hbar\beta\Delta_b/2) = \sin\varphi\,\langle\tau_z\rangle_{eq},
\end{aligned}
\tag{21.85}
$$

where $\langle\tau_z\rangle_{eq}$ is the thermal equilibrium value in the eigenbasis of H_{TSS},

$$
\langle\tau_z\rangle_{eq} = \tanh(\hbar\beta\Delta_b/2),
\tag{21.86}
$$

and where $\tan\varphi = \Delta/\epsilon$. As distinguished from the NIBA expressions (21.83), the particle is not confined in reality to one of the two wells at $T = 0$. Rather Eq. (21.85) yields $\langle\sigma_z\rangle_{eq\,|T=0} = \epsilon/\Delta_b$, and $\langle\sigma_x\rangle_{eq}$ satisfies the bound $|\langle\sigma_x\rangle_{eq\,|T=0}| \leqslant 1$. This indicates that the NIBA fails for a biased system in the regime $k_B T \lesssim \hbar\Delta_b$. The appropriate treatment of this regime within the dynamical approach is given in Sec. 22.3.

The NIBA expressions (21.83) match the exact weak-damping expressions (21.85) in the temperature regime $T \gtrsim \hbar\Delta_b/k_B$. This points to the fact that the NIBA gives the behavior of both the expectation values $\langle\sigma_j\rangle_t$ and the equilibrium correlation function $C_z^{\pm}(t)$ correctly for all times at temperatures above $\hbar\Delta_b/k_B$.

21.4 The interacting-blip chain approximation (IBCA)

A systematic improvement on the NIBA consists in taking into account, in addition to the intra-blip correlations, the correlations between adjacent blips and the phase correlations between neighboring sojourn-blip pairs. Diagrammatically, we then have a chain of blips in which the nearest-neighbor interblip correlations are fully included. A pictorial representation illustrating the contribution of three blips to $\langle\sigma_z\rangle_t$ is sketched in Fig. 21.2. Because of the chain-like structure, this approximation has been dubbed "interacting-blip chain approximation" (IBCA) [464].

In the IBCA, the bath influence function $\mathcal{F}^{(n)}$ in Eq. (21.31) is approximated as

$$
\mathcal{F}_{IBCA}^{(n)} = \exp\left\{-\sum_{j=1}^{n}\left(Q'_{2j,2j-1} - i\,\xi_j\eta_{j-1}X_{j,j-1}\right) - \sum_{j=2}^{n}\xi_j\xi_{j-1}\Lambda_{j,j-1}\right\}.
\tag{21.87}
$$

The nearest-neighbor blip-blip and sojourn-blip interactions read

$$
\begin{aligned}
\Lambda_{j,j-1} &= Q'_{2j,2j-3} + Q'_{2j-1,2j-2} - Q'_{2j,2j-2} - Q'_{2j-1,2j-3}, \\
X_{j,j-1} &= Q''_{2j,2j-1} + Q''_{2j-1,2j-2} - Q''_{2j,2j-2}.
\end{aligned}
\tag{21.88}
$$

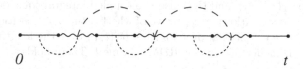

Figure 21.2: Three-blip contribution to $\langle\sigma_z\rangle_t$ in the IBCA. The solid line represents a sojourn and the curly line a blip interval. The blip-blip correlations $\Lambda_{j,j-1}$ are symbolically sketched by a dashed curve and the sojourn-blip correlations by a dotted curve. The intrablip interactions and the individual interactions in $\Lambda_{j,j-1}$ are not displayed.

To set up the dynamical equations for the RDM in the IBCA, it is expedient to introduce propagating functions $R_\pm(t,\tau)$ in which the particle is released from the sojourn state $\eta_0 = +1$ at time zero, then makes any number of sojourn-blip-sojourn transitions, and finally hops at time $t-\tau$ into the final blip state $\xi_f = \pm 1$, and remains there until time t. The quantity $R_\pm(t,\tau)$ is made up of three different components:

- An amplitude $A_\pm(\tau)$, which represents the first sojourn-blip pair. For an initial state of class B, followed by a hop to the blip $k = \pm$ of length τ, the amplitude is

$$A_\pm(\tau) = \mp i\,(\Delta/2)\,e^{-Q'(\tau)\pm i Q''(\tau)}\,. \tag{21.89}$$

- An amplitude $B_\pm(t,\tau)$ accounting for the bias force. For a time-dependent bias and stay in the blip state $\xi = \pm 1$ from time $t-\tau$ until time t, the amplitude is

$$B_\pm(t,\tau) = \exp\left(\pm i \int_{t-\tau}^{t} dt_1\,\epsilon(t_1)\right). \tag{21.90}$$

- A kernel $Y_{\xi_2,\xi_1}(\tau_2,s_1,\tau_1)$, which represents the two-step transition from the blip state ξ_1 of length τ_1 via both intermediate sojourn states of length s_1 to the blip state ξ_2 of length τ_2. The kernel, augmented with the bath correlations, reads

$$Y_{\xi_2,\xi_1}(\tau_2,s_1,\tau_1) = -\tfrac{1}{2}\,\xi_2\,\xi_1\,\Delta^2\,e^{-Q'(\tau_2)-\xi_2\xi_1\Lambda_{2,1}(\tau_2,s_1,\tau_1)}\cos[X_{2,1}(\tau_2,s_1)]\,. \tag{21.91}$$

The leading terms of the iteration of the elementary sojourn-blip sequence are

$$
\begin{aligned}
R_\pm(t,\tau) \;=\;& B_\pm(t,\tau)\bigg\{A_\pm(\tau) \\
&+ \sum_{k=\pm}\int_0^{t-\tau} ds \int_0^{t-\tau-s} d\tau_1\, Y_{\pm,k}(\tau,s,\tau_1)\,B_k(t-\tau-s,\tau_1)A_k(\tau_1) + \cdots\bigg\}.
\end{aligned}
\tag{21.92}
$$

The infinite series can be summed to the set of IBCA integral equations

$$
\begin{aligned}
R_\pm(t,\tau) \;=\;& B_\pm(t,\tau)\bigg[A_\pm(\tau) \\
&+ \sum_{k=\pm}\int_0^{t-\tau} ds \int_0^{t-\tau-s} d\tau_1\, Y_{\pm,k}(\tau,s,\tau_1)\,R_k(t-\tau-s,\tau_1)\bigg].
\end{aligned}
\tag{21.93}
$$

The kernels $Y_{\pm,k}(\tau, s, \tau_1)$, and thus also the coupled integral equations (21.93), are not in the form of convolutions, irrespective of whether the bias is static or dynamical.

Integration of the conditional probabilities $R_{\pm}(t, \tau)$ over the length τ of the final blip state gives the coherences of the RDM at time t. There holds

$$
\begin{aligned}
\langle \sigma_x \rangle_t &= \int_0^t d\tau \, [\, R_+(t, \tau) + R_-(t, \tau) \,] \,, \\
\langle \sigma_y \rangle_t &= i \int_0^t d\tau \, [\, R_+(t, \tau) - R_-(t, \tau) \,] \,.
\end{aligned}
\tag{21.94}
$$

and the population $\langle \sigma_z \rangle_t$ results from integration of the relation (21.4),

$$
\langle \sigma_z \rangle_t = 1 - \Delta \int_0^t ds \, \langle \sigma_y \rangle_s \,,
\tag{21.95}
$$

The integration over s takes into account that the final step back to the diagonal may occur at any time s in the interval $0 \leqslant s \leqslant t$.

In conclusion, the dynamical problem of finding the reduced density matrix is solved, except for quadrature, once the conditional quantities $R_{\pm}(s, \tau)$ are known in the interval $0 \leqslant \tau \leqslant s \leqslant t$. An efficient numerical algorithm consists in solving (21.93) by iteration on an equidistant grid in time [464].

The method presented here differs from the iterative solution of the GME (21.69) (cf. Ref. [462]). To include nearest-neighbor blip correlations in the GME, the respective kernel has to be considered at least in order Δ^4. However, iteration of the GME does not lead to linked blip clusters of higher order than those included in the kernel. Furthermore, the GME (21.69) is in the form of a convolution in the absence of time-dependent deterministic forces, while the dynamical equation (21.93) is generally in non-convolutive form.

As the nearest-neighbor blip correlations constitute the most relevant corrections to the NIBA, the IBCA is a valid approximation for longer propagation times than the NIBA. The IBCA is most suitable for moderate-to-strong damping and short to intermediate times.

Systematic improvement of the IBCA is possible along two lines of development. For weak-to-moderate damping, we may insert, in analogy with the proceeding in Subsection 22.3, all possible tunneling events of the undamped system into the intervals of the chain-links. For higher damping, the first step to do would be to include all next-to-nearest-neighbor interblip correlations. The relevant kernels $Y_{\xi, \xi', \xi''}$ would then depend on three blip labels and on five time intervals, namely the lengths of three blips and of two sojourns in between. Upon book-keeping more and more time intervals in the kernels, the range in which the bath correlations are taken into account exactly is systematically enlarged. The corresponding generalization of the numerical algorithm is clear, but the numerical costs increase drastically with each step.

22 Two-state dynamics: sundry topics

22.1 Symmetric TSS in the NIBA

22.1.1 Ohmic scaling limit

This subsection deals with the symmetric TSS ($\epsilon = 0$) in the Ohmic scaling limit within the NIBA. With the scaling form (18.53) of $Q'(\tau)$, the NIBA kernel can be calculated in analytic form. In the range $0 < K < 1$, one finds

$$\hat{K}_z^{(s)}(\lambda) \equiv g(\lambda) = \Delta^2 \cos(\pi K) \int_0^\infty d\tau\, e^{-\lambda\tau}\, e^{-Q'(\tau)}\,, \qquad (22.1)$$

$$g(\lambda) = \Delta_{\text{eff}} \left(\frac{\hbar\beta\Delta_{\text{eff}}}{2\pi}\right)^{1-2K} \frac{h(\lambda)}{K + \hbar\beta\lambda/2\pi}\,; \quad h(\lambda) = \frac{\Gamma(1 + K + \hbar\beta\lambda/2\pi)}{\Gamma(1 - K + \hbar\beta\lambda/2\pi)}\,, \qquad (22.2)$$

where Δ_{eff} is the effective tunneling amplitude, $\Delta_{\text{eff}}^{2-2K} = \Gamma(1 - 2K)\cos(\pi K)\Delta_{\text{r}}^{2-2K}$. The Laplace transform $\langle\sigma_z(\lambda)\rangle$ of the population $\langle\sigma_z\rangle_t$ reads

$$\langle\sigma_z(\lambda)\rangle = 1/[\lambda + g(\lambda)]\,. \qquad (22.3)$$

Zero temperature

In the limit $T \to 0$, the expression (22.2) becomes $g(\lambda) = \Delta_{\text{eff}}(\Delta_{\text{eff}}/\lambda)^{1-2K}$, and thus

$$\langle\sigma_z(\lambda)\rangle = 1/[\lambda + \Delta_{\text{eff}}(\Delta_{\text{eff}}/\lambda)^{1-2K}]\,. \qquad (22.4)$$

This is the Laplace transform of the Mittag-Leffler function $E_\nu(z) \equiv E_{\nu,1}(z)$ [441, a], of which the integral representation is given in Eq. (7.50) [269]. In real time, there is

$$\langle\sigma_z\rangle_t = E_{2-2K}[-(\Delta_{\text{eff}}t)^{2-2K}] = \sum_{m=0}^\infty \frac{(-1)^m}{\Gamma[1 + (2 - 2K)m]} (\Delta_{\text{eff}}t)^{(2-2K)m}\,. \qquad (22.5)$$

As $K \to 0$, we recover from Eq. (22.5) the persistent oscillation $\langle\sigma_z\rangle_t = \cos(\Delta t)$, whereas for $K = \frac{1}{2}$ we get pure exponential relaxation $\langle\sigma_z\rangle_t = e^{-\Delta_{\text{eff}}(1/2)t}$ with the rate $\Delta_{\text{eff}}(K = \frac{1}{2}) = \pi\Delta^2/2\omega_c$, as follows from Eq. (18.35). In addition, as the NIBA is exact in order Δ^2, the true short-time evolution of $\langle\sigma_z\rangle_t$ is found from Eq. (22.5) as $\langle\sigma_z\rangle_t = 1 - (\Delta_{\text{eff}}t)^{2-2K}/\Gamma(3 - 2K) + \mathcal{O}[(\Delta_{\text{eff}}t)^{4-4K}]$ for $\Delta_{\text{eff}}t \ll 1$.

The singularities of $\langle\sigma_z(\lambda)\rangle$ determine the behavior of $\langle\sigma_z\rangle_t$ at times $t \gtrsim 1/\Delta_{\text{eff}}$. In the regime $0 < K < \frac{1}{2}$, the singularities of $\langle\sigma_z(\lambda)\rangle$ are:

(1) A complex conjugate pair of simple poles on the principal sheet at

$$\lambda_{0,\pm} \equiv -\gamma_0 \pm i\Omega_0 = \Delta_{\text{eff}} \exp[\pm i\,\tfrac{\pi}{2(1-K)}]\,. \qquad (22.6)$$

(2) A branch point at $\lambda = 0$. The complex λ-plane is cut along the negative real axis, and in the cut plane the integrand is single-valued.

According to this, we divide $P(t) = \langle \sigma_z \rangle_t$ into a coherent and an incoherent part,

$$P(t) = P_0(t) + P_1(t) . \tag{22.7}$$

The residues of the complex conjugate poles at $\lambda = \lambda_{0,\pm} = -\gamma_0 \pm i\Omega_0$ yield the coherent part

$$P_0(t) = \cos(\Omega_0 t)\, e^{-\gamma_0 t}/(1 - K) . \tag{22.8}$$

The oscillation frequency Ω_0 and the decoherence rate γ_0 are given by

$$\Omega_0 = \Delta_{\text{eff}} \cos\left[\tfrac{\pi K}{2(1-K)} \right] , \quad \text{and} \quad \gamma_0 = \Delta_{\text{eff}} \sin\left[\tfrac{\pi K}{2(1-K)} \right] . \tag{22.9}$$

The quality factor of the oscillation Q_0 at $T = 0$ is a function of the parameter K,

$$Q_0 \equiv \Omega_0/\gamma_0 = \cot\left[\tfrac{\pi K}{2(1-K)} \right] . \tag{22.10}$$

Recently, it was argued by Lesage and Saleur (LS) [465] by using a generalization of the boundary conditions changing operators of conformal field theory [466] that the leading long-time behavior of $P(t)$ is a damped oscillation as in Eq. (22.8). They found that the quality factor of the oscillation in the regime $0 < K < \tfrac{1}{2}$ matches the NIBA form (22.10). Only the LS frequency scale differs from the NIBA scale Δ_{eff}. In our notation, the expressions for Ω_0 and γ_0 obtained by Lesage and Saleur read

$$\Omega_{0,\text{LS}} = \Delta_{\text{LS}} \cos\left[\tfrac{\pi K}{2(1-K)} \right] , \quad \text{and} \quad \gamma_{0,\text{LS}} = \Delta_{\text{LS}} \sin\left[\tfrac{\pi K}{2(1-K)} \right] , \tag{22.11}$$

with the frequency scale

$$\Delta_{\text{LS}} = \frac{\sin\left[\tfrac{\pi K}{2(1-K)} \right]}{\sqrt{\pi}} \frac{\Gamma\!\left(\tfrac{K}{2(1-K)}\right)}{\Gamma\!\left(\tfrac{1}{2(1-K)}\right)} \left(\frac{\Gamma(\tfrac{1}{2} + K)\Gamma(1-K)}{\sqrt{\pi}} \right)^{1/[2(1-K)]} \Delta_{\text{eff}} . \tag{22.12}$$

There holds $\lim_{K \to 0} \Delta_{\text{LS}}/\Delta_{\text{eff}} = 1 + \mathcal{O}(K^2)$, and $\lim_{K \to 1/2} \Delta_{\text{LS}}/\Delta_{\text{eff}} = 1 + \mathcal{O}(1 - 2K)$. Thus, the frequency scale Δ_{LS} coincides with the NIBA scale Δ_{eff} in these limits. For intermediate values of K, the scale Δ_{LS} differs from Δ_{eff} by only a few percent; e.g., for $K = \tfrac{1}{4}$ we find from Eq. (22.12) $\Delta_{\text{LS}} \approx 1.038\, \Delta_{\text{eff}}$. From this we conclude that the coherent contribution $P_0(t)$ is well described by the NIBA.

In the range $0 < K < \tfrac{1}{2}$, the cut contribution $P_1(t)$ has the negative initial value $P_1(0) = -K/(1 - K)$, and the asymptotic expansion

$$P_1(t) = \sum_{n=1}^{\infty} \frac{(-1)^{n-1}}{\Gamma[1 - (2 - 2K)n]} \frac{1}{(\Delta_{\text{eff}} t)^{(2-2K)n}} , \qquad \Delta_{\text{eff}} t \gg 1 . \tag{22.13}$$

The leading term is negative and decays as $(\Delta_{\text{eff}} t)^{2K-2}/\Gamma(2K - 1)$. For $K = 0$ and $K = \tfrac{1}{2}$, the branch cut is absent, and hence $P_1(t) = 0$ for these special cases.

In the regime $\tfrac{1}{2} < K < 1$, the poles of the Laplace transform (22.4) are not on the principal sheet. Thus, $P(t)$ is fully determined by the branch-cut contribution $P_1(t)$ with the asymptotic series (22.13). This function decays monotonically with growing time and has the initial value $P_1(0) = 1$. For $K = 3/4$, the function $P(t)$ can be written as $P(t) = e^{\Delta_{\text{eff}} t} \text{erfc}(\sqrt{\Delta_{\text{eff}} t})$, where $\text{erfc}(z)$ is the incomplete error function.

The cut contribution in the NIBA decays asymptotically as $(\Delta_{\text{eff}} t)^{-2(1-K)}$. Hence the nonequilibrium correlation function $P_1(t)$ exhibits asymptotically slower decay than the Ohmic equilibrium correlation function $S_z(t) \propto 1/t^2$. This is in contradiction with the fluctuation-dissipation theorem and therefore indicates an unphysical artifact of the NIBA. As opposed to the sluggish decay predicted in the NIBA, the nonequilibrium correlation function $P_1(t)$ decays in reality exponentially fast, as proved in Subsec. 22.6.3 for K slightly below and slightly above the Toulouse point $K = \frac{1}{2}$. See also the discussion in Subsecs. 22.7.1 and 22.7.2.

Finite temperature

At finite T, $P_1(t)$ resolves into an infinite sequence of exponentials,

$$P_0(t) = \sum_{\xi=\pm} A_\xi \exp(\lambda_\xi t) , \qquad P_1(t) = \sum_{n=1}^{\infty} A_n \exp(\lambda_n t) . \tag{22.14}$$

The λ_ξ and the λ_n are solutions of the pole equation $\lambda + g(\lambda) = 0$, where $g(\lambda)$ is defined in Eq. (22.2). With the dimensionless variables θ and x_j ($j = \xi, n$),

$$\theta = 2\pi k_B T / \hbar \Delta_{\text{eff}} , \qquad x_j = (\hbar/2\pi k_B T)\lambda_j , \tag{22.15}$$

the pole equation reads

$$x(x + K) + \frac{1}{\theta^{2-2K}} \frac{\Gamma(1 + K + x)}{\Gamma(1 - K + x)} = 0 . \tag{22.16}$$

The amplitudes resulting from Eq. (22.3) are

$$A_j = 1 / \left\{ 1 + x_j [\psi(1 - K + x_j) - \psi(K + x_j)] \right\} , \tag{22.17}$$

and according to the initial condition $P(0) = 1$ they satisfy the sum rule

$$\sum_{\xi=\pm} A_\xi + \sum_{n=1}^{\infty} A_n = 1 . \tag{22.18}$$

With regard to $P_1(t)$, the pole condition yields $x_n = -n - v_n$, where v_n lies to linear order in K in the interval $-K \leqslant v_n \leqslant K$.[1] Each contribution to $P_1(t)$ has a negative amplitude for $K < 1/2$. These add up to $P_1(0) = -(K/\theta) \operatorname{Im} \psi'(1 - i/\theta)$. Hence, in the regime $K \ll 1$, the incoherent part $P_1(t)$ is negligibly small.

Consider next the oscillatory component $P_0(t)$. To linear order in K, one finds iteratively from Eq. (22.16) for $-\gamma \pm i\Omega = \Delta_{\text{eff}} \theta x_\pm$ for any T

$$\Omega = \Delta_{\text{eff}} \left\{ 1 + K[\operatorname{Re} \psi(i/\theta) + \ln \theta] \right\} , \qquad \gamma = \tfrac{1}{2}\pi K \Delta_{\text{eff}} \coth(\pi/\theta) \tag{22.19}$$

The decoherence rate is in Korringa form. The temperature dependence of Eq. (22.19) has been observed in a variety of physical systems involving electron-hole excitations, such as interstitials in metals [103] and rare impurities in metals [467]. The generalization of Eq. (22.19) to the biased case is given in Section 22.3.

---

[1]Beyond NIBA, the λ_n are equipped with an additional negative shift analogous to the one discussed in Subsec. 22.6.3 for K slightly below and slightly above 1/2.

The leading low temperature behavior of $x_{\pm}$ is found from Eq. (22.16) using the asymptotic expansion of the gamma function. We find $\gamma = \gamma_0[1 - \kappa\,\theta^2 + \mathcal{O}(\theta^4)]$ and $\Omega = \Omega_0[1 + \kappa\,\theta^2 + \mathcal{O}(\theta^4)]$, where $\kappa = K(1-2K)/12$, and thus

$$Q(\theta) = \Omega/\gamma = Q_0\left[1 + 2\,\kappa\,\theta^2 + \mathcal{O}(\theta^4)\right],\qquad (22.20)$$

Hence the quality factor increases at very low T. Since κ is of order K, the enhancement is not an artefact of the NIBA. As T is increased, Q reaches a maximum at $T = T_{\mathrm{m}}$, and then rapidly decreases above T_{m}. For $K \ll 1$, T_{m} is near $\hbar\Delta_{\mathrm{eff}}/2\pi k_{\mathrm{B}}$.

For $T > T_0 = 2T_{\mathrm{m}}$, or rather $\Theta > \Theta_0$, we may expand the r.h.s. in Eq. (22.16) about $x = 0$. Then the pole condition becomes a quadratic equation in x,

$$x^2 + [\,K + g_1(K)/\Theta\,]\,x + 1/\Theta = 0\,,\qquad \Theta \equiv \frac{\Gamma(1-K)}{\Gamma(1+K)}\theta^{2-2K}\,.\qquad (22.21)$$

where $g_1(K) = 1/K - \pi\cot(\pi K)$. One finds that the solutions of this equation

$$x_{\pm} = \frac{1}{2\Theta}\left(-[\,K\Theta + g_1(K)\,] \pm \sqrt{[\,K\Theta + g_1(K)\,]^2 - 4\Theta}\,\right)\qquad (22.22)$$

are very close to the numerical solutions of Eq. (22.16) for $\Theta > \Theta_0$ or $T > T_0$. In the damping regime $K > 1/2$, the roots (22.22) are consistently real for $\Theta > \Theta_0$.

In the range $0 < K < 1/2$, the roots $x_{\pm}$ are complex conjugate in the coherent regime $T < T^*(K)$ and real in the incoherent regime $T > T^*(K)$ [468]. The temperature $T^*(K)$ at which the transition between the two "phases" occurs is

$$T^* = (\hbar\Delta_{\mathrm{eff}}/2\pi k_{\mathrm{B}})\left(\frac{\Gamma(1+K)}{\Gamma(1-K)}\Theta^*\right)^{\frac{1}{2-2K}}\,,\qquad \Theta^* = [\,1 + \sqrt{\pi K \cot(\pi K)}\,]^2/K^2\,.\qquad (22.23)$$

The formula (22.23) has a relative error of less than 0.4% as compared with the bound calculated numerically from Eq. (22.3) by Garg [469, 86]. The transition temperature $T^*(K)$ is plotted in Fig. 22.1. It monotonically increases when K is decreased. For weak damping, there is

$$T^*(K) = \frac{1}{\pi K}\left(\frac{2}{K}\right)^{\frac{K}{1-K}}\frac{\hbar\Delta_{\mathrm{r}}}{k_{\mathrm{B}}}\,,\qquad K \ll 1\,.\qquad (22.24)$$

In the coherence regime $\Theta < \Theta^*$, the roots (22.22) are conveniently written as

$$x_{\pm} = \frac{1}{2\Theta}\left(K(\Theta^* - \Theta) - 2\sqrt{\Theta^*} \pm i\sqrt{4(\Theta^* - \Theta)[K\sqrt{\Theta^*} - 1] - K^2(\Theta^* - \Theta)^2}\,\right).$$

With the limiting value $\lim_{K\to 1/2} T^* = \hbar\Delta_{\mathrm{eff}}/k_{\mathrm{B}}\pi$, or rather $\lim_{K\to 1/2}\Theta^* = 4$, and thus $\lim_{K\to 1/2} K\sqrt{\Theta^*} \to 1$, we see from this expression that the coherent regime ceases to exist at the Toulouse point $K = 1/2$ for all T. Additional results in the Ohmic scaling limit within the framework of the NIBA are presented in Ref. [468].

We refrain from giving results in the NIBA for a biased TSS at low T because of the weakness of the NIBA in this parameter regime, as pointed out in Sec. 21.3. In the two subsequent sections, I shall outline the systematic analysis of the biased TSS (i) in the limit of memoryless noise correlations relevant at elevated temperature, and (ii) in the limit of weak-coupling for any T and any spectral coupling.

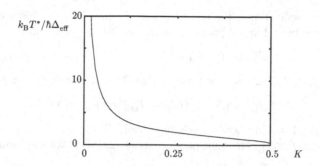

Figure 22.1: The temperature T^* at which the transition from the coherent to the incoherent phase occurs, given in Eq. (22.23), is plotted versus K.

22.1.2 The super-Ohmic case

We now study the behavior of $\langle\sigma_z\rangle_t$ for a symmetric system with a super-Ohmic spectral density, Eq. (18.44) with $s > 2$. Our analysis is based on the expression[2]

$$\langle\sigma_z(\lambda)\rangle = \frac{\lambda}{\lambda^2 + \tilde{\Delta}^2 + \lambda\Sigma_z(\lambda)} . \tag{22.25}$$

Here, $\tilde{\Delta}$ is the polaron-dressed tunneling amplitude (20.96). The NIBA self-energy is

$$\hbar\Sigma_z^{(\text{blip})}(\lambda) = \hbar\tilde{\Delta}^2 \int_0^\infty d\tau\, e^{-\lambda\tau} \left(e^{-[Q'(\tau)-X_1]} \cos[Q''(\tau)] - 1 \right) . \tag{22.26}$$

For low-to-moderate temperature, the poles of Eq. (22.25) are found from the relation

$$\lambda^2 + \lambda\Sigma_z^{(\text{blip})}(\pm i\,\tilde{\Delta}) + \tilde{\Delta}^2 = 0 . \tag{22.27}$$

The shift of the oscillation frequency in leading order of $\Sigma_z^{(\text{blip})}$ is determined by the imaginary part of the self-energy.[3] Since generally $\text{Im}\,\Sigma_z^{(\text{blip})}(\pm i\tilde{\Delta}) \ll \text{Re}\,\Sigma_z^{(\text{blip})}(\pm i\tilde{\Delta})$ for $s > 2$ [472], the imaginary part of the self-energy can be disregarded.

Consider now the rate $\Upsilon \equiv \text{Re}\,\Sigma_z^{(\text{blip})}(\pm i\,\tilde{\Delta})$. With the parallel shift of the integration path, which has led us from Eq. (20.30) to the expression (20.31), we find

$$\Upsilon = \tilde{\Delta}^2 \cosh\left(\tfrac{1}{2}\beta\hbar\tilde{\Delta}\right) \int_0^\infty dt\, \cos(\tilde{\Delta}t) \left(e^{X_2(t)} - 1\right) , \tag{22.28}$$

where $X_2(t)$ is given in Eq. (18.50). Expansion of the integrand in powers of $X_2(t)$ gives the multi-phonon series for the rate Υ. For $s = 3$, this is a hypergeometric series for the variable $\phi = (T/T'_{\text{ph}})^2$, where T'_{ph} is defined in Eq. (18.49)

$$\Upsilon = \frac{\tilde{\Delta}}{2\pi} \coth\left(\tfrac{1}{2}\beta\hbar\tilde{\Delta}\right) \left(\frac{\hbar\tilde{\Delta}}{k_{\text{B}}T}\right)^2 \phi\; {}_2F_2\left(1 + i\frac{\beta\hbar\tilde{\Delta}}{2\pi}, 1 - i\frac{\beta\hbar\tilde{\Delta}}{2\pi}; 2, \frac{3}{2}; \phi\right) . \tag{22.29}$$

---

[2]In this subsection we put for simplicity $\Sigma_z^{(s)}(\lambda) = \Sigma_z(\lambda)$.
[3]The shift due to the one-phonon process is given in Eq. (22.93) [cf. also Eq. (19.26)].

For later use we split Υ into the contributions Υ_- and Υ_+, which are odd and even in the number n of exchanged phonons, which is just the power of ϕ,

$$\Upsilon = \Upsilon_- + \Upsilon_+ . \tag{22.30}$$

At temperature $T \lesssim \hbar\tilde{\Delta}/k_B$, the leading contribution is one-phonon exchange,

$$\Upsilon_{-,1} = (\tilde{\Delta}/2\pi)\,(\hbar\tilde{\Delta}/k_B T'_{ph})^2 \coth(\tfrac{1}{2}\beta\hbar\tilde{\Delta}) . \tag{22.31}$$

As T is increased, multi-phonon exchange gradually develops. In the temperature regime $\hbar\tilde{\Delta}/k_B \ll T \ll T_D$, the multiphonon series (22.29) takes the form

$$\Upsilon = \frac{\tilde{\Delta}}{\pi}\,\frac{\hbar\tilde{\Delta}}{k_B T}\sum_{n=1}^{\infty}\frac{\Gamma(\tfrac{3}{2})}{n\Gamma(n+\tfrac{1}{2})}\,\phi^n , \qquad \text{with} \qquad \phi = (T/T'_{ph})^2 . \tag{22.32}$$

The asymptotic limit of the series (22.32) is found by calculation of the integral (22.28) in SDA in the limit $\beta\hbar\tilde{\Delta} \to 0$, but $\beta\hbar\omega_D \gg 1$. The limiting value is given by the expression (20.104) with $\epsilon = 0$ and $\kappa = 0$, and with $\lambda_1(s,0)$ and $\lambda_2(s,0)$ given in Eq. (20.68). This yields in the regime $s > 1$

$$\Upsilon_\pm = \frac{\Upsilon}{2} = \frac{\hbar\Delta^2\,e^{-2B_s}}{2\,k_B T'_{ph}}\sqrt{\frac{\pi}{\lambda_2(s,0)}}\left(\frac{T'_{ph}}{T}\right)^{(s+1)/2}\exp\left\{\lambda_1(s,0)\left(\frac{T}{T'_{ph}}\right)^{s-1}\right\} , \tag{22.33}$$

and for the special case $s = 3$

$$\Upsilon_\pm = \frac{\Upsilon}{2} = \frac{1}{4\sqrt{\pi}}\,\frac{\hbar\Delta^2\,e^{-2B_3}}{k_B T'_{ph}}\left(\frac{T'_{ph}}{T}\right)^2\exp\left(\frac{2T^2}{3T'^2_{ph}}\right) . \tag{22.34}$$

In the underdamped regime, the poles of Eq. (22.25) are at $\lambda_\pm = -\gamma \pm i\,\Omega$ with

$$\gamma = \tfrac{1}{2}\Upsilon , \qquad \Omega = \sqrt{\tilde{\Delta}^2 - \tfrac{1}{4}\Upsilon^2} , \tag{22.35}$$

and the populations undergo damped oscillations

$$\langle\sigma_z\rangle_t = \cos(\Omega t + \varphi)\,e^{-\gamma t}/\cos\varphi , \qquad \varphi = \arctan(\gamma/\Omega) . \tag{22.36}$$

The temperature T^*, at which the transition from underdamped to overdamped motion occurs, is determined by the equality $\Upsilon = 2\tilde{\Delta}$, as follows from Eq. (22.35). With Eqs. (20.96) and (22.33), one finds for T^* the transcendental equation

$$T^* = T'_{ph}\left(\frac{\ln\left[\Delta\,e^{-B_s}/A_s(T^*)\right]}{\lambda_1(s,0) + d_s(0)}\right)^{1/(s-1)} , \tag{22.37}$$

where $A_s(T)$ is the pre-exponential factor in the expression (22.33).

For T above T^*, we have overdamped relaxation

$$\langle\sigma_z\rangle_t = \frac{1}{(\gamma_1 - \gamma_2)}\left(\gamma_1\,e^{-\gamma_1 t} - \gamma_2\,e^{-\gamma_2 t}\right) , \qquad \gamma_{1/2} = \tfrac{1}{2}\Upsilon \pm \sqrt{\tfrac{1}{4}\Upsilon^2 - \tilde{\Delta}^2} . \tag{22.38}$$

For $T \gg T^*$, there is $\gamma_1 \gg \gamma_2$. As a result, the amplitude of the second term is small compared to that of the first term, and $\langle \sigma_z \rangle_t$ exhibits mono-exponential decay,

$$\langle \sigma_z \rangle_t = e^{-\Upsilon t} . \tag{22.39}$$

Until now, the study of the super-Ohmic case has been restricted to the NIBA. Beyond that, an additional approximate self-energy expression including all orders of Δ^2 can be derived, in which the intra-sojourn correlations are fully taken into account, but all other correlations are disregarded [472]. In Laplace space, each blip-sojourn sequence contributes a factor $-\Sigma_z^{(\mathrm{soj})}(\lambda)/\lambda$. The factor $1/\lambda$ stems from the blip integral, and $\hbar \Sigma_z^{(\mathrm{soj})}(\lambda)$ is the self-energy term of a single sojourn. Because of the ξ_j sum in Eq. (21.72), the odd multi-phonon contributions cancel out, and thus

$$\hbar \Sigma_z^{(\mathrm{soj})}(\lambda) = \hbar \tilde{\Delta}^2 \int_0^\infty ds\, e^{-\lambda s} \Big(\cosh[\, Q'(s) - X_1 \,] \cos[\, Q''(s)\,] - 1 \Big) . \tag{22.40}$$

The sum of all blip-sojourn sequences is a geometrical series. Finally, one obtains

$$\hbar \Sigma_z(\lambda) = \hbar \Sigma_z^{(\mathrm{blip})}(\lambda) - \frac{\hbar \tilde{\Delta}^2}{\lambda} \frac{\Sigma_z^{(\mathrm{soj})}(\lambda)/\lambda}{1 + \Sigma_z^{(\mathrm{soj})}(\lambda)/\lambda} . \tag{22.41}$$

With the self-energy (22.41) the expression (22.25) takes the form

$$\langle \sigma_z(\lambda) \rangle = \frac{\lambda + \Sigma_z^{(\mathrm{soj})}(\lambda)}{[\,\lambda + \Sigma_z^{(\mathrm{soj})}(\lambda)\,][\,\lambda + \Sigma_z^{(\mathrm{blip})}(\lambda)\,] + \tilde{\Delta}^2} . \tag{22.42}$$

The iteration of the pole condition in Eq. (22.42) leads to the expression

$$\langle \sigma_z(\lambda) \rangle = \frac{\lambda + \Upsilon_+}{(\lambda + \Upsilon_+)(\lambda + \Upsilon_+ + \Upsilon_-) + \tilde{\Delta}^2} , \tag{22.43}$$

Here we have put $\mathrm{Re}\,\Sigma_z^{(\mathrm{soj})}(\lambda = \pm i\,\tilde{\Delta}) = \Upsilon_+$ and $\mathrm{Re}\,\Sigma_z^{(\mathrm{blip})}(\lambda = \pm i\,\tilde{\Delta}) = \Upsilon_+ + \Upsilon_-$, and the respective imaginary parts are disregarded.

The inverse transform of (22.43) shows in the regime $T < T^*$ damped oscillations,

$$\langle \sigma_z \rangle_t = \cos(\Omega t + \phi)\, e^{-\gamma t}/\cos(\phi) , \qquad \phi = \arctan(\Upsilon_-/2\Omega) ,$$
$$\gamma = \Upsilon_+ + \tfrac{1}{2}\Upsilon_- , \qquad \Omega = \sqrt{\tilde{\Delta}^2 - \Upsilon_-/4} . \tag{22.44}$$

The transcendental equation for the transition temperature T^* is as in (22.37), but now with an extra factor 2 in the argument of the logarithm.

In the overdamped regime $T > T^*$, $\langle \sigma_z \rangle_t$ shows bi-exponential relaxation,

$$\langle \sigma_z \rangle_t = \frac{\gamma_1 - \Upsilon_+}{\gamma_1 - \gamma_2} e^{-\gamma_1 t} + \frac{\Upsilon_+ - \gamma_2}{\gamma_1 - \gamma_2} e^{-\gamma_2 t} ,$$
$$\gamma_{1/2} = \tfrac{1}{2}\Upsilon_- + \Upsilon_+ \pm \sqrt{\tfrac{1}{4}\Upsilon_-^2 - \tilde{\Delta}^2} . \tag{22.45}$$

For T well above T^*, γ_1 is near Υ, and γ_2 is near Υ_+. Hence in this regime, the expression (22.45) coincides with the NIBA expression (22.39). As the temperature is increased further, the crossover to the classical Marcus regime takes place. This has been discussed already in some detail in Subsection 20.2.8.

In conclusion, the above treatment of interblip correlations is not systematic. In Eq. (22.41), a specially selected interblip correlation term is summed in all orders in $\tilde{\Delta}^2$. The characteristic features of the dynamics are found in qualitative agreement with those of the NIBA.

22.2 White-noise regime

22.2.1 Power spectrum of the stochastic force

With a view to the regime of weak damping and moderate-to-high temperature, consider the normalized power spectrum of dimension frequency

$$S_\xi(\omega) \equiv (q_0/\hbar)^2 \, \tilde{\mathcal{K}}(\omega)/2 = (\pi/2) \, G(\omega) \coth(\beta\hbar\omega/2) . \qquad (22.46)$$

Here $\tilde{\mathcal{K}}(\omega)$ is given in Eq. (2.14). The second form follows with use of Eqs. (3.27) and (3.148). The relation (22.46) is a ramification of the fluctuation-dissipation theorem. For Ohmic spectral density $G(\omega) = 2K\omega$, the power spectrum is white in the classical regime $\beta\hbar|\omega| \ll 1$,

$$S_{\xi,\mathrm{wn}} \equiv \vartheta = 2\pi K k_\mathrm{B} T/\hbar . \qquad (22.47)$$

The quantity ϑ is a thermal frequency scaled with the Ohmic damping parameter K.

In the associated time regime $|t|/\hbar\beta \gg 1$, the Ohmic scaling form (18.53) of the pair interaction $Q(t)$ reduces to the "white noise" correlation function

$$Q_\mathrm{wn}(t) = Q_\mathrm{adia} + \vartheta |t| + i\pi K \operatorname{sgn}(t) ,$$

$$Q_\mathrm{adia} \equiv \int_{\nu_1}^{\omega_c} d\omega \, \frac{G_\mathrm{ohm}(\omega)}{\omega^2} = 2K \ln\left(\frac{\hbar\omega_c}{2\pi k_\mathrm{B} T}\right) . \qquad (22.48)$$

This expression is in accordance with the memoryless force correlation (2.2), as follows with the relation $\ddot{Q}(t) = q_0^2 L(t)/\hbar$ and with Eq. (5.25). The pair interaction $Q_\mathrm{wn}(t)$ establishes in effect the Markovian dynamics in the charge representation (21.31). The term Q_adia with the lower bound $\nu_1 = 2\pi/\hbar\beta$ is the adiabatic bath contribution which induces the temperature dependent dressed tunneling amplitude

$$\Delta_T \equiv \Delta \, e^{-Q_\mathrm{adia}/2} = \Delta_\mathrm{r}\left(\frac{2\pi k_\mathrm{B} T}{\hbar\Delta_\mathrm{r}}\right)^K = \Delta_\mathrm{eff}\left(\frac{2\pi k_\mathrm{B} T}{\hbar\Delta_\mathrm{eff}}\right)^K \sqrt{h(0)} . \qquad (22.49)$$

Here Δ_r and Δ_eff are given in (18.34) and (18.35), and $h(0) = \Gamma(1+K)/\Gamma(1-K)$. There is $(\Delta_\mathrm{eff}/\Delta_\mathrm{r})^{1-K} \sqrt{h(0)} = 1 + \mathcal{O}(K^2)$. The dynamical part $\vartheta |t| + i\pi K \operatorname{sgn}(t)$ of $Q_\mathrm{wn}(t)$ results from the nonadiabatic modes with frequencies $\omega < 2\pi/\hbar\beta$.

The white-noise expression (22.48) practically holds for temperatures in the range

$$T_\mathrm{b} \lesssim T \ll \hbar\omega_c/k_\mathrm{B} , \qquad T_\mathrm{b} \equiv \hbar(\Delta_\mathrm{eff}^2 + \epsilon^2)^{1/2}/k_\mathrm{B} . \qquad (22.50)$$

22.2.2 Symmetric Ohmic TSS at moderate-to-high temperature

In this subsection, we study the case of weak Ohmic dissipation, $K \ll 1$, in the temperature regime (22.50). The temperature regime below it, $0 \leqslant T \lesssim T_b$, is discussed in Sec. 22.3. The weak-damping regime $K \ll 1$ applies, e.g., to interstitial tunneling in metals. In addition, this regime is of fundamental importance with regard to "Macroscopic Quantum Coherence" (MQC) and qubit operation, as the lower the value of K the higher is the temperature until which quantum coherence persists [see Fig. 22.1 on page 405 and Eq. (22.24)].

In the temperature regime (22.50), the fluctuations of the collective bath mode $\mathfrak{E}(t) = q_0 \xi(t)$ are virtually memory-less, Eq. (22.47), and the pair correlation function takes the form $Q_{\mathrm{wn}}(t)$ given in Eq. (22.48). Use of the form (22.48) instead of the full expression (18.53) in the breathing mode integral of the kernels (21.80) results in the substitution $h(\lambda) \to h(0)$ in Eq. (22.2). This is the appropriate approximation when the dynamical poles $\lambda_{1/2}$ of $\langle \sigma_z(\lambda) \rangle$ satisfy the condition $\hbar \beta |\lambda_{1/2}| \ll 1$.

With the white-noise form (22.48), the interblip interactions cancel each other out exactly, $\Lambda_{jk} = 0$. Therefore, the noninteracting-blip assumption is *exact* in the regime (22.50). In addition, the kernel (22.1) takes the simple form

$$\hat{K}_z^{(s)}(\lambda) = \Delta_T^2 / [\lambda + \vartheta] = \Delta_T^2 / [\lambda + 2\pi K/\hbar\beta] . \tag{22.51}$$

Consider first the *symmetric* TSS for which Eq. (21.75) with (22.51) takes the from

$$\langle \sigma_z(\lambda) \rangle = \frac{\lambda + \vartheta}{\lambda(\lambda + \vartheta) + \Delta_T^2} = \frac{\lambda - \lambda_1 - \lambda_2}{(\lambda - \lambda_1)(\lambda - \lambda_2)} . \tag{22.52}$$

The positions $\lambda_{1/2}$ of the poles of $\langle \sigma_z(\lambda) \rangle$ are complex-conjugate for $T < T^*$, and real-valued for $T > T^*$. The transition temperature $T^*(K)$ separating the coherent from the incoherent "phase" is given in Eq. (22.24). For small K, it is inversely proportional to K. In the coherent regime $\hbar \Delta_{\mathrm{eff}}/k_B \lesssim T \leqslant T^*(K)$, we have $\lambda_{1/2} = -\gamma \pm i\, \Omega$ with

$$\Omega(T) = \Delta_T \sqrt{1 - (T/T^*)^{2-2K}} \,; \qquad \gamma(T) = \vartheta/2 = \pi K\, k_B T/\hbar . \tag{22.53}$$

The population $\langle \sigma_z \rangle_t$ performs damped oscillations,

$$\langle \sigma_z \rangle_t = [\cos(\Omega t - \varphi)/\cos\varphi]\, e^{-\gamma t}, \qquad \varphi = \arctan(\gamma/\Omega) . \tag{22.54}$$

Because of the absence of inter-blip correlations, the symmetrized correlation function $S_z(t) = \mathrm{Re} \langle \sigma_z(t)\sigma_z(0) \rangle_\beta$ coincides with $\langle \sigma_z \rangle_{|t|}$ for zero bias. The Fourier transform is the spectral function $\tilde{S}_z(\omega)$. This function turns out from Eq. (22.54) as a superposition of two Lorentzians centered at $\omega = \pm\Omega(T)$ with line width $2\gamma(T)$,

$$\tilde{S}_z(\omega) = \frac{\gamma + (\Omega - \omega)\tan\varphi}{(\omega - \Omega)^2 + \gamma^2} + \frac{\gamma + (\Omega + \omega)\tan\varphi}{(\omega + \Omega)^2 + \gamma^2} . \tag{22.55}$$

We now have, together with the findings in Subsection 22.1.1, a complete picture of the unbiased Ohmic TSS. In the regime $T \lesssim \hbar\Delta_r/k_B$ the expressions (22.19) for $\Omega(T)$

and $\gamma(T)$ apply. They smoothly match near $T = \hbar\Delta_r/k_B$ with the expressions (22.53). The oscillation frequency increases $\propto T^2$ at low temperatures,[4] as we can see from Eq. (22.19). For $T \gtrsim \hbar\Delta_r/k_B$, the form (22.53) applies. Observing that $\Delta_T \propto T^K$, one finds that the oscillation frequency $\Omega(T)$ has a maximum at $T = K^{1/(2-2K)}T^*$. Above this temperature, $\Omega(T)$ decreases monotonously and approaches zero for $T = T^*$. The coherent quantum oscillations "dephase" on the time scale $1/\gamma(T)$.

For T above T^*, the two poles of $\langle\sigma_z(\lambda)\rangle$ in Eq. (22.52) are situated on the real negative λ-axis at $\lambda = -\gamma_1$ and $\lambda = -\gamma_2$. The rates are

$$\gamma_{1,2} = \pi K\,k_B T/\hbar \pm \sqrt{(\pi K\,k_B T/\hbar)^2 - \Delta_T^2}\,. \tag{22.56}$$

Instead of the oscillatory behavior (22.54), there is pure bi-exponential relaxation

$$\langle\sigma_z\rangle_t = \left(\gamma_1\,e^{-\gamma_2 t} - \gamma_2\,e^{-\gamma_1 t}\right)/(\gamma_1 - \gamma_2)\,. \tag{22.57}$$

As T is increased further, the poles move in opposite direction along the negative real λ-axis. The rate γ_1 increases with temperature and approaches linear dependence on T in the regime $T \gg T^*$, $\gamma_1 = 2\pi K k_B T/\hbar$. In contrast, the rate γ_2 decreases with increasing temperature and takes for $T \gg T^*(K)$ the asymptotic form

$$\gamma_2 \equiv \gamma_r(T) = \frac{\Delta_r}{K}\left(\frac{\hbar\Delta_r}{2\pi k_B T}\right)^{1-2K}\,. \tag{22.58}$$

For T fairly above T^*, there is $\gamma_2 \ll \gamma_1$, and thus the amplitude of the second exponential is negligibly small. Then, the spectral function $\tilde{S}_z(\omega)$ consists of a single quasi-elastic Lorentzian, and $\langle\sigma_z\rangle_t$ decays mono-exponentially,

$$\tilde{S}_z(\omega) = \frac{2\gamma_r}{\omega^2 + \gamma_r^2}\,, \qquad \langle\sigma_z\rangle_t = e^{-\gamma_r t}\,, \qquad T \gg T^*\,. \tag{22.59}$$

The relaxation rate γ_r is the sum of the forward and backward "Golden Rule" tunneling rates. The rate (22.58) is twice the expression (20.80) in the regime $K \ll 1$. Since the tunneling rate varies with temperature as T^{2K-1}, the width of the Lorentzian decreases with increasing T in the regime $K < 1/2$. The anomalous T^{2K-1} law has been predicted by Kondo [393, 138]. According to this characteristic feature, the temperature regime well above T^* is referred to as Kondo regime.

22.2.3 Biased Ohmic TSS at moderate-to-high temperature

The above analysis is easily extended to the *biased* system. Upon inserting the white-noise correlation $Q_{wn}(t)$, Eq. (22.48), in Eqs. (21.80), we directly obtain [459]

$$
\begin{aligned}
\hat{K}_z^{(s)}(\lambda) &= \Delta_T^2(\lambda + \vartheta)/[(\lambda + \vartheta)^2 + \epsilon^2]\,, \\
\hat{K}_z^{(a)}(\lambda) &= \Delta_T^2\pi K\epsilon/[(\lambda + \vartheta)^2 + \epsilon^2]\,, \\
\hat{K}_x^{(s)}(\lambda) &= (\pi K/\Delta)\hat{K}_z^{(s)}(\lambda)\,, \\
\hat{K}_x^{(a)}(\lambda) &= \hat{K}_z^{(a)}(\lambda)/(\pi K\Delta)\,.
\end{aligned}
\tag{22.60}
$$

---

[4] The T^2-law for $\Omega(T)$ gives a linear contribution to the specific heat (cf. Subsection 19.2.1).

With these kernels, the expressions (21.75) and (21.76) take the forms

$$\langle \sigma_z(\lambda) \rangle = \frac{K\pi\Delta_T^2 \epsilon + \lambda[\epsilon^2 + (\vartheta + \lambda)^2]}{\lambda N(\lambda)} , \qquad (22.61)$$

$$\langle \sigma_x(\lambda) \rangle = \frac{\Delta_T[\epsilon\lambda + K\pi(\Delta_T^2 + \lambda(\vartheta + \lambda))]}{\lambda N(\lambda)} , \qquad (22.62)$$

where

$$N(\lambda) = \lambda[(\lambda + \vartheta)^2 + \epsilon^2] + (\lambda + \vartheta)\Delta_T^2 . \qquad (22.63)$$

Hence the dynamical behaviors of $\langle \sigma_z \rangle_t$ and $\langle \sigma_x \rangle_t$ are guided by four singularities,

- a simple pole at $\lambda = 0$, the residua of which are the equilibrium values,

$$\langle \sigma_z \rangle_{\mathrm{eq}} = \frac{\hbar\epsilon}{2k_{\mathrm{B}}T} ; \qquad \langle \sigma_x \rangle_{\mathrm{eq}} = \frac{\Delta_T}{\Delta} \frac{\hbar\Delta_T}{2k_{\mathrm{B}}T} , \qquad (22.64)$$

- three simple poles which are located at the zeros of the cubic equation $N(\lambda) = 0$.

The characteristics of the cubic equation $N(\lambda) = 0$ is as follows [468]. In the bias range $|\epsilon| < \epsilon_c \equiv \Delta_T/\sqrt{8}$, there are three temperature regimes with qualitatively different behaviors. These are separated by the transition temperatures

$$T_{2/1} = \frac{1}{4\sqrt{2}\pi K k_{\mathrm{B}}} \frac{\hbar}{|\epsilon|} \sqrt{\Delta_T^4 + 20\epsilon^2\Delta_T^2 - 8\epsilon^4 \pm \Delta_T(\Delta_T^2 - 8\epsilon^2)^{3/2}} , \qquad (22.65)$$

For $T < T_1$ and $T > T_2$, one of the zeros of $N(\lambda)$ is negative real, and the other two are complex conjugate,

$$\lambda_1 = -\gamma_{\mathrm{r}} , \qquad \lambda_{2/3} = -\gamma \pm i\Omega . \qquad (22.66)$$

The former entails exponential relaxation and the latter damped oscillations.

In the intermediate temperature regime $T_1 < T < T_2$, all zeros are negative real, and thus the system undergoes triple-exponential relaxation.

At the critical bias strength $|\epsilon| = \epsilon_c$, the transition temperatures T_1 and T_2 coincide, $T_1(\epsilon_c) = T_2(\epsilon_c) = 3\sqrt{3/2}\,\hbar\Delta_T/(4\pi K k_{\mathrm{B}})$.

For $|\epsilon| \ll \epsilon_c$, we may expand the expression (22.65) about $\epsilon = 0$. This yields

$$T_1 = \frac{1}{2\pi K} \frac{2\Delta_T^2 - \epsilon^2}{\Delta_T} , \qquad T_2 = \frac{1}{4\pi K} \frac{\Delta_T^2 + 2\epsilon^2}{\epsilon} . \qquad (22.67)$$

When $|\epsilon|$ exceeds the critical bias ϵ_c, the pole condition $N(\lambda) = 0$ has one real root and two complex conjugate ones, as in Eq. (22.66), in the entire temperature range.

In the oscillatory regime, the Vieta relations read

$$\gamma_{\mathrm{r}} + 2\gamma = 2\vartheta , \qquad (22.68)$$

$$\gamma_{\mathrm{r}}(\gamma^2 + \Omega^2) = \Delta_T^2\,\vartheta , \qquad (22.69)$$

$$\gamma^2 + 2\gamma\gamma_{\mathrm{r}} + \Omega^2 = \Delta_T^2 + \epsilon^2 + \vartheta^2 .$$

Here, the polaronic effects are in the dressed amplitude Δ_T, and the dynamical impact from the bath depends via ϑ on the product of damping strength and temperature. For temperatures in the range $T_b \lesssim T \lesssim T_1$ we find (disregarding higher orders in ϑ)

$$\gamma_r = \frac{\Delta_T^2}{\Delta_T^2 + \epsilon^2}\vartheta - \frac{\Delta_T^2 \epsilon^4}{(\Delta_T^2 + \epsilon^2)^2}\vartheta^3 ,$$

$$\gamma = \frac{\Delta_T^2 + 2\epsilon^2}{2(\Delta_T^2 + \epsilon^2)}\vartheta + \frac{\Delta_T^2 \epsilon^4}{2(\Delta_T^2 + \epsilon^2)^2}\vartheta^3 , \qquad (22.70)$$

$$\Omega^2 = \Delta_T^2 + \epsilon^2 - \frac{\Delta_T^2(\Delta_T^2 + 4\epsilon^2)}{4(\Delta_T^2 + \epsilon^2)^2}\vartheta^2 .$$

At these low temperatures, the dynamics is dominated by the system's Hamiltonian, and the environmental coupling is a perturbation. In leading (one-phonon exchange) order, the relaxation rate γ_r and the decoherence rate γ grow linearly with ϑ,

$$\gamma_r = \sin^2\varphi\,\vartheta , \qquad \text{and} \qquad \gamma = \tfrac{1}{2}\gamma_r + \cos^2\varphi\,\vartheta , \qquad (22.71)$$

where $\tan\varphi = \Delta_T/\epsilon$.[5] The trigonometric factors $\sin^2\varphi$ and $\cos^2\varphi$ determine the weights of the transverse and longitudinal coupling. On the other hand, the oscillation frequency decreases with growing ϑ. Since $\vartheta \propto KT$, perturbative treatment of the damping breaks down even for small K, when T is sufficiently large. As temperature is raised, multi-phonon exchange becomes increasingly important.

In the Kondo regime $T \gtrsim T_2$, Eqs. (22.68) yield (disregarding higher orders in $1/\vartheta$)

$$\gamma_r = \frac{\Delta_T^2}{\vartheta} + \Delta_T^2(\Delta_T^2 - \epsilon^2)\frac{1}{\vartheta^3} ,$$

$$\gamma = \vartheta - \frac{\Delta_T^2}{2\vartheta} - \frac{\Delta_T^2(\Delta_T^2 - \epsilon^2)}{2}\frac{1}{\vartheta^3} , \qquad (22.72)$$

$$\Omega^2 = \epsilon^2 + \frac{\Delta_T^2(4\epsilon^2 - \Delta_T^2)}{4\epsilon^2}\frac{1}{\vartheta^2} .$$

In this regime, the dynamics is dominated by the noise force. In addition, the dephasing and relaxation rates show opposite behaviors. The dephasing rate γ continues growing linearly with ϑ in leading order of ϑ, whereas the relaxation rate γ_r drops inversely with ϑ. With the polaronic effect $\Delta_T \propto T^K$, the relaxation rate behaves Kondo-like, $\gamma_r \propto T^{2K-1}$. Thus, decoherence is much faster than relaxation. In addition, the oscillation frequency approaches the bias frequency as T is increased. The crossover betweensystem-controlled behavior (22.70) and noise-controlled behavior (22.72) is determined by the zeros of the discriminant of $N(\lambda) = 0$, Eq. (22.63).

Laplace inversion of the expression (22.61) for $\langle\sigma_z(\lambda)\rangle$ yields in the time regime

$$\langle\sigma_z\rangle_t = a_r\,e^{-\gamma_r t} + \left[(1 - a_r - \langle\sigma_z\rangle_{eq})\cos\Omega t + a_s \sin\Omega t\right]e^{-\gamma t} + \langle\sigma_z\rangle_{eq} , \qquad (22.73)$$

---

[5]In difference to Eq. (3.139) here we define φ in terms of the dressed tunneling coupling Δ_T.

with the initial condition $\langle\sigma_z\rangle_{t=0} = 1$ and $\langle\dot\sigma_z\rangle_{t=0} = 0$. The amplitudes are given by

$$
\begin{aligned}
a_{\mathrm{r}} &= [(\Omega^2 + \gamma^2)(1 - \langle\sigma_z\rangle_{\mathrm{eq}}) - \Delta_T^2]/D\,, \\
a_{\mathrm{s}} &= [(\gamma_{\mathrm{r}} - \gamma)a_{\mathrm{r}} + \gamma\,(1 - \langle\sigma_z\rangle_{\mathrm{eq}})]/\Omega\,,
\end{aligned}
\tag{22.74}
$$

where $D = \Omega^2 + (\gamma - \gamma_{\mathrm{r}})^2$. Similarly, we obtain from Eq. (22.62)

$$
\langle\sigma_x\rangle_t = b_{\mathrm{r}}\,e^{-\gamma_{\mathrm{r}}t} + \left[-(b_{\mathrm{r}} + \langle\sigma_x\rangle_{\mathrm{eq}})\cos\Omega t + b_{\mathrm{s}}\sin\Omega t\right]e^{-\gamma t} + \langle\sigma_x\rangle_{\mathrm{eq}}\,,
\tag{22.75}
$$

with $\langle\sigma_x\rangle_{t=0} = 0$ and $\langle\dot\sigma_x\rangle_{t=0} = \pi K\Delta_T^2/\Delta$, and with the amplitudes

$$
\begin{aligned}
b_{\mathrm{r}} &= [\Delta_T^2(\epsilon + \pi K(\gamma - \gamma_{\mathrm{r}}/2))/\Delta - (\Omega^2 + \gamma^2)\langle\sigma_x\rangle_{\mathrm{eq}}]/D\,, \\
b_{\mathrm{s}} &= [\pi K\Delta_T^2/\Delta + (\gamma_{\mathrm{r}} - \gamma)\,b_{\mathrm{r}} - \gamma\langle\sigma_x\rangle_{\mathrm{eq}}]/\Omega\,.
\end{aligned}
\tag{22.76}
$$

The expressions (22.70) – (22.76) specify the dynamics of the expectation values $\langle\sigma_z\rangle_t$ and $\langle\sigma_x\rangle_t$ for $K \ll 1$ in the Markov or white-noise regime $T \gtrsim T_{\mathrm{b}}$.

Consider next the spin correlation function $S_z(t)$. In the NIBA, the correlation term $R(t)$ in Eq. (21.52) is zero. Further, the function $S_{z,\mathrm{unc}}(t)$ differs from $\langle\sigma_z\rangle_t$ by an extra factor $\langle\sigma_z\rangle_{\mathrm{eq}}$ in the antisymmetric terms of $\langle\sigma_z\rangle_t$, as follows from Eq. (21.52). In Fourier space one then finds with the NIBA kernels (21.80)

$$
\tilde S_z(\omega) = 2\,\mathrm{Re}\,\frac{-i\omega - \langle\sigma_z\rangle_\infty \hat K_z^{(\mathrm{a})}(-i\omega)}{-i\omega\,[-i\omega + \hat K_z^{(\mathrm{s})}(-i\omega)]}\,,
\tag{22.77}
$$

The same form has been found in Ref. [470] using a perturbative Liouville relaxation method. The dynamical susceptibility $\tilde\chi_z(\omega)$ in the NIBA is given below in Eq. (23.18).

For temperatures in the Markov regime (22.50), the NIBA is correct for all frequencies. Inserting the kernels (22.60) into Eq. (22.77) and decomposing the resulting expression into partial fractions, one obtains the spectral spin correlation function as a superposition of three Lorentzians,

$$
\tilde S_z(\omega) = c_{\mathrm{r}}\frac{2\gamma_{\mathrm{r}}}{\omega^2 + \gamma_{\mathrm{r}}^2} + \sum_{\zeta=\pm 1}\frac{(1 - c_{\mathrm{r}} - \langle\sigma_z\rangle_{\mathrm{eq}}^2)\,\gamma + c_{\mathrm{s}}(\Omega + \zeta\omega)}{(\Omega + \zeta\omega)^2 + \gamma^2}\,,
\tag{22.78}
$$

$$
\begin{aligned}
c_{\mathrm{r}} &= [\Omega^2 + \gamma^2 - \Delta_T^2 - (\Omega^2 + \gamma^2)\langle\sigma_z\rangle_{\mathrm{eq}}^2]/D\,, \\
c_{\mathrm{s}} &= [(\gamma_{\mathrm{r}} - \gamma)c_{\mathrm{r}} + \gamma\,(1 - \langle\sigma_z\rangle_{\mathrm{eq}}^2)]/\Omega\,.
\end{aligned}
\tag{22.79}
$$

The quasi-elastic peak at $\omega = 0$ arises from incoherent relaxation to thermal equilibrium on the time scale $1/\gamma_{\mathrm{r}}$. The inelastic peaks of width γ centered at $\omega = \pm\Omega$ are signatures of coherent tunneling at frequency Ω with dephasing time $1/\gamma$. There follows from the relation (21.21) that the inelastic peaks of the spectral function $\tilde C_z^\pm(\omega)$ at $\omega = \zeta\Omega$ have additional weight factors $\kappa_{\mathrm{inel}}^\pm(\zeta) = 1 \pm \zeta\tanh(\frac{1}{2}\beta\hbar\Omega)$. These factors determine the relative magnitudes of the absorption and emission lines.

In the regime $\vartheta \ll \Delta_T$, the spectral correlation function $\tilde C_z^\pm(\omega)$ takes the form

$$\tilde{C}_z^\pm(\omega) = \frac{\Delta_T^2}{\Omega^2} \sum_{\zeta=\pm 1} \frac{[1 \pm \zeta \tanh(\frac{1}{2}\beta\hbar\Omega)]\gamma}{(\omega - \zeta\Omega)^2 + \gamma^2} + \left(\frac{\epsilon^2}{\Omega^2} - \langle\sigma_z\rangle_{\rm eq}^2 \right) \frac{2\gamma_{\rm r}}{\omega^2 + \gamma_{\rm r}^2} . \qquad (22.80)$$

The inelastic peaks of $\tilde{S}_z(\omega)$ and $\tilde{C}_z^\pm(\omega)$ merge near $T = T_1$ with the quasi-elastic peak. In the Kondo regime, in which T is well above T_2, the spectral functions are reduced to a single central Lorentzian [441], as follows from Eq. (22.78) with (22.72),

$$\tilde{C}_z^\pm(\omega) = \tilde{S}_z(\omega) = \frac{2\gamma_r}{\omega^2 + \gamma_r^2} . \qquad (22.81)$$

According to Eq. (22.72), the width γ_r decreases as temperature is increased.

At low T and weak damping, the weight of the quasielastic peak is found from Eq. (22.79) as $c_r = (\epsilon/\Omega)^2 - \langle\sigma_z\rangle_{\rm eq}^2$. With the NIBA expression (21.83) for $\langle\sigma_z\rangle_{\rm eq}$, one then has $c_r = (\epsilon/\Omega)^2 - \tanh^2(\hbar\beta\epsilon/2)$. Hence the weight of the quasielastic peak turns into negative territory as temperature is decreased. This behavior is qualitatively incorrect and unphysical. The flaw is remedied, however, by considering the one-boson self-energy beyond the NIBA. The corresponding analysis is given in the next section.

Finally, one remark concerning the scaling limit (18.56) is appropriate. We see from the above results that the expectation value $\langle\sigma_z\rangle_t$ is universal in the sense specified in Subsection 18.2.2. In contrast, the expectation values $\langle\sigma_x\rangle_t$ and $\langle\sigma_y\rangle_t$ are non-universal since they have an overall factor $\Delta_r/\Delta = (\Delta_r/\omega_c)^K$. Hence, the off-diagonal elements of the reduced density matrix vanish in the scaling limit (18.56).

22.3 Weak quantum noise in the biased TSS

For weak damping and temperature below the Markov regime (22.50), $T \lesssim T_{\rm b}$, the white noise form (22.48) for $Q(\tau)$ is not valid anymore, and therefore the noninteracting-blip assumption is potentially questionable. We have set out already in item (a) of Subsection 21.3.1 and after Eq. (21.83), when this may occur. In the sequel, we study the self-energy for weak damping with all interblip correlations taken into account.

22.3.1 The one-boson self-energy

Interestingly enough, the self-energies (21.77) resulting from the irreducible terms of the series (21.72) and (21.74) can be summed in analytic form for weak damping with a strategy similar to that presented in Subsection 19.1.4 [471, 462, 459]. In one-phonon approximation, there are two contributions to $\Sigma_z^{(\rm s)}(\lambda)$. The first includes the internal correlations of a single blip of length τ, and the second covers the correlations between two blips of length τ_1 and τ_2 at distance s. In the interval s, the TSS undergoes uncorrelated transitions between the four states of the RDM, as given in Eq. (21.6),

$$P_0(s) \equiv \langle\sigma_z\rangle_s^{(0)} = \frac{\epsilon^2}{\Delta_{\rm b}^2} + \frac{\Delta_{\rm eff}^2}{\Delta_{\rm b}^2} \cos(\Delta_{\rm b}s) . \qquad (22.82)$$

With this insertion in the interval s, the one-phonon contribution to the self-energy $\hbar\Sigma_z^{(s)}(\lambda)$ takes the form

$$\Sigma_z^{(s)}(\lambda) = -\Delta^2 \int_0^\infty d\tau\, e^{-\lambda\tau} \cos(\epsilon\tau)\, Q'(\tau)$$
$$- \Delta^4 \int_0^\infty d\tau_2\, ds\, d\tau_1\, e^{-\lambda(\tau_2+s+\tau_1)} \sin(\epsilon\tau_2)\, P_0(s)\sin(\epsilon\tau_1) \qquad (22.83)$$
$$\times \left[\, Q'(\tau_1+\tau_2+s) + Q'(s) - Q'(\tau_1+s) - Q'(\tau_2+s)\,\right].$$

The analogous expression for the self-energy $\hbar\Delta\Sigma_x^{(a)}(\lambda)$ is found from Eq. (21.74) as

$$\Sigma_x^{(a)}(\lambda) = -\Delta \int_0^\infty d\tau\, e^{-\lambda\tau} \sin(\epsilon\tau)\, Q'(\tau)$$
$$+ \Delta^3 \int_0^\infty d\tau_2\, ds\, d\tau_1\, e^{-\lambda(\tau_2+s+\tau_1)} \cos(\epsilon\tau_2)\, P_0(s)\sin(\epsilon\tau_1) \qquad (22.84)$$
$$\times \left[\, Q'(\tau_1+\tau_2+s) + Q'(s) - Q'(\tau_1+s) - Q'(\tau_2+s)\,\right].$$

In contrast, the one-phonon contributions to the selfenergies $\hbar\Sigma_z^{(a)}(\lambda)$ and $\hbar\Delta\Sigma_x^{(s)}(\lambda)$ are determined by the correlations of the initial sojourn with the last blip,

$$\Sigma_z^{(a)}(\lambda) = \Delta^2 \int_0^\infty d\tau\, e^{-\lambda\tau} \sin(\epsilon\tau)\, Q''(\tau) - \Delta^4 \int_0^\infty d\tau_2\, ds\, d\tau_1\, e^{-\lambda(\tau_2+s+\tau_1)}$$
$$\times \sin(\epsilon\tau_2)\, P_0(s) \cos(\epsilon\tau_1) \left[\, Q''(\tau_1+\tau_2+s) - Q''(\tau_1+s)\,\right],$$
$$\Sigma_x^{(s)}(\lambda) = \Delta \int_0^\infty d\tau\, e^{-\lambda\tau} \cos(\epsilon\tau)\, Q''(\tau) - \Delta^3 \int_0^\infty d\tau_2\, ds\, d\tau_1\, e^{-\lambda(\tau_2+s+\tau_1)} \qquad (22.85)$$
$$\times \cos(\epsilon\tau_2)\, P_0(s) \cos(\epsilon\tau_1) \left[\, Q''(\tau_1+\tau_2+s) - Q''(\tau_1+s)\,\right],$$

With the spectral representation (18.42) for $Q(\tau)$ in Eq. (22.83), the time integrals are easily done, yielding [471]

$$\Sigma_z^{(s)}(\lambda) = -\frac{\Delta^2}{2\Delta_b^2} \frac{2\Delta^2\lambda^3 u(\lambda) + \epsilon^2\lambda_{b-}\lambda_{b+}[\lambda_{b-}u(\lambda_{b+}) + \lambda_{b+}u(\lambda_{b-})]}{(\lambda^2+\epsilon^2)^2}, \qquad (22.86)$$

where $\lambda_{b\pm} = \lambda \pm i\,\Delta_b$, and where

$$u(z) = \int_0^\infty d\omega\, \frac{G(\omega)}{\omega^2+z^2}\, \coth(\tfrac{1}{2}\beta\hbar\omega). \qquad (22.87)$$

The function $u(z)$ is related to the function $v(y)$ occurring in the imaginary-time approach [cf. Eq. (19.27)] by analytic continuation, $u(z) = v(y=-iz)$.

With analogous procedure in the first expression of Eq. (22.85) one gets

$$\Sigma_z^{(a)}(\lambda) = i\, \frac{\epsilon}{2} \frac{\Delta^2(\lambda^2+\Delta_b^2)}{\Delta_b^2\,(\lambda^2+\epsilon^2)} \left[w(\lambda+i\,\Delta_b) - w(\lambda-i\,\Delta_b)\right], \qquad (22.88)$$

where

$$w(z) = \Delta_b \int_0^\infty d\omega\, \frac{G(\omega)}{\omega(\omega^2+z^2)}. \qquad (22.89)$$

Finally, the self-energies $\hbar\Delta\Sigma_x^{(a)}(\lambda)$ and $\hbar\Delta\Sigma_x^{(s)}(\lambda)$ can be cast into forms which are similar to Eqs. (22.86) and (22.88), respectively.

22.3.2 Populations and coherences (super-Ohmic and Ohmic)

For weak damping, the poles of $\langle \sigma_z(\lambda) \rangle$ in Eq. (21.75) can be found iteratively. Therefore, we can limit ourselves to the self-energies at $\lambda = 0$ and at $\lambda = \pm i \, \Delta_{\mathrm{b}}$. For these particular values of λ, the expressions (22.86) and (22.88) significantly simplify,

$$\Sigma_z^{(\mathrm{s})}(0) = \frac{\Delta^2}{\epsilon^2} \, S_\xi(\Delta_{\mathrm{b}}, \beta) \,, \qquad \Sigma_z^{(\mathrm{a})}(0) = \frac{\Delta^2}{\epsilon \, \Delta_{\mathrm{b}}} \frac{\pi}{2} \, G(\Delta_{\mathrm{b}}) \,, \qquad (22.90)$$

$$\begin{aligned} \operatorname{Re} \Sigma_z^{(\mathrm{s})}(\pm i \, \Delta_{\mathrm{b}}) &= S_\xi(\Delta_{\mathrm{b}}, \beta) + 2 \frac{\epsilon^2}{\Delta^2} \lim_{\omega \to 0} S_\xi(\omega, \beta) \,, \\ \operatorname{Im} \Sigma_z^{(\mathrm{s})}(\pm i \, \Delta_{\mathrm{b}}) &= \pm \Delta_{\mathrm{b}} \int_0^\infty d\omega \, \frac{G(\omega)}{\omega^2 - \Delta_{\mathrm{b}}^2} \left(1 + \frac{2}{e^{\hbar \beta \omega} - 1} \right) \,. \end{aligned} \qquad (22.91)$$

Here, $S_\xi(\omega, \beta)$ is the power spectrum of the stochastic force introduced in Eq. (22.46),

$$S_\xi(\omega, \beta) = \frac{\pi}{2} \, G(\omega) \coth \left(\frac{\hbar \beta \omega}{2} \right) \,. \qquad (22.92)$$

Similar to the discussion in Subsec. 19.1.4, the first term of $\operatorname{Im} \Sigma_z^{(\mathrm{s})}(\pm i \, \Delta_{\mathrm{b}})$ yields the one-phonon contribution to the dressed transition amplitude Δ_{eff} defined in Eqs. (18.33) and (18.35) respectively. It is pertinent, to include this term by consistently replacing Δ by Δ_{eff}. With the one-boson exchange taken into account, the squared oscillation frequency Ω^2, relaxation rate γ_{r} and dephasing rate γ read

$$\begin{aligned} \Omega^2 &= (\Delta_{\mathrm{eff}}^2 + \epsilon^2) \left(1 - \frac{\Delta_{\mathrm{eff}}^2}{\Delta_{\mathrm{b}}^2} \int_0^\infty d\omega \, \frac{G(\omega)}{\omega^2 - \Delta_{\mathrm{b}}^2} \frac{2}{e^{\hbar \beta \omega} - 1} \right) \,, \\ \gamma_{\mathrm{r}} &= \frac{\epsilon^2}{\Delta_{\mathrm{b}}^2} \Sigma_z^{(\mathrm{s})}(0) = \frac{\Delta_{\mathrm{eff}}^2}{\Delta_{\mathrm{b}}^2} S_\xi(\Delta_{\mathrm{b}}, \beta) = \sin^2 \varphi \, S_\xi(\Delta_{\mathrm{b}}, \beta) \,, \qquad (22.93) \\ \gamma &= \frac{1}{2} \operatorname{Re} \Sigma_z^{(\mathrm{s})}(\pm i \, \Delta_{\mathrm{b}}) = \frac{1}{2} \gamma_{\mathrm{r}} + \frac{\epsilon^2}{\Delta_{\mathrm{b}}^2} S_\xi(0, \beta) = \frac{1}{2} \gamma_{\mathrm{r}} + \cos^2 \varphi \, \gamma_{\mathrm{pd}} \,. \end{aligned}$$

The factors $\sin^2 \varphi$ in γ_{r} and $\cos^2 \varphi$ in γ represent the weights of the transverse ($\propto \sin \varphi$) and of the longitudinal ($\propto \cos \varphi$) coupling in the Hamiltonian (3.146). The relaxation rate γ_{r} is the inverse time scale for relaxation of diagonal states of the RDM to the thermal equilibrium state. It is proportional to $\sin^2 \varphi$ and to the spectral power of the random force at the level splitting frequency. The decoherence rate γ characterizes the decay of the off-diagonal elements (coherences) of the RDM. It represents the inverse time scale for loss of phase coherence between the two states, and is a combination of the relaxation contribution $\frac{1}{2} \gamma_{\mathrm{r}}$ and the "pure-dephasing" contribution due to longitudinal noise at zero frequency,

$$\gamma_{\mathrm{pd}} \equiv S_\xi(0, \beta) = \begin{cases} \dfrac{\pi}{2} \lim_{\omega \to 0} G(\omega) \,, & T = 0 \,, \\[2mm] \pi \dfrac{k_{\mathrm{B}} T}{\hbar} \lim_{\omega \to 0} \dfrac{G(\omega)}{\omega} \,, & T > 0 \,. \end{cases} \qquad (22.94)$$

The longitudinal coupling $\propto \cos\varphi$ entails fluctuations of the eigen energies. These lead to a random phase between the two eigen states. Since there is no energy transferred in this process, the power spectrum at $\omega = 0$ appears in the pure dephasing rate. Clearly, the dephasing rate $\gamma_{\rm pd}$ is only meaningful if the noise power neither diverges nor drops to zero in the limit $\omega \to 0$. Since $\gamma_{\rm pd}(T = 0) = 0$ for $s > 0$, pure dephasing is generally slower than exponential decay at $T = 0$. At finite T, by contrast, pure dephasing is exponential for Ohmic damping, and slower (faster) than exponential decay in the super-Ohmic (sub-Ohmic) case. These conclusions are substantiated in the following section. Finally, the level splitting $\hbar\Omega$ undergoes a temperature-dependent shift due to transverse noise, which is a kind of Lamb shift.

For Ohmic dissipation, the expressions (22.93) take the form

$$
\begin{aligned}
\Omega^2 &= \Delta_{\rm b}^2 \left\{ 1 + 2K\sin^2\varphi \left[{\rm Re}\,\psi(i\,\hbar\Delta_{\rm b}/2\pi k_{\rm B}T) - \ln(\hbar\Delta_{\rm b}/2\pi k_{\rm B}T) \right] \right\} , \\
\gamma_{\rm r} &= \pi K\sin^2\varphi \coth(\hbar\Delta_{\rm b}/2k_{\rm B}T)\,\Delta_{\rm b} , \\
\gamma &= \tfrac{1}{2}\gamma_{\rm r} + \cos^2\varphi\,\gamma_{\rm pd} , \\
\gamma_{\rm pd} &= \vartheta \equiv 2\pi K k_{\rm B}T/\hbar .
\end{aligned}
\tag{22.95}
$$

The expression for Ω^2 coincides with the result (19.38) of the imaginary-time approach. In addition, the expressions for γ and $\gamma_{\rm r}$ smoothly map in the Markov regime $k_{\rm B}T \gtrsim \hbar\Delta_{\rm b}$ on the white-noise expressions (22.70).

In Eq. (21.75), the residuum of the pole at $\lambda = 0$ is the equilibrium value $\langle\sigma_z\rangle_{\rm eq}$,

$$
\langle\sigma_z\rangle_{\rm eq} = \frac{\Sigma_z^{(a)}(0)}{\Sigma_z^{(s)}(0)} = \frac{\epsilon}{\Delta_{\rm b}}\tanh\left(\frac{\hbar\Delta_{\rm b}}{2k_{\rm B}T}\right) = \cos\varphi\,\langle\tau_z\rangle_{\rm eq} ,
\tag{22.96}
$$

which is exactly the weak-damping form anticipated above in Eq. (21.85).

The Laplace transform (21.75) is easily inverted to obtain $\langle\sigma_z\rangle_t$. With use of the expressions (22.90) and (22.91) one readily gets

$$
\begin{aligned}
\langle\sigma_z\rangle_t &= \left[\epsilon^2/\Delta_{\rm b}^2 - \langle\sigma_z\rangle_\infty\right] e^{-\gamma_{\rm r}t} + \langle\sigma_z\rangle_\infty + (\Delta_{\rm eff}^2/\Delta_{\rm b}^2)\cos(\Omega t)\,e^{-\gamma t} \\
&\quad + \left[(\gamma_{\rm r}\epsilon^2 + \gamma\Delta_{\rm eff}^2)/\Delta_{\rm b}^3 - \gamma_{\rm r}\langle\sigma_z\rangle_\infty/\Delta_{\rm b} \right]\sin(\Omega t)\,e^{-\gamma t} .
\end{aligned}
\tag{22.97}
$$

The spectral spin correlation function $\tilde{S}_z(\omega)$ is found in the form

$$
\tilde{S}_z(\omega) = \left(\frac{\epsilon^2}{\Delta_{\rm b}^2} - \langle\sigma_z\rangle_{\rm eq}^2\right)\frac{2\gamma_{\rm r}}{\omega^2 + \gamma_{\rm r}^2} + \sum_{\zeta=\pm 1}\frac{(\Delta_{\rm eff}^2/\Delta_{\rm b}^2)\gamma + c_{\rm s}(\Omega + \zeta\omega)}{(\Omega + \zeta\omega)^2 + \gamma^2} ,
\tag{22.98}
$$

where $c_{\rm s} = (\gamma_{\rm r}\epsilon^2 + \gamma\Delta_{\rm eff}^2)/\Delta_{\rm b}^3 - (\gamma_{\rm r}/\Delta_{\rm b})\langle\sigma_z\rangle_{\rm eq}^2$. With the expression (22.96) for $\langle\sigma_z\rangle_{\rm eq}$, the weight of the quasi-elastic peak is $(\epsilon^2/\Delta_{\rm b}^2)/\cosh^2(\hbar\Delta_{\rm b}/2k_{\rm B}T)$. Therefore, the NIBA flaw, namely that the weight of the quasi-elastic peak becomes negative at low T, as discussed below Eq. (22.77), is dissolved by the interblip correlations.

Following similar lines, it is also straightforward to calculate $\langle\sigma_x\rangle_t$. We obtain

$$\langle \sigma_x \rangle_t = \left(\frac{\epsilon \Delta_{\mathrm{eff}}^2}{\Delta \Delta_{\mathrm{b}}^2} - \langle \sigma_x \rangle_{\mathrm{eq}} \right) e^{-\gamma_{\mathrm{r}} t} - \frac{\epsilon \Delta_{\mathrm{eff}}^2}{\Delta \Delta_{\mathrm{b}}^2} \cos(\Omega t)\, e^{-\gamma t} + \langle \sigma_x \rangle_{\mathrm{eq}}$$
$$+ \left(\frac{\Delta_{\mathrm{eff}}^2}{\Delta \Delta_{\mathrm{b}}} [\pi K + \epsilon (\gamma_{\mathrm{r}} - \gamma)/\Delta_{\mathrm{b}}^2] - \frac{\gamma_{\mathrm{r}}}{\Delta_{\mathrm{b}}} \langle \sigma_x \rangle_{\mathrm{eq}} \right) \sin(\Omega t)\, e^{-\gamma t} . \tag{22.99}$$

The equilibrium value of $\langle \sigma_x \rangle_{\mathrm{eq}}$ is

$$\langle \sigma_x \rangle_{\mathrm{eq}} = \frac{\Delta_{\mathrm{eff}}^2}{\Delta \Delta_{\mathrm{b}}} \tanh \left(\frac{\hbar \Delta_{\mathrm{b}}}{2 k_{\mathrm{B}} T} \right) . \tag{22.100}$$

The above expressions for $\langle \sigma_z \rangle_t$ and $\langle \sigma_x \rangle_t$ match in the Ohmic case with the expressions (22.70) – (22.76) at temperatures $T \gtrsim T_{\mathrm{b}}$. Thus, we have found, together with the results of Subsec. 22.2.3, analytic expressions for the dynamics of the biased, weakly damped Ohmic TSS in the range $0 \leqslant T \ll \hbar \omega_{\mathrm{c}}/k_{\mathrm{B}}$. Interestingly, the above dynamical approach yields the correct equilibrium values reached at asymptotic time. These coincide with the expressions (21.85) found by a simple thermostatic consideration. As a final remark, with a view at the amplitudes (22.74), (22.76), and (22.79), it seems natural to replace Δ_{b} by Ω in the above expressions (22.96) – (22.100).

An insightful quantity is the difference of the populations of the ground state and the excited state. In the eigenbasis of H_{TSS} we have $\langle \tau_z \rangle \equiv \langle |g><g|\rangle - \langle |e><e|\rangle$. By means of the rotations (3.141) and (3.138) we may relate $\langle \tau_z \rangle_t$ to $\langle \sigma_x \rangle_t$ and $\langle \sigma_z \rangle_t$ as

$$\langle \tau_z \rangle_t = \sin \varphi \langle \sigma_x \rangle_t + \cos \varphi \langle \sigma_z \rangle_t . \tag{22.101}$$

With the above results for the expectation values $\langle \sigma_x \rangle_t$ and $\langle \sigma_z \rangle_t$ we get

$$\langle \tau_z \rangle_t = \left(\epsilon/\Omega - \langle \tau_z \rangle_{\mathrm{eq}} \right) e^{-\gamma_{\mathrm{r}} t} + \langle \tau_z \rangle_{\mathrm{eq}} + A_{\mathrm{s}} \sin(\Omega t)\, e^{-\gamma t} , \tag{22.102}$$

where the oscillation amplitude is $\mathcal{O}(K)$, $A_{\mathrm{s}} = \pi K \Delta_{\mathrm{eff}}^2/\Delta\,\Omega + \gamma_{\mathrm{r}} \epsilon/\Omega^2 - \langle \tau_z \rangle_{\mathrm{eq}} \gamma_{\mathrm{r}}/\Omega$. This expression describes relaxation from the initial value ϵ/Ω to the equilibrium value $\langle \tau_z \rangle_{\mathrm{eq}} = \tanh(\hbar\Omega/2 k_{\mathrm{B}} T)$, and points out again the role of the rate γ_{r}.

Finally, let us take a look at the decay of the off-diagonal elements or coherences of the TSS density matrix in energy representation. These are the expectation values of the flip operators $\tau_{\pm}$. The latter are expressed in terms of the σ-matrices by the relation (3.143). With the expressions (22.97), (22.99) and (21.4) we find

$$\langle \tau_{\pm} \rangle_t \equiv \frac{1}{2} \left(\cos \varphi \langle \sigma_x \rangle_t \pm i \langle \sigma_y \rangle_t - \sin \varphi \langle \sigma_z \rangle_t \right) = -\frac{\Delta}{2\Omega} e^{\mp i \Omega t} e^{-\gamma t} , \tag{22.103}$$

where amplitudes of order K are disregarded. Interestingly, the relaxation contributions $\propto e^{-\gamma_{\mathrm{r}} t}$ cancel out. This confirms that $1/\gamma$ is the time scale for dephasing.

The analysis of the dynamics can be generalized to the case of time-dependent external fields coupled to σ_z and to σ_x by employing the substitutions given in Subsection 23.1.1. The dynamics of $\langle \sigma_z \rangle_t$ under influence of a monochromatic high-frequency field coupled to σ_z has been studied in Ref. [462]. Alternative approaches based on second-order perturbation in the TSS-bath coupling have been frequently employed in this parameter regime. In these treatments, the adiabatic renormalization of the bare tunneling matrix element is usually disregarded. For a discussion of $\langle \sigma_z \rangle_t$ and $\langle \tau_z \rangle_t$ under monochromatic low-frequency driving, we refer to Refs. [473, 474].

22.4 Pure dephasing

The dynamics induced by the spin-boson Hamiltonian (18.15) is easily solved if we disregard the tunneling term. For $\Delta = 0$, we have $\tau_j = \sigma_j$, where $j = x, y, z$, and the observables of interest are the coherences. These are the expectation values of the flip operators $\sigma_\pm = \frac{1}{2}(\sigma_x \pm i\,\sigma_y)$,

$$\langle \sigma_\pm \rangle_t = \text{tr}\,[\sigma_\pm(t)\hat{\rho}(0)]\,. \tag{22.104}$$

In the absence of tunneling, the operator $\sigma_\pm(t)$ obeys the equation of motion

$$\dot{\sigma}_\pm(t) = (i/\hbar)[H_{\text{SB}}, \sigma_\pm(t)]_{\Delta=0} = \mp i\,[\epsilon + \mathfrak{E}(t)/\hbar]\sigma_\pm(t)\,, \tag{22.105}$$

where $\mathfrak{E}(t) = \sum_\alpha \hbar\lambda_\alpha[b_\alpha\,e^{-i\omega_\alpha t} + b_\alpha^\dagger\,e^{i\omega_\alpha t}]$ is the polarization energy. Hence we have

$$\begin{aligned}
\langle \sigma_\pm \rangle_t &= \langle \sigma_\pm \rangle_0\,e^{\mp i\epsilon t} \left\langle \exp\left\{ \mp \frac{i}{\hbar} \int_0^t dt'\,\mathfrak{E}(t') \right\} \right\rangle \\
&= \langle \sigma_\pm \rangle_0\,e^{\mp i\epsilon t} \exp\left\{ -\frac{1}{\hbar^2} \int_0^t dt' \int_0^{t'} dt''\,\langle \mathfrak{E}(t')\mathfrak{E}(t'') \rangle \right\}\,.
\end{aligned} \tag{22.106}$$

The second form holds for Gaussian statistics of the fluctuating polarization energy $\mathfrak{E}(t)$. With the thermal averages (3.67) and the definition (18.16) for the spectral density $G(\omega)$ one finds that, apart from the minus sign, the exponent is just the spectral representation (18.42) of the pair correlation function $Q(t)$ discussed in Section 18.2. Hence the off-diagonal elements of the RDM decay exactly as

$$\langle \sigma_\pm \rangle_t = \langle \sigma_\pm \rangle_0\,e^{\mp i\epsilon t}\,e^{-Q(t)}\,. \tag{22.107}$$

An analytic expression for $Q(t)$ for general spectral bath parameter s, general T and all times is given in Eq. (18.45). The corresponding zero temperature expression is stated in Eq. (20.55).

The loss of coherence process is characterized by three different time regimes:
(i) the core regime $0 < t < 1/\omega_c$, in which $Q(t)$ depends on the particular choice of the cutoff in the spectral density $G(\omega)$.
(ii) the quantum regime $1/\omega_c \ll t \ll \hbar\beta$, in which $Q(t)$ is essentially determined by the ground state of the reservoir $[\coth(\beta\hbar\omega/2) \approx 1]$, i.e, $Q(t) = Q_{T=0}(t)$.
(iii) the thermal regime $t \gg \hbar\beta$, in which $Q(t)$ is ruled by the classical state of the reservoir $[\coth(\beta\hbar\omega/2) \approx 2/\beta\hbar\omega]$.

Therefore, $Q(t)$ is equipped with an additional factor t in the thermal regime, as compared with $Q_{T=0}(t)$. Hence the loss of coherence is always faster in the time regime $t > \hbar\beta$ in relation to the loss in the regime $t < \hbar\beta$ [475]. At zero temperature, the quantum regime persists until time infinity.

Consider now first the case of Ohmic friction, $s = 1$. With the Ohmic form (20.55) for the pair correlation function $Q(t)$ we find in the quantum regime

$$|\langle \sigma_\pm \rangle_t| = |\langle \sigma_\pm \rangle_0| \frac{\cos[2K \arctan(\omega_c t)]}{[1 + (\omega_c t)^2]^K}\,, \qquad 0 \leqslant t \ll \hbar\beta \tag{22.108}$$

with the limiting power-law drop $|\langle\sigma_\pm\rangle_t| = |\langle\sigma_\pm\rangle_0| \cos(\pi K) (\omega_c t)^{-2K}$ outside the core regime, $1/\omega_c \ll t \ll \hbar\beta$. The asymptotic behavior is characterized by exponential decay, as we infer from Eq. (22.48) with Eq. (22.47),

$$|\langle\sigma_\pm\rangle_t| = |\langle\sigma_\pm\rangle_0| \cos(\pi K) e^{-Q_{\text{adia}}} e^{-\gamma_{\text{pd}}t}, \qquad t \gg \hbar\beta . \qquad (22.109)$$

Here $\gamma_{\text{pd}} = S_\xi(0) = \vartheta$ is the pure dephasing rate introduced in Eq. (22.95). Thus the off-diagonal states of the RDM decay algebraically in the quantum regime, and the power of the decay law depends on the coupling strength. As $T \to 0$, the pure dephasing rate γ_{pd} drops to zero. As we now see from Eq. (22.108), the off-diagonal states dephase anyhow, but with slower than exponential, viz. algebraic drop.

In the super-Ohmic range $s > 1$ we find from Eq. (18.42) the limiting forms

$$
\begin{aligned}
\operatorname{Re} Q(t) &= Q_{\text{adia}} - 2\delta_s \frac{\Gamma(s)\sin(\frac{1}{2}\pi s)}{(s-1)} \frac{1}{(\omega_{\text{ph}}t)^{s-1}} , & 1/\omega_c \ll t \ll \hbar\beta , \\
\operatorname{Re} Q(t) &= Q_{\text{adia}} + 4\delta_s \frac{\Gamma(s-2)\cos(\frac{1}{2}\pi s)}{\beta\hbar\omega_{\text{ph}}}(\omega_{\text{ph}}t)^{2-s} , & t \gg \hbar\beta ,
\end{aligned}
\qquad (22.110)
$$

where $Q_{\text{adia}} = 2\delta_s \left[\Gamma(s)/(s-1)\right](\omega_c/\omega_{\text{ph}})^{s-1}$.

The reduction of the off-diagonal states $\propto e^{-Q_{\text{adia}}}$ by adiabatic dressing is quickly built up in the initial core regime $t \lesssim 1/\omega_c$. From the viewpoint of quantum state engineering, this is already a substantial loss of phase coherence. The drop depends strongly on the cutoff frequency ω_c. In the subsequent time regime $1/\omega_c < t < \hbar\beta$ coherent oscillations persist in which the amplitude roughly maintains a plateau value.

In the range $1 < s < 2$ there is finally in the thermal regime $t \gg \hbar\beta$ decay $\propto e^{-\text{const}\times T(\omega_{\text{ph}}t)^{2-s}}$. In contrast, in the range $s > 2$ the plateau regime persists in practice until time infinity. The preliminary result (22.94) for the pure dephasing rate γ_{pd} is consistent with these findings. Since for $s > 1$ the decay is actually slower than exponential, γ_{pd} vanishes in this regime.

Let us finally discuss the sub-Ohmic regime $0 < s < 1$. In the limit $\omega_c \to \infty$ the core regime becomes arbitrarily narrow, and there is no reduction of the coherences in this regime. In the subsequent time regimes we have

$$
\begin{aligned}
\operatorname{Re} Q(t) &= 2\delta_s \frac{\Gamma(s)\sin(\frac{1}{2}\pi s)}{(1-s)}(\omega_{\text{ph}}t)^{1-s} , & 0 \ll t \ll \hbar\beta , \\
\operatorname{Re} Q(t) &= 4\delta_s \frac{\Gamma(s-2)\cos(\frac{1}{2}\pi s)}{\beta\hbar\omega_{\text{ph}}}(\omega_{\text{ph}}t)^{2-s} , & t \gg \hbar\beta .
\end{aligned}
\qquad (22.111)
$$

The expressions (22.94) for γ_{pd} are consistent with the behaviors (22.111). The pure dephasing rate γ_{pd} vanishes for $T = 0$, since the actual decay $\propto e^{-\text{const}\times(\omega_{\text{ph}}t)^{1-s}}$ in the quantum regime is slower than exponential decay $\propto e^{-\gamma_{\text{pd}}t}$ would be. On the other hand, the pure dephasing rate γ_{pd} diverges for finite T, because the true decay in the thermal regime $\propto e^{-\text{const}\times(T/\omega_{\text{ph}})(\omega_{\text{ph}}t)^{2-s}}$ is actually faster than decay under $e^{-\gamma_{\text{pd}}t}$.

22.5 1/f noise and decoherence

Experiments with superconducting qubits revealed that the relaxation dynamics is dominated by Ohmic high frequency noise at ω near Ω, while the dominant source of dephasing at low T is slow flicker noise with roughly a $1/f$ power spectrum [119, 120]. The $1/f$ noise may have different origin depending on the system. A widely-used assumption is that the qubit is affected by impurities which essentially act as an ensemble of two-level systems [118]. The discrete nature of a TLS environment was observed before, e.g., in metals and in single-electron devices. There, the random or coherent switching causes conductance fluctuations and $1/f$ current noise. In charge qubits, the discrete fluctuators add to the fluctuations of the gate charge regulating the qubit.

Basically, a TLS environment induces non-Gaussian statistics, which in general can not simply be shaped with a Gaussian influence functional. Rather the full qubit-TLS dynamics may be relevant, which becomes apparent in the transient qubit dynamics and in saturation effects [476, 477]. Gaussian statistics is found to hold definitely only in the limit of very weak qubit-TLS coupling. Slow flicker noise becomes effective when there is a whole ensemble of thermal or non-thermal fluctuators with a roughly log-uniform distribution of flip rates [118, 478].

Let us now take a closer look to the effects of a TLS environment in Gaussian approximation. We start with the observation that the TLS coupling induces in the qubit Hamiltonian (3.146) the random polarization energy

$$\mathfrak{E}(t) = \sum_\alpha v_\alpha \, \sigma_{z,\alpha}(t) \, . \tag{22.112}$$

The power spectrum of the related noise, normalized as in Eq. (22.46), then is

$$S_{\text{TLS}}(\omega) = \frac{1}{2\hbar^2} \sum_\alpha v_\alpha^2 \tilde{C}_z^+(\omega) \, , \tag{22.113}$$

where $\tilde{C}_z^+(\omega)$ is the transform of the correlation function $C_z^+(t)$ defined in (21.13).

Consider now two different scenarios: the slow noise comes (i) from a randomly switching and (ii) from a coherent TLS environment.

22.5.1 1/f noise from fluctuating background charges

One possibility is that the two-level systems are fluctuating background charges (BC) switching randomly between two states [118]. In the low-frequency regime of interest the spectral correlation function of the BCs is given by the quasi-elastic Lorentzian (22.81). The power spectrum resulting from the back ground charges then is

$$S_{\text{BC}}(\omega) = \frac{1}{2\hbar^2} \sum_\alpha v_\alpha^2 \frac{2\gamma_{\text{r},\alpha}}{\omega^2 + \gamma_{\text{r},\alpha}^2} \, , \tag{22.114}$$

where $\gamma_{\text{r},\alpha}$ is the switching rate of two-state fluctuator α. Since in general there are many fluctuators, one must average over the distribution of coupling strengths and switching rates,

$$S_{\mathrm{BC}}(\omega) \propto \int dv\, d\gamma_{\mathrm{r}}\, P(v, \gamma_{\mathrm{r}})\, v^2\, \frac{2\gamma_{\mathrm{r}}}{\omega^2 + \gamma_{\mathrm{r}}^2}. \tag{22.115}$$

Various forms for the distribution function $P(v, \gamma_{\mathrm{r}})$ with uncorrelated v and γ_{r} have been analyzed in Refs. [478, 479]. Generally one should distinguish between results for specific samples and those averaged over a statistical ensemble of samples. This is particularly important when the qubit is influenced by a small number of BCs, which results in sizeable sample-to-sample fluctuations. There is also the possibility that the bulk of the weights in v and in γ_{r} comes from different BCs.

Consider now the simple case in which the switching rates are dispersed with a smooth distribution on a log scale, as in the standard tunneling model of glasses [121, 122, 123]. Putting $\gamma_{\mathrm{r}} \propto e^{-\ell/\ell_0}$, where ℓ may be thought of as the tunneling distance (or any power of it), and assuming uniform distribution of ℓ over a range substantially larger than ℓ_0, we obtain precisely $1/f$ noise,

$$S_{\mathrm{BC}}(\omega) \propto \overline{v^2} \int d\ell\, \frac{2\gamma_{\mathrm{r}}(\ell)}{\omega^2 + \gamma_{\mathrm{r}}^2(\ell)} \propto \overline{v^2} \int \frac{d\gamma_{\mathrm{r}}}{\gamma_{\mathrm{r}}}\, \frac{2\gamma_{\mathrm{r}}}{\omega^2 + \gamma_{\mathrm{r}}^2} \propto \frac{\overline{v^2}}{|\omega|}. \tag{22.116}$$

Here $\overline{v^2}$ is the average squared coupling. This may be compared with the power spectrum resulting from an oscillator bath. Eq. (22.46) yields in the regime $\omega \ll k_{\mathrm{B}}T/\hbar$

$$S_\xi(\omega) = \frac{\pi k_{\mathrm{B}} T}{\hbar}\, \frac{G(|\omega|)}{|\omega|}. \tag{22.117}$$

The expression (22.117) reproduces the $1/f$ power spectrum (22.116) precisely in the extreme sub-Ohmic limit $s \to 0$, in which the spectral coupling becomes constant,

$$G(\omega) \quad \to \quad G_{\mathrm{BC}}(\omega) = 2\delta_0\, \omega_{\mathrm{ph}}\, \Theta(\omega - \omega_{\mathrm{ir}}). \tag{22.118}$$

Here we have subjoined an intrinsic low-frequency cut-off ω_{ir} for the $1/f$ noise spectrum. Since the slow $1/f$ noise sources are largely out of equilibrium, the bath temperature in Eq. (22.117) should be considered as a fit parameter T_{b}. Upon introducing a suitable frequency scale $\omega_{1/f}$, we then may write

$$S_{\mathrm{BC}}(\omega) = \frac{\omega_{1/f}^2}{|\omega|}\, \Theta(\omega - \omega_{\mathrm{ir}}) \quad \text{with} \quad \omega_{1/f} = \sqrt{2\pi\delta_0\, \omega_{\mathrm{ph}} k_{\mathrm{B}} T_{\mathrm{b}}/\hbar}. \tag{22.119}$$

22.5.2 $1/f$ noise from coherent two-level systems

The case in which the noise comes from a set of coherent two level systems is equally interesting. Now the dynamics of the dissipative TSS is embodied in the form (22.80) of the spectral spin correlation function $\tilde{C}_z^+(\omega)$. In addition, we take a continuous distribution $P(v, \epsilon, \Delta)$ of the coupling parameter v and the TSS parameters ϵ and Δ. At low T, the emission line at $\omega = -\Omega$ is irrelevant since it is exponentially suppressed by a factor $e^{-\beta\hbar\Omega}$. Then the power spectrum takes the form

$$S_{\text{TLS}}(\omega) \quad \propto \quad \int dv\, d\epsilon\, d\Delta\, P(v, \epsilon, \Delta)$$
$$\times \left(\frac{2\gamma \sin^2 \varphi}{(\omega - \Omega)^2 + \gamma^2} + \frac{\cos^2 \varphi}{\cosh^2(\frac{1}{2}\beta\hbar\Omega)} \frac{2\gamma_r}{\omega^2 + \gamma_r^2} \right), \tag{22.120}$$

where $\Omega = \sqrt{\Delta^2 + \epsilon^2}$. It is natural to assume that the distribution of v is not correlated with that of Δ and ϵ. When the width γ_r of the quasielastic peak is small compared to the Rabi frequency Ω, the power spectrum (22.120) splits into low-frequency noise resulting from the quasielastic peak and high-frequency noise stemming from the inelastic peak.

If we assume a log-uniform distribution of level splittings Δ or switching rates $\gamma_r \propto \Delta^2$, the low-frequency behavior resulting from Eq. (22.120) is again $1/f$ noise as in Eq. (22.116). It is self-evident to choose again a log uniform distribution of level splittings, $P_\Delta(\Delta) \propto 1/\Delta$, for the inelastic contribution in Eq. (22.120). Taking the integral over Δ, we then obtain

$$S_{\text{TLS}}(\omega) \propto \frac{1}{\omega} \int_0^\omega d\epsilon\, P_\epsilon(\epsilon). \tag{22.121}$$

The choice $P_\epsilon(\epsilon) \propto \epsilon$ readily yields Ohmic power spectrum $S_{\text{TLS}}(\omega) \propto \omega$ at high frequency. Most interestingly, the integral over ϵ in the low-frequency contribution $S(\omega) \propto 1/|\omega|$ with the same assumption $P_\epsilon(\epsilon) \propto \epsilon$ yields a weight factor T^2.

In sum, with the distribution $P(\epsilon, \Delta) \propto \epsilon/\Delta$ the expression (22.120) simultaneously yields the limiting behaviors [120]

$$S_{\text{TLS}}(\omega) = \begin{cases} a\left(\dfrac{k_{\text{B}}T}{\hbar}\right)^2 \dfrac{1}{\omega} & \text{for} \quad 0 < \omega \ll \Omega, \\ a\,\omega & \text{for} \quad \omega \approx \Omega. \end{cases} \tag{22.122}$$

The different spectral power-law forms with the same prefactor a intersect at $\omega = k_{\text{B}}T/\hbar$. A connection between the low-frequency $1/f$ noise power relevant for dephasing and the Ohmic high-frequency noise power responsible for relaxation was observed in experiments with Josephson devices [482]. The above T^2 dependence of the strength of the $1/f$ noise is consistent with these experiments. This indicates relevance of a coherent TLS environment in Josephson qubit devices.

There are still open questions about the statistics of the low-frequency noise, and whether the average over the TLS parameters is universal or sample-dependent.

22.5.3 Decoherence from $1/f$ noise

Consider now the transient coherent dynamics of the dissipative TSS in which the power spectrum of the BC ensemble is characterized by the form (22.119). In this case, the expression (22.94) for the pure dephasing rate is not valid. In fact, the decay of coherences is given in generalization of Eq. (22.107) by the expression

$$\langle \sigma_\pm \rangle_t = \langle \sigma_\pm \rangle_0\, e^{\mp i\Omega t}\, e^{-\gamma_r t/2}\, e^{-\cos^2 \varphi\, \text{Re}\, Q(t)}. \tag{22.123}$$

In one-boson exchange approximation, the relaxation rate reads $\gamma_r = \sin^2\varphi\, S_{BC}(\Omega)$. With the power spectrum (22.119), and with the relation (22.46), the bath correlation function (18.42) takes in the time regime $\hbar\beta \ll t \ll 1/\omega_{ir}$ the form

$$\mathrm{Re}\, Q(t) \;=\; (\omega_{1/f} t)^2 \,|\ln(\omega_{ir} t)|/\pi \;. \tag{22.124}$$

For longitudinal noise, $\varphi = 0$, the expression (22.123) with (22.124) yields for the coherence decay quadratic time-dependence in the exponent (see also Ref. [480]),

$$|\langle\sigma_\pm\rangle_t| \;=\; |\langle\sigma_\pm\rangle_0|\, e^{-(\gamma^* t)^2} \quad\text{with}\quad \gamma^* = \omega_{1/f}\sqrt{\ln(\omega_{1/f}/\omega_{ir})/\pi}\;. \tag{22.125}$$

In contrast, transverse noise, $\varphi = \pi/2$, leads to exponential decay

$$|\langle\sigma_\pm\rangle_t| \;=\; |\langle\sigma_\pm\rangle_0|\, e^{-\gamma t} \quad\text{with}\quad \gamma = \kappa\,\omega_{1/f}^2/\Omega\;. \tag{22.126}$$

The one-phonon exchange contribution gives $\gamma = \frac{1}{2}\gamma_r = \frac{1}{2}\omega_{1/f}^2/\Omega$, and thus $\kappa = \frac{1}{2}$. The analysis of multi-phonon contributions to the rate shows that these are sensitive to the divergent low-frequency power spectrum and thus may dominate. It was found with a diagrammatic self-energy method that by consideration of higher-order contributions the factor κ is actually moved with logarithmic accuracy to [481]

$$\kappa \;\approx\; \ln(\omega_{1/f}^2/\omega_{ir}\Omega)/\pi\;. \tag{22.127}$$

Hence the ratio γ/γ_r is actually much larger than $\frac{1}{2}$. In the experiments reported in Ref. [125] for the ratio γ/γ_r the numerical value $\kappa \approx 3$ was found.

22.6 The Ohmic TSS at and close to the Toulouse point

For the special case $K = \frac{1}{2}$, the Ohmic TSS can be mapped on the Toulouse model, as we have discussed already in Subsec. 19.2.2. The equations of motion for the imaginary-time Green's function of the d-level are solved without difficulties and yield the expression (19.67). The real-time equilibrium correlation functions of interest can be expressed in terms of this function, as we shall see below in Subsec. 22.6.6. Alternatively, we can directly sum the exact series expressions given in Subsecs. 21.2.3 and 21.2.4 [223]. As an advantage over the fermionic approach, the latter method facilitates the calculation of nonequilibrium conditional probabilities.

22.6.1 Grand-canonical sums of collapsed blips and sojourns

The reason that the path summation is possible in analytic form in the particular case $K = \frac{1}{2}$ can most easily be understood using the concept of *collapsed* blips and *collapsed* sojourns. In the related charge picture, neighboring charges of opposite sign form *collapsed* dipoles which have zero dipole length and zero dipole moment, and thus are not interacting with other charges. The corresponding calculation is done for $K = \frac{1}{2} - \kappa$, where $\kappa \ll 1$, and after all the limit $\kappa \to 0$ is performed. In the scaling limit

(18.56) with (18.53), the bath influence factors $F_m^{(\pm)}$ and $F_{m,n}$ are given in Eqs. (21.51) and (21.55). The crucial point now is that the factor $\cos(\pi K) = \sin(\pi\kappa)$ occurring in $F_m^{(\pm)}$ and $F_{m,n}$ drops to zero $\propto \kappa$, as $\kappa \to 0$. Thus, to obtain a non-vanishing contribution, each $\cos(\pi K)$-factor must come along with a divergent factor $1/\kappa$. This factor is exactly provided by the short-distance singular behavior of the breathing mode integral of a dipole with intra-dipole interaction $\lim_{\tau\to 0} e^{-Q'(\tau)} \approx (\omega_c\tau)^{-1+2\kappa}$. The dipole length τ may represent either a sojourn or a blip interval. The contribution of a blip (sojourn) dipole together with the associated $\cos(\pi K)$-factor is

$$J_{b\,(s)}(K = \tfrac{1}{2}) \equiv \lim_{\kappa\to 0} \frac{\Delta^2}{2} \sin(\pi\kappa) \int_0^\infty d\tau \, \frac{1}{(\omega_c\tau)^{1-2\kappa}} B_{b\,(s)}(\tau) f_{b\,(s)}(\tau) \,. \qquad (22.128)$$

The bias factor for a blip dipole is $B_b(\tau) = e^{\pm i\,\epsilon\tau}$, and is unity for a sojourn dipole, $B_s(\tau) = 1$. The function $f_{b\,(s)}(\tau)$ represents the interaction factor of the blip (sojourn) dipole with the other charges. Observing that a blip (sojourn) dipole with zero dipole length does not interact with other charges, we have $f_{b\,(s)}(0) = 1$. Thus we obtain

$$J_{b\,(s)}(K = \tfrac{1}{2}) = \lim_{\kappa\to 0} \frac{\Delta^2}{2\omega_c} B_{b\,(s)}(0) f_{b\,(s)}(0) \sin(\pi\kappa)\Gamma(2\kappa) = \frac{\pi}{4}\frac{\Delta^2}{\omega_c} \equiv \frac{\gamma}{2} \,. \qquad (22.129)$$

From this we see that a collapsed blip dipole, as well as a collapsed sojourn dipole, depends neither on the bias strength nor on temperature. The expression $J(K = \tfrac{1}{2})$ is half the effective frequency $\Delta_{\text{eff}}(K = \tfrac{1}{2}) = \gamma$ introduced in Eq. (19.52).

Of particular importance now is that collapsed dipoles are noninteracting. Therefore, the grand-canonical sum of these entities can be performed without difficulty.

During a wide sojourn interval the system may make visits of duration zero to either of the two blip states. We refer to these "blitz" visits of blip states as collapsed blips (CB). All the possibilities of these short visits within a wide sojourn interval form a grand-canonical ensemble of these entities. Taking into account the minus sign associated with each factor Δ^2, and the multiplicity factor two of the possible intermediate blip states,[6] the grand-canonical sum of noninteracting collapsed blips in the sojourn interval of length s adds up to the exponential factor

$$U_{\text{CB}}(s) = e^{-\gamma s} \,, \qquad (22.130)$$

which we refer to as CB form factor. Equivalently, during an extended stay in a blip state the system may make any number of visits of duration zero to a sojourn state. The ensemble of these short visits may be viewed as a noninteracting gas of collapsed sojourns (CS). The grand-canonical sum of these entities in the blip interval of length τ, together with the CS multiplicity factor one, yields the CS form factor

$$U_{\text{CS}}(\tau) = e^{-\gamma\tau/2} \,. \qquad (22.131)$$

With these preliminaries, the path or charge sum is facilitated substantially.

---

[6]A dipole is formed by two neighbouring ξ-charges of opposite sign. A blip dipole is formed by a visit of a blip state from a sojourn state with return to the same or the other sojourn state (See footnote on page 385). A sojourn dipole is formed by a visit of a sojourn state from a blip state with return to the same blip state. Thus, blip dipoles have a multiplicity factor 2 compared to sojourn dipoles.

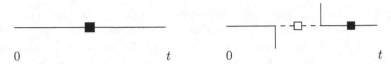

Figure 22.2: The diagrams for $P_s(t)$ (left) and $P_a(t)$ (right). The full (dashed) lines represent sojourns (blips). A full box represents the insertion of a CB form factor U_{CB} within a sojourn interval, and the empty box stands for a CS form factor U_{CS} inserted in a blip interval. The downward and upward spikes symbolize the remaining two solitary charges forming an extended blip dipole.

22.6.2 The function $\langle \sigma_z \rangle_t$ for $K = \frac{1}{2}$

Consider first the contribution $P_s(t)$ to $\langle \sigma_z \rangle_t$ in Eq. (21.41), which is symmetric under bias inversion. In order Δ_{2m}, there are m $\cos(\pi K)$-factors. These must be combined with m collapsed blips, which is the maximum number in this order. Hence $P_s(t)$ is pictorially represented by a single sojourn of length t being dressed by a CB form factor, as sketched diagrammatically in Fig. 22.2 (left). Thus $P_s(t)$ decays exponentially with temperature- and bias-independent relaxation rate $\gamma = \pi \Delta^2 / 2\omega_c$,

$$P_s(t) = U_{CB}(t) = e^{-\gamma t} . \tag{22.132}$$

Consider next the antisymmetric contribution $P_a(t)$ to $\langle \sigma_z \rangle_t$. We see from Eq. (21.51) that the influence function $F_m^{(-)}$ has one $\cos(\pi K)$-factor less than $F_m^{(+)}$. Hence in each order there is a single wide blip. Because of the factor ξ_1 in $F_m^{(-)}$ and summation over $\xi_1 = \pm 1$ is implied, the product $F_m^{(-)} B_m^{(a)}$ in Eq. (21.41) with Eq. (21.42) must provide the bias factor $\sin(\xi_1 \epsilon \tau_1)$. Thus, the first blip is the wide one. The picture we now have is as sketched in Fig. 22.2 (right). The contribution $P_a(t)$ is pictorially represented by two wide sojourns of lengths s_1 and s_2 which are separated by a single wide blip of length $\tau_1 = \tau$. The initial sojourn is bare. In contrast, the final sojourn is dressed with a CB form factor, and the intermediate wide blip with a CS form factor. The intervals of these three components add up to the total time, $s_1 + \tau + s_2 = t$. Upon putting together the various pieces, and taking into account the interaction between the remaining two solitary charges, one obtains

$$P_a(t) = \Delta^2 \int_0^\infty ds_1 \, d\tau \, ds_2 \, \delta(t - s_1 - \tau - s_2) \sin(\epsilon \tau) \, e^{-Q'(\tau)} \, e^{-\gamma \tau/2} \, e^{-\gamma s_2} . \tag{22.133}$$

Using Eqs. (22.132), (22.133), and (18.53), one finally gets

$$\langle \sigma_z \rangle_t = e^{-\gamma t} + \frac{2}{\hbar \beta} \int_0^t d\tau \, \frac{\sin(\epsilon \tau)}{\sinh(\pi \tau / \hbar \beta)} \left[e^{-\gamma \tau/2} - e^{-\gamma t} \, e^{\gamma \tau/2} \right] . \tag{22.134}$$

The integral is a linear combination of hypergeometric functions.

The canonical equilibrium state reached at asymptotic time has the analytic form

$$\langle \sigma_z \rangle_{eq} = \frac{2}{\hbar \beta} \int_0^\infty d\tau \, \frac{\sin(\epsilon \tau) \, e^{-\gamma \tau/2}}{\sinh(\pi \tau / \hbar \beta)} = \frac{2}{\pi} \, \mathrm{Im} \, \psi \left(\frac{1}{2} + \frac{\hbar \gamma}{4\pi k_B T} + i \frac{\hbar \epsilon}{2\pi k_B T} \right) , \tag{22.135}$$

where $\psi(z)$ is the digamma function. As $T \to 0$, this expression simplifies to

$$\langle \sigma_z \rangle_{\mathrm{eq}, T=0} = (2/\pi) \arctan(2\epsilon/\gamma) . \tag{22.136}$$

Thus, there persists nonzero occupation of the higher (left) state at zero temperature. At $T = 0$, the leading time dependence at asymptotic time $\sqrt{\gamma^2/4 + \epsilon^2}\, t \gg 1$ is

$$\langle \sigma_z \rangle_t = \langle \sigma_z \rangle_{\mathrm{eq}, T=0} - \frac{8}{\pi} \frac{\gamma^2}{\gamma^2 + 4\epsilon^2} \frac{\sin(\epsilon t)\, e^{-\gamma t/2}}{\gamma t} . \tag{22.137}$$

Here, terms of order $e^{-\gamma t}$ are disregarded. At low temperature, $k_{\mathrm{B}} T \ll \gamma$, the asymptotic time dependence is expressed with the Matsubara frequency $\nu_1 = 2\pi/\hbar\beta$ as

$$\langle \sigma_z \rangle_t = \langle \sigma_z \rangle_{\mathrm{eq}} - \frac{16}{\hbar\beta\gamma} \frac{\gamma^2}{\gamma^2 + 4\epsilon^2} \sin(\epsilon t)\, e^{-(\gamma+\nu_1)t/2} . \tag{22.138}$$

Both expressions exhibit a term describing a damped oscillation. The quality of the oscillation increases with increasing bias and decreases with increasing temperature.

The expectation values $\langle \sigma_x \rangle_t$ and $\langle \sigma_y \rangle_t$ are non-universal, as we have dicussed at the end of Subsection 22.2.3, Hence they are zero in the universality limit (18.56).

22.6.3 The case $K = \frac{1}{2} - \kappa$; coherent-incoherent crossover

In the NIBA, the time evolution of $\langle \sigma_z \rangle_t$ for a symmetric TSS at $T = 0$ is described by the Mittag-Leffler-function $E_{2-2K}[-(\Delta_{\mathrm{eff}} t)^{2-2K}]$, as discussed in Subsec. 22.1.1. This function predicts (i) that the coherent-incoherent transition occurs precisely at $K = 1/2$, and (ii), that the leading asymptotic cut contribution is $\propto 1/(\Delta_{\mathrm{eff}} t)^{2-2K}$. For $K = \frac{1}{2} - \kappa$, where $|\kappa| \ll 1$, the frequency scale Δ_{eff} is related to the frequency scale $\gamma = \pi\Delta^2/2\omega_c$ holding at the Toulouse point $K = \frac{1}{2}$ by $\Delta_{\mathrm{eff}}/\gamma = (1 - 2C_{\mathrm{E}} \kappa) (\omega_c/\gamma)^{2\kappa}$.

Near the Toulouse point, the oscillation frequency Ω, damping rate Γ and cut contribution $P_{\mathrm{inc}}(t)$ are found from Eqs. (22.9) and (22.13) as

$$\Omega = \Delta_{\mathrm{eff}} [2\pi\kappa + \mathcal{O}(\kappa^2)] , \qquad \Gamma = \Delta_{\mathrm{eff}} [1 + \mathcal{O}(\kappa^2)] , \tag{22.139}$$

$$P_{\mathrm{inc}}(t) = -2\kappa/(\Delta_{\mathrm{eff}} t)^{1+2\kappa} , \tag{22.140}$$

Alas, since $P_{\mathrm{inc}}(t)$ decays slower than $1/t^2$, this sluggish decrease contradicts the fluctuation-dissipation theorem, and therefore is a bug of the NIBA.

Examine now the coherent-incoherent transition for zero bias beyond NIBA. First, we observe that away from the Toulouse point the NIBA kernel is an extended blip. Beyond NIBA, this blip is dressed by a formfactor. For K slightly below 1/2, i.e., in the range $0 < \kappa \ll 1$, the form factor is a CS formfactor built up by collapsed sojourns, as given in Eq. (22.131). With this modification, the kernel now reads [483]

$$\hat{K}_z(\lambda) = \Delta^2 \pi\kappa \int_0^\infty d\tau \frac{e^{-(\lambda+\Delta_{\mathrm{eff}}/2)\tau}}{(\omega_c \tau)^{1-2\kappa}} = \Delta_{\mathrm{eff}} (\lambda/\Delta_{\mathrm{eff}} + 1/2)^{-2\kappa} . \tag{22.141}$$

With this kernel, the pole condition in the NIBA expression (22.4)

$$\Delta_{\text{eff}} + \lambda(\lambda/\Delta_{\text{eff}})^{2\kappa} = 0 \,, \tag{22.142}$$

is changed into

$$\Delta_{\text{eff}} + \lambda\left(1/2 + \lambda/\Delta_{\text{eff}}\right)^{2\kappa} = 0 \,. \tag{22.143}$$

The resulting oscillation frequency and damping rate are

$$\Omega = \Delta_{\text{eff}}\left[2\pi\kappa + \mathcal{O}(\kappa^2)\right], \qquad \Gamma = \Delta_{\text{eff}}\left[1 + 2\kappa\ln 2 + \mathcal{O}(\kappa^2)\right]. \tag{22.144}$$

These expressions coincide in leading order of κ with the NIBA forms (22.139). Thus, $K = \frac{1}{2}$ is exactly the critical damping strength at which the coherent-incoherent transition occurs for $T = 0$ in the scaling limit, as already predicted by the NIBA.

As was to be expected, the NIBA bug (22.140) is dissolved by use of the improved kernel (22.141). We see from the pole/cut condition (22.143), that the branch point of $\langle\sigma_z(\lambda)\rangle$ is not anymore at $\lambda = 0$, but is shifted to $\lambda = -\Delta_{\text{eff}}/2$. As a result, the spurious algebraic long-time tail (22.140) is eliminated. For $\kappa > 0$, the leading contribution of the asymptotic series at times $t \gg \Delta_{\text{eff}}^{-1}$ is found as

$$P_{\text{inc}}(t) = -2\kappa\,\frac{\exp(-\Delta_{\text{eff}}t/2)}{(\Delta_{\text{eff}}t)^{1+2\kappa}} \,, \tag{22.145}$$

while for $\kappa < 0$, i.e., for K slightly above $\frac{1}{2}$, it is given by

$$P_{\text{inc}}(t) = 8|\kappa|\,\frac{\exp(-\Delta_{\text{eff}}t/2)}{(\Delta_{\text{eff}}t)^{1+2|\kappa|}} \,. \tag{22.146}$$

Thus, the unphysical algebraic long-time tail (22.140) predicted by the NIBA is in reality suppressed by an exponential decay factor. It is straightforward to see that for $\kappa > 0$ ($K < \frac{1}{2}$) the cut contribution is negative for all times, while it is positive for $\kappa < 0$ ($K > \frac{1}{2}$). For $K \geqslant \frac{1}{2}$, the expectation value $\langle\sigma_z\rangle_t$ is decaying monotonously from the initial value $+1$ to zero, which means that the dynamics is fully incoherent. In marked contrast to the NIBA result (22.140), the power of the algebraic decay factor does not depend on the sign of κ.

In the following, we consider various equilibrium correlation functions at $K = \frac{1}{2}$.

22.6.4 σ_z equilibrium correlation function and response function

The function $S_z(t) = \text{Re}\langle\sigma_z(t)\sigma_z(0)\rangle$

Consider first the symmetric σ_z autocorrelation function $S_z(t)$. Starting out from the expression (21.52) and using results achieved in Subsec. 22.6.2, we get

$$S_{z\,\text{unc}}(t) = e^{-\gamma|t|} + \langle\sigma_z\rangle_{\text{eq}}\left[P_{\text{a}}(|t|) - \langle\sigma_z\rangle_{\text{eq}}\right], \tag{22.147}$$

where $P_{\text{a}}(t)$ and $\langle\sigma_z\rangle_{\text{eq}}$ are given in Eqs. (22.133) and (22.135), respectively.

Next we turn to the correlation contribution $R(t)$. Because of the factor $\xi_1\xi_{n+1}\sin^2(\pi K)$ in the expression (21.55) for $F_{m,n}$, the correlation term $R(t)$, Eq. (21.53), consists of two extended blips, the one being the first blip in the negative-time branch, and the other the first blip in the positive-time branch. The two blips of

Figure 22.3: The left diagram represents the correlation contribution $R(t)$ to $S_z(t)$, and the right diagram pictorially gives the response function $\chi_z(t)$. The symbols are as in Fig. 22.2.

length τ_i, $i = 1,\, 2$, are dressed with a CS form factor $e^{-\gamma\tau_i/2}$, as explained in Subsec. 22.6.1. The first extended blip is followed by a sojourn of length $s_1 + s_2$, in which the section s_1 is in the negative, and the section s_2 in the positive time branch. The interval s_1 is dressed with a CB form factor $e^{-\gamma s_1}$, while the interval s_2 is bare. The final sojourn of length $s_3 = t - s_2 - \tau_2$ is dressed again with a CB form factor $e^{-\gamma s_3}$. The resulting correlation term $R(t)$ is pictorially sketched in Fig. 22.3 (left).

For $K = \frac{1}{2}$, the blip interaction factor G_2 [cf. Eq. (21.31)] for two blips of type ξ and $\pm\xi$ with flip times t_1, t_2 and t_3, t_4 in chronological order $t_1 < t_2 < t_3 < t_4$, can be decomposed as [cf. the similar decomposition (19.49) for imaginary times]

$$
\begin{aligned}
G_2(\xi, \xi) &= e^{-Q'_{21}}\, e^{-Q'_{43}} + e^{-Q'_{32}}\, e^{-Q'_{41}} \,, \\
G_2(\xi, -\xi) &= e^{-Q'_{21}}\, e^{-Q'_{43}} - e^{-Q'_{31}}\, e^{-Q'_{42}} \,.
\end{aligned}
\tag{22.148}
$$

On the ξ summation, the respective first contributions in (22.148) cancel with the subtraction term in Eq. (21.53). The correlation term $R(t)$ is readily found to read

$$
\begin{aligned}
R(t) &= -\frac{\Delta^4}{2} \int_0^\infty d\tau_1\, d\tau_2\, ds_1\, ds_2\, \Theta(|t| - \tau_2 - s_2)\, e^{-\gamma(\tau_1+\tau_2)/2}\, e^{-\gamma s_1}\, e^{-\gamma(|t|-\tau_2-s_2)} \\
&\quad \times \Big\{ \cos[\epsilon(\tau_1 + \tau_2)]\, e^{-Q'(s_1+s_2)}\, e^{-Q'(\tau_1+s_1+s_2+\tau_2)} \\
&\qquad + \cos[\epsilon(\tau_1 - \tau_2)]\, e^{-Q'(\tau_1+s_1+s_2)}\, e^{-Q'(s_1+s_2+\tau_2)} \Big\} \,.
\end{aligned}
\tag{22.149}
$$

Here the negative-time segment $\{s_1, \tau_1\}$ is correlated with the positive-time segment $\{s_2, \tau_2\}$ via inter-dipole charge interactions.

The expression (22.149) is most appropriately processed by first choosing the arguments of the function $Q'(\tau)$ as new integration variables, and then executing all the other integrations. Upon combining the resulting form for $R(t)$ with the expression (22.147), the symmetrized correlation function takes the analytic form

$$
S_z(t) = e^{-\gamma|t|} - F_1^2(t) - F_2^2(t) \,,
\tag{22.150}
$$

$$
\begin{aligned}
F_1(t) &= \frac{\Delta^2}{2\gamma} \int_0^\infty d\tau\, \sin(\epsilon\tau)\, e^{-Q'(\tau)} \left[e^{-\gamma|t-\tau|/2} + e^{-\gamma|t+\tau|/2} \right] , \\
F_2(t) &= \frac{\Delta^2}{2\gamma} \int_0^\infty d\tau\, \cos(\epsilon\tau)\, e^{-Q'(\tau)} \left[e^{-\gamma|t-\tau|/2} - e^{-\gamma|t+\tau|/2} \right] .
\end{aligned}
\tag{22.151}
$$

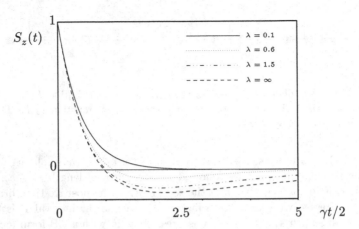

Figure 22.4: The epression (22.150) is plotted for zero bias as a function of $\gamma t/2$ for different values of the scaled inverse temperature $\lambda = \hbar\beta\gamma/2\pi$. The high temperature curve ($\lambda = 0.1$) for $S_z(t)$ cannot be resolved from the function $\langle\sigma_z\rangle_t = e^{-\gamma t}$.

The analytic expression (22.150) with (22.151) is exact for all T, ϵ, and t. In obtaining this concise expression, we have used the relation $F_1(0) = \langle\sigma_z\rangle_{\rm eq}$.

There holds $F_1(t) = F_1(-t)$ and $F_2(t) = -F_2(-t)$. Evaluation of the integrals (22.151) for zero temperature and asymptotic time $|t| \gg 1/\gamma$ yields the expressions

$$F_1(t) = {\rm sgn}(\epsilon)\, e^{-\gamma|t|/2} + \frac{4}{\pi}\,\frac{\gamma^2}{\gamma^2 + 4\epsilon^2}\,\frac{\sin(\epsilon t)}{\gamma t}\,, \qquad F_2(t) = \frac{4}{\pi}\,\frac{\gamma^2}{\gamma^2 + 4\epsilon^2}\,\frac{\cos(\epsilon t)}{\gamma t}\,. \quad (22.152)$$

Here we have kept the exponentially decaying term in $F_1(t)$, and we have disregarded in the algebraic tails terms of orders $\mathcal{O}[1/(\gamma t)^2]$.

Consider now first the unbiased case. The expression (22.150) yields with the terms (22.152) at zero temperature besides the exponential term an algebraic long-time tail with negative prefactor,

$$S_z(t) = e^{-\gamma|t|} - \frac{16}{\pi^2}\,\frac{1}{(\gamma t)^2}\,, \quad (22.153)$$

At very low T, the algebraic law (22.153) holds in the intermediate time region $\gamma^{-1} \ll |t| \ll \hbar\beta$, while in the asymptotic regime $|t| \gg \hbar\beta$ the algebraic term is replaced by exponential decay with the rate given by the lowest Matsubara frequency $\nu_1 = 2\pi/\hbar\beta$ [223],

$$S_z(t) = e^{-\gamma|t|} - \left(\frac{8}{\hbar\beta\gamma}\right)^2 e^{-\nu_1|t|}\,. \quad (22.154)$$

The time behavior of $S_z(t)$ is shown for different temperatures in Fig. 22.4. The overshot into the negative domain is owed to the correlation term $R(t)$ of $S_z(t)$.

For nonzero bias and time $\gamma|t| \gg 1$, the term $e^{-\gamma|t|}$ and a term $-e^{-\gamma|t|}$ resulting from $-F_1(t)^2$ cancel each other out in (22.150). The resulting expression for $T = 0$ is

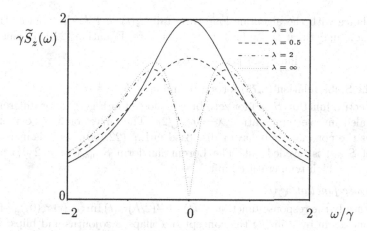

Figure 22.5: The spectral function $\gamma \tilde{S}_z(\omega)$ for zero bias is plotted versus ω/γ for different values of the scaled inverse temperature $\lambda = \hbar\beta\gamma/2\pi$.

$$S_z(t) = -\frac{8}{\pi} \frac{\gamma^2}{\gamma^2 + 4\epsilon^2} \frac{\sin(|\epsilon|t)}{\gamma t} e^{-\gamma|t|/2} - \frac{16}{\pi^2} \frac{\gamma^4}{(\gamma^2 + 4\epsilon^2)^2} \frac{1}{(\gamma t)^2} . \tag{22.155}$$

Here, the first term describes again damped oscillations, as found for $\langle \sigma_z \rangle_t$ in the same limit in Eq. (22.137), and even the prefactors match.

The leading algebraic decay of $S_z(t)$ can be written with the static susceptibility $\overline{\chi}_z$ in the universal form

$$S_z(t) = -\left(\frac{\hbar\overline{\chi}_z}{2}\right)^2 \frac{1}{t^2}, \qquad \overline{\chi}_z = \frac{8}{\pi\hbar\gamma} \frac{\gamma^2}{\gamma^2 + 4\epsilon^2} . \tag{22.156}$$

The static susceptibility $\overline{\chi}_z$ at $T = 0$ may be calculated either by use of Eq. (21.60) with Eq. (22.136), or alternatively from Eq. (19.76). The expression (22.156) is consistent with the expression (6.78) for $s = 1$ at the Toulouse point $\delta_1 = \frac{1}{2}$.

Next, we turn to the spectral function $\tilde{S}_z(\omega)$, which is the Fourier transform of the expression (22.150). Use of the expressions (22.151) yields for any T and ϵ

$$\tilde{S}_z(\omega) = \coth\left(\frac{\hbar\omega}{2k_B T}\right) \frac{2}{\pi} \frac{\gamma}{\omega[\omega^2 + \gamma^2]} \left(\omega\,\Phi''(\omega) - \gamma\,\Phi'(\omega)\right), \tag{22.157}$$

where $\Phi'(\omega)$ and $\Phi''(\omega)$ are the real and imaginary parts of the complex function

$$\Phi(\omega) = \psi(x_+) + \psi(x_-) - \psi(x_+ - i\hbar\beta\omega/2\pi) - \psi(x_- - i\hbar\beta\omega/2\pi) . \tag{22.158}$$

Here $\psi(x)$ is the digamma function, and $x_\pm = \frac{1}{2} + (\frac{1}{2}\gamma \pm i\epsilon)\hbar\beta/2\pi$. At $T = 0$, this is

$$\begin{aligned}
\Phi'(\omega) &= -\tfrac{1}{2}\ln\{[\gamma^2 + 4(\omega+\epsilon)^2][\gamma^2 + 4(\omega-\epsilon)^2]/[\gamma^2 + 4\epsilon^2]^2\}, \\
\Phi''(\omega) &= \arctan[2(\omega+\epsilon)/\gamma] + \arctan[2(\omega-\epsilon)/\gamma] .
\end{aligned} \tag{22.159}$$

431

In accordance with the algebraic long-time tail $S_z(t) \propto t^{-2}$, the Fourier transform $\tilde{S}_z(\omega)$ is non-analytic at $\omega = 0$ for zero temperature. Equation (22.157) yields

$$\lim_{\omega \to 0} \tilde{S}_z(\omega)/|\omega| = \pi(\hbar\bar{\chi}_z/2)^2 . \tag{22.160}$$

This is the Shiba relation (6.79) for $s = 1$ and $\delta_1 = K = \frac{1}{2}$.

The spectral function $\tilde{S}_z(\omega)$ for zero bias is plotted in Fig. 22.5 for different values of the scaled inverse temperature $\lambda = \hbar\beta\gamma/2\pi$. The curve for $\lambda = \infty$ in Fig. 22.5 illustrates the nonanalytic behavior disclosed in Eq. (22.160). As T is increased, the trough of $\tilde{S}_z(\omega)$ is levelled out. The Lorentzian form $\tilde{S}_{z,\,\text{unc}}(\omega) = 2\gamma/(\omega^2 + \gamma^2)$ is reached in the high temperature limit $\lambda \to 0$.

The response function $\chi_z(t)$

Next, we study the response function $\chi_z(t) = -(2/\hbar)\Theta(t) \,\text{Im} \langle \sigma_z(t)\sigma_z(0)\rangle_\beta$. The analysis is done again by utilizing the concept of collapsed sojourns and blips. First, we recognize from the formally exact expression for $\chi_z(t)$, Eq. (21.56), that the system is in a blip state at time zero. Secondly, we see from the form (21.51) for the influence factor $F_m^{(-)}$ that there is only one extended blip for $K = \frac{1}{2}$. Evidently, the system hops into the extended blip state at some negative time and leaves this state only at some positive time. The extended blip is dressed in both time segments with a CS form factor each. The extended blip is followed by an extended sojourn being dressed with a CB form factor. The associated diagram is depicted in Fig. 22.3 (right) on page 429. In mathematical terms, we have

$$\chi_z(t) = \frac{2}{\hbar}\Theta(t)\Delta^2 \int_0^\infty d\tau_1\, d\tau_2\, ds_2\, \delta(t - \tau_2 - s_2)$$
$$\times \cos[\epsilon(\tau_1 + \tau_2)]\, e^{-Q'(\tau_1+\tau_2)}\, e^{-\gamma(\tau_1+\tau_2)/2}\, e^{-\gamma s_2} . \tag{22.161}$$

Here, $\tau_1 + \tau_2$ is the overall length of the blip and s_2 is the remaining sojourn length in the positive branch. Upon introducing $\tau = \tau_1 + \tau_2$ as new integration variable and executing the other integrals, the resulting expression can be written as

$$\chi_z(t) = (4/\hbar)\,\Theta(t)\,F_2(t)\,e^{-\gamma t/2} , \tag{22.162}$$

where the function $F_2(t)$ is defined in Eq. (22.151). This is the analytic expression for the response function $\chi_z(t)$ for any bias and temperature.

With the asymptotic form (22.152) for $F_2(t)$, one finds that the response function at $T = 0$ decays asymptotically as

$$\chi_z(t) \approx \Theta(t)\frac{16}{\pi\hbar}\frac{\gamma^2}{\gamma^2 + 4\epsilon^2}\frac{\cos(\epsilon t)\,e^{-\gamma t/2}}{\gamma t} , \qquad \gamma t \gg 1 . \tag{22.163}$$

Fourier transformation of Eq. (22.162) yields an analytic expression for the dynamical susceptibility holding for all T and ϵ,

$$\tilde{\chi}_z(\omega) = \frac{2}{\pi}\frac{1}{\hbar\omega}\frac{\gamma}{\omega + i\gamma}\,\Phi(\omega) , \tag{22.164}$$

where $\Phi(\omega)$ is the complex function defined in Eq. (22.158).

At this point, three remarks are appropriate. First, the independently derived expressions (22.157) and (22.164) satisfy the fluctuation-dissipation theorem

$$\tilde{S}_z(\omega) = \hbar \coth(\beta\hbar\omega/2)\,\tilde{\chi}_z''(\omega)\,. \tag{22.165}$$

Secondly, the static susceptibility $\overline{\chi}_z = \tilde{\chi}_z(\omega = 0)$ found from Eq. (22.164) coincides with the expression (19.76) obtained within the thermodynamic approach and normalized as in Eq. (19.4). The expression (19.76) also complies with the result found from the dispersion relation

$$\overline{\chi}_z = \frac{2}{\pi}\int_0^\infty d\omega\,\frac{\tilde{\chi}_z''(\omega)}{\omega}\,. \tag{22.166}$$

Thirdly, we may calculate the static susceptibility directly from the expression (21.58). For $K = \frac{1}{2}$, this series adds up to a single extended blip with a CS formfactor followed by a sojourn with a CB form factor,

$$\overline{\chi}_z = \frac{2\Delta^2}{\hbar}\int_0^\infty d\tau\,\tau\cos(\epsilon\tau)\,e^{-\gamma\tau/2}\,e^{-Q'(\tau)}\int_0^\infty ds\,e^{-\gamma s}\,. \tag{22.167}$$

This integral yields again the analytic expression (19.76).

22.6.5 σ_x equilibrium correlation function and response function

The function $S_x(t) = \mathrm{Re}\langle\sigma_x(t)\sigma_x(0)\rangle$

With the concept of collapsed blips and collapsed sojourns explained in Subsection 22.6.1, also the path sums (21.64) – (21.67) for the coherence correlation and response functions can be executed in the scaling limit for $K = \frac{1}{2}$.

Consider first the symmetrized correlation function $S_x(t)$. Upon assigning a factor $\cos(\pi K)$ to each potentially collapsing dipole, there is a surplus of one $\cos(\pi K)$ factor in each term of the series (21.65) for $S_x^B(t)$. As a result, the contribution $S_x^B(t)$ vanishes as $K \to \frac{1}{2}$. In the contribution $S_x^A(t)$, the system dwells in the initial sojourn state η_0 throughout the negative-time branch. At time zero it then hops into the blip state $\xi_1 = -\eta_0$. Afterwards, it stays there until time t except for flash visits of sojourn states, which altogether sum up to a CS form factor $e^{-\gamma t/2}$. The contribution $S_x^A(t)$ is sketched diagrammatically in Fig. 22.6. With the omission of the charges at times zero and t, all charges merge into collapsed dipoles. Hence, $S_x(t)$ does not depend on temperature, and it undergoes damped oscillation

$$S_x(t) = \cos(\epsilon t)\,e^{-\gamma|t|/2}\,. \tag{22.168}$$

The response function $\chi_x(t)$

Next, turn to the response function $\chi_x(t)$. The contribution $\chi_x^A(t)$ is sketched in Fig. 22.7 (left). In the negative-time branch, there is a single extended dressed blip followed by a dressed sojourn. At time zero, the system hops again into a dressed blip state, and stays there until time t. In mathematical terms, one then has

Figure 22.6: The diagram describing $S_x^A(t)$ with symbols as in Fig. 22.2. The bullets mark transitions which are free of bath correlations because of the modified influence functional.

$$\chi_x^A(t) = \Theta(t) \frac{2}{\hbar} \sin(\epsilon t)\, e^{-\gamma t/2} \Delta^2 \int_0^\infty d\tau \int_0^\infty ds\, \sin(\epsilon\tau)\, e^{-Q'(\tau)}\, e^{-\gamma\tau/2}\, e^{-\gamma s}\,. \qquad (22.169)$$

Observe that the double integral times the factor Δ^2 is exactly $P_a(t \to \infty) = \langle\sigma_z\rangle_{\mathrm{eq}}$, as follows from Eq. (22.135) or from Eq. (22.133). Thus one obtains [223]

$$\chi_x^A(t) = \Theta(t)\,(2/\hbar)\,\langle\sigma_z\rangle_{\mathrm{eq}}\,\sin(\epsilon t)\, e^{-\gamma t/2}\,. \qquad (22.170)$$

Consider finally $\chi_x^B(t)$. First, observe that the term $\cos^2(\pi K)$ in the square bracket of Eq. (21.67) vanishes as $(K-\frac{1}{2})^2$ in the limit $K \to \frac{1}{2}$, while the first term is unity in this limit. The related diagram is depicted in Fig. 22.7 (right). The TSS hops at negative time $-\tau$ from a sojourn into a blip, and it stays there until time zero, where it returns to a sojourn state. At time $s > 0$, it hops again into a blip and dwells in this state until time t. The two extended blips have opposite ξ-labels intervals, and they are dressed with CS form factors, as discussed above. However, because of the factor ξ_{n+1} in the square bracket in Eq. (21.67) and the summation of the terms with $\xi_{n+1} = \pm 1$, the extended sojourn in the positive-time branch is free of collapsed blips. The existing dipole of length $\tau + s$ introduces correlations between the negative- and positive-time branches. All these attributes are condensed in the formula

$$\chi_x^B(t) = \Theta(t)\frac{\Delta^2}{\hbar}\int_0^\infty d\tau \int_0^t ds\, e^{-\gamma(t+\tau-s)/2}\, e^{-Q'(\tau+s)}\cos[\epsilon(t-\tau-s)]\,. \qquad (22.171)$$

Introducing the dipole length $\tau + s$ as new integration variable, performing the other integrations, and combining the resulting expression with Eq. (22.170), we find

$$\chi_x(t) = (2/\hbar)[\sin(\epsilon t)F_1(t) + \cos(\epsilon t)F_2(t)]\,, \qquad (22.172)$$

where the functions $F_1(t)$ and $F_2(t)$ are defined above in Eq. (22.151).

The function $\chi_x(t)$ describes the linear response of the system to a variation of the tunneling splitting Δ. Using Eq. (22.152), we find asymptotically algebraic decay,

$$\chi_x(t) \approx \frac{8}{\pi\hbar}\frac{\gamma^2}{\gamma^2 + 4\epsilon^2}\frac{1}{\gamma t} \qquad \text{for} \qquad t \gg 1/\gamma\,. \qquad (22.173)$$

Next, we turn to the spectral representations. Taking the Fourier transform of $S_x(t)$ given in Eq. (22.168), we obtain

$$\tilde{S}_x(\omega) = \gamma\frac{\omega^2 + \epsilon^2 + \gamma^2/4}{(\omega^2 + \epsilon^2 + \gamma^2/4)^2 - 4\epsilon^2\omega^2}\,. \qquad (22.174)$$

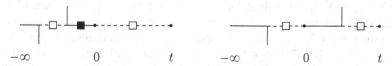

Figure 22.7: The contribution of group A (left) and group B (right) to $\chi_x(t)$. The symbols are as in Fig. 22.2. Each diagram has only one extended dipole.

On the other hand, we may calculate $\tilde{\chi}''_x(\omega)$ from Eq. (22.172). We then discover that the resulting expression is in accordance with the fluctuation-dissipation theorem

$$\hbar\tilde{\chi}''_x(\omega) = \tanh(\beta\hbar\omega/2)\,\tilde{S}_x(\omega)\,. \tag{22.175}$$

Finally, the real part of the dynamical susceptibility in the unbiased case is found as

$$\tilde{\chi}'_x(\omega) = \frac{2\gamma}{\pi\hbar}\frac{1}{\omega^2+\gamma^2/4}\,\mathrm{Re}\left\{\psi\!\left(\frac{1}{2}+\frac{\hbar\gamma}{4\pi k_\mathrm{B}T}\right)-\psi\!\left(\frac{1}{2}+i\frac{\hbar\omega}{2\pi k_\mathrm{B}T}\right)\right\}\,. \tag{22.176}$$

This yields that the static susceptibility $\overline{\chi}_x = \tilde{\chi}'_x(\omega=0)$ for zero bias diverges logarithmically with inverse temperature as $T\to 0$, $\lim_{T\to 0}\overline{\chi}_x(T) = (8/\pi\hbar\gamma)\ln(\hbar\gamma/k_\mathrm{B}T)$

22.6.6 Correlation and response functions in the Toulouse model

Additional insights are gained by calculating the equilibrium tunneling and coherence correlation functions in the original fermionic model (19.56). The equivalence relations of the TSS operators with the fermionic operators of the d level are

$$\sigma_z = 2d^\dagger d - 1\,, \qquad \sigma_x = d^\dagger + d\,, \qquad \sigma_y = i(d-d^\dagger)\,. \tag{22.177}$$

To proceed, we rewrite the Matsubara representation (19.67) of the single-particle Euclidean Green function $\mathcal{G}(\tau)$ as a contour integral. Analytic continuation gives the real-time Green function $G(t) \equiv \langle T_t d^\dagger(t)d(0)\rangle_\beta$ of the d-level in the form

$$G(\pm t) = \mathcal{G}(\tau=\pm i t) = \tfrac{1}{2}e^{\mp i\epsilon t}\left[F_1(t)\mp i\,F_2(t)\pm e^{-\gamma|t|/2}\right]\,, \tag{22.178}$$

where the functions $F_1(t)$ and $F_2(t)$ are defined by the integral representations

$$F_1(t) = \frac{\gamma}{2\pi}\int_{-\infty}^{\infty}d\omega\,\frac{\cos(\omega t)}{\omega^2+\gamma^2/4}\tanh\left(\frac{\hbar(\omega+\epsilon)}{2k_\mathrm{B}T}\right)\,,$$
$$F_2(t) = \frac{\gamma}{2\pi}\int_{-\infty}^{\infty}d\omega\,\frac{\sin(\omega t)}{\omega^2+\gamma^2/4}\tanh\left(\frac{\hbar(\omega+\epsilon)}{2k_\mathrm{B}T}\right)\,. \tag{22.179}$$

The key relation linking fermionic representations in 1D to bosonic ones is

$$\tanh(\hbar\omega/2k_\mathrm{B}T) = (2\omega_\mathrm{c}/\pi)\int_0^{\infty}d\tau\,\sin(\omega\tau)\,e^{-Q'_{K=1/2}(\tau)}\,, \tag{22.180}$$

where $Q'_{K=1/2}(\tau)$ is the scaling form (18.53) for $K = \frac{1}{2}$. With use of the relation (22.180), the "fermionic" expressions (22.179) can directly be transformed into the bosonic forms (22.151). Thus, an important intermediate result is that the expressions (22.151) and (22.179) are different integral representations of the same functions.

Next, we insert the form (22.177) for σ_z into Eq. (21.13), use Eq. (21.14) for the index $j = z$, and observe that the two-particle Green function factorizes into products of one-particle Green functions. We readily obtain

$$
\begin{aligned}
S_z(t) &= -4\,\mathrm{Re}\,\{G(t)G(-t)\}\,, \\
\chi_z(t) &= (8/\hbar)\Theta(t)\,\mathrm{Im}\,\{G(t)G(-t)\}\,.
\end{aligned}
\tag{22.181}
$$

Similarly, the correlation function $S_x(t) = \mathrm{Re}\,\langle\sigma_x(t)\sigma_x(0)\rangle_\beta$ and the response function $\chi_x(t) = -(2/\hbar)\Theta(t)\,\mathrm{Im}\,\langle\sigma_x(t)\sigma_x(0)\rangle_\beta$ can be expressed with the function $G(t)$,

$$
\begin{aligned}
S_x(t) &= \mathrm{Re}\,\{G(t) - G(-t)\}\,, \\
\chi_x(t) &= -\Theta(t)\,(2/\hbar)\,\mathrm{Im}\,\{G(t) - G(-t)\}\,.
\end{aligned}
\tag{22.182}
$$

With the relation (22.178) between $G(\pm t)$ and the functions $F_{1,2}(t)$ one finds full agreement of the correlation and response functions of the fermionic Touloude model with the corresponding functions of the spin-boson model at the Toulouse point. The expressions (22.181) are actually transformed into the forms (22.150) and (22.162), and the expressions (22.182) into the forms (22.168) and (22.172). Because σ_x in Eq. (22.177) is the bare tunneling operator, we have worked in the representation (21.24). Hence the Toulouse Hamiltonian (19.56), and the resonant-level Hamiltonian (19.95), directly correspond to the polaron-transformed Hamiltonian $\tilde{H}$ given in Eq. (18.38). This concludes the discussion of the exactly solvable case $K = \frac{1}{2}$.

22.7 Long-time behavior at $T = 0$ in the regime $0 < K < 1$: general

We have seen in the previous section that in the special case $K = \frac{1}{2}$ the grand-canonical sum of collapsed dipoles within an extended dipole generates an exponential form factor which affects the effective length of the dipole. For K near to $\frac{1}{2}$ the collapsed dipoles are widened, and hence they are interacting with each other. As a result, the grand-canonical sum of these dipoles can not be done in analytic form any more. Nevertheless, it seems quite natural to assume that at long time the alternating series of these entities add up to an effective form factor which cuts off the respective time interval at a length of order $1/\Delta_{\mathrm{eff}}$. With these preliminaries, the asymptotic dynamics can be understood by means of two rules:

(I) Every time interval which is free of a form factor for $K = \frac{1}{2}$ is free of a form factor also for $K \neq \frac{1}{2}$. We shall refer to such intervals as bare intervals.

(II) The grand-canonical alternating sum of extended dipoles placed between bare intervals forms an effective neutral cluster. This acts as a form factor which limits the respective interval to an effective length of order $1/\Delta_{\mathrm{eff}}$.

In the series expressions given in Subsecs. 21.2.4 and 21.2.5, those sojourn states n which have an influence phase factors $\sin(\phi_{n,m})$ [which is $\xi_{n+1}\sin(\pi K)$ in the scaling limit] play a particular role. Rule I directly applies to those sojourns, as the subsequent $\{\xi_j\}$ summation leads to cancellations among the interactions stretching over the sojourn interval s_n For this reason, e.g., the first sojourn in the positive-time regime of the series expression (21.53) for $R(t)$ remains bare. We now apply these rules to correlation functions of the TSS.

22.7.1 The populations

The noninteracting-blip approximation does not describe the long-time dynamics of the population $\langle \sigma_z \rangle_t$ correctly, as pointed out on page 403. In the Ohmic regime $K < 1$ and for zero bias, the expectation value $\langle \sigma_z \rangle_t$ is found to decay asymptotically $\propto (\Delta_{\text{eff}} t)^{-(2-2K)}$, as follows from Eq. (22.13). The algebraic decay in the NIBA originates from a branch point of the Laplace-transformed kernel $\hat{K}_z^{(s)}(\lambda) \propto \lambda^{2K-1}$ at $\lambda = 0$ [cf. Eq. (22.4)]. The asymptotic behavior is changed qualitatively by the interblip correlations, as we have shown for K near the Toulouse point, $K = \frac{1}{2} - \kappa$ with $|\kappa| \ll 1$, in Subsec. 22.6.3. It is difficult to work out the effects of the interblip correlations quantitatively for general K since many different frequency scales are involved, as is well-known from the closely related Kondo problem in the antiferromagnetic sector. The resummation of the alternating series (21.72) of blip sequences results in kernels $\hat{K}_z^{(s,a)}(\lambda)$ which are regular at $\lambda = 0$ (see Subsec. 22.8.1). Thus $\langle \sigma_z \rangle_t$, or the envelope function of it, approaches the equilibrium value exponentially fast, as follows by Laplace inversion of Eq. (21.75). Further discussion of relaxation and decoherence, and several exact analytical results at $T = 0$ are given in Sec. 22.8.

The exponential decay towards the equilibrium distribution $\langle \sigma_z \rangle_{\text{eq}}$ also follows directly from the above two rules. As in the diagrams in Fig. 22.2, there is (at long time) only a single cluster both for $P_s(t)$ and $P_a(t)$, and hence exponential decay.

22.7.2 The population correlations and Shiba relation

With rules (I) and (II), we see from Fig 22.3 (right) that the response function $\chi_z(t)$ is diagrammatically represented by a single neutral cluster surrounding the origin of the time axis. Hence, $\chi_z(t)$ decays exponentially at long times.

Similarly, the contribution $S_z^{(\text{unc})}(t)$ to $S_z(t)$ given in Eq. (22.147) decays exponentially, as follows with the results obtained in Subsection 22.7.1. Employing rules (I, II) to the diagram in Fig. 22.3 (left), we find that the correlation term $R(t)$ is represented by two neutral clusters, the one in the negative time branch near the origin, the other in the positive time branch near t. Therefore, at asymptotic times, the clusters are separated by an interval of length t. Since the interval is bare, the clusters are interacting with the unscreened dipole-dipole interaction $\ddot{Q}'(t)$, which is $-2K/t^2$ in the Ohmic case for times $t \ll \hbar\beta$. Hence we should expect that $S_z(t)$ drops to zero as $1/t^2$ in the time regime $\Delta_{\text{eff}}^{-1} \ll t \ll \hbar\beta$.

To put this argument in concrete form, we expand the blip interaction factor G_{n+m} in Eq. (21.55) under the assumption that the effective lengths of the two clusters are small compared to the interval t between the clusters. We then have

$$F_{m,n} \rightarrow F_m^{(-)}(\{\tau_k\}) F_n^{(-)}(\{\tau_j\}) \left(1 - \ddot{Q}'(t) \sum_{j=1}^{n} \sum_{k=n+1}^{n+m} \xi_j \xi_k \tau_j \tau_k \right). \qquad (22.183)$$

Here $F_n^{(-)}$ and $F_m^{(-)}$ are the influence functions (21.51) for the clusters with n blips in the negative and m blips in the positive time branch, and $\{\tau_k\}$ and $\{\tau_j\}$ are the sets of charge intervals within these clusters. Next, we insert Eq. (22.183) into Eq. (21.53). Because of the $\{\xi_j\}$ summations, those terms which are odd in the blip labels ξ_j for the individual time branches cancel out. The remaining contribution can be written in the symmetric form

$$\begin{aligned}
R(t) &= \ddot{Q}'(t) \lim_{t_q \to \infty} \sum_{m=1}^{\infty} (-1)^m \int_0^{t_q} \mathcal{D}_{2m,0}\{t_j\} \frac{1}{2^m} \sum_{\{\xi_j = \pm 1\}} B_m^{(s)} F_m^{(-)} \left(\sum_{k=1}^{m} \xi_k \tau_k \right) \\
&\times \lim_{t_p \to -\infty} \sum_{n=1}^{\infty} (-1)^n \int_{t_p}^{0} \mathcal{D}_{0,2n}\{t_j\} \frac{1}{2^n} \sum_{\{\xi_j = \pm 1\}} B_n^{(s)} F_n^{(-)} \left(\sum_{\ell=1}^{n} \xi_\ell \tau_\ell \right).
\end{aligned}$$

This expression describes two neutral clusters which are interacting with each other via the dipole-dipole interaction $\ddot{Q}'(t)$. Next, we observe that the series expansion for each cluster can be identified with the expression (21.58) for the static susceptibility $\overline{\chi}_z$ (apart from a missing factor $2/\hbar$). With this identification, we obtain for $S_z(t)$ the asymptotic behavior [cf. Eq. (6.78) for s=1 and $\delta_1 = K$]

$$S_z(t) = \left(\frac{\hbar \overline{\chi}_z}{2} \right)^2 \ddot{Q}'(t) \overset{T \to 0}{=} -2K \left(\frac{\hbar \overline{\chi}_{z,0}}{2} \right)^2 \frac{1}{t^2}, \qquad t \gg 1/\Delta_{\text{eff}}, \qquad (22.184)$$

where $\overline{\chi}_{z,0} = \overline{\chi}_z(T = 0)$. The $1/t^2$ tail of $S_z(t)$ also holds for finite T in the time region $1/\Delta_{\text{eff}} \ll t \ll \hbar\beta$. At larger time, $t \gg \hbar\beta$, the function $S_z(t)$ drops to zero exponentially fast $\propto e^{-\nu_1 t}$, with the rate given by the Matsubara frequency $\nu_1 = 2\pi/\hbar\beta$.

The expression (22.184) applies in the regime $K < 1$ for any bias. The intrinsic properties of the TSS are in the linear or nonlinear static susceptibility. For zero bias, one obtains in the zero damping limit from Eq. (19.4) with (22.96) the relation $\overline{\chi}_{z,0} \rightarrow 2/\hbar\Delta$, while for $K = 1/2$ one has $\overline{\chi}_{z,0} = 8/(\pi\hbar\gamma)$, where $\gamma = \pi\Delta^2/2\omega_c$, as follows from Eq. (22.156). Making use of the correspondence with the Kondo model [143, 86], one obtains in the damping regime $1 - K \ll 1$ with Eq. (19.94) $\overline{\chi}_{z,0} = [2(1 - K)]^{K/(1-K)}/\hbar\Delta_r$, where $\Delta_r = (\Delta/\omega_c)^{K/(1-K)}\Delta$.

From Eq. (22.184) we can also derive the behavior in frequency space near $\omega = 0$, known as the *Shiba relation*,[7]

$$\lim_{\omega \to 0} \tilde{S}_z(\omega \to 0)/|\omega| = 2\pi K(\hbar\overline{\chi}_{z,0}/2)^2, \qquad (22.185)$$

---

[7]For the normalization of $\overline{\chi}_z$, see Eq. (19.4) and footnote on page 323, and Eq. (19.93).

or equivalently

$$\lim_{\omega \to 0} \hbar \tilde{\chi}''_{z,0}(\omega)/\omega = 2\pi K (\hbar \overline{\chi}_{z,0}/2)^2 . \tag{22.186}$$

These relations are analogous to a relation proven by Shiba [260] for the Anderson model. While Shiba's derivation is essentially based upon a particle number conservation law, the proof given here is based upon the Coulomb gas representation in complex time. With the oscillator-spin correspondence $\chi_{n,0} = \overline{\chi}_{z,0}$, the relations (22.185) and (22.186) are in correspondence with the expressions (6.79).

In the strong damping regime $K > 1$, the Kondo temperature $T_K \equiv 2/(\pi k_B \overline{\chi}_{z,0})$ is zero in the scaling limit. Hence the condition $t \gg \hbar/k_B T_K$ for the above asymptotic analysis fails, but rather the high-temperature expansion applies down to $T = 0$.

Generalized Shiba relation for non-Ohmic spectral density

The exact formal expression (21.58) for the static susceptibility is well-defined in the limit $T \to 0$ also in the super-Ohmic regime $s > 1$. This we may infer, for instance, indirectly from the expression (21.60) with (22.96). In the sub-Ohmic weak-damping regime (18.36), the renormalized tunneling matrix element Δ_r at $T = 0$ is nonzero, as follows from Eq. (18.31). Hence the static susceptibility at zero temperature is nonzero as well. As a result, the reasoning leading to $S_z(t) = (\hbar \overline{\chi}_{z,0}/2)^2 \ddot{Q}'(t)$, applies for these regimes, just as for the Ohmic case. We then find with the expression (20.55) for $Q(t)$ algebraic decay $\propto |t|^{-1-s}$ at asymptotic time,

$$S_z(t) = -2\delta_s \Gamma(1 + s) \sin(\pi s/2) \omega_{\rm ph}^{1-s} (\hbar \overline{\chi}_{z,0}/2)^2 |t|^{-1-s} , \tag{22.187}$$

and alternatively in frequency space

$$\lim_{\omega \to 0^\pm} \tilde{S}_z(\omega)/|\omega|^s = \lim_{\omega \to 0^\pm} \hbar \, {\rm sgn}(\omega) \tilde{\chi}''_{z,0}(\omega)/|\omega|^s = 2\pi \delta_s \omega_{\rm ph}^{1-s} (\hbar \overline{\chi}_{z,0}/2)^2 . \tag{22.188}$$

The Shiba relation (22.185), and the respective non-Ohmic generalization (22.187) and (22.188) have been derived first in Ref. [224]. Recently, this relation has been corroborated by diverse other methods. It has been verified numerically by employing the correspondence with the anisotropic Kondo model [484, 485], and by numerical integration of the flow equations resulting from a continuous sequence of infinitesimal unitary transformations [405, 406]. The relation has been confirmed also analytically by using nonperturbative methods derived from a Bethe ansatz [486].

22.7.3 The coherence correlation function

Finally, consider the asymptotic dynamics of the coherence correlation function in the Ohmic case at $T = 0$. Using rules (I) and (II) given on page 436, one finds that both $S_x^A(t)$ and $\chi_x^A(t)$ have a single dressed blip of length t in the positive-time branch [cf. Figs. 22.6 and 22.7 (left)]. Hence, both $S_x^A(t)$ and $\chi_x^A(t)$ decay exponentially fast. In group B, we have one blip cluster in the negative-time and one blip cluster in the positive-time branch. Since in each branch the initial sojourn is bare, the two clusters are situated near the origin and near t, as depicted in Fig. 22.7

(right). The two clusters have opposite unit charge. Hence they are interacting with the unscreened charge-charge interaction $e^{-Q'(t)} \propto t^{-2K}$. Evidently, this interaction directly determines the algebraic long-time tails of $S_x(t)$ and $\chi_x(t)$ for $0 < K < 1$,

$$S_x(t) = c_S(K)(\Delta_{\text{eff}}t)^{-2K}, \qquad \chi_x(t) = c_\chi(K)(\Delta_{\text{eff}}t)^{-2K}. \tag{22.189}$$

Thus, the coherence correlation function $S_x(t)$ and the response function $\chi_x(t)$ decay with a power law in which the exponent depends on the damping strength.

Taking the Fourier transforms of the expressions (22.189), one gets the relation

$$\lim_{\omega \to 0^+} \tilde{S}_x(\omega)/\tilde{\chi}_x''(\omega) = 2\tan(\pi K)c_S(K)/c_\chi(K). \tag{22.190}$$

The ratio $c_S(K)/c_\chi(K)$ can be fixed by reconciling the relation (22.190) with the fluctuation-dissipation theorem (22.175) for $T = 0$, $\hbar\tilde{\chi}_x''(\omega) = \text{sgn}(\omega)\tilde{S}_x(\omega)$. This yields $c_S(K)/c_\chi(K) = \hbar\cot(\pi K)/2$, and after all in the time regime

$$\lim_{t \to \infty} S_x(t)/\chi_x(t) = \tfrac{1}{2}\hbar\cot(\pi K). \tag{22.191}$$

We have shown in Subsec. 22.6.5 that at $K = \frac{1}{2}$ and $T = 0$ the function $S_x(t)$ decays exponentially and the function $\chi_x(t)$ inversely with t. These different behaviors are consistent with the relation (22.191) because the $\cot(\pi K)$-factor is zero at $K = \frac{1}{2}$.

We recall that the logarithmic divergence of the static susceptibility $\overline{\chi}_x(T)$ for zero bias in the limit $T \to 0$ [cf. Eq. (22.176)] comes along with the asymptotic behavior $\chi_x(t, T = 0) \propto 1/t$ given in Eq. (22.173). It is quite natural to assume that the static susceptibility for zero bias behaves differently in the limit $T \to 0$ for K below and above $\frac{1}{2}$. For $K < \frac{1}{2}$, the slow decay $\chi_x(t) \propto t^{-2K}$ implies that the static susceptibility for zero bias diverges algebraically, $\lim_{T \to 0} \overline{\chi}_x(T) \propto T^{2K-1}$. On the other hand, the decay of $\chi_x(t)$ for $K > \frac{1}{2}$ is sufficiently fast so that $\overline{\chi}_x(T)$ remains finite, as $T \to 0$. Recently, these obvious characteristics have been confirmed numerically [487].

In summary, the decay of the population correlation function $S_z(t)$ at long times is determined by the unscreened dipole interaction, whereas the decay of the coherence correlation function $S_x(t)$ is determined by the unscreened charge interaction.

22.8 From weak to strong tunneling: relaxation and decoherence

22.8.1 Incoherent tunneling beyond the nonadiabatic limit

In the time-local limit $K_z^{(s/a)}(t-t') \propto \delta(t-t')$, the generalized master equation (GME) (21.69) simplifies to the rate equation

$$d\langle\sigma_z\rangle_t/dt = k^+ - k^- - (k^+ + k^-)\langle\sigma_z\rangle_t, \tag{22.192}$$

where $k^\pm$ is the forward/backward rate from the state $\sigma = \pm 1$ to the state $\sigma = \mp 1$,

$$k^\pm = \frac{1}{2}\int_0^\infty d\tau\left[K_z^{(s)}(\tau) \pm K_z^{(a)}(\tau)\right] = \frac{1}{2}\left[\hat{K}_z^{(s)}(\lambda = 0) \pm \hat{K}_z^{(a)}(\lambda = 0)\right]. \tag{22.193}$$

Strictly speaking, the total transfer rate $k = k^+ + k^-$ is given by the solution of the equation $k - \hat{K}_z^{(s)}(\lambda = -k) = 0$ with smallest positive real value. When k is small compared to the other frequencies of the system, one obtains the expression (22.193).

Equation (22.192) describes the population dynamics in the incoherent regime. The solution of this equation with initial value $\langle\sigma_z\rangle_0$ is

$$\langle\sigma_z\rangle_t = \langle\sigma_z\rangle_{eq} + [\langle\sigma_z\rangle_0 - \langle\sigma_z\rangle_{eq}]e^{-(k^+ + k^-)t}. \tag{22.194}$$

The equilibrium value

$$\langle\sigma_z\rangle_{eq} = \frac{k^+ - k^-}{k^+ + k^-} \tag{22.195}$$

agrees with the exact formal expression given in Eq. (21.78).

Writing the series (21.72) for the kernels of the GME in terms of the blip and sojourn intervals (21.30) and switching to the Laplace transform, we obtain the power series in Δ^2 for the forward/backward rate as

$$k^\pm = \sum_{n=1}^{\infty} (-1)^{n-1} \frac{\Delta^{2n}}{2^{n+1}} \int_0^\infty d\tau_n \left[\prod_{j=1}^{n-1} ds_j\, d\tau_j\right] \sum_{\{\xi_j = \pm 1\}} \left(\tilde{F}_n^{(+)} B_n^{(s)} \pm \tilde{F}_n^{(-)} B_n^{(a)}\right). \tag{22.196}$$

Here, the $\tilde{F}_n^{(\pm)}$ are the irreducible influence clusters introduced in Eq. (21.70). The rate expression is in the form of a series in the number of bounces (blips). The $n = 1$ term is the nonadabatic rate discussed in Sec. 20.2. The terms with $n > 1$ represent the adiabatic corrections. They would vanish if the inter-blip interactions were disregarded. In the strong-tunneling regime, the full series must be summed up.

A graph-theoretic approach to the computation of adiabatic corrections in the real-time path integral formulation has been given by Stockburger and Mak [488]. The linked cluster sum considered by these authors is the graphical equivalent of the series expression (22.196) with the irreducible influence functions $\tilde{F}_n^{(\pm)}$. They presented an approximate summation of the multi-blip contributions under the assumption that the separation of blips is larger than $1/\omega_c$. The numerical studies show two competing adiabatic effects. (i) at high temperature, the rate is reduced (compared to the nonadiabatic rate) because of correlated re-crossings of the Landau-Zener region. (ii) for low enough temperature, the rate is enhanced above the nonadiabatic rate because of an effective reduction of the reaction barrier by the electronic coupling and by nuclear tunneling. The exact summation of the series expressions (21.72) and (22.196) in the weak-damping limit is given above in Section 22.3.

Exact solution in analytic form at $T = 0$ for general K

Interestingly, in the Ohmic scaling limit at $T = 0$, the incoherent rate can be determined in analytic form in all orders of Δ for any K. The analysis of the Coulomb gas representation (22.196) performed in Ref. [489] reveals a close relationship between the rate contribution of order Δ^{2n} in the TSS with the rate k_n^+ from site 0 to site n of the same order in the Schmid model, (see page 525 in Sec. 29.1). With the explicit form of the rate k_n^+ given below in Eq. (29.6), one then obtains the weak-tunneling series of the TSS (forward) rate in the analytic form (see Ref. [489] for details)

$$k_{\mathrm{wt}}^+(K, \epsilon) = \frac{\epsilon}{2\sqrt{\pi}} \sum_{m=1}^{\infty} \frac{1}{m!} \frac{\Gamma(Km)\,[\,1 - \cos(2\pi Km)\,]}{\Gamma[\frac{3}{2} + (K-1)m]} \left(\frac{\epsilon_{\mathrm{sb}}}{\epsilon}\right)^{(2-2K)m}. \qquad (22.197)$$

The Kondo-like frequency ϵ_{sb} is defined in terms of the parameters of the TSS by

$$\epsilon_{\mathrm{sb}}^{2-2K} = \frac{\Gamma^2(1-K)}{2^{2K}} \frac{\Delta^2}{\omega_c^{2K}} = \frac{1}{[2\sin(\pi K)]^2} \epsilon_0^{2-2K}. \qquad (22.198)$$

The second relation connects the Kondo frequency of the spin-boson model with the Kondo frequency of the Schmid model introduced in Eq. (28.120). For rational K, the series (22.197) is a linear combination of hypergeometric functions.

The critical bias

$$\epsilon_{\mathrm{cr}}(K) = \sqrt{|1-K|}\, K^{K/[2(1-K)]}\, \epsilon_{\mathrm{sb}} \qquad (22.199)$$

determines the convergence radius of the series (22.197). In the regime $K < 1$, the weak-tunneling series (22.197) is absolutely converging when the bias ϵ exceeds $\epsilon_{\mathrm{cr}}(K)$, whereas it converges in the regime $K > 1$ when the bias ϵ is below $\epsilon_{\mathrm{cr}}(K)$.

The $m = 1$ term in Eq. (22.197) is the nonadiabatic rate (20.78) apart from the factor $e^{-\epsilon/\omega_c}$. Expressed in terms of ϵ_{sb}, the nonadiabatic rate takes the form

$$k_{\mathrm{na}}^+(K, \epsilon) = \sin^2(\pi K) \frac{\Gamma(K)}{\Gamma(\frac{1}{2} + K)} \frac{\epsilon}{\sqrt{\pi}} \left(\frac{\epsilon_{\mathrm{sb}}}{\epsilon}\right)^{2-2K}. \qquad (22.200)$$

Next, we wish to derive the strong-tunneling series from the weak-tunneling series (22.197). This is achieved upon performing transformations analogously to those executed below in Subsec. 28.5.3. With these, the series (22.197) is transformed into the integral representation

$$k^+(\epsilon) = \operatorname{Re} \frac{\epsilon}{2\pi i} \int_C \frac{dz}{z} \left\{ \sqrt{z - 1 - z^K u_2} - \sqrt{z - 1 - z^K u_1} \right\}, \qquad (22.201)$$

where $u_1 = (\epsilon/\epsilon_{\mathrm{sb}})^{2K-2}$ and $u_2 = e^{i2\pi K} u_1$. The contour C starts at the origin, encircles the branch point in counter-clockwise sense, and returns to the origin. The representation (22.201) converges not only in the regime where the series (22.197) does, but for all values of K and of $\epsilon/\epsilon_{\mathrm{SB}}$, . This allows us to determine the asymptotic series.

Consider first the regime $K < 1$. Upon changing variable z to $y_1 = z^{1-K}/u_1$ and to $y_2 = z^{1-K}/u_2$, respectively, and expanding the ensuing integrands in powers of $\epsilon/\epsilon_{\mathrm{sb}}$, the asymptotic (strong-tunneling) series is found to read [489],

$$k_{\mathrm{st}}^+(K, \epsilon) = \frac{\epsilon_{\mathrm{sb}}}{2\sqrt{\pi}} \sum_{n=0}^{\infty} b_n(K) \frac{1}{n!} \frac{\Gamma[(\frac{1}{2} - n)\frac{K}{1-K}]}{(\frac{1}{2} - n)\Gamma[(\frac{1}{2} - n)\frac{1}{1-K}]} \left(\frac{\epsilon}{\epsilon_{\mathrm{sb}}}\right)^{2n}, \qquad (22.202)$$

with

$$b_n(K) = \begin{cases} 2\sin^2\left[\frac{\pi K}{2(1-K)}(1-2n)\right], & \text{for} \qquad K < \frac{1}{3}, \\[2mm] 1, & \text{for} \qquad \frac{1}{3} < K < 1. \end{cases} \qquad (22.203)$$

The leading term of the asymptotic series (22.202) is independent of the bias and is

$$k_{as}^+(K, \epsilon) = b_0(K) \frac{2\Gamma[\frac{K}{2(1-K)}]}{\Gamma[\frac{1}{2(1-K)}]} \frac{\epsilon_{sb}}{2\sqrt{\pi}} . \qquad (22.204)$$

There are two simple checks testing the two sectors in Eq. (22.203):

(1) For very weak damping $K \ll 1$, both the weak-tunneling series (22.197) and the strong-tunneling series (22.202) are summed to the simple form ($\epsilon_{sb} = \Delta_{eff}$)

$$k_{wt}^+(K, \epsilon) = k_{st}^+(K, \epsilon) = \pi K \Delta_{eff}^2 / \sqrt{\Delta_{eff}^2 + \epsilon^2} , \qquad \text{for} \qquad K \ll 1 . \quad (22.205)$$

This expression agrees with the earlier result (22.95) for $T = 0$, $k^+ = \gamma_r(T = 0)$.

(2) At the Toulouse point $K = \frac{1}{2}$, only the leading term (22.200) of the series (22.197) and the leading term (22.204) of the series (22.202) are nonzero. Most satisfactorily, the expressions (22.200) and (22.204) coincide for $K = \frac{1}{2}$ and yield $k_{na}^+(\frac{1}{2}, \epsilon) = k_{as}^+(\frac{1}{2}, \epsilon) = \pi \Delta^2 / 2\omega_c$, which is the exact rate for $K = \frac{1}{2}$.

In the narrow regime $1 - K \ll 1$, we get from Eq. (22.204)

$$k^+(0) = \sqrt{2(1 - K)/\pi}\, \epsilon_{sb} \quad \text{with} \quad \epsilon_{sb} = \left(\frac{1}{2(1-K)}\right)^{\frac{1}{1-K}} \Delta_r = \frac{2}{\pi} \frac{k_B T_K}{\hbar} . \quad (22.206)$$

The second form relates ϵ_{sb} to the Kondo temperature T_K of the anisotropic Kondo model in the corresponding regime $\rho J_\perp \ll \rho J_\parallel$ [see Eq. (19.94)].

Consider next the asymptotic expansion for $K > 1$ and large bias. This is found from the contour integral (22.201) upon changing from variable z to $t_{1,2} = e^{-i\pi} u_{1,2} z^K$. We then find in the section $K = p + \kappa$, where $p = 1, 2, \cdots$, and $0 \leqslant \kappa < 1$,

$$k^+(\epsilon) = \frac{\epsilon}{\sqrt{\pi}} \sum_{m=1}^{\infty} \frac{(-1)^m}{m!} \frac{\Gamma(\frac{m}{K}) \sin(\frac{1+p}{K} m\pi) \sin(\frac{p}{K} m\pi)}{K \Gamma[\frac{3}{2} + (\frac{1}{K} - 1)m]} \left(\frac{\epsilon}{\epsilon_{sb}}\right)^{(2/K-2)m} , \quad (22.207)$$

From this we see that the rate is zero for a symmetric system. This confirms that there is a transition to self-trapping at $K = 1$ in the Ohmic scaling limit (18.56), as is known also from the anisotropic Kondo model. The localization transition has been discussed by Chakravarty [144], Bray and Moore [145], and by Hakim et al. [146].

In Fig. 22.8 (left) the normalized rate k^+/k_{na}^+ is plotted versus the scaled bias $x = (\Delta_r/\epsilon)^{1-K}$ for various values of K in the regime $K < 1$. The horizontal line represents the particular case $K = \frac{1}{2}$. For $K < \frac{1}{4}$, the full rate is always lower than the golden rule rate. Hence the numerous multi-bounce contributions interfere destructively in this regime. For $\frac{1}{2} < K < 1$, the multi-bounce contributions interfere constructively for all x so that the full rate is always above the golden rule rate. In the regime $\frac{1}{4} < K < \frac{1}{2}$, the rate goes through a maximum as tunneling is increased, and finally falls below the golden rule rate. This reflects constructive interference at small and intermediate $x = (\Delta_r/\epsilon)^{1-K}$, and destructive interference at large x.

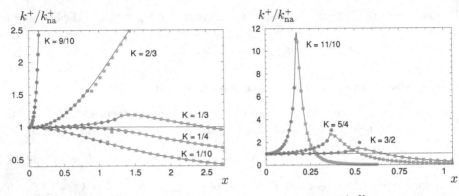

Figure 22.8: The scaled rate k^+/k_{na}^+ is plotted versus $x = (\Delta_{\mathrm{r}}/\epsilon)^{1-K}$ for various values of K, $K < 1$ (left) and $K > 1$ (right). The circles are calculated from the weak-tunneling series, and the squares from the asymptotic strong-tunneling series. The full curve is the respective hypergeometric function expression.

Fig. 22.8 (right) shows plots of the normalized rate k^+/k_{na}^+ versus the scaled bias $x = (\epsilon/\Delta_{\mathrm{r}})^{K-1}$ for $K > 1$. The normalized rate goes through a maximum which is shifted to higher x when K is increased. At large enough x, the rate k^+ falls below the golden rule rate. Hence there is constructive interference of the tunneling terms at small and intermediate x, and destructive interference in the strong-tunneling regime.

22.8.2 Decoherence at zero temperature: analytic results

One might guess from the known special cases that there is a close relationship between the relaxation rate $k^+(\epsilon)$ and the decoherence rate $\gamma(\epsilon)$. The former describes relaxation to the ground state of the TSS, and the latter loss of phase coherence, which reveals itself in the damping of the oscillatory dynamics as given, e.g., in the expression (22.97). In the weak damping limit, we simply have $\gamma(\epsilon) = \frac{1}{2}k^+(\epsilon)$ for $T = 0$, as we see from Eq. (22.95).

In Ref. [489] it was conjectured that the strong-tunneling series for the decoherence rate at zero temperature in the damping regime $0 < K \leqslant \frac{1}{2}$ reads

$$\gamma_{\mathrm{st}}(K,\epsilon) = \frac{\epsilon_{\mathrm{sb}}}{4\sqrt{\pi}} \sum_{n=0}^{\infty} \tilde{b}_n(K) \frac{1}{n!} \frac{\Gamma[(\frac{1}{2}-n)\frac{K}{1-K}]}{(\frac{1}{2}-n)\Gamma[(\frac{1}{2}-n)\frac{1}{1-K}]} \left(\frac{\epsilon}{\epsilon_{\mathrm{sb}}}\right)^{2n}, \qquad (22.208)$$

with

$$\tilde{b}_n(K) = 2\sin^2\left[\frac{\pi K}{2(1-K)}(1-2n)\right], \qquad 0 < K \leqslant \frac{1}{2}. \qquad (22.209)$$

The function $\tilde{b}_n(K)$ coincides with the function $b_n(K)$ in the regime $0 < K \leqslant \frac{1}{3}$, while it is twice as large as $b_n(K)$ for $K = \frac{1}{2}$. As a result, the ratio $\gamma_{\mathrm{st}}(K,\epsilon)/k_{\mathrm{st}}^+(K,\epsilon)$ is $\frac{1}{2}$ for $0 < K \leqslant \frac{1}{3}$, and larger for $\frac{1}{3} < K < \frac{1}{2}$, and it approaches unity as $K \to \frac{1}{2}$.

The strong-tunneling series (22.208) with (22.209) coincides with exact expressions for the decoherence rate known in special cases. A relevant cheque is the asymptotic strong-tunneling limit $\epsilon \to 0$. In this limit, one obtains from Eq. (22.208)

$$\gamma_{st}(K, 0) = \frac{1}{\sqrt{\pi}} \sin^2 \left[\frac{\pi K}{2(1-K)} \right] \frac{\Gamma[\frac{K}{2(1-K)}]}{\Gamma[\frac{1}{2(1-K)}]} \epsilon_{sb} . \tag{22.210}$$

The decoherence rate of the unbiased Ohmic TSS has been calculated by Lesage and Saleur within the framework of integrable QFT [465]. The resulting expression (22.11) with (22.12) coincides indeed with the expression (22.210) in the regime $0 < K \leqslant \frac{1}{2}$.

The series (22.208) matches also the known exact expressions for the decoherence rate $\gamma(K, \epsilon)$ in the regimes $K \ll 1$ and K close to $\frac{1}{2}$. These matches strongly support validity of the asymptotic series (22.208). Assuming that the conjecture is correct, the decoherence rate (22.208) represents the exact bound for minimal decoherence in the SB model in the scaling limit in the regime $0 < K \leqslant \frac{1}{2}$. For given $K < 1/2$, Δ_r, and ϵ, the bound is saturated at $T = 0$.

22.9 Thermodynamics from dynamics

In computation of dynamical correlation functions, often the imaginary-time correlation function is calculated by standard Monte Carlo simulations. The imaginary-time data are then continued to real time by using, e.g., Padé approximant methods or image reconstruction techniques like Max-Ent [490]. However, the analytic continuation of a function which is blurred by numerical noise is ill-defined: small errors in the input data can lead to exponentially enhanced errors in the output data.

Here we disseminate that sometimes the opposite course, i.e., computation of thermodynamics from the asymptotic dynamics may have considerable advantages.

In the standard thermodynamic approach, first of all the partition function is computed, which then provides the basis to calculate the static susceptibility, specific heat, etc. This proceeding has two major disadvantages. First, it is difficult to perform approximations in the Coulomb gas representation for the partition function which preserve the symmetry of the imaginary-time interaction, $W(\tau) = W(\hbar\beta - \tau)$. Secondly, the fugacity expansion for Z is often found to converge badly. Remarkably, the fugacity expansion for $\ln Z$ or the free energy (cumulant expansion) usually converges much faster [cf. the discussion in Ref. [491] for an unordered Coulomb gas].

Instead of using the imaginary-time approach for the partition function Z, we can find the thermodynamic properties of an ergodic open system within a dynamical approach by studying the equilibrium state which the system takes at asymptotic time. To proceed, we first note that the equilibrium distribution $\langle \sigma_z \rangle_{eq}$ is related to the partition function Z by $\langle \sigma_z \rangle_{eq} = (2/\hbar\beta)\partial \ln Z/\partial\epsilon = -(2/\hbar)\partial F/\partial\epsilon$ [cf. Eq. (19.3)]. Thus, we directly obtain the free energy if we integrate the formal expression for $\langle \sigma_z \rangle_{eq}$ with respect to the bias. Since the integration constant does not depend on Δ and β, it may be set equal to zero without any restriction. It is just in this way that one obtains directly an exact formal series expression for $\ln Z$ or the free energy.[8] We now briefly sketch this approach, and then use it to show that the damped system has a surprising universal behavior at low temperature (see also [224]).

Based on the formal solution (21.46) for $\langle \sigma_z \rangle_{\text{eq}}$, we obtain upon integration with respect to the bias the series expansion of the free energy in the form

$$F(T, \epsilon) = \lim_{t \to \infty} \frac{\hbar}{2} \sum_{m=1}^{\infty} (-1)^{m-1} \int_0^t \mathcal{D}_{2m,0}\{t_j\} \frac{1}{2^m} \sum_{\{\xi_j\}} F_m^{(-)} A_m^{(s)}, \tag{22.211}$$

where

$$A_m^{(s)} = \cos\left(\epsilon \sum_{j=1}^{m} \xi_j \tau_j\right) \bigg/ \sum_{j=1}^{m} \xi_j \tau_j. \tag{22.212}$$

We are not allowed to interchange in Eq. (22.211) $\lim_{t \to \infty}$ with the summation of the series since the limit $t \to \infty$ would give a divergent result for each individual term.

Secondly, we use the fact that the static (nonlinear) susceptibility $\overline{\chi}_z$ is related to $\langle \sigma_z \rangle_{\text{eq}}$ by $\overline{\chi}_z = (2/\hbar)\partial\langle \sigma_z \rangle_{\text{eq}}/\partial\epsilon$ [cf. Eq. (19.4)]. Using this relation, we obtain from the series expression (21.46) for $\langle \sigma_z \rangle_{\text{eq}}$ the exact formal solution for $\overline{\chi}_z$ in the form

$$\overline{\chi}_z(T, \epsilon) = \lim_{t \to \infty} \frac{2}{\hbar} \sum_{m=1}^{\infty} (-1)^{m-1} \int_0^t \mathcal{D}_{2m,0}\{t_j\} \frac{1}{2^m} \sum_{\{\xi_j\}} F_m^{(-)} C_m^{(s)}, \tag{22.213}$$

where

$$C_m^{(s)} = \cos\left(\epsilon \sum_{j=1}^{m} \xi_j \tau_j\right) \sum_{j=1}^{m} \xi_j \tau_j. \tag{22.214}$$

Let us now determine the asymptotic low temperature expansion of F and $\overline{\chi}_z$. First, we observe that T enters into the expressions (22.211) and (22.213) only through the pair interaction $Q'(\tau)$ occurring via the factor G_m in $F_m^{(-)}$, Eq. (21.43) with (21.31). Secondly, we write $Q'(\tau) = Q_0'(\tau) + Q_1'(\tau)$, where $Q_0'(\tau)$ is the pair interaction for zero temperature, and $Q_1'(\tau)$ is the finite temperature contribution. The leading correction at low T is easily computed for the form (18.44) of $G(\omega)$ from Eq. (18.45), and is

$$Q_1'(\tau) = \kappa_s (\hbar\beta\omega_{\text{ph}})^{1-s} (\tau/\hbar\beta)^2 \left\{1 + \mathcal{O}[(\tau/\hbar\beta)^2]\right\}, \tag{22.215}$$

where

$$\kappa_s = 2\delta_s \Gamma(1+s)\zeta(1+s), \tag{22.216}$$

and where $\Gamma(z)$ and $\zeta(z)$ are the gamma and Riemann zeta function. With this expression for $Q_1'(\tau)$, the low temperature expansion of the blip correlation factor G_m defined in Eq. (21.31) takes the form

$$G_m = G_m^{(0)}\left\{1 - \kappa_s \omega_{\text{ph}}^{1-s}\left(\frac{k_B T}{\hbar}\right)^{1+s}\left(\sum_{j=1}^{m} \xi_j \tau_j\right)^2 + \mathcal{O}\left(T^{3+s}\right)\right\}, \tag{22.217}$$

---

[8]The imaginary-time approach yields a series expression for the partition function (cf. Chap. 19).

where $G_m^{(0)}$ is the blip interaction factor for zero temperature. Inserting now (15.96) into the expressions (15.92) and (15.93), and observing that the resulting extra factor $(\sum_j \xi_j \tau_j)^2$ can be generated by double differentiation with respect to the bias, one immediately finds for F and $\overline{\chi}_z$ the low temperature expressions

$$F(T, \epsilon) = F(0, \epsilon) - \frac{1}{4}\kappa_s \left(k_B T / \hbar \omega_{ph}\right)^{s-1} (k_B T / \hbar)^2 \overline{\chi}_z(0, \epsilon) , \qquad (22.218)$$

$$\overline{\chi}_z(T, \epsilon) = \overline{\chi}_z(0, \epsilon) + \kappa_s \left(k_B T / \hbar \omega_{ph}\right)^{s-1} (k_B T / \hbar)^2 \overline{\chi}_z''(0, \epsilon) , \qquad (22.219)$$

where terms of order T^{3+s} are disregarded. Here, $F(0, \epsilon)$ and $\overline{\chi}_z(0, \epsilon)$ are the free energy and static nonlinear susceptibility for $T = 0$, and $\overline{\chi}_z''(0, \epsilon) = \partial^2 \overline{\chi}_z(0, \epsilon)/(\partial \epsilon)^2$. The expressions (22.218) and (22.219) are the asymptotic expansions of F and $\overline{\chi}_z$ for the two-state system described by a spectral density of the form (18.44) under the limitations discussed below. The leading temperature dependence both of F and $\overline{\chi}_z$ is T^{1+s}. Interestingly, the bias enters only through the nonlinear susceptibility at zero temperature. Observing now that the specific heat $c(T) = \partial U(T)/\partial T$ can be expressed in terms of the free energy $F(T)$ as

$$c(T) = -T \partial^2 F(T)/(\partial T)^2 , \qquad (22.220)$$

we immediately see that the leading dependence of $c(T)$ on temperature is T^s. Thus it is natural to adjust the definition of the Wilson ratio, Eq. (19.41), to non-Ohmic spectral density. As anticipated in Eq. (6.118), the generalized Wilson ratio is

$$R_s \equiv \lim_{T \to 0} \frac{4c(T)/k_B}{\overline{\chi}_z(0, \epsilon) (\hbar \omega_{ph})^{1-s} (k_B T)^s} . \qquad (22.221)$$

With use of the expression (22.218) we finally obtain

$$R_s = 2s\Gamma(2+s)\zeta(1+s)\delta_s . \qquad (22.222)$$

These results for the free energy, the specific heat, and the Wilson ratio are generally valid provided that the static susceptibility is nonzero at $T = 0$. This generally holds for $s > 0$ when the system is *biased*, i.e., the static susceptibility is nonlinear.

The linear susceptibility is finite in the super-Ohmic case for all δ_s, and in the Ohmic case for K in the regime $0 < K < 1$. Thus in these cases the relation (22.222) is also valid for the symmetric TSS. For $s = 1$ and $K > 1$, and in the sub-Ohmic case $s < 1$ and any $\delta_s > 0$, the limit $\lim_{\epsilon \to 0} \overline{\chi}_z(T = 0, \epsilon)$ is divergent. Hence in these parameter regimes, the above analysis is not applicable. Instead of that, in these regimes the high temperature expansions for $\overline{\chi}(T, \epsilon = 0)$ and $F(T)$, which are the expansions in powers of Δ^2, Eq. (22.211) and Eq. (22.213), are valid down to $T = 0$. In the Ohmic case, the Wilson ratio simplifies to the form

$$R_1 = 2K\pi^2/3 . \qquad (22.223)$$

Using the correspondence relation (19.91) for K, we see that the Wilson ratio (22.223) of the Ohmic TSS coincides with the Wilson ratio (19.93) of the s-d model. The Wilson ratio has been studied numerically for the s-d model, and excellent agreement with the formula (22.223) has been found [485].

23 The driven two-state system

With the advance of laser technology and the possibility for experimental time resolution in the sub-picosecond regime, there has been growing interest in the study of the dynamics of quantum systems that are driven by strong time-dependent external fields. Quantum dynamics of explicitly time-dependent Hamiltonians shows a variety of novel effects, such as the phenomenon of coherent destruction of tunneling [492], stabilization of localized states which would otherwise decay [493], and the possibility of controlling the tunneling dynamics with pulsed monochromatic light [494, a] [495] and sinusoidal fields [494, b] [495]. For a review, see Refs. [496, 497]. Here we restrict the attention to a two-state system which is simultaneously exposed to a fluctuating force by the surroundings and to deterministic time-dependent forces.

23.1 Time-dependent external fields

We generalize the spin-boson Hamiltonian (18.15) by taking into account the interaction with external time-dependent fields. We assume that the external fields couple to the system's operators σ_z and σ_x only, and have no effect on the bath.

23.1.1 Diagonal and off-diagonal driving

The coupling of an external field to σ_z describes, e.g., an electric field coupled to the dipole moment of the TSS, or in the rf SQUID device discussed in Subsection 3.2.2 a time-dependent magnet flux threading the ring. This coupling leads to a temporal modulation of the bias which is superimposed on the static bias. This entails in the Hamiltonian (18.15) the substitution

$$\epsilon \quad \Longrightarrow \quad \epsilon(t) = \epsilon_0 + \epsilon_1(t) . \tag{23.1}$$

Here, ϵ_0 is the bias frequency related to the intrinsic static strain field, and $\epsilon_1(t)$ is a bias modulation due to the externally applied time-dependent force. Regarding electron transfer in a solvent, it is conceivable to control charge tunneling by application of strong continuous laser fields. For charge transfer in nano-structured devices, we may think of regulating the dynamics by turning on microwave irradiation or a high-frequency voltage. In pump-probe set-ups, the relevant system is subject to a pulse-shaped driving force. The modulation of the bias energy leads to modified bias phases. The accumulated phase for a path with a sequence of m blips is given by

$$\varphi_m = \sum_{j=1}^{m} \xi_j \vartheta(t_{2j}, t_{2j-1}) , \quad \text{where} \quad \vartheta(t_2, t_1) = \int_{t_1}^{t_2} dt' \, \epsilon(t') . \tag{23.2}$$

The bias factors, which replace the forms (21.42), read of φ_m as

$$B_m^{(s)} = \cos\varphi_m, \qquad B_m^{(a)} = \sin\varphi_m. \tag{23.3}$$

In contrary, a "quadrupole"-like time-dependent coupling induces a temporal variation of the barrier opacity. Barrier modulation can be realized, e.g., in a superconducting loop with two Josephson junctions [498]. The quadrupole coupling leads to a *multiplicative* modification of the tunneling coupling. For harmonic pulsation of the barrier, we have in the Hamiltonian (18.15) the substitution [499]

$$\Delta \quad \Longrightarrow \quad \Delta(t) = \Delta \exp[\mu\cos(\nu t)], \tag{23.4}$$

where ν is the angular frequency and μ a suitable dimensionless amplitude.

23.1.2 Exact formal solution

In the exact formal series expressions obtained in Chapter 21, the time-dependent tunneling matrix element is accounted for by the substitution [e.g., in Eq. (21.36)]

$$\Delta^m \quad \longrightarrow \quad \Delta_m(\{t_j\}) = \prod_{j=1}^{m} \Delta(t_j). \tag{23.5}$$

In the non-convolutive case, the Laplace transform of the master equation (21.69) for $P(t) = \langle\sigma_z\rangle_t$ takes the form[1]

$$\lambda\hat{P}(\lambda) = 1 + \int_0^\infty dt\, e^{-\lambda t}\left[\hat{K}_\lambda^{(a)}(t) - \hat{K}_\lambda^{(s)}(t)P(t)\right] \tag{23.6}$$

with the kernel $\hat{K}_\lambda^{(s,a)}(t) = \int_0^\infty d\tau\, e^{-\lambda\tau} K_z^{(s,a)}(t+\tau, t)$.

Consider now first the case of a periodic bias modulation with period $\mathcal{T} = 2\pi/\omega$,

$$\epsilon(t) = \epsilon(t + n\mathcal{T}). \tag{23.7}$$

Then the kernel $\hat{K}_\lambda^{(s,a)}(t)$ is time-periodic and may be written as Fourier series,

$$\hat{K}_\lambda^{(s,a)}(t) = \sum_{m=-\infty}^{\infty} k_m^{(s,a)}(\lambda)\, e^{-i\,m\omega t}. \tag{23.8}$$

Use of the Fourier series (23.8) in Eq. (23.6) allows for the calculation of $P(t)$ in a recursive manner [500]. The resulting expression for the population $P(t)$ consists of a transient part $P^{(\mathrm{tr})}(t)$ and a stationary part $P^{(\mathrm{st})}(t)$ representing persistent oscillations. The transient behavior is governed by the zeros of the characteristic equation

$$\lambda + im\omega + k_0^{(s)}(\lambda + im\omega) = 0. \tag{23.9}$$

---

[1]We restrict the attention to the discussion of $\langle\sigma_z\rangle_t$. The corresponding study of $\langle\sigma_x\rangle_t$ is without difficulty and left to the reader.

23 The driven two-state system

The root λ_0 of Eq. (23.9) for $m = 0$ with negative real part of smallest absolute value γ_0 determines the time scale $\tau_0 = 1/\gamma_0$ for the decay of the primary transient. The roots of Eq. (23.9) for $m = \pm 1, \pm 2, \ldots$, are obtained from λ_0 by shifts with a multiple of the driving frequency in imaginary direction, $\lambda_m = \lambda_0 - i\,m\omega$. The related transients drop to zero on the same time scale as the primary transient.

At times $t \gg \tau_0$, the transients have already receded, and $P(t)$ oscillates persistently with period $\mathcal{T}$. This stationary regime is described by the Fourier series

$$P^{(\mathrm{st})}(t) = \sum_{m=-\infty}^{\infty} p_m e^{-i\,m\omega t} . \tag{23.10}$$

The Fourier coefficient p_m is found from Eq. (23.6) as the residuum of the pole of $\hat{P}(\lambda)$ at $\lambda = -i\,m\omega$. The coefficients obey the recursive relations [500] (see also Ref. [501])

$$
\begin{aligned}
p_0 &= \frac{k_0^{(\mathrm{a})}(0)}{k_0^{(\mathrm{s})}(0)} - \sum_{m\neq 0} \frac{k_m^{(\mathrm{s})}(0)}{k_0^{(\mathrm{s})}(0)} p_m , \\
p_m &= \frac{1}{-i\,m\omega}\left(k_m^{(\mathrm{a})}(-i\,m\omega) - \sum_n k_{m-n}^{(\mathrm{s})}(-i\,m\omega)\,p_n \right) .
\end{aligned}
\tag{23.11}
$$

In the NIBA, the kernels are of order Δ^2. For monochromatic modulation of the bias

$$\epsilon(t) = \epsilon_0 + \epsilon_1 \cos(\omega t) , \tag{23.12}$$

the Fourier series of the bias factors are

$$
\begin{aligned}
B_m^{(\mathrm{s})}(\tau + t, t) &\equiv \cos[\vartheta(\tau + t, t)] = \textstyle\sum_m A_m^{(\mathrm{s})}(\tau)\,e^{-i\,m\omega t} , \\
B_m^{(\mathrm{a})}(\tau + t, t) &\equiv \sin[\vartheta(\tau + t, t)] = \textstyle\sum_m A_m^{(\mathrm{a})}(\tau)\,e^{-i\,m\omega t} ,
\end{aligned}
\tag{23.13}
$$

with the Fourier coefficients

$$
\begin{aligned}
A_{2m}^{(\mathrm{s})}(\tau) &= (-1)^m \cos(\epsilon_0\tau) J_{2m}[(2\epsilon_1/\omega)\sin(\omega\tau/2)]\,e^{-i\,m\omega\tau} , \\
A_{2m}^{(\mathrm{a})}(\tau) &= (-1)^m \sin(\epsilon_0\tau) J_{2m}[(2\epsilon_1/\omega)\sin(\omega\tau/2)]\,e^{-i\,m\omega\tau} , \\
A_{2m+1}^{(\mathrm{s})}(\tau) &= (-1)^{m+1} \sin(\epsilon_0\tau) J_{|2m+1|}[(2\epsilon_1/\omega)\sin(\omega\tau/2)]\,e^{-i\,(m+1/2)\omega\tau} , \\
A_{2m+1}^{(\mathrm{a})}(\tau) &= (-1)^m \cos(\epsilon_0\tau) J_{|2m+1|}[(2\epsilon_1/\omega)\sin(\omega\tau/2)]\,e^{-i\,(m+1/2)\omega\tau} ,
\end{aligned}
\tag{23.14}
$$

where $J_n(z)$ is a J-Bessel function. The Fourier coefficients of the NIBA kernels are

$$
\begin{aligned}
k_m^{(\mathrm{s})}(\lambda) &= \Delta^2 \int_0^{\infty} d\tau\, e^{-\lambda\tau} e^{-Q'(\tau)} \cos[Q''(\tau)] A_m^{(\mathrm{s})}(\tau) , \\
k_m^{(\mathrm{a})}(\lambda) &= \Delta^2 \int_0^{\infty} d\tau\, e^{-\lambda\tau} e^{-Q'(\tau)} \sin[Q''(\tau)] A_m^{(\mathrm{a})}(\tau) .
\end{aligned}
\tag{23.15}
$$

For zero static bias, $\epsilon_0 = 0$, there follow from the expressions (23.14) the relations $k_{2m}^{(a)}(\lambda) = 0$ and $k_{2m+1}^{(s)}(\lambda) = 0$. With these, the set of equations (23.11) yield the selection rule[2]

$$p_{2m} = 0 \,. \tag{23.16}$$

Thus in the absence of a static bias only terms with odd multiples of the fundamental frequency contribute to the stationary dynamics (23.10) of $P(t)$.

23.1.3 Linear response

When the driving amplitude ϵ_1 is small compared to the driving frequency ω, we may expand the kernels (23.15) in a power series in ϵ_1. Within linear response, the relevant kernels are of order unity and of order ϵ_1. We then find in the Fourier series (23.10) the limitation $|m| \leqslant 1$. The term $k_0^{(s,a)}(\lambda)$ is independent of ϵ_1, and $k_{\pm 1}^{(s,a)}(\lambda) = \mathcal{O}(\epsilon_1)$. As a result, $P(t)$ behaves in the stationary regime as

$$P^{(st)}(t) = P_\infty + \hbar\epsilon_1 \left[\tilde{\chi}(\omega)\,e^{-i\omega t} + \tilde{\chi}(-\omega)\,e^{i\omega t} \right]/4 \,, \tag{23.17}$$

where $P_\infty = \lim_{\epsilon_1 \to 0} k_0^{(a)}(0)/k_0^{(s)}(0) = \tanh(\tfrac{1}{2}\hbar\beta\epsilon_0)$ [cf. Eq. (21.83)]. The evaluation of the recursive relations (23.11) for $|m| \leqslant 1$ gives

$$\tilde{\chi}(\omega) = \lim_{\epsilon_1 \to 0} \frac{4}{\hbar\epsilon_1 \left[-i\,\omega + k_0^{(s)}(-i\,\omega) \right]} \left[k_1^{(a)}(-i\,\omega) - P_\infty k_1^{(s)}(-i\,\omega) \right] \,. \tag{23.18}$$

This is the dynamical susceptibility in the NIBA. The reader may easily convince himself that the expression (23.18) is directly connected with the spectral correlation function (22.77) by the fluctuation-dissipation theorem (22.165).

23.1.4 The Ohmic case with Kondo parameter $K = \tfrac{1}{2}$

For an Ohmic heat bath with damping strength $K = \tfrac{1}{2}$ and $\hbar\beta\omega_c \gg 1$, the dynamics can be solved exactly, up to quadratures, in the presence of diagonal and off-diagonal driving. In generalization of the expression (22.134), we obtain [501, 499]

$$P(t) = \exp\left(-\int_0^t d\tau\,\gamma(\tau) \right) + P_a(t) \,, \qquad \gamma(\tau) = \frac{\pi}{2}\frac{\Delta^2(\tau)}{\omega_c} \,,$$

$$P_a(t) = \int_0^t dt_2\,\Delta(t_2) \exp\left(-\int_{t_2}^t d\tau\,\gamma(\tau) \right) \tag{23.19}$$

$$\times \int_0^{t_2} dt_1\,\Delta(t_1)\sin[\vartheta(t_2,t_1)]\,e^{-Q'(t_2-t_1)} \exp\left(-\frac{1}{2}\int_{t_1}^{t_2} d\tau'\,\gamma(\tau') \right) \,,$$

where $\vartheta(t_2,t_1)$ is given in Eq. (23.2). The asymptotic dynamics is described by $P_a(t)$.

Consider now the case of time-independent tunneling coupling in some more detail. We find from Eq. (23.19) for the Fourier amplitudes p_m the exact expression

$$p_m(\omega,\epsilon_1) = \frac{\Delta^2}{-i\,m\omega + \gamma} \int_0^\infty d\tau\, e^{i\,m\omega\tau - \gamma\tau/2 - Q'(\tau)} A_m^{(a)}(\tau) \,. \tag{23.20}$$

---

[2] $\langle \sigma_y \rangle_t^{(st)}$ obeys the same selection rule [cf. Eq. (21.4)], and $\langle \sigma_x \rangle_t^{(st)}$ has only even harmonics.

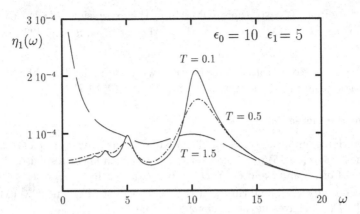

Figure 23.1: The spectral amplitude η_1 is plotted versus ω for different temperatures. See text for details. Frequencies and temperature are in units of $\gamma = \pi\Delta^2/2\omega_c$.

The spectral amplitude of the fundamental frequency, $\eta_1(\omega, \epsilon_1) = 4\pi|p_1(\omega, \epsilon_1)/\hbar\epsilon_1|^2$ [cf. Eq. (23.41)], is plotted versus ω for various temperatures in Fig. 23.1.[3] At high T, $\eta_1(\omega)$ is peaked at $\omega = 0$, and there is only little structure. As the temperature is decreased, resonances are formed at fractional values of the static bias, $\omega = \epsilon_0/n$ ($n = 1, 2, \ldots$). At these frequencies, the driving-induced coherent motion is amplified.

23.2 Markovian regime

When the system's characteristic motion is slow on the time scale τ on which the kernels $K_z^{(\mathrm{s,a})}(t, t-\tau)$ decay, the incoherent dynamics at long times is described by a time-local master equation. The GME (21.69) simplifies to the form

$$\dot{P}(t) = \mathcal{M}_z^{(\mathrm{a})}(t) - \mathcal{M}_z^{(\mathrm{s})}(t)P(t), \qquad (23.21)$$

with

$$\mathcal{M}_z^{(\mathrm{s})}(t) = \int_0^\infty d\tau\, K_z^{(\mathrm{s})}(t, t-\tau) = \Delta^2 \int_0^\infty d\tau\, e^{-Q'(\tau)} \cos[Q''(\tau)] \cos[\vartheta(t, t-\tau)],$$

$$\mathcal{M}_z^{(\mathrm{a})}(t) = \int_0^\infty d\tau\, K_z^{(\mathrm{a})}(t, t-\tau) = \Delta^2 \int_0^\infty d\tau\, e^{-Q'(\tau)} \sin[Q''(\tau)] \sin[\vartheta(t, t-\tau)].$$

The respective second forms are the NIBA expressions, and $\vartheta(t_2, t_1)$ is given in Eq. (23.2). The master equation (23.21) can be solved directly. The solution with initial value $P(0) = 1$ is

$$P(t) = \exp\left(-\int_0^t dt'\, \mathcal{M}_z^{(\mathrm{s})}(t')\right) + \int_0^t dt'\, \mathcal{M}_z^{(\mathrm{a})}(t') \exp\left(-\int_{t'}^t dt''\, \mathcal{M}_z^{(\mathrm{s})}(t'')\right).$$

---

[3]Figs. 23.1 – 23.4 are by courtesy of M. Grifoni, P. Hänggi and L. Hartmann.

In the adiabatic limit we may put $\vartheta(t, t - \tau) = \epsilon(t)\,\tau$. With this form, the kernels are found as linear combinations of the forward/backward Golden Rule rates $k^{\pm}$ introduced in Subsection 20.2.1,

$$\mathcal{M}_z^{(s,a)}(t) \;=\; k^{+}[\epsilon(t)] \pm k^{-}[\epsilon(t)]\,, \tag{23.22}$$

In these expressions the time-dependence is solely in the bias $\epsilon(t)$. The adiabatic rates obey detailed balance with respect to the current bias $\epsilon(t)$.

23.3 High-frequency regime

When the driving frequency ω is very large compared to the characteristic frequencies of the damped TSS, the system is sluggish on the time scale $\mathcal{T}_{\omega} = 2\pi/\omega$. Then a coarse-grained description, in which the dynamics is averaged over the period $\mathcal{T}_{\omega} = 2\pi/\omega$, is expedient. In this spirit, we replace the kernel $\hat{K}_{\lambda}^{(s,a)}(t)$ in Eq. (23.6) with the time-averaged kernel

$$\hat{K}_{\lambda}^{(s,a)}(t) \;\Longrightarrow\; \langle \hat{K}_{\lambda}^{(s,a)} \rangle \equiv \frac{1}{\mathcal{T}_{\omega}} \int_0^{\mathcal{T}_{\omega}} dt\, \hat{K}_{\lambda}^{(s,a)}(t) \;=\; k_0^{(s,a)}(\lambda)\,. \tag{23.23}$$

Then the averaged dynamics is determined by the $m = 0$ Fourier component of the kernels in the series (23.8), and Eq. (23.6) with the time-averaged kernels $k_0^{(s,a)}(\lambda)$ can be easily solved for $\hat{P}(\lambda)$. The resulting expression is analogous in form to Eq. (21.75),

$$\hat{P}(\lambda) \;=\; \frac{1 + k_0^{(a)}(\lambda)/\lambda}{\lambda + k_0^{(s)}(\lambda)}\,. \tag{23.24}$$

This expression establishes the averaged transient and the long-time behavior.

The characteristics of the transient dynamics is determined by the zeros of the equation $\lambda + k_0^{(s)}(\lambda) = 0$. With use of the decomposition

$$J_0(2z \sin\alpha) \;=\; \sum_{n=-\infty}^{\infty} J_n^2(z) \cos(2n\alpha) \tag{23.25}$$

in the expression for $k_0^{(s)}(\lambda)$ given in Eq. (23.15) with (23.14), the pole condition of Eq. (23.24) is found to read

$$\lambda + \frac{1}{2} \sum_{n=-\infty}^{\infty} J_n^2\left(\tfrac{\epsilon_1}{\omega}\right) \Big\{ g[\lambda + i\,(n\omega + \epsilon_0)] + g[\lambda - i\,(n\omega + \epsilon_0)] \Big\} \;=\; 0\,, \tag{23.26}$$

where the kernel $g(\lambda)$ is given in Eq. (22.1).

In the incoherent regime, we keep the relaxation pole at $\lambda = -\gamma_{\rm r} \equiv -k_0^{(s)}(\lambda = 0)$ and disregard all other poles since their residua are very small. Thus we obtain

$$P(t) \;=\; P_{\infty} + [1 - P_{\infty}]e^{-\gamma_{\rm r} t}\,, \tag{23.27}$$

453

where $P_\infty = k_0^{(\mathrm{a})}(0)/k_0^{(\mathrm{s})}(0)$ is the time-averaged equilibrium value. The relaxation rate is given by the series[4]

$$\gamma_{\mathrm{r}}(\epsilon_0; \omega, \epsilon_1) = \sum_{n=-\infty}^{\infty} J_n^2\left(\frac{\epsilon_1}{\omega}\right)\left[k^+(\epsilon_0 + n\omega) + k^-(\epsilon_0 + n\omega)\right]. \tag{23.28}$$

The formula (23.28) describes the inclusive relaxation rate γ_{r} as a sum over all possible decay channels. The individual channels represent tunneling processes with simultaneous emission and absorption of a fixed number of quanta with fundamental frequency ω, and the weight factor for n quanta is $J_n^2(\epsilon_1/\omega)$. The partial rates in Eq. (23.28) are given in terms of the relaxation rates $k^\pm$ for a static bias $\epsilon_n = \epsilon_0 + n\omega$ for the dissipative mechanism under consideration. These rates have been discussed in Section 20.2. The inclusive relaxation rate γ_{r} sensitively depends on the parameters of the driving field. We see in Fig. 23.2 that the rate is enhanced compared to the static case when the resonance condition $\epsilon_0 = \pm n\omega$ is met and when the weight does not fall on a zero of the respective Bessel function. Upon tuning ϵ_1/ω to a zero of one of the Bessel functions in Eq. (23.28), the corresponding decay channel is missing.

The striking dependence of γ_{r} on the external field parameters may be utilized, e.g., to control electron transfer rates in chemical reaction processes.

The fast ac field leads to an asymptotic population $P_\infty = k_0^{(\mathrm{a})}(0)/k_0^{(\mathrm{s})}(0)$ which may differ drastically from the static case, $P_\infty = \tanh(\hbar\beta\epsilon_0/2)$. Even inverted population can be reached by means of a suitable choice of the parameters ϵ_1 and ω, as predicted, e.g., in the works in Ref. [503].

Next, consider a symmetric Ohmic TSS. Using for $g(\lambda)$ the NIBA form (22.2), the pole condition (23.26) can be written in the regime $\omega \gg 2\pi K k_{\mathrm{B}}T/\hbar$ as [499]

$$\lambda + g(\lambda)\left\{ J_0^2(\epsilon_1/\omega) + \left(\frac{2\pi K + \hbar\beta\lambda}{\hbar\beta\omega}\right)^2 \sum_{n\neq 0} \frac{J_n^2(\epsilon_1/\omega)}{n^2}\right\} = 0. \tag{23.29}$$

The first term in the curly bracket dominates for large ω when ϵ_1/ω is sufficiently distant from a zero of the J_0-Bessel function. Then the pole condition simply reads

$$\lambda + g(\lambda)J_0^2(\epsilon_1/\omega) = 0. \tag{23.30}$$

Thus the results presented in Subsections 22.2.2 and 22.2.3 directly apply to the present case if we replace Δ by a driving-renormalized tunneling matrix element,

$$\Delta \longrightarrow J_0(\epsilon_1/\omega)\Delta. \tag{23.31}$$

There is weakly-damped coherent dynamics in the transient regime when $K \ll 1$.

The coherent dynamics is suppressed when ϵ_1/ω is tuned to a zero of the Bessel function $J_0(\epsilon_1/\omega)$. This phenomenon of an undamped or weakly damped driven TSS

---

[4]Representations of the form (23.28) for transport quantities have a long tradition since the pioneering work by Tien and Gordon [502].

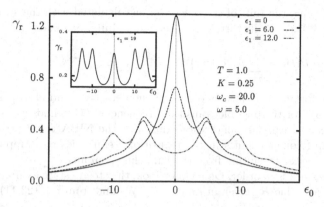

Figure 23.2: The inclusive rate γ_r is shown as a function of ϵ_0 for different amplitudes ϵ_1. The frequency parameters are given in units of $\Delta/2$. In the inset, the amplitude ϵ_1 is tuned on a zero of the $n = 1$ Bessel function, $J_1(\epsilon_1/\omega) = 0$ so that the first side band is absent.

is termed "coherent destruction of tunneling" [504, 505]. At the destruction point $J_0(\epsilon_1/\omega) = 0$, the system relaxes incoherently with the rate

$$\gamma_r \;=\; g(0) \left(\frac{2\pi K}{\hbar\beta\omega} \right)^2 \sum_{n\neq 0} \frac{J_n^2(\epsilon_1/\omega)}{n^2} \;. \tag{23.32}$$

Consider now the coherent regime for $K \ll 1$ and $\epsilon_1/\omega \ll 1$. For large ω, the leading effects of the driving force are taken into account by expanding the Bessel functions in Eq. (23.26) up to terms of order $(\epsilon_1/\omega)^2$. In this approximation, the *undamped* driven system is characterized by three bias frequencies $\mu_1 = \epsilon_0$, $\mu_2 = \epsilon_0 + \omega$, and $\mu_3 = \epsilon_0 - \omega$ and three tunneling frequencies ν_j. The squared ν_j are the solutions of a cubic equation in ν^2,

$$\nu^6 - a_4\nu^4 + a_2\nu^2 - a_0 = 0 \;, \tag{23.33}$$

with coefficients

$$a_0 \;=\; \mu_1^2\mu_2^2\mu_3^2 + [\,\Delta_1^2\mu_2^2\mu_3^2 + \text{cycl.}\,] \;,$$
$$a_2 \;=\; \Delta_1^2(\mu_2^2 + \mu_3^2) + \mu_2^2\mu_3^2 + \text{cycl.} \;, \tag{23.34}$$
$$a_4 \;=\; \mu_1^2 + \Delta_1^2 + \text{cycl.} \;,$$

where $\Delta_1^2 = (1 - \epsilon_1^2/2\omega^2)\Delta^2$ and $\Delta_2^2 = \Delta_3^2 = (\epsilon_1^2/2\omega^2)\Delta^2$. The undamped system performs a superposition of coherent oscillations

$$P_{\text{undamped}}(t) \;=\; p_0 + \sum_{j=1}^{3} p_j \cos(\nu_j t) \;, \tag{23.35}$$

with the amplitudes (p_2, p_3 cycl.)

$$p_1 \;=\; \frac{(\nu_1^2 - \mu_1^2)(\nu_1^2 - \mu_2^2)(\nu_1^2 - \mu_3^2)}{\nu_1^2(\nu_1^2 - \nu_2^2)(\nu_1^2 - \nu_3^2)} \;, \qquad p_0 \;=\; \prod_{j=1}^{3} \frac{\mu_j^2}{\nu_j^2} \;. \tag{23.36}$$

For weak damping, the shift of the poles can be calculated with the strategy set out in Section 22.3. Disregarding irrelevant frequency shifts, one finds

$$P(t) = \sum_{j=1}^{3} p_j \cos(\nu_j t)\, e^{-\gamma_j t} + (p_0 - P_\infty)\, e^{-\gamma_0 t} + P_\infty\,. \tag{23.37}$$

The rates are linear combinations of one-phonon emission and absorption processes [cf. Eq. (22.93)] for the possible transition frequencies.[5] One finds again by comparison of the exact weak-damping expressions with the NIBA results that the NIBA disregards the frequency shifts in the transition frequencies and amplitudes of the one-phonon processes resulting from the tunneling coupling.

One final remark is in order. On a closer look, the steady state exhibits oscillations about the mean value P_∞ given in Eq.(23.27). We find from Eq. (23.11) for large ω

$$P^{(\mathrm{st})}(t) = P_\infty + \sum_{m \neq 0} \frac{1}{-i\,m\omega} \left(k_m^{(\mathrm{a})}(-i\,m\omega) - k_m^{(\mathrm{s})}(-i\,m\omega)P_\infty \right) e^{-i m\omega t}\,. \tag{23.38}$$

With increasing driving-frequency, the oscillations of $P^{(\mathrm{st})}(t)$ about P_∞ become suppressed since the $m \neq 0$ Fourier components get less important.

23.4 Quantum stochastic resonance

The stochastic resonance phenomenon refers to the amplification of the response to an applied periodic signal at a certain optimal value of the noise strength and is a cooperative effect of friction, noise, and periodic driving in a bistable system. This phenomenon has been dubbed *stochastic resonance* (SR) since classically the maximal output signal occurs when the thermal hopping rate is in resonance with the frequency of the driving force. It has been argued that this phenomenon is of fundamental importance in biological evolution. Several comprehensive reviews on classical stochastic resonance are available [506]. Qualitatively new signatures of stochastic resonance appear in the quantum regime [507, 508].

When the condition (3.135) is met, the double well system is well described by the discrete two-state system. Then the quantity of interest in SR is the power spectrum

$$S(\nu) = \int_{-\infty}^{\infty} d\tau\, e^{i\nu\tau}\, \overline{C}^{(\mathrm{st})}(\tau)\,, \tag{23.39}$$

where $\overline{C}^{(\mathrm{st})}(\tau)$ is the time-averaged steady-state population correlation function,

$$\overline{C}^{(\mathrm{st})}(\tau) \equiv \lim_{\tau \to \infty} \frac{\omega}{2\pi} \,\mathrm{Re} \int_0^{2\pi/\omega} dt\, \langle\, \sigma_z(t+\tau)\sigma_z(t)\, \rangle = \sum_{m=-\infty}^{\infty} |p_m(\omega, \epsilon_1)|^2\, e^{-i m\omega\tau}\,.$$

The set of amplitudes $\{p_m(\omega, \epsilon_1)\}$ obeys the recursion relations (23.11). The power spectrum takes the form

---

[5]The various rate expressions are given in Ref. [462].

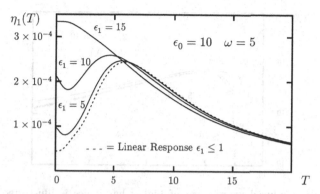

Figure 23.3: Depiction of QSR in the deep quantum regime for Ohmic damping $K = \frac{1}{2}$. The power amplitude η_1 of the fundamental frequency is plotted as a function of temperature for different driving amplitudes. See text for details. The units are the same as in Fig. 23.1.

$$S(\nu) = 2\pi \sum_{m=-\infty}^{\infty} |p_m(\omega, \epsilon_1)|^2 \, \delta(\nu - m\omega). \tag{23.40}$$

A quantitative study of the power amplitude

$$\eta_m(\omega, \epsilon_1) = 4\pi \, |p_m(\omega, \epsilon_1)/\hbar\epsilon_1|^2 \tag{23.41}$$

can be performed by first calculating numerically the kernels $k_m^{(s,a)}(\lambda)$ for the environment of interest and subsequently solving the recursive relations (23.11).

The findings in the Ohmic scaling limit are as follows. In classical SR, the spectral amplification is maximal for a *symmetric* system [506], while in the deep quantum regime and $K < 1$, QSR is only effective in the presence of a static bias. QSR is most striking when the static bias ϵ_0 is larger than the driving amplitude ϵ_1. For $K = \frac{1}{2}$, the power amplitudes are given by the expression (23.20). In Fig. 23.3, the power amplitude $\eta_1(T)$ for $K = \frac{1}{2}$ is plotted versus temperature for different driving amplitudes. When $\epsilon_1 > \epsilon_0$, the power amplitude decreases monotonously with increasing T. As ϵ_1 is decreased, a minimum at low T followed by a QSR maximum at $T \approx \hbar\omega/k_B$ is formed. For $\epsilon_1 \lesssim 5\gamma$, the QSR can be studied within linear response theory.

In the linear response regime, we find from Eqs. (23.17) and (23.41) the relation $p_1 = -\hbar\epsilon_1\tilde{\chi}(\omega)/4$. In the regime $\hbar\omega < 2\pi\alpha k_B T$ and $k_B T > \hbar\Delta$ and/or $K > 1$, the NIBA form (23.18) is correct. From this we obtain in the Lorentzian approximation for the quasi-elastic peak centered at $\omega = 0$ the expression

$$\eta_1(\omega) = \frac{\pi}{(2k_B T)^2} \frac{1}{\cosh^4(\hbar\epsilon_0/2k_B T)} \frac{\gamma_r^2}{\omega^2 + \gamma_r^2}, \tag{23.42}$$

where $\gamma_r = k^+ + k^- = [1 + \exp(-\hbar\beta\epsilon)]k^+$ is the width, and where $k^+(T, \epsilon_0)$ is the forward tunneling rate (20.74). Since the quasi-elastic peak reflects exponential

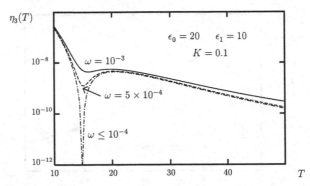

Figure 23.4: Noise-induced suppression of higher harmonics is illustrated for the third power amplitude η_3. Frequencies and temperature are given in units of Δ_{eff}.

relaxation, it is not surprising that the same form is also found in the classical case, e.g. for the bistable double well, in which γ_r is the thermal relaxation rate [506].

For large amplitude ϵ_1 of the driving force, the response of higher harmonics may become significant. One finds from the recursive relations (23.11) that quantum noise can enhance or suppress higher harmonics in $P^{(\text{as})}(t)$. In Fig. 23.4, the temperature dependence of the power amplitude η_3 is shown for different ω. As the driving frequency is decreased, noise-induced suppression of this amplitude occurs at a temperature where the fundamental power amplitude η_1 has a maximum.

23.5 Driving-induced symmetry breaking

As a final neat example we show that driving-induced localization can occur when both bias and tunneling coupling (TC) energy are harmonically modulated [509].

Consider the cooperative effect of monochromatic fields modulating both the bias of a TSS with zero static bias and the tunneling coupling,

$$\epsilon(t) = \epsilon_1 \sin(\omega t), \qquad \Delta(t) = \Delta \exp[\mu \sin(\nu t)]. \qquad (23.43)$$

If either the bias or the coupling energy is modulated, the left-right symmetry is dynamically broken. However, on average over a period, one finds asymptotically again equal occupation of both states.

When both parameters are modulated with commensurable frequencies,

$$\omega = m\,\Omega \quad \text{and} \quad \nu = n\,\Omega \quad \text{with} \quad m, n \text{ integer}, \qquad (23.44)$$

the TSS Hamiltonian has a discrete time translation symmetry. Interestingly, when n and m are odd, the left-right symmetry is broken even on average. The symmetry breaking is maximal when $n = m$, i.e., $\nu = \omega$. This case is pictorially sketched and

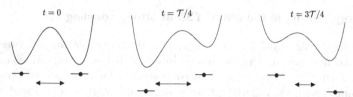

Figure 23.5: Bistable potential (top) and related TSS (bottom) for the case $\nu = \omega$. At time $t = 0$, the TSS is symmetric. At time $t = \mathcal{T}/4$, the left state is lower and the TC has the maximum value. At time $t = 3\mathcal{T}/4$, the right state is lower, but the TC has the minimum value. Thus, in the presence of damping, relaxation towards the left state is preferred.

qualitatively explained in Fig. 23.5. We expect from the illustrative presentation that at long time the occupation of the left state is preferred on average.

The averaged dynamics is again established by the expression (23.24) for $\hat{P}(\lambda)$ in which the kernels $k_0^{(s,a)}(\lambda)$ are calculated for time-periodic bias and tunneling coupling. For simplicity, we now restrict the attention to the averaged equilibrium state

$$P_\infty = \frac{k_0^{(a)}(0)}{k_0^{(s)}(0)} = \frac{\gamma^+ - \gamma^-}{\gamma^+ + \gamma^-} . \tag{23.45}$$

Here, $\gamma^\pm$ are the inclusive forward/backward tunneling rates. We start out from the time-dependent NIBA kernel

$$K_z^{(s,a)}(t, t - \tau) = \Delta(t)\Delta(t - \tau) e^{-Q'(\tau)} \begin{cases} \cos[Q''(\tau)] \cos\vartheta(t, t - \tau) \\ \sin[Q''(\tau)] \sin\vartheta(t, t - \tau) \end{cases} , \tag{23.46}$$

with time-dependent bias and tunneling coupling

$$e^{\pm i\vartheta(t, t - \tau)} = e^{\mp i(\epsilon_1/\omega) \cos(\omega t)} e^{\pm i(\epsilon_1/\omega) \cos[\omega(t - \tau)]} , \tag{23.47}$$

$$\Delta(t)\Delta(t - \tau) = e^{\mu \sin(\omega t)} e^{\mu \sin[\omega(t - \tau)]} , \tag{23.48}$$

The time-averaged Laplace transform of the kernel (23.46)

$$k_0^{(s,a)}(\lambda) = \frac{\omega}{2\pi} \int_0^{2\pi/\omega} dt \int_0^\infty d\tau \, e^{-\lambda\tau} K_z^{(s,a)}(t, t - \tau) . \tag{23.49}$$

can be calculated in analytic form by expanding the bias phase term (23.47) in a double series in J-Bessel functions, and the oscillatory tunneling coupling (23.48) in a double series in I-Bessel functions. Readily we obtain the inclusive rates $\gamma^\pm$ as

$$\gamma^\pm = \sum_{\ell,m,n=-\infty}^{\infty} I_\ell(\mu) I_m(\mu) J_n(\epsilon_1/\omega) J_{\ell+n-m}(\epsilon_1/\omega) \, k^\pm[(n-m)\omega] , \tag{23.50}$$

where $k^\pm(\epsilon)$ is the exclusive forward/backward rate for a static bias ϵ given in Eq. (20.28). The expression (23.50) is the generalization of Eq. (23.28) to an additional periodic modulation of the tunneling coupling. It reveals that the forward/backward symmetry is dynamically broken, $\gamma^+ \neq \gamma^-$, when both the driving amplitude of the bias ϵ_1 and the modulation amplitude of the tunneling coupling μ are nonzero.

23.6 Energy transfer in the driven TSS at strong coupling

In the externally driven TSS, there is perpetual energy exchange between system, reservoir and interaction. The convoluted interplay between the various competitive dissipative channels can be comprehensively studied for arbitrary spectral coupling of any strength with the tailored influence functional method introduced in Sec. 5.6.

23.6.1 Energy transferred to reservoir and interaction

Consider now the mean energy $\langle \mathcal{E}_{\kappa_{\mathrm{f}}}(t) \rangle_\kappa = \langle H_{\mathrm{R}}(t) + \kappa_{\mathrm{f}} H_{\mathrm{I}}(t) \rangle_\kappa$, where $\kappa_{\mathrm{f}} = 0,\, 1$, and where $\kappa = 0,\, 1$ refers to the unshifted and shifted thermal initial state (21.11). We put for simplicity the initial state $p_{\mathrm{i}}(\eta_{\mathrm{i}}) = \delta_{\eta_{\mathrm{i}},1}$. With the path relations $r(s) = \frac{1}{2} q_0\, \eta(s)$ and $y(s) = q_0\, \xi(s)$, where $\eta(s)$ and $\xi(s)$ are the kink-like spin paths (21.29), the functional integral (5.127) for $\langle \mathcal{E}_{\kappa_{\mathrm{f}}}(t) \rangle_\kappa$ reduces to a form which is similar to the spin path representation (21.37) for the propagating function $J(\eta, t; \eta_{\mathrm{i}}, 0)$ [246],

$$\langle \mathcal{E}_{\kappa_{\mathrm{f}}}(t) \rangle_\kappa = \sum_{m=1}^{\infty} \frac{(-1)^m}{2^m} \int_0^t \mathcal{D}_{2m}\{t_j\} \sum_{\{\xi_j = \pm 1\}} B_m \sum_{\{\eta_j = \pm 1\}''} \mathcal{F}_m(\kappa)\, \mathfrak{E}_m(\kappa_{\mathrm{f}}, \kappa) \,. \qquad (23.51)$$

Here, $B_m = e^{i\varphi_m}$ is the bias phase factor with the phase (23.2), $\mathcal{F}_m(\kappa) = G_m\, H_m(\kappa)$ is the influence function (21.31), and the double prime in $\{\eta_j = \pm 1\}''$ indicates summation over the intermediate sojourn states only. The function $\mathfrak{E}_m(\kappa_{\mathrm{f}}, \kappa)$ governs the energy transfer to the dissipative channels. After summation over the two final sojourn states, the energy functional (5.129) can be written as follows,

$$\mathfrak{E}_m(\kappa_{\mathrm{f}}, \kappa) = \sum_{k=0}^{m-1} [\mathcal{U}_k(t_{2m}) - (1 - \kappa_{\mathrm{f}})\mathcal{U}_k(t)]\, \eta_k + i \sum_{j=1}^{m} [\mathcal{V}_j(t_{2m}) - (1 - \kappa_{\mathrm{f}})\mathcal{V}_j(t)]\, \xi_j$$
$$- \kappa\, \hbar \left[\dot{Q}''(t_{2m}) - (1 - \kappa_{\mathrm{f}})\dot{Q}''(t) \right], \qquad (23.52)$$

with the correlation functions

$$\begin{aligned}
\mathcal{U}_k(s) &= \hbar[\dot{Q}''(s - t_{2k}) - \dot{Q}''(s - t_{2k+1})]\,, \\
\mathcal{V}_j(s) &= \hbar[\dot{Q}'(s - t_{2j}) - \dot{Q}'(s - t_{2j-1})]\,.
\end{aligned} \qquad (23.53)$$

The amount of energy transferred to the reservoir $\langle E_{\mathrm{R}}(t) \rangle_\kappa$ is given by the series expression (23.51) with the energy functional $\mathfrak{E}_{\mathrm{R},m}(\kappa) = \mathfrak{E}_m(0, \kappa)$. The amounts of energy transferred to reservoir-plus-interaction, $\langle E_{\mathrm{RI}}(t) \rangle_\kappa$, and to the interaction alone, $\langle E_{\mathrm{I}}(t) \rangle_\kappa$ are also given by the series (23.51), but with the energy functionals

$$\begin{aligned}
\mathfrak{E}_{\mathrm{RI},m}(\kappa) &= \sum_{k=0}^{m-1} \mathcal{U}_k(t_{2m})\eta_k + i \sum_{j=1}^{m} \mathcal{V}_j(t_{2m})\xi_j - \kappa\, \hbar\, \dot{Q}''(t_{2m})\,, \\
\mathfrak{E}_{\mathrm{I},m}(\kappa) &= \sum_{k=0}^{m-1} \mathcal{U}_k(t)\eta_k + i \sum_{j=1}^{m} \mathcal{V}_j(t)\xi_j - \kappa\, \hbar\, \dot{Q}''(t)\,.
\end{aligned} \qquad (23.54)$$

23.6.2 Energy balance relation

The function $\mathfrak{E}_{\mathrm{RI},m}(\kappa) = \mathfrak{E}_m(1,\kappa)$ does not depend on the end time t. The key point now is that it can be generated by differentiation of the influence function $\mathcal{F}_m(\kappa)$,

$$\mathfrak{E}_{\mathrm{RI},m}(\kappa)\,\mathcal{F}_m(\kappa) = i\,\hbar\,\xi_m \partial \mathcal{F}_m(\kappa)/\partial t_{2m}\,. \tag{23.55}$$

With this, the breathing integral of the final sojourn can be integrated by parts,

$$
\begin{aligned}
\int_{t_{2m-1}}^{t} dt_{2m}\,\mathfrak{E}_{\mathrm{RI},m}(\kappa) B_m \mathcal{F}_m(\kappa) &= \left. i\,\hbar\,\xi_m B_m \mathcal{F}_m(\kappa)\right|_{t_{2m}=t_{2m-1}}^{t_{2m}=t} \\
&\quad + \hbar \int_{t_{2m-1}}^{t} dt_{2m}\,\epsilon(t_{2m}) B_m \mathcal{F}_m(\kappa)\,.
\end{aligned}
\tag{23.56}
$$

There are three observations: (i) The lower boundary of the first term does not contribute because of the subsequent ξ-summation in the expression (23.51). (ii) The series (23.51) with the upper boundary term is the expection $\langle \sigma_x(t)\rangle_\kappa$, apart from a constant. (iii) The series (23.52) with the last term of Eq. (23.56) is essentially an integral over $\langle \dot\sigma_z(s)\rangle_\kappa$ weighted with the bias. More precisely, there results

$$\langle \mathcal{E}_{\mathrm{RI}}(t)\rangle_\kappa = \frac{\hbar\Delta}{2}\langle \sigma_x(t)\rangle_\kappa + \frac{\hbar}{2}\int_0^t ds\,\epsilon(s)\frac{\partial \langle \sigma_z(s)\rangle_\kappa}{\partial s}\,. \tag{23.57}$$

This can be written by virtue of integration by parts as the energy balance relation

$$\langle E_{\mathrm{S}}(t)\rangle_\kappa + \langle E_{\mathrm{R}}(t)\rangle_\kappa + \langle E_{\mathrm{I}}(t)\rangle_\kappa = \langle E_{\mathrm{exc}}(t)\rangle_\kappa\,, \tag{23.58}$$

where $\langle E_{\mathrm{S}}(t)\rangle_\kappa$ is the TSS energy, and $\langle E_{\mathrm{exc}}(t)\rangle_\kappa$ is the excess energy pumped into or out of the TSS,

$$
\begin{aligned}
\langle E_{\mathrm{S}}(t)\rangle_\kappa &= -\frac{\hbar\Delta}{2}\langle \sigma_x(t)\rangle_\kappa - \frac{\hbar\epsilon(t)}{2}\left[\langle \sigma_z(t)\rangle_\kappa - \langle \sigma_z(0)\rangle_\kappa\right], \\
\langle E_{\mathrm{exc}}(t)\rangle_\kappa &= -\frac{\hbar}{2}\int_0^t ds\,\dot\epsilon(s)\langle \sigma_z(s)\rangle_\kappa\,.
\end{aligned}
\tag{23.59}
$$

This verifies as an important consistency check that the exact formal expression (23.51) with (23.52) directly implies the energy balance relation (23.58).

23.6.3 Energy transfer in the driven TSS at strong coupling

The formal series expressions for the individual contributions of the balance equation (23.58) hold for general linear dissipation. Summation in analytic form is possible in particular limits. In the Ohmic strong coupling regime at the Toulouse point $K = \frac{1}{2}$, the path sums are feasible with techniques reported in Refs. [224, 501], and above in Sec. 22.6 and Subsec. 23.1.2. Following these lines $\langle \sigma_z(t)\rangle$ and $\langle \sigma_x(t)\rangle$ are found as

$$
\begin{aligned}
\langle \sigma_z(t)\rangle &= e^{-\gamma t} + \Delta^2 \int_0^t d\tau\, e^{-Q'(\tau)} e^{-\gamma\tau/2} \int_0^{t-\tau} ds\, e^{-\gamma s} \sin[\vartheta(t-s,t-s-\tau)]\,, \\
\langle \sigma_x(t)\rangle &= \Delta \int_{1/\omega_c}^t d\tau\, e^{-Q'(\tau)} e^{-\gamma\tau/2} \cos[\vartheta(t,t-\tau)]\,,
\end{aligned}
\tag{23.60}
$$

where $Q'(\tau) = \ln[(\hbar\beta\omega_c/\pi)\sinh(\pi\tau/\hbar\beta)]$, and $\gamma = \pi\Delta^2/2\omega_c$. With these expressions, the mean energy $\langle E_{RI}(t)\rangle$ can be found from the balance equation (23.57). The remaining energy contribution $\langle E_I(t)\rangle$ is also accessible in analytic form, and is

$$
\langle E_I(t)\rangle = \frac{\hbar}{2}\Delta^2 \int_0^t d\tau\, e^{-Q'(\tau)} e^{-\gamma\tau/2} \int_{1/\omega_c}^{t-\tau} ds\, e^{-\gamma s}
\tag{23.61}
$$
$$
\times \cos[\vartheta(t-s, t-s-\tau)]\left[\dot{Q}'(s) - \dot{Q}'(\tau-s)\right].
$$

For harmonic drive $\epsilon(t) = \epsilon_0 + \epsilon_1\cos(\omega t)$, the energy transfer contribution $\langle \mathcal{E}_j(t)\rangle$ in the stationary time regime, where $j = $ S, R, I, exc, and the corresponding time-averaged energies $\langle \overline{\mathcal{E}_j(t)}\rangle$, are

$$
\langle \mathcal{E}_j(t)\rangle = P_j(\omega)\,t + \sum_n E_{j,n}(\omega)\, e^{-in\omega t},
$$
$$
\langle \overline{\mathcal{E}_j(t)}\rangle = P_j(\omega)\,t + E_{j,0}(\omega).
\tag{23.62}
$$

The linearly growing term represents the energy draining into channel j at constant power P_j. There follows from Eq. (23.59) with (23.13) and (23.14)

$$
P_S = P_I = 0,
$$
$$
P_R(\omega) = P_{\text{exc}}(\omega) = \hbar\epsilon_1 \frac{\omega}{2}\Delta^2 \int_0^\infty d\tau\, e^{-Q'(\tau)} e^{-\gamma\tau/2} \cos(\epsilon_0\tau)
\tag{23.63}
$$
$$
\times J_1[(2\epsilon_1/\omega)\sin(\omega\tau/2)] \int_0^\infty ds\, e^{-\gamma s} \sin[\omega(\tau/2 + s)],
$$

where $J_1(z)$ is a Bessel function of the first kind. The term P_{exc} is the constant part of the power injected into the TSS. The findings show that the steadily growing amount of energy $P_{\text{exc}}t$ is fully transferred to the reservoir, as one would expect directly. The constant contributions $E_{I,0}$ and $E_{S,0}$ are found to depend logarithmically on ω_c,

$$
E_{I,0} = (\gamma/2\pi)\ln^2(\omega_c/\gamma),
$$
$$
E_{S,0} = -(\gamma/\pi)\ln(\omega_c/\gamma),
\tag{23.64}
$$

whereas $E_{\text{exc},0}$ is independent of the cutoff,

$$
E_{\text{exc},0}(\omega) = -\hbar\epsilon_1 \frac{\omega}{2}\Delta^2 \int_0^\infty d\tau\, \tau\, e^{-Q'(\tau)} e^{-\gamma\tau/2} \cos(\epsilon_0\tau)
\tag{23.65}
$$
$$
\times J_1[(2\epsilon_1/\omega)\sin(\omega\tau/2)] \int_0^\infty ds\, e^{-\gamma s} \sin[\omega(\tau/2 + s)].
$$

Finally, the balance equation (23.58) yields $E_{R,0}(\omega) = E_{\text{exc},0}(\omega) - E_{I,0} - E_{S,0}$. One directly sees from Eq. (23.62) with (23.63) that in the long run $\langle \mathcal{E}_R(t)\rangle$ dominates over $\langle \mathcal{E}_I(t)\rangle$. However at intermediate times, they are of comparable magnitude. For harmonic drive, the energy flow into the reservoir shows distinct quantum stochastic resonance characteristics in the quantities $P_R(\omega)$ and $E_{R,0}(\omega)$. Pronounced resonances occur also in the harmonics $E_{R,1}(\omega)$ and $E_{I,1}(\omega)$, and in harmonics of higher order at $\omega = \epsilon_0/n$, $n = 1, 2, \cdots,$. The absolute values of $E_{R,1}(\omega)$ and $E_{I,1}(\omega)$ are about the same, but the phases differ by π. Details are given in Ref. [246].

PART V:

THE DISSIPATIVE MULTI-STATE SYSTEM

A quantum Brownian particle in a multi-well potential coupled to a dissipative environment is archetypal for many problems in physics and chemistry. Examples include super-ionic conductors, atoms on surfaces, and interstitials in dielectrics and metals. Also the current-voltage characteristics of a Josephson junction, and charge transport in a quantum wire hindered by impurity scattering are described by this model. At high temperature, the particle moves forward or backward from well to well by incoherent tunneling or thermally activated transitions. As the temperature is lowered, coherent tunneling across many wells may become significant, and the competing different tunneling paths may interfere constructively or destructively. The model has a profound and powerful duality symmetry between the weak-binding and strong-binding representation. In view of the broad area of applications, the understanding of quantum transport in multi-well systems is a central issue.

463

24 Quantum Brownian particle in a washboard potential

24.1 Introduction

In the preceding part, I have considered the dynamics of a damped quantum system which is effectively restricted to a two-dimensional Hilbert space. Many of the concepts and approximation schemes developed in Part IV can be generalized for a dissipative system with N tight-binding sites. A system with three sites, $N = 3$, for instance, is of particular interest for the study of the ultrafast primary electron transfer in bacterial photosynthesis [510]. In many systems in chemical and biological physics, the transfer of a particle or a charge from a donor to an acceptor state occurs via a bridge of few or many intermediate tight-binding states. The RDM for a discrete N-site system can be visualized as a square lattice with $N \times N$ lattice sites. The path sum for the conditional propagating function (5.12) covers all paths on this lattice for given boundary sites. Each path consists of a sequence of segments in which the path dwells for some time on a lattice site and then undertakes a sudden flip to a neighboring site. A general path can be divided into a sequence of clusters: each cluster is a path section between two successive visits of a diagonal state. The sum of all clusters with the same final diagonal state for a given initial diagonal state constitute the associated kernel. A natural generalization of the noninteracting-blip approximation to the case of N states is the "noninteracting-cluster" approximation (NICA) developed in Ref. [511]. In the NICA, all intra-cluster correlations are taken into account while the inter-cluster and thus the inter-kernel correlations are disregarded. The infinite series resulting from the iteration of the kernels can be cast into master equations for the populations, as given below in Eq. (25.23).

Quantum Brownian motion (QBM) in a periodic potential is a key model for many transport phenomena in condensed matter [512]. Early work was mainly concerned with the significance of polaronic effects [134]–[136]. Recent interest has been focussed on the influence of frequency-independent damping. For instance, the electron-hole drag of charged particles in metals, as well as quasiparticle tunneling in Josephson junctions [513, 154] give rise to Ohmic dissipation (cf. Subsections 4.2.8 and 4.2.10). A duality symmetry between the weak- and strong-binding representations of the QBM model was put forward by Schmid [147]. The duality in the QBM model corresponds to the charge-phase duality in the Josephson junction model [154]. Transport of charge through impurities in quantum wires [514] and tunneling of edge currents through constrictions in fractional quantum Hall devices [515] are also described by the QBM model. The collective excitations of the correlated fermions away from the barrier manifest themselves as density fluctuations. In a theoretical description they are represented by a Luttinger harmonic liquid, and they corresponds to the thermal reservoir in the QBM model. Short-range electron interaction is equivalent to Ohmic damping in the related QBM model, whereas unscreened long-range Coulomb repulsion [516] corresponds to a sub-Ohmic reservoir coupling [517]. Many other physical and chemical systems involve transport of charge through barriers under Ohmic and super-Ohmic dissipation [86, 179].

24.2 Weak- and tight-binding representation

Consider a quantum Brownian particle moving in a tilted corrugated potential. The dynamics is described by the translational-invariant global model (3.12) with a bilinear coordinate coupling of the particle to a bath of harmonic oscillators,

$$
H_{\mathrm{WB}} = \frac{P^2}{2M} + V(X,t) + \sum_\alpha \left[\frac{p_\alpha^2}{2m_\alpha} + \frac{m_\alpha \omega_\alpha^2}{2} \left(x_\alpha - \frac{c_\alpha}{m_\alpha \omega_\alpha^2} X \right)^2 \right]. \tag{24.1}
$$

The label WB indicates the "weak binding" representation. We choose a trigonometric form for the corrugation with period X_0 and a global tilting force $F = 2\pi V_{\mathrm{tilt}}/X_0 = \hbar \epsilon_{\mathrm{WB}}/X_0$. To study the particle's response in thermal equilibrium, we assume that this force is acting only for $t > 0$. We then have [cf. Eq.(17.43)]

$$
V(X,t) = -V_0 \cos\left(2\pi X/X_0\right) - \Theta(t)FX. \tag{24.2}
$$

There are many different physical situations in which a discrete translational symmetry is assigned to an underlying periodic potential. Clearly, the model (24.1) with (24.2) provides an idealized description. Despite its simplicity, it involves substantial complexity and shows various interesting nontrivial behaviors. In real systems, there may arise complications due to perturbations of the periodic order by impurities or local strain fields. These may potentially be included by taking coupling and bias parameters as random or noisy.

In the tight-binding regime $k_{\mathrm{B}}T/\hbar\omega_0$, $FX_0/\hbar\omega_0 \ll 1 \ll V_0/\hbar\omega_0$, where $\hbar\omega_0$ is the frequency of small oscillations about the potential minima, only the lowest state in each well is occupied. Then the smooth potential system is effectively reduced to a single-band tight-binding lattice with lattice points $q_n = nq_0$, where q_0 is the lattice constant, and n is integer. The Hilbert space of the isolated system is spanned by the set of localized ground states $\{|n\rangle\}$ of the individual wells positioned at the lattice points. In TB representation, the position and space translation operators may be written as

$$
\begin{aligned}
\hat{q} &= \hat{q}_0 \sum_n n \, |n\rangle\langle n| = q_0 \sum_n n \, a_n^\dagger a_n, \\
\hat{T}(q) &\equiv e^{i q_0 \hat{p}/\hbar} = \sum_n |n\rangle\langle n+1| = \sum_n a_n^\dagger a_{n+1}.
\end{aligned} \tag{24.3}
$$

The respective second forms are expressed with operators $a_n^\dagger$ and a_n creating and annihilating a particle at site n, and obeying the commutator relation $[a_n, a_m^\dagger] = \delta_{m,n}$.

The localized states are weakly coupled by a transfer or tunneling amplitude Δ due to quantum mechanical overlap of nearest-neighbor localized states. The tight-binding (TB) system-plus-reservoir model with translational invariant coupling is

$$H_{\mathrm{TB}} = H_{\mathrm{S}} + H_{\mathrm{RI}} + H_{\mathrm{tilt}} ,$$

$$H_{\mathrm{S}} = -\frac{\hbar\Delta}{2} \sum_n \left(a_n a_n^\dagger + a_n^\dagger a_n\right) ,$$

$$H_{\mathrm{RI}} = \sum_\alpha \left[\frac{\pi_\alpha^2}{2M_\alpha} + \frac{M_\alpha \Omega_\alpha^2}{2}\left(u_\alpha - \frac{d_\alpha}{M_\alpha \Omega_\alpha^2}\hat{q}\right)^2\right] ,$$

$$H_{\mathrm{tilt}} = -\Theta(t)F\hat{q} .$$

(24.4)

It is convenient to express the tilting force F in terms of the potential drop $\hbar\epsilon$ between neighboring sites. For the above two models with site distance X_0 and q_0, respectively, we have

$$F = \hbar\,\epsilon_{\mathrm{WB}}/X_0 = \hbar\,\epsilon_{\mathrm{TB}}/q_0 .$$

(24.5)

As far as we are solely interested in properties of the WB and TB particle, the set of coupling constants $\{c_\alpha\}$ and $\{d_\alpha\}$, and the parameters of the reservoirs occur only in the respective spectral densities of the system-bath coupling. It is obvious to define the spectral densities of the coupling for the two models as in Eq. (3.25),

$$J_{\mathrm{WB}}(\omega) = \frac{\pi}{2}\sum_\alpha \frac{c_\alpha^2}{m_\alpha \omega_\alpha}\delta(\omega - \omega_\alpha) , \qquad J_{\mathrm{TB}}(\omega) = \frac{\pi}{2}\sum_\alpha \frac{d_\alpha^2}{M_\alpha \Omega_\alpha}\delta(\omega - \Omega_\alpha) .$$

(24.6)

At this point I would like to anticipate that the tight-binding and weak-binding models are related by a duality symmetry which becomes an exact self-duality in the Ohmic scaling limit. This property will open the possibility to perform explicit computations in the one model and then transfer the results obtained to the dual model upon using the duality tranformation. The discussion on such issue is given in Chapter 28. With this property in mind, the subsequent studies are mostly done for simplicity in the TB representation.

25 Multi-state dynamics

25.1 Quantum transport and quantum-statistical fluctuations

The reduced density matrix describes all properties pertaining to the Brownian particle. In the TB limit, the originally continuous coordinate q is discrete, $q = nq_0$ ($n = 0, \pm 1, \pm 2, \cdots$). Correspondingly, the density matrix is discrete and conveniently labelled with integer indices which number the wells. We now study the dynamics in TB representation for two different kinds of initial states.

25.1.1 Product initial state

The first kind is a product initial state (pis) in which the state of the system and the thermal state of the reservoir are in factorized form as in Eq. (5.9). Suppose that the particle was prepared to start out at time zero from the site $n = 0$ of the discrete lattice. The dynamical quantity of interest is then the probability $P_n(t)$ for finding the particle at site n at a later time $t > 0$. To formulate the evolution of $P_n(t)$, we use again the real-time influence functional method for a product initial state of the system-plus-reservoir complex discussed in Section 5.2. The populations $P_n(t)$ are the diagonal elements of the reduced density matrix. They can be written as a double path integral for the propagating function analogous to the expression (5.12),

$$P_n(t) = J(nn, t; 00, 0) = \int \mathcal{D}q(\cdot) \int \mathcal{D}q'(\cdot) \, \mathcal{A}[q(\cdot)] \mathcal{A}^*[q'(\cdot)] \mathcal{F}[q(\cdot), q'(\cdot)] . \quad (25.1)$$

Here, $q(\cdot)$, $q'(\cdot)$ are discontinuous paths propagating vertically and horizontally in steps of length q_0 in the (q, q')-plane [see Eq. (26.1)] with the boundary conditions

$$q(0) = q'(0) = 0 , \quad \text{and} \quad q(t) = q'(t) = n \, q_0 . \quad (25.2)$$

The functional $\mathcal{A}[q(\cdot)]$ is the probability amplitude that the particle propagates along the path $q(\cdot)$, and $\int \mathcal{D}q(\cdot) \int \mathcal{D}q'(\cdot)$ means summation of all paths on the TB lattice with boundary points (25.2). Again, all the influences of the coupling to the reservoir are contained in the influence functional $\mathcal{F}[q(\cdot), q'(\cdot)]$ which conveys self-interactions of the paths $q(\cdot)$ and $q'(\cdot)$ and interactions between the paths $q(\cdot)$ and $q'(\cdot)$. The expression (25.1) will be processed in Chap. 26.

25.1.2 Characteristic functions of moments and cumulants

The Fourier transform of the populations is the *characteristic function*

$$\mathcal{Z}(\rho, t) = \sum_{n=-\infty}^{\infty} e^{i \rho n q_0} P_n(t) = \langle e^{i \rho q(t)} \rangle = \sum_{m=0}^{\infty} \frac{(i \rho)^m}{m!} \langle q^m(t) \rangle . \quad (25.3)$$

The function $\mathcal{Z}(\rho, t)$ is the *moment generating function* (MGF), in which ρ is the counting field. The Nth derivative of the MGF at $\rho = 0$ gives the Nth moment

$$\langle q^N(t) \rangle \equiv q_0^N \sum_{n=-\infty}^{\infty} n^N P_n(t) = \left(-i \frac{\partial}{\partial \rho} \right)^N \mathcal{Z}(\rho, t) \Big|_{\rho=0}. \tag{25.4}$$

The reducible moments of $P_n(t)$ can be written in terms of the irreducible moments or cumulants $\langle q^N(t) \rangle_c$ as

$$\begin{aligned}
\langle q(t) \rangle &= \langle q(t) \rangle_c, \\
\langle q^2(t) \rangle &= \langle q(t) \rangle_c^2 + \langle q^2(t) \rangle_c, \\
\langle q^3(t) \rangle &= \langle q(t) \rangle_c^3 + 3 \langle q^2(t) \rangle_c \langle q(t) \rangle_c + \langle q^3(t) \rangle_c.
\end{aligned} \tag{25.5}$$

The cumulant generating function (CGF) $\ln \mathcal{Z}(\rho, t)$ yields the cumulant expansion

$$\ln \mathcal{Z}(\rho, t) = \sum_{m=1}^{\infty} \frac{(i\rho)^m}{m!} \langle q^m(t) \rangle_c. \tag{25.6}$$

Evidently, the Nth cumulant can be found from the CGF by differentiation,

$$\langle q^N(t) \rangle_c = \left(-i \frac{\partial}{\partial \rho} \right)^N \ln \mathcal{Z}(\rho, t) \Big|_{\rho=0}. \tag{25.7}$$

The cumulant generating function $\ln \mathcal{Z}(\rho, t)$ carries all statistical properties of the quantum transport process. The transport dynamics is referred to as diffusive when $\ln \mathcal{Z}(\rho, t)$, and hence all irreducible moments or cumulants, grow linearly with t at long time. This is the case when the dynamics comes about by incoherent tunneling events between the system's diagonal states (see Section 25.2).

The first cumulant gives the average current. From this one may deduce the mobility or conductance. The second cumulant yields the diffusion constant or, in connection with charge transport, the (nonequilibrium) dc current noise. Higher cumulants are zero when the statistics is Gaussian. The third cumulant, referred to as skewness, gives information about the leading asymmetric deviation from the Gaussian distribution. Experimentally, it can be discriminated from Gaussian noise by inverting the current. The fourth cumulant, called curtosis or sharpness, is a measure for the flatness of the distribution compared to the standard distribution. When the curtosis is positive, the distribution is sharp, and when it is negative, the distribution is flat.

25.1.3 Thermal initial state and correlation functions

The second kind of initial state is a thermal initial state: the system is prepared by letting the particle equilibrate with the reservoir, and then at time $t = 0$ a measurement of the observable of interest of the particle is performed. The measurement leads to a reduction of the canonical density operator according to $\hat{W}_0 = \hat{P} \hat{W}_\beta \hat{P}$, where $\hat{W}_\beta$ is the equilibrium density operator of the global system, and the operator

$\hat{P}$ projects onto the measured interval of the observable in question. Upon measuring the same observable again at a later time $t > 0$, we obtain information about the equilibrium autocorrelations of this observable. Here we restrict our attention to position correlation functions. These can be deduced from the generating function

$$\mathscr{Z}_{\text{th}}(\rho, \kappa, \mu; t) \equiv \langle e^{i\rho q(0)} e^{i\kappa q(t)} e^{i\mu q(0)} \rangle_{\beta} , \qquad (25.8)$$

where $\langle \cdots \rangle_{\beta}$ means thermal average of the states of the global system. We are mainly interested in the evolution of the average position $\langle q(t) \rangle_{\beta}$, the mean square displacement

$$D_{\text{th}}(t) \equiv \langle [\, q(t) - q(0) \,]^2 \rangle_{\beta} , \qquad (25.9)$$

and the antisymmetrized correlation function

$$A(t) \equiv \frac{1}{2i} \langle [\, q(t)q(0) - q(0)q(t) \,] \rangle_{\beta}|_{F=0} . \qquad (25.10)$$

These functions are directly found from $\mathscr{Z}_{\text{th}}$ upon differentiation,

$$\langle q(t) \rangle_{\beta} = -i \frac{\partial}{\partial \kappa} \mathscr{Z}_{\text{th}} \Big|_{\rho=\kappa=\mu=0} ,$$

$$D_{\text{th}}(t) = \left[-\frac{\partial^2}{\partial \kappa^2} - \frac{\partial}{\partial \rho} \frac{\partial}{\partial \mu} + \frac{\partial}{\partial \kappa} \frac{\partial}{\partial \mu} + \frac{\partial}{\partial \kappa} \frac{\partial}{\partial \rho} \right] \mathscr{Z}_{\text{th}} \Big|_{\rho=\kappa=\mu=0} , \qquad (25.11)$$

$$A(t) = \frac{i}{2} \left[\frac{\partial}{\partial \kappa} \frac{\partial}{\partial \mu} - \frac{\partial}{\partial \kappa} \frac{\partial}{\partial \rho} \right] \mathscr{Z}_{\text{th}} \Big|_{\rho=\kappa=\mu=0, F=0} .$$

The generating function $\mathscr{Z}_{\text{th}}$ can be expressed in terms of three-time conditional propagating functions. For an ergodic open system, these can be written as real-time path integrals which represent propagation since the infinite past (see Subsecs. 5.3.3 and 21.2.2). Suppose that the particle has been prepared at some negative time t_{p} in the diagonal state $(0,0)$. Then the particle will have relaxed at time zero to the thermal equilibrium state if we send t_p to the infinite past. Performing now at time $t = 0$ a measurement of the observable of interest, the desired thermal correlated initial state is prepared. We have

$$\mathscr{Z}_{\text{th}}(\rho, \kappa, \mu; t) = \lim_{t_{\text{p}} \to -\infty} \sum_{n,m,r} J(n\, n, t;\, m\, r, 0;\, 0\, 0, t_{\text{p}})\, e^{iq_0(\rho m + \kappa n + \mu r)} . \qquad (25.12)$$

The propagating function may be expressed as a double path integral of the form (25.1) in which the paths on the (q, q')-lattice are constrained as

$$\begin{aligned} q(t_{\text{p}}) &= 0 , & q(0) &= m\, q_0 , & q(t) &= n\, q_0 , \\ q'(t_{\text{p}}) &= 0 , & q'(0) &= r\, q_0 , & q'(t) &= n\, q_0 . \end{aligned} \qquad (25.13)$$

The effects of different initial preparation are now easily visualized. For the product initial state discussed in Subsec. 25.1.1, the dynamics of the particle at negative times is quenched. For the thermal initial state discussed here, the particle undergoes dynamics since the infinite past, and the effects of the correlations at time zero are represented by the interactions between the negative- and positive-time branches.

25.2 Poissonian quantum transport

25.2.1 Incoherent nearest-neighbor transitions (weak tunneling)

If the temperature is high and/or the bath coupling is strong, the particle tunnels incoherently between neighboring TB states.[1] This implies that the quantum particle occupies a diagonal state of the RDM after every second transition it makes. Hence the paths on the (q, q')-lattice are restricted to visits of sites of a tridiagonal matrix. Each path of a walker on the RDM consists of sequences of three elementary two-step (order Δ^2) processes from a diagonal to a diagonal state:

(1) The walker may step forward to the next diagonal state

(2) The walker may walk back to the preceding diagonal state

(3) The walker may go off-diagonal and then return to the same diagonal state

The elementary processes (1) and (2) have each two variants, whereas the third has four variants, as counted from the number of possible intermediate off-diagonal states. On the assumption that the above elementary processes are noninteracting and thus statistically independent, the nearest-neighbor tunneling transitions are incoherent. We denote the weights of the double jumps per unit time by w^+, w^-, and w, respectively. The weights are of order Δ^2. They are directly related to the non-adiabatic forward and backward tunneling rates in the two-state system discussed in Section 20.2. We have $w^+ = k^+$, $w^- = k^-$, and $w = -k$, where $k = k^+ + k^-$. As the walker finally reaches the diagonal state $(n\,n)$, there is an excess of n k^+-transitions over the k^--transitions. Thus the grand-canonical sum of the statistically independent elementary two-step processes is

$$P_n(t) = \sum_{j,\ell,m=0}^{\infty} \delta_{j,\,\ell+n} \frac{1}{j!\,\ell!\,m!} (k^+)^j (k^-)^\ell (-k)^m \, t^{j+\ell+m} \,, \qquad (25.14)$$

which can be summed upon using the detailed balance relation (20.34) to the form

$$P_n(t) = \exp(n\beta\hbar\epsilon/2 - kt) \, I_{|n|}\big[\, kt/\cosh(\beta\hbar\epsilon/2)\,\big] \,, \qquad (25.15)$$

where $I_n(z)$ is a modified Bessel function. A simple interpretation of this result may be obtained by observing that the nearest-neighbor hopping dynamics is alternatively described in terms of the master equation [440, 512]

$$\dot{P}_n(t) = k^+ P_{n-1}(t) + k^- P_{n+1}(t) - k P_n(t) \,. \qquad (25.16)$$

Upon using a recursion formula of the modified Bessel function, it is straightforward to see that the expression (25.15) solves the master equation.

Next, we discover that the Fourier series (25.3) with the populations (25.15) can be summed up to the generating function of the modified Bessel functions. The resulting expression for $\mathcal{Z}(\rho, t)$ is

---

[1]The precise conditions for this weak-tunneling regime are discussed (in connection with the dissipative two-state system) in Chapters 20 and 21, and in Sections 28.4 and 28.5.

$$\mathcal{Z}(\rho, t) = \exp\left[\cos(\rho q_0) - 1\right] k t + i \sin(\rho q_0)[k^+ - k^-] t\} . \tag{25.17}$$

The linear time-dependence in the exponent tells us that the dynamics is purely diffusive. With this form, we find from Eq. (25.7) the first and second cumulants as

$$\langle q(t)\rangle_c = q_0 \left[k^+(\epsilon) - k^-(\epsilon)\right] t , \qquad \langle q^2(t)\rangle_c = q_0^2 k(\epsilon) t . \tag{25.18}$$

Next, we employ the definining expressions for the nonlinear mobility $\mu(T, \epsilon)$ and the diffusion coefficient $D(T, \epsilon)$,

$$\mu = \frac{q_0}{\hbar\epsilon} \lim_{t\to\infty} \frac{\langle q(t)\rangle}{t} , \qquad D = \lim_{t\to\infty} \frac{\langle q^2(t)\rangle_c}{2t} . \tag{25.19}$$

With these definitions, we finally find from the forms (25.18) and the defining expressions (25.19) the nonlinear mobility and the diffusion coefficient as

$$\mu(\epsilon) = q_0^2 \left[k^+(\epsilon) - k^-(\epsilon)\right]/\hbar\epsilon ,$$
$$D(\epsilon) = q_0^2 \left[k^+(\epsilon) + k^-(\epsilon)\right]/2 . \tag{25.20}$$

If the forward/backward rates are related by detailed balance, Eq. (20.34), there holds

$$k^+(\epsilon) - k^-(\epsilon) = \tanh(\beta\hbar\epsilon/2)\left[k^+(\epsilon) + k^-(\epsilon)\right] . \tag{25.21}$$

one obtains with use of Eq. (25.20) the relation

$$D(T, \epsilon) = k_{\mathrm{B}}T \frac{\beta\hbar\epsilon/2}{\tanh(\beta\hbar\epsilon/2)} \mu(T, \epsilon) \tag{25.22}$$

The nonadiabatic forward/backward tunneling rates $k^\pm(T, \epsilon)$ in a TB double-well system have been discussed in Section 20.2. With these results we then get explicit expressions for $\mu(T, \epsilon)$ and $D(T, \epsilon)$. The resulting expression for the nonlinear mobility in the Ohmic scaling limit is given below in Subsection 27.1.

25.2.2 The general case (strong tunneling)

We now generalize the discussion to a TB model in which the transport of mass or charge between the sites $n = 0, \pm 1, \pm 2, \cdots$ takes place via direct forward and backward transitions by ℓ sites, $\ell = 1, 2, \cdots$. We denote the respective transitions weights ("rates") by $k_\ell^\pm$. Assuming statistically independent transitions, the dynamics of the population probability $P_n(t)$ of site n is governed by the master equation

$$\dot{P}_n(t) = \sum_{\ell=1}^{\infty} \left[k_\ell^+ P_{n-\ell}(t) + k_\ell^- P_{n+\ell}(t) - (k_\ell^+ + k_\ell^-)P_n(t)\right] . \tag{25.23}$$

The moment generating function defined in Eq. (25.3) is found from this equation as

$$\mathcal{Z}(\rho, t) = \prod_{n=1}^{\infty} \exp\left\{ t(e^{i\rho q_0 n} - 1)k_n^+ + t(e^{-i\rho q_0 n} - 1)k_n^- \right\} . \tag{25.24}$$

The cumulants are obtained from the cumulant generating function $\ln \mathcal{Z}(\rho, t)$ by differentiation. With use of Eq. (25.7) we get

$$\langle q^N(t) \rangle_c = q_0^N \sum_{n=1}^{\infty} n^N [k_n^+ + (-1)^N k_n^-] t . \qquad (25.25)$$

The characteristic function is conveniently written in terms of partial forward/backward currents $I_n^{\pm} = n k_n^{\pm}$ as

$$\mathcal{Z}(\rho, t) = \prod_{n=1}^{\infty} Z_n^+(\rho, t) \, Z_n^-(\rho, t) , \qquad (25.26)$$

where

$$Z_n^{\pm}(\rho, t) = \sum_{\ell=0}^{\infty} e^{\pm i \rho q_0 n \ell} \frac{(t \, I_n^{\pm}/n)^\ell}{\ell!} \, e^{-t I_n^{\pm}/n} = \exp[t(e^{\pm i \rho q_0 n} - 1) I_n^{\pm}/n] . \qquad (25.27)$$

The physical meaning of this expression is elucidated as follows. Suppose that the mass or charge transferred per unit time in forward direction were the results of a Poisson process for particles of unit mass propagating via uncorrelated nearest-neighbor forward transitions k_1^+ contributing a current $I_1^+ = k_1^+$, plus a Poisson process of uncorrelated forward moves via next-to-nearest-neighbor transitions contributing a current $I_2^+ = 2 k_2^+$, etc. Suppose also that the total backward current were the result of independent Poisson processes with partial backward currents $I_n^- = n k_n^-$, $n = 1, 2, \cdots$. The final form of the characteristic function would then be the expression (25.26) with (25.27).

The transport model (25.23) finds application, e.g., to charge transport through a weak link or through a quantum impurity in a 1D quantum wire, and to tunneling of edge currents in the fractional quantum Hall regime [cf. Chapter 31]. In these cases, the interpretation would be that the expression (25.26) with (25.27) describes independent Poisson processes for particles of charge one going across the impurity's barrier in forward/backward direction, contributing a current $I_1^{\pm}$, plus a Poisson process for particles of charge two (or for joint transport of a pair of particles of charge one) contributing a forward/backward current $I_2^{\pm}$, etc.

When the tunneling entities are coupled to a thermal reservoir, then the forward and backward transition weights are connected by detailed balance. Assuming that the potential drop per lattice period in forward direction is $\hbar \epsilon$, we have

$$k_n^-(\epsilon) = e^{-n \beta \hbar \epsilon} k_n^+(\epsilon) . \qquad (25.28)$$

The nonequilibrium relation (25.22) between diffusion and mobility does not hold when transfer rates $k_n^{\pm}$ with $n > 1$ contribute to the transport process, since they have different detailed balance factors, $e^{-n \beta \hbar \epsilon}$.

In classical transport, the transition weights $k_n^{\pm}$, $n = 1, 2, \cdots$ have the usual meaning of rates, since they are all positive. When the Poisson processes come about quantum mechanically, the sign of transition weights can be negative (see, e.g., explicit results in Chapter 29, in which the sign of $k_n^{\pm}$ alternates as a function of n). This does not spoil conservation of probability, since the master equation (25.23) provides $\sum_n \dot{P}_n(t) = 0$ regardles of the particular form chosen for the set $\{k_n^{\pm}\}$. Nevertheless, each population $P_n(t)$, $n = 1, 2, \cdots$, must be non-negative at any time.

Below we shall see that the quantum mechanical transition weights $k_n^{\pm}(\epsilon)$ can be given in analytic form for particular models in ample regions of the parameter space.

26 Exact formal expressions for current and current noise

In the tight-binding model (24.4), the quantum particle makes sudden transitions between neighboring states of the density matrix with a probability amplitude $\pm i\Delta/2$ per unit time. The sequence of states of the density matrix visited in succession can be visualized in terms of a paths of a walker on an infinite square lattice spanned by the sites of the discrete (q, q') double path. The walker starts on some site on the principal diagonal, say at $(0,0)$, and then randomly makes horizontal and vertical steps along the q- and q'-axis, respectively. At each site of the square lattice there are four possible directions to proceed further. Starting out from the initial diagonal state (n, n) and ending at the diagonal state (m, m) requires an even number of steps with the minimum number $2|n - m|$.

A TB path on the square lattice is conveniently parametrized in terms of charges $u_j = \pm 1$ and $v_j = \pm 1$. A path with k steps in $q(\tau)$ and ℓ steps in $q'(\tau)$ is then given by

$$q^{(k)}(\tau) \;=\; q_0 \sum_{j=1}^{k} u_j \Theta(\tau - t_j)\,, \qquad q'^{(\ell)}(\tau) \;=\; q_0 \sum_{i=1}^{\ell} v_i \Theta(\tau - t_i')\,. \tag{26.1}$$

A double path from the state $(0,0)$ to the state (n,n) is subject to the constraints

$$\sum_{j=1}^{k} u_j \;=\; n\,, \qquad \sum_{i=1}^{\ell} v_i \;=\; n\,, \tag{26.2}$$

where $k, \ell \geqslant n$.

The environmental coupling manifests itself in complex-valued interactions between the charges. The influence action is conveniently split into the self-interactions of the paths $q(\tau)$ and $q'(\tau)$ denoted by Φ_1 and Φ_1^*, and into the interaction between the paths $q(\tau)$ and $q'(\tau)$, denoted by Φ_2. Upon inserting the tight-binding paths (26.1) into Eq. (5.21) with (5.22), the influence function then takes the form

$$\mathcal{F}[q^{(k)}, q'^{(\ell)}] \;=\; \exp\left(\Phi_1[q^{(k)}] + \Phi_1^*[q'^{(\ell)}] + \Phi_2[q^{(k)}, q'^{(\ell)}]\right)\,, \tag{26.3}$$

with the interactions

$$\Phi_1[q^{(k)}] \;=\; \sum_{j=2}^{k} \sum_{i=1}^{j-1} u_j u_i Q(t_j - t_i)\,,$$

$$\Phi_1^*[q'^{(\ell)}] \;=\; \sum_{j=2}^{\ell} \sum_{i=1}^{j-1} v_j v_i Q^*(t_j' - t_i')\,, \tag{26.4}$$

$$\Phi_2[q^{(k)}, q'^{(\ell)}] \;=\; -\sum_{j=1}^{\ell} \sum_{i=1}^{k} v_j u_i Q(t_j' - t_i)\,,$$

where $Q(t)$ is the pair correlation function introduced in Sec. 18.2. For computational reasons it is again useful to introduce symmetric and antisymmetric flip or spin paths,

$$\eta(\tau) \equiv [\,q(\tau) + q'(\tau)\,]/q_0\,, \qquad \xi(\tau) = [\,q(\tau) - q'(\tau)\,]/q_0\,. \tag{26.5}$$

The path $\eta(\tau)$ describes propagation on the (q, q')-lattice in direction parallel to the principal diagonal, whereas the path $\xi(\tau)$ measures moves in perpendicular (off-diagonal) direction. A general path with $2m$ time-ordered steps on the square lattice, which is released at time $t_{\rm p}$ from the diagonal state $\xi = 0$, $\eta = 0$, is represented as

$$\eta^{(2m)}(\tau) = \sum_{k=1}^{2m} \eta_k\,\Theta(\tau - t_k)\,, \qquad \xi^{(2m)}(\tau) = \sum_{k=1}^{2m} \xi_k\,\Theta(\tau - t_k)\,, \tag{26.6}$$

where $t_k > t_{\rm p}$ for all k. The charges $\xi_k = \pm 1$ and $\eta_k = \pm 1$ label the four possibilities to move at time t_k instantaneously from a site to a neighboring site. Every individual u- or v-charge is associated with a $\{\eta, \xi\}$-pair. We have the correspondences

$$\begin{aligned} u_j &= \pm 1 &\longleftrightarrow&\quad \{\eta_j, \xi_j\} = \{\pm 1,\, \pm 1\}\,, \\ v_k &= \pm 1 &\longleftrightarrow&\quad \{\eta_k, \xi_k\} = \{\pm 1,\, \mp 1\}\,. \end{aligned} \tag{26.7}$$

In time-ordered (η, ξ)-representation, the influence function (26.3) for $2m$ steps reads under constraint (26.2)

$$\mathcal{F}^{(2m)} = G_{2m} H_{2m}\,, \qquad \text{where} \quad \begin{cases} G_{2m} = \exp\left(\displaystyle\sum_{j=2}^{2m}\sum_{k=1}^{j-1} \xi_j \xi_k Q'(t_j - t_k)\right)\,, \\[4mm] H_{2m} = \exp\left(-i\displaystyle\sum_{k=1}^{2m-1} \eta_k\,\chi_{k,2m}\right)\,, \end{cases} \tag{26.8}$$

with the influence phase

$$\chi_{k,2m} = -\sum_{j=k+1}^{2m} \xi_j Q''(t_j - t_k)\,. \tag{26.9}$$

The filter function G_{2m} is the charge representation of the noise action factor $e^{-\mathcal{S}^{(\rm N)}/\hbar}$ in Eq. (5.32). It suppresses far and long excursions away from the diagonal of the RDM. The phase factor H_{2m} corresponds to the friction action factor $e^{-i\mathcal{S}^{(\rm F)}/\hbar}$.

For nonzero tilting force at time $t > 0$ in the Hamiltonian (24.4), $F = \Theta(t)\hbar\epsilon/q_0$, the path (26.6) at positive time is weighted with a bias phase factor. For k transitions at negative times and $j = 2m - k$ transitions at positive times, the bias factor reads

$$B_{j,k} = \exp(i\,\varphi_{j,k}) \qquad \varphi_{j,k} = \epsilon \int_0^t d\tau\, \xi^{(k+j)}(\tau) = \epsilon \sum_{\ell=k+1}^{k+j} \xi_\ell(t - t_i)\,. \tag{26.10}$$

With these preliminaries, it is straightforward to write down the path sum for all dynamical quantities of interest.

One remark seems appropriate. We shall see that the (η, ξ)-representation is expedient in actual computations. On the other hand, the (q, q')-representation (26.3) is advantageous if one aims at derivation of fluctuation-dissipation theorems, e.g., detailed balance and the Einstein relation (see Subsection 26.3.3).

26.1 Product initial state

Consider now first the case of a product initial state of the global system. The path sum $\int \mathcal{D}q \int \mathcal{D}q'$ in Eq. (25.1) is expressed for the discrete system as follows. First, $P_n(t)$ is represented as an expansion in even numbers of steps or tunneling transitions. Secondly, in a given order of steps, say $2m$, we have to sum over all possible arrangements of labels $\{\xi_j = \pm 1\}$ and $\{\eta_j = \pm 1\}$ which fulfill the boundary conditions (25.2). This entails the constraints

$$\sum_{j=1}^{2m} \eta_j = 2n \,, \qquad \sum_{j=1}^{2m} \xi_j = 0 \,. \qquad (26.11)$$

Thirdly, we must take into account that the individual steps may occur at any time. To save space, we compactly write time-ordered integrations for ℓ steps in the negative time-branch and k steps in the positive-time branch as

$$\int_{t_p}^t \mathcal{D}_{k,\ell}\{t_j\} \times \cdots \equiv \int_0^t dt_{\ell+k} \cdots \int_0^{t_{\ell+2}} dt_{\ell+1} \int_{t_p}^0 dt_\ell \cdots \int_{t_p}^{t_2} dt_1 \, \Delta^{k+\ell} \times \cdots \,, \qquad (26.12)$$

where $t_p < 0$. We have included in the definition the product of the tunneling amplitudes. The various components are combined to yield for $P_n(t)$ the form

$$P_n(t) = \sum_{m=|n|}^\infty (-1)^{m-n} \frac{1}{2^{2m}} \int_0^t \mathcal{D}_{2m,0}\{t_j\} \sum_{\{\xi_j\}'} B_{2m,0} \, G_{2m} \sum_{\{\eta_j\}'} H_{2m} \,. \qquad (26.13)$$

The prime in $\{\xi_j\}'$ and $\{\chi_j\}'$ denotes summation under the constraints (26.11). With this form, the characteristic function (25.3) for a product initial state reads

$$\mathcal{Z}(\rho,t) = \sum_{m=0}^\infty (-1)^m \int_0^t \mathcal{D}_{2m,0}\{t_j\} \sum_{\{\xi_j\}'} B_{2m,0} \, G_{2m} F_{2m}(\rho) \,. \qquad (26.14)$$

The weighted average over the populations $P_n(t)$ is in the function

$$\begin{aligned}
F_{2m}(\rho) &= \frac{1}{2^{2m}} \sum_{n=-m}^{+m} (-1)^n e^{i\rho n q_0} \sum_{\{\eta_j\}'} H_{2m} \\
&= \sin\left(\frac{\rho q_0}{2}\right) \prod_{j=1}^{2m-1} \sin\left(\frac{\rho q_0}{2} - \chi_{j,2m}\right) \,.
\end{aligned} \qquad (26.15)$$

To obtain the second form, we have used the Fourier representation for the Kronecker delta which brings in the constraint $\sum_j \eta_j = 2n$, and then performed the sum over n and $\{\eta_j = \pm 1\}$. Observe that every individual path together with its charge-conjugate counter part gives a real contribution to the generating function.

It is now easy to find the expressions for the moments of the probability distribution. For the most interesting cases $N = 1$ and $N = 2$ we obtain using Eq. (25.4)

$$\langle q(t) \rangle = \frac{q_0}{2} \sum_{m=1}^{\infty} \int_0^t \mathcal{D}_{2m,0}\{t_j\} \sum_{\{\xi_j\}'} b_{2m} \, G_{2m} \sin \varphi_{2m} \,, \qquad (26.16)$$

$$\langle q^2(t) \rangle = \frac{q_0^2}{2} \sum_{m=1}^{\infty} \int_0^t \mathcal{D}_{2m,0}\{t_j\} \sum_{\{\xi_j\}'} c_{2m} \, G_{2m} \cos \varphi_{2m} \,, \qquad (26.17)$$

where

$$b_{2m} \equiv b_{2m,0} = (-1)^{m-1} \prod_{\ell=1}^{2m-1} \sin(\chi_{\ell,2m}) \,, \qquad (26.18)$$

$$c_{2m} \equiv c_{2m,0} = (-1)^{m-1} \sum_{k=1}^{2m-1} \cos(\chi_{k,2m}) \prod_{\ell=1,\ell \neq k}^{2m-1} \sin(\chi_{\ell,2m}) \,. \qquad (26.19)$$

Eqs. (26.13)–(26.19) are exact expressions adapted to a product initial state. The correlations resulting from the friction action are in the phase functions $\chi_{k,2m}$. The correlations due to the noise action are captured by the filter function G_{2m}.

We also give the corresponding Laplace transforms. These are conveniently expressed in terms of the intervals the system spends in a particular state, $\tau_\ell = t_{\ell+1} - t_\ell$. We introduce for the corresponding integrations the compact symbol

$$\int_0^\infty \hat{\mathcal{D}}_{k,0}(\lambda, \{\tau_\ell\}) \times \cdots \equiv \Delta^k \int_0^\infty \prod_{\ell=1}^{k-1} d\tau_\ell \, e^{-\lambda \tau_\ell} \times \cdots \,, \qquad \tau_\ell = t_{\ell+1} - t_\ell \,. \quad (26.20)$$

The Laplace transforms of $\langle q(t) \rangle$ and $\langle q^2(t) \rangle$ are then given by

$$\langle \hat{q}(\lambda) \rangle = \frac{1}{\lambda^2} \frac{q_0}{2} \sum_{m=1}^{\infty} \int_0^\infty \hat{\mathcal{D}}_{2m,0}(\lambda, \{\tau_\ell\}) \sum_{\{\xi_j\}'} b_{2m} \, G_{2m} \sin \varphi_{2m} \,, \qquad (26.21)$$

$$\langle \hat{q}^2(\lambda) \rangle = \frac{1}{\lambda^2} \frac{q_0^2}{2} \sum_{m=1}^{\infty} \int_0^\infty \hat{\mathcal{D}}_{2m,0}(\lambda, \{\tau_\ell\}) \sum_{\{\xi_j\}'} c_{2m} \, G_{2m} \cos \varphi_{2m} \,. \qquad (26.22)$$

The bias phase $\varphi_{2m,0}$ is conveniently expressed in terms of the cumulated charge

$$p_\ell = \sum_{k=1}^{\ell} \xi_k = - \sum_{k=\ell+1}^{2m} \xi_k \,. \qquad (26.23)$$

The second form holds because of the constraint (26.11). The cumulated charge p_ℓ measures how far the system is off-diagonal after ℓ transitions. When the systems dwells during the interval τ_{2k} in a diagonal state or sojourn, the associated cumulative charge p_{2k} is zero. Accordingly, the cumulative charge p_{2k} is zero, when the systems dwells in a diagonal state. In order Δ_{2m}, the sequence of $2m$ transitions can be divided into a product of kernels. Here, a kernel is the section between two successive sojourns. In the charge picture, a kernel is an irreducible neutral charge cluster with total charge zero, i,e., it cannot be divided into neutral subclusters. The meaning of

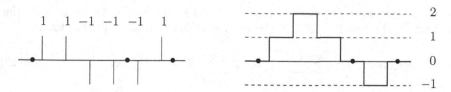

Figure 26.1: A sequence of six steps falling into two kernels or irreducible neutral charge clusters separated by a sojourn ($\bullet$) is sketched. In the charge picture (left), the spikes or charges $\{\xi_j = \pm 1\}$ give the direction of the moves perpendicular to the diagonal of the RDM. In the respective state representation (right), the cumulated charge p_j measures distance from the main diagonal after j steps.

the charges $\{\xi_j\}$ and the cumulated charges $\{p_j\}$ is illustrated in Fig. 26.1. With the cumulated charges $\{p_j\}$ and with $\sum_i \xi_i = 0$, the bias phase (26.10) can be written as

$$\varphi_{j,k} = \epsilon p_k(t_{k+1} - t) + \epsilon \sum_{\ell=k+1}^{k+j-1} p_\ell \tau_\ell, \qquad \varphi_{2m} \equiv \varphi_{2m,0} = \epsilon \sum_{\ell=1}^{2m-1} p_\ell \tau_\ell. \qquad (26.24)$$

The diffusive dynamics to be reached at long time can now be traced from the residua of the singularities at $\lambda = 0$ in the expressions (26.21) and (26.22).

26.2 Thermal initial state

To derive the dynamical expressions for a thermal initial state, we follow the discussion in Section 5.3.3 and Subsection 21.2.2. We then obtain the propagating function, which is a constituent of the expression (25.12), in the form

$$J(n\,n, t; m\,r, 0; 0\,0, t_{\rm p}) = \sum_{\substack{k=|m|+|r| \\ j=|n-m|+|n-r|}}^{\infty} (-1)^n \left(\frac{i}{2}\right)^{j+k}$$

$$\times \int_{t_{\rm p}}^{t} \mathcal{D}_{j,k}\{t_\ell\} \sum_{\{\xi_\ell\}''} B_{j,k} G_{j+k} \sum_{\{\eta_\ell\}''} H_{j+k}. \qquad (26.25)$$

The double prime in $\{\xi_\ell\}''$ and $\{\eta_\ell\}''$ is to indicate the constraints

$$\sum_{\ell=1}^{k} \eta_\ell = m + r, \qquad \sum_{\ell=k+1}^{k+j} \eta_\ell = 2n - (m + r),$$

$$\sum_{\ell=1}^{k} \xi_\ell = m - r, \qquad \sum_{\ell=k+1}^{k+j} \xi_\ell = r - m. \qquad (26.26)$$

The summations $\{\xi_\ell = \pm 1\}''$ and $\{\eta_\ell = \pm 1\}''$ create all possible paths on the (q, q')-lattice to go in k steps in the negative time interval $-|t_{\rm p}| < t' < 0$ from $(0, 0)$ to

(m, r) and in j steps at positive times $0 < t' < t$ from (m, r) to (n, n). Substituting the form (26.25) into Eq. (25.12), the generating function $\mathcal{Z}_{\text{th}}$ takes the form

$$\mathcal{Z}_{\text{th}}(\rho, \kappa, \nu; t) = \sum_{j=0}^{\infty} \sum_{k=0}^{\infty} \int_{-\infty}^{t} \mathcal{D}_{j,k}\{t_\ell\} \sum_{\{\xi_\ell\}'} B_{j,k} G_{j+k} F_{j,k}(\rho, \kappa, \nu), \tag{26.27}$$

where the prime in $\{\xi_\ell\}'$ denotes again summation over all charge sequences obeying the neutrality condition (26.11). The function $F_{j,k}(\rho, \kappa, \nu)$ is given by

$$F_{j,k}(\rho, \kappa, \nu) = \left(\frac{1}{2}\right)^{j+k} \sum_{n, m, r = -\infty}^{+\infty} (-1)^{(j+k-2n)/2} e^{i q_0 (n\kappa + m\rho + r\nu)}$$

$$\times \Theta(k - |m| - |r|)\, \theta(j - |n - m| - |n - r|) \sum_{\{\eta_\ell\}''} H_{j+k}. \tag{26.28}$$

Following the strategy which has led us to the second form in Eq. (26.15), it is again easy to perform the constrained sums. The resulting expression is

$$F_{j,k}(\rho, \kappa, \nu) = i^{j+k} \exp\left(i \frac{q_0}{2}(\rho - \nu)p_k\right) \sin\left(\frac{\kappa q_0}{2}\right) \tag{26.29}$$

$$\times \prod_{\ell=1}^{k} \sin\left(\frac{q_0}{2}(\kappa + \rho + \nu) - \chi_{\ell, j+k}\right) \prod_{m=k+1}^{j+k-1} \sin\left(\frac{q_0}{2}\kappa - \chi_{m, j+k}\right).$$

Finally, performing the differentiations of $\mathcal{Z}_{\text{th}}$ given in Eq. (25.11), we find for $t > 0$

$$\langle q(t) \rangle_{\text{th}} = \langle q(t) \rangle + R_1(t), \tag{26.30}$$

$$D_{\text{th},+}(t) = \langle q^2(t) \rangle + R_2(t). \tag{26.31}$$

The implications of the different initial conditions $\langle \cdots \rangle$ and $\langle \cdots \rangle_{\text{th}}$ are captured by the functions $R_1(t)$ and $R_2(t)$. These are formally given by the series expressions

$$R_1(t) = \frac{q_0}{2} \sum_{j=0}^{\infty} \sum_{k=0}^{\infty} \int_{-\infty}^{t} \mathcal{D}_{2j+1, 2k+1}\{t_\ell\} \sum_{\{\xi_\ell\}'} b_{2j+1, 2k+1} G_{2j+2k+2} \sin\varphi_{2j+1, 2k+1},$$

$$R_2(t) = \frac{q_0^2}{2} \sum_{j=0}^{\infty} \sum_{k=0}^{\infty} \int_{-\infty}^{t} \mathcal{D}_{2j+1, 2k+1}\{t_\ell\} \sum_{\{\xi_\ell\}'} c_{2j+1, 2k+1} G_{2j+2k+2} \cos\varphi_{2j+1, 2k+1}, \tag{26.32}$$

with the coefficients

$$b_{j,k} = (-1)^{(j+k)/2 - 1} \prod_{\ell=1}^{j+k-1} \sin(\chi_{\ell, j+k}),$$

$$c_{j,k} = (-1)^{(j+k)/2 - 1} \sum_{m=k+1}^{j+k-1} \cos(\chi_{m, j+k}) \prod_{\ell=1, \ell \neq m}^{j+k-1} \sin(\chi_{\ell, j+k}). \tag{26.33}$$

Further, the response function $\chi(t)$, which is related to the antisymmetric correlation function $A(t)$ by the relation (6.15), is found as

$$\chi(t) = \Theta(t) \frac{q_0^2}{2\hbar} \sum_{j=0}^{\infty} \sum_{k=0}^{\infty} \int_{-\infty}^{t} \mathcal{D}_{2j+1,2k+1}\{t_\ell\} \sum_{\{\xi_\ell\}'} \times p_{2k+1} b_{2j+1,2k+1} G_{2j+2k+2} \cos\varphi_{2j+1,2k+1} . \tag{26.34}$$

Finally, consider the Laplace transforms of $R_1(t)$ and $R_{2,+}(t) = \Theta(t)\,R_2(t)$. In generalization of the expression (26.20), we define

$$\int_0^{\infty} \hat{\mathcal{D}}_{j,k}(\lambda, \{\tau_\ell\}) \times \cdots \equiv \Delta^{j+k} \prod_{i=k+1}^{k+j-1} \int_0^{\infty} d\tau_i\, e^{-\lambda\tau_i} \prod_{\ell=1}^{k-1} \int_0^{\infty} d\tau_\ell\, e^{-0^+\tau_\ell} \times \cdots . \tag{26.35}$$

The additional two intervals not included in Eq. (26.35) are

$$r_k = -t_k ; \qquad s_k = t_{k+1} . \tag{26.36}$$

The interval r_k is the time passed between the last step in the negativ-time branch and time zero, and s_k is the interval between time zero and the first step at positive time. The Laplace transforms of the functions $R_1(t)$ and $R_{2,+}(t)$ are

$$\hat{R}_1(\lambda) = \frac{1}{\lambda}\frac{q_0}{2} \sum_{j=0}^{\infty} \sum_{k=0}^{\infty} \int_0^{\infty} \hat{\mathcal{D}}_{2j+1,2k+1}(\lambda, \{\tau_\ell\}) \times \int_0^{\infty} dr_k\, ds_k\, e^{-(\lambda s_k + 0^+ r_k)} \sum_{\{\xi_\ell\}'} b_{2j+1,2k+1}\, G_{2j+2k+2} \sin\varphi_{2j+1,2k+1} ,$$

$$\hat{R}_{2,+}(\lambda) = \frac{1}{\lambda}\frac{q_0^2}{2} \sum_{j=0}^{\infty} \sum_{k=0}^{\infty} \int_0^{\infty} \hat{\mathcal{D}}_{2j+1,2k+1}(\lambda, \{\tau_\ell\}) \times \int_0^{\infty} dr_k\, ds_k\, e^{-(\lambda s_k + 0^+ r_k)} \sum_{\{\xi_\ell\}'} c_{2j+1,2k+1}\, G_{2j+2k+2} \cos\varphi_{2j+1,2k+1} , \tag{26.37}$$

and the Laplace transform of the response function for zero bias is

$$\hat{\chi}(\lambda) = \frac{1}{\lambda}\frac{q_0^2}{2\hbar} \sum_{j=0}^{\infty} \sum_{k=0}^{\infty} \int_0^{\infty} \hat{\mathcal{D}}_{2j+1,2k+1}(\lambda, \{\tau_\ell\}) \times \int_0^{\infty} dr_k\, ds_k\, e^{-(\lambda s_k + 0^+ r_k)} \sum_{\{\xi_\ell\}'} p_{2k+1}\, b_{2j+1,2k+1}\, G_{2j+2k+2} . \tag{26.38}$$

The functions $R_1(t)$ and $R_{2,+}(t)$ describe the dynamical effects of the correlated initial state for $\langle q(t)\rangle_{\text{th}}$ and $D_{\text{th}}(t)$ as distinguished from the product initial state.

26.3 Mobility and Diffusion

26.3.1 Exact formal series expressions for transport coefficients

The mobility and the diffusion coefficient are found from the limits $\lim_{\lambda \to 0} \lambda^2 \langle \hat{q}(\lambda) \rangle$ and $\lim_{\lambda \to 0} \lambda^2 \langle \hat{q}^2(\lambda) \rangle$ in the expressions (26.21) and (26.22).

In the charge picture, an off-diagonal (blip) interval with cumulated charge p_k separates a sequence of charges with total charge zero into two sections with charge $\pm p_k$. When the blip length τ_k is large compared to the effective length of the sections, the blip interaction factor is $\exp\left[-p_k^2 Q'(\tau_k)\right]$. For the form (18.44) of the spectral density $G(\omega)$, the function $Q(\tau)$ behaves for nonzero T at asymptotic time as

$$Q'(\tau) \propto \tau^{2-s}, \qquad Q''(\tau) \propto \tau^{1-s}, \qquad \text{for} \qquad \tau \gg 1/\omega_c, \hbar\beta, \qquad (26.39)$$

as we see from Eq. (18.45). Thus, the interaction factor G_{2m} of a neutral cluster of $2m$ ξ-charges (cf. Fig. 26.1 for the notion of a neutral cluster) ties the cluster together so that the breathing integrals over the internal intervals τ_k, $k = 1, \cdots, 2m - 1$, are convergent in the regime $s < 2$ even if the limit $\lambda \to 0$ is performed. In contrast, the dipole-dipole interaction between two neutral clusters at large distance σ,

$$G_{\text{dipole}}(\sigma) = \exp\left[-(\textstyle\sum_{j,\text{left}} p_j \tau_j)(\sum_{i,\text{right}} p_i \tau_i) \ddot{Q}'(\sigma)\right], \qquad (26.40)$$

can not bind the clusters together in the limit $\lambda \to 0$. Instead, the leading interaction between two neutral clusters arises from the phase correlations between the intermediate sojourn and the subsequent cluster. These are contained in the factor b_{2m} given in Eq. (26.18). For two clusters consisting of $2k$ and $(2m - 2k)$ charges and an intermediate sojourn of length τ_{2k}, which is very large compared to the lengths of the two surrounding clusters, this factor is split up as follows,

$$b_{2m} = b_{2k} \otimes \{-\dot{Q}''(\tau_{2k})\} \otimes b_{2m-2k} \sum_{\ell=2k+1}^{2m-1} p_\ell \tau_\ell, \qquad \tau_{2k} \gg \Delta_{\text{r}}^{-1}, \qquad (26.41)$$

where Δ_{r}^{-1} is the effective length, apart from a numerical factor, of a cluster at low T. The generalisation to a sequence of three or more clusters is clear. As a result, the moment $\langle \hat{q}(\lambda) \rangle$ emerges as a sequence of clusters. Observing that the phase factor b_{2k} is odd under charge conjugation, the initial cluster is antisymmetric in the bias, while all subsequent clusters are even in the bias. The associated kernels are

$$\hat{k}_{\text{ini}}(\lambda) = \sum_{m=1}^{\infty} \int_0^{\infty} \hat{\mathcal{D}}_{2m,0}(\lambda, \{\tau_\ell\}) \frac{1}{2} \sum_{\{\xi_j\}_{\text{cl}}} b_{2m} G_{2m} \sin \varphi_{2m},$$

$$\hat{k}(\lambda) = \sum_{m=1}^{\infty} \int_0^{\infty} \hat{\mathcal{D}}_{2m,0}(\lambda, \{\tau_\ell\}) \sum_{\{\xi_j\}_{\text{cl}}} b_{2m} G_{2m} \cos \varphi_{2m} \times (\textstyle\sum_j p_j \tau_j). \qquad (26.42)$$

The kernel $\hat{k}_{\text{ini}}(\lambda)$ has dimension frequency, while the kernel $\hat{k}(\lambda)$ is dimensionless. The symbol $\{\xi_j\}_{\text{cl}}$ denotes the constraint (26.11) with the addition that neutral subclusters are excluded. The breathing integral of the sojourn between the clusters is

$$\hat{f}(\lambda) = \int_{\Delta_{\mathrm r}^{-1}}^{\infty} d\sigma \, \dot{Q}''(\sigma) \, e^{-\lambda \sigma} \,, \tag{26.43}$$

The entirety of the kernel sequences is a geometrical series which is summed to

$$\langle \hat{q}(\lambda) \rangle = \frac{q_0}{\lambda^2} \frac{\hat{k}_{\mathrm{ini}}(\lambda)}{1 + \hat{f}(\lambda)\,\hat{k}(\lambda)} \,. \tag{26.44}$$

The regime $\lambda \ll \Delta_{\mathrm r}$ determines the current at asymptotic time. In the linear response regime, there holds $\hat{k}_{\mathrm{ini}}(\lambda) = (\epsilon/2)\,\hat{k}(\lambda)|_{\epsilon=0}$. Thus we obtain from Eq. (26.44)

$$\langle \hat{q}(\lambda) \rangle_{\mathrm{lin}} = \frac{\epsilon\, q_0}{2} \frac{1}{\lambda^2\, \hat{f}(\lambda, s)} \frac{\hat{f}(\lambda, s)\,\hat{k}(\lambda)|_{\epsilon=0}}{1 + \hat{f}(\lambda, s)\,\hat{k}(\lambda)|_{\epsilon=0}} \,. \tag{26.45}$$

With the expression $\dot{Q}''(\sigma) = 2\delta_s \Gamma(s)\, \omega_{\mathrm{ph}}^{1-s}\, \mathrm{Im}\,(i\,\sigma + 1/\omega_{\mathrm c})^{-s}$ resulting from Eq. (18.45), the leading terms of $\hat{f}(\lambda)$ in the regime $s < 2$ and $\omega_{\mathrm c} \to \infty$ are

$$\hat{f}(\lambda, s) = \frac{\pi\,\delta_s}{\sin(\frac{1}{2}\pi s)} \left(\frac{\omega_{\mathrm{ph}}}{\Delta} \right)^{1-s} \left[\left(\frac{\lambda}{\Delta} \right)^{s-1} - \frac{1}{\Gamma(2-s)} + \frac{1-s}{\Gamma(3-s)} \frac{\lambda}{\Delta} \right] \,. \tag{26.46}$$

In the Ohmic limit $s \to 1$, this expression drops to zero. This indicates that for Ohmic spectral coupling ($s = 1$) the leading phase correlation of a sojourn with the subsequent kernel is of order $1/\omega_{\mathrm c}$. More specifically, there holds

$$\hat{f}(\lambda, 1) = 2K\,[\,\Delta + (C_{\mathrm E} - 1)\lambda + \lambda \log(\lambda/\Delta)\,]/\omega_{\mathrm c} \,. \tag{26.47}$$

With these findings, consider now the various sectral coupling regimes.

26.3.2 The sub- and super-Ohmic regimes

Sub-Ohmic regime

In the sub-Ohmic regime $s < 1$ at nonzero T, the charge interaction factor G_{2m} is determined by Poissonian multiphonon exchange. This ensures with the asymptotic form (26.39) that large intracluster lengths are exponentially suppressed at $\omega_{\mathrm{ph}}\tau_k \gg 1$,

$$G_{2m} = \prod_{k=1}^{2m-1} \exp\left(-p_k^2 \frac{2\pi\delta_s}{\sin(\frac{1}{2}\pi s)\Gamma(3-s)\,\hbar\beta\omega_{\mathrm{ph}}} (\omega_{\mathrm{ph}}\tau_k)^{2-s} \right) \,. \tag{26.48}$$

Hence all the τ-integrals in Eq. (26.42) remain convergent in the limit $\lambda \to 0$, and thus the kernel $\hat{k}(\lambda, s)$ is regular in this limit. In addition, the kernel $\hat{k}(\lambda = 0, s)|_{\epsilon=0}$ is nonzero at finite T. On the other hand, the function $\hat{f}(\lambda, s)$ diverges in the limit $\lambda \to 0$ due to the first term in the square bracket in Eq. (26.46). As a result, the expression (26.45) can be expanded in inverse powers of the kernel $\hat{k}(0, s)$,

$$\langle \hat{q}(\lambda) \rangle_{\mathrm{lin}} = \frac{\epsilon\, q_0}{2} \frac{1}{\lambda^2\, \hat{f}(\lambda, s)} \left(1 - \frac{1}{\hat{f}(\lambda, s)\,\hat{k}(0, s)|_{\epsilon=0}} + \mathcal{O}(1/\hat{f}^2(\lambda, s)) \right) \,. \tag{26.49}$$

One then finds upon inverse Laplace transformation in the asymptotic time regime

$$\langle q(t)\rangle_{\text{lin}} = q_0 \frac{\sin(\frac{1}{2}\pi s)}{2\pi\delta_s} \frac{\epsilon}{\omega_{\text{ph}}} \frac{(\omega_{\text{ph}}t)^s}{\Gamma(1+s)} \left\{ 1 - \frac{s\Gamma(s)/\Gamma(2s)}{\pi\delta_s \, \hat{k}(0,s)_{|_{\epsilon=0}}} (\omega_{\text{ph}}t)^{s-1} + \cdots \right\} . \quad (26.50)$$

This result is quite remarkable. First, it is nonperturbative with regard to the kernel. Second, and most importantly, $\langle q(t)\rangle_{\text{lin}}$ loses asymptotically dependence on temperature and on the TB lattice, and actually exhibits subdiffusive free Brownian motion at asymptotic time. Writing the linear-response relation (7.24) in the form

$$\chi(t) = (q_0/\hbar\epsilon) \langle \dot{q}(t)\rangle_{\text{lin}} , \quad (26.51)$$

the response function at asymptotic time is found from Eq. (26.50) to read

$$\chi(t) = \frac{\sin(\frac{1}{2}\pi s)}{\Gamma(s)} \frac{q_0^2}{2\pi\hbar\,\delta_s} (\omega_{\text{ph}}t)^{s-1} = \frac{\sin(\frac{1}{2}\pi s)}{\Gamma(s)} \frac{1}{M\gamma_s} (\omega_{\text{ph}}t)^{s-1} . \quad (26.52)$$

This form coincides with that of the sub-Ohmic free Brownian particle, Eq. (7.55).

The direct calculation of the second moment for asymptotic time is more difficult because of the $\cos(\chi_{k,2m})$-term in Eq. (26.19). Fortunately, this can be circumvented by relying (i) on the fact that for ergodic systems $\lim_{t\to\infty}\langle q^2(t)\rangle_{\text{pis}}/D_{\text{th}}(t) = 1$, and (ii) on the version (7.21) of the Einstein relation. We then get with (26.52)

$$\langle q^2(t)\rangle_{\epsilon=0} = \frac{2}{\beta} \int_0^t dt' \, \chi(t') = \frac{\hbar\sin(\frac{1}{2}\pi s)}{M\gamma_s} \frac{2}{\beta\hbar\omega_{\text{ph}}} \frac{(\omega_{\text{ph}}t)^s}{\Gamma(1+s)} . \quad (26.53)$$

With the relation $\hat{\mu}(\lambda) = (q_0/\hbar\epsilon)\lambda^2\langle\hat{q}(\lambda)\rangle_{\text{lin}}$, the linear ac mobility $\tilde{\mu}(\omega) = \hat{\mu}(-i\,\omega)$ at low frequency is found from Eq. (26.49),

$$\tilde{\mu}_{\text{lin}}(\omega, s) = \frac{\sin(\frac{1}{2}\pi s)}{M\gamma_s} \left(\frac{-i\,\omega}{\omega_{\text{ph}}}\right)^{1-s} \left\{ 1 - \frac{\sin(\frac{1}{2}\pi s)}{\pi\delta_s\,\hat{k}(0,s)} \left(\frac{-i\,\omega}{\omega_{\text{ph}}}\right)^{1-s} + \cdots \right\} . \quad (26.54)$$

The leading term is the spectral mobility of a free sub-Ohmic particle, Eq. (7.59).

For $s < 1$ and $T = 0$, the kernel $\hat{k}(\lambda = 0, s)$ vanishes in the limit $\epsilon \to 0$ with an essential singularity, which in order Δ^2 is $\propto \exp[-a(s)\,\delta_s^{1/s}\,(\omega_{\text{ph}}/\epsilon)^{(1-s)/s}]$, where $a(s) > 0$ is a numerical coefficient. This follows from an analysis analogous to that given in Subsec. 20.2.3 [cf. Eq. (20.60)]. As a result, the mobility $\tilde{\mu}_{\text{TB}}(\omega, s)$ drops to zero faster than any power of ω as $\omega \to 0$, and the particle becomes strictly localized.

Super-Ohmic regime

In the super-Ohmic regime $1 < s < 2$ and nonzero T, large intra-cluster lengths are exponentially suppressed by the interaction factor G_{2m} given in Eq. (26.48). As a result, the kernels $\hat{k}_{\text{ini}}(\lambda, s)$ and $\hat{k}(\lambda, s)$ remain regular in the limit $\lambda \to 0$. As a function of the bias, the kernels exhibit a maximum at intermediate bias values due to multiphon-exchange processes. As to the function $\hat{f}(\lambda, s)$, the second term in Eq. (26.46), which is independent of λ, is the leading one in the limit $\lambda \to 0$. Thus the particle moves ahead along the TB lattice with constant mean velocity at asymptotic time,

$$\langle \hat{q}(\lambda) \rangle = \langle v \rangle / \lambda^2 \,, \qquad\qquad \langle q(t) \rangle = \langle v \rangle t \,. \qquad (26.55)$$

In the most general case, the mean velocity is found from Eq. (26.44),

$$\langle v \rangle = q_0 \frac{\hat{k}_{\text{ini}}(0, s)}{1 + \hat{f}(0, s)\, \hat{k}(0, s)} \,. \qquad (26.56)$$

The Ohmic case

For Ohmic spectral coupling, $s = 1$, the kernels $\hat{k}_{\text{ini}}(\lambda, s)$ and $\hat{k}(\lambda, s)$ are regular in the limit $\lambda \to 0$. In addition, the sojourn-cluster phase correlation function $\hat{f}(\lambda)$ becomes constant in this limit, $\hat{f}(0, 1) = 2K\Delta/\omega_c$, as follows from Eq. (26.47).

With this, we obtain from Eq. (26.44) at asymptotic time diffusive behavior,

$$\langle q(t) \rangle = q_0\, \hat{k}_{\text{ini}}(0, 1)\,[\,1 - 2K(\Delta/\omega_c)\hat{k}(0, 1) + \cdots\,]\,t \,. \qquad (26.57)$$

In the Ohmic scaling limit, the expression (26.47) yields $\lim_{\omega_c \to \infty} \hat{f}(\lambda, 1) = 0$. Thus, the cluster-sojourn-cluster contribution drops to zero, so that the current in the Ohmic scaling limit is exactly determined by the single-cluster contribution,

$$\langle \hat{q}(\lambda) \rangle = q_0\, \hat{k}_{\text{ini}}(\lambda, 1)/\lambda^2 \,, \qquad \text{Ohmic scaling limit}\,. \qquad (26.58)$$

With the definition (25.19) for the nonlinear mobility μ, one finds upon using Eq. (26.42) the exact formal expression

$$\mu(\epsilon, T) = \lim_{\lambda \to 0^+} \frac{q_0^2}{\hbar\epsilon} \frac{1}{2} \sum_{m=1}^{\infty} \int_0^{\infty} \hat{\mathcal{D}}_{2m,0}(\lambda, \{\tau_\ell\}) \sum_{\{\xi_\ell\}'} b_{2m} G_{2m} \sin\left(\epsilon \textstyle\sum_j p_j \tau_j\right) . \qquad (26.59)$$

The diffusion coefficient is defined by the limiting expression

$$D(T, \epsilon) = \lim_{t \to \infty} \frac{1}{2t}[\,\langle \hat{q}^2(t) \rangle - \langle \hat{q}(t) \rangle^2\,] \,, \qquad (26.60)$$

This is written in Laplace space with the expressions (26.22) and (26.58) as

$$D(T, \epsilon) = \frac{q_0^2}{2} \lim_{\lambda \to 0^+} \left(\frac{1}{2} \sum_{m=1}^{\infty} \int_0^{\infty} \hat{\mathcal{D}}_{2m,0}(\lambda, \{\tau_\ell\}) \sum_{\{\xi_j\}'} c_{2m}\, G_{2m} \cos\left(\epsilon \textstyle\sum_j p_j \tau_j\right) \right.$$
$$\left. - \frac{2}{\lambda}\, \hat{k}_{\text{ini}}^2(0, 1) \right) \,. \qquad (26.61)$$

With the scaling form (18.53) of the bath correlation function $Q(t)$, the influence phase (26.9) expressed with the cumulative charge p_ℓ, Eq. (26.23), takes the form $\chi_{\ell,2m} = \pi K p_\ell$. With this, the phase coefficients in Eqs. (26.59) and (26.61) become

$$b_{2m} \equiv (-1)^{m-1} \prod_{\ell=1}^{2m-1} \sin(\pi K p_\ell) , \qquad (26.62)$$

$$c_{2m} \equiv (-1)^{m-1} \sum_{k=1}^{2m-1} \cos(\pi K p_k) \prod_{\ell=1,\ell \neq k}^{2m-1} \sin(\pi K p_\ell) . \qquad (26.63)$$

An intermediate sojourn, say $\ell = 2j$, has cumulative charge $p_{2j} = 0$. Thus, the coefficient b_{2m} ensures that paths with an interim visit of a sojourn do not contribute to the mobility. In contrast, path with a single visit of the same sojourn state contribute to the diffusion coefficient because of the $\cos(\pi K p_{2j})$-factor in c_{2m}. The corresponding charge configuration consists of two neutral clusters. The breathing integral of the intermediate sojourn diverges in leading order as $1/\lambda$, since the dipole-dipole interaction (26.40) can not bind the clusters.[1] This divergent contribution in the first term of the round bracket in Eq. (26.61) and the second divergent term in the round bracket cancel each other out exactly. As a result, the expression (26.61) is regular in the limit $\lambda \to 0$.

The exact formal expressions (26.59) and (26.61) holding in the Ohmic scaling limit will be analyzed in subsequent sections.

26.3.3 Einstein relation

The Einstein relation links the diffusion coefficient for $\epsilon = 0$ to the linear mobility,

$$D(T,0) = k_{\mathrm{B}}T \mu_{\mathrm{lin}}(T) . \qquad (26.64)$$

As emphasized in the preceeding subsection, the static transport coefficients D and μ_{lin} in this relation are well-defined in the parameter regime $1 \leqslant s < 2$. We have seen already in Section 7.3 for the case of free Brownian motion that the relation (26.64) is a special version of the fluctuation-dissipation theorem. The relation was proven rigorously for Brownian motion in a corrugated potential in the classical regime [518]. There have been raised concerns about the general validity of the Einstein relation in the quantum regime [222]. The questions are related to possible problems with the definition of μ_{lin} since the stationary state in which the particle moves with constant velocity is not normalizable and therefore the standard Green-Kubo method might not be applicable. On the other hand, if one defines the mobility as in Eq. (25.19), it is not obvious a priori whether the Einstein relation is valid in all orders in Δ^2.

To prove the relation (26.64) for the TB model [374], we rewrite μ_{lin} and D as moments of the distribution function, and return to the parametrization (26.1),

$$\mu_{\mathrm{lin}} = q_0^2 \sum_n n X_n , \qquad D = \tfrac{1}{2} q_0^2 \sum_n n^2 Y_n , \qquad (26.65)$$

$$X_n = \sum_{k,\ell=|n|}^{\infty} i^{\ell-k} \left(\frac{\Delta}{2}\right)^{\ell+k} \int_0^\infty \prod_{j=1}^{k-1} d\rho_j \prod_{i=1}^{\ell-1} d\rho_i' \int_{-\infty}^{\infty} d\tau \sum_{\{u_k,v_\ell\}'} \frac{i\varphi_{k,\ell}}{\hbar\epsilon} \mathcal{F}[q^{(k)}, q'^{(\ell)}] ,$$

---

[1] For details, see also the expression (27.58) and subsequent text on page 496.

$$Y_n = \sum_{k,\ell=|n|}^{\infty} i^{\ell-k} \left(\frac{\Delta}{2}\right)^{\ell+k} \int_0^{\infty} \prod_{j=1}^{k-1} d\rho_j \prod_{i=1}^{\ell-1} d\rho_i' \int_{-\infty}^{\infty} d\tau \sum_{\{u_k, v_\ell\}'} \mathcal{F}[q^{(k)}, q'^{(\ell)}],$$

where $\rho_j = t_{j+1} - t_j$ and $\rho_i' = t_{i+1}' - t_i'$ are the intervals in the paths $q(\tau)$ and $q'(\tau)$, and $\tau = t_1' - t_k$. The bias phase is $\varphi_{k,\ell} = \epsilon \left(\sum_i v_i t_i' - \sum_j u_j t_j \right)$. The ρ- and ρ'-integrals preserve charge ordering in the individual sets $\{u_k\}$ and $\{v_\ell\}$, respectively. The unbounded τ-integral introduces all possibilities of mixing up the set of charges $\{u_k\}$ with the set of charges $\{v_\ell\}$.

To proceed, we assign to the double path $\{q^{(k)}(\tau), q'^{(\ell)}(\tau)\}$ a conjugate double path $\{q^{(\ell)}(\tau), q'^{(k)}(\tau)\}$ in which the charges are in reverse order. Since the function $Q(z)$ is analytic in the strip $0 \geqslant \text{Im}\, z > -\hbar\beta$, the τ-integration contour can be shifted parallel in this strip, as explained below Eq. (20.29).[2] If we perform a translation of the contour by $-i\hbar\beta$, $\tilde{\tau} = \tau - i\hbar\beta$. we may use the reflection symmetry (18.43) of $Q(z)$. As a result, the influence function $\tilde{\mathcal{F}}$ calculated for a conjugate pair of double paths on the shifted contour $\tilde{\tau}$ has the property

$$\tilde{\mathcal{F}}[q^{(k)}, q'^{(\ell)}] + \tilde{\mathcal{F}}[q^{(\ell)}, q'^{(k)}] = \mathcal{F}^*[q^{(k)}, q'^{(\ell)}] + \mathcal{F}^*[q^{(\ell)}, q'^{(k)}], \qquad (26.66)$$

and the bias phase for the shifted contour is related to that for the original one by

$$\tilde{\varphi}_{k,\ell} = \varphi_{k,\ell} - i\beta\hbar\epsilon n. \qquad (26.67)$$

Writing X_n for the shifted contour, then using the relations (26.66) and (26.67), and comparing the resulting form with Y_n, one finds in the regime $1 \leqslant s < 2$ [374]

$$X_n = -X_n + n\beta Y_n. \qquad (26.68)$$

This relation is valid in all orders in Δ^2. Substituting Eq. (26.68) into the expressions (26.65), we find the Einstein relation (26.64) (see also the discussion in Ref. [519]).

Since a conjugate pair of double paths visit the same sites, the relation (26.68) and hence the Einstein relation holds also in the presence of disorder.

---

[2] In the absence of a bias, there must hold $\lim_{|z|\to\infty} \exp[-Q(z)] = 0$. This gives the bound $s < 2$.

27 The Ohmic case

In the scaling limit, it is expedient to scale the mobility with that of a free Brownian partice, $\mu_0 = 1/\eta = q_0^2/2\pi\hbar K$ [cf. Eq. (7.29)]. The normalized mobility is defined as

$$\mathcal{M}(T,\epsilon) \equiv \mu(T,\epsilon)/\mu_0 . \tag{27.1}$$

Using the scaling form (18.53) for $Q(\tau)$ and the expression (26.59), the normalized mobility in the TB model can be written

$$\mathcal{M}_{\mathrm{TB}}(T,\epsilon,K) = \frac{\pi K}{\epsilon} \sum_{m=1}^{\infty} (-1)^{m-1} \Delta^{2m} \sum_{\{\xi_j = \pm 1\}'} \int_0^\infty d\tau_1 \, d\tau_2 \cdots d\tau_{2m-1} \tag{27.2}$$

$$\times \sin\left(\sum_{j=1}^{2m-1} \epsilon \, p_j \tau_j \right) \prod_{k=1}^{2m-1} \sin(\pi K p_k) \prod_{j>k=1}^{2m} \left[\frac{\beta\hbar\omega_{\mathrm{c}}}{\pi} \sinh\left(\frac{\pi\tau_{jk}}{\hbar\beta} \right) \right]^{2K\xi_j\xi_k} ,$$

where $\tau_{jk} \equiv t_j - t_k = \sum_{\ell=k}^{j-1} \tau_\ell$, and $K = \eta q_0^2/2\pi\hbar$ is the dimensionless Kondo coupling.

In the charge picture, the expression (27.2) is the grand-canonical sum of all charge sequences which can not be separated into neutral sub-clusters, i.e., in each summand the cumulated charges p_j, $j = 1, \cdots, 2m-1$, have the same sign. The double product is the charge interaction factor G_{2m}, Eq. (26.8), which represents the $2m(2m-1)/2$ interactions of the $2m$ unit charges $\xi_j = \pm 1$ of the neutral charge cluster.

27.1 Weak-tunneling limit

The weak-tunneling or Golden Rule limit is the $m = 1$ term of the series (27.2),

$$\mathcal{M}_{\mathrm{TB}} = \frac{2\pi K}{\epsilon} \sin(\pi K) \, \Delta^2 \int_0^\infty d\tau \, \frac{\sin(\epsilon\tau)}{[(\hbar\beta\omega_{\mathrm{c}}/\pi) \sinh(\pi\tau/\hbar\beta)]^{2K}} . \tag{27.3}$$

The integral can be calculated directly. Alternatively, we may combine previous results, i.e., the relation $k^- = e^{-\beta\hbar\epsilon} k^+$ with the expression (20.74) for the rate k^+, and the expression (25.20) for the mobility, In either way, one gets

$$\mathcal{M}_{\mathrm{TB}}(T,\epsilon,K) = \frac{\pi^2 K}{\Gamma(2K)} \left| \Gamma\left(K + i\frac{\hbar\epsilon}{2\pi k_{\mathrm{B}}T} \right) \right|^2 \frac{\sinh(\hbar\epsilon/2k_{\mathrm{B}}T)}{\hbar\epsilon/2k_{\mathrm{B}}T} \left(\frac{\hbar\Delta_{\mathrm{r}}}{2\pi k_{\mathrm{B}}T} \right)^{2-2K} . \tag{27.4}$$

From this we find for the nonlinear mobility at zero temperature

$$\mathcal{M}_{\mathrm{TB}}(0,\epsilon,K) = \frac{\pi^2 K}{\Gamma(2K)} \left(\frac{\Delta_{\mathrm{r}}}{\epsilon} \right)^{2-2K} , \tag{27.5}$$

and for the linear mobility at finite temperature

$$\mathcal{M}_{\text{TB, lin}}(T, K) \equiv \mathcal{M}(T, 0, K) = \frac{\pi^2\sqrt{\pi}}{2}\frac{\Gamma(1+K)}{\Gamma(1/2+K)}\left(\frac{\hbar\Delta_r}{\pi k_{\text{B}}T}\right)^{2-2K}. \tag{27.6}$$

These expressions describe the mobility for an Ohmic environment in the nearest-neighbor tunneling regime discussed in Subsec. 25.2.1.

The diffusion coefficient follows by use of expression (27.4) in the relation (25.22),

$$D_{\text{TB}}(T, \epsilon, K) = \frac{q_0^2 \Delta_r}{4\Gamma(2K)}\left(\frac{\hbar\Delta_r}{2\pi k_{\text{B}}T}\right)^{1-2K}\left|\Gamma\left(K+i\frac{\hbar\epsilon}{2\pi k_{\text{B}}T}\right)\right|^2 \cosh\left(\frac{\hbar\epsilon}{2k_{\text{B}}T}\right). \tag{27.7}$$

Thus, for zero bias, the diffusion exhibits Kondo-type behavior $D(T, 0, K) \propto T^{2K-1}$, that is, the diffusion coefficient increases with decreasing temperature in the damping regime $K < 1/2$ [see the discussion below Eq. (20.80)].

27.2 Weak-damping limit

When damping is very weak, $K \ll 1$, the charge cluster is dilute, so that the average distance between charges is large compared to the thermal time $\hbar\beta$. On this condition, the noise correlation $Q'(\tau)$ takes the "white noise" form [cf. Eq. (22.48)]

$$Q'(\tau) = 2K[\pi|\tau|/\hbar\beta + \ln(\beta\hbar\omega_c/2\pi)]. \tag{27.8}$$

With the correlation function (27.8), the noise filter (26.8) combined with the bias filter $e^{i\varphi_{2m}}$ provides an exponential filter function for each individual time interval,

$$G_{2m}\,e^{i\varphi_{2m}} = \left(\frac{2\pi}{\beta\hbar\omega_c}\right)^{2Km}\prod_{k=1}^{2m-1}\exp\left\{-\left(2\pi K\frac{p_k}{\hbar\beta}-i\,\epsilon\right)p_k\tau_k\right\}. \tag{27.9}$$

In the regime $\beta\hbar\epsilon \ll 2\pi pK$, where p is a characteristic cumulated charge, the path weight is ruled by the noise filter function G_{2m}. For $\beta\hbar\epsilon \gg 2\pi pK$, conversely, the filter G_{2m} is ineffective, and the path weight is dominated by the bias factor $e^{i\varphi_{2m}}$.

With the product form (27.9), the time integrals in the Laplace transform $\langle\hat{q}(\lambda)\rangle$ are uncorrelated, and each integral is a simple definite integral of an exponential function. Eventually, the mobility is found from Eq. (27.2) as

$$\mathcal{M}_{\text{TB}}(T, \epsilon) = \frac{2\pi K}{\beta\hbar\epsilon}\sum_{m=1}^{\infty}(-1)^{m-1}u^m\sum_{\{\xi_j=\pm1\}'}\text{Im}\prod_{k=1}^{2m-1}\frac{\sin(\pi K p_k)}{\pi K p_k[p_k - i\,\beta\hbar\epsilon/2\pi K]}, \tag{27.10}$$

where

$$u \equiv (\beta\hbar\Delta_T/2)^2, \qquad \Delta_T = \Delta_r\,(2\pi k_{\text{B}}T/\hbar\Delta_r)^K. \tag{27.11}$$

The adiabatic Franck-Condon prefactor $(2\pi/\beta\hbar\omega_c)^K$ in Eq. (27.9) is included in the frequency scale Δ_T [see also Eq. (22.49)]. The series (27.10) resembles the grand-canonical partition function of a one-dimensional Coulomb gas discussed first by Lenard [520]. The summation over all arrangements of charges $\xi_j = \pm1$ satisfying the neutrality condition (26.11) leads to the continued fraction expression [520, 513]

$$\mathcal{M}_{\mathrm{TB}}(T,\epsilon) \;=\; \frac{4\pi K}{\beta\hbar\epsilon}\;\mathrm{Im}\;\frac{a_1 u}{1 + \dfrac{a_1 a_2 u}{1 + \dfrac{a_2 a_3 u}{1 + \cdots}}} \tag{27.12}$$

with coefficients

$$a_n \;=\; \frac{\sin(\pi K n)}{\pi K n}\;\frac{1}{n - i\,\beta\hbar\epsilon/2\pi K}\,. \tag{27.13}$$

For $K \ll 1$, the sine function in Eq. (27.13) can be linearized. This yields

$$a_n \;=\; 1/(n - i\nu) \qquad \text{with} \qquad \nu \equiv \beta\hbar\epsilon/2\pi K\,. \tag{27.14}$$

With this form, the expression (27.12) matches the continued fraction (9.1.73) in Ref. [90] which can be written in terms of modified Bessel functions $I_\mu(z)$ with complex order μ [521]. With adjustment of the parameters, we obtain the mobility in analytic form for all T and ϵ as

$$\mathcal{M}_{\mathrm{TB}}(T,\epsilon) \;=\; \frac{\beta\hbar\Delta_T}{\nu}\;\mathrm{Im}\left(\frac{I_{1-i\nu}(\beta\hbar\Delta_T)}{I_{-i\nu}(\beta\hbar\Delta_T)}\right)\,. \tag{27.15}$$

With recursion relations of the modified Bessel function, this can be written as

$$\mathcal{M}_{\mathrm{TB}}(T,\epsilon) \;=\; 1 - \frac{\sinh(\pi\nu)}{\pi\nu}\;\frac{1}{|I_{i\nu}(\beta\hbar\Delta_T)|^2}\,. \tag{27.16}$$

Taking the limit $\epsilon \to 0$ either in (27.15) or in (27.16), yields the linear mobility

$$\mathcal{M}_{\mathrm{TB,lin}}(T) \;=\; 1 - \frac{1}{I_0^2(\beta\hbar\Delta_T)}\,. \tag{27.17}$$

From this we see that the linear mobility at zero temperature coincides with the mobility of a free Brownian particle, $\mu_{\mathrm{TB,lin}}(T=0) = \mu_0$.

Consider next the nonlinear mobility at $T = 0$. Now, thermal noise is absent and quantum noise is so weak in the regime $K \ll 1$ that the only effect of the filter function G_{2m} in the series expression (27.2) is to regularize the time integrals, $\lim_{\kappa\to 0^+}\int_0^\infty d\tau_j\, e^{-(\kappa - i\,\epsilon p_j)\tau_j}$, $j = 1,\cdots,2m-1$. If we put $\sin(\pi K p_k) \to \pi K p_k$ in addition, the cumulated charges $\{p_j\}$ drop out in the expression (27.10). As a result, for fixed number of charges in a neutral cluster, every individual sequence of charges yields the same contribution.

Altogether, the perturbative series for the nonlinear mobility takes the form

$$\mathcal{M}_{\mathrm{TB}}(0,\epsilon) \;=\; \sum_{m=1}^\infty \left(\frac{\pi K\Delta}{\epsilon}\right)^{2m} \sum_{\{\xi_j = \pm 1\}'} 1\,. \tag{27.18}$$

The number of possibilities for 2m charges with total charge zero to form a single irreducible cluster, i.e. all $p_j \neq 0$, is $2^{2m-1}\Gamma(m - 1/2)/\sqrt{\pi}\,\Gamma(m+1)$. Thus we find

$$\mathcal{M}_{\mathrm{TB}}(0,\epsilon) \;=\; \frac{1}{2\sqrt{\pi}}\sum_{m=1}^\infty \frac{\Gamma(m-1/2)}{\Gamma(m+1)}\left(\frac{2\pi K\Delta}{\epsilon}\right)^{2m}\,. \tag{27.19}$$

This series is convergent in the regime

$$\epsilon > \epsilon_0 \equiv 2\pi K \Delta \,, \tag{27.20}$$

and is summed to the square root expression

$$\mathcal{M}_{\mathrm{TB}}(0,\epsilon) = 1 - \sqrt{1 - (\epsilon_0/\epsilon)^2}\,. \tag{27.21}$$

Two remarks seem expedient. First, the expression (27.21) is also found from the continued fraction expression (27.12) if we substitute $a_n = i/\nu$ and put $u = (\hbar\beta\Delta/2)^2$. Secondly, in the zero temperature limit of the expression (27.15), or of the expression (27.16), we may employ the uniform asymptotic expansion of the modified Bessel function for large complex order [cf. Eq. (9.7.7) in Ref. [90]]. Taking into account the asymptotically leading term and disregarding the adiabatic Franck-Condon dressing factor, we then arrive again at the expression (27.21).

27.3 Exact solution in the Ohmic scaling limit at $K = \frac{1}{2}$

The quantum transport problem of the Ohmic TB particle for the particular case $K = \frac{1}{2}$ can be analytically solved in two different ways. In the first, one employs in the Coulomb gas representation of the TB model presented in Chapter 26 the concept of collapsed dipoles, which has been introduced in Subsection 22.6.1. In the second way, one passes on to a representation of noninteracting fermionic quasiparticles by following the treatment given in Subsections 19.2.2 and 22.6.6.

27.3.1 Current and mobility

In the Ohmic scaling limit, the TB series (27.2) for the mobility and the series (26.21) for $\langle \hat{q}(\lambda) \rangle_{\mathrm{pis}}$ can be calculated in analytic form for the coupling value $K = \frac{1}{2}$ using the concept of collapsed dipoles developed in Subsection 22.6.1 [522, 374]. The calculation is done by putting $K = \frac{1}{2} - \kappa$ and eventually taking the limit $\kappa \to 0$. In this limit, the phase factors $\sin(\pi K p_k)$ are reduced to factors $1, 2\pi\kappa, -1, -4\pi\kappa, \cdots$ for $p_k = 1, 2, 3, 4, \cdots$. On the other hand, the breathing mode integral of a dipole diverges as $1/\kappa$ in the limit $\kappa \to 0$. The singularity originates again from the $\tau^{-1+2\kappa}$ short-distance behavior of the intra-dipole interaction $\exp[-Q'(\tau)]$. Altogether, only those charge sequences, in which the occurring κ-factors are compensated by $1/\kappa$-singularities, contribute to the mobility. These are just the formations which have cumulated charge sequence $p_k = 1, 2, 1, 2, \cdots, 2, 1$ and $p_k = -1, -2, -1, -2, \cdots, -2, -1$, or pictorially

$$+ \; (+ \,-) \; (+ \,-) \; \cdots \; (+ \,-) \; (+ \,-) \; - \;,$$
$$- \; (- \,+) \; (- \,+) \; \cdots \; (- \,+) \; (- \,+) \; + \;. \tag{27.22}$$

Here, the round brackets indicate the charge pairs which form collapsed dipoles. Thus, to the path sum for the mobility only paths contribute which are limited to the pentadiagonal matrix of the (q, q')-lattice.

With regard to the equivalent fermionic representation discussed below and in Subsec. 28.2.1, we now use the symmetric Kondo frequency scale $\tilde{\epsilon}_0$ introduced for general K below in Eq. (28.121). At $K = \frac{1}{2}$, there is

$$\tilde{\epsilon}_0 = 2\gamma = \pi\Delta^2/\omega_c , \tag{27.23}$$

where $\gamma = \Delta_{\text{eff}}(K = \frac{1}{2})$ is the frequency scale familiar from the spin-boson model at the Toulouse point $K = \frac{1}{2}$ (see Subsection 19.2.2 and Section 22.6).

The grand-canonical sum of collapsed noninteracting dipoles in the interval τ between the two peripheral charges yields the form factor (see Subsection 22.6.1)

$$U_{\text{CD}}(\tau) = e^{-\tilde{\epsilon}_0\tau} . \tag{27.24}$$

The exponent in Eq. (27.24) differs from the exponent in the CS form factor (22.131) by a factor 4. A factor 2 stems from the phase factor $\sin(2\pi\kappa)$ of the collapsed dipole compared with the blip phase factor $\sin(\pi\kappa)$ in Eq. (22.128), and an additional multiplicity factor 2 exists because there are 2 off-diagonal states with cumulative charge $p = 2$ accessible from an offdiagonal state with $p = 1$.

Accordingly, the dressed extended $+-$ and $-+$ dipoles depicted in (27.22) determine the mean particle current and the mobility, $\langle I \rangle = (\hbar\epsilon/q_0)\,\mu = q_0(\epsilon/\pi)\,\mathcal{M}$. The breathing mode integral of these extended dipoles readily gives

$$\langle I \rangle = \lim_{t\to\infty}\langle q(t)\rangle/t = q_0\Delta^2\int_0^\infty d\tau\, e^{-Q'(\tau)-\tilde{\epsilon}_0\tau}\sin(\epsilon\tau) , \tag{27.25}$$

which in essence is an integral representation of the digamma function,

$$\langle I \rangle = q_0\frac{\tilde{\epsilon}_0}{\pi}\,\text{Im}\,\psi\left(\frac{1}{2} + \frac{\beta\hbar\tilde{\epsilon}_0}{2\pi} + i\,\frac{\beta\hbar\epsilon}{2\pi}\right) . \tag{27.26}$$

This yields for the scaled nonlinear mobility of the TB model the analytic expression

$$\mathcal{M}_{\text{TB}} = \frac{\tilde{\epsilon}_0}{\epsilon}\,\text{Im}\,\psi\left(\frac{1}{2} + \frac{\beta\hbar\tilde{\epsilon}_0}{2\pi} + i\,\frac{\beta\hbar\epsilon}{2\pi}\right) . \tag{27.27}$$

At high temperature, $\hbar\tilde{\epsilon}_0/2\pi k_{\text{B}}T \ll 1$, we get in agreement with expression (27.4)

$$\mathcal{M}_{\text{TB}} = \frac{\pi\tilde{\epsilon}_0}{2\epsilon}\tanh\left(\frac{\beta\hbar\epsilon}{2}\right) . \tag{27.28}$$

In the opposite limit $T \to 0$, we obtain the equivalent forms

$$\mathcal{M}_{\text{TB}} = \frac{\tilde{\epsilon}_0}{\epsilon}\arctan\left(\frac{\epsilon}{\tilde{\epsilon}_0}\right) = \frac{\tilde{\epsilon}_0}{\epsilon}\left\{\frac{\pi}{2} - \arctan\left(\frac{\tilde{\epsilon}_0}{\epsilon}\right)\right\} . \tag{27.29}$$

Thus we find, as in the limit $K \to 0$ in Subsection 27.2, $\mu_{\text{lin}}(T = 0) = \mu_0$. We shall see in Subsection 28.4.1 that this relation generally holds in the regime $K < 1$.

Fermionic representation

It has been recognized by Guinea that the TB model (24.4) in the Ohmic scaling limit at $K = \frac{1}{2}$ can be mapped on free fermions [523] (see also Ref. [514], Sec. VIII). The correspondence is analogous to that of the spin-boson model, which maps at $K = \frac{1}{2}$ on the Toulouse point of the anisotropic Kondo model (cf. Subsec. 19.4.1) and on the resonance level model with vanishing repulsive contact interaction (cf. Subsec 19.4.2). The equivalence of the bosonic Coulomb gas representation (presented in Sec. 26.1) at $K = \frac{1}{2}$ with a fermionic representation directly follows with use of the inverse Fourier transform of the expression (22.180),

$$e^{-Q'(\tau, K=\frac{1}{2})} = \frac{i}{2\omega_c} \int_{-\infty}^{\infty} d\omega \, \tanh\left(\frac{\beta\hbar\omega}{2}\right) e^{-i\omega\tau} . \tag{27.30}$$

With this representation the integral (27.25) is transformed into the frequency integral

$$\langle I \rangle = 2q_0 \int_0^\infty \frac{d\omega}{2\pi} \, \mathcal{T}(\omega)[\mathcal{N}_+(\omega, \epsilon) - \mathcal{N}_-(\omega, \epsilon)] , \tag{27.31}$$

where

$$\mathcal{N}_\pm(\omega, \epsilon) = f(\omega \mp \epsilon)[1 - f(\omega \pm \epsilon)] , \tag{27.32}$$

and where $f(\omega) = 1/(e^{\beta\hbar\omega} + 1)$ is the Fermi function,

The interpretation of this expression is straightforward. The quantity $\mathcal{N}_+(\omega, \epsilon)$ represents the spectral weight for right-moving fermions for an occupied state on the left side at frequency $\omega - \epsilon$ and an empty state on the right side of the barrier at frequency $\omega + \epsilon$. Accordingly, the quantity $\mathcal{N}_-(\omega, \epsilon)$ represents the spectral weight for left-moving fermions for an occupied state on the right side at frequency $\omega + \epsilon$ and an empty state on the left side of the barrier at frequency $\omega - \epsilon$.

The functions $\mathcal{T}(\omega)$ and $\mathcal{R}(\omega)$ represent the spectral transmission and reflection probability of the fermionic entity at frequency ω , $\mathcal{T}(\omega) + \mathcal{R}(\omega) = 1$,

$$\mathcal{T}(\omega) = \frac{\tilde{\epsilon}_0^2}{\tilde{\epsilon}_0^2 + \omega^2} , \qquad \mathcal{R}(\omega) = \frac{\omega^2}{\tilde{\epsilon}_0^2 + \omega^2} . \tag{27.33}$$

The expression (27.31) describes the net current across the barrier.

It is convenient to introduce the linear combinations

$$F_1(\omega, \epsilon) \equiv \mathcal{N}_+(\omega, \epsilon) - \mathcal{N}_-(\omega, \epsilon) = f(\omega - \epsilon) - f(\omega + \epsilon) ,$$
$$F_2(\omega, \epsilon) \equiv \mathcal{N}_+(\omega, \epsilon) + \mathcal{N}_-(\omega, \epsilon) = \coth(\hbar\epsilon/k_B T) \, F_1(\omega, \epsilon) . \tag{27.34}$$

In the second line, we have used the explicit form of the Fermi function $f(\omega)$. In terms of these functions, the first cumulant (27.31) may be written as

$$\langle I \rangle = 2q_0 \int_0^\infty \frac{d\omega}{2\pi} \, \mathcal{T}(\omega) F_1(\omega, \epsilon) = 2q_0 \frac{\epsilon}{2\pi} - 2q_0 \int_0^\infty \frac{d\omega}{2\pi} \mathcal{R}(\omega) F_1(\omega, \epsilon) . \tag{27.35}$$

The first form expresses the current in terms of the transmission through the barrier. The first term in the second form is the maximum current, which appears in the absence of the barrier. The second term diminishes the maximum current because of the possibility that the particle is reflected at the barrier.

One remark about small deviations from $K = \frac{1}{2}$ is in order. For $K = \frac{1}{2} - \kappa$, $|\kappa| \ll 1$, a leading log summation of all diagrams contributing to the current can be performed [524]. The resulting expression for the mobility is again of the form (27.27), though with a bias- and temperature-dependent renormalization of the frequency scale,

$$\tilde{\epsilon}_0 \rightarrow \tilde{\epsilon}(\epsilon, T) = \overline{\epsilon} [(\epsilon/\overline{\epsilon})^2 + (2\pi k_{\mathrm{B}} T/\hbar\overline{\epsilon})^2]^{-\kappa} \tag{27.36}$$

with $\overline{\epsilon} = \tilde{\epsilon}_0 (\omega_c/\tilde{\epsilon}_0)^{2\kappa}$. We remark that the expression (27.27) with (27.36) is consistent with the scaling forms at $K = \frac{1}{2} - \kappa$ of the weak- and strong-tunneling expansions of the mobility discussed below in Section 28.4.

27.3.2 Diffusion and skewness

The calculation of the series (26.22) for the second moment $\langle \hat{q}^2(\lambda) \rangle_{\mathrm{pis}}$ is more intricate because of the cosine factor in the phase term (26.19). This factor allows the particle to make a single virtual visit of diagonal state at some intermediate time. The combined analysis of zeros from phase factors and short-distance singularities from the charge interaction shows that the paths which contribute to the second cumulant are limited to the heptadiagonal matrix of the (q, q')-lattice. In the equivalent charge picture, the charge sequences are restricted to configurations in which the cumulated charges p_j are in the range $|p_j| \leqslant 3$.

For zero bias, the resulting "bosonic" integral expression at long times reads [374]

$$\langle q^2(t) \rangle_c = t q_0^2 \frac{\tilde{\epsilon}_0}{2} \left\{ 1 - \frac{4\tilde{\epsilon}_0 \omega_c^2}{\pi^2} \int_0^\infty d\tau_1 \, d\tau_2 \, d\tau_3 \; e^{-\tilde{\epsilon}_0(\tau_1+\tau_3)} \; e^{-Q'(\tau_1+\tau_2)-Q'(\tau_2+\tau_3)} \right\} .$$

Evaluation of the triple integral yields for the diffusion coefficient the analytic form

$$D(T,0) = (\hbar\tilde{\epsilon}_0/2\pi) \, \psi' \left(1/2 + \hbar\tilde{\epsilon}_0/2\pi k_{\mathrm{B}} T \right) \mu_0 . \tag{27.37}$$

Evidently, the diffusion coefficient (27.37) and the linear mobility resulting from Eq. (27.27) meet the Einstein relation (26.64), $D(T,0) = k_{\mathrm{B}} T \mu_{\mathrm{lin}}(T)$.

The diffusion coefficient vanishes for zero bias and $T = 0$. This is an indication that the dispersion is in fact sub-diffusive in this limit. We see in Subsection 27.4.2 that the second moment for zero bias and zero temperature spreads out at long time only logarithmically with time, $\langle q_c^2(t \rightarrow \infty) \rangle = (2/\pi^2) q_0^2 \ln(\tilde{\epsilon}_0 t)$, and that the logarithmic spread of the position dispersion at long time generally holds for all translational-invariant systems with Ohmic dissipation which have nonzero linear mobility at $T = 0$ (see also Subsection 7.3.2). A study of the diffusion coefficient in the Coulomb gas representation for the biased case is reported in Ref. [525].

The bosonic forms of the cumulants, extracted from the formally exact series (26.14) [e.g. Eq. (26.17)], may be transformed with use of the integral representation (27.30) into fermionic representation. The resulting expressions for the second cumulant (diffusion) and third cumulant (skewness) in the biased case are

$$\langle q^2(t)\rangle_c \;=\; t\,(2q_0)^2 \int_0^\infty \frac{d\omega}{2\pi}\,\left[\,\mathcal{T}(\omega)F_2(\omega,\epsilon) \,-\, \mathcal{T}^2(\omega)F_1^2(\omega,\epsilon)\,\right], \tag{27.38}$$

$$\langle q^3(t)\rangle_c \;=\; t\,(2q_0)^3 \int_0^\infty \frac{d\omega}{2\pi}\,\mathcal{T}(\omega)F_1(\omega,\epsilon) \tag{27.39}$$

$$\times\left[\,1 - 3\mathcal{T}(\omega)F_2(\omega,\epsilon) + 2\mathcal{T}^2(\omega)F_1^2(\omega,\epsilon)\,\right].$$

Upon employing the expressions (27.34) and (27.33), the cumulants (27.38) and (27.39) can be expressed in terms of cumulants of lower order. One finds

$$\langle q^2(t)\rangle_c \;=\; q_0\left\{\,\coth(\beta\hbar\epsilon)\,\tilde{\epsilon}_0\frac{d}{d\tilde{\epsilon}_0} + \left(2 - \tilde{\epsilon}_0\frac{d}{d\tilde{\epsilon}_0}\right)\frac{1}{\hbar\beta}\frac{d}{d\epsilon}\,\right\}\langle q(t)\rangle_c, \tag{27.40}$$

$$\langle q^3(t)\rangle_c \;=\; q_0\left\{\,\coth(\beta\hbar\epsilon)\,\tilde{\epsilon}_0\frac{d}{d\tilde{\epsilon}_0} + \left(2 - \frac{1}{2}\tilde{\epsilon}_0\frac{d}{d\tilde{\epsilon}_0}\right)\frac{1}{\hbar\beta}\frac{d}{d\epsilon}\,\right\}\langle q^2(t)\rangle_c$$

$$-\, q_0^2\left\{\,\coth(\beta\hbar\epsilon)\,\frac{1}{\hbar\beta}\frac{d}{d\epsilon} + \frac{1}{\sinh^2(\beta\hbar\epsilon)}\,\right\}\tilde{\epsilon}_0\frac{d}{d\tilde{\epsilon}_0}\langle q(t)\rangle_c. \tag{27.41}$$

With use of the analytic expression (27.26) for the current, it is now straightforward to derive analytic expressions for diffusion and skewness, e.g., the generalisation of the diffusion coefficient (27.37) to nonzero bias.

27.4 The effects of a thermal initial state

27.4.1 Mean position and variance

We have seen in Section 26.2 that the difference between $\langle q(t)\rangle_{\text{th}}$ (thermal initial state) and $\langle q(t)\rangle$ (product initial state) is described by the function $R_1(t)$ given in Eq. (26.32). Under conditions specified in Subsection 26.3.1 all the integrals in the expression for $\hat{R}_1(\lambda)$ in Eq. (26.37) are convergent in the limit $\lambda \to 0$. Hence, because of the extra $1/\lambda$-factor, the function $R_1(t)$ approaches a constant at asymptotic time, $R_1(t \to \infty) = R_{1,\infty}$. As a result, the leading asymptotic behaviors of the mean values $\langle q(t)\rangle_{\text{th}}$ and $\langle q(t)\rangle$ coincide, while sub-leading terms are different,

$$\lim_{t\to\infty}\left[\langle q(t)\rangle_{\text{th}} - \langle q(t)\rangle\right] \;=\; R_{1,\infty}. \tag{27.42}$$

In the presence of drift due to a bias, the variances for $t > 0$ are defined by

$$\sigma^2(t) \equiv \langle q^2(t)\rangle - \langle q(t)\rangle^2, \quad \text{and} \quad \sigma_{\text{th}}^2(t) \equiv D_{\text{th},+}(t) - \langle q(t)\rangle_{\text{th}}^2. \tag{27.43}$$

Consider the term of order Δ^{2m} in Eq. (26.22) with (26.19). The leading contribution to $\langle \hat{q}^2(\lambda)\rangle$ in the limit $\lambda \to 0$ comes from the terms in which the factor $\cos(\chi_{k,2m})$ in Eq. (26.19) is associated with a sojourn interval τ_k ($p_k = 0$). The respective τ_k-integration yields a factor $1/\lambda$. The two branches to the left and right of this interval are easily identified as terms of the series expansion

$$\lambda \langle \hat{q}(\lambda) \rangle = F\mu/\lambda + q_\infty + \mathcal{O}(\lambda) . \tag{27.44}$$

Besides, the diffusion contribution $2D/\lambda^2$ to $\langle \hat{q}^2(\lambda) \rangle$ results from the summands in which the phase factor $\cos(\chi_{k,2m})$ falls on a blip state $(p_k \neq 0)$. In the end we find

$$\lim_{t \to \infty} \langle q^2(t) \rangle = (F\mu t)^2 + 2(F\mu q_\infty + D)t . \tag{27.45}$$

With this form and with the subtraction of $\langle q(t) \rangle^2$, we obtain for the variance in the diffusive regime

$$\lim_{t \to \infty} \sigma^2(t) = 2Dt . \tag{27.46}$$

The effects of the different initial conditions are in the function

$$\sigma_{\mathrm{th}}^2(t) - \sigma^2(t) = R_2(t) - R_1^2(t) - 2\langle q(t) \rangle R_1(t) , \tag{27.47}$$

as follows from Eqs. (27.43), (26.30), and (26.31). In the limit $\lambda \to 0$, the leading contribution in order Δ^{i+j} to $\hat{R}_2(\lambda)$ comes again from terms with the factor $\cos(\chi_{k,i+j})$ in which τ_k is a sojourn. The branch to the left of this interval can be identified with a contribution of $\hat{R}_1(\lambda)$, and the branch to the right is a term of the series expression for $\langle \hat{q}(\lambda) \rangle$. Readily we find that in the combined expression $R_2(t) - 2\langle q(t) \rangle R_1(t)$ the asymptotically leading term of $R_2(t)$, which grows linearly with t, is cancelled. The remaining term approaches asymptotically a constant value,

$$\lim_{t \to \infty} [\, R_2(t) - 2\langle q(t) \rangle R_1(t) \,] = R_{2,\infty} . \tag{27.48}$$

Therefore, the leading behavior at long times of $\sigma_{\mathrm{th}}^2(t)$ coincides with that of $\sigma^2(t)$, Eq. (27.46), whereas sub-leading terms are different,

$$\lim_{t \to \infty} [\, \sigma_{\mathrm{th}}^2(t) - \sigma^2(t) \,] = R_{2,\infty} - R_{1,\infty}^2 . \tag{27.49}$$

Thus in the diffusive regime, the variances $\sigma_{\mathrm{th}}^2(t)$ and $\sigma^2(t)$ differ by a constant at asymptotic time.

27.4.2 Linear response

In this subsection, we focus the attention on the linear response to an external force $F = \hbar\epsilon/q_0$ switched on at time $t = 0$. One may think that the definition of the mobility within Kubo's linear-response theory is problematic [222] since the stationary state at asymptotic times is not normalizable. We now draw our attention to this question.

The Kubo formalism yields for the mean velocity at time t the expression

$$\langle v(t) \rangle_{\mathrm{lin}} = -(2\epsilon/q_0)\,\Theta(t)\,A(t) = \hbar\epsilon\,\chi(t)/q_0 . \tag{27.50}$$

Here, $A(t)$ is the antisymmetric equilibrium correlation function (25.10) at zero bias, and $\chi(t)$ is the response function. Alternatively, we may define $\langle v(t) \rangle_{\mathrm{lin}}$ as

$$\langle v(t) \rangle_{\mathrm{lin}} = \epsilon \lim_{\epsilon \to 0} \langle \dot{q}(t) \rangle_{\mathrm{th}}/\epsilon . \tag{27.51}$$

It is now interesting to see, using Eqs. (26.30) and (26.34), whether the two expressions coincide. Equating the expressions (27.50) and (27.51) and switching to the Laplace transforms, we obtain

$$\lim_{\epsilon \to 0} [\langle \hat{q}(\lambda) \rangle + \hat{R}_1(\lambda)]/\epsilon = (\hbar/q_0)\,\hat{\chi}(\lambda)/\lambda . \tag{27.52}$$

To prove the validity of this relation, we use for the time integrals occurring in the series expression for $\hat{R}_1(\lambda)$ and $\hat{\chi}(\lambda)$ the integral identities

$$\int_0^\infty dr_k\,ds_k\,e^{-\lambda s_k} f(r_k + s_k) = \int_0^\infty d\tau\, f(\tau)\,(1 - e^{-\lambda\tau})/\lambda ,$$

$$\int_0^\infty dr_k\,ds_k\,s_k\,e^{-\lambda s_k} f(r_k + s_k) = \int_0^\infty d\tau\, f(\tau)\,[\,(1 - e^{-\lambda\tau})/\lambda^2 - \tau e^{-\lambda\tau}/\lambda\,] ,$$

where $\tau = r_k + s_k$, and where the factor $f(\tau)$ contains the interactions between the two sets of charges to the left and right of the interval τ. With these forms, it is straightforward to see that the relation (27.52) is satisfied for all λ. Thus, the expressions (27.50) and (27.51) agree.

Using Eqs. (27.44) and (27.52), we find that the *linear* mobility is found from the absorptive part of the dynamical susceptibility as

$$\mu_{\text{lin}} = \lim_{\omega \to 0} \omega\,\text{Im}\,\tilde{\chi}(\omega) = \lim_{\omega \to 0} \omega\,\text{Im} \int_0^\infty dt\,\chi(t)\,e^{i\omega t} . \tag{27.53}$$

Consider next the mean square displacement. In the absence of the bias, $\epsilon = 0$, the variance $\sigma_{\text{th}}^2(t)$ coincides with the equilibrium correlation function $D_{\text{th},+}(t)$. Use of the expression (7.23) and the relation (27.53) yields

$$\lim_{\lambda \to 0} \lambda^2\,\hat{D}_{\text{th},+}(\lambda) = 2k_\text{B}T \lim_{\omega \to 0} \omega\,\text{Im}\,\tilde{\chi}(\omega) = 2k_\text{B}T\,\mu_{\text{lin}} . \tag{27.54}$$

On the other hand, we get in the diffusive regime with the findings of Subsec. 27.4.1

$$\lim_{\lambda \to 0} \lambda^2\,\hat{D}_{\text{th},+}(\lambda) = \lim_{\lambda \to 0} \lambda^2 \langle \hat{q}^2(\lambda) \rangle = 2D . \tag{27.55}$$

Equating Eq. (27.54) with Eq. (27.55), we arrive at the Einstein relation (26.64).

Now assume that the zero frequency limit in Eq. (27.53) for zero temperature, and hence the linear mobility $\mu_{\text{lin},0} = \mu_{\text{lin}}(T = 0)$ is nonzero. Then the integral expression (7.22) with a low-frequency cut-off at $1/t_0$ yields for zero temperature

$$\lim_{\lambda \to 0} \hat{D}_{\text{th},+}(\lambda) = -(2\hbar/\pi)\,\mu_{\text{lin},0}\,\ln(\lambda t_0)/\lambda , \tag{27.56}$$

and hence in the time domain

$$\lim_{t \to \infty} D_{\text{th},+}(t) = (2\hbar/\pi)\mu_{\text{lin},0}\,\ln(t/t_0) , \tag{27.57}$$

Since we have not referred to a particular system in the derivation, the logarithmic spread of $D_{\text{th}}(t)$ at long time holds at zero temperature for any system of which the linear mobility at $T = 0$ is nonzero. Actually this is the case for the TB model in the regime $K < 1$, as we shall see below in Subsection 28.4.1.

We can also study the spread of the second moment at long time directly from the exact formal expression (26.22). To this, consider now the case in which the cosine-term in the phase factor c_{2m} falls on a sojourn with cumulative charge $p_{2k} = 0$ and length τ_{2k}. Then the $2m$ charges are arranged into two neutral clusters with $2k$ and $2m - 2k$ charges. On the assumption that the average sojourn length is very large compared to the internal lengths of the clusters, there results the decomposition

$$c_{2m}\, G_{2m}\cos(\varphi_{2m}) = \sum_{k=1}^{m-1} \prod_{\ell=1, \ell \neq 2k}^{2m-1} b_{2m-2k}\, G_{2m-2k}\, b_{2k} G_{2k}$$

$$\times\left(\sin(\varphi_{2m-2k})\sin(\varphi_{2k}) \right. \tag{27.58}$$

$$\left. + \cos(\varphi_{2m-2k})\sum_{i,\text{left}}(p_i\tau_i)\cos(\varphi_{2k})\sum_{j,\text{right}}(p_j\tau_j)\,\ddot{Q}'(\tau_{2k})\right).$$

The first term in the round bracket represents two noninteracting clusters. Thus, in Laplace space, the intermediate sojourn integral yields a factor $1/\lambda$. The second term includes the dipole-dipole interaction of the clusters. It is immediately clear that all contributions of the first type add up in the time regime to the term $\langle q(t)\rangle^2$ in the second relation of Eq. (25.5).

There follows from the Einstein relation that the diffusion coefficient for $\epsilon = 0$ is zero in the limit $T \to 0$. Hence in this limit, $\langle q^2(t)\rangle$ grows subdiffusively. This behavior originates from the dipole interaction between the two neutral clusters. Each dipole represents for $\epsilon = 0$ the series (26.59) for the linear mobility. Thus we obtain for $T = 0$ and zero bias with the interaction $\ddot{Q}'(\tau) = -2K/\tau^2$

$$\lim_{\lambda\to 0}\lambda^2\langle\hat{q}^2(\lambda)\rangle = 2\frac{\hbar^2}{q_0^2}\mu_{\text{lin},0}^2\int_{t_0}^{\infty}d\tau\,\frac{2K}{\tau^2}\left(e^{-\lambda\tau} - 1\right). \tag{27.59}$$

The lower bound of the integral is equated with the reference time t_0 introduced in Eq. (27.56). Further, in the round bracket the term unity is subtracted in order to suppress in the limit $\lambda \to 0$ the short-time regime of the dipole interaction. Thus we get using the relation $q_0^2 = 2\pi\hbar K\,\mu_0$

$$\lim_{\lambda\to 0}\lambda^2\langle\hat{q}^2(\lambda)\rangle = -(2\hbar/\pi)\,(\mu_{\text{lin},0}^2/\mu_0)\,\lambda\,\ln(\lambda\,t_0). \tag{27.60}$$

The inverse Laplace transform then yields in the time regime

$$\lim_{t\to\infty}\langle q^2(t)\rangle = (2\hbar/\pi)\,(\mu_{\text{lin},0}^2/\mu_0)\,\ln(t/t_0). \tag{27.61}$$

Since the Ohmic TSS is ergodic (cf. Subsec. 3.1.10), the prefactor of the logarithmic law (27.57) must coincide with that in (27.61). The surprising result is that the linear mobility at $T = 0$ coincides with the mobility in the absence of the cosine potential,

$$\mu_{\text{lin},0} = \mu_0. \tag{27.62}$$

Thus, in the zero temperature limit the TB potential is renormalized to zero. This result is in correspondence with the findings in Subsection 28.4.1.

27.4.3 The exactly solvable case $K = \frac{1}{2}$

In the Ohmic scaling limit at the Toulouse point $K = \frac{1}{2}$, the function $\langle q(t) \rangle_{\text{th}}$ and the correlation functions $D_{\text{th}}(t)$ and $\chi(t)$ can be calculated in analytic form [525] using again the concept of collapsed blips explained in Subsection 22.6.1.

The response function $\chi(t)$ is diagrammatically represented by a single extended dipole of length $\tau_- + \tau_+$, where τ_- is the length in the negative and τ_+ the length in the positive time branch. Both intervals are dressed with a grand-canonical sum of collapsed dipoles resulting in a combined form factor $U_{\text{CD}}(\tau_- + \tau_+) = e^{-2\gamma(\tau_- + \tau_+)}$ [see Eq. (27.24) with (27.23)]. The resulting expression is

$$
\begin{aligned}
\hat{\chi}(\lambda) &= \frac{q_0^2 \gamma}{\pi \hbar} \frac{2\omega_c}{\lambda} \int_0^\infty d\tau_- \int_0^\infty d\tau_+ \, e^{-\lambda \tau_+} \, e^{-2\gamma(\tau_- + \tau_+)} \, e^{-Q'(\tau_- + \tau_+)} \\
&= \frac{q_0^2 \gamma}{\pi \hbar} \frac{2}{\lambda^2} \left\{ \psi\left(\frac{1}{2} + \frac{\beta \hbar (2\gamma + \lambda)}{2\pi} \right) - \psi\left(\frac{1}{2} + \frac{2\beta \hbar \gamma}{2\pi} \right) \right\},
\end{aligned}
\tag{27.63}
$$

where $\psi(z)$ is the digamma function.

The dynamical susceptibility is found from Eq. (27.63) as $\tilde{\chi}(\omega) = \hat{\chi}(-i\omega)$. This yields for the absorptive part $\tilde{\chi}''(\omega) = \text{Im}\,\tilde{\chi}(\omega)$ with the substitution $\tau_- + \tau_+ = \tau$

$$
\begin{aligned}
\tilde{\chi}''(\omega) &= \frac{q_0^2}{\pi \hbar} \frac{2\gamma \omega_c}{\omega^2} \int_0^\infty d\tau \, e^{-Q'(\tau) - 2\gamma \tau} \sin(\omega \tau) \\
&= \frac{q_0^2}{\pi \hbar} \frac{2\gamma}{\omega^2} \, \text{Im}\, \psi\left(\frac{1}{2} + \frac{\beta \hbar (2\gamma + i\omega)}{2\pi} \right).
\end{aligned}
\tag{27.64}
$$

The calculation of $\hat{D}_{\text{th},+}(\lambda)$ is more difficult than that of $\hat{\chi}(\lambda)$ since there are contributions from paths which cross the diagonal. These occur on account of the cosine factor in the phase terms (26.19) and (26.33). The resulting integral expression in Laplace space for $\epsilon = 0$ is [525]

$$
\hat{D}_{\text{th},+}(\lambda) = \frac{\hbar}{\pi \lambda} \frac{q_0^2}{\pi \hbar} 2\gamma \omega_c \int_{-\infty}^\infty d\omega \, \frac{\coth(\beta \hbar \omega / 2)}{\omega^2 + \lambda^2} \int_0^\infty d\tau \, e^{-Q'(\tau) - 2\gamma \tau} \sin(\omega \tau).
\tag{27.65}
$$

Noting the expression (27.64) for $\tilde{\chi}''(\omega)$, one directly sees that the expression (27.65) can be written in the form (7.22). This yields in the time regime [cf. Eq. (7.19)]

$$
D_{\text{th},+}(t) = \frac{\hbar}{\pi} \int_{-\infty}^\infty d\omega \, \tilde{\chi}''(\omega) \coth(\beta \hbar \omega / 2) \left[1 - \cos(\omega t) \right].
\tag{27.66}
$$

In the limit $t \to \infty$, the expression (27.66) with (27.64) yields

$$
\lim_{t \to \infty} \frac{D_{\text{th},+}(t)}{t} = 2D(T, 0) \lim_{t \to \infty} \frac{1}{\pi t} \int_{-\infty}^\infty d\omega \, \frac{1 - \cos(\omega t)}{\omega^2} = 2D(T, 0),
\tag{27.67}
$$

where the diffusion coefficient $D(T, 0)$ is given in Eq. (27.37). This verifies that $D_{\text{th},+}(t)$ coincides with $\langle q^2(t) \rangle$ at asymptotic time.

28 Duality symmetry

Duality is a recognized concept by now. In electromagnetism, the simplest form of duality is the invariance of the source free Maxwell equations under interchange of electric and magnetic field, $\mathbf{B} \to \mathbf{E}$, $\mathbf{E} \to -\mathbf{B}$. In the presence of sources, the product of electric and magnetic charges obeys the Dirac quantization condition. In general, duality maps a theory with strong coupling to one with weak coupling. Thus, if a duality symmetry exists, one can study the strong-coupling regime via the perturbative analysis of the weak-coupling regime. A theory is self-dual when there is an exact map between the strong- and the weak-coupling sector of the same theory.

28.1 Duality for general spectral density

The model of the damped Brownian particle in a cosine potential has an intrinsic duality symmetry. The weak-binding representation of this model can be mapped exactly on its tight-binding representation, and vice versa. Schmid [147] demonstrated in an imaginary-time path sum approach by comparing the perturbative series in the corrugation strength V_0 with the perturbative series in the tunneling matrix element Δ (multi-kink expansion) that the equilibrium density matrix at $T = 0$ is self-dual in the Ohmic scaling limit.[1] The analysis by Schmid led to a transformation for the dc mobility in which diffusive and localized behavior are interchanged. The duality was generalized by Fisher and Zwerger in Ref. [512] to real time and finite temperature. They showed within the influence functional method that in the duality transformation the strict Ohmic spectral density in the WB model (24.1) is mapped onto an Ohmic spectral density with a Drude cut-off in the dual TB model (24.4).

About ten years later, it was shown that the Hamiltonians (24.1) and (24.4) are exactly related by a duality transformation for general density $J(\omega)$ with power $s < 2$ at low frequencies [526]. In the duality, the power s maps on the power $2 - s$ and vice versa. Hence super-Ohmic and sub-Ohmic friction are interchanged, while Ohmic friction maps on Ohmic, but with the coupling strength transformed into the inverse of it. This matter is discussed in the following subsection.

We shall expound in Section 28.3 that a corresponding duality symmetry holds between the charge and phase representation for a Josephson junction. An interesting correspondence between impurity scattering in a Tomonaga-Luttinger liquid and Brownian motion in a periodic potential is discussed in Chapter 31. In this case, there is an exact self-duality map between weak and strong backscattering.

---

[1]The universality or scaling limit is discussed in Subsection 18.2.2.

28.1.1 The map between the TB and WB Hamiltonian

We begin with demonstrating that the Hamiltonians (24.1) and (24.4) can directly be mapped onto each other. For simplicity, we confine ourselves to the case $F = 0$.[2] In the first step, we perform the canonical transformation

$$
\begin{aligned}
p_\alpha &\rightarrow -m_\alpha \omega_\alpha x_\alpha\,, & x_\alpha &\rightarrow p_\alpha/m_\alpha \omega_\alpha + p\,c_\alpha/\kappa m_\alpha \omega_\alpha^2\,, \\
X &\rightarrow p/\kappa\,, & P &\rightarrow -\kappa q + \sum_\alpha c_\alpha x_\alpha/\omega_\alpha\,.
\end{aligned}
\tag{28.1}
$$

To map the cosine potential in (24.1) on the hopping term in (24.4), we choose

$$
V_0 = \hbar\Delta\,, \qquad \kappa = 2\pi\hbar/X_0 q_0\,.
\tag{28.2}
$$

With this transformation the Hamiltonian (24.1) takes the form

$$
\begin{aligned}
H &= -\hbar\Delta\cos\left(\frac{q_0 p}{\hbar}\right) + \frac{\kappa^2 q^2}{2M} - \frac{\kappa q}{M}\sum_\alpha \frac{c_\alpha x_\alpha}{\omega_\alpha} + H_{\mathrm{R}}\,, \\
H_{\mathrm{R}} &= \sum_\alpha\left(\frac{p_\alpha^2}{2m_\alpha} + \frac{m_\alpha \omega_\alpha^2 x_\alpha^2}{2}\right) + \frac{1}{2M}\left(\sum_\alpha \frac{c_\alpha x_\alpha}{\omega_\alpha}\right)^2\,.
\end{aligned}
\tag{28.3}
$$

Now, in the transformed reservoir Hamiltonian H_{R} the modes of the environment are coupled with each other. In the next step, we diagonalize H_{R}. This leads to new canonical variables π_α and u_α, and new parameters M_α and Ω_α. By the transformation, the interaction term $-(\kappa q/M)\sum_\alpha c_\alpha x_\alpha/\omega_\alpha$ is converted into $-q\sum_\alpha d_\alpha u_\alpha$. The mapping is completed by imposing spatial invariance on the system-bath coupling, which requires that the q^2 term in Eq. (28.3) becomes the counter term of the model (24.4). This condition is implemented by the constraint

$$
\kappa^2/M = \sum_\alpha d_\alpha^2/M_\alpha \Omega_\alpha^2\,.
\tag{28.4}
$$

A relation between $J_{\mathrm{WB}}(\omega)$ and $J_{\mathrm{TB}}(\omega)$ is found with manipulations of the dynamical matrix $A_{\alpha\beta}$ of the reservoir Hamiltonian H_{R}. We have

$$
A_{\alpha\beta}(\omega) = B_{\alpha\beta}(\omega) + c_\alpha c_\beta/M\omega_\alpha\omega_\beta\,, \qquad B_{\alpha\beta}(\omega) = \delta_{\alpha\beta}m_\alpha(\omega_\alpha^2 - \omega^2)\,.
\tag{28.5}
$$

Since the coupling term of the bath modes is in the form of an exterior vector product, the ratio of the determinants of the matrices $\mathbf{A}(\omega)$ and $\mathbf{B}(\omega)$ gives

$$
\frac{\det\mathbf{A}(\omega)}{\det\mathbf{B}(\omega)} = 1 + \frac{1}{M}\sum_\alpha \frac{c_\alpha^2}{m_\alpha\omega_\alpha^2(\omega_\alpha^2 - \omega^2)} = 1 + i\,\frac{\tilde{\gamma}_{\mathrm{WB}}(\omega)}{\omega}\,.
\tag{28.6}
$$

In the last form, we have introduced the spectral damping function $\tilde{\gamma}(\omega)$ for the weak binding model. In the continuum limit, we have

---

[2]The case $F \neq 0$ is more intricate since the duality transformation changes a force coupled to a coordinate into a force coupled to a momentum. In the Ohmic scaling limit, the final result is as discussed in Subsection 28.1.3.

$$\tilde{\gamma}(\omega) = \lim_{\varepsilon \to 0} \frac{-i\,\omega}{M} \frac{2}{\pi} \int_0^\infty d\omega' \frac{J(\omega')}{\omega'(\omega'^2 - \omega^2 - i\varepsilon\,\mathrm{sgn}\,\omega)} \,. \tag{28.7}$$

On the other hand, we may resolve $A_{\alpha\beta}$ for $B_{\alpha\beta}$ and then switch to the unitarily equivalent form in which $\mathbf{A}$ is diagonal. We then get

$$\tilde{B}_{\alpha\beta}(\omega) = \tilde{A}_{\alpha\beta}(\omega) - (M/\kappa^2)\,d_\alpha d_\beta\,, \qquad \tilde{A}_{\alpha\beta}(\omega) = \delta_{\alpha\beta}M_\alpha(\Omega_\alpha^2 - \omega^2)\,, \tag{28.8}$$

$$\frac{\det \mathbf{B}(\omega)}{\det \mathbf{A}(\omega)} = 1 - \frac{M}{\kappa^2} \sum_\alpha \frac{d_\alpha^2}{M_\alpha(\Omega_\alpha^2 - \omega^2)} = -i\frac{M^2}{\kappa^2}\,\omega\tilde{\gamma}_{\mathrm{TB}}(\omega)\,. \tag{28.9}$$

To obtain the second form, we have used Eq. (28.4), and we have introduced the damping function $\tilde{\gamma}_{\mathrm{TB}}(\omega)$ of the TB model. Combining Eq. (28.6) with Eq. (28.9), we find an exact relation between the damping functions of the two models [526],

$$\tilde{\gamma}_{\mathrm{TB}}(\omega)\,[\tilde{\gamma}_{\mathrm{WB}}(\omega) - i\,\omega] = \kappa^2/M^2\,. \tag{28.10}$$

Using $J(\omega) = M\omega\,\mathrm{Re}\,\tilde{\gamma}(\omega)$, we then have

$$J_{\mathrm{WB}}(\omega) = (\kappa^2/M^2)J_{\mathrm{TB}}(\omega)/|\tilde{\gamma}_{\mathrm{TB}}(\omega)|^2\,, \tag{28.11}$$

$$J_{\mathrm{TB}}(\omega) = (\kappa^2/M^2)J_{\mathrm{WB}}(\omega)/|\tilde{\gamma}_{\mathrm{WB}}(\omega) - i\,\omega|^2\,. \tag{28.12}$$

Thus, the spectral density of the one model can be calculated for any form of the spectral density of the other model.

Consider now the WB model with the spectral power-law form

$$J_{\mathrm{WB}}(\omega) = M\gamma_s\omega_{\mathrm{ph}}(\omega/\omega_{\mathrm{ph}})^s\,, \qquad 0 < s < 2\,. \tag{28.13}$$

Here we have introduced for $s \neq 1$ a phononic reference frequency ω_{ph}, as in Eq. (3.38). The corresponding spectral damping function is [cf. the first term in Eq. (3.49)]

$$\tilde{\gamma}_{\mathrm{WB}}(\omega) \equiv \tilde{\gamma}(\omega, s) = \lambda_s(-i\,\omega/\omega_{\mathrm{ph}})^{s-1}\,, \qquad \lambda_s \equiv \gamma_s/\sin(\pi s/2)\,. \tag{28.14}$$

Let us now check the consistency of the transformation. We see from Eq. (28.9) that the constraint (28.4) is satisfied if $\omega\tilde{\gamma}_{\mathrm{TB}}(\omega)$ vanishes in the limit $\omega \to 0$. This in turn implies that $\lim_{\omega \to 0} \tilde{\gamma}_{\mathrm{WB}}(\omega)/\omega$ must diverge, which is the case in the parameter range $0 < s < 2$, as we see with use of Eq. (28.14). For $s \geqslant 2$, we have at low frequency in leading order $\tilde{\gamma}_{\mathrm{WB}}(\omega) \propto \omega$, as we catch from Eq. (3.53). This term leads to mass renormalization [88], and not to friction, as we have discussed in Subsec. 3.1.3. Moreover, with the form $\tilde{\gamma}(\omega) \propto \omega$ the condition (28.4) cannot be satisfied.

Thus we have shown that there is an exact duality between the weak corrugation model (24.1) and the tight-binding model (24.4) in the regime $s < 2$. In the mapping, the continuous coordinate X in the WB model is identified, up to a scale factor $1/\kappa$, with the quasimomentum p in the dual TB model. Hence the duality is a sort of a Fourier transformation between real and momentum space [513]. A non-zero system-bath coupling is essential since otherwise the scale factor $1/\kappa$ is infinity. Strictly speaking, the mapping holds under the condition $\lim_{\omega \to 0} \omega^2/J(\omega) \to 0$. There are no restrictions on the form of $J(\omega)$ at finite frequencies except those enforced on physical grounds. There follows from Eq. (28.11) or Eq. (28.12) that the spectral density $J_{WB}(\omega) \propto \omega^s$ of the weak corrugation model maps on the spectral density $J_{TB}(\omega \to 0) \propto \omega^{2-s}$ of the dual TB model. Thus the power s in the spectral density is mapped on the power $2 - s$. This means that sub-Ohmic and super-Ohmic friction are interchanged in the transformation, while Ohmic friction is mapped on Ohmic.

So far, the scales X_0 and q_0 are independent. However, the relation (28.10) is simplified with the particular choice

$$M\lambda_s X_0 q_0 / 2\pi\hbar \equiv M\lambda_s/\kappa = 1 . \tag{28.15}$$

Then the standard dimensionless coupling strengths

$$K_{WB} \equiv M\lambda_s X_0^2 / 2\pi\hbar , \quad \text{and} \quad K_{TB} \equiv M\lambda_s q_0^2 / 2\pi\hbar , \tag{28.16}$$

are related by

$$X_0/q_0 = K_{WB} = 1/K_{TB} . \tag{28.17}$$

With this, there follows from the condition $F = \hbar\epsilon_{TB}/q_0 = \hbar\epsilon_{WB}/X_0$ the relations

$$\epsilon_{TB} = \epsilon_{WB}/K_{WB} , \quad \text{and} \quad \epsilon_{WB} = \epsilon_{TB}/K_{TB} . \tag{28.18}$$

Upon inserting the expressions (28.13) and (28.14) into Eq. (28.12), the spectral density of the dual TB model obtains the power law ω^{2-s} with an algebraic cut-off

$$J_{TB}(\omega) = M\lambda_s^2 \frac{\omega_{ph}}{\gamma_s} \frac{(\omega/\omega_{ph})^{2-s}}{1 + \left[\cot(\pi s/2) - (\omega_{ph}/\gamma_s)(\omega/\omega_{ph})^{2-s}\right]^2} . \tag{28.19}$$

In the Ohmic case, $s = 1$, the density $J_{TB}(\omega)$ reduces to the Drude form [512]

$$J_{TB}(\omega) = \frac{M\gamma\omega}{1 + (\omega/\omega_c)^2} \quad \text{with} \quad \omega_c = \gamma . \tag{28.20}$$

Let us briefly pause and dwell on the issue why $J_{TB}(\omega)$ has a soft cut-off while the cutoff is missing in $J_{WB}(\omega)$. For simplicity, we consider the Ohmic case $s = 1$. First, we observe that the trajectories in the TB lattice are step-like paths as given in Eq. (26.1) or in Eq. (26.6), while those in the cosine potential are rounded on the timescale γ^{-1}. The rounding of the Heaviside step function is described by the function $h(\tau) = \Theta_{WB}(\tau)$, which is a particular solution of the differential equation $-\ddot{h}(\tau) + \gamma\dot{h}(\tau) = \gamma\delta(\tau)$. The smeared Θ-function is [512]

$$\Theta_{WB}(\tau) = e^{\gamma t} \Theta(-\tau) + \Theta(\tau) . \tag{28.21}$$

In the WB model, the (η, ξ)-path analogous to Eq. (26.6) is smoothed out by replacing the Heaviside function $\Theta(\tau)$ with the smeared function $\Theta_{WB}(\tau)$. It is straightforward to see that the influence function (26.8) for the smoothed (η, ξ)-path with the straight spectral density $J_{WB}(\omega) = M\gamma\omega$ is the same as that for the steplike path (26.6), but with the modified spectral density (28.20), which has a soft cutoff at frequency γ.

In other words, the algebraic cutoff in the spectral density $J_{TB}(\omega)$ is due to the sharp tight-binding trajectory (26.1) instead of the smoothed path in the WB model. Formally, the two influence functions are reconciled by the change of the spectral densities which enter the respective kernels $Q(\tau)$, as given by the relations (28.11) and (28.12). Physically, the inertia of the particle with a finite mass provides a natural cutoff in the coupling to high-frequency oscillators of the reservoir.

Finally, we note that the dual spectral functions $\tilde{\gamma}_{WB}(\omega)$ and $\tilde{\gamma}_{TB}(\omega)$ are symmetric in form at low frequency, as we see from Eqs. (28.14) and (28.10) with (28.15),

$$\tilde{\gamma}_{WB}(\omega) = \lambda_s(-i\omega/\omega_{ph})^{s-1}, \qquad \tilde{\gamma}_{TB}(\omega) = \lambda_s(-i\omega/\omega_{ph})^{1-s}. \qquad (28.22)$$

28.1.2 Frequency-dependent linear mobility

Now we turn to the linear ac mobilities of the WB and TB model. First, we observe that by the canonical transformation (28.1) a coordinate autocorrelation function of the WB model is transformed into a momentum autocorrelation function of the associated TB model. Thus we have for the linear mobility of the WB model

$$\tilde{\mu}_{WB}(\omega) \equiv -i\omega\,\frac{i}{\hbar}\int_0^\infty dt\,e^{i\omega t}\,\langle[X(t),X(0)]\rangle_\beta = -i\omega\,\tilde{\Pi}_{TB}(\omega)/\kappa^2, \qquad (28.23)$$

where $\tilde{\Pi}_{TB}(\omega)$ is the Fourier transform of the retarded momentum response function of the TB model, $\Pi_{TB}(t) = (i/\hbar)\,\Theta(t)\langle[p(t),p(0)]\rangle_\beta$. On the other hand, the ac mobility of the TB model is related to the Fourier transform $\tilde{Y}_{TB}(\omega)$ of the respective retarded coordinate response function $Y_{TB}(t) = (i/\hbar)\Theta(t)\langle[q(t),q(0)]\rangle_\beta$ by

$$\tilde{\mu}_{TB}(\omega) = -i\omega\tilde{Y}_{TB}(\omega). \qquad (28.24)$$

To connect $\tilde{\Pi}_{TB}(\omega)$ with $\tilde{Y}_{TB}(\omega)$, we use the equations of motion resulting from the Hamiltonian (24.4). We then obtain the relation

$$\omega^2\tilde{\Pi}_{TB}(\omega) = iM\omega\tilde{\gamma}_{TB}(\omega) - M^2\omega^2\tilde{\gamma}_{TB}^2(\omega)\tilde{Y}_{TB}(\omega). \qquad (28.25)$$

Using this form, and Eqs. (28.23), (28.24) and (28.10), we find the exact relations

$$\tilde{\mu}_{WB}(\omega) = \frac{1}{M[\tilde{\gamma}_{WB}(\omega) - i\omega]} - \frac{\tilde{\gamma}_{TB}(\omega)}{\tilde{\gamma}_{WB}(\omega) - i\omega}\tilde{\mu}_{TB}(\omega), \qquad (28.26)$$

$$\tilde{\mu}_{TB}(\omega) = \frac{1}{M\tilde{\gamma}_{TB}(\omega)} - \frac{\tilde{\gamma}_{WB}(\omega) - i\omega}{\tilde{\gamma}_{TB}(\omega)}\tilde{\mu}_{WB}(\omega). \qquad (28.27)$$

The expressions (28.10)–(28.12), (28.26) and (28.27) are the central duality relations. Most remarkably, they hold for spectral coupling $\propto \omega^s$ with s in the range $0 < s < 2$. Here, the correspondence is shown for the frequency-dependent linear mobility. Clearly, the mapping can be extended to different kinds of dynamical variables.

In the low-frequency regime $\omega/\tilde{\gamma}(\omega,s) \ll 1$, the relations (28.26) and (28.27) become symmetric. With use of Eqs. (28.14) and (28.22) we obtain the concise form

$$M\tilde{\gamma}(\omega, s)\tilde{\mu}_{\mathrm{WB}}(\omega; s) = 1 - M\tilde{\gamma}(\omega, 2 - s)\tilde{\mu}_{\mathrm{TB}}(\omega; 2 - s) . \tag{28.28}$$

In the Ohmic scaling limit we have $\tilde{\gamma}_{\mathrm{WB}}(\omega) = \tilde{\gamma}_{\mathrm{TB}}(\omega) = \gamma$, and the mobility of the free Brownian particle is $\mu_0 = 1/M\gamma$. Then the duality relation reads

$$\tilde{\mu}_{\mathrm{WB}}(\omega; K) = \mu_0 - \tilde{\mu}_{\mathrm{TB}}(\omega; 1/K) . \tag{28.29}$$

The mobility is labelled with WB or TB, to indicate dependence on V_0 and Δ, respectively. With the universal frequency scale $\tilde{\epsilon}_0$ given in Eq. (28.121) the mobility loses distinction between WB and TB, and the relation (28.29) takes the self-dual form

$$\tilde{\mu}(\omega, \tilde{\epsilon}_0, K) = \mu_0 - \tilde{\mu}(\omega, \tilde{\epsilon}_0, 1/K) . \tag{28.30}$$

This relation equally applies within the TB or WB model, and between these models.

28.1.3 Nonlinear static mobility

For an external force F, the potential drop per lattice constant X_0 in the continuous WB potential (24.2) is $\hbar\epsilon_{\mathrm{WB}} = FX_0$, while the potential drop in the dual TB model per lattice constant q_0 is $\hbar\epsilon_{\mathrm{TB}} = Fq_0$. Using the relation (28.17), the drops of both models are related by $\epsilon_{\mathrm{WB}} = \epsilon_{\mathrm{TB}}/K_{\mathrm{TB}}$. Thus, in generalization of Eq. (28.29), the nonlinear static mobilities in the Ohmic scaling limit are related for all T and ϵ by

$$\mu_{\mathrm{WB}}(V_0/\hbar, T, \epsilon, K) = \mu_0 - \mu_{\mathrm{TB}}(\Delta, T, \epsilon/K, 1/K) . \tag{28.31}$$

Again we may eliminate the explicit dependence on the TB and WB frequency scales Δ and $V_0/\hbar$ by introducing the universal frequency scale $\tilde{\epsilon}_0$ given in Eq. (28.121),

$$\mu(\tilde{\epsilon}_0, T, \epsilon, K) = \mu_0 - \mu(\tilde{\epsilon}_0, T, \epsilon/K, 1/K) . \tag{28.32}$$

At this point, one remark is expedient. The self-duality relation (28.32) is free of explicit dependence on the particular model. In the mapping, both the Ohmic viscosity η and the biasing force stay constant. Because the lattice constants transform as $X_0 \leftrightarrow q_0/K$, where $K = \eta q_0^2/2\pi\hbar$, we have in the duality $K \leftrightarrow 1/K$ and $\epsilon \leftrightarrow \epsilon/K$. In addition, due to the correspondence $\omega_{\mathrm{c}} \triangleq \gamma$, the scaling limit in the TB model is mapped on the high-friction limit in the WB model, $\hbar\gamma \gg V_0, \hbar\epsilon_{\mathrm{WB}}$. The mapping described by the self-duality relation (28.32) is discussed in Section 28.4.

For $T = 0$, it is convenient to eliminate the change of the bias $\epsilon \leftrightarrow \epsilon/K$ by employing the modified frequency scale ϵ_0 given in Eqs. (28.119) and (28.120), as explained in Subsection 28.5.1. With the scaled bias $\upsilon = \epsilon/\epsilon_0$, we then have

$$\mu(\upsilon, K) = \mu_0 - \mu(\upsilon, 1/K) . \tag{28.33}$$

As an example, we now transform the TB mobility (27.16) holding for $K \ll 1$ into the mobility of the WB model. We see from the definition $K_{\mathrm{WB}} = M\gamma X_0^2/2\pi\hbar$ and the duality $K_{\mathrm{WB}} = 1/K_{\mathrm{TB}}$ that the condition $K_{\mathrm{TB}} \to 0$ corresponds to the classical limit $K_{\mathrm{WB}} \to \infty$ in the dual WB model.

Using the self-duality relation (28.31), the scaled mobility of the WB model in the classical limit $K_{WB} \to \infty$ is found from the TB expression (27.16) to read ($\epsilon = \epsilon_{TB}/K$)

$$\mathcal{M}_{WB}(T, \epsilon) = \frac{\sinh(\beta\hbar\epsilon/2)}{\beta\hbar\epsilon/2} \frac{1}{|I_{i\beta\hbar\epsilon/2\pi}(\beta V_0)|^2}. \tag{28.34}$$

This expression reduces in the special cases $\epsilon = 0$ and $T = 0$, respectively, to

$$\mathcal{M}_{WB}(T, 0) = 1/I_0^2(\beta V_0), \tag{28.35}$$

$$\mathcal{M}_{WB}(0, \epsilon) = \Theta(\epsilon - \epsilon_0)\sqrt{1 - (\epsilon_0/\epsilon)^2}. \tag{28.36}$$

Here we have put the Kondo scale (28.119). In the classical limit, this is

$$\epsilon_0 = 2\pi V_0/\hbar. \tag{28.37}$$

The results (28.34)–(28.36) coincide with those of the Smoluchowski diffusion equation (11.19) (see Ref. [318]). For $\epsilon \leqslant \epsilon_0$ or equivalently $V_{tilt} \equiv F X_0/2\pi \leqslant V_0$, the mobility is zero in the classical limit because the creeping motion of the overdamped particle comes to a stop at a point with zero slope and cannot move anymore since there are neither quantal nor thermal fluctuations. For $V_{tilt} > V_0$, the potential (24.2) is sloping downwards everywhere, so that the particle is not trapped. This results in a nonzero mobility at $T = 0$ in the classical limit for $V_{tilt} > V_0$.

As already emphasized, the mobility is most easily calculated in the TB representation. A major advantage of the duality now is that one can shift the numerical problem of calculating or simulating the dynamics of the WB model to the discrete TB model. For instance in quantum Monte Carlo simulations of the dynamics, the discrete variables of the latter model significantly reduce the relevant configuration space subject to Monte Carlo sampling compared to the continuous model, and meaningful real-time simulations in the interesting non-perturbative low-temperature regime are possible [528]. In many interesting cases, e. g. charge transfer in chemical reactions, the density $J_{WB}(\omega)$ is not in the simple power-law form (28.13), and also the band width ω_c may be of the same order as the other frequencies. Nevertheless, the associated density $J_{TB}(\omega)$ is given by (28.12). Therefore, one may take advantage of performing numerical simulations for the equivalent TB model. Since the mapping is exact, the entire regime from quantum tunneling to thermal hopping across the barrier can be studied in the equivalent TB model.

In conclusion, we have discussed an exact duality symmetry between quantum Brownian motion in a continuous cosine potential and in a discrete tight-binding lattice. Because excited states at each lattice site are disregarded in the TB limit, one might think that a general mapping between the two models is impossible. However, the TB model is fully sensitive to all aspects of the quantum and thermal hopping dynamics of the continuous model. In fact, equivalence of the discrete TB model with the continuous WB model is reached by taking into account the change of the spectral density of the coupling as given in Eq. (28.12).

28.2 Self-duality in the exactly solvable cases $K = \frac{1}{2}$ and $K = 2$

In Section 27.3 we have discussed the three lowest cumulants of the tight-binding transport model (24.4) in the Ohmic scaling regime at $K = \frac{1}{2}$. Now, we first extend the discussion to the full counting statistics. Then we demonstrate, upon employing self-duality, that the resulting cumulant generating function can be transformed into the CGF of the TB model at $K = 2$. Finally, we show that the CGFs also describe the full-counting statistics of the WB model (24.1) in corresponding regimes.

28.2.1 Full counting statistics at $K = \frac{1}{2}$

In Subsection 27.3.1 we have seen that the current of the bosonic transport problem at $K = \frac{1}{2}$ can be mapped on one of free fermions with the transmission probability (27.33). Therefore, it seems reasonable that the cumulant generating function is given in fermionic representation by the Levitov-Lesovik formula [529]

$$
\ln \mathcal{Z}_{\mathrm{F}}(\rho, t) = t \int_0^\infty \frac{d\omega}{2\pi} \ln\Big\{ 1 + \mathcal{T}(\omega) \big[\mathcal{N}_+(\omega, \epsilon) \, (e^{i \, 2q_0\rho} - 1) \\
+ \mathcal{N}_-(\omega, \epsilon) \, (e^{-i \, 2q_0\rho} - 1) \big] \Big\}
\tag{28.38}
$$

with the spectral transmission probability at frequency ω [see Eq. (27.33)]

$$
\mathcal{T}(\omega) = \frac{\tilde{\epsilon}_0^2}{\tilde{\epsilon}_0^2 + \omega^2} .
\tag{28.39}
$$

In fact, the CGF (28.38) reproduces the cumulants $\langle q(t) \rangle_{\mathrm{c}}$, $\langle q^2(t) \rangle_{\mathrm{c}}$ and $\langle q^3(t) \rangle_{\mathrm{c}}$, as given in Eqs. (27.35), (27.38) and (27.39).

In the remainder of this subsection, we consider transport and noise at zero temperature. In this limiting case, up-hill particle flow is absent, $\mathcal{N}_+(\omega, \epsilon) = \Theta(\epsilon - |\omega|)$, and $\mathcal{N}_-(\omega, \epsilon) = 0$, and the CGF (28.38) yields for the cumulants the concise form

$$
\langle q^N(t) \rangle_{\mathrm{c}} = t \, 2q_0^N \int_0^\epsilon \frac{d\omega}{2\pi} \, \mathcal{T}_N(\omega) .
\tag{28.40}
$$

Here

$$
\begin{aligned}
\mathcal{T}_1(\omega) &= \mathcal{T}(\omega) , \\
\mathcal{T}_2(\omega) &= 2\mathcal{T}(\omega)[1 - \mathcal{T}(\omega)] , \\
\mathcal{T}_3(\omega) &= 4\mathcal{T}(\omega)[1 - \mathcal{T}(\omega)][1 - 2\mathcal{T}(\omega)] ,
\end{aligned}
\tag{28.41}
$$

etc. The function $\mathcal{T}_N(\omega)$ can be generated by differentiation,

$$
\mathcal{T}_N(\omega) = [\tilde{\epsilon}_0(d/d\tilde{\epsilon}_0)]^{N-1} \mathcal{T}(\omega) .
\tag{28.42}
$$

This allows to express every single cumulant in terms of derivatives of the current,

$$
\langle q^N(t) \rangle_{\mathrm{c}} = \left(q_0 \, \tilde{\epsilon}_0 \frac{d}{d\tilde{\epsilon}_0} \right)^{N-1} \langle q(t) \rangle_{\mathrm{c}} = t \left(q_0 \, \tilde{\epsilon}_0 \frac{d}{d\tilde{\epsilon}_0} \right)^{N-1} \langle I \rangle .
\tag{28.43}
$$

On the other hand, the CGF can be written in the form (25.24). Since backward flow is absent at $T = 0$, this is simplified to the expression

$$\ln \mathcal{Z}(\rho, t) = t \sum_{n=1}^{\infty} \left(e^{i \rho q_0 n} - 1 \right) k_n^+ .$$
(28.44)

The resulting cumulants are [cf. Eq. (25.25)]

$$\langle q^N(t) \rangle_c = t q_0^N \sum_{n=1}^{\infty} n^N k_n^+ .$$
(28.45)

Apparently, the form (28.45) can be reconciled with the expression (28.43) for arbitrary N if and only if the transition weight k_n^+ is of the particular order $\tilde{\epsilon}_0^n$, i.e., the transition weight k_n^+ is established by the paths with the minimum number $2n$ of steps. This specific property allows us to extract all transition weights k_n^+ from the power series in $\tilde{\epsilon}_0$ of one of the cumulants, e.g. from the series of the mobility. The power series in $\tilde{\epsilon}_0$ of the second form in Eq. (27.29) yields for the transition weight k_n^+ at zero temperature and $K = \frac{1}{2}$ the analytic expression[3]

$$k_n^+ = \frac{\epsilon}{\pi} \frac{(-1)^{n-1}}{n} \frac{\sin[\frac{\pi}{2}(n-1)]}{n-1} \left(\frac{\tilde{\epsilon}_0}{\epsilon} \right)^n .$$
(28.46)

The rate k_n^+ is zero for odd n, except for the case $n = 1$, The non-vanishing rates are

$$k_1^+ = \frac{\tilde{\epsilon}_0}{2} , \qquad k_{2n}^+ = \frac{\epsilon}{\pi} \frac{(-1)^n}{2n} \frac{1}{2n-1} \left(\frac{\tilde{\epsilon}_0}{\epsilon} \right)^{2n} , \qquad n = 1, 2, \cdots .$$
(28.47)

Hence the sign of the transition weights is not strictly positive. This is the interference phenomenon addressed in Subsec. 25.2.2 and discussed in greater depth in Chap. 29.

The integral (28.38) at $T = 0$ and the sum (28.44) with (28.46) can be done in analytic form. The resulting CGFs actually coincide, $\ln \mathcal{Z}_F(\rho, t) = \ln \mathcal{Z}(\rho, t)$.

The expressions given here have lost explicit dependence on the particular representation. In application to the TB model (24.4), the length q_0 is the lattice constant, ϵ is the bias frequency and the renormalized tunneling amplitude is [cf. Eq. (27.23)]

$$\tilde{\epsilon}_0 = \pi \frac{\Delta^2}{\omega_c} .$$
(28.48)

Passage to the WB representation (24.1) with lattice period X_0 and bias frequency ϵ_{WB} implies the substitutions $q_0 \to X_0/2$ and $\epsilon \to \epsilon_{WB}/2$, as follows from the relation (28.17). Further, we see from the relation (28.121) at $K = \frac{1}{2}$ that the dressed tunneling amplitude $\tilde{\epsilon}_0$ is expressed in terms of the WB parameters as

$$\tilde{\epsilon}_0 = \frac{1}{2\pi} \frac{\omega_c^2}{V_0/\hbar} .$$
(28.49)

With these assignments, the above expressions for the CGF and the cumulants are equally valid both for the WB and the TB model at $K = \frac{1}{2}$. The perturbative series (28.45) represents both the weak-tunneling series of the TB model and the strong-backscattering series of the WB model.

---

[3]The rate (28.46) coincides with the rate (29.6) for $K = 1/2$ and $\tilde{\epsilon}_0 = \epsilon_0/2$ [cf. Eq. (28.121)].

28.2.2 Full counting statistics at $K = 2$

In the preceding subsection, we have seen that the full counting statistics at the Toulouse point $K = \frac{1}{2}$ is available in analytic form. Equipped with this solution, the case $K = 2$ can now be tackled without splendid calculation upon employing self-duality of the model under consideration. The dual mapping $K = \frac{1}{2} \leftrightarrow K = 2$ in the TB representation implies

$$\tilde{\epsilon}_0 \leftrightarrow \tilde{\epsilon}_0 , \qquad \text{and} \qquad \epsilon \leftrightarrow \epsilon/2 , \tag{28.50}$$

as follows from the relations (28.18) and (28.32).

Curent and mobility:

Employing the mapping (28.32) for $K = \frac{1}{2}$, we find from the expression (27.27) that the normalized mobility (27.1) at $K = 2$ takes the form

$$\mathcal{M} = 1 - \frac{2\tilde{\epsilon}_0}{\epsilon} \operatorname{Im} \psi \left(\frac{1}{2} + \frac{\beta\hbar\tilde{\epsilon}_0}{2\pi} + i \frac{\beta\hbar\epsilon}{4\pi} \right) , \tag{28.51}$$

and the mean particle current is

$$\langle I \rangle = q_0 \frac{\epsilon}{4\pi} - q_0 \frac{\tilde{\epsilon}_0}{2\pi} \operatorname{Im} \psi \left(\frac{1}{2} + \frac{\beta\hbar\tilde{\epsilon}_0}{2\pi} + i \frac{\beta\hbar\epsilon}{4\pi} \right) . \tag{28.52}$$

Here, q_0 and $\epsilon = F q_0/\hbar$ are the lattice constant and bias frequency in the TB representation (24.4). The renormalized scattering amplitude $\tilde{\epsilon}_0$ is connected with the TB parameters Δ and ω_c by the relation

$$\tilde{\epsilon}_0 = \frac{1}{2\pi} \frac{\omega_c^2}{\Delta} , \tag{28.53}$$

as follows from Eq. (28.121) at $K = 2$. Because $\tilde{\epsilon}_0 \propto 1/\Delta$, the weak-tunneling series of current and mobility emerge from the asymptotic expansion of the digamma function in the expressions (28.51) and (28.52). The leading term of the weak tunneling series resulting from the expression (28.51) coincides indeed with the weak-tunneling expression (27.4) at $K = 2$,

$$\mathcal{M} = \frac{\pi^2}{3} \frac{\Delta^2}{\omega_c^4} \left(\epsilon^2 + \left(\frac{2\pi}{\hbar\beta} \right)^2 \right) . \tag{28.54}$$

Passage to the WB model implies the substitutions $q_0 \to 2X_0$ and $\epsilon \to 2\epsilon_{\text{WB}}$, and the relation of the dressed amplitude with the WB parameters is

$$\tilde{\epsilon}_0 = \pi \frac{(V_0/\hbar)^2}{\omega_c} . \tag{28.55}$$

With these assignments, we may switch easily between the dual representations.

In the limit $T \to 0$, the expression (28.52) is simplified to the equivalent forms

$$\langle I \rangle / q_0 = \frac{\epsilon}{4\pi} - \frac{\tilde{\epsilon}_0}{2\pi} \arctan \left(\frac{\epsilon}{2\tilde{\epsilon}_0} \right) = \frac{\epsilon}{4\pi} - \frac{\tilde{\epsilon}_0}{4} + \frac{\tilde{\epsilon}_0}{2\pi} \arctan \left(\frac{2\tilde{\epsilon}_0}{\epsilon} \right) . \tag{28.56}$$

The Taylor expansion of the first form yields the weak-tunneling series of the current for the TB model, and the strong-backscattering series for the WB model. The Taylor expansion of the second form gives the strong-tunneling and weak-backscsattering series of the TB and WB model, respectively.

By taking up the line pursued in Subsection 27.3.1, we may transform the expression (28.52) into fermionic representation. The resulting forms are

$$\langle I \rangle = q_0 \frac{\epsilon}{4\pi} - q_0 \int_0^\infty \frac{d\omega}{2\pi} \mathcal{R}(\omega) \, F_1(\omega, \epsilon/2) = q_0 \int_0^\infty \frac{d\omega}{2\pi} \overline{\mathcal{T}}(\omega) \, F_1(\omega, \epsilon/2) \,, \quad (28.57)$$

where $\overline{\mathcal{T}}(\omega)$ and $\overline{\mathcal{R}}$ are the spectral transition and reflexion probabilities at $K = 2$,

$$\overline{\mathcal{T}}(\omega) = \frac{\omega^2}{\omega^2 + \tilde{\epsilon}_0^2} \,, \qquad \overline{\mathcal{R}}(\omega) = \frac{\tilde{\epsilon}_0^2}{\omega^2 + \tilde{\epsilon}_0^2} \,. \quad (28.58)$$

Comparison of the expressions (28.57) with the forms (27.35) shows that in the mapping $K = \frac{1}{2} \leftrightarrow K = 2$ transmission and reflexion perform role reversal

$$\mathcal{T}(\omega) = \overline{\mathcal{R}}(\omega) \,, \qquad \text{and} \qquad \mathcal{R}(\omega) = \overline{\mathcal{T}}(\omega) \,. \quad (28.59)$$

The meaning of the expression (28.57) is as in the case $K = \frac{1}{2}$. The first term in the first form is the maximum current in the absence of the barrier. The second term diminishes the current because of the possibility that the particle is reflected at the barrier. The second form expresses the current in terms of the barrier transmission.

Full counting statistics:

With the relations (28.50) and the substitution $\mathcal{T}(\omega) \rightarrow \overline{\mathcal{T}}(\omega)$, we may also deduce the cumulant generating function at $K = 2$ from the CGF expression at $K = \frac{1}{2}$, Eq. (28.38). The resulting expression is

$$\ln \mathcal{Z}(\rho, t) = t \int_0^\infty \frac{d\omega}{2\pi} \ln \left\{ 1 + \overline{\mathcal{T}}(\omega) \left[\mathcal{N}_+(\omega, \epsilon/2) \, (e^{i q_0 \rho} - 1) \right. \right.$$
$$\left. \left. + \, \mathcal{N}_-(\omega, \epsilon/2) \, (e^{-i q_0 \rho} - 1) \right] \right\} \,. \quad (28.60)$$

The CGF yields for the second and third cumulant the fermionic frequency integrals analogous to the expressions (27.38) and (27.39),

$$\langle q^2(t) \rangle_{\mathrm{c}} = t q_0^2 \int_0^\infty \frac{d\omega}{2\pi} \left[\overline{\mathcal{T}}(\omega) F_2(\omega, \epsilon/2) - \overline{\mathcal{T}}^2(\omega) F_1^2(\omega, \epsilon/2) \right] \,,$$

$$\langle q^3(t) \rangle_{\mathrm{c}} = t q_0^3 \int_0^\infty \frac{d\omega}{2\pi} \overline{\mathcal{T}}(\omega) F_1(\omega, \epsilon/2) \quad (28.61)$$
$$\times \left[1 - 3\overline{\mathcal{T}}(\omega) F_2(\omega, \epsilon/2) + 2\overline{\mathcal{T}}^2(\omega) F_1^2(\omega, \epsilon/2) \right] \,.$$

Again, these forms can be rewritten in terms of derivatives of lower-order cumulants,

$$\langle q^2(t) \rangle_{\rm c} = \frac{q_0}{2} \left\{ -\coth\left(\tfrac{1}{2}\beta\hbar\epsilon\right) \tilde{\epsilon}_0 \frac{d}{d\tilde{\epsilon}_0} + \left(2 + \tilde{\epsilon}_0 \frac{d}{d\tilde{\epsilon}_0}\right) \frac{2}{\hbar\beta} \frac{d}{d\epsilon} \right\} \langle q(t) \rangle_{\rm c} , \qquad (28.62)$$

$$\langle q^3(t) \rangle_{\rm c} = \frac{q_0}{2} \left\{ -\coth\left(\tfrac{1}{2}\beta\hbar\epsilon\right) \tilde{\epsilon}_0 \frac{d}{d\tilde{\epsilon}_0} + \left(2 + \frac{\tilde{\epsilon}_0}{2} \frac{d}{d\tilde{\epsilon}_0}\right) \frac{2}{\hbar\beta} \frac{d}{d\epsilon} \right\} \langle q^2(t) \rangle_{\rm c} \qquad (28.63)$$

$$+ \frac{q_0^2}{4} \left\{ \coth\left(\tfrac{1}{2}\beta\hbar\epsilon\right) \frac{2}{\hbar\beta} \frac{d}{d\epsilon} + \frac{1}{\sinh^2(\tfrac{1}{2}\beta\hbar\epsilon)} \right\} \tilde{\epsilon}_0 \frac{d}{d\tilde{\epsilon}_0} \langle q(t) \rangle_{\rm c} .$$

At $T = 0$, we have $\mathcal{N}_+(\omega, \epsilon/2) = \Theta(\epsilon/2 - |\omega|)$ and $N_-(\omega, \epsilon/2) = 0$. Eq. (28.60) yields a relation which expresses cumulants again in terms of derivatives of the mean current,

$$\langle q^N(t) \rangle_{\rm c} = t \left(-\frac{q_0}{2} \tilde{\epsilon}_0 \frac{d}{d\tilde{\epsilon}_0} \right)^{N-1} \langle I \rangle . \qquad (28.64)$$

The weak-tunneling representations of the CGF and the cumulants can be written as given in Eqs. (28.44) and (28.45), respectively, but now the transition weights are[4]

$$\tilde{k}_n^+ = \frac{\epsilon}{4\pi} \frac{(-1)^{n-1}}{n(2n+1)} \left(\frac{\epsilon}{2\tilde{\epsilon}_0} \right)^{2n} . \qquad (28.65)$$

Like in the case $K = \tfrac{1}{2}$, Eq. (28.47), the sign of the transition weights k_n^+ alternates.

We conclude the discussion of the cases $K = \tfrac{1}{2}$ and $K = 2$ with an outlook. First we anticipate that the full counting statistics at $T = 0$ can be found in analytic form for general K (see Chap. 29). The relations (28.43) and (28.64) are consistent with the general functional relations (29.5), and the rates (28.46) and (28.65) are special cases of the general rate expression (29.6). The varying sign of the transition weights is a signature of quantum interference of the contributing decay channels.

28.3 Duality and supercurrent in Josephson junctions

28.3.1 Charge-phase duality

In Subsection 3.4.3 we have introduced the model Hamiltonian (3.248) for the Josephson junction. For weak Josephson coupling, $E_{\rm J} \ll E_{\rm c}$, the charge on the junction is nearly sharp and, therefore, the charge representation (3.158) is appropriate. The fluctuations of the phase jump across the junction are described by the phase correlation function $Q_\psi(t)$ defined in Eq. (3.249).

In the opposite limit of large Josephson coupling, $E_{\rm J} \gg E_{\rm c}$, the phase jump across the junction is nearly sharp. The relevant Hamiltonian

$$H = -V_0 \cos\left(\frac{2\pi \bar{Q}}{2e} \right) + \sum_\alpha \left[\frac{\bar{q}_\alpha^2}{2C_\alpha} + \left(\frac{\hbar}{2e} \right)^2 \frac{1}{2\bar{L}_\alpha} (\bar{\psi} - \bar{\psi}_\alpha)^2 \right], \qquad (28.66)$$

---

[4]The rate (28.65) coincides with the rate (29.8) for $K = 1/2$ and $\tilde{\epsilon}_0 = \epsilon_0/4$ [cf. Eq. (28.121)].

is conveniently represented in the basis of discrete phase states $|n\rangle$,

$$
\begin{aligned}
\bar{\psi} &= 2\pi \sum_n n a_n^\dagger a_n = 2\pi \sum_n n |n\rangle\langle n| , \\
e^{i 2\pi \bar{Q}/2e} &= \sum_n a_n^\dagger a_{n+1} = |n\rangle\langle n+1| .
\end{aligned}
\tag{28.67}
$$

The Hamiltonians (3.248) and (28.66) correspond to the WB and TB Hamiltonians, Eq. (24.1) and Eq. (24.4), respectively.

Following the discussion given in Subsec. 28.1.1, we now show that the Hamiltonian (3.248) can be transformed into the form (28.66). First, we put $E_J \to U_0$, and we perform the canonical transformation [we put $\lambda_\alpha = (\hbar/2e)\sqrt{C_\alpha/L_\alpha}$],

$$
\begin{aligned}
q_\alpha &\to -\lambda_\alpha \psi_\alpha , & \psi_\alpha &\to q_\alpha/\lambda_\alpha + (\pi/e)\bar{Q} , \\
\psi &\to \pi\bar{Q}/e , & Q &\to -(e/\pi)\bar{\psi} + \sum_\alpha \lambda_\alpha \psi_\alpha .
\end{aligned}
\tag{28.68}
$$

In the transformed reservoir Hamiltonian H_{R} the electromagnetic modes are coupled,

$$
\begin{aligned}
H &= -V_0 \cos\left(\frac{2\pi\bar{Q}}{2e}\right) + \left(\frac{e}{\pi}\right)^2 \frac{\bar{\psi}^2}{2C} - \frac{e\bar{\psi}}{\pi C} \sum_\alpha \lambda_\alpha \psi_\alpha + H_{\mathrm{R}} , \\
H_{\mathrm{R}} &= \sum_\alpha \left[\frac{q_\alpha^2}{2C_\alpha} + \left(\frac{\hbar}{2e}\right)^2 \frac{\psi_\alpha^2}{2L_\alpha}\right] + \frac{1}{2C}\left(\sum_\alpha \lambda_\alpha \psi_\alpha\right)^2 .
\end{aligned}
\tag{28.69}
$$

The mapping is completed by equating the $\bar{\psi}^2$ terms in Eqs. (28.69) and (28.66),

$$
\left(\frac{e}{\pi}\right)^2 \frac{1}{C} = \left(\frac{\hbar}{2e}\right)^2 \sum_\alpha \frac{1}{L_\alpha} ,
\tag{28.70}
$$

and by switching to the diagonal basis of the bath modes. The ratio of the determinants of the dynamical matrices $\mathbf{A}(\omega)$ and $\mathbf{B}(\omega)$,

$$
B_{\alpha\beta}(\omega) = \delta_{\alpha\beta} C_\alpha (\hbar/2e)^2 (\omega_\alpha^2 - \omega^2) , \qquad A_{\alpha\beta}(\omega) = B_{\alpha\beta}(\omega) + \lambda_\alpha\lambda_\beta/C , \tag{28.71}
$$

where $\omega_\alpha = 1/\sqrt{L_\alpha C_\alpha}$, reads

$$
\frac{\det \mathbf{A}(\omega)}{\det \mathbf{B}(\omega)} = 1 + \frac{1}{C} \sum_\alpha \frac{1}{L_\alpha} \frac{1}{\omega_\alpha^2 - \omega^2} = \frac{-i\,\omega C + Y^*(\omega)}{-i\,\omega C} .
\tag{28.72}
$$

In the second form we have introduced the admittance $Y(\omega)$ given in Eq. (3.222). On the other hand, in the eigen basis of the matrix $\mathbf{A}(\omega)$ we have

$$
\begin{aligned}
\bar{A}_{\alpha\beta} &= \delta_{\alpha\beta} \bar{C}_\alpha \left(\frac{\hbar}{2e}\right)^2 (\bar{\omega}_\alpha^2 - \omega^2) , \\
\bar{B}_{\alpha\beta} &= \bar{A}_{\alpha\beta} - \frac{\pi^2 C}{e^2}\left(\frac{\hbar}{2e}\right)^2 \bar{\lambda}_\alpha \bar{\omega}_\alpha \bar{\lambda}_\beta \bar{\omega}_\beta .
\end{aligned}
\tag{28.73}
$$

In this representation we obtain with use of Eq. (28.70)

$$\frac{\det \mathbf{B}(\omega)}{\det \mathbf{A}(\omega)} = -i\,\omega C \left(\frac{\pi\hbar}{2e^2}\right)^2 \sum_\alpha \frac{1}{\bar{L}_\alpha} \frac{-i\,\omega}{\bar{\omega}_\alpha^2 - \omega^2} = -i\,\omega C\, R_Q^2 \bar{Y}^*(\omega)\,, \qquad (28.74)$$

where $R_Q = h/4e^2$ is the resistance quantum for Cooper pairs. Equating the inverse of Eq. (28.74) with the expression (28.72), we find a simple relation between the admittance $Y(\omega)$ of the model for weak Josephson coupling and the admittance $\bar{Y}(\omega)$ of the strong-coupling model,

$$\bar{Y}(\omega)\,[\,i\,\omega C + Y(\omega)\,] = 1/R_Q^2\,. \qquad (28.75)$$

Using this correspondence, we can express quantitites of the one model in terms of those of the other. The phase-charge duality transformations are [154, 157]

$$\psi \longleftrightarrow \pi Q/e\,, \qquad \frac{1}{R_Q[\,i\,\omega C + Y(\omega)\,]} \longleftrightarrow R_Q Y(\omega)\,. \qquad (28.76)$$

The phase autocorrelation function $Q_\psi(t)\langle[\,\psi(0) - \psi(t)\,]\psi(0)\rangle_\beta$ takes the form (3.249). The spectral density of the coupling reads [cf. Eq. (3.250) with (3.229)]

$$G_\psi(\omega) = 2\omega R_Q \operatorname{Re}\bar{Y}(\omega) = 2\omega \operatorname{Re}\frac{1}{R_Q[\,i\,\omega C + Y(\omega)\,]}\,. \qquad (28.77)$$

In the strong-coupling limit $E_J \gg E_c$, the influence of the environment is described by the charge autocorrelation function $Q_Q(t) \equiv \langle[\,Q(0) - Q(t)\,]Q(0)\rangle_\beta$. Using correspondence (28.76), we find the function $Q_Q(t)$ in the form (3.249) with the spectral density

$$G_Q(\omega) = 2(e/\pi)^2 R_Q \operatorname{Re} Y(\omega)\,\omega\,. \qquad (28.78)$$

In the Ohmic scaling limit, we then obtain with

$$\rho = R/R_Q \qquad (28.79)$$

the forms

$$G_\psi(\omega) = 2\rho\omega\,, \qquad \text{and} \qquad G_Q(\omega) = (e/\pi)^2\,(2/\rho)\,\omega\,, \qquad (28.80)$$

which provide exact self-duality between the charge and phase representation. These results can directly be transferred to electron tunneling through a normal junction with the substitution $2e \to e$ and $R_Q \to R_K$. The respective spectral densities are in correspondence with the earlier results (3.226) with (3.227) and (3.230) with (3.231).

Nano-electronic devices based on low-capacitance Josephson junctions appear to be suitable for large-scale integration and physical realization of quantum bits. In the regime $E_c \gg E_J$, the Coulomb blockade effect allows the controll of individual charges [116, 530]. Single- and two-bit operations may be performed upon applying sequences of gate voltages. When the gate voltage is chosen such that neighboring charge states are degenerate, the system reduces to two states which are weakly coupled by E_J [cf. Subsection 3.2.1]. One may also think of quantum-logic Josephson elements with flux states instead of charge states. The effects of the electromagnetic environment are described by the phase correlation function (3.249) with (3.250).

In Subsection 20.3.4 we have studied the supercurrent through an ultrasmall Josephson junction with Josephson coupling energy E_J in the weak-tunneling limit, in which $I \propto E_J^2$. In the region $\rho < 1$, we found divergence of the expression (20.162) for $\rho < 1$ in the low-energy regime $k_B T$, $eU \ll E_J (E_J/E_c)^{\rho/(1-\rho)}$. This indicates that tunneling processes of higher order in E_J become relevant in this parameter regime. As we shall see, the higher-order terms smooth out the singularity in question.

Interestingly enough, in the particular cases $\rho \ll 1$, $\rho = \frac{1}{2}$ and $\rho = 2$, the tunneling dynamics of Cooper pairs can be solved in analytic form in all orders in E_J.

28.3.2 Supercurrent-voltage characteristics for $\rho \ll 1$

By virtue of the correspondence relations (20.157) between tunneling of bosons under Ohmic friction and tunneling of Cooper pairs in a resistive electromagnetic environment, we can translate the Ivanchenko-Zil'berman expression (27.15) into the supercurrent-voltage characteristics of a small capacitance Josephson junction [527],

$$I(U) = \frac{U}{R} \frac{\beta E_J^*}{\nu} \, \mathrm{Im} \left(\frac{I_{1-i\nu}(\beta E_J^*)}{I_{-i\nu}(\beta E_J^*)} \right), \qquad \rho \ll 1, \qquad (28.81)$$

where $\nu = \beta eU/\pi\rho$. Furthermore, E_J^* is the equivalent of $\hbar\Delta_T$ defined in Eq. (27.11),

$$E_J^* = E_J \, e^{-\rho\zeta} \left(\frac{2\pi^2\rho}{\beta E_c} \right)^\rho, \qquad (28.82)$$

where ζ is given in Eq. (20.160) with (20.133). The formula (28.81) can be rewritten as [see Eqs. (27.15) and (27.16)]

$$I(U) = \frac{U}{R} \left(1 - \frac{\sinh(\pi\nu)}{\pi\nu} \frac{1}{|I_{i\nu}(\beta E_J^*)|^2} \right), \qquad \rho \ll 1. \qquad (28.83)$$

The $I(U)$-characteristics (28.83) covers the full range from weak to strong Cooper pair tunneling. It shows a peak at small voltage, as explained below Eq. (20.162).

The expression (28.83) has three important limits:

(1) The order E_J^2 coincides with the weak-tunneling expression (20.165).

(2) The linear superconductance at general T is found to read

$$G_{\mathrm{lin}}(T) \equiv \lim_{\nu \to 0} \frac{I(U)}{U} = \frac{1}{R} \left(1 - \frac{1}{I_0^2(\beta E_J^*)} \right). \qquad (28.84)$$

(3) The supercurrent-voltage characteristics at zero temperature is

$$I(U) = \Theta(U - RI_c) \frac{U}{R} \left(1 - \sqrt{1 - (RI_c/U)^2} \right). \qquad (28.85)$$

Here we have used the relation $\beta E_J/\nu = RI_c/U$.

By means of the duality map it is straightforward to translate the expression (28.83) into the *I-V* characteristics of a large capacitance Josephson junction ($E_J \gg E_c$) coupled to a very strong resistive environment, $R/R_Q \gg 1$. The relevant Hamiltonian is given in Eq. (28.66), and the spectral density weighting the charge fluctuations is $G_Q(\omega) = (e/\pi)^2 2(R_Q/R)\,\omega$. The changeover to the latter case is analogous to the passage from the expression (27.16) to the form (28.34). In the classical limit $R/R_Q \to \infty$, the dc voltage drop U at the current-biased junction with bias current $I_{\text{ext}} = I_b$ due to the backscattering current is (we put $\sigma = e\beta R I_b/\pi$) [531]

$$U(I_b) = RI_b\left(1 - \frac{\sinh(\pi\sigma)}{\pi\sigma}\frac{1}{|I_{i\sigma}(\beta V_0)|^2}\right). \qquad (28.86)$$

As $T \to 0$, this reduces to

$$U(I_b) = \Theta(I_b - \pi V_0/eR)\,RI_b\left(1 - \sqrt{1 - (\pi V_0/eRI_b)^2}\right). \qquad (28.87)$$

28.3.3 Supercurrent-voltage characteristics at $\rho = \frac{1}{2}$

With the spadework set out in Section 27.3, it is only a small step to give the supercurrent-voltage characteristics for the case $\rho = \frac{1}{2}$, or equivalently $R = \pi\hbar/4e^2$. Employing the correspondence relations (20.157)–(20.159), the renormalized tunneling amplitude (28.48) is expressed in terms of the device parameters as

$$\tilde{\epsilon}_0 = \frac{\pi^2}{2\hbar}\frac{E_J^2\,e^{-\varsigma}}{E_c} = \frac{e\pi}{\hbar}\frac{E_J\,e^{-\varsigma}}{E_c}RI_c. \qquad (28.88)$$

Putting $q_0 \hat{=} 2e$, the expression (27.26) transforms into the supercurrent

$$I(U) = I_c\frac{\hbar\tilde{\epsilon}_0}{\pi E_J}\,\text{Im}\,\psi\left(\frac{1}{2} + \frac{\beta\hbar\tilde{\epsilon}_0}{2\pi} + i\frac{\beta eU}{\pi}\right). \qquad (28.89)$$

At zero temperature, the current-voltage characteristics takes the form [532]

$$I(U) = I_c\frac{\hbar\tilde{\epsilon}_0}{\pi E_J}\,\arctan\left(\frac{2eU}{\hbar\tilde{\epsilon}_0}\right) = I_c\frac{\hbar\tilde{\epsilon}_0}{2E_J} - I_c\frac{\hbar\tilde{\epsilon}_0}{\pi E_J}\,\arctan\left(\frac{\hbar\tilde{\epsilon}_0}{2eU}\right). \qquad (28.90)$$

In contrast to the regime $\rho < \frac{1}{2}$, this expression does not show a peak structure anymore. The first term in the second form is the plateau value. It is equal to the Coulomb blockade expression (20.163) at $\rho = \frac{1}{2}$. This term is the leading one for weak Cooper pair tunneling, i.e., when the bias energy $2eU$ is large. The opposite regime of strong Cooper pair tunneling is captured, when the bias energy $2eU$ is small compared to $\hbar\tilde{\epsilon}_0$.

28.3.4 Supercurrent-voltage characteristics at $\rho = 2$

At $\rho = 2$, which corresponds to $R = \pi\hbar/e^2$, the renormalized frequency (28.53) with the correspondence relations (20.157)–(20.159) takes the form

$$\tilde{\epsilon}_0 = \frac{e}{\hbar}\left(\frac{E_c\,e^\varsigma}{4\pi^2 E_J}\right)^2 RI_c \,. \tag{28.91}$$

The supercurrent-voltage characteristics dual to the expression (28.89) is

$$I(U) = \frac{U}{R} - \left(\frac{E_c\,e^\varsigma}{4\pi^2 E_J}\right)^2 I_c\,\operatorname{Im}\psi\left(\frac{1}{2} + \frac{\beta\hbar\tilde{\epsilon}_0}{2\pi} + i\frac{\beta eU}{2\pi}\right). \tag{28.92}$$

In the zero temperature limit, this expressions simplifies to [532]

$$I(U) = \frac{U}{R} - \left(\frac{E_c\,e^\varsigma}{4\pi^2 E_J}\right)^2 I_c\,\arctan\left[\left(\frac{4\pi^2 E_J}{E_c\,e^\varsigma}\right)^2\frac{U}{RI_c}\right]. \tag{28.93}$$

The power series in U represents the weak-tunneling series. The leading term $\propto U^3$ matches exactly the Coulomb blockade expression (20.163) at $\rho = 2$.

In the opposite limit of large U, we have strong Cooper pair tunneling. The corresponding weak-backscattering series is found from the supercurrent-voltage relation

$$I(U) = \frac{U}{R} - \frac{\pi}{2}\left(\frac{E_c\,e^\varsigma}{4\pi^2 E_J}\right)^2 I_c + \left(\frac{E_c\,e^\varsigma}{4\pi^2 E_J}\right)^2 I_c\,\arctan\left(\left(\frac{E_c\,e^\varsigma}{4\pi^2 E_J}\right)^2\frac{RI_c}{U}\right), \tag{28.94}$$

which is complementary to the expression (28.93). Interestingly, the leading backscattering term is independent of the applied voltage.

With the correspondences (20.157)–(20.159), it is easy to translate the expressions for higher order cumulants and the CGF, which for $K = \frac{1}{2}$ are given in Eqs. (27.40), (27.41), and (28.38), and for $K = 2$ in Eqs. (28.62), (28.63) and (28.60), into the corresponding expressions for Cooper pair tunneling at $\rho = \frac{1}{2}$ and at $\rho = 2$.

28.4 Self-duality in the Ohmic scaling limit

In the WB model (24.1), the scaling limit (18.56) corresponds to the high-friction (Smoluchowski) limit $\gamma = \omega_c \gg V_0/\hbar$, ϵ_{WB}. Without the potential, the normalized mobility (27.1) is $\mathcal{M}_{WB} = 1$. With the potential, the mobility is reduced because of backscattering in the corrugated potential. The backscattering contribution of the model (24.1) can be written as a perturbative series in the corrugation strength V_0. It is analogous in form to the TB series (27.2) and may be written as [147, 512]

$$\mathcal{M}_{WB}(T,\epsilon,K) = 1 - \frac{\pi}{\epsilon}\sum_{m=1}^{\infty}(-1)^{m-1}\left(\frac{V_0}{\hbar}\right)^{2m}$$

$$\times \int_0^\infty d\tau_1\,d\tau_2\cdots d\tau_{2m-1}\sum_{\{\xi_j\}'}G_{2m}\left(\{\tau_j\}; 1/K, \{\xi_j\}\right) \tag{28.95}$$

$$\times \sin\left(\sum_{i=1}^{2m-1}\epsilon\,p_i\tau_i/K\right)\prod_{k=1}^{2m-1}\sin(\pi p_k/K),$$

Comparison of the WB series (28.95) with the TB series (27.2) for the mobility in the Ohmic scaling limit confirms validity of the duality relation (28.31).

28.4.1 Linear mobility at finite T

We see from the TB series (27.2) by simply counting dimensions that in the zero bias limit the temperature T is combined with the frequencies Δ and ω_c in the form $\Delta^2 T^{2K-2} \omega_c^{-2K}$, and there is no other dependence on these quantities. On the other hand, taking into account the mapping (28.17), $K_{\text{WB}} = 1/K$, we find from the WB series (28.95) that V_0 is combined with T and ω_c as $V_0^2 T^{2/K-2} \omega_c^{-2/K}$. From this we see that there is a Kondo-type temperature which is related to the parameters of the TB and WB model according to

$$k_{\text{B}}T_0 = a_{\text{TB}}(K)(\Delta/\omega_c)^{K/(1-K)}\hbar\Delta = a_{\text{WB}}(K)(V_0/\hbar\omega_c)^{1/(K-1)}V_0 , \qquad (28.96)$$

where the function $a_{\text{TB}}(K)$ depends only on K, and where

$$a_{\text{WB}}(K) = a_{\text{TB}}(1/K) . \qquad (28.97)$$

Upon introducing the Kondo-scaled temperature

$$\vartheta = T/T_0 , \qquad (28.98)$$

the TB series (27.2) and the WB series (28.95) for the dimensionless linear mobility $\mathcal{M}_{\text{lin}}(\vartheta, K) \equiv \mu_{\text{lin}}/\mu_0$ can be rewritten in the self-dual scaling forms

$$\mathcal{M}_{\text{lin}}(\vartheta, K) = \sum_{m=1}^{\infty} d_m(K)\,\vartheta^{2(K-1)m} , \qquad (28.99)$$

$$\mathcal{M}_{\text{lin}}(\vartheta, K) = 1 - \sum_{m=1}^{\infty} d_m(1/K)\,\vartheta^{2(1/K-1)m} . \qquad (28.100)$$

In the Coulomb gas representation (27.2), the dimensionless coefficient $d_m(K)$ is given in terms of a $(2m-1)$-fold integral.

Self-duality means that the expressions (28.99) and (28.100) are the Taylor and the asymptotic expansion of the same mobility function. By the self-duality transformation $K \to 1/K$, the two expansions are interchanged. Equating Eq. (28.99) with Eq. (28.100), we obtain the nonperturbative self-duality

$$\mathcal{M}_{\text{lin}}(\vartheta, K) = 1 - \mathcal{M}_{\text{lin}}(\vartheta, 1/K) , \qquad (28.101)$$

holding for all K. By this relation, the Taylor expansion is mapped on the asymptotic expansion and vice versa. With the series expressions (28.99) and (28.100), the mapping holds term by term. Observe that the expressions (28.99)–(28.101) have lost explicit dependence on the TB and WB model. The expressions (28.99) and (28.100) describe the same physical system in complementary regions of the parameter space.

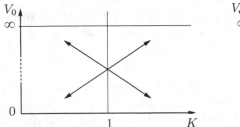
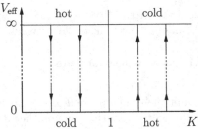

Figure 28.1: Regarding the TB lattice as the large barrier limit of the WB model, both models can be represented in a single diagram of V_0 versus K. The duality maps a WB model with small V_0 and Kondo parameter $K \lesssim 1$ to a TB model with small Δ (large V_0) and $K \gtrsim 1$ (left diagram). In the right diagram, the flow of the effective barrier height with decreasing temperature is sketched for the linear mobility. For $K < 1$ the system flows towards a vanishing barrier whereas it becomes localized for $K > 1$ at zero temperature. For the nonlinear mobility at $T = 0$, the flows as a function of the bias are similar.

The radius of convergence of the power series (28.99) and (28.100) is conveniently expressed in terms of the temperature scale

$$T_{\rm cr}(K) \equiv \lim_{m \to \infty} |d_m(K)|^{1/[2(1-K)m]} T_0 = T_{\rm cr}(1/K) . \qquad (28.102)$$

The equality is a consequence of self-duality [cf. the validation of the corresponding relation (28.118) with the expression (28.117)]. The scale $T_{\rm cr}$ is a crossover temperature analogous to the Kondo temperature in Kondo models. For $K < 1$, the series (28.99) absolutely converges for $T > T_{\rm cr}$, and the series (28.100) for $T < T_{\rm cr}$. For $K > 1$, the regions of convergence of these two series are interchanged.

For $K < 1$, we have the following picture holding both for the TB and WB model. At $T \gg T_{\rm cr}$, we are in the perturbative regime of the series (28.99), the $m = 1$ term being the leading one. As T is lowered, the barrier gets effectively weaker so that higher-order terms of the series become increasingly important. Physically, this means that coherent tunneling through two, three, $\cdots$, many barriers give relevant contributions to the mobility. At $T = T_{\rm cr}$, the region of convergence of the series (28.99) is left. For $T < T_{\rm cr}$, the asymptotic series (28.100) applies. As temperature is lowered further, convergence of the series is improved. At $T = 0$, all terms $m \geq 1$ have dropped to zero, and we are left with $\mu_{\rm lin} = \mu_0$. Thus by cooling down the system from high to zero temperature, the originally strong barrier gradually fades away.

In passing from $K < 1$ to $K > 1$, the role of high and low T are interchanged. Thus, as $T \to 0$, the dressed barrier becomes infinitely high and the particle is localized. The pertinent flows of the effective barrier height are depicted in Fig. 28.1 (right).

28.4.2 Nonlinear mobility at $T = 0$

Similar scaling behavior is found for the nonlinear mobility at $T = 0$. In this case, the bias takes the role of temperature. By counting dimensions in the series expressions (27.2) and (28.95), we see that the Kondo energy scale is given in terms of the parameters of the TB and WB model as

$$\epsilon_0 = b_{\mathrm{TB}}(K)\,\Delta^{1/(1-K)}\omega_c^{-K/(1-K)} = b_{\mathrm{WB}}(K)(V_0/\hbar)^{K/(K-1)}\omega_c^{-1/(K-1)}\,. \qquad (28.103)$$

It is now convenient to introduce the Kondo-scaled dimensionless bias frequency

$$v = \epsilon/\epsilon_0\,. \qquad (28.104)$$

Since v is chosen invariant, while ϵ maps on ϵ/K in the duality transfromation (28.31), we must have in difference to the relation (28.97)

$$b_{\mathrm{WB}}(K) = K\,b_{\mathrm{TB}}(1/K)\,. \qquad (28.105)$$

Explicit expressions for the prefactors $b_{\mathrm{TB}}(K)$ and $b_{\mathrm{WB}}(K)$ satisfying the relation (28.105) can be extracted from the relations given below in Eqs. (28.119) and (28.120).

With the scaled bias v, the TB series (27.2) for the dimensionless nonlinear mobility at zero temperature, $\mathcal{M}(v,K) \equiv \mu(T=0,\epsilon,K)/\mu_0$, takes the scaling form

$$\mathcal{M}(v,K) = \sum_{m=1}^{\infty} c_m(K)\,v^{2(K-1)m}\,. \qquad (28.106)$$

Likewise, the WB series (28.100) at $T = 0$ can be written as

$$\mathcal{M}(v,K) = 1 - \sum_{m=1}^{\infty} c_m(1/K)\,v^{2(1/K-1)m}\,. \qquad (28.107)$$

Thus we find again nonperturbative self-duality, but now of the form

$$\mathcal{M}(v,K) = 1 - \mathcal{M}(v,1/K)\,. \qquad (28.108)$$

In the Coulomb gas representation, the coefficient $c_m(K)$ is again given in terms of a $(2m-1)$-fold integral. The convergence radius of the expansions (28.106) and (28.107) is determined by the critical frequency

$$\epsilon_{\mathrm{cr}}(K) \equiv \lim_{m\to\infty} |c_m(K)|^{1/[\,2(1-K)m]}\epsilon_0 = \epsilon_{\mathrm{cr}}(1/K)\,. \qquad (28.109)$$

For $K < 1$, the series (28.106) converges absolutely for $\epsilon > \epsilon_{\mathrm{cr}}$, while the series (28.107) converges for $\epsilon < \epsilon_{\mathrm{cr}}$. For $K > 1$, the regions of convergence are interchanged.

The dual series expressions (28.106) and (28.107) are again the Taylor and asymptotic expansion of the same physical quantity, and vice versa, depending on whether K is smaller or larger than one and whether the bias ϵ is larger or smaller than ϵ_{cr}. They hold both for the TB and WB model. Upon utilizing the self-duality property of the model, we shall determine in Subsec. 28.5.1 the coefficients $c_m(K)$ for general parameter K. There, we also relate the dressed frequencies ϵ_0 and ϵ_{cr} to the parameters of the original WB and TB model. Finally, the relation (28.103) allows to express the parameters of the one model in terms of those of the other [see Eq. (28.122)].

28.5 Scaling function at $T = 0$ for arbitrary K

28.5.1 Construction of the self-dual scaling solution

The dual expansions (28.106) and (28.107) can be derived from the contour integral representation [533]

$$\mathcal{M}(\upsilon, K) = \frac{1}{2\pi i} \int_C dz \, P(z) F(z) \, \upsilon^{2z} . \tag{28.110}$$

Here we design the function $P(z)$ such that it provides the analytic structure of the power series (28.106) and (28.107),

$$P(z) = \frac{1}{z} \Gamma\left(1 + \frac{zK}{K-1}\right) \Gamma\left(1 + \frac{z}{1-K}\right) . \tag{28.111}$$

The function $P(z)$ is analytic over the entire complex plane save for the points

$$
\begin{aligned}
z &= 0 \,, \\
z &= z_n^{(1)} \equiv (1/K - 1)n \,, & n &= 1, 2, \ldots \,, \\
z &= z_m^{(2)} \equiv (K - 1)m \,, & m &= 1, 2, \ldots \,,
\end{aligned}
\tag{28.112}
$$

where it possesses simple poles. The function $F(z)$ serves as an adaptive function. We assume that $F(z)$ is an analytic function on the entire z-plane. The contour C starts at infinity, then circles the origin such that the pole at $z = 0$ and the set of poles $\{z_n^{(1)}\}$ lie to the left, and the set of poles $\{z_m^{(2)}\}$ lie to the right of the integration path, and finally returns to infinity (cf. Fig. 28.2). The contour is closed upon requiring that the integrand in Eq. (28.110) tends to zero faster than $1/|z|$ on the chosen semicircle.

(1) In the regime $(\epsilon_{\mathrm{cr}}/\epsilon)^{2(1-K)} < 1$, the contour is closed such that the set of poles $\{z_m^{(2)}\}$ is circled. The resulting series is in the form (28.106) with coefficients

$$c_m(K) = (-1)^{m-1} \Gamma(Km + 1) \, F[(K - 1)m]/\Gamma(m + 1) . \tag{28.113}$$

This case applies in the regimes $\epsilon \lesssim \epsilon_{\mathrm{cr}}$ for $K \gtrsim 1$.

(2) For $(\epsilon_{\mathrm{cr}}/\epsilon)^{2(1-K)} > 1$, the contour is closed in the reverse direction of rotation, so that the pole at $z = 0$ and the set of poles $\{z_n^{(1)}\}$ are circled. With the normalization $F(0) = 1$, the resulting series is exactly in the form (28.107). The coefficients for $m \geq 1$ are given again by Eq. (28.113), but with $1/K$ substituted for K. This case applies in the regimes $\epsilon \gtrsim \epsilon_{\mathrm{cr}}$ for $K \gtrsim 1$.

The entire analytic structure of the integral representation (28.110) required by the self-duality property is carried by the function $P(z)$. The only additional requirement imposed by self-duality is that the yet unknown function $F(z)$ is invariant under the substitution $K \to 1/K$. It is tempting to assume that self-duality at zero temperature is provided by a "minimal theory" in the sense that the whole dependence on the

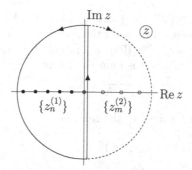

Figure 28.2: Sketch of the integration path C in the contour integral (28.110) for $K > 1$. Closing the contour in the counter-clockwise and clockwise sense gives the series (28.107) and (28.106), respectively. For $K < 1$, the respective poles ensue by reflection at the origin while the contours are left unchanged.

coupling constant K is determined by the analytic properties of the function $P(z)$. Thus, in a minimal theory, the function $F(z)$ does not depend on K at all. On that condition it is possible to determine $F(z)$ provided that the series (28.106) or (28.107) is known in analytic form for a particular value of K different from unity. Fortunately, the TB mobility is known in analytic form for the case $K = \frac{1}{2}$ and in the limit $K \to 0$, as discussed in Subsec. 27.3.1 and Sec. 27.2. The latter case is equivalent to the classical limit $K \to \infty$ in the WB model. Indeed, each one of these special cases can be used for the determination of the function $F(z)$.

To obtain $F(z)$, we equate the terms of the series (27.19) with those of the series (28.106) for $K = 0$ and use Eq. (28.113). Alternatively, we may match the power series of (28.36) with the series (28.107) for $K \to \infty$. In both ways, we unambiguously get

$$F(z) = \Gamma\left(\tfrac{3}{2}\right) / \Gamma\left(z + \tfrac{3}{2}\right) . \tag{28.114}$$

Actually, the same function would have resulted, if we had matched the series (28.106) with (28.113) for $K = \frac{1}{2}$ with the scaled mobility (27.29).

The matching function (28.114) is indeed an analytic function in the entire z-plane, as postulated. Thus we have established an exact contour integral representation of the scaling function for general K at zero temperature [533],

$$\mathcal{M}(v, K) = \frac{1}{2\pi i} \int_C dz \frac{1}{z} \Gamma\left(1 + \frac{zK}{K-1}\right) \Gamma\left(1 + \frac{z}{1-K}\right) \frac{\Gamma(\tfrac{3}{2})}{\Gamma(z + \tfrac{3}{2})} v^{2z} . \tag{28.115}$$

This is a Mellin-Barnes integral [377], and is closely related to generalized hypergeometric functions. With the form (28.114), the series coefficient (28.113) is found as

$$c_m(K) = \frac{(-1)^{m-1}}{m!} \frac{\sqrt{\pi}}{2} \frac{\Gamma(Km+1)}{\Gamma[(K-1)m + \tfrac{3}{2}]} . \tag{28.116}$$

The series expressions (28.106) and (28.107) with (28.116) constitute the exact solution for the mobility for general values of the parameters K and v at zero temperature.

Readily the crossover scale (28.109) delimiting the regions of convergence of the expansions (28.106) and (28.107) is found to be given by

$$\epsilon_{cr}(K) = \sqrt{|1 - K|} \, K^{K/[2(1-K)]} \, \epsilon_0 \,. \tag{28.117}$$

This expression satisfies for fixed ϵ_0 the relation

$$\epsilon_{cr}(K) = \epsilon_{cr}(1/K) \,. \tag{28.118}$$

One obtains, e.g., in the limit $K \to \infty$ the radius $\epsilon_{cr} = 2\pi V_0/\hbar$. This matches the convergence radius of the power series resulting from the square root expression (28.36).

So far we have discussed the nonlinear mobility detached from the particular model. To relate the frequency scale ϵ_0 of the scaling solution to the parameters of the WB model, we equate the $m = 1$ term of the series (28.107) with the $m = 1$ term of the series (28.95) at zero temperature. We then obtain

$$\epsilon_0^{2-2/K} = K^{2-2/K} 2^{2-2/K} [\pi/\Gamma(1/K)]^2 (V_0/\hbar)^2 \omega_c^{-2/K} \,. \tag{28.119}$$

Likewise, we can express ϵ_0 in terms of the parameters of the TB model. Equating the $m = 1$ term of the series (28.106) with the expression (27.5), we find

$$\epsilon_0^{2-2K} = 2^{2-2K} [\pi/\Gamma(K)]^2 \Delta^2 \omega_c^{-2K} \,. \tag{28.120}$$

The contour integral (28.115) with (28.119) and with (28.120) is an exact representation of the nonlinear dc-mobility of the WB model and TB model, respectively, holding at $T = 0$ in the scaling limit for general bias ϵ and general friction K.

The expressions (28.119) and (28.120) are symmetrical to one another under substitution $K \to 1/K$ except for the factor $K^{2-2/K}$ in (28.119). This factor appears because of the correspondence $\epsilon \leftrightarrow K\epsilon$ in the duality relation (28.31) [see relation (28.105)]. A frequency scale $\tilde{\epsilon}_0$, which is fully symmetric under inversion $K \to 1/K$ is $\tilde{\epsilon}_0 = a K^{1/(K-1)} \epsilon_0$, where a is an arbitrary positive constant. With the convenient choice $a = 1/8$ we obtain the relations of $\tilde{\epsilon}_0$ with the TB and WB model in the form

$$\tilde{\epsilon}_0 = 2^{-3} K^{\frac{1}{K-1}} \epsilon_0 \,,$$

$$\tilde{\epsilon}_0^{2-2K} = \frac{\pi^2 \, 2^{4K-4}}{\Gamma^2(1+K)} \frac{\Delta^2}{\omega_c^{2K}} \,, \qquad \tilde{\epsilon}_0^{2-\frac{2}{K}} = \frac{\pi^2 \, 2^{\frac{4}{K}-4}}{\Gamma^2(1+\frac{1}{K})} \frac{(V_0/\hbar)^2}{\omega_c^{2/K}} \,. \tag{28.121}$$

At this point we wish to emphasize a number of conclusions [533]:

(1) With use of the relations (28.119) and (28.120) [or equivalently with the relations (28.121)] and elimination of $\tilde{\epsilon}_0$, we can express Δ in terms of V_0,

$$\Delta = \Gamma(1 + K)[\Gamma(1 + 1/K)]^K (\omega_c/\pi)^{1+K} (V_0/\hbar)^{-K} \,. \tag{28.122}$$

Resolution of this relation for V_0 gives the same functional form except that $1/K$ is substituted for K. It is interesting to compare the expression (28.122) with the corresponding relation (17.76) obtained by use of the bounce method for large K. Employing Stirling's asymptotic formula for $\Gamma(K)$ and identifying the cutoff ω_c with the damping frequency γ, the expression (28.122) indeed coincides with the result (17.76) of the bounce method. However the form (28.122) is universal since it holds for all K. Interestingly enough, here we have found the single-bounce (and all multi-bounce contributions) to the TB mobility solely by use of duality and analytic properties, without calculating bounce actions and fiddly fluctuation determinants.

(2) Besides duality and analytic properties, we have used in the construction of the scaling solution only knowledge of the classical mobility. From this we infer that the entire quantum regime of this transport problem is fully determined by self-duality.

(3) One should expect that also the scaling function $\mathcal{M}_{\mathrm{lin}}(\vartheta, K)$ for the *linear* mobility can be found from the above "minimal" assumptions. However, a formidable practical complication arises: an ansatz equivalent to (28.110) does not work since one finds from the $m = 1$ term that the respective adaptive function $F(z)$ is non-analytic in the complex z-plane.

(4) The property of self-duality is a direct consequence of integrability in the equivalent boundary sine-Gordon model [534]. Integrability determines the low-energy properties of this model and shows itself in the above scaling behavior.

28.5.2 Supercurrent-voltage characteristics at $T = 0$ for arbitrary ρ

We have seen in Section 28.3 that the charge-phase duality in the Josephson junction dynamics is an exact self-duality in the Ohmic scaling limit. Thus, with the correspondence relations (20.157)–(20.159), the findings of the previous subsection can directly be applied to Cooper pair tunneling in a resistive environment. The supercurrent-voltage characteristics at $T = 0$ and arbitrary $\rho = R/R_Q$ takes the scaling-invariant form $I(U) = \mathcal{M}(v, \rho)\, U/R$ [532]. Here, $\mathcal{M}(v, \rho)$ is the scaling function (28.106), but now it represents the normalized conductance of the Josephson contact. The various energy scales of the system are aggregated in the scaling variable

$$v = \frac{eU}{\pi E_{\mathrm{J}}} \left[\Gamma(\rho) \left(\frac{e^{\zeta}}{\pi^2 \rho} \frac{E_{\mathrm{c}}}{E_{\mathrm{J}}} \right)^{\rho} \right]^{\frac{1}{1-\rho}} . \tag{28.123}$$

The current resulting from weak Cooper-pair-tunneling has the scaling form

$$I(U) = c_1(\rho)\, v^{2\rho-2}\, U/R . \tag{28.124}$$

This expression matches exactly with the Coulomb blockade current (20.163).

Likewise, the current resulting from the leading order of the dual weak-backscattering series (28.107),

$$I(U) = [\,1 - c_1(1/\rho)\, v^{2/\rho-2}\,]\, U/R , \tag{28.125}$$

exactly matches with the current (17.87) of a voltage-biased Josephson junction.

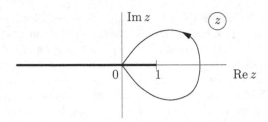

Figure 28.3: The path of integration C in Eq. (28.126) starts at the origin, circles around the branch point at $z = 1$ in the counter-clockwise sense, and returns to the origin.

28.5.3 Connection with Seiberg-Witten theory

Now, an interesting connection of the duality with supersymmetric Seiberg-Witten theory [535] is established. Consider the series (28.107) with (28.116) and express the ratio of gamma functions in the coefficient $c_m(1/K)$ as a contour integral taken along the path in the complex z-plane as sketched in Fig. 28.3 [536],

$$\frac{\Gamma(x)}{\Gamma(x+y)} = \Gamma(1-y)\frac{1}{2\pi i}\int_C dz\, z^{x-1}\,(z-1)^{y-1}\,. \qquad (28.126)$$

Since the series (28.107) is absolutely convergent within the circle of convergence, we can interchange the order of integration and summation. This yields

$$\mathcal{M}(v,K) = \frac{1}{4\sqrt{\pi}\,i}\int_C dz\,\frac{1}{\sqrt{z-1}}\left\{\sum_{n=0}^{\infty}\frac{(-1)^n}{n!}\Gamma(n+\tfrac{1}{2})\left(\frac{z^{1/K}v^{2/K-2}}{z-1}\right)^n\right\}\,. \qquad (28.127)$$

The series in the curly bracket can be summed to a square root. With the substitution $z = t/v^2$, we finally obtain for the scaling function the integral representation [534]

$$\mathcal{M}(v,K) = \frac{1}{4v\,i}\int_{C'} dt\,\frac{1}{\sqrt{t+t^{1/K}-v^2}}\,. \qquad (28.128)$$

The expression (28.128) is analytic in K and therefore holds for all K. One can now prove with this form directly the self-duality relation (28.108) without recourse to the series expansions. With the substitution $t^K = x$ in Eq. (28.128), we obtain

$$M(v,1/K) = \frac{1}{4v\,i}\int_{C'} dx\,\frac{1}{\sqrt{x+x^{1/K}-v^2}}\left[\left(\frac{x^{1/K-1}}{K}+1\right)-1\right]\,, \qquad (28.129)$$

where C' changes accordingly. The first term of the integrand is a total derivative and provides the term unity in Eq. (28.108), while the residual integrand yields $-\mathcal{M}(v,K)$. Thus, the integral (28.128) satisfies the self-duality relation (28.108).

The representation (28.128) bears resemblance with representations for mass gaps in SU(2) supersymmetric Yang-Mills theory [535]. The parameter v corresponds to an order parameter related to the expectation value of the Higgs field. A geometrical picture in terms of tori for the integral (28.128) is discussed in Ref. [537].

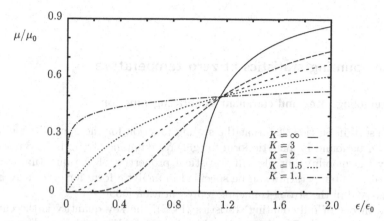

Figure 28.4: The normalized mobility μ/μ_0 at $T = 0$ is plotted as a function of ϵ/ϵ_0 for different K. In the regime $\epsilon < \epsilon_{cr}$ and $K > 1$, the weak-tunneling (TB) series (28.106) with coefficient (28.116) converges, whereas for $\epsilon > \epsilon_{cr}$ and $K > 1$ the strong-tunneling (WB) series (28.107) with coefficient (28.116) converges. The curve for $K = \infty$ shows the square root singularity of the classical case, Eq. (28.36). For finite K, the mobility is nonzero in the bias regime $\epsilon/\epsilon_0 < 1$ due to quantum mechanical tunneling.

28.5.4 Special limits

At $K = \frac{1}{2}$, the frequency scale is $\epsilon_0 = 2\tilde{\epsilon}_0 = 2\pi\Delta^2/\omega_c$. The scaling function $\mathcal{M}(v, \frac{1}{2})$ reproduces previous results given in Subsec. 27.3.1. For $K = \frac{1}{2}$, the series (28.107) is summed to the first form given in Eq. (27.29), whereas the dual series (28.106) gives the second form in Eq. (27.29). In the narrow regime $K = 1/2 - \kappa$, with $|\kappa| \ll 1$, the scaling function is consistent with the former result (27.36) at $T = 0$.

For $K = 1 + \kappa$ with $|\kappa| \ll 1$, the series (28.107) is summed to the expression

$$\mathcal{M}_{\mathrm{WB}}(\epsilon, K = 1 + \kappa) = 1/[1 + (\pi V_0/\hbar\omega_c)^2(2\omega_c/\epsilon)^{2\kappa}], \qquad (28.130)$$

where we have used the relation (28.119). Hence the normalized mobility in the limit $\epsilon \to 0$ is unity for $K < 1$ and zero for $K > 1$. The form (28.130) agrees with a leading-log summation in a related impurity scattering problem (cf. Sec. 31.1) with Luttinger parameter $g = 1 - \kappa$, where $g \widehat{=} 1/K$ [538].

The normalized mobility μ/μ_0 is plotted in Fig. 28.4 as a function of ϵ/ϵ_0 for different values of the damping parameter K. The mobility shows a smooth transition from the square root singular behavior (28.36) in the classical limit $K \to \infty$ via the kink-like shape $\mu/\mu_0 = 1 - (\epsilon_0/2\epsilon) \arctan(2\epsilon/\epsilon_0)$ at $K = 2$ to a plateau-like behavior slightly above $K = 1$. The curves for $K < 1$ are found from the self-duality relation (28.108). All curves cross the line $\mu/\mu_0 = 1/2$ in the interval $\epsilon/\epsilon_0 = 2/\sqrt{3} \pm 0.01$. As $K \to 1^+$, the frequency scale ϵ_0 drops to zero and μ/μ_0 becomes unity. Because of the scale employed in Fig. 28.4, this behavior does not show up in the figure.

29 Full counting statistics at zero temperature

29.1 Tunneling rates and cumulant generating function

We have seen in Section 28.5 that the scaling function for the nonlinear mobility at $T = 0$ can be found in analytic form for general damping strength K. Surprisingly, not only the mobility but also all statistical properties of the quantum transport process at zero temperature can be specified in analytic form for general K both in the weak- and in the strong-tunneling representation [539].

In the theory of full counting statistics (FCS), the key quantity is the cumulant generating function $\ln \mathcal{Z}(\rho, t)$ introduced in Eq. (25.6). For weak tunneling or strong backscattering, the TB model (24.4) applies. Since at $T = 0$ there is no transition from the lower to the higher state, $k_n^- = 0$, the expression (25.24) reduces to

$$\ln \mathcal{Z}(\rho, t) = t \sum_{n=1}^{\infty} \left(e^{i \rho q_0 n} - 1 \right) k_n^+(K) . \tag{29.1}$$

This form can be discovered indeed from a systematic cluster decomposition of the series expression (26.14) in the limit of very large t. The clusters are irreducible path segments between diagonal states of the RDM.[1] In Laplace space, an irreducible cluster becomes independent of the Laplace variable λ, as $\lambda \to 0$. By definition, the clusters are noninteracting. Hence each time interval which separates neighboring clusters gives a factor $1/\lambda$. Path segments with intermediate visit of a diagonal state have a reducible component which factorizes into two clusters of lower order times a factor $1/\lambda$. After subtraction of the reducible component an irreducible part remains. The transition rates $k_n^{\pm}$ can be identified as the sum of all irreducible clusters which interpolate between the (arbitrary) diagonal state m and the diagonal state $m \pm n$. The clusters are formally given in terms of a power series in the number of tunneling transitions. In the Ohmic scaling limit at $T = 0$, the tunneling matrix element Δ is merged with the bias ϵ and the cut-off ω_c into the dimensionless expansion parameter

$$u = \frac{\Delta}{\epsilon} \left(\frac{\epsilon}{\omega_c} \right)^K \propto v^{K-1} . \tag{29.2}$$

The rate k_n^+ is given in terms of the series

$$k_n^+ = \sum_{\ell=n}^{\infty} k_{n,\ell}^+ \quad \text{with} \quad k_{n,\ell}^+ = R_{n,\ell}^+ u^{2\ell} . \tag{29.3}$$

The partial rate $k_{n,\ell}^+ \propto \Delta^{2\ell}$ covers all irreducible contributions with 2ℓ moves, or 2ℓ charges in the charge representation.

---

[1] The notion "irreducibility" is explained in Subsection 21.2.6.

The irreducible Coulomb integrals have been analyzed in the limit $T \to 0$ in Ref. [489]. It was found that there are formidable cancellations among the individual Coulomb integrals of the same order: only those charge sequences contribute to the partial rates in which the charges $\{u_j\}$ and $\{v_j\}$ introduced in Eq. (26.1) are all positive. As a result, the partial rate $k_{n,\ell}^+$ is zero when $\ell > n$. Accordingly, the rate k_n^+ is fully determined by the direct paths from well 0 to well n which have $2n$ moves, i.e., the minimal number of moves required in order to move forward n sites,

$$k_n^+ = k_{n,n}^+ \propto \Delta^{2n}. \tag{29.4}$$

This property allows us to generate the cumulant $\langle q^N(t) \rangle_c / t = q_0^N \sum_{n>0} n^N k_n^+$ from the current $\langle I \rangle = q_0 \sum_{n>0} n \, k_n^+$ by differentiation with respect to Δ or v,

$$\langle q^N(t) \rangle_c / t = \left(q_0 \frac{\Delta}{2} \frac{\partial}{\partial \Delta} \right)^{N-1} \langle I \rangle = \left(q_0 \frac{v}{2(K-1)} \frac{\partial}{\partial v} \right)^{N-1} \langle I \rangle, \tag{29.5}$$

In addition, the central property (29.4) allows us to determine the transition rates k_n^+ unambiguously from one of the cumulants, e.g., from the current.

Equating the series for the current $\langle I \rangle = \langle q(t) \rangle / t = q_0 \sum_{n>0} n k_n^+$ with the expression $\langle I \rangle = (\epsilon q_0 / 2\pi K) \mathcal{M}(v, K)$, where $\mathcal{M}(v, K)$ is given by the series (28.106), the rate to advance forward n sites in a single tunneling event is found to be given by

$$k_n^+(K) = \frac{(-1)^{n-1}}{n!} \frac{\Gamma(\frac{3}{2}) \Gamma(Kn)}{\Gamma[\frac{3}{2} + (K-1)n]} \frac{\epsilon}{2\pi} \left(\frac{\epsilon}{\epsilon_0} \right)^{(2K-2)n}. \tag{29.6}$$

The expression (29.1) with (29.6) is the weak-tunneling series of the full counting statistics for general damping parameter K. For classical Poissonian transport, all the individual transition rates k_n^+ in Eq. (29.1) are positive. Certainly, the rate k_1^+ is positive for all K, and the cotunneling rates k_n^+, $n > 1$, are positive in the regime $K < 1/2n$. The fine distinction to a classical Poisson process, however, is that the cotunneling rates (preferably weights) (29.6) change sign at $K = m/2n$, where $m = 1, 3, \cdots, 2n - 3$. As a result, in the entire regime $K > 1 - 3/2n$, the weight k_n^+ is positive when n is odd, and negative, when n is even. Negative sign of transition weights imply destructive superposition of the respective transport channel.

In the opposite limit of weak backscattering or strong tunneling, the series expression for $\ln \mathcal{Z}(\rho, t)$ at $T = 0$ reads

$$\ln \mathcal{Z}(\rho, t) = t \left(i \, \rho q_0 \frac{\epsilon}{2\pi K} + \sum_{n=1}^{\infty} \left(e^{-i \rho q_0 n / K} - 1 \right) \tilde{k}_n^+(K) \right), \tag{29.7}$$

where

$$\tilde{k}_n^+(K) = \frac{(-1)^{n-1}}{n!} \frac{\Gamma(\frac{3}{2}) \Gamma(n/K)}{\Gamma[\frac{3}{2} + (1/K - 1)n]} \frac{\epsilon}{2\pi} \left(\frac{\epsilon}{\epsilon_0} \right)^{(2/K-2)n}. \tag{29.8}$$

These expressions together with the relation (28.119) entail that the cumulant relation analogous to Eq. (29.5) in the strong-tunneling representation is

$$\langle q^N(t) \rangle_c / t = \left(-\frac{q_0}{K} \frac{V_0}{2} \frac{\partial}{\partial V_0} \right)^{N-1} \langle I \rangle = \left(-q_0 \frac{v}{2(1-K)} \frac{\partial}{\partial v} \right)^{N-1} \langle I \rangle, \tag{29.9}$$

where the weak-backscattering current is $\langle I \rangle = (q_0 \, \epsilon / 2\pi K) \left[1 - (2\pi/\epsilon) \sum_{n>0} n \tilde{k}_n^+ \right]$.

The strong-tunneling series (29.7) is quite similar to the weak-tunneling series (29.1), but there are subtle differences. The first term in the expression (29.7) represents the current in the absence of the barrier. The exponential factor $e^{-i \rho q_0 n / K}$ indicates that now we have tunneling over distance q_0 / K and multiples thereof, and the minus sign in the exponent means that the tunneling contribution diminishes the current instead of building it up as in the weak-tunneling limit.

The rate $\tilde{k}_n^+$ is positive when $K > 2n$. Therefore, the perception of the backscattering dynamics as a classical Poisson process is quite appropriate for transition weights with modest n, when K is sufficiently large. In the classical limit $K \to \infty$, each particular transition weight is positive and hence amenable to classical interpretation. In this regime one finds with use of the rate expression (29.8)

$$ \ln \mathcal{Z}(\rho, t) = i t \rho \frac{q_0 \, \epsilon}{2\pi K} \left[1 - \sum_{n=1}^{\infty} \frac{\Gamma(n - \frac{1}{2})}{2\sqrt{\pi} n!} \left(\frac{\epsilon_0}{\epsilon} \right)^{2n} \right] = i t \rho \frac{q_0 \, \epsilon}{2\pi K} \sqrt{1 - \left(\frac{\epsilon_0}{\epsilon} \right)^2} \,, \quad (29.10) $$

where $\epsilon > \epsilon_0$ is assumed. This yields for the mobility the expression (28.36), but the quantum fluctuations $\langle q^n(t) \rangle_c$ with $n > 1$ drop to zero, as one reaches the classical limit $K \to \infty$. In this limit, the Kondo scale ϵ_0 is related to the corrugation strength V_0 of the weak-binding model (24.1) with (24.2) by the relation $\epsilon_0 = 2\pi V_0 / \hbar$. For $\epsilon < \epsilon_0$, the particle is trapped in a well, as described on page 504, and thus $\ln \mathcal{Z}(\rho, t) = 0$.

29.2 Leading low temperature contribution

For general damping strength K and general temperature, the evaluation of the Coulomb gas integral representation given in Sec. 26.1 is not possible. Progress may be achieved by employment of thermodynamic Bethe ansatz techniques in the related BSG model [540, 541]. A general procedure for the calculation of cumulants of any order is sketched and the third moment is explicitly calculated in Ref. [542].

Interestingly, the leading low-temperature contribution to the zero temperature cumulant generating function can be calculated in analytic form. To this aim, we insert the charge interaction with the leading thermal contribution

$$ Q'(\tau) = Q'_0(\tau) + K \frac{\pi^2}{3} \left(\frac{\tau}{\hbar \beta} \right)^2 + \mathcal{O}\left[(\tau/\hbar\beta)^4 \right] \,, \quad (29.11) $$

where $Q'_0(\tau) = 2K \ln(\omega_c \tau)$ is the interaction at $T = 0$, into the charge interaction factor G_{2m} given in Eq. (26.8). The resulting expression up to terms of order T^2 is

$$ G_{2m} = G_{2m}^{(0)} \left\{ 1 - K \frac{\pi^2}{3} \left(\frac{k_B T}{\hbar} \right)^2 \left(\sum_{j=1}^{2m-1} p_j \tau_j \right)^2 \right\} \,, \quad (29.12) $$

where $G_{2m}^{(0)}$ is the full interaction factor at $T = 0$. The decisive point now is that the sum $\sum_j p_j \tau_j$ in the curly bracket can be generated by differentiation of the bias factor $B_{2m} = \prod_{j=1}^{2m-1} e^{i \, \epsilon p_j \tau_j}$ with respect to the bias. As a result, the asymptotic low-temperature expression of the term $G_{2m} B_{2m}$ can be written as

$$G_{2m}B_{2m} = G_{2m}^{(0)} \left\{ 1 + K \frac{\pi^2}{3} \left(\frac{k_B T}{\hbar} \right)^2 \frac{\partial^2}{\partial \epsilon^2} \right\} B_{2m} . \tag{29.13}$$

Thus, according to the relation (29.13), the T^2-contribution to the cumulant generating function can be universally written in terms of the second derivative of the zero temperature expression with respect to the bias. One finds that subleading contributions $\propto T^4, T^6, \cdots$ are nonuniversal, and therefore can not be expressed in simple terms. Substituting the form (29.13) into the series expression (26.14), we readily obtain the cumulant generating function in the TB representation in the form (29.1), in which the rate k_n^+ up to order T^2 is given by

$$k_n^+(\epsilon, T) = \left\{ 1 + K \frac{\pi^2}{3} \left(\frac{k_B T}{\hbar} \right)^2 \frac{\partial^2}{\partial \epsilon^2} \right\} k_n^+(\epsilon, 0) . \tag{29.14}$$

Observing from the form (29.6) that $k_n^+(\epsilon, 0) \propto \epsilon^{(2K-2)n+1}$, we readily get

$$k_n^+(\epsilon, T) = \left\{ 1 + (2K-2)n[1 + (2K-2)n] \frac{\pi^2 K}{3} \left(\frac{k_B T}{\hbar \epsilon} \right)^2 \right\} k_n^+(\epsilon, 0) . \tag{29.15}$$

The dual weak-backscattering or strong-tunneling representation of the CGF to order T^2 is found to read

$$\ln \mathcal{Z}(\rho, t) = t \left[i \rho \frac{q_0 \epsilon}{2\pi K} - \frac{\rho^2}{2} \frac{q_0^2}{2\pi K} \frac{2 k_B T}{\hbar} + \sum_{n=1}^{\infty} \left(e^{-i \rho n q_0 / K} - 1 \right) \tilde{k}_n^+(\epsilon, T) \right] , \tag{29.16}$$

$$\tilde{k}_n^+(\epsilon, T) = \left\{ 1 + (2/K - 2)n[1 + (2/K - 2)n] \frac{\pi^2}{3K} \left(\frac{k_B T}{\hbar \epsilon} \right)^2 \right\} \tilde{k}_n^+(\epsilon, 0) . \tag{29.17}$$

In Eq. (29.16), the first and second term are related by the Einstein relation (26.64).

Altogether, the expression (29.1) with (29.15), and the expression (29.16) with (29.17), represent the dual weak- and strong-tunneling series representations of the cumulant generating function $\mathcal{Z}(\rho, t)$ with the T^2-dependence included. It is straightforward to derive from these forms the corresponding series of all cumulants.

The T^2-contribution to the full counting statistics is a distinctive signature of Ohmic dissipation inherent in the quantum transport process. A related phenomenon is the universal T^2-behavior observed at low temperatures in other open quantum systems subject to Ohmic dissipation, e.g. the T^2-enhancement in macroscopic quantum tunneling (cf. Section 17.2). It is also the origin of the universal Wilson ratio occurring e.g. in Kondo systems and in the related Ohmic two-state system (cf. Subsection 19.2.1 and Section 22.9). Other examples are the T^2 enhancement of the free energy and static susceptibility of the two-state system [cf. Eqs. (22.218) and (22.219) for $s = 1$]. In these cases, the prefactor is determined by the susceptibility at zero temperature and its second derivative, respectively. The physical origin of the T^2 enhancement is the low-frequency thermal noise at Ohmic dissipation [373].

We finally remark that with the same strategy one may calculate also finite temperature corrections for non-Ohmic damping, $s \neq 1$, e.g. for the nonlinear mobility. It is straightforward to see that in such case the power-law of the leading thermal enhancement is T^{1+s} [224].

30 Twisted partition function and nonlinear mobility

Analytic evaluation of the grand-canonical sum of the real-time Coulomb gas (27.2) for any temperature, bias and Kondo parameter K is still unresolved. A route has been proposed in which the problem is solved crabwise [491, 544]. There it was shown that the grand-canonical sum in imaginary time can be solved in analytic form for twisted partition functions. From these the nonlinear mobility is found with a conjecture relying on analytical continuation of winding numbers to the real physical bias. Here we show how the method works. In addition, we confirm results of this approach obtained in particular regions of the parameter space. The regime $K \ll 1$ for general T and ϵ, and the regime $T = 0$ for general ϵ and K have been studied in Ref. [545]. Here we present the method for the TB model (24.4). Translation to the WB model (24.1) is straightforward along the lines given above.

30.1 Solving the imaginary-time Coulomb gas with Jack polynomials

The perturbative series of the partition function is equivalent to the grand-canonical sum of a one-dimensional gas of positive and negative unit charges with overall neutrality. The charges represent the forward and backward moves in the TB lattice. While in the imaginary-time Coulomb gas for the TSS the charges are lined up alternatingly, in the multi-state TB system with infinitely many states the charges are unordered. In the term of order Δ^{2n}, we have $2n$ charges, and we must integrate over the positions $\{\tau_i'\}$ and $\{\tau_j\}$ of the n negative and n positive charges, respectively, in the period of length $\hbar\beta$. The effects of the environmental coupling are carried by the exponential charge interaction factor $\mathrm{e}^{\pm \mathcal{W}(\tau)}$, where the pair interaction $\mathcal{W}(\tau)$ is given in Eq. (18.55). The grand-canonical series of the TB partition function is

$$
\begin{aligned}
\mathcal{Z}_{\mathrm{TB}} = {} & 1 + \sum_{n=1}^{\infty} \frac{(\Delta/2)^{2n}}{(n!)^2} \int_0^{\hbar\beta} \prod_{i=1}^{n} \mathrm{d}\tau_i \int_0^{\hbar\beta} \prod_{j=1}^{n} \mathrm{d}\tau_j' \, \exp\left\{ -\epsilon \sum_{k=1}^{n} (\tau_k - \tau_k') \right\} \\
& \times \exp\left\{ \sum_{m>k=1}^{n} [\mathcal{W}(\tau_m - \tau_k) + \mathcal{W}(\tau_m' - \tau_k')] - \sum_{m,k=1}^{n} \mathcal{W}(|\tau_m - \tau_k'|) \right\} .
\end{aligned}
$$

(30.1)

The change of integration variables, $u_i = 2\pi\, \tau_i/\hbar\beta$, maps the Coulomb gas (30.1) with (18.55) on the unit circle. The perturbative series of the partition function (30.1) may be written as a power series in the effective dimensionless fugacity

$$
x = (2\pi/\hbar\beta\omega_{\mathrm{c}})^K \, \hbar\beta\Delta/2 = \pi(\hbar\beta\Delta_{\mathrm{r}}/2\pi)^{1-K} .
$$

(30.2)

The resulting series for the partition function is

$$\mathcal{Z}_{\mathrm{TB}}(x,p) = 1 + \sum_{n=1}^{\infty} x^{2n} \mathcal{I}_{2n}(p) , \tag{30.3}$$

with the $2n$-fold Coulomb gas integral

$$\mathcal{I}_{2n}(p) = \frac{2^{-2Kn}}{(n!)^2} \int_0^{2\pi} \prod_{i=1}^{n} \left(\frac{du_i}{2\pi} \frac{du_i'}{2\pi}\right) \left| \frac{\prod_{i<j} \sin(\frac{u_i-u_j}{2}) \sin(\frac{u_i'-u_j'}{2})}{\prod_{i,j} \sin(\frac{u_i-u_j'}{2})} \right|^{2K} e^{ip\sum_i (u_i-u_i')} , \tag{30.4}$$

where p represents the analytically continued bias,

$$p = iq \quad \text{with} \quad q = \hbar\epsilon/(2\pi k_{\mathrm{B}} T) . \tag{30.5}$$

Changing integration variables $z_i = e^{iu_i}$ and $z_i' = e^{iu_i'}$ we get

$$\mathcal{I}_{2n}(p) = \frac{1}{(n!)^2} \oint \prod_{i=1}^{n} \left(\frac{dz_i}{2i\pi z_i} \frac{dz_i'}{2i\pi z_i'}\right) \left(\frac{z_1 \cdots z_n}{z_1' \cdots z_n'}\right)^p \frac{[\Delta(z)\overline{\Delta(z)}\,\Delta(z')\overline{\Delta(z')}]^K}{\prod_{i,k}[(1-z_i\bar{z}_k')(1-z_k'\bar{z}_i)]^K} , \tag{30.6}$$

where $\Delta(z) = \prod_{i<k}(z_i - z_k)$ is the n-variable Vandermonde determinant.

The multiple integrals in (30.6) are regular in the regime $K < \frac{1}{2}$. Unfortunately, they can not be evaluated for general K and for general complex p. To advance, we must put K rational and p integer. Integer p may be regarded as a winding number due to a magnetic charge located at the origin. For rational K, we may expand the integrand in terms of Jack polynomials [491, 546]

$$\prod_{i,j} \frac{1}{(1-r_i s_j)^K} = \sum_{\lambda} b_{\lambda}(K) P_{\lambda}(r, K) P_{\lambda}(s, K) . \tag{30.7}$$

Here, the function $P_{\lambda}(r, K)$ is a symmetric polynomial in the set of variables $(r_1, r_2, \cdots, r_n)$, and $\lambda = (\lambda_1, \lambda_2, \cdots, \lambda_n)$ is a partition of an integer with order $\lambda_1 \leqslant \lambda_2 \leqslant \cdots \leqslant \lambda_n$. For positive integer p, we have

$$(z_1 \cdots z_n)^p P_{\lambda}(z, K) = P_{\lambda+p}(z, K) . \tag{30.8}$$

Here $\lambda + p$ means the partition λ, whereby p columns of length n have been added in the respective Young tableau. The multiple integrals in (30.6) can be executed by use of the orthogonality relation of the Jack polynomials holding because the Vandermonde determinant is involved. In the end, one arrives at the totally ordered n-fold series expression

$$\mathcal{I}_{2n}(p) = \sum_{m_n=0}^{\infty} \sum_{m_{n-1}=0}^{m_n} \cdots \sum_{m_1=0}^{m_2} \prod_{j=1}^{n} e_j(m_j) \tag{30.9}$$

with

$$e_j(m) = \frac{1}{\Gamma^2(K)} \frac{\Gamma(jK+m)}{\Gamma(1-K+jK+m)} \frac{\Gamma(jK+p+m)}{\Gamma(1-K+jK+p+m)} . \tag{30.10}$$

It has been suggested to use (30.3) with (30.9) and (30.10) to define $\mathcal{Z}_{\mathrm{TB}}(x,p)$ for complex p, and conjectured that this is the unique analytic continuation [544]. Below, the conjecture is verified in various limits where analytic expressions are available by different methods. Especially conclusive is the limit $|p| \to \infty$ in which the free energy should prove to have proper low-temperature behavior in all orders of Δ.

Consider now the cumulant expansion

$$\mathcal{F}_{\mathrm{TB}}(x,p) = -k_{\mathrm{B}}T \ln \mathcal{Z}_{\mathrm{TB}}(x,p) = -k_{\mathrm{B}}T \sum_{n=1}^{\infty} x^{2n} C_{2n}(p) . \tag{30.11}$$

The first three cumulant coefficients are

$$\begin{aligned}
C_2(p) &= \mathcal{I}_2(p) , \\
C_4(p) &= \mathcal{I}_4(p) - \tfrac{1}{2}\mathcal{I}_2(p)^2 , \\
C_6(p) &= \mathcal{I}_6(p) - \mathcal{I}_2(p)\,\mathcal{I}_4(p) + \tfrac{1}{3}\mathcal{I}_2(p)^3 .
\end{aligned} \tag{30.12}$$

In the analysis of the cumulant expressions it proves to be decisive to rearrange the multiple sums in $C_{2n}(p)$ into totally ordered sums, as already given for the functions $\mathcal{I}_{2n}(p)$ in Eq. (30.9). This leads to the split-up

$$C_{2n}(p) = \sum_{m=1}^{n} C_{2n}^{(m)}(p) , \tag{30.13}$$

where $C_{2n}^{(m)}(p)$ is an m-fold ordered sum. The first cumulant is

$$C_2(p) = \sum_{j=0}^{\infty} e_1(j) . \tag{30.14}$$

The second cumulant has two contributions,

$$\begin{aligned}
C_4^{(1)}(p) &= \frac{1}{2} \sum_{j=0}^{\infty} e_1^2(j) , \\
C_4^{(2)}(p) &= \sum_{j=0}^{\infty} \sum_{k=0}^{j} \left[e_2(j) - e_1(j) \right] e_1(k) .
\end{aligned} \tag{30.15}$$

The multiple integral in Eq. (30.4) diverges at short distances when $K \geqslant 1/2$. Accordingly, the multiple series (30.9) for the coefficient $\mathcal{I}_{2n}(p)$ diverges in the regime $K \geqslant 1/2$ also. Interestingly, some divergences of the Coulomb integrals cancel each other out when taking the connected part $C_{2n}(p)$. As a result, the coefficient $C_{2n}(p)$ in the cumulant series (30.11) is regular for K-values at which $\mathcal{I}_{2n}(p)$ is singular. The analysis of the ordered m-fold sums $C_{2n}^{(m)}(p)$ in Eq. (30.13) reveals that the cumulant coefficient $C_{2n}(p)$ is nonsingular in the range $0 < K < 1 - \frac{1}{2n}$.

30.2 Nonlinear mobility

In Ref. [544] a conjecture was made which relates the nonlinear mobility of the quantum Brownian particle or the nonlinear conductance in quantum impurity models[1] directly to the partition function $\mathcal{Z}(x, p)$ or to the free energy $\mathcal{F}(x, p)$. The conjecture for the normalized mobility of the TB model (24.4) is

$$
\begin{aligned}
\mathcal{M}_{\mathrm{TB}}(x, q) &= \frac{K}{2p} x \frac{d}{dx} \ln \left(\frac{\mathcal{Z}_{\mathrm{TB}}(x, -p)}{\mathcal{Z}_{\mathrm{TB}}(x, p)} \right) \bigg|_{p=iq} \\
&= \frac{K}{2p} \frac{x}{k_{\mathrm{B}}T} \frac{d}{dx} \left(\mathcal{F}_{\mathrm{TB}}(x, p) - \mathcal{F}_{\mathrm{TB}}(x, -p) \right) \bigg|_{p=iq},
\end{aligned}
\tag{30.16}
$$

where $q = \beta\hbar\epsilon/2\pi$. The conjecture is based on the fact that the free energy is real only for integer p, and that the continuation of $C_{2n}(p)$ is not an even function of p. Hence the free energy acquires an imaginary part when p is complex. The imaginary part of the free energy then yields the mobility according to the relation (30.16).

Use of the perturbative cumulant expansion (30.11) in the relation (30.16) yields the weak-tunneling series of the normalized conductance,

$$
\mathcal{M}_{\mathrm{TB}}(x, q) = \sum_{n=1}^{\infty} \mathcal{M}_n(x, q) .
\tag{30.17}
$$

The perturbative tunneling contribution of order x^{2n} to the nonlinear mobility is

$$
\mathcal{M}_n(x, q) = \frac{K}{iq} n [C_{2n}(-iq) - C_{2n}(iq)] x^{2n} .
\tag{30.18}
$$

We now study the conjecture in diverse regimes of the parameter space.

30.2.1 Strong barrier limit

The series for $C_2(p) = \mathcal{I}_2(p)$ found from Eq. (30.9) can be summed in analytic form,

$$
\begin{aligned}
C_2(p) &\equiv \frac{1}{\Gamma^2(K)} \sum_{j=0}^{\infty} \frac{\Gamma(K+j)\Gamma(K+p+j)}{\Gamma(1+j)\Gamma(1+p+j)} \\
&= \sin(\pi K)\Gamma(1-2K) \frac{\Gamma(K+p)}{\pi\Gamma(1-K+p)} .
\end{aligned}
\tag{30.19}
$$

With this, the conjecture (30.18) yields for the mobility in order Δ^2

$$
\mathcal{M}_1 = K \left(\frac{\beta\hbar\Delta}{2} \right)^2 \left(\frac{2\pi}{\beta\hbar\omega_c} \right)^{2K} \frac{\sinh(\beta\hbar\epsilon/2)}{\beta\hbar\epsilon/2} \frac{|\Gamma(K + i\,\beta\hbar\epsilon/2\pi)|^2}{\Gamma(2K)} .
\tag{30.20}
$$

The expression (30.20) coincides indeed with the result (27.4) obtained above in the non-equilibrium real-time approach.

---

[1] Charge transport in quantum impurity systems and Brownian particle transport in a cosine potential are closely related. This is discussed in Chapter 31.

Next, we study the conjecture (30.16) in order x^2 in some detail. First of all, consider the real- and imaginary-time breathing mode or noise integrals

$$J_1(K,p) \equiv \frac{1}{\pi} \int_0^\infty du \, \frac{e^{-2pu}}{(2\sinh u)^{2K}} \, , \tag{30.21}$$

$$H_1(K,p) \equiv \frac{1}{\pi} \int_0^\pi dv \, \frac{e^{2ipv}}{(2\sin v)^{2K}} \, . \tag{30.22}$$

Analytic evaluation of the integrals yields

$$J_1(K,p) = \frac{\Gamma(1-2K)\Gamma(K+p)}{2\pi\,\Gamma(1-K+p)} \, , \tag{30.23}$$

$$H_1(K,p) = 2\,e^{ip\pi}\sin[\pi(p+K)]\,J_1(K,p) \, . \tag{30.24}$$

The relation (30.24) can be understood as follows. The function $e^{-2pz}/(\sinh z)^{2K}$ of the complex variable z is analytic in the half-strip $0 < \mathrm{Re}\,z < \infty$, $0 > \mathrm{Im}\,z > -\pi$ of the complex plane $z = u - iv$. Accordingly, the closed contour integral along the edges of the half-strip vanishes. This entails that the noise integrals $H_1(K,p)$ and $J_1(K,p)$ are related as given in Eq. (30.24). For integer p, the relation (30.24) is simplified to

$$H_1(K,p) = 2\sin(\pi K)\,J_1(K,p) \, , \qquad p \in \text{Integers} \, . \tag{30.25}$$

Thus, the function $H_1(K,p)$ coincides for integer p with the funtion $\mathcal{C}_2(p)$, Eq, (30.19). As a result, the conjectured relation (30.16) with (30.3) yields for the mobility in order x^2 with integer p [cf. Eq. (30.18) for $n=1$ and $q = -ip$]

$$\mathcal{M}_1 = (K/p)\,x^2\,[\,H_1(K,-p) - H_1(K,p)\,] \, . \tag{30.26}$$

Thus, for integer p, we can equally write instead of Eq. (30.26)

$$\mathcal{M}_1 = (K/p)\,x^2\,2\sin(\pi K)\,[\,J_1(K,-p) - J_1(K,p)\,] \, . \tag{30.27}$$

On the other hand, if we choose $p = i\,\beta\hbar\epsilon/2\pi$ in the expression (30.27), it exactly matches with the mobility (27.4). From this we infer that the analytic continuation of the "thermodynamic" expression (30.26) from winding number p to the physical bias, $p = i\,\beta\hbar\epsilon/2\pi$, is unique and yields in fact the correct mobility in order x^2.

Unfortunately, the real-time approach is not practicable for calculation of terms $\mathcal{M}_n(x,q)$ with $n \geqslant 2$ in the series (30.17), since the relevant multiple Coulomb integrals for charges distributed on the forward/backward paths cannot be carried out in analytic form. The loophole now is to take the imaginary time-route: First one calculates the free energy for integer p, as outlined in Section 30.1. Then one employs the conjectured relation (30.16) and performs analytical continuation to complex p. Let us now see whether this line of argument is valid beyond the term $\mathcal{M}_1(x,q)$.

30.2.2 Weak-damping limit

For weak damping $K \ll 1$, the leading contribution to the coefficient $\mathcal{I}_{2n}(p)$ in the multiple series (30.9) is the term with $m_1 = m_2 = \cdots = m_n = 0$. Furthermore, for $K \ll 1$ the coefficient $e_j(0)$ given in (30.10) reduces to $e_j(0) = 1/[j\,(j + p/K)]$. With these reductions, the series coefficient $\mathcal{I}_{2n}(p)$ is simplified to the concise form

$$\mathcal{I}_{2n}(p) = \prod_{j=1}^{n} e_j(0) = \frac{1}{n!}\frac{\Gamma(1 + p/K)}{\Gamma(1 + n + p/K)} . \tag{30.28}$$

With the expression (30.28), the series (30.3) can be summed in closed form [545],

$$\mathcal{Z}_{\mathrm{TB}}(x,p) = \frac{\Gamma(1 + p/K)}{x^{p/K}} \sum_{n=0}^{\infty} \frac{x^{2n+p/K}}{n!\Gamma(1 + n + p/K)} = \frac{\Gamma(1 + p/K)}{x^{p/K}} I_{p/K}(2x) . \tag{30.29}$$

The function $I_\nu(z)$ is a modified Bessel function with index ν. With functional relations of the Bessel functions [90] the tunneling susceptibility $\mathrm{d}\mathcal{F}/\mathrm{d}x$ is found as

$$\frac{\mathrm{d}}{\mathrm{d}x}\mathcal{F}_{\mathrm{TB}}(x,p) = -k_{\mathrm{B}}T\frac{I_{1+p/K}(2x)}{I_{p/K}(2x)} . \tag{30.30}$$

With this form the conjecture (30.16) yields for the nonlinear mobility the expression

$$\mathcal{M}(x,q) = \frac{K}{q}\,2x\,\mathrm{Im}\,\frac{I_{1-\mathrm{i}q/K}(2x)}{I_{-\mathrm{i}q/K}(2x)} , \qquad q = \frac{\hbar\epsilon}{2\pi k_{\mathrm{B}}T} , \tag{30.31}$$

which is in agreement with the result of an elaborate real-time calculation, Eq. (27.15).

30.2.3 Zero temperature limit

The limit $T \to 0$ corresponds to the limit $|p| \to \infty$ and thus is a pivotal test whether the conjecture about the analytic continuation from integer p to complex p in the expression (30.16) yields the correct mobility for general K in all orders of x [545].

As $|p| \to \infty$, the ordered sums in $\mathcal{I}_{2n}(p)$ turn into ordered integrals. With the substitution $k \to pu$, we have the mapping

$$\sum_{k} \cdots \to p\int \mathrm{d}u \cdots , \tag{30.32}$$

and the asymptotic expansion of $e_j(pu)$ yields

$$\lim_{|p|\to\infty} e_j(pu) = \frac{p^{2K-2}}{\Gamma^2(K)}\left[u\,(1 + u)\right]^{K-1}\left\{1 + \mathcal{O}(1/p)\right\} . \tag{30.33}$$

With the relations (30.32) and (30.33) one finds that $\mathcal{I}_{2n}(p)$ behaves asymptotically as $p^{(2K-1)n}$. On passing from $\mathcal{I}_{2n}(p)$ to the connected part $\mathcal{C}_{2n}(p)$, the $n-1$ leading orders in $1/p$ cancel each other out, so that $\lim_{|p|\to\infty}\mathcal{C}_{2n}(p)/\mathcal{I}_{2n}(p) \propto p^{1-n}$, and thus $\mathcal{C}_{2n}(p) \propto p^{2(K-1)n+1}$ in leading order in $1/p$. Actually, one obtains

$$C_{2n}(p) = \frac{(-1)^{n-1}}{n!} \frac{\Gamma(nK)}{\Gamma^{2n}(K)} \frac{\Gamma(2n-1-2nK)}{\Gamma(n-nK)} p^{2(K-1)n+1} . \tag{30.34}$$

The fraction $r_n(K) = \Gamma(2n-1-2nK)/\Gamma(n-nK)$ in the cumulant $C_{2n}(p)$ has simple poles at $K = 1 - \frac{1}{2n} + \frac{m}{n}$, where $m = 0, 1, 2, \cdots$. On the other hand, the analytic continuation $p = i\,q$ in the mobility coefficient (30.18) with (30.34) yields a trigonometric factor $\cos[(\pi(1-K)n]$, whose zeros just compensate the poles of $r_n(K)$. Thus, the mobility contribution (30.18) is regular for all $K > 0$. In addition, temperature cancels out in the combined factor $(q/2)^{2(K-1)n}x^{2n}$. There is

$$(q/2)^{2(K-1)n}x^{2n} = \Gamma^{2n}(K)(\epsilon/\epsilon_0)^{2(K-1)n} . \tag{30.35}$$

Here ϵ_0 is the universal frequency scale defined in Eq. (28.120). With the expressions (30.34) and (30.35) the contribution (30.18) takes the form

$$\mathcal{M}_n = \frac{(-1)^{n-1}}{n!} \frac{\Gamma(\frac{3}{2})\Gamma(1+nK)}{\Gamma(\frac{3}{2}-n+nK)} \left(\frac{\epsilon}{\epsilon_0}\right)^{2(K-1)n} . \tag{30.36}$$

The normalized mobility (30.17) with (30.36) coincides indeed with the scaling solution (28.106) with (28.104) and (28.116). This confirms validity of the conjecture (30.16) in the zero temperature limit.

Let us finally address the question whether the imaginary-time approach can also render the full counting statistics at nonzero T. To this, we write the mobility contribution $\mathcal{M}_n$ by relying on the partial rate expansion (29.3) as

$$\mathcal{M}_n = \frac{2\pi K}{\epsilon} \sum_{m=1}^{n} m \left(k_{m,n}^+ - k_{m,n}^-\right) . \tag{30.37}$$

Here, $k_{m,n}^{\pm}$ is the rate contribution of order x^{2n} for direct forward/backward transitions by m sites. With this we can solve the relation (30.16) in order x^{2n},

$$\mathcal{M}_n = \frac{K\beta}{q} \frac{\partial}{\partial x} \mathrm{Im}\, \mathcal{F}_n(i\,q) , \tag{30.38}$$

for the free energy contribution $\mathcal{F}_n$. The resulting expression is

$$\mathrm{Im}\, \mathcal{F}_n(i\,q) = \frac{\hbar}{2} \sum_{m=1}^{n} \frac{m}{n}(k_{m,n}^+ - k_{m,n}^-) . \tag{30.39}$$

Evidently, the free energy $\mathcal{F}_n$ does not provide information about individual tunneling rates. Rather $\mathrm{Im}\, \mathcal{F}_n(i\,q)$ covers a particular linear combination of partial rates of same order x^{2n}, whereas the full counting statistics requires knowledge of all individual partial rates $k_{m,n}^{\pm}$, $m = 1, 2, \cdots, n$. Only for $T = 0$, where $k_{m,n}^+ = \delta_{m,n}\, k_{m,m}^+$ and $k_{m,n}^- = 0$, the full-counting statistics is available, as discussed in Chapter 29.

Thus, on the route towards complete information of transport and noise statistics one has to overcome the hurdle of handling the real-time Coulomb gas.

31 Charge transport in quantum impurity systems

The physics of interacting particles in one dimension is drastically different from the physics of interacting particles in two or three dimensions. The theoretical methods and techniques relevant to quantum physics in one dimension have been reviewed in compendia by Gogolin, Nersesyan, and Tsvelik [547], and by Giamarchi [548]. Signatures of many-body correlations have attracted a great deal of interest in recent years. Many investigations have been focussed on one-dimensional (1D) electron systems, in which the usual Fermi liquid behavior is destroyed by the interaction. The generic features of many 1D interacting fermion systems are well described in terms of the Tomanaga-Luttinger liquid (TLL) model [549, 415]. In the TLL model, the effects of the electron-electron interaction are captured by a dimensionless parameter g. A sensitive experimental probe of a Luttinger liquid state is the tunneling conductance through a point contact in a 1D quantum wire, as observed in Ref. [514]. Of interest are also the dc nonequilibrium current noise [550] and higher cumulants. The generic model is a quantum impurity embedded in a Luttinger liquid environment (QI-TLL model). Tunneling of edge currents in the fractional quantum Hall (FQH) regime provides another realization of a Luttinger phase. As shown by Wen [515], the edge state excitations are described by a (chiral) Luttinger liquid with Luttinger parameter $g = \nu$, where ν is the fractional filling parameter.

31.1 Generic models for transmission of charge through barriers

In this section, we discuss the conductance of 1D interacting spinless electrons in the presence of a barrier. First, we consider a pure interacting electron gas and give the two-terminal conductance. Then we discuss transport through a single barrier. We approach this problem perturbatively in two limits: a very weak barrier (strong tunneling) and a very large barrier (weak tunneling). We find that the QI-TLL model is closely related to the model of an Ohmic Brownian particle in a tilted washboard potential, discussed in the preceding sections. The model also describes a one-channel coherent conductor in a resistive electromagnetic environment, as we shall see in Subsection 31.1.4.

31.1.1 The Tomonaga-Luttinger liquid

The characteristic feature of a Fermi liquid is the discontinuity of the momentum density of states at the Fermi surface. In one dimension, the electron-electron interaction is so strong that the discontinuity is dissolved and instead a power law essential singularity at the Fermi points arises. The ground state of the interacting 1D electron gas is a Tomonaga-Luttinger liquid which is distinguished by a gapless collective sound mode [415]. At low energy and long wave length, the electron interaction is regarded as a contact interaction covered by a single dimensionless parameter g.

The low-energy modes of the 1D interacting electron liquid are conveniently treated in the framework of bosonization [549, 415, 547]. This approach is appropriate for low temperature at which the excitations are located near the Fermi points. The creation operator for spinless fermions can equivalently be expressed in terms of boson phase fields $\theta(x)$ and $\phi(x)$, which obey the equal-time commutation relation

$$[\,\phi(x,t),\theta(x',t)\,] \;=\; -(i/2)\,\mathrm{sgn}(x-x')\,. \tag{31.1}$$

Thus $\partial_x\phi(x)$ is the canonically conjugate momentum density to $\theta(x)$, and $\partial_x\theta(x)$ is conjugate to $\phi(x)$. With the fields $\theta(x)$ and $\phi(x)$, the fermion field operator can be written as

$$\psi^\dagger(x) \propto \sum_{n\,\mathrm{odd}} \exp\{i\,n[k_\mathrm{F}x + \sqrt{\pi}\theta(x)]\}\exp[i\,\sqrt{\pi}\phi(x)]\,. \tag{31.2}$$

At long wavelengths, the terms $n = \pm 1$ are most essential. These constitute the right- and left-moving components of the electron field. The boson representation of the electron density operator is then given by

$$\rho(x) \;=\; k_\mathrm{F}/\pi + \partial_x\theta(x)/\sqrt{\pi} + k_\mathrm{F}\cos[2k_\mathrm{F}x + 2\sqrt{\pi}\theta(x)]/\pi\,, \tag{31.3}$$

where $\hbar k_\mathrm{F}$ is the Fermi momentum. The first term is the background charge, the second term represents the density fluctuations of the right- and left-movers, and the last term describes interference between right- and left-movers.

The electron interaction in boson representation emerges from a combination of the hard-core condition of the bosons and the electron-electron interaction of the original fermions. Assuming electron-electron contact interactions and disregarding backscattering, the TLL liquid is described by the generic harmonic Hamiltonian

$$H_\mathrm{L}(\theta,\phi) \;=\; \frac{\hbar v}{2}\int dx\,[(\partial_x\theta)^2/g + g(\partial_x\phi)^2]\,, \tag{31.4}$$

where g is the interaction parameter, and $v = v_\mathrm{F}/g$ is the sound velocity. For the noninteracting Fermi gas, we have $g = 1$, and the case $g < 1$ corresponds to repulsive interaction. We assume a sharp cutoff ω_c in the band width for the linear dispersion relation implicit in H_L and take ω_c as the largest frequency of the problem.

By switching from the Hamiltonian to the Lagrangian, we get the equivalent θ- and ϕ-representations of the Luttinger liquid. The corresponding Euclidean actions are

$$\mathbb{S}_\mathrm{L} \;=\; \begin{cases} \dfrac{\hbar v}{2g}\displaystyle\int dx\,d\tau\left((\partial_x\theta)^2 + \dfrac{1}{v^2}(\partial_\tau\theta)^2\right), \\[2ex] \dfrac{\hbar v g}{2}\displaystyle\int dx\,d\tau\left((\partial_x\phi)^2 + \dfrac{1}{v^2}(\partial_\tau\phi)^2\right). \end{cases} \tag{31.5}$$

It is well-known that the "two-terminal" conductance for a single-channel Fermi liquid is $\mathbb{G}_0(g{=}1) = e^2/2\pi\hbar$. For interacting electrons, the conductance may be calculated from the current-current correlation function in the zero frequency limit, where the current is $J = ie\dot{\theta}/\sqrt{\pi}$. One then finds that the conductance of a single-channel (1D) quantum wire is renormalized by the electron-electron interaction [514],[1]

$$\mathbb{G}_0(g) = ge^2/2\pi\hbar . \tag{31.6}$$

Thus the parameter g is a measure of the conductance of the pure Luttinger liquid.

31.1.2 Charge transport through a single weak barrier

The weak impurity is modelled by a barrier Hamiltonian $H_{\mathrm{sc}} = \int dx\, V(x)\psi^\dagger(x)\psi(x)$, where $V(x)$ is a scattering potential [514]. Omitting multiple-electron backscattering processes, one finds for a short-ranged impurity potential, which is suppposed to be centered at $x = 0$, the form

$$H_{\mathrm{sc}}(\bar{\theta}) = -V_0 \cos(2\sqrt{\pi}\,\bar{\theta}) , \tag{31.7}$$

where $\bar{\theta}(t) \equiv \theta(x = 0, t)$, and where V_0 is the Fourier transform of $V(x)$ at momentum $2k_{\mathrm{F}}$. The scattering Hamiltonian originates from the interference between left- and right-movers and represents $2k_{\mathrm{F}}$-backscattering. An applied voltage drop V_{a} at the impurity gives rise to the contribution

$$H_{\mathrm{U}}(\bar{\theta}) = eU\,\bar{\theta}/\sqrt{\pi} . \tag{31.8}$$

The weak barrier or θ model is then given by

$$H_{\bar{\theta}}(\theta, \phi) = H_{\mathrm{L}}(\theta, \phi) + H_{\mathrm{sc}}(\bar{\theta}) + H_{\mathrm{U}}(\bar{\theta}) . \tag{31.9}$$

In the model (31.9), the nonlinear tunneling degree of freedom $\bar{\theta}(t)$ is coupled to a harmonic field which represents the modes in the leads away from $x = 0$. This is the convenient starting point if one wishes to study backscattering off a weak impurity, in particular with regard to computations of conductance and statistical fluctuations.

Since the backscattering term $H_{\mathrm{sc}}(\bar{\theta})$ operates at $x = 0$ only, we may integrate out the fluctuations of $\theta(x)$ for all x away from the origin. This can be done exactly because of the harmonic nature of the pure Luttinger liquid action (31.5). With the Fourier ansatz $\theta(x, \tau) = (1/\hbar\beta) \sum_n \theta(x, \nu_n) e^{i\nu_n\tau}$, where $\nu_n = (2\pi/\hbar\beta)n$, the Euclidean action is minimized when $\theta(x, \nu_n) = \bar{\theta}_n \exp(-|\nu_n|x|/v)$. Here, $\bar{\theta}_n$ is the Fourier coefficient of $\bar{\theta}(\tau) = \theta(0, \tau)$. The resulting Euclidean influence action [514] includes all effects of the modes in the leads on the tunneling degree of freedom $\bar{\theta}(\tau)$,

$$\mathbb{S}_{\mathrm{infl}}[\bar{\theta}]/\hbar = \frac{1}{g}\frac{1}{\hbar\beta}\sum_n |\nu_n|\,|\bar{\theta}_n|^2 . \tag{31.10}$$

This is analogous to the influence action (4.43) of a harmonic oscillator bath.

It is convenient to consider the nonlinear conductance given in units of the conductance $\mathbb{G}_0(g)$ of a 1D quantum wire,

---

[1] The conductance considered here is a low-frequency microwave conductance. A two-terminal setup with reservoirs held at fixed chemical potentials would lead to modifications [551].

$$\mathcal{G}(T, U, g) \equiv \frac{I(T, U, g)}{U\mathcal{G}_0(g)} . \tag{31.11}$$

The normalized conductance $\mathcal{G}$ in charge transport directly corresponds to the normalized mobility $\mathcal{M}$ in particle transport discussed in the preceding chapters.

For a weak barrier, the θ-representation (31.7) is appropriate. Then the conductance is determined by

$$\mathcal{G}_\theta(T, U, g) = \lim_{t \to \infty} \frac{e}{\sqrt{\pi}\, U\mathcal{G}_0(g)} \langle \dot{\bar{\theta}}(t) \rangle_\beta , \tag{31.12}$$

where $\langle \cdots \rangle_\beta$ denotes the thermal average of all modes of H_L away from the impurity. As we have just seen, this means average with the weight function $\exp\{-\mathcal{S}_{\mathrm{infl}}[\bar{\theta}]/\hbar\}$.

An analytical expression for $\mathcal{G}_\theta$ can be derived as follows. First, we expand the formal path integral expression into a power series in V_0^2. Then, in each term of the series, we integrate out the Gaussian field $\theta(x, \tau)$ away from $x = 0$. The resulting expression has the form of a statistical, grand-canonical ensemble of interacting discrete charges, analogous to the corresponding expressions given in Chapter 26. Because of the analogy of the interaction term (31.7) with the interaction (24.2), the charge conditions are as specified in Section 26.1. As a result of the elimination of the harmonic modes in the leads, the charges are interacting with each other. The pair interaction in imaginary time $\mathcal{W}_\theta(\tau)$ is related to the correlator of $\bar{\theta}(\tau)$ by

$$\langle \mathcal{T} e^{i 2\sqrt{\pi}[\bar{\theta}(\tau) - \bar{\theta}(0)]} \rangle_\beta = e^{-\mathcal{W}_\theta(\tau)} . \tag{31.13}$$

Here, the thermal average $\langle \cdots \rangle$ denotes again average with the weight function $\exp[-\mathcal{S}_{\mathrm{infl}}[\bar{\theta}]/\hbar]$. Upon completing the square, we get

$$\mathcal{W}_\theta(\tau) = g\, \frac{2\pi}{\hbar\beta} \sum_{n \neq 0} \frac{1}{|\nu_n|} \left[1 - e^{i\nu_n\tau} \right] . \tag{31.14}$$

We see from the expression (4.77) with (3.83), or from Eqs. (4.211) - (4.213), that $\mathcal{W}_\theta(\tau)$ is the charge interaction for an Ohmic spectral density $G_\theta(\omega) = 2g\,\omega$. The analytically continued real-time charge interaction $Q_\theta(t) = \mathcal{W}_\theta(\tau = it)$ has the integral representation (18.42) and takes the Ohmic scaling form [see Eq. (18.53)]

$$Q_\theta(t) = 2g \ln\left[\frac{\hbar\beta\omega_c}{\pi} \sinh\left(\frac{\pi|t|}{\hbar\beta} \right) \right] + i\pi g\, \mathrm{sgn}(t) . \tag{31.15}$$

The strong-tunneling series for the normalized conductance is found to read

$$\mathcal{G}_\theta(T, U, g) = 1 - \frac{\pi\hbar}{eU} \sum_{m=1}^{\infty} (-1)^{m-1} \left(\frac{V_0}{\hbar} \right)^{2m} \int_0^\infty d\tau_1\, d\tau_2 \cdots d\tau_{2m-1} \tag{31.16}$$

$$\times \sum_{\{\xi_j = \pm 1\}'} G_{2m}(\{\tau_j\}; g, \{\xi_j\}) \sin\left(\sum_{i=1}^{2m-1} g\, eU\, p_i \tau_i/\hbar \right) \prod_{k=1}^{2m-1} \sin(\pi p_k g) ,$$

where p_i is the cumulated charge defined in Eq. (26.23). The interaction factor $G_{2m}(\{\tau_j\}; g, \{\xi_j\})$ is defined in Eq. (26.8) with the interaction (31.15).

31.1.3 Charge transport through a single strong barrier

In the opposite limit of a large barrier or weak link, there are two disconnected semi-infinite Luttinger leads in zeroth order. To account for electron tunneling through the barrier, a tunneling Hamiltonian describing punctual tunneling is added,

$$H_{\mathrm{T}} \propto \psi^{\dagger}(x = 0^+)\psi(x = 0^-) + \text{h.c.} . \tag{31.17}$$

This term depends via Eq. (31.2) on the phase jump $\bar{\phi} \equiv \frac{1}{2}[\phi(x = 0^+) - \phi(x = 0^-)]$ of the ϕ-field at the point-like barrier. A voltage drop U at the barrier induces an additional phase jump $eUt/\hbar$. These together establish the tunneling Hamiltonian

$$H_{\mathrm{T}} = -\hbar\Delta\cos[2\sqrt{\pi}\bar{\phi} + eUt/\hbar] . \tag{31.18}$$

With addition of the harmonic liquids in the left $(-)$ and right $(+)$ leads described by Eq. (31.4) we arrive at the Hamiltonian of the weak link problem (ϕ-model)

$$H_{\bar{\phi}}(\theta, \phi) = H_{\mathrm{L},+}(\theta, \phi) + H_{\mathrm{L},-}(\theta, \phi) + H_{\mathrm{T}}(\bar{\phi}) . \tag{31.19}$$

This model describes again a nonlinear tunneling degree of freedom coupled to a harmonic field. We may again integrate out the harmonic modes away from the barrier, this time in the ϕ-representation of the lead action (31.5). Taking the route similar to that towards the action (31.10), the Euclidean influence action of the tunneling mode $\bar{\phi}$ resulting from the Luttinger liquid environment is found to read

$$\mathbb{S}_{\mathrm{infl}}[\bar{\phi}]/\hbar = g\frac{1}{\hbar\beta}\sum_n |\nu_n||\bar{\phi}_n|^2 . \tag{31.20}$$

Because of the substitution $g \to 1/g$, when switching from the θ- to the ϕ-representation, the action (31.20) is the dual of the action (31.10).

With use of the Heisenberg equation of motion for $\bar{\theta}(t)$, we may express the normalized conductance (31.12) in terms of the ϕ-field. The resulting expression is

$$\mathcal{G}_{\phi}(T, U, g) = \lim_{t\to\infty} \frac{e\Delta}{U\mathbb{G}_0(g)}\langle \sin[2\sqrt{\pi}\bar{\phi}(t) + eUt/\hbar]\rangle_{\beta} , \tag{31.21}$$

where $\langle \cdots \rangle_{\beta}$ now means average with the weight function $\exp[-\mathbb{S}_{\mathrm{infl}}[\bar{\phi}]/\hbar]$.

The proceeding is as in the preceeding subsection. We formally expand the path integral in powers of Δ^2 and, in view of the formal similarity of the (31.18) with (31.7), we introduce again a charge representation. The effects of the Luttinger liquid modes are included in the charge interaction, which is again a pair correlation function, but now for $\bar{\phi}(t)$. Following the lines (31.13)–(31.15), we get for the real-time correlator

$$\langle \mathcal{T}e^{i2\sqrt{\pi}[\bar{\phi}(t)-\bar{\phi}(0)]}\rangle_{\beta} = \exp\left\{ -\frac{2}{g}\ln\left[\frac{\hbar\beta\omega_c}{\pi}\sinh\left(\frac{\pi|t|}{\hbar\beta}\right)\right] - i\frac{\pi}{g}\mathrm{sgn}(t)\right\} . \tag{31.22}$$

The weak-tunneling series for the normalized conductance is readily found as the dual of the series (31.16),

$$G_\phi(T, U, g) = \frac{\pi\hbar}{g\,eU} \sum_{m=1}^{\infty} (-1)^{m-1} \Delta^{2m} \int_0^{\infty} d\tau_1 \, d\tau_2 \cdots d\tau_{2m-1} \tag{31.23}$$

$$\times \sum_{\{\xi_j = \pm 1\}'} G_{2m}(\{\tau_j\}; 1/g, \{\xi_j\}) \sin\left(\sum_i^{2m-1} eU p_i \tau_i/\hbar\right) \prod_{k=1}^{2m-1} \sin(\pi p_k/g) \,.$$

31.1.4 Coherent conductor in a resistive Ohmic environment

A mesoscopic conductor embedded in an electromagnetic environment forms a quantum system violating Ohm's law. The electrons in the conductor induce electromagnetic modes in the electrical circuit and hence undergo inelastic scattering processes. By this, the current at low voltage is reduced, as we have discussed already for weak tunneling in Subsection 20.3.1. This picture of dynamical Coulomb blockade changes in the opposite limit of a good conductor. The description of tunneling of discrete charges is then no longer valid. The Luttinger parameter of the coherent conductor in the absence of the electromagnetic environment is $g = 1$. According to convenience, we may use either the θ- or the ϕ-representation of the action (31.5).

An Ohmic environment can simulate the electron interactions in a coherent conductor, as we may infer from the study of single charge tunneling in Sec. 20.3. We now extend the analogy of a one-channel conductor in a resistive electromagnetic environment with coupling $\alpha = R/R_K$ to impurity scattering in a Luttinger liquid [552].

Let us take the *weak-tunneling* (wt) Hamiltonian (3.238) as a starting-point. For spinless electrons and real tunneling amplitude T_T of electrons, it can be written as

$$H_{\mathrm{wt}} = H_1 + H_2 + H_{\mathrm{env}}(\mathcal{Q}, \varphi, \{\varphi_\alpha\}) + T_T[\psi_2^\dagger(0^+)\psi_1(0^-)\,e^{-i[\varphi(t)+\delta\varphi(t)]} + \mathrm{h.c.}] \,. \tag{31.24}$$

Here, $H_{1,2}$ is the electron part for the left/right electrode with the voltage drop U across the barrier included in H_1, as specified in Eq. (3.237). The term H_{env} describes the resistive electromagnetic environment (3.217) with spectral coupling of the Drude form (20.126), and U is the applied voltage drop at the barrier. The last term couples the phase fluctuations $\varphi(t) + \delta\varphi(t)$ induced by the impedance to the local electronic fields $\psi_{1,2}(0)$ at the edges of the barrier. The phase correlations are found upon elimination of the set of modes $\{\varphi_\alpha\}$ as given in Eq. (3.230) with (3.231). We choose that $\varphi(t)$ fluctuates dynamically, and $\delta\varphi(t)$ adiabatically, as specified in Eq. (20.133),

$$\langle\, [\varphi(0) - \varphi(t)]\, \varphi(0)\,\rangle_\beta = Q_{\mathrm{ohm}}(t; \alpha) \,,$$
$$\langle\, [\delta\varphi(0) - \delta\varphi(t)]\, \delta\varphi(0)\,\rangle_\beta = \Delta Q_{\mathrm{em}}(\alpha) + \mathcal{O}[1/(\omega_R t)^2] \,. \tag{31.25}$$

The term $\Delta Q_{\mathrm{em}}(\alpha)$ provides adiabatic dressing of the tunneling amplitude.

For a point-like barrier, electron tunneling is effectively one-dimensional. With the bosonization of the electronic entity discussed in Subsection 31.1.1, the tunneling term in Eq. (31.24) can be written as [cf. Eq. (31.18)]

$$H_T = -\tfrac{1}{2}\hbar\Delta\left(e^{i(2\sqrt{\pi}\bar{\phi} - \varphi)} + \mathrm{h.c.}\right) \,. \tag{31.26}$$

The relation of the tunneling amplitude Δ with the electronic counterpart $T_{\rm T}$, and with the tunneling resistance $R_{\rm T}$ is given in Eq. (20.145) with (20.117).

The autocorrelation functions for the electronic phase jump at the barrier $2\sqrt{\pi}\bar{\phi}(t)$ and for the combined phase $\chi(t) = 2\sqrt{\pi}\bar{\phi}(t) + \varphi(t)$ are [cf. Eqs. (31.22) and (20.132)]

$$4\pi \langle [\bar{\phi}(0) - \bar{\phi}(t)]\,\bar{\phi}(0) \rangle_\beta = Q_{\rm ohm}(t; \alpha = 1)\,,$$
$$\langle [\chi(0) - \chi(t)]\chi(0) \rangle_\beta = Q_{\rm ohm}(t; 1 + \alpha)\,. \tag{31.27}$$

Thus, $H_{\rm wt}$ is equivalent to the impurity Hamiltonian $H_{\bar{\phi}}$, Eq. (31.19), if we equate

$$1 + \alpha = 1/g\,, \tag{31.28}$$

as one infers from Eq. (31.22). The correspondence holds in all orders of Δ^2 in the series expression (31.23) for the nonlinear conductance. The respective contribution of lowest order to the weak-tunneling current has been discussed above in Subsec. 20.3.2.

In the opposite limit of strong tunneling, the tunneling term $\propto T_{\rm T}$ is replaced by the scattering potential $H_{\rm sc}(\bar{\theta})$ given in bosonized form in Eq. (31.7). The electromagnetic environment $H_{\rm env}(\mathcal{Q}, \varphi)$, Eq. (3.217), induces a fluctuating voltage $\hbar\dot{\varphi}(t)/e$ which is coupled to the scattering mode $\bar{\theta}$ analogous to the voltage U in $H_{\rm U}$, Eq. (31.8). With the electronic Hamiltonian $H_{{\rm L},g=1}$ added, the Hamiltonian dual to (31.24) thus is

$$H_{\rm st} = H_{{\rm L},g=1}(\theta, \phi) + H_{\rm sc}(\bar{\theta}) + (eU + \hbar\dot{\varphi})\bar{\theta}/\sqrt{\pi} + H_{\rm env}(\mathcal{Q}, \varphi, \{\varphi_\alpha\})\,. \tag{31.29}$$

It is straightforward to integrate out the modes $\{\varphi_\alpha\}$ of the electromagnetic environment. The resulting Euclidean action in Fourier representation for an Ohmic environment, $\hat{Y}(|\nu_n|) = 1/R$, is analogous to the second term in Eq. (4.208) and reads

$$S_{\rm em}[\bar{\theta}, \varphi]/\hbar = \frac{1}{\hbar\beta} \sum_n \left\{ \frac{1}{4\pi} \frac{R_{\rm K}}{R} |\nu_n||\varphi_n|^2 - i\,\nu_n \varphi_n \frac{\bar{\theta}_n}{\sqrt{\pi}} \right\}\,, \tag{31.30}$$

where φ_n is the Matsubara component of the field $\varphi(\tau)$. In the next step, the phase components φ_n are integrated out after completion of the square. This yield the electromagnetic contribution to the Euclidean influence action ($\alpha = R/R_{\rm K}$)

$$S_{\rm infl, em}[\bar{\theta}]/\hbar = \alpha \frac{1}{\hbar\beta} \sum_n |\nu_n|\,|\bar{\theta}_n|^2\,. \tag{31.31}$$

The Fermi liquid fluctuations in the leads ($g = 1$) give rise to the action (31.10) with $g = 1$. Thus, the combined influence action of the resistive environment and the fermionic leads is

$$S_{\rm infl, em}[\bar{\theta}]/\hbar + S_{\rm infl, el}[\bar{\theta}/]/\hbar = (1 + \alpha)\frac{1}{\hbar\beta} \sum_n |\nu_n|\,|\bar{\theta}_n|^2\,, \tag{31.32}$$

Comparison with the expression (31.10) shows that there is again a formal equivalence to backscattering off an impurity in a TLL, this time in the weak-backscattering limit, and the effective coupling parameter $1 + \alpha$ is again related to the fictitious TLL parameter g by $1 + \alpha = 1/g$, just as in the weak-tunneling case (31.28). The correspondence of the model (31.29) with the QI-TLL model (31.9) is again on the level of the effective action.

31.1.5 Equivalence with quantum transport in a cosine potential

In the preceding subsections, we have established a formal equivalence between the models of a quantum impurity in a TLL and a coherent one-channel conductor in a resistive environment. Now we discuss the correspondence of the QI-TLL model with that of a Brownian particle in a tilted cosine potential.

First, we note that the θ model (31.9) corresponds to the WB model (24.1). The equivalence can be shown by employing canonical transformations and by a study of the equations of motion of the coordinate and momentum autocorrelation functions (cf. the discussion in Subsecs. 28.1.1 and 28.1.2). The tunneling degree of freedom $\bar{\theta}$ corresponds to $\sqrt{\pi}X/X_0$. Ohmic damping of the mode $\bar{\theta}$ is provided by excitation of the TLL liquid away from the barrier, and the parameter g is related to the viscosity η by $g = \eta X_0^2/2\pi\hbar$. In the correspondence, the cutoff frequency ω_c of the liquid modes is identified with η/M. The equivalence becomes exact when the inertia force is negligibly small compared with the friction force, i.e., when ω_c is the largest frequency of the problem. For $k_B T \ll V_0$ and $eU \ll V_0$, this means $\eta^2 \gg MV_{WB}''(0)$, or equivalently $\hbar\omega_c \gg 2\pi gV_0$. As a result of the mapping, the nonlinear conductance in the θ-model is directly related to the nonlinear mobility in the WB model.

Similarly, the high-barrier ϕ-model (31.19) is equivalent to the TB model (24.4), as follows again by unitary transformations.

Alternatively, the correspondences of the θ- and ϕ-model with the dissipative WB and TB model are directly visible from the exact formal series expressions for the conductance and mobility, respectively. The strong-tunneling series (31.16) coincides with the series (28.95) for the WB mobility, and the weak-tunneling series (31.23) with the TB mobility (27.2) under the assignments

$$g \leftrightarrow 1/K, \quad \text{and} \quad eU/\hbar \leftrightarrow \epsilon. \tag{31.33}$$

The mapping relations (31.33) hold on the level of the effective actions. Hence they apply not only for the conductance, but also for the full counting statistics.

31.2 Self-duality between weak and strong tunneling

The results of Subsec. 31.1.5 can be condensed into a match of the normalized mobility of the Brownian particle with the normalized conductance in the QI-TLL model,

$$\mathcal{G}(\tilde{\epsilon}_0, T, U, g) = \mu(\tilde{\epsilon}_0, T, \epsilon = eU/\hbar, K = 1/g)/\mu_0, \tag{31.34}$$

which relates the θ-model to the WB model, and the ϕ-model to the TB model. An important difference however is that, for repulsive electron interaction, the Luttinger parameter g is restricted to the domain $g < 1$, whereas in the related Brownian particle model the parameter regime is $0 < K < \infty$.

Using the equality (31.34), the self-duality relation (28.32) is converted into

$$\mathcal{G}(\tilde{\epsilon}_0, T, U, g) = 1 - \mathcal{G}(\tilde{\epsilon}_0, T, gU, 1/g). \tag{31.35}$$

With the symmetric frequency scale $\tilde{\epsilon}_0$ given in Eq. (28.121), this relation has lost explicit dependence on the ϕ- or θ-model, and therefore holds regardless of the particular model. It equally applies for the conductance within the ϕ-, and within the θ-model, and for the conductance between these models. The entire expansions around weak and strong backscattering are in fact related to each other term by term.

The correspondence relation (31.34) allows to transfer the analytical results obtained in Chapt. 28.4 for the Brownian particle to charge transport across a quantum impurity in a Luttinger liquid. The exact scaling solutions at $T = 0$, Eqs. (28.106) and (28.107) with Eq. (28.116), agree with expressions derived by Fendley et al. [540]. These authors utilized a suitable basis of interacting quasiparticles in which the model is integrable, and they employed sophisticated thermodynamic Bethe-ansatz (TBA) technology to calculate the non-Fermi distribution function and the density of states of the quasiparticles. These quantities determine the conductance by a Boltzmann-type rate expression. The Kondo frequency $\epsilon_0 = 8K^{1/(1-K)}\tilde{\epsilon}_0$ directly corresponds to the temperature scale T_B' used in Ref. [540], $T_B' = \hbar\epsilon_0/k_B$. The simple derivation given in Sec. 28.5 sheds additional light on the underlying symmetries of these models.

31.3 Full counting statistics of charge transfer

The full counting statistics of charge transport through an impurity is included in the moment generating function $\chi(\rho, t)$. This is the Fourier transform of the probability distribution $P(Q, t)$ of the charge Q crossing the impurity during time t, $\chi(\rho, t) = \sum_Q e^{i\rho Q} P(Q, t)$, where ρ is the counting field. The function $\chi(\rho, t)$ generates moments of the charge $Q(t) = \int_0^t dt' \, I(t')$ transferred during time t. At long time, there is

$$\chi(\rho, t) = \sum_k \frac{(i\rho)^k}{k!} \langle Q^k(t) \rangle = \exp\left\{ t \sum_k \frac{(i\rho)^k}{k!} \langle \delta^k Q \rangle \right\}. \tag{31.36}$$

Here, $t \langle \delta^k Q \rangle$ is the kth cumulant of the distribution, and $I(t)$ is the time-dependent current through the scattering region.

31.3.1 Charge transport at low temperature for arbitrary g

An important limiting case is the zero temperature regime in which the strong- and weak-backscattering expansions of all cumulants can be found in analytic form [539]. There are clear physical pictures in these different limits.

Weak tunneling:

For strong backscattering or weak tunneling, the ϕ-model (31.19) applies, which is the equivalent of the TB model (24.4) in the Ohmic scaling limit. Here, the true ground state is that of two completely disconnected leads. Then, evidently, only quasiparticles with integer unit charge, i.e. electrons, may tunnel through the barrier between the leads. Since at $T = 0$ there are no transition from the energetically lower to the higher lead, we have in correspondence with the expression (29.1)

$$\ln \chi(\rho, t) = t \sum_{n=1}^{\infty} \left(e^{i \rho e n} - 1 \right) \frac{I_n^+}{n}, \tag{31.37}$$

where $I_n^+ = n k_n^+$. The physical meaning of this expression is quite illuminating. Suppose that k_n^+ is the probability per unit time to transfer a particle of charge ne through the impurity barrier. Then the charge transferred in the time interval t is the result of a Poisson process for particles of charge e crossing the barrier, contributing a current $e I_1^+$, plus a Poisson process for particles of charge $2e$ contributing a current $e I_2^+$, etc. All these Poisson processes are represented by $\ln \chi(\rho, t)$.

Observing that the correspondence of the TB model (24.4) with the ϕ-model (31.19) does hold not only for the conductance but also for all cumulants, we immediately get from the expression (29.6) for the partial current $e I_n^+$ at $T = 0$

$$e I_n^+(\epsilon, 0) = e \frac{(-1)^{n-1}}{\Gamma(n)} \frac{\Gamma(\frac{3}{2}) \Gamma(n/g)}{\Gamma[\frac{3}{2} + (1/g - 1)n]} \frac{\epsilon}{2\pi} \left(\frac{\epsilon}{\epsilon_0} \right)^{(2/g - 2)n}. \tag{31.38}$$

This yields for the first cumulant or mean current the expression

$$\langle \delta Q \rangle = \langle I(U) \rangle = e \sum_{n=1}^{\infty} I_n^+(eU/\hbar, 0). \tag{31.39}$$

While in a classical Poisson process all the partial currents $e I_n^+$ in Eq. (31.37) are positive, the subtle point now is that the partial currents (31.38) are not. Certainly, the odd current contribution $e I_{2n-1}^+$ are positive, but the joint tunneling of pairs of electrons, and multiples thereof, $n = 2, 4, \cdots$, come with a negative sign and, thus interfere destructively with the odd current contributions, $n = 1, 3, \cdots$.

As pointed out in Sec. 29.2, the leading thermal dependence is included in $\ln \chi(\rho, t)$, if we add in the expression (31.37) the leading thermal correction to the partial currents $e I_n^+$. There follows from Eq. (29.15) with $K = 1/g$

$$e I_n^+(\epsilon, T) = \left\{ 1 + (2/g - 2)n \left[1 + (2/g - 2)n \right] \frac{\pi^2}{3g} \left(\frac{k_B T}{\hbar \epsilon} \right)^2 \right\} e I_n^+(\epsilon, 0). \tag{31.40}$$

Strong tunneling:

In the strong-tunneling limit, a collective state between the edges with elementary excitations of charge ge is formed. The resulting expression for $\ln \chi(\rho, t)$ is

$$\ln \chi(\rho, t) = t \left(i \rho g e \frac{\epsilon}{2\pi} - \frac{\rho^2}{2} \frac{g e^2}{2\pi} \frac{2 k_B T}{\hbar} + \sum_{n=1}^{\infty} \left(e^{-i \rho g e n} - 1 \right) \frac{\tilde{I}_n^+}{n} \right). \tag{31.41}$$

There holds up to terms of second order in $k_B T / \hbar \epsilon$

$$\tilde{I}_n^+(\epsilon, T) = \left\{ 1 + (2g - 2)n \left[1 + (2g - 2)n \right] \frac{\pi^2 g}{3} \left(\frac{k_B T}{\hbar \epsilon} \right)^2 \right\} \tilde{I}_n^+(\epsilon, 0), \tag{31.42}$$

$$\tilde{I}_n^+(\epsilon, 0) = \frac{(-1)^{n-1}}{\Gamma(n)} \frac{\Gamma(\frac{3}{2}) \Gamma(ng)}{\Gamma[\frac{3}{2} + (g - 1)n]} \frac{g \epsilon}{2\pi} \left(\frac{\epsilon}{\epsilon_0} \right)^{(2g - 2)n}.$$

The expression (31.41) with (31.42) corresponds to Eq. (29.16) with (29.17).

The form (31.41) is quite similar to Eq. (31.37), but there are subtle differences. The first two terms in the expression (31.41) represent the current and noise of quasi-particles with charge ge in the absence of the barrier. The exponential factor $e^{-i\rho g e n}$ indicates that now we have tunneling of quasiparticles of charge ge and of multiples thereof, and the minus sign in the exponent means that the tunneling diminishes the current instead of building it up as in the strong-backscattering limit. Since the partial currents $ge\bar{I}_n$ are positive for $g < 1/2n$, the perception of a classical Poisson process, in which clusters of quasiparticles with charge ge are tunneling independently, is quite appropriate for bundles with modest n, when g is small. The mean current at zero temperature is

$$\langle \delta Q \rangle = \langle I \rangle = g\frac{U}{R_K} - ge\sum_{n=1}^{\infty} \tilde{I}_n^+(eU/\hbar, 0) \,. \tag{31.43}$$

where $R_K = 2\pi\hbar/e^2$.

As g goes to zero, all the partial currents become positive, but the quantum fluctuations $\langle \delta^n Q \rangle$ with $n > 1$ disappear at $T = 0$, as one reaches the classical limit,

$$\ln \chi(\rho, t) = it\rho g e\frac{\epsilon}{2\pi}\left[1 - \sum_{n=1}^{\infty} \frac{\Gamma(n-\frac{1}{2})}{2\sqrt{\pi}n!}\left(\frac{\epsilon_0}{\epsilon}\right)^{2n}\right] = it\rho g e\frac{\epsilon}{2\pi}\sqrt{1 - \left(\frac{\epsilon_0}{\epsilon}\right)^2}, \tag{31.44}$$

where $\epsilon_0 = 2\pi V_0/\hbar$. The first term is the current in the absence of the barrier, and the second is the sum of all partial backscattering currents.

The T^2-contribution to the cumulants vanishes at $g = 1$. Hence T^2-variation of the current and of its fluctuations is a distinctive signature of 1D interacting electrons.

Finally, we remark that cumulant relations analogous to the expressions (29.5) and (29.9) hold. The corresponding relation in the weak-tunneling representation is

$$\langle \delta^N Q \rangle = \left(e\frac{\Delta}{2}\frac{\partial}{\partial\Delta}\right)^{N-1}\langle \delta Q \rangle \,, \tag{31.45}$$

whereas in the strong-tunneling representation

$$\langle \delta^N Q \rangle = \left(-ge\frac{V_0}{2}\frac{\partial}{\partial V_0}\right)^{N-1}\langle \delta Q \rangle \,. \tag{31.46}$$

These relations help us to reduce every single cumulant to a calculation of derivatives of the current $\langle \delta Q \rangle$ with respect to the tunneling coupling Δ or to the corrugation stingrength V_0, respectively. They are in agreement with the findings from the integrable approach to the BSG model [539].

In the application of the model to a fractional quantum Hall bar, the collective excitations of the harmonic reservoir are Laughlin quasiparticles, the weak or strong impurity or barrier corresponds to a point contact, and the interaction parameter g represents the filling fraction, $g = \nu = \frac{1}{3}, \frac{1}{5}, \cdots$. In the weak-backscattering limit, Laughlin quasiparticles with fractional charge νe are tunneling. In the strong-backscattering limit, the system consists of two different Hall devices. These are weakly coupled by the interaction (31.18), and only electrons (integer charge) can tunnel. Strict duality means that the entire expansions around weak and strong backscattering are related. In the FQHE system, the crossover from weak to strong backscattering comes along with a crossover from Laughlin quasiparticle tunneling to electron tunneling.

For general g and general temperature, the analytical calculation of higher-order Coulomb integrals in the bosonic representation is not possible. Then one may resort to thermodynamic Bethe ansatz techniques in the related BSG model [540, 541]. In Ref. [542] a general scheme for the calculation of cumulants of any order is proposed.

31.3.2 Full counting statistics at $g = \frac{1}{2}$ and general temperature

In the weak-tunneling limit (strong impurity), the mean current at $T = 0$ depends on the applied voltage as $U^{2/g-1}$, while the current $I_{\rm B}$ backscattered from a weak barrier (weak impurity) varies as U^{2g-1}. This we can see from the leading orders of the series (31.39) and (31.43). Hence the backscattered current becomes independent of the applied voltage at $g = \frac{1}{2}$ in leading order. This indicates that $g = \frac{1}{2}$ is a special point. According to the mapping (31.28), $1+\alpha = 1/g$, the case $g = \frac{1}{2}$ corresponds to a coherent conductor (e.g. a quantum dot) with series resistance $R = R_{\rm K}$ or $\alpha = 1$. An open point contact is a realization of such series resistance, as discussed in Ref. [553].

Employing the nonequilibrium Keldysh formalism and refermionization techniques, which map a Luttinger liquid at $g = \frac{1}{2}$ on noninteracting fermions, the cumulant generating function has been calculated in Ref. [553]. With a transformation set out in Appendix C of Ref. [554], the expression of the CGF given in Ref. [553] can be converted into a form analogous to Eq. (28.60),

$$
\ln \chi(\rho, t) = t \int_0^\infty \frac{d\omega}{2\pi} \ln \left\{ 1 + \overline{\mathcal{T}}(\omega) \left[\mathcal{N}_+(\omega, eU/2\hbar) \left(e^{i e\rho} - 1 \right) \right. \right.
$$
$$
\left. \left. + \mathcal{N}_-(\omega, eU/2\hbar) \left(e^{-i e\rho} - 1 \right) \right] \right\}, \tag{31.47}
$$

with the spectral transition probability $\overline{\mathcal{T}}(\omega) = \omega^2/(\omega^2 + \tilde{\epsilon}_0^2)$. The functions $\mathcal{N}_\pm(\omega, \epsilon)$ are given in Eqs. (27.32). The renormalized frequency $\tilde{\epsilon}_0$ is related to the bare impurity strength V_0 by

$$
\tilde{\epsilon}_0 = \pi \left(V_0/\hbar \right)^2/\omega_{\rm c}, \tag{31.48}
$$

as follows from Eq. (28.121) with the correspondence $1/g = K = 2$.

The CGF (31.47) of the QI-TLL model (31.9) at $g = \frac{1}{2}$ also describes the FCS of a coherent conductor in series with an Ohmic resistor of resistance $R = R_{\rm K}$. The expression (31.47) is in agreement with the CGF (28.60) of the quantum Brownian particle model (24.1) at $K = 2$, and with the CGF of Cooper pair tunneling in a resistive environment with resistance $R = 2R_{\rm Q}$. Finally, the leading cumulants are easily found from the expressions (28.61)–(28.63).

32 Nonlinear quantum Brownian duet as work-to work converter

32.1 Introduction

Thermodynamic machines transform different forms of energy into one another. According to the laws of thermodynamics, the efficiency for conversion of heat Q_h into work W, $\eta = W/Q_h$ is bounded from above by Carnot's efficiency $\eta_C = 1 - T_c/T_h$, where T_c and T_h are the temperatures of the cold and hot heat reservoirs. As Carnot cyclic processes require quasi-static conditions, it has been generally assumed that the Carnot efficiency comes with a vanishing output power. This view has been contested by Benenti et al. [555] who demonstrated that, if time-reversal symmetry is broken, Carnot efficiency at finite power is not forbidden. Since then, various attempts to get close to Carnot efficiency upon retaining finite power output have been made [556]-[558]. These works taken together indicated that optimization of power and efficiency comes along with large power fluctuations. This phenomenon is closely related to a thermodynamic uncertainty relation discovered by Barato and Seifert [559], which describes a tradeoff between entropy production (cost), power (yield) and power fluctuations (precision) of an arbitrary current. A general proof of the uncertainty relation was given for Markov jump processes satisfying a local detailed balance condition [560].

Recently, it was shown that in heat engines large power output, operation close to Carnot efficiency, and small fluctuations in the output are not compatible. In fact, these three characteristics satisfy a universal tradeoff criterion, a bound constraining power, efficiency and large power fluctuations [561]. The bound was derived for steady-state engines described by Markovian dynamics on a discrete state space and for overdamped Langevin dynamics. Bounds on entropy production in terms of mean current and current fluctuation in periodically driven classical Markov systems were discussed in Refs. [562]–[564]. Mitigation of the tradeoff bound has been found in ballistic multi-terminal transport [565], and in coherent electron transport through single- and double-dot junctions without [566] and with electron interactions [567].

With two time-dependent external drives, the QBP model (24.4) forms a quantum Brownian duet and acts as an iso-thermal work-to-work converter. It is now quite interesting to investigate whether the tradeoff criterion also holds for such system when quantum tunneling is predominant.

The tradeoff bound is introduced in the next section. The subsequent sections deal with the quantum Brownian duet and give results both in the linear and nonlinear response regimes. It is found that the tradeoff bound can be undercut down to zero for sufficiently low temperature and weak damping where the non-Markovian quantum nature prevails [568].

32.2 Universal tradeoff between power, efficiency and relative uncertainty

32.2.1 Power and power fluctuations

Consider now the energy in- and output of a quantum Brownian particle (QBP) moving in a 1D lattice model TB, as described by the Hamiltonian (24.4). Here assume that the QBP is subjected to a time periodic force $\hat{H}_{\text{tilt}}(t) = -\hbar\epsilon(t)\hat{q}/q_0$, where $\epsilon(t) = \epsilon(t + 2\pi/\omega)$. When the QBP is subjected to two independent drives,

$$\epsilon(t) = \epsilon_1(t) + \epsilon_2(t) , \tag{32.1}$$

as discussed for a classical setting in Refs. [556, 557], it can operator as a work-to-work converter. When the two harmonic drives have different frequency, the work rates are clearly distinguishable. But then the converter could not operate in the *linear* regime, as different frequencies would not couple (cf. Subsec. 32.3). In case of common frequency, one may question whether both powers and variances can be experimentally distinguished. This is possible in fact, when the two forces are independent, e.g., when they operate spatially separated, as in the device discussed in Ref. [568].

The power $P_j(t)$ for the drive $\epsilon_j(t)$ is related to the particle's velocity $\dot{q}(t)$ as

$$P_j(t) = \hbar\epsilon_j(t)\,\dot{q}(t)/q_0 , \qquad j = 1, 2 . \tag{32.2}$$

The exact formal series representations of position and position spread of the QBP have been discussed for a constant tilt in Chap. 26.

Consider now first in some detail the order Δ^2, which is the leading one in the weak-tunneling regime. It describes transport via nearest-neighbour transitions, and single-electron transport in the related fermionic model,

$$\langle q(t)\rangle = q_0 \int_0^t dt_2 \int_0^{t_2} dt_1 \, k_{\text{P}}(t_2 - t_1) \sin[\vartheta(t_2, t_1)] , \tag{32.3}$$

where

$$k_{\text{P}}(\tau) = \Delta^2 \sin[Q''(\tau)]e^{-Q'(\tau)} \tag{32.4}$$

includes the bath correlations, and $\vartheta(t_2, t_1)$ is the total bias phase accumulated from the two drives during the time period from t_1 to t_2,

$$\vartheta(t_2, t_1) = \int_{t_1}^{t_2} dt' \, [\epsilon_1(t') + \epsilon_2(t')] . \tag{32.5}$$

At times much larger than the decay time of $k_{\text{P}}(\tau)$, the transient dynamics is gone, and the power $\langle P_j(t)\rangle = \hbar\epsilon_j(t)\langle\dot{q}(t)\rangle/q_0$ is a periodic function of t,

$$\langle P_j(t)\rangle = \hbar \int_0^\infty d\tau \, k_{\text{P}}(\tau)\,\epsilon_j(t)\sin[\vartheta(t, t-\tau)] = \sum_n P_{j,n}\, e^{-in\omega t} . \tag{32.6}$$

The time-averaged mean power $\langle \overline{P(t)}\rangle = P_{j,0}$ emerges with the expression (32.3) as

$$P_{j,0} = \hbar \int_0^\infty d\tau \, k_\mathrm{P}(\tau) \frac{\omega}{2\pi} \int_0^{2\pi/\omega} dt \, \epsilon_j(t) \sin[\vartheta(t, t - \tau)] \, . \tag{32.7}$$

Consider next the power variance, which is defined as

$$\langle D_j(t) \rangle = \frac{1}{t} \int_0^t dt_2 \int_0^t dt_1 \, \langle (P_j(t_2) - \langle P_j(t_2) \rangle)(P_j(t_1) - \langle P_j(t_1) \rangle) \rangle \, . \tag{32.8}$$

This yields in the stationary regime in order Δ^2

$$\langle D_j(t) \rangle = \hbar^2 \int_0^\infty d\tau \, k_\mathrm{D}(\tau) \epsilon_j(t) \epsilon_j(t - \tau) \cos[\vartheta(t, t - \tau)] = \sum_n D_{j,n} \, e^{-in\omega t} \, , \tag{32.9}$$

where $k_\mathrm{D}(\tau) = \cot[Q''(\tau)] \, k_\mathrm{P}(\tau)$. The time-averaged power variance then is

$$D_{j,0} = \hbar^2 \int_0^\infty d\tau \, k_\mathrm{D}(\tau) \frac{\omega}{2\pi} \int_0^{2\pi/\omega} dt \, \epsilon_j(t) \epsilon_j(t - \tau) \cos[\vartheta(t, t - \tau)] \, . \tag{32.10}$$

32.2.2 Trade-off relation

When the resulting mean powers have opposite sign, $P_{1,0} P_{2,0} < 0$, the QBP entity is acting as work-to-work converter with the positive mean power being the load, and the negative mean power being the yield. If $P_{2,0}$ is the load and $P_{1,0}$ is the yield, the efficiency of the converter is $\eta \equiv |P_{1,0}|/P_{2,0} \leqslant 1$. Optimal performance is characterized by maximal efficiency at a given load. However, optimization of the converter should also conform to power fluctuations as low as possible. The latter may be rated with the estimate of relative uncertainty

$$\Sigma_{1,0} = \sqrt{D_{1,0}/P_{1,0}^2} \, . \tag{32.11}$$

It has been argued and proven for a huge class of steady-state heat engines with internal classical states that there is a tradeoff between large power, high efficiency and low relative uncertainty, being expressed by the joint bound [561]

$$Q_1 \equiv \beta |P_{1,0}| \, (1/\eta - 1) \, \Sigma_{1,0}^2 \geqslant 2 \, . \tag{32.12}$$

If efficiency is close to unity with considerable yield, the bound implies that the power fluctuations are quite large. Conversely, if the bound is broken, simultaneous attainment of maximal efficiency, sizeable yield and low power fluctuations are within reach. This can happen indeed, as shown below.

32.2.3 Harmonic Brownian duet

Consider mean power and power variance for isochromatic driving with a phase shift $\varphi = \arctan \alpha$ in the input channel,

$$\begin{aligned}
\epsilon_1(t) &= f_1 \sin(\omega t) \, , \\
\epsilon_2(t) &= f_2 \cos(\omega t - \varphi) \, .
\end{aligned} \tag{32.13}$$

The overall bias $\epsilon(t) = \epsilon_1(t) + \epsilon_2(t)$ can be written as

$$
\begin{aligned}
\epsilon(t) &= f\sin(\omega t) + \phi\,, \\
f &= \text{sgn}\,(f_1 + f_2\sin\varphi)\,\sqrt{f_1^2 + f_2^2 + 2f_1 f_2\sin\varphi]}\,, \qquad (32.14)\\
\phi &= \arctan[f_2\cos\varphi/(f_1 + f_2\sin\varphi)]\,.
\end{aligned}
$$

With the drives (32.13) and (32.14), the Fourier coefficients of $e^{i\vartheta(t,t-\tau)}$ are Bessel functions $J_n(z)$ times phasefactors. The resulting time-averaged powers are

$$
\begin{aligned}
P_{1,0} &= \hbar\int_0^\infty d\tau\, k_P(\tau)\, f_1\, J_1[A(\tau)]\cos(\phi - \omega\tau/2)\,, \\
&\qquad\qquad\qquad\qquad\qquad\qquad\qquad\qquad (32.15)\\
P_{2,0} &= \hbar\int_0^\infty d\tau\, k_P(\tau)\, f_2\, J_1[A(\tau)]\sin(\phi + \varphi - \omega\tau/2)\,.
\end{aligned}
$$

where $A(\tau) = (2f/\omega)\sin(\omega\tau/2)$.

The mean variances of the powers (32.10) are found as

$$
\begin{aligned}
D_{1,0} &= \frac{\hbar^2}{2}\int_0^\infty d\tau\, k_D(\tau)\, f_1^2\left[J_0[A(\tau)]\cos(\omega\tau) - J_2[A(\tau)]\cos(2\phi)\right]\,, \\
&\qquad\qquad\qquad\qquad\qquad\qquad\qquad\qquad (32.16)\\
D_{2,0} &= \frac{\hbar^2}{2}\int_0^\infty d\tau\, k_D(\tau)\, f_2^2\left[J_0[A(\tau)]\cos(\omega\tau) + J_2[A(\tau)]\cos(2\phi + 2\varphi)\right]\,.
\end{aligned}
$$

The expressions (32.15) and (32.16) are exact in the weak-tunneling limit for arbitrary strength of the driving amplitudes f_1 and f_2.

32.3 Linear response

In linear thermodynamics, the fluxes depend linearly on the forces F_j, and the mean powers are quadratic forms of the forces,

$$
P_{j,0} = F_j\sum_{k=1,2}\mathcal{L}_{j,k}\,F_k\,. \qquad (32.17)
$$

The matrix $\mathcal{L}$ is referred to as Onsager matrix in literature. Expanding the general expression (32.7) up to terms quadratic in $F_1 = \hbar f_1/q_0$ and $F_2 = \hbar f_2/q_0$, one has

$$
P_{j,0} = \hbar\int_0^\infty d\tau\, k_P(\tau)\,\frac{\omega}{2\pi}\int_0^{2\pi/\omega} dt\,\epsilon_j(t)\vartheta(t,t-\tau)\,. \qquad (32.18)
$$

Using expression (32.5), the Onsager matrix can be extracted from Eq. (32.18) as

$$
\mathcal{L}_{j,k} = \frac{\hbar}{F_1 F_2}\int_0^\infty d\tau\, k_P(\tau)\,\frac{\omega}{2\pi}\int_0^{2\pi/\omega} dt\,\epsilon_j(t)\int_{t-\tau}^t dt'\,\epsilon_k(t')\,. \qquad (32.19)
$$

For the drive (32.13), we then obtain, upon writing the phase shift as $\varphi = \arctan\alpha$,

$$
\mathcal{L} = \frac{q_0^2}{\hbar} L_s(\omega) \begin{pmatrix} 1 & \dfrac{\alpha + \kappa}{\sqrt{1+\alpha^2}} \\ \dfrac{\alpha - \kappa}{\sqrt{1+\alpha^2}} & 1 \end{pmatrix},
$$

(32.20)

where $\kappa = L_c(\omega)/L_s(\omega)$. The functions $L_s(\omega)$ and $L_c(\omega)$ are

$$
\begin{aligned}
L_s(\omega) &= \frac{1}{2\omega} \int_0^\infty d\tau\, k_P(\tau) \sin(\omega\tau), \\
L_c(\omega) &= \frac{1}{2\omega} \int_0^\infty d\tau\, k_P(\tau) \left[1 - \cos(\omega\tau)\right].
\end{aligned}
$$

(32.21)

The mean power variance $D_{1,0}$ to second order in the force F_1 is given by

$$
\begin{aligned}
D_{1,0} &= \hbar^2 \int_0^\infty d\tau\, k_D(\tau) \frac{\omega}{2\pi} \int_0^{2\pi/\omega} dt\, \epsilon_1(t)\epsilon_1(t-\tau) \\
&= \frac{1}{2} F_1^2 q_0^2 \int_0^\infty d\tau\, k_D(\tau) \cos(\omega\tau) = F_1^2 q_0^2\, \omega \coth(\hbar\beta\omega/2)\, L_s(\omega),
\end{aligned}
$$

(32.22)

Here we have used the relation

$$
\int_0^\infty d\tau\, k_P(\tau) \sin(\omega\tau) = \tanh(\hbar\beta\omega/2) \int_0^\infty d\tau\, k_D(\tau) \cos(\omega\tau),
$$

(32.23)

which is based on the reflection property (18.43) of the function $Q(z)$ and the resulting detailed balance relations (20.34) and (25.21).

In the Ohmic scaling limit, Eq. (18.53), the integrals (32.21) can be calculated in analytic form. The resulting expressions for $L_s(\omega)$ and $\kappa(\omega) = L_c(\omega)/L_s(\omega)$ are

$$
\begin{aligned}
L_s(\omega) &= \frac{1}{2\hbar\beta\omega} \left(\frac{\hbar\beta\Delta_r}{2\pi}\right)^{2-2K} \sin(2\pi K)\, \Gamma(1-2K) \\
&\quad \times \sinh\left(\frac{\hbar\beta\omega}{2}\right) \Gamma\left(K - i\frac{\hbar\beta\omega}{2\pi}\right) \Gamma\left(K + i\frac{\hbar\beta\omega}{2\pi}\right),
\end{aligned}
$$

(32.24)

$$
\kappa(\omega) = \frac{\tan(\pi K)}{\sinh(\frac{1}{2}\hbar\beta\omega)} \left[\frac{\Gamma(K)^2}{\Gamma(K - i\frac{\hbar\beta\omega}{2\pi})\Gamma(K + i\frac{\hbar\beta\omega}{2\pi})} - \cosh\left(\frac{\hbar\beta\omega}{2}\right)\right].
$$

(32.25)

The Onsager matrix (32.20) administers the interplay of phase tuning of the driving force and exchange of energy between bath and QBP. In the limit $\alpha \to \infty$ ($\varphi \to \pi/2$), the Onsager matrix is symmetric. Then, the converter operates with time reversal symmetry. As α is lowered, this symmetry is broken because of anti-symmetric admixtures. In the limit $\alpha \to 0$ ($\varphi \to 0$), the Onsager matrix is anti-symmetric. Upon tuning the phase parameter α, one moves forth or back between these limits.

32.3.1 Maximum output power

For the drive (32.13), the maximal yield $|\overline{P}_{1,0}| = |P_{1,0}(\overline{F}_1, F_2)|$ is at $F_1 = \overline{F}_1$, where

$$
\begin{aligned}
\overline{F}_1 &= -\frac{\alpha + \kappa}{2\sqrt{1 + \alpha^2}} F_2 \,, \\
\overline{P}_{1,0} &= -(q_0^2/\hbar) L_{\rm s}(\omega) \overline{F}_1^2 \,.
\end{aligned}
\tag{32.26}
$$

Consider next the relation

$$
P_{1,0}(F_1, F_2) = P^\star \overline{P}_{1,0} \,,
\tag{32.27}
$$

where $P^\star$ is a parameter in the range $0 < P^\star \leqslant 1$. The equality (32.27) has two roots, which are $F_{1,\pm} = (1 \pm \sqrt{1 - P^\star}) \, \overline{F}_1$. Correspondingly, efficiency $\eta = |P_{1,0}|/P_{2,0}$ and power fluctuations $\Sigma_1 = \sqrt{D_{1,0}/P_{1,0}^2}$ at $F_1 = F_{1,\pm}$ have two branches,

$$
\begin{aligned}
\eta_\pm &= \frac{P^\star}{2} \frac{X}{1 + 2/Y \mp \sqrt{1 - P^\star}} \,, \\
\Sigma_{1,\pm} &= \frac{2\hbar}{q_0 F_2} \frac{\sqrt{1 + \alpha^2}}{\alpha + \kappa} \sqrt{\frac{\omega \coth(\beta\omega/2)}{L_{\rm s}(\omega)} \frac{(1 \pm \sqrt{1 - P^\star})}{P^\star}} \,,
\end{aligned}
\tag{32.28}
$$

where

$$
\begin{aligned}
X &= (\alpha + \kappa)/(\alpha - \kappa) \,, \\
Y &= (\alpha^2 - \kappa^2)/(1 + \kappa^2) \,.
\end{aligned}
\tag{32.29}
$$

In Fig. 32.1, the two branches of the efficiency $\eta_\pm$ (left) and the power fluctuations $\Sigma_{1,\pm}$ (right) are plotted versus $P^\star$. The behaviors are qualitatively different for $\alpha > \kappa$ and $\alpha < \kappa$. The left panel shows that high efficiency can be reached on the $(+)$-branch when $\alpha > \kappa$, and on the $(-)$-branch when $\alpha < \kappa$. In contrast, low power fluctuations arise only in branch $(-)$ when $\alpha < \kappa$ (right panel). Hence high efficiency is compatible with low power fluctuations when $\alpha < \kappa$, i.e., when the anti-symmetric off-diagonal parts of the Onsager matrix outweigh the symmetric ones.

32.3.2 Maximal efficiency

To find out optimal working conditions, we now focus on the maximal efficiency (ME) at notable power yield. The efficiency $\eta(F_1)$ for fixed F_2 is maximal at $F_1 = F_{1,\rm ME}$, where $F_{1,\rm ME} = -F_2(\sqrt{1 + \kappa^2} - \sqrt{1 + \alpha^2})/(\kappa - \alpha)$, and the maximal efficiency is

$$
\eta_{\rm ME} = \frac{\sqrt{1 + \alpha^2} - \sqrt{1 + \kappa^2}}{\sqrt{1 + \alpha^2} + \sqrt{1 + \kappa^2}} \frac{\alpha + \kappa}{\alpha - \kappa} \,.
\tag{32.30}
$$

Fig. 32.2 (a) shows $\eta_{\rm ME}$ versus $\hbar\beta\omega$ for different interaction strength K. With decreasing K, $\eta_{\rm ME}$ is strictly increasing. In the asymptotic non-Markovian low temperature regime $\hbar\beta\omega \gg 1$, the function (32.25) takes the form

$$
\kappa(\omega) = \tan(\pi K) \left[(\hbar\beta\omega/2\pi)^{1-2K} \Gamma(K)^2/\pi - 1 \right] \,, \qquad \hbar\beta\omega \gg 1 \,,
\tag{32.31}
$$

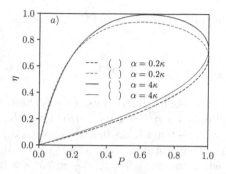
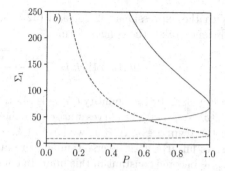

Figure 32.1: Efficiency and power fluctuations in the LR regime versus P^*. Solid curve is $\alpha = 4\kappa$ and dashed curve $\alpha = 0.2\kappa$. In a) the upper branches are η_+ (solid) and η_- (dashed). In b) both upper branches represent $\Sigma_{1,+}$. Other parameters are $\hbar\beta\omega = 6$, $\omega = 5\Delta_r$, $\Delta_r = 1$, $K = 0.1$, and $f_2 = 0.1\,\omega$.

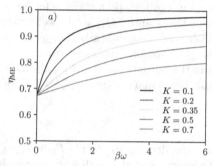
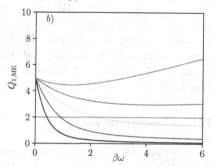

Figure 32.2: Maximum efficiency η_{ME} and tradeoff criterion $Q_{1,\mathrm{ME}}$ versus $\beta\omega$ ($\hbar = 1$) in the LR regime for $\alpha = 5$ and different K. Curves from up to down with increasing K in Panel a), and with decreasing K in Panel b). Shortfall of the bound 2 occurs for $K \lesssim 0.4$. The gradual increase of $Q_{1,\mathrm{ME}}$ in the range $1/4 < K \lesssim 0.4$ takes place at higher $\beta\omega$ than shown in Panel b).

Hence the function $\kappa(\omega)$ diverges in the limit $\hbar\beta\omega \to \infty$ as $(\hbar\beta\omega)^{1-2K}$ for $K < 1/2$ and becomes a positive constant in the range $1/2 < K < 1$. Thus, η_{ME} reaches unity in the former, and a value less than unity in the latter case. For $K \ll 1/2$, the prefactor of the term $(\hbar\beta\omega/2\pi)^{1-2K}$ is $1/K$. As a result, the ME efficiency dwells close to unity in a considerably wide temperature range for weak Ohmic damping, or large repulsive Coulomb interaction in the associated fermionic transport model.

The mean output power and relative uncertainty at maximal efficiency are

$$P_{1,0\,\mathrm{ME}} = -F_2^2 L_s(\omega)\sqrt{(1+\kappa^2)/(1+\alpha^2)}\,\eta_{\mathrm{ME}}\,,$$

$$\Sigma_{1,\mathrm{ME}} = \sqrt{L_s\omega \coth(\hbar\beta\omega/2)}\,F_{1\,\mathrm{ME}}/P_{1,0\,\mathrm{ME}}\,.$$

$$(32.32)$$

With the expressions (32.30) and (32.32) the tradeoff criterion (32.12) at maximal efficiency takes the concise form

$$Q_{1,\mathrm{ME}} = 2\hbar\beta\omega \coth(\hbar\beta\omega/2)\frac{\sqrt{1+\alpha^2}}{(\alpha+\kappa)^2}\left(\frac{1-\alpha\kappa}{\sqrt{1+\kappa^2}} + \sqrt{1+\alpha^2}\right). \tag{32.33}$$

In Fig. 32.2 (b) the quantity $Q_{1,\mathrm{ME}}$ is plotted versus $\beta\omega$ for different values of K. Since $\kappa(\beta\omega \to 0) \to 0$, the curves start out at the value $4\sqrt{1+\alpha^2}(1+\sqrt{1+\alpha^2})/\alpha^2$ for all K. In the regime $\frac{1}{2} < K < 1$, we have $\kappa(\beta\omega \to \infty) = -\tan(\pi K) > 0$, and hence $Q_{1,\mathrm{ME}}$ grows linearly with inverse temperature at low temperatures, whereas $|\langle P_{1,\mathrm{ME}}\rangle|$ and $\Sigma_{1,\mathrm{ME}}$ become constant in this limit. In contrast, in the range $0 < K < \frac{1}{2}$, $\kappa(\beta\omega \to \infty)$ diverges asymptotically as $(\beta\omega)^{1-2K}$. Thus, the yield $|P_{1,0\mathrm{ME}}|$ grows as $(\beta\omega)^{1-2K}$, the relative uncertainty $\Sigma_{1,\mathrm{ME}}$ drops to zero as $(\beta\omega)^{2K-1}$, and the quantity $Q_{1,\mathrm{ME}}$ varies as $(\beta\omega)^{4K-1}$ in this limit. As a result, $Q_{1,\mathrm{ME}}$ diverges in the range $\frac{1}{4} < K < \frac{1}{2}$, stays flat below 2 for $K = \frac{1}{4}$, and drops to zero when K is in the range $0 < K < \frac{1}{4}$, as $\beta\omega \to \infty$. Hence the QBP work converter has optimal performance for weak damping $0 < K < \frac{1}{4}$. With decreasing temperature the quantity $Q_{1,\mathrm{ME}}$ falls well below the classical bound 2, and eventually drops to zero, as the non-Markovian quantum regime is reached. High efficiency and small power fluctuations are in fact compatible. The impact of an added n-fold frequency drive in the output, $\delta\epsilon_1(t) = F_1\gamma_n \sin(n\omega t)$, changes the behaviors shown in Figs. 32.1 and 32.2 only marginally for $|\gamma_n| < 1$ [568].

32.4 Nonlinear response

The above results of the LR regime apply when the driving amplitude $F_2 = \hbar f_2/q_0$ is sufficiently small, $f_2 \ll \alpha\omega$. For larger F_2, the interplay of the nonlinear drive with the bath correlations becomes significant. Then the ME analysis must start with the original expressions (32.15) and (32.16). The ME point $F_1 = F_{1,\mathrm{ME}}$ can be found as numerical root of the extremal condition $d\eta/dF_1 = 0$. With this, numerical NLR computation of $P_{1,\mathrm{ME}}$, $\eta_{1,\mathrm{ME}}$, $\Sigma_{1,\mathrm{ME}}$ and $Q_{1,\mathrm{ME}}$ is fairly straightforward.

The characteristic behaviors of $P_{1,0\mathrm{ME}}$, $\eta_{1,\mathrm{ME}}$, and $\Sigma_{1,\mathrm{ME}}$ versus F_2/ω in NLR are shown in Fig. 32.3 (a)-(c) for $\beta\omega = 6$ (blue) and $\beta\omega = 2$ (red) for $K = 0.1$. Clear deviations from the LR behaviors occur in (a), (b), and (c), as F_2/ω is increased. The yield $|P_{1,\mathrm{ME}}|$ reaches a maximum near $F_2/\omega = 7.5$ for both temperatures. By contrast, the NLR power fluctuations run through a flat minimum located near $F_2/\omega = 2$ for $\beta\omega = 6$ and near $F_2/\omega = 4$ for $\beta\omega = 2$ and spanning a broad amplitude range. In this area, the work-to-work converter has sizeable power yield with simultaneous low power fluctuations and efficiency still close to unity, only slightly smaller than in LR. This may indicate that the NLR regime is promising for finding best compromise between large power yield, low power fluctuations and high efficiency. Panel (d) shows the tradeoff criterion $Q_{1,\mathrm{ME}}$ versus $\beta\omega$ for different values of F_2/ω. Interestingly, the quantity $Q_{1,\mathrm{ME}}$ falls below 2 for F_2/ω below 5.5 and sufficiently low temperature. On the contrary, it consistently stays above 2 for larger F_2/ω and arbitrarily low

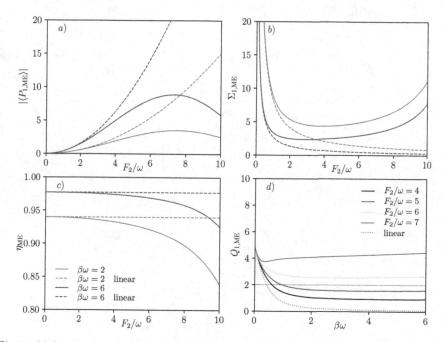

Figure 32.3: Comparison of nonlinear (solid) with linear (dashed) regime at maximal efficiency. Panels a) - c) show $P_{1,0,\text{ME}}$, $\Sigma_{1,\text{ME}}$ and η_{ME} versus F_2/ω ($\hbar = q_0 = 1$). Upper curves in a) and c) for $\beta\omega = 6$ and in b) for $\beta\omega = 3$. Panel d): Trade-off $Q_{1,\text{ME}}$ versus $\beta\omega$ for different F_2/ω. Curves from up to down with decreasing F_2/ω. The parameters are $K = 0.1$, $\alpha = 5$, $\omega = 5\Delta_\text{r}$, and $\Delta_\text{r} = 1$.

temperatures. In the former case, the power fluctuations are in the flat minimum of panel (b), thereby facilitating shortfall of the tradeoff bound in the NLR.

So far, mean powers and power dispersion of the QBP converter have been studied in order Δ^2. Higher orders of Δ^2 in the series expressions (26.16) and (26.17) may be significant at sufficiently low T. The leading correction results from direct next-to-nearest-neighbour transitions in the TB lattice, or rather coherent transport of two charges in the associated fermionic transport model. For moderate damping, continuation to asymptotic low temperatures is possible upon including form factors of width $1/\Delta_\text{r}$, analogous to the method discussed in Sec. 27.3, being exact at the Toulouse point $K = \frac{1}{2}$. The conclusions of the numerical analysis are that the higher-order tunneling terms yield marginal contributions up to inverse temperature $\beta\hbar\omega = 6$ for $\Delta_\text{r} = 1$, and that the above weak-tunneling results are qualitatively correct down to much lower temperatures.

In conclusion, the hitherto mostly unregarded nonlinear response regime proves to be a new promising operation field for isothermal machines.

Bibliography

[1] G. Baym, *Lectures on Quantum Mechanics* (Benjamin, Reading, 1969).

[2] D. Chandler, *Introduction to Modern Statistical Mechanics* (Oxford University Press, New York, 1987).

[3] R. P. Feynman and A. R. Hibbs, *Quantum Mechanics and Path Integrals* (Mc Graw-Hill, New York, 1965).

[4] R. P. Feynman, *Statistical Mechanics* (Benjamin, Reading, Mass., 1972).

[5] L. S. Schulman, *Techniques and Applications of Path Integration* (Wiley, 1981).

[6] H. Kleinert, *Path Integrals in Quantum Mechanics, Statistics, Polymer Physics, and Financial Markets*, 3rd edition (World Scientific, Singapore, 2004).

[7] S. Chandrasekhar, Rev. Mod. Phys. **15**, 1 (1943).

[8] M. S. Green, J. Chem. Phys. **20**, 1281 (1952);
R. Kubo, Rep. Progr. Phys. (London) **29**, 255 (1966).

[9] P. Caldirola, Il Nuovo Cim. **18**, 393 (1941).

[10] E. Kanai, Progr. Theor. Phys. **3**, 440 (1948).

[11] W. H. Louisell, *Quantum Statistical Properties of Radiation* (Wiley, N.Y., 1973).

[12] D. Schuch, Phys. Rev. A **55**, 935 (1997).

[13] H. Dekker, Phys. Rev. A **16**, 2116 (1977).

[14] M. D. Kostin, J. Chem. Phys. **57**, 3589 (1972).

[15] K. Yasue, Ann. Phys. (N.Y.) **114**, 479 (1978).

[16] E. Nelson, Phys. Rev. **150**, 1079 (1966).

[17] S. Nakajima, Progr. Theor. Phys. **20**, 948 (1958).

[18] R. Zwanzig, J. Chem. Phys. **33**, 1338 (1960);
R. Zwanzig, in: *Lectures in Theoretical Physics* (Boulder), Vol. 3, ed. by W. E. Brittin, B. W. Downs, and J. Down (Interscience, New York, 1961).

[19] J. Prigogine and P. Resibois, Physica **27**, 629 (1961).

[20] J. R. Senitzky, Phys. Rev. **119**, 670 (1960).

[21] G. W. Ford, M. Kac, and P. Mazur, J. Math. Phys. **6**, 504 (1965).

[22] H. Mori, Progr. Theor. Phys. **33**, 423 (1965).

[23] F. Haake, in *Quantum Statistics in Optics and Solid State Physics*, Springer Tracts in Modern Physics, Vol. 66, ed. by G. Höhler (Springer, Berlin, 1973).

[24] H. Haken, Rev. Mod. Phys. **47**, 67 (1975).

[25] H. Spohn, Rev. Mod. Phys. **52**, 569 (1980).

[26] H. Dekker, Phys. Rep. **80**, 1 (1981).

[27] H. Grabert, *Projection Operator Techniques in Nonequilibrium Statistical Mechanics*, Springer Tracts in Modern Physics, Vol. 95 (Springer, 1982).

[28] K. Blum, *Density Matrix Theory and Applications* (Plenum Press, 1981).

[29] P. Talkner, Ann. Phys. (N.Y.) **167**, 390 (1986).

[30] R. Alicki and K. Lendi, in *Quantum Dynamical Semigroups and Applications*, Lecture Notes in Physics Vol. 286, ed. H. Araki *et al.* (Springer, Berlin, 1987).

[31] C. W. Gardiner, *Quantum Noise* (Springer, Berlin, 1991).

[32] A.G. Redfield, IBM J. Res. Dev. **1**, 19 (1957); Adv. Magn. Reson. **1**, 1 (1965).

[33] G. C. Schatz and M. A. Ratner, *Quantum Mechanics in Chemistry* (Prentice Hall, Englewood Ciffs, New Jersey, 1993).

[34] M. Mehring, *Principles of High-Resolution NMR in Solids* (Springer, 1983).

[35] R. R. Ernst, G. Bodenhausen, and A. Wokaun, *Principles of Nuclear Magnetic Resonance in One and Two Dimensions* (Clarendon Press, Oxford, 1990).

[36] C. P. Slichter, *Principles of Magnetic Resonance* (Springer, Berlin, 1990).

[37] L. Allen and Z. H. Eberly, *Optical Resonance and Two-Level Atoms* (Wiley, New York, 1975).

[38] Y. R. Shen, *The Principles of Nonlinear Optics* (Wiley, New York, 1984).

[39] J. M. Jean, J. Chem. Phys. **101**, 10464 (1994);
A. K. Felts, W. T. Pollard, and R. A. Friesner, J. Phys. Chem. **99**, 2929 (1995);
J. M. Jean and G. R. Fleming, J. Chem. Phys. **103**, 2092 (1995).

[40] O. Kühn, V. May, and M. Schreiber, J. Chem. Phys.**101**, 10 404 (1994).

[41] V. Sidis, Adv. Chem. Phys. **82**, 73 (1992).

[42] T. Pacher, L. S. Cederbaum, and H. Köppel, Adv. Chem. Phys. **84**, 293 (1993).

[43] W. Domcke and G. Stock, Adv. Chem. Phys. **100**, 1 (1997).

[44] G. Lindblad, Commun. Math. Phys. **48**, 119 (1976).

[45] A. Isar *et al.*, Int. J. Mod. Phys. E **3**, 635 (1994).

[46] Sh. Gao, Phys. Rev. Lett. **79**, 3101 (1997).

[47] a.: Ph. Pechukas, Phys. Rev. Lett. **73**, 1060 (1994);
b.: A. Suárez, R. Silbey, and I. Oppenheim, J. Chem. Phys. **97**, 5101 (1992);
c.: W. J. Munro and C. W. Gardiner, Phys. Rev. A **53**, 2633 (1996).

[48] R. Karrlein and H. Grabert, Phys. Rev. E **55**, 153 (1997).

[49] E. Fick and G. Sauermann, *The Quantum Statistics of Dynamic Processes*, Springer Series in Solid-State Sciences, Vol. 88 (Springer, Berlin, 1990).

[50] C. W. Gardiner, IBM J. Res. Dev. **32**, 127 (1988).

[51] G. W. Ford and M. Kac, J. Stat. Phys. **46**, 803 (1987).

[52] G. W. Ford. J. T. Lewis, and R. F. O'Connell, Phys. Rev. A **37**, 4419 (1988).

[53] R. Araújo, S. Wald, and M. Henkel, J. Stat. Mech. (2019) 053101.

[54] R. Benguria and M. Kac, Phys. Rev. Lett. **46**, 1 (1981).

[55] A. Schmid, J. Low Temp. Phys. **49**, 609 (1982).

[56] U. Eckern, W. Lehr, A. Menzel-Dorwarth, F. Pelzer, and A. Schmid, J. Stat. Phys. **59**, 885 (1990).

[57] R. H. Koch, D. J. van Harlingen, and J. Clarke, Phys. Rev. Lett. **45**, 2132 (1980); *ibid.* **47**, 1216 (1981).

[58] D. Giulini, E. Joos, C. Kiefer, J. Kupsch, I.-O. Stamatescu, and H. D. Zeh, *Decoherence and the Appearance of a Classical World in Quantum Theory* (Springer, Berlin, 1996).

[59] M. H. A. Davis, *Markov Models and Optimization* (Chapman, London, 1993).

[60] N. Gisin, Phys. Rev. Lett. **52**, 1657 (1984).

[61] N. Gisin and I. C. Percival, J. Phys. A **25**, 5677 (1992); I. C. Percival, Proc. R. Soc. London A **447**, 189 (1994).

[62] I. Percival, *Quantum State Diffusion* (Cambridge Univ. Press, 1998).

[63] H. P. Breuer and F. Petruccione, J. Phys. A: Math. Gen. **31**, 33 (1998).

[64] N. G. van Kampen, *Stochastic Processes in Physics and Chemistry* (North-Holland, Amsterdam, 1992).

[65] L. Diósi, J. Phys. Math. Gen. **21**, 2885 (1988).

[66] J. Dalibard, Y. Castin, and K. Mølmer, Phys Rev. Lett. **68**, 580 (1992).

[67] R. Dum, P. Zoller, and H. Ritsch, Phys. Rev. A **45**, 4879 (1992); C. W. Gardiner, A. S. Parkins, and P. Zoller, *ibid.* A **46**, 4363 (1992); R. Dum, A. S. Parkins, P. Zoller, and C. W. Gardiner, *ibid.* A **46**, 4382 (1992).

[68] H. J. Carmichael, S. Singh, R. Vyas, and P. R. Rice, Phys. Rev. A **39**, 1200 (1989); H. J. Carmichael, *An Open System Approach to Quantum Optics* (Springer, Berlin, 1993).

[69] M. Naraschewski and A. Schenzle, Z. Physik A **25**, 5677 (1992).

[70] K. Mølmer, Y. Castin, and J. Dalibard, J. Opt. Soc. Am. B **10**, 524 (1993).

[71] H. M. Wiseman and G. J. Milburn, Phys. Rev A **47**, 1652 (1992).

[72] H. P. Breuer and F. Petruccione, Phys. Rev. Lett. **74**, 3788 (1995); Phys. Rev. E **51**, 4041 (1995); *ibid.* E **52**, 428 (1995).

[73] H.-P. Breuer and F. Petruccione, *The Theory of Open Quantum Systems* (Oxford University Press, Oxford, 2007).

[74] R. J. Cook, *Quantum Jumps* in Progress in Optics, Vol. XXVIII, ed. by E. Wolf (Elsevier, Amsterdam, 1990).

[75] W. Nagourney, J. Sandberg, and H. Dehmelt, Phys. Rev. Lett.**56**, 2797 (1986); Th. Sauter, W. Neuhauser, R. Blatt, and P. E. Toschek, Phys. Rev. Lett. **57**, 1699 (1986); Th. Basché, S. Kummer, and C. Bräuchle, Nature **373**, 132 (1995).

[76] P. Ullersma, Physica (Utrecht) **32**, 27, 56, 74, 90 (1966).

[77] R. Zwanzig, J. Stat. Phys. **9**, 215 (1973).

[78] A. O. Caldeira and A. J. Leggett, Phys. Rev. Lett. **46**, 211 (1981); and Ann. Phys. (N.Y.) **149**, 374 (1983); *ibid.* **153**, 445(E) (1983).

[79] K. H. Stevens, J. Phys. C **16**, 5765 (1983).

[80] V. Ambegaokar and U. Eckern, Z. Physik B **69**, 399 (1987).

[81] V. B. Magalinskiǐ, Sov. Phys.–JETP **9**, 1381 (1959).

[82] R. J. Rubin, J. Math. Phys. **1**, 309 (1960); *ibid.* **2**, 373 (1961).

[83] A. J. Leggett, Phys. Rev. B **30**, 1208 (1984).

[84] H. Grabert and U. Weiss, Z. Physik B **56**, 171 (1984).

[85] P. Hänggi, in *Stochastic Dynamics*, Lecture Notes in Physics, Vol.**484**, ed. by L. Schimansky-Geier and Th. Pöschel (Springer, Berlin, 1997), p. 15.

[86] A. J. Leggett, S. Chakravarty, A. T. Dorsey, M. P. A. Fisher, A. Garg, and W. Zwerger, Rev. Mod. Phys. **59**, 1 (1987); *i*bid. **67**, 725 (1995) [erratum].

[87] J.-D. Bao, J. Stat. Phys. **114**, 503 (2004).

[88] H. Grabert, P. Schramm, and G.-L. Ingold, Phys. Rep. **168**, 115 (1988); P. Schramm and H. Grabert, J. Stat. Phys. **49**, 767 (1987).

[89] I. S. Gradshteyn and I. M. Ryzhik, *Tables of Integrals, Series and Products* (Academic Press, London, 1965).

[90] M. Abramowitz and I. Stegun, *Handbook of Mathematical Functions* (Dover, New York, 1971).

[91] B. Spreng, G.-L. Ingold, Phys. Scr. **T 165**, 014028 (2015).

[92] A. I. Saichev and G. M. Zaslavsky, Chaos **7**, 753 (1997).

[93] R. Metzler and J. Klafter, Phys. Rep. **339**, 1 (2000).

[94] E. Lutz, Phys. Rev. E **64**, 051106 (2001).

[95] R. J. Rubin, Phys. Rev. **131**, 964 (1963).

[96] M. Maekawa and K. Wada, Phys. Lett. A **80**, 293 (1980).

[97] A.V. Mokshin, R. M. Yulmetyev, and P. Hänggi, New J. Phys. **7**, 9 (2005); Phys. Rev. Lett. **95**, 200601 (2005).

[98] H. Spohn, *Dynamics of Charged Particles and Their Radiation Field* (Cambridge University Press, Cambridge, 2004).

[99] G. W. Ford, J. T. Lewis, and R. F. O'Connell, Phys. Rev. Lett. **55**, 2273 (1985).

[100] J.-D. Bao, P. Hänggi, and Yi-Zh. Zhuo, Phys. Rev. E **72**, 061107 (2005).

[101] H. Wipf, D. Steinbinder, K. Neumaier, P. Gutsmiedl, A. Magerl, and A. J. Dianoux, Europhys. Lett. **4**, 1379 (1989); D. Steinbinder, H. Wipf, A. Magerl, A. D. Dianoux, and K.Neumaier, *ibid.* **6**, 535 (1988); *ibid.* **16**, 211 (1991). See also H. Grabert and H. Wipf, in: Festkörperprobleme/Advances in Solid State Physics, Vol. 30, p. 1, ed. by U. Rössler (Vieweg, Braunschweig, 1990).

[102] G. M. Luke *et al.*, Phys. Rev. B **43**, 3284 (1991); O. Hartmann *et al.*, Hyperfine Interactions **64**, 641 (1990), and references therein; I. S. Anderson, Phys. Rev. Lett. **65**, 1439 (1990).

[103] *Hydrogen in Metals III*, Topics in Applied Physics, Vol. 73, ed. by H. Wipf (Springer, Berlin, 1997).

[104] A. Würger, *From Coherent Tunneling to Relaxation*, Springer Tracts in Modern Physics, Vol. 135 (Springer, Berlin, 1997).

[105] K. Chun and N. O. Birge, Phys. Rev. B **48**, 11500 (1993); B **54**, 4629 (1996).

[106] B. Golding, N. M. Zimmermann, and S. N. Coppersmith, Phys. Rev. Lett. **68**, 998 (1992).

[107] R. Marcus, J. Chem. Phys. **24**, 966 (1956).

[108] R. A. Marcus and N. Sutin, Biochim. Biophys. Acta **811**, 265 (1985).

[109] S. Coleman, Phys. Rev. D **15**, 2929 (1977); S. Coleman, in *The Whys of Subnuclear Physics*, ed. by A. Zichichi (Plenum, New York, 1979), p. 805.

[110] U. Weiss and W. Häffner, Phys. Rev. D **27**, 2916 (1983).

[111] A. Barone and G. Paterno, *Physics and Application of the Josephson Effect* (Wiley, New York, 1982).

[112] J. R. Friedman *et al.*, Nature (London) **406**, 43 (2000).

[113] C. H. van der Wal *et. al.*, Science **290**, 773 (2000).

[114] J. E. Mooij *et al.*, Science **285**, 1036 (1999).

[115] M. V. Feigelman *et al.*, J. Low Temp. Phys. **118**, 805 (2000).

[116] Yu. Makhlin, G. Schön, and A. Shnirman, Rev. Mod. Phys. **73**, 357 (2001).

[117] Y. Nakamura, Yu. A. Pashkin, and J. S. Tsai, Nature **398**, 786 (1999).

[118] E. Paladino, L. Faoro, G. Falci, R. Fazio, Phys. Rev. Lett. **88**, 228304 (2002).

[119] G. Ithier, E. Collin, P. Joyez, P. J. Meeson, D. Vion, D. Esteve, F. Chiarello, A. Shnirman, Yu. Makhlin, J. Schriefl, and G. Schön, Phys. Rev. B **72**, 134519 (2005).

[120] A. Shnirman, G. Schön, I. Martin, and Yu. Makhlin, Phys. Rev. Lett. **94**, 127002 (2005).

[121] P. W. Anderson, B. I. Halperin, and C. M. Varma, Phil. Mag. **25**, 1 (1972).

[122] W. A. Phillips, J. Low Temp. Phys. **7**, 351 (1972).

[123] J. L. Black, in *Glassy Metals I*, Topics in Applied Physics, Vol. 46, ed. by H.-J. Güntherodt and H. Beck (Springer, Berlin, 1981).

[124] Yu. Makhlin, and A. Shnirman, Phys. Rev. Lett. **92**, 178301 (2004).

[125] D. Vion *et al.*, Science **296**, 886 (2002).

[126] J. M. Martinis *et al.*, Phys. Rev. Lett. **89**, 117901 (2002).

[127] I. Chiorescu, Y. Nakamura, C. Harmans, J. E. Mooij, Science **299**, 1869 (2003).

[128] A. Wallraff *et al.*, Nature **431**, 162 (2004).

[129] A. O. Niskanen, K. Harrabi, F. Yoshihara, Y. Nakamura, S. Lloyd, and J. S. Tsai, Science **316**, 723 (2007).

[130] J. Yamashita and T. Kurosawa, J. Chem. Solids **5**, 34 (1958).

[131] a.: T. Holstein, Ann. Phys. (N.Y.) **8**, 325 (1959); b.: *ibid.* **8**, 343 (1959).

[132] H. B. Shore and L. M. Sander, Phys. Rev. B **12**, 1546 (1975).

[133] M. Wagner, J. Phys. A **18**, 1915 (1986).

[134] C. P. Flynn and A. M. Stoneham, Phys. Rev. B **1**, 3966 (1970).

[135] Y. Kagan and M. I. Klinger, J. Phys. C **7**, 2791 (1974).

[136] H. Teichler and A. Seeger, Phys. Lett. **82** A, 91 (1981).

[137] H. Fröhlich, Adv. Phys. **3**, 325 (1954).

[138] J. Kondo, in *Fermi Surface Effects*, Vol. 77 of Springer Series in Solid State Sciences, eds. J. Kondo and A. Yoshimori (Springer, Berlin, 1988).

[139] T. Regelmann, L. Schimmele, and A. Seeger, Z. Physik B **95**, 441 (1994).

[140] G. D. Mahan, Many-Particle Physics (Plenum Press, New York, 1981).

[141] G. D. Mahan, in *Fermi Surface Effects*, Vol. 77 of Springer Series in Solid State Sciences, eds. J. Kondo and A. Yoshimori (Springer, Berlin, 1988).

[142] K. Ohtaka and Y. Tanabe, Rev. Mod. Phys. **62**, 929 (1990).

[143] A. M. Tsvelik and P. B. Wiegmann, Adv. Phys. **32**, 453 (1983).

[144] S. Chakravarty, Phys. Rev. Lett. **49**, 681 (1982).

[145] A. J. Bray and M. A. Moore, Phys. Rev. Lett. **49**, 1546 (1982).

[146] V. Hakim, A. Muramatsu, and F. Guinea, Phys. Rev. B **30**, 464 (1984).

[147] A. Schmid, Phys. Rev. Lett. **51**, 1506 (1983).

[148] S. Bulgadaev, Sov. Phys.–JETP Lett. **39**, 314 (1985).

[149] F. Guinea, V. Hakim, and A. Muramatsu, Phys. Rev. Lett. **54**, 263 (1985).

[150] P. G. de Gennes, *Superconductivity of Metals and Alloys* (Wesley, N.Y., 1989).

[151] D. J. Scalapino, in *Superconductivity*, Vol. 1, ed. by R. D. Parks (Marcel Dekker, New York, 1969).

[152] a.: V. Ambegaokar, U. Eckern, and G. Schön, Phys. Rev. Lett. **48**, 1745 (1982); b.: U. Eckern, G. Schön, and V. Ambegaokar, Phys. Rev. B **30**, 6419 (1984).

[153] A. I. Larkin and Yu. N. Ovchinnikov, Phys. Rev. B **28**, 6281 (1983).

[154] G. Schön and A. D. Zaikin, Phys. Rep. **198**, 237 (1990).

[155] Special Issue on *Single Charge Tunneling*, Z. Physik B **85** (3), 317-468 (1991).

[156] *Single Charge Tunneling*, ed. by H. Grabert and M. H. Devoret, NATO ASI Series B: Physics Vol. 294 (Plenum Press, New York, 1992).

[157] G.-L. Ingold and Yu. V. Nazarov, in Ref. [156], p. 21-107.

[158] D. V. Averin and K. K. Likharev, J. Low Temp. Phys. **62**, 345 (1986).

[159] *Quantum Tunneling of Magnetization - QTM 94*, ed. by L. Gunther and B. Barbara (Kluwer, Dordrecht, 1995).

[160] P. C. E. Stamp in Ref. [161].

[161] *Tunneling in Complex Systems*, Proceedings from the Institute for Nuclear Theory, Vol. 5, ed. by S. Tomsovic (World Scientific, Singapore, 1998).

[162] J. L. van Hemmen and A. Suto, Europhys. Lett. **1**, 481 (1986); and in Ref. [159].

[163] M. Enz and R. Schilling, J. Phys. C **19**, L 711 and 1765 (1986); and in Ref. [159].

[164] D. Gatteschi, R. Sessoli, and J. Villain, *Molecular Nanomagnets* (Oxford University Press, Oxford, 2006).

[165] I. S. Tupitsyn, N. V. Prokof'ev and P. C. E. Stamp, Intl. J. Mod. Phys. B **11**, 2901 (1997).

[166] A. O. Caldeira, A. H. Castro Neto, and T. O. de Carvalho, Phys. Rev. B **48**, 13 974 (1993).

[167] A. Cuccoli *et al.*, Physical Rev. E **64**, 066124 (2001).
H. Kohler and F. Sols, New. J. Phys., **8**, 149 (2006).
J.Ankerhold and E. Pollak, Physical Review E **75**, 041103 (2007).
G. Rastelli, New. J. Phys. **18**, 053033 (2016); D. Maille, S. Andergassen, and G. Rastelli, Physical Review Research **2**, 013226 (2020).

[168] A. Widom and T.D. Clark, Phys. Rev. B 30, 1205 (1984).

[169] G. Mahler and V. A. Weberruß, *Quantum Networks* (Springer, Berlin, 1998).

[170] J. Allinger and U. Weiss, Z. Physik B **98**, 289 (1995).

[171] D. Cohen, Phys. Rev. E **55**, 1422 (1997); Phys. Rev. Lett. **78**, 2878 (1997).

[172] S. Doniach and E. H. Sondheimer, *Green's Functions for Solid State Physicists* (Benjamin, Reading, 1974).

[173] J. P. Sethna, Phys. Rev. B **24**, 698 (1981); *ibid.* B **25**, 5050 (1982).

[174] A. A. Louis and J. P. Sethna, Phys. Rev. Lett. **74**, 1363 (1995).

[175] H. Sugimoto and Y. Fukai, Phys. Rev. B **22**, 670 (1980);
A. Klamt and H. Teichler, Phys. Stat. Sol. (B) **134**, 103 (1986).

[176] A. Sumi and Y. Toyozawa, J. Phys. Soc. Jpn. **35**, 137 (1973).

[177] F. M. Peeters and J. T. Devreese, Phys. Rev. B **32**, 3515 (1985).

[178] B. Gerlach and H. Löwen, Rev. Mod. Phys. **63**, 63 (1991).

[179] Yu. Kagan, J. Low Temp. Phys. **87**, 525 (1992).

[180] *Quantum Tunnelling in Condensed Media*, ed. by Yu. Kagan and A. J. Leggett (Elsevier Publishers, Amsterdam, 1992).

[181] Yu. Kagan and N. V. Prokov'ev, in Ref. [180].

[182] F. Napoli, M. Sassetti, and U. Weiss, Physica B **202**, 80 (1994).

[183] J. A. Stroscio and D. M. Eigler, Science **254**, 1319 (1991).

[184] D. M. Eigler, C. P. Lutz, and W. E. Rudge, Nature **352**, 600 (1991).

[185] M.F. Crommie *et al.*, Nature (London) **363** 524 (1993); Science **262**, 218 (1993).

[186] J. E. Artacho and L. M. Falicov, Phys. Rev. B **47**, 1190 (1993).

[187] M. Sassetti, E. Galleani d'Agliano, and F. Napoli, Physica B **154**, 359 (1989).

[188] E. G. d'Agliano, P. Kumar, W. Schaich, H. Suhl, Phys. Rev. B **11**, 2122 (1975).

[189] L.-D. Chang and S. Chakravarty, Phys. Rev. B **31**, 154 (1985).

[190] P. Nozières and C. De Dominicis, Phys. Rev. **178**, 1097 (1969).

[191] F. Guinea, V. Hakim, and A. Muramatsu, Phys. Rev. B **32**, 4410 (1985).

[192] F. Sols and F. Guinea, Phys. Rev. B **36**, 7775 (1987).

[193] K. Schönhammer, Phys. Rev. B **43**, 11323 (1991).

[194] P. W. Anderson, Phys. Rev. Lett. **18**, 1049 (1967).

[195] K. Yamada and K. Yosida, Progr. Theor. Phys. **68**, 1504 (1982); K. Yamada, A. Sakurai, S. Miyazima, and H. S. Wang, Progr. Theor. Phys. **75**, 1030 (1986).

[196] A. Oguchi and K. Yosida, Progr. Theor. Phys. **75**, 1048 (1986);
T. Kitamura, A. Oguchi and K. Yosida, Progr. Theor. Phys. **78**, 583 (1987).

[197] N. R. Wertheimer, Phys. Rev. **147**, 255 (1966).

[198] G. Schön and A. D. Zaikin, Phys. Rev. B **40**, 5231 (1989).

[199] F. Guinea and G. Schön, Physica B **152**, 165 (1988).

[200] F. W. J. Hekking, L. I. Glazman, K. A. Matveev, and R. I. Shekhter, Phys. Rev. Lett. **70**, 4138 (1993).

[201] F. W. J. Hekking and Yu V. Nazarov, Phys. Rev. Lett. **71**, 1625 (1993).

[202] E. Pollak, Chem. Phys. Lett. **127**, 178 (1986).

[203] W. P. Schleich, *Quantum Optics in Phase Space* (Wiley-VCH, 2001).

[204] H. Weyl, Z. Physik **46**, 1 (1927).

[205] M. Hillery, R. F. O'Connell, M. O. Scully, and E. P. Wigner, Phys. Rep. **106**, 122 (1984).

[206] E. Wigner, Phys. Rev. **40**, 749 (1932).

[207] U. Weiss, Z. Physik B **30**, 429 (1978).

[208] R. P. Feynman and F. L. Vernon, Ann. Phys. (N.Y.) **24**, 118 (1963).

[209] A. O. Caldeira and A. J. Leggett, Physica **121 A**, 587 (1983).

[210] A. Stern, Y. Aharonov, and Y. Imry, Phys. Rev. A **41**, 3436 (1990).

[211] D. Loss and K. Mullen, Phys. Rev. B **43**, 13 252 (1991).

[212] B. d'Espagnat, *Conceptual Foundations of Quantum Mechanics* (Benjamin, Reading, 1976).

[213] P. Grigolini, *Quantum Mechanical Irreversibility and Measurement* (World Scientific, 1993).

[214] S. Dattagupta, *Relaxation Phenomena in Condensed Matter Physics* (Academic Press, New York, 1987).

[215] J. Schwinger, J. Math. Phys. **2**, 407 (1961).

[216] L. P. Kadanoff and G. Baym, *Quantum Statistical Mechanics* (Benjamin, 1962).

[217] L. V. Keldysh, Sov. Phys.–JETP **20**, 1018 (1965).

[218] A. Kamenev, *Many-body theory of nonequilibrium systems*, in *Nanophysics: coherence and transport*, ed. by H. Bouchiat et al. (Elsevier, Amsterdam, 2005), [arXiv: condmat/0412296v2 (2005)].

[219] K. C. Chou, Z. B. Su, B. L. Hao, and L. Yu, Phys. Rep. **118**, 1 (1985).

[220] J. Rammer and H. Smith, Rev. Mod. Phys. **58**, 323 (1986).

[221] J. Rammer, *Quantum Field Theory of Nonequilibrium States* (Cambridge University Press, Cambridge, 2007).

[222] Y.-C. Chen, J. L. Lebowitz, and C. Liverani, Phys. Rev. B **40**, 4664 (1989).

[223] M. Sassetti and U. Weiss, Phys. Rev. A **41**, 5383 (1990).

[224] M. Sassetti and U. Weiss, Phys. Rev. Lett. **65**, 2262 (1990).

[225] H. Svensmark and K. Flensberg, Phys. Rev. A **47**, R23 (1993).

[226] K. S. Chow, D. A. Browne, and V. Ambegaokar, Phys. Rev. B **37**, 1624 (1988); K. S. Chow and V. Ambegaokar, Phys. Rev. B **38**, 11 168 (1988).

[227] D. S. Golubev, J. König, H. Schoeller, G. Schön, and A. D. Zaikin, Phys. Rev. B **56**, 15 782 (1997).

[228] W. T. Strunz, L. Diósi, and N. Gisin, Phys. Rev. Lett. **82**, 1801 (1999).

[229] J. T. Stockburger and C. H. Mak, Phys. Rev. Lett. **80**, 2657 (1998); J. Chem. Phys. **110**, 4983 (1999).

[230] J. T. Stockburger and H. Grabert, Phys. Rev. Lett. **88**, 170407 (2002).

[231] J. T. Stockburger, Chem. Phys. **296**, 159 (2003).

[232] J .H. Van Vleck, J. Math. Natl. Acad. Sci. U.S.A. **14**, 178 (1928).

[233] M. C. Gutzwiller, *Chaos in Classical and Quantum Mechanics*, Interdisciplinary Applied Mathematics, Vol. 1 (Springer, Berlin, 1990).

[234] M. C. Gutzwiller, J. Math. Phys. **8**, 1979 (1967).

[235] E. J. Heller, J. Chem. Phys. **75**, 2923 (1981).

[236] M. F. Herman and E. Kluk, Chem. Phys. **91**, 27 (1984).

[237] K. G. Kay, Chem. Phys. **322**, 3 (2006); G. Hochman and K. G. Kay, Phys. Rev. A **73**, 064102 (2006).

[238] W. Koch, F. Großmann, J. T. Stockburger, and J. Ankerhold, Phys. Rev. Lett. **100**, 230402 (2008).

[239] S. Zhang and E. Pollak, Phys. Rev. Lett. **91**, 190201 (2003).

[240] M. Campisi, P. Talkner, and P. Hänggi, Phys. Rev. Lett. **102**, 210401 (2009).

[241] M. Esposito, M. A. Ochoa, and M. Galperin, Phys. Rev. B **92**, 235440 (2015).

[242] U. Seifert, Phys. Rev. Lett. **116**, 020601 (2016).

[243] M. Esposito, U. Harbola, and S. Mukamel, Rev. Mod. Phys. **81**, 1665 (2009).

[244] S. Gasparinetti, P. Solinas, A. Braggio, and M. Sassetti, New J. Phys. **16**, 115001 (2014).

[245] M. Carrega, P. Solinas, A. Braggio, M. Sassetti, and U. Weiss, New J. Phys. **17**, 045030 (2015).

[246] M. Carrega, P. Solinas, M. Sassetti, and U. Weiss, Phys. Rev. Lett. **116**, 240403 (2016).

[247] H. Grabert, U. Weiss, and P. Talkner, Z. Physik B **55**, 87 (1984).

[248] F. Haake and R. Reibold, Phys. Rev. A **32**, 2462 (1985).

[249] R. Jung, G.-L. Ingold, and H. Grabert, Phys. Rev. A **32**, 2510 (1985).

[250] H. Metiu and G. Schön, Phys. Rev. Lett. **53**, 13 (1984).

[251] C. Aslangul, N. Pottier, and D. Saint-James, J. Stat. Phys. **40**, 167 (1985).

[252] P. S. Riseborough, P. Hänggi, and U. Weiss, Phys. Rev. A **31**, 471 (1985).

[253] K. Lindenberg and B. J. West, Phys. Rev. A **30**, 568 (1984).

[254] B. L. Hu, J. P. Paz, and Y. Zhang, Phys. Rev. D **45**, 2843 (1992).

[255] A. Einstein, Ann. Phys. (Leipzig) **17**, 549 (1905).

[256] J. B. Johnson, Phys. Rev. **32**, 97 (1928).

[257] H. Nyquist, Phys. Rev. **32**, 110 (1928).

[258] H. B. Callen and T. A. Welton, Phys. Rev. **83**, 34 (1951).

[259] P. Talkner, Z. Physik B **41**, 365 (1981).

[260] H. Shiba, Progr. Theor. Phys. **54**, 967 (1975).

[261] G. W. Ford, J. T. Lewis, and R. F. O'Connell, Ann. of Physics **185**, 270 (1988).

[262] G.-L. Ingold, Eur. Phys. J. B **85**, 30 (2012).

[263] R. Adamietz, G.-L. Ingold, and U. Weiss, Eur. Phys. J. B **87**, 90 (2014).

[264] A. Hanke and W. Zwerger, Phys. Rev. E **52**, 6875 (1995).

[265] S. A. Adelman, J. Chem. Phys. **64**, 124 (1976).

[266] H. A. Kramers, Physica (Utrecht) **7**, 284 (1940).

[267] A. Sandulescu and H. Scutaru, Ann. Phys. (N.Y.) **173**, 277 (1987).

[268] V. Hakim and V. Ambegaokar, Phys. Rev. A **32**, 423 (1985).

[269] A. Erdélyi, *Higher Transcendental Functions*, Vol. 3 (McGraw-Hill, N.Y., 1955).

[270] P. Hänggi, G.-L. Ingold, and P. Talkner, New. J. Phys. **10**, 115008 (2008).

[271] B. Spreng, G.-L. Ingold, and U. Weiss, EPL **103**, 60007 (2013).

[272] R. P. Feynman, Phys. Rev. **97**, 660 (1955).

[273] R. Giachetti and V. Tognetti, Phys. Rev. Lett. **55**, 912 (1985); Phys. Rev. B **33**, 7647 (1986).

[274] R. P. Feynman and H. Kleinert, Phys. Rev. A **34**, 5080 (1986).

[275] H. Leschke, in *Path Summations: Achievements and Goals*, ed. by S. Lundquist *et al.* (World Scientific, Singapore, 1987).

[276] W. Janke, in *Path Integrals from meV to MeV*, ed. by V. Sa-yakanit *et al.* (World Scientific, Singapore, 1989).

[277] R. Giachetti, V. Tognetti, and R. Vaia, in *Path Summations: Achievements and Goals*, ed. by S. Lundquist *et al.* (World Scientific, Singapore, 1987).

[278] A. Cuccoli, V. Tognetti, and R. Vaia, in *Quantum Fluctuations in Mesoscopic and Macroscopic Systems*, ed. by H. A. Cerdeira, F. Guinea Lopez, and U. Weiss (World Scientific, Singapore, 1991).

[279] R. Giachetti and V. Tognetti, Phys. Rev. A **36**, 5512 (1987); R. Giachetti, V. Tognetti, R. Vaia, Phys. Rev. A **37**, 2165 (1988); A **38**, 1521, 1638 (1988).

[280] G. Falci, R. Fazio, and G. Giaquinta, Europhys. Lett. **14**, 145 (1991); S. Kim and M. Y. Choi, Phys. Rev. B **42**, 80 (1990).

[281] A. Cuccoli *et al.*, Phys. Rev. A **45**, 8418 (1992); A. Cuccoli, R. Giachetti, V. Tognetti, R. Vaia, and P. Verrucchi, J. Phys.: Condens. Matter **7**, 7891 (1995).

[282] H. Kleinert, Phys. Lett. A **174**, 332 (1992).

[283] H. Kleinert, W. Kürzinger, and A. Pelster, J. Phys. A **31**, 8307 (1998).

[284] A. Cuccoli, A. Rossi, V. Tognetti, and R. Vaia, Phys. Rev. E **55**, 4849 (1997).

[285] D. M. Larsen, Phys. Rev. B **32**, 2657 (1985), B **33**, 799 (1986); S. N. Gorshkov, A. V. Zabrodin, C. Rodriguez, V. K. Fedyanin, Theor. Math. Phys. **62**, 205 (1985); K. M. Broderix, N. Heldt, H. Leschke, Z. Physik B **66**, 507 (1987).

[286] J. T. Devreese and F. Brosens, Phys. Rev. B **45**, 6459 (1992).

[287] W. H. Zurek, Phys. Today **44** (10), 36 (1991).

[288] S. Chakravarty and A. Schmid, Phys. Rep. **140**, 193 (1986).

[289] S. Washburn and R. A. Webb, Adv. Phys. **35**, 375 (1986).

[290] P. Mohanty, E. M. Q. Jariwala, R. A. Webb, Phys. Rev. Lett. **78**, 3366 (1997).

[291] P. Mohanty and R. A. Webb, Phys. Rev. B **55**, R13 452 (1997).

[292] L. Saminadayar, P. Mohanty, R. A. Webb, P. Degiovanni, and C. Bäuerle, Physica E **40**, 12 (2007).

[293] I. L. Aleiner, B. L. Altshuler, and M. E. Gershenson, Waves in Random Media **9**, 201 (1999); Phys. Rev. Lett. **82**, 3190 (1999); I. L. Aleiner, B. L. Altshuler, and M. G. Vavilov, J. Low Temp. Phys. **126**, 1377 (2002).

[294] D.S. Golubev and A. D. Zaikin, Phys. Rev. B **59**, 9195 (1999).

[295] D. S. Golubev and A. D. Zaikin, Phys. Rev. B **62**, 14061 (2000).

[296] D. S. Golubev, A. D. Zaikin, G. Schön, J. Low Temp. Phys. **126**, 1355 (2002).

[297] J. von Delft, Intl. J. Mod. Phys. B **22**, 727 (2008).

[298] F. Marquardt, J. von Delft, R. A. Smith, and V. Ambegaokar, Phys. Rev. B **76**, 19 5331 (2007).

[299] D. S. Golubev and A. D. Zaikin, J. Phys.: Conf. Ser. **129**, 012016 (2008).

[300] F. Guinea, Phys. Rev. B **65**, 205317 (2002).

[301] B. Altshuler, A. G. Aronov, and D. E. Khmelnitskii, J. Phys. C **15**, 7367 (1982).

[302] D. Cohen, J. von Delft, F. Marquardt, and Y. Imry, Phys. Rev. B **80**, 245410 (2009).

[303] D. Cohen, J. Phys. A **31**, 8199 (1998); D. Cohen and Y. Imry, Phys. Rev. B **59**, 11 143 (1999).

[304] D. Golubev and A. D. Zaikin, Phys. Rev. Lett. **81**, 1074 (1998).

[305] R. Schuster *et al.*, Nature (London) **385**, 417 (1997).

[306] I. L. Aleiner, N. S. Wingreen, and Y. Meir, Phys. Rev. Lett. **79**, 3740 (1997).

[307] K. A. Eriksen, P. Hedegård, and H. Bruus, Phys. Rev. B **64**, 195 327 (2001).

[308] S. Arrhenius, Z. Phys. Chem. (Leipzig) **4**, 226 (1889).

[309] A. J. Leggett, Contemp. Phys. **25**, 583 (1984).

[310] A. J. Leggett, in *Directions in Condensed Matter Physics*, Vol. 1, ed. by G. Grinstein and G. Mazenko (World Scientific, Singapore, 1986), p. 187.

[311] H. Grabert, P. Olschowski, and U. Weiss, Phys. Rev. B **36**, 1931 (1987).

[312] P. Hänggi, P. Talkner, and M. Borkovec, Rev. Mod. Phys. **62**, 251 (1990).

[313] J. Ankerhold, *Quantum Tunneling in Complex Systems*, Springer Tracts in Modern Physics, Vol. 224 (Springer Verlag, Berlin, 2007).

[314] A. N. Cleland, J. M. Martinis, and J. Clarke, Phys. Rev. B **37**, 5950 (1988).

[315] D. M. Brink, J. M. Neto, and H. A. Weidenmüller, Phys. Lett. B **80**, 170 (1979).

[316] K. Möhring and U. Smilansky, Nucl. Phys. A **338**, 227 (1980).

[317] D. Emin and T. Holstein, Ann. Phys. (N.Y.) **53**, 439 (1969).

[318] H. Risken, *The Fokker-Planck Equation* (Springer Verlag, Berlin, 1984).

[319] F. Hund, Z. Physik **43**, 805 (1927).

[320] J. R. Oppenheimer, Phys. Rev. **31**, 80 (1928).

[321] G. Gamow, Z. Physik **51**, 204 (1928).

[322] R. W. Gurney and E. U. Condon, Nature (London) **122**, 439 (1928).

[323] E. P. Wigner, Z. Phys. Chem. B **19**, 203 (1932).

[324] W. H. Miller, J. Chem. Phys. **62**, 1899 (1975).

[325] W. H. Miller, S. D. Schwartz, J. W. Tromp, J. Chem. Phys. **79**, 4889 (1983).

[326] W. H. Miller, J. Chem. Phys. **61**, 1823 (1974).

[327] T. Yamamoto, J. Chem. Phys. **33**, 281 (1960).

[328] E. Pollak and J.-L. Liao, J. Chem. Phys. **108**, 2733 (1998); G. Gershinsky and E. Pollak, J. Chem. Phys. **108**, 2756 (1998).

[329] F. Matzkies and U. Manthe, J. Chem. Phys. **106**, 2646 (1997).

[330] W. H. Thompson and W. H. Miller, J. Chem. Phys. **102**, 7409 (1995); *ibid.* **106**, 142 (1997).

[331] F. J. Mc Lafferty and Ph. Pechukas, Chem. Phys. Lett. **27**, 511 (1974).

[332] F. Haake, *Quantum Signatures of Chaos* (Springer, Berlin, 2nd edition, 2000).

[333] I. Affleck, Phys. Rev. Lett. **46**, 388 (1981).

[334] J. S. Langer, Ann. Phys. (N.Y.) **41**, 108 (1967); *ibid.* **54**, 258 (1969).

[335] J. S. Langer, in *Systems far from Equilibrium*, Lecture Notes in Physics, Vol. 132, ed. by L. Garrido (Springer, Berlin, 1980), p. 12.

[336] C. G. Callan and S. Coleman, Phys. Rev. D **16**, 1762 (1977).

[337] M. Stone, Phys. Lett. **67 B**, 186 (1977).

[338] T. Nakamura, A. Ottewill, and S. Takagi, Ann. Phys. (N.Y.) **260**, 9 (1997).

[339] R. F. Dashen, B. Hasslacher, and A. Neveu, Phys. Rev. D **10**, 4114 (1974).

[340] R. P. Bell, *The Tunnel Effect in Chemistry* (Chapman and Hall, London, 1980).

[341] V. I. Goldanskii, Dokl. Acad. Nauk SSSR **124**, 1261 (1959); **127**, 1037 (1959).

[342] P. Reimann, M. Grifoni, and P. Hänggi, Phys. Rev. Lett. **79**, 10 (1997).

[343] S. Keshavamurthy and W. H. Miller, Chem. Phys. Lett. **218**, 189 (1994).

[344] N. T. Maitra and E. J. Heller, Phys. Rev. Lett. **78**, 3035 (1997).

[345] M. J. Gillan, J. Phys. C **20**, 3621 (1987); see also P. G. Wolynes, J. Chem. Phys. **87**, 6559 (1987).

[346] G. A. Voth, D. Chandler, and W. H. Miller, J. Chem. Phys. **91**, 7749 (1989).

[347] J. Cao and G. A. Voth, J. Chem. Phys. **105**, 6856 (1996).

[348] D. Makarov and M. Topaler, Phys. Rev. E **52**, 178 (1995).

[349] M. C. Gutzwiller, J. Math. Phys. **12**, 343 (1971); M. C. Gutzwiller, Physica D (Utrecht) **5**, 183 (1982).

[350] T. Banks, C. M. Bender, and T. T. Wu, Phys. Rev. D **8**, 3346 (1973).

[351] R. F. Grote and J. T. Hynes, J. Chem. Phys. **73**, 2715 (1980); P. Hänggi and F. Mojtabai, Phys. Rev. A **29**, 1168 (1982).

[352] P. G. Wolynes, Phys. Rev. Lett. **47**, 968 (1981); see also: V. I. Mel'nikov and S. V. Meshkov, Sov. Phys.–JETP Lett. **60**, 38 (1983).

[353] H. Grabert, U. Weiss, and P. Hänggi, Phys. Rev. Lett. **52**, 2193 (1984).

[354] H. Grabert and U. Weiss, Phys. Rev. Lett. **53**, 1787 (1984).

[355] A. I. Larkin and Yu. N. Ovchinnikov, Sov. Phys.–JETP Lett. **37**, 382 (1983).

[356] A. I. Larkin and Yu. N. Ovchinnikov, Sov. Phys.–JETP **59**, 420 (1984).

[357] E. Pollak, J. Chem. Phys. **85**, 865 (1986); Phys. Rev. A **33**, 4244 (1986).

[358] H. Grabert, Phys. Rev. Lett. **61**, 1683 (1988).

[359] E. Pollak, H. Grabert, and P. Hänggi, J. Chem. Phys. **91**, 4073 (1989).

[360] V. I. Mel'nikov and S. V. Meshkov, J. Chem. Phys. **85**, 1018 (1986).

[361] E. Hershkovitz and E. Pollak, J. Chem. Phys. **106**, 7678 (1997).

[362] P. Hänggi, H. Grabert, G.-L. Ingold, and U. Weiss, Phys. Rev. Lett. **55**, 761 (1985). For an experimental confirmation of Eq. (15.16) see J. B. Bouchaud, E. Cohen de Lara, and R. Kahn, Europhys. Lett. **17**, 583 (1992).

[363] J. Ankerhold, Ph. Pechukas, and H. Grabert, Phys. Rev. Lett. **87**, 086802 (2001).

[364] W. T. Coffey, Yu. P. Kalmykov, S. V. Titov, and B. P. Mulligan, J. Phys. A: Math. Theor. **40**, F91 (2007); *ibid.* **40**, 12505 (2007).

[365] L. Machura *et al.*, Phys. Rev. E **70**, 031107 (2004); J. Luczka, R. Rudnicki, and P. Hänggi, Physica A 351, 60 (2005).

[366] D. Waxman and A. J. Leggett, Phys. Rev. B **32**, 4450 (1985).

[367] P. Hänggi and W. Hontscha, J. Chem. Phys. **88**, 4094 (1988); Ber. Bunsenges. Phys. Chem. **95**, 379 (1991).

[368] A. Schmid, Ann. Phys. (N.Y.) **170**, 333 (1986).

[369] U. Eckern and A. Schmid, in Ref. [180].

[370] H. Grabert, P. Olschowski, and U. Weiss, Z. Physik B **68**, 193 (1987).

[371] E. Freidkin, P. S. Riseborough, and P. Hänggi, Z. Physik B **64**, 237 (1986).

[372] H. Grabert and U. Weiss, Z. Physik B **56**, 171 (1984).

[373] J. M. Martinis and H. Grabert, Phys. Rev. B **38**, 2371 (1988).

[374] U. Weiss, M. Sassetti, Th. Negele, M. Wollensak, Z. Physik B **84**, 471 (1991).

[375] S. Washburn, R. A. Webb, R. F. Voss, and S. M. Faris, Phys. Rev. Lett. **54**, 2712 (1985); D. B. Schwartz, B. Sen, C. N. Archie, and J. E. Lukens, **55**, 1547 (1985); A. N. Cleland, J. M. Martinis, J. Clarke, Phys. Rev. B **36**, 58 (1987).

[376] S. E. Korshunov, Sov. Phys.–JETP **65**, 1025 (1987).

[377] A. Erdélyi, *Higher Transcendental Functions*, Vol. 1 (McGraw-Hill, N.Y., 1955).

[378] L.-D. Chang and S. Chakravarty, Phys. Rev. B **29**, 130 (1984); *ibid.* B **30**, 1566(E) (1984).

[379] M. H. Devoret, D. Esteve, C. Urbina, J. Martinis, A. N. Cleland, and J. Clarke, in Ref. [180].

[380] S. Takagi, *Macroscopic Quantum Tunneling* (Cambridge University Press, Cambridge, 2002).

[381] G. Careri and G. Consolini, Ber. Bunsenges. Phys. Chem. **95**, 376 (1991).

[382] W. Kleemann, V. Schönknecht, D. Sommer, Phys. Rev. Lett. **66**, 762 (1991).

[383] F. Bruni, G. Consolini, and G. Careri, J. Chem. Phys. **99**, 538 (1993).

[384] W. Wernsdorfer and R. Sessoli, Science **284**, 133 (1999).

[385] M. N. Leuenberger and D. Loss, Nature **410**, 789 (2001).

[386] B. Golding, J. E. Graebner, A. B. Kane, and J. L. Black, Phys. Rev. Lett. **41**, 1487 (1978).

[387] J. L. Black and P. Fulde, Phys. Rev. Lett. **43**, 453 (1979).

[388] G. Weiss, W. Arnold, K. Dransfeld, and H.-J. Güntherodt, Solid State Comm. **33**, 111 (1980).

[389] J. Stockburger, U. Weiss, and R. Görlich, Z. Physik B **84**, 457 (1991).

[390] J. Stockburger, M. Grifoni, M. Sassetti, U. Weiss, Z. Physik B **94**, 447 (1994).

[391] P. Esquinazi, R. König, and F. Pobell, Z. Physik B **87**, 305 (1992).

[392] *Tunneling Systems in Amorphous and Crystalline Solids*, ed. by P. Esquinazi (Springer Verlag, Berlin, 1998).

[393] J. Kondo, Physica **125** B, 279 (1984); *ibid.* **126** B, 377 (1984).

[394] G. Cannelli, R. Cantelli, F. Cordero, and F. Trequattrini, in Ref. [392].

[395] H. Grabert and H. R. Schober, in Ref. [103].

[396] C. D. Tesche, Ann. N. Y. Acad. Sci. **480**, 36 (1986); S. Chakravarty, *ibid.* **480**, 25 (1986).

[397] K. Huang and A. Rhys, Proc. Roy. Soc. A **204**, 406 (1950); M. Lax, J. Chem. Phys. **20**, 1752 (1952); R. Kubo and Y. Toyozawa, Progr. Theor. Phys. **13**, 160 (1955); V. G. Levich and R. R. Dogonadze, Coll. Czech. Chem. Comm. **26**, 193 (1961). V. G. Levich in *Advances in Electrochemistry and Electrochemical Engineering*, ed. by P. Delahay and C. W. Tobias, Vol. 4, 249 (Interscience, 1965).

[398] H. Spohn and R. Dümcke, J. Stat. Phys. **41**, 389 (1985).

[399] U. Weiss, H. Grabert, P. Hänggi, P. Riseborough, Phys. Rev. B **35**, 9535 (1987).

[400] S. Chakravarty and S. Kivelson, Phys. Rev. B **32**, 76 (1985).

[401] A. T. Dorsey, M. P. A. Fisher, and M. Wartak, Phys. Rev. A **33**, 1117 (1986).

[402] R. Silbey and R. A. Harris, J. Chem. Phys. **80**, 2615 (1983); J. Phys. Chem. **93**, 7062 (1989).

[403] F. Wegner, Ann. Physik (Leipzig) **3**, 77 (1994).

[404] S. Kehrein, *The Flow Equation Approach to Many-Particle Systems*, Springer Tracts in Modern Physics, Vol. 217 (Springer Verlag, Berlin, 2006).

[405] S. K. Kehrein, A. Mielke, and P. Neu, Z. Physik B **99**, 269 (1996).

[406] S. K. Kehrein and A. Mielke, Ann. Physik (Leipzig) **6**, 90 (1997); J. Stat. Phys. **90**, 889 (1997).

[407] S. K. Kehrein and A. Mielke, Phys. Lett. A **219**, 313 (1996).

[408] R. Görlich and U. Weiss, Phys. Rev. B **38**, 5254 (1988).

[409] R. Görlich and U. Weiss, Il Nuovo Cim. **11** D, 123 (1989).

[410] P. B. Vigmann and A. M. Finkel'steĭn, Sov. Phys.–JETP **48**, 102 (1978).

[411] P. Nozières, J. Low Temp. Phys. **17**, 31 (1974).

[412] P. Schlottmann, J. Magn. Mater. **7**, 72 (1978); Phys. Rev. B **25**, 4805 (1982).

[413] P. W. Anderson, Phys. Rev. Lett. **18**, 1049 (1967).

[414] K. D. Schotte and U. Schotte, Phys. Rev. **182**, (1969);
K. Schönhammer, Z. Phys. B **45**, 23 (1981).

[415] V. J. Emery, in *Highly Conducting One-Dimensional Solids*, ed. by J. T. Devreese *et al.* (Plenum, New York, 1979);
F. D. M. Haldane, Phys. Rev. Lett. **47**, 1840 (1981).

[416] D. L. Cox and A. Zawadowski, Adv. Phys. **47**, 599 (1998).

[417] G. Yuval and P. W. Anderson, Phys. Rev. B **1**, 1522 (1970).

[418] P. W. Anderson and G. Yuval, J. Phys. C **4**, 607 (1971).

[419] J. Cardy, Journ. of Physics A **14**, 1407 (1981).

[420] S. Chakravarty and J. Rudnick, Phys. Rev. Lett. **75**, 501 (1995).

[421] K. Völker, Phys. Rev. B **58**, 1862 (1998).

[422] H. Spohn, Comm. Math. Phys. **123**, 277 (1989).

[423] B. Carmeli and D. Chandler, J. Chem. Phys. **82**, 3400 (1985);
D. Chandler in *Liquids, Freezing and Glass Transition*, ed. by D. Levesque, J. P. Hansen, and J. Zinn-Justin (Elsevier Science Publishers, 1990).

[424] a.: J. Ulstrup, *Charge Transfer in Condensed Media* (Springer, 1979);
b.: B. Fain, *Theory of Rate Processes in Condensed Media* (Springer, 1980).

[425] P. Ao and J. Rammer, Phys. Rev. Lett. **62**, 3004 (1989).

[426] A. Garg, J. N. Onuchic, and V. Ambegaokar, J. Chem. Phys. **83**, 4491 (1985).

[427] I. Rips and J. Jortner, J. Chem. Phys. **87**, 2090 (1987);
M. Sparpaglione and S. Mukamel, J. Chem. Phys. **88**, 3263 (1987).

[428] J. N. Gehlen and D. Chandler, J. Chem. Phys. **97**, 4958 (1992); J. N. Gehlen, D. Chandler, H. J. Kim, and J. T. Hynes, J. Phys. Chem. **96**, 1748 (1992).

[429] X. Song and A. A. Stuchebrukhov, J. Chem. Phys. **99**, 969 (1993);
A. A. Stuchebrukhov and X. Song, J. Chem. Phys. **101**, 9354 (1994).

[430] J. Cao, C. Minichino, and G. A. Voth, J. Chem. Phys. **103**, 1391 (1995).

[431] M. H. Devoret, D. Esteve, H. Grabert, G.-L. Ingold, H. Pothier, and C. Urbina, Phys. Rev. Lett. **64**, 1824 (1990); Physica B **165 & 166**, 977 (1990);
S. M. Girvin, L. I. Glazman, M. Jonson, D. R. Penn, and M. D. Stiles, Phys. Rev. Lett. **64**, 3183 (1990).

[432] R. Bruinsma and P. Bak, Phys. Rev. Lett. **56**, 420 (1986).

[433] M. Sassetti and U. Weiss, Europhys. Lett. **27**, 311 (1994).

[434] J. Jortner, J. Chem. Phys. **64**, 4860 (1975).

[435] P. Minnhagen, Phys. Lett. A **56**, 327 (1976).

[436] G. Falci, V. Bubanja, and G. Schön, Z. Physik B **85**, 451 (1991).

[437] K. Ando, J. Chem. Phys. **106**, 116 (1997).

[438] J. S. Bader, R. A. Kuharski, and D. Chandler, J. Chem. Phys. **93**, 230 (1990).

[439] P. Siders and R. A. Marcus, J. Am. Chem. Soc. **103**, 741 (1981).

[440] U. Weiss and H. Grabert, Phys. Lett. **108** A, 63 (1985).

[441] a.: H. Grabert and U. Weiss, Phys. Rev. Lett. **54**, 1605 (1985).
b.: M. P. A. Fisher and A. T. Dorsey, Phys. Rev. Lett. **54**, 1609 (1985).

[442] C. Aslangul, N. Poitier, and D. Saint-James, J. Phys. (Paris) **47**, 1671 (1986).

[443] R. Egger, C. H. Mak, and U. Weiss, J. Chem. Phys. **100**, 2651 (1994).

[444] H. Grabert, Phys. Rev. B **46**, 12 753 (1992).

[445] A. Würger, in Ref. [392]; Phys. Rev. Lett. **78**, 1759 (1997).

[446] Q. Niu, J. Stat. Phys. **65**, 317 (1991).

[447] H. Grabert, U. Weiss, and H. R. Schober, Hyperfine Interactions **31**, 147 (1986).

[448] R. Pirc and P. Gosar, Phys. Kondens. Mater. **9**, 377 (1969).

[449] A. Würger, Physics Letters A **236**, 571 (1998).

[450] A. Würger, Solid State Comm. **106**, 63 (1998).

[451] M. Sassetti, F. Napoli, and U. Weiss, Phys. Rev. B **52**, 11 213 (1995).

[452] H. Schoeller and G. Schön, Phys. Rev. B **50**, 18 436 (1994);
J. König, J. Schmid, H. Schoeller, and G. Schön, Phys. Rev. B **54**, 16 820 (1996); H. Schoeller, in Ref. [453].

[453] *Mesoscopic Electron Transport*, ed. by L. L. Sohn, L. P. Kouwenhoven, and G. Schön, NATO ASI Series E, Vol. 345 (Kluwer, Dordrecht, 1997).

[454] J. C. Cuevas and E. Scheer, *Molecular Electronics*, Series in Nanoscience and Nanotechnology – Vol. 1 (World Scientific, Singapore, 2010).

[455] G.-L. Ingold, H. Grabert, and U. Eberhardt, Phys. Rev. B **50**, 395 (1994).

[456] R. D. Coalson, D. G. Evans, and A. Nitzan, J. Chem. Phys. **101**, 436 (1994).

[457] A. Lucke, C. H. Mak, R. Egger, J. Ankerhold, J. Stockburger, and H. Grabert, J. Chem. Phys. **107**, 8397 (1997).

[458] C. Cohen-Tannoudji, B. Diu, and F. Laloë, *Quantum Mechanics*, Vol. 1 (Wiley, New York).

[459] M. Grifoni, M. Winterstetter, and U. Weiss, Phys. Rev. E **56**, 334 (1997).

[460] F. Guinea, Phys. Rev. B **32**, 4486 (1985).

[461] G. Lang, E. Paladino, and U. Weiss, Europhys. Lett. **43**, 117 (1998); Phys. Rev. E **58**, 4288 (1998).

[462] M. Grifoni, M. Sassetti, and U. Weiss, Phys. Rev. E **53**, R2033 (1996).

[463] H. Dekker, Phys. Rev. A **35**, 1436 (1987).

[464] M. Winterstetter and U. Weiss, Chem. Phys. **217**, 155 (1997).

[465] F. Lesage and H. Saleur, Phys. Rev. Lett. **80** 4370 (1998).

[466] J. Cardy, Nucl. Phys. B **324**, 581 (1989); I. Affleck and A. W. W. Ludwig, Nucl. Phys. B **360**, 641 (1991); *ibid.* B **428**, 545 (1994).

[467] P. Fulde and I. Peschel, Adv. Phys. **21**, 1 (1972).

[468] U. Weiss and H. Grabert, Europhys. Lett. **2**, 667 (1986); U. Weiss, H. Grabert, and S. Linkwitz, J. Low Temp. Phys. **68**, 213 (1987).

[469] A. Garg, Phys. Rev. B **32**, 4746 (1985).

[470] S. Dattagupta, H. Grabert, and R. Jung, J. Phys.: Cond. Mat. **1**, 1405 (1989).

[471] U. Weiss and M. Wollensak, Phys. Rev. Letters **62**, 1663 (1989); R. Görlich, M. Sassetti, and U. Weiss, Europhys. Lett. **10**, 507 (1989).

[472] A. Würger, Phys. Rev. B **57**, 347 (1998).

[473] D. A. Parshin, Z. Physik B **91**, 367 (1993).

[474] J. Stockburger, M. Grifoni, and M. Sassetti, Phys. Rev. B **51**, 2835 (1995).

[475] W. G. Unruh, Phys. Rev. A **51**, 992 (1995).

[476] E. Paladino, M. Sassetti, G. Falci, and U. Weiss, Phys. Rev. B **77**, 041303(RC) (2008).

[477] P. Nägele, G. Campagnano, and U. Weiss, New. J. Phys. **10**, 115010 (2008); P. Nägele and U. Weiss, Physica E **42**, 622 (2010).

[478] Y. M. Galperin, B. L. Altshuler, J. Bergli, and D. V. Shantsev, Phys. Rev. Lett. **96**, 097009 (2006); Y. M. Galperin, B. L. Altshuler, and D. V. Shantsev, ArXiv:cond-mat/0312490v1 (2003).

[479] J. Schriefl, Yu. Makhlin, A. Shnirman, and G. Schön, New. Journ. of Physics **8**, 1 (2006).

[480] G. Falci, A. D'Arrigo, A. Mastellone, and E. Paladino, Phys. Rev. Lett. **94**, 167002 (2005).

[481] Yu. Makhlin, G. Schön, and A. Shnirman, in *New Directions in Mesoscopic Physics (Towards Nanoscience)*, R. Fazio, V. F. Gantmakher, and Y. Imry (Eds.) (Kluwer Academic Publishers, 2003) [ArXiv:cond-mat/0309049v1].

[482] O. Astafiev, Yu. A. Pashkin, Y. Nakamura, T. Yamamoto, and J. S. Tsai, Phys. Rev. Lett. **93**, 267007 (2004).

[483] R. Egger, H. Grabert, and U. Weiss, Phys. Rev. E **55**, R3809 (1997).

[484] T. A. Costi and C. Kieffer, Phys. Rev. Lett. **76**, 1683 (1996).

[485] T. A. Costi, Phys. Rev. B **55**, 3003 (1997); Phys. Rev. Lett. **80**, 1038 (1998).

[486] F. Lesage and H. Saleur, Nucl. Phys. B **490**, 543 (1997).

[487] S. P. Strong, Phys. Rev. E **55**, 6636 (1997).

[488] J. T. Stockburger and C. H. Mak, J. Chem. Phys. **105**, 8126 (1996).

[489] H. Baur, A. Fubini, and U. Weiss, Phys. Rev. B **70**, 024302 (2004).

[490] R. N. Silver, D. S. Sivia, and J. E. Gubernatis in: *Quantum Simulations of Condensed Matter Phenomena*, ed. by J. D. Doll and J. E. Gubernatis (World Scientific, Singapore, 1990).

[491] P. Fendley, F. Lesage, and H. Saleur, J. Stat. Phys. **79**, 799 (1995).

[492] F. Grossmann, T. Dittrich, and P. Hänggi, Phys. Rev. Lett. **67**, 516 (1991).

[493] N. Makri, J. Chem. Phys. **106**, 2286 (1997); N. Makri and L. Wei, Phys. Rev. E **55**, 2475 (1997).

[494] a.: D. G. Evans, R. D. Coalson, H. J. Kim, and Yu. Dakhnovskii, Phys. Rev. Lett. **75**, 3649 (1995).
b.: M. Morillo and R. I. Cukier, Phys. Rev. B **54**, 13 962 (1996).

[495] M. Grifoni, L. Hartmann, and P. Hänggi, Chem. Phys. **217**, 167 (1997).

[496] Special issue on *Dynamics of Driven Quantum Systems*, ed. by W. Domcke, P. Hänggi, and D. Tannor, Chem. Phys. **217** (2, 3), 117-416 (1997).

[497] M. Grifoni and P. Hänggi, *Driven Quantum Tunneling*, Phys. Rep. **304**, 229 (1998).

[498] S. Han, J. Lapointe, and J. E. Lukens, Phys. Rev. Lett. **66**, 810 (1991); Phys. Rev. B **46**, 6338 (1992).

[499] M. Grifoni, Phys. Rev. E **54**, R3086 (1996).

[500] M. Grifoni, M. Sassetti, P. Hänggi, and U. Weiss, Phys. Rev. E **52**, 3596 (1995).

[501] M. Grifoni, M. Sassetti, J. Stockburger, U. Weiss, Phys. Rev. E **48**, 3497 (1993).

[502] P. K. Tien and J. P. Gordon, Phys. Rev. **129**, 647 (1963).

[503] Yu. Dakhnovskii, Phys. Rev. B **49**, 4649 (1994); Yu. Dakhnovskii and R. D. Coalson, J. Chem. Phys. **103**, 2908 (1995); I. A. Goychuk, E. G. Petrov, and V. May, Chem. Phys. Lett. **253**, 428 (1996). See also A. Lück, M. Winterstetter, U. Weiss, and C.H. Mak, Phys. Rev. E **58**, 5565 (1998).

[504] F. Grossmann and P. Hänggi, Europhys. Lett. **18**, 571 (1992).

[505] J. M. Gomez-Llorente, Phys. Rev. A **45**, R6958 (1992); erratum Phys. Rev. E **49**, 3547 (1994).

[506] P. Jung, Phys. Rep. **234**, 175 (1993); Proc. NATO Workshop on *Stochastic Resonance in Physics and Biology*, F. Moss *et al.* (Eds.), J. Stat. Phys. **70**, 1 (1993); L. Gammaitoni, P. Hänggi, P. Jung, and F. Marchesoni, Rev. Mod. Phys. **70**, 223 (1998).

[507] R. Löfstedt and S. N. Coppersmith, Phys. Rev. Lett. **72**, 1947 (1994); Phys. Rev. E **49**, 4821 (1994).

[508] M. Grifoni and P. Hänggi, Phys. Rev. Lett. **76**, 1611 (1996); Phys. Rev. E **54**, 1390 (1996).

[509] H. Adam, M. Winterstetter, M. Grifoni, and U. Weiss, Phys. Rev. Lett. **83**, 252 (1999).

[510] *The Photosynthetic Reaction Center*, Vol. 1 and 2, ed. by J. Deisenhofer and J. R. Norris (Academic Press, New York, 1993).

[511] R. Egger, C. H. Mak, and U. Weiss, Phys. Rev. E **50**, R655 (1994).

[512] M. P. A. Fisher and W. Zwerger, Phys. Rev. B **32**, 6190 (1985).

[513] W. Zwerger, Phys. Rev. B **35**, 4737 (1987).

[514] C. L. Kane and M. P. A. Fisher, Phys. Rev. Lett. **68**, 1220 (1992); Phys. Rev. B **46** 15 233 (1992).

[515] X. G. Wen, Phys. Rev. B **41**, 12 838 (1990); **43** 11 025 (1991); **44**, 5708 (1991).

[516] H. J. Schulz, Phys. Rev. Lett. **71**, 1864 (1993).

[517] M. Fabrizio, A. O. Gogolin, and S. Scheidl, Phys. Rev. Lett. **72**, 2235 (1994).

[518] H. Rodenhausen, J. Stat. Phys. **55**, 1065 (1989).

[519] Y.-C. Chen, J. Stat. Phys. **65**, 761 (1991).

[520] A. Lenard, J. Math. Phys. **2**, 682 (1961).

[521] Yu. M. Ivanchenko and L. A. Zil'berman, Sov. Phys.–JETP **28**, 1272 (1969).

[522] U. Weiss and M. Wollensak, Phys. Rev. B **37**, 2729 (1988).

[523] F. Guinea, Phys. Rev. B **32**, 7518 (1985).

[524] U. Weiss, R. Egger, and M. Sassetti, Phys. Rev. B **52**, 16 707 (1995).

[525] M. Sassetti, M. Milch, and U. Weiss, Phys. Rev. A **46**, 4615 (1992).

[526] M. Sassetti, H. Schomerus, and U. Weiss, Phys. Rev. B **53**, R2914 (1996).

[527] H. Grabert, G.-L. Ingold, and B. Paul, Europhys. Lett. **44**, 360 (1998).

[528] K. Leung, R. Egger, and C. H. Mak, Phys. Rev. Lett. **75**, 3344 (1995).

[529] L. S. Levitov and G. B. Lesovik, JETP Lett. **58**, 230 (1993).

[530] D. V. Averin, Solid State Comm. **105**, 659 (1998).

[531] I. S. Beloborodov, F. W. J. Hekking and F. Pistolesi, in *New Directions in Mesoscopic Physics*, ed. by R. Fazio et al. (Kluwer Academic Publisher, 2003).

[532] G.-L. Ingold and H. Grabert, Phys. Rev. Lett. **83**, 3721 (1999).

[533] U. Weiss, Solid State Comm. **100**, 281 (1996);
U. Weiss, in *Tunneling and Its Implications*, ed. by D. Mugnai, A. Ranfagni, and L. S. Schulman, p. 134 (World Scientific, Singapore, 1997).

[534] P. Fendley and H. Saleur, Phys. Rev. Lett. **81**, 2518 (1998).

[535] N. Seiberg and E. Witten, Nucl. Phys. B **426**, 19 (1994); B **431**, 484 (1994).

[536] N. M. Temme, *Special Functions* (Wiley, New York, 1996), p. 49.

[537] L. Álvarez-Gaumé and F. Zamora, ArXiv:hep-th/9709180v3 (1997).

[538] K. A. Matveev, D. Yue, and L. I. Glazman, Phys. Rev. Lett. **71**, 3351 (1993).

[539] H. Saleur and U. Weiss, Phys. Rev. B **63**, 201302(R) (2001).

[540] P. Fendley, A. W. W. Ludwig, and H. Saleur, Phys. Rev. B **52**, 8934 (1995).

[541] P. Fendley and H. Saleur, Phys. Rev. B **54**, 10845 (1996).

[542] A. Komnik and H. Saleur, Phys. Rev. Lett. **96**, 216406 (2006).

[543] Y.-C. Chen and J. L. Lebowitz, Phys. Rev. B **46**, 10743 (1992).

[544] P. Fendley, F. Lesage, and H. Saleur, J. Stat. Phys. **85**, 211 (1996).

[545] J. Honer and U. Weiss, Chem. Phys. **375**, 265 (2010).

[546] R.P. Stanley, Adv. Math. **77**, 76 (1989).

[547] A. O. Gogolin, A. A. Nersesyan, and A. M. Tsvelik, *Bosonization and Strongly Correlated Systems* (Cambridge University Press, Cambridge, 1998).

[548] T. Giamarchi, *Quantum Physics in One Dimension* (Clarendon, Oxford, 2004).

[549] A. Luther and I. Peschel, Phys. Rev. B **9**, 2911 (1974).

[550] P. Fendley, A. W. W. Ludwig, and H. Saleur, Phys. Rev. Lett. **75**, 2196 (1995).

[551] R. Egger and H. Grabert, Phys. Rev. Lett. **77**, 538 (1996), and Ref. 10 therein.

[552] I. Safi and H. Saleur, Phys. Rev. Lett. **93**, 126602 (2004).

[553] M. Kindermann and B. Trauzettel, Phys. Rev. Lett. **94**, 166803 (2005).

[554] A. O. Gogolin and A. Komnik, Phys. Rev. B **73**, 195301 (2006).

[555] G. Benenti, K. Saito, and G. Casati, Phys. Rev. Lett. **106**, 230602 (2011).

[556] K. Proesmans, B. Cleuren, and C. Van den Broeck, Phys. Rev. Lett. **116**, 220601 (2016).

[557] K. Proesmans and C. Van den Broeck, Chaos **27**, 104601 (2017).

[558] V. Holubec and A. Ryabov, Phys. Rev. Lett. **121**, 120601 (2018).

[559] A. C. Barato and U. Seifert, Phys. Rev. Lett. **114**, 158101 (2015).

[560] T.R. Gingrich, J. M. Horowitz, N. Perunov and J. L. England, Phys. Rev. Lett. **116**, 120601 (2016).

[561] P. Pietzonka and U. Seifert, Phys. Rev. Lett. **120**, 190602 (2018).

[562] T. Koyuk, U. Seifert, and P. Pietzonka, J. Phys. A **52**, 02LT02 (2019).

[563] T. Koyuk and U. Seifert, Phys. Rev. Lett. **122**, 230601 (2019).

[564] T. Koyuk and U. Seifert, Phys. Rev. Lett. **125**, 260604 (2020).

[565] K. Brandner, T. Hanazato, and K. Saito, Phys. Rev. Lett. **120**, 090601 (2018).

[566] B. K. Agarwalla and D. Segal, Phys. Rev. B **98**, 155438 (2018).

[567] K. Ptaszyński, Phys. Rev. B **98**, 085425 (2018).

[568] M. Carrega, M. Sassetti, and U. Weiss, Physical Review A **99**, 062111 (2019).

Index

Printed in the United States
by Baker & Taylor Publisher Services